T0312163

The Biomarker Guide
Second Edition
Volume 1

The second edition of *The Biomarker Guide* is a fully updated and expanded version of this essential reference. Now in two volumes, it provides a comprehensive account of the role that biomarker technology plays both in petroleum exploration and in understanding Earth history and processes.

Biomarkers and Isotopes in the Environment and Human History details the origins of biomarkers and introduces basic chemical principles relevant to their study. It discusses analytical techniques, and applications of biomarkers to environmental and archeological problems.

The Biomarker Guide is an invaluable resource for geologists, petroleum geochemists, biogeochemists, environmental scientists, and archeologists.

KENNETH E. PETERS is currently Senior Research Geologist at the US Geological Survey in Menlo Park, California, where he is involved in three-dimensional petroleum system modeling for the North Slope of Alaska, the San Joaquin Basin, and elsewhere. He has 25 years of research experience with Chevron, Mobil, and ExxonMobil. Ken taught formal courses in petroleum geochemistry and thermal modeling both in industry and at various universities. He was Chair of the Organic Geochemistry Division of the Geochemical Society (2001–2004).

CLIFFORD C. WALTERS is Senior Research Associate with the ExxonMobil Research and Engineering Company, where he models oil generation and reservoir transformations, geomicrobiology, and processes of solids formation. He has over 20 years of industrial experience, including research at Sun Exploration and Production Company, and Mobil.

J. MICHAEL MOLDOWAN is Professor (Research) in the Department of Geological and Environmental Sciences at Stanford University. He joined Chevron's Biomarker Group in 1974, which, under the leadership of the late Dr. Wolfgang K. Seifert, is largely credited with pioneering the application of biological marker technology to petroleum exploration.

The Biomarker Guide

Second Edition
I. Biomarkers and Isotopes in the Environment and Human History

K. E. Peters
US Geological Survey, Menlo Park, CA, USA

C. C. Walters
ExxonMobil Research & Engineering Co.,
Corporate Strategic Research, Annandale, NJ, USA

J. M. Moldowan
Department of Geological and Environmental Sciences,
School of Earth Sciences, Stanford, CA, USA

CAMBRIDGE
UNIVERSITY PRESS

CAMBRIDGE UNIVERSITY PRESS
Cambridge, New York, Melbourne, Madrid, Cape Town, Singapore, São Paulo

Cambridge University Press
The Edinburgh Building, Cambridge CB2 8RU, UK

Published in the United States of America by Cambridge University Press, New York

www.cambridge.org
Information on this title: www.cambridge.org/9780521781589

First published 1993 by Prentice Hall, Inc.
Second edition published 2005 by Cambridge University Press
Reprinted with corrections 2005
This digitally printed version 2007

A catalogue record for this publication is available from the British Library

Library of Congress Cataloguing in Publication data

Peters, Kenneth E.
The biomarker guide. – 2nd ed. / K. E. Peters, C. C. Walters, and J. M. Moldowan.
    p.   cm.
Includes bibliographical references and index.
Contents: 1. Biomarkers in the environment and human history – 2. Biomarkers in petroleum systems and Earth history.
ISBN 0 521 78158 2
1. Petroleum – Prospecting.   2. Biogeochemical prospecting.   3. Biochemical markers.   I. Walters, C. C. (Clifford C.)
II. Moldowan, J. M. (J. Michael), 1946–   III. Title.
TN271.P4P463   2004
622′.1828–dc22   2003065416

ISBN 978-0-521-78158-9 hardback vol. 1
ISBN 978-0-521-78697-3 paperback vol. 1

ISBN 978-0-521-83762-0 hardback vol. 2
ISBN 978-0-521-03998-7 paperback vol. 2

Dedicated to
Vanessa, Brent, and Miwok
Johnet
Mary

Contents

About the authors   *page* ix
Preface   xi
Purpose   xvi
Acknowledgments   xvii

PART I BIOMARKERS AND ISOTOPES IN THE ENVIRONMENT AND HUMAN HISTORY

1 **Origin and preservation of organic matter**   3
   Introduction to biomarkers   3
   Domains of life   3
   Primary productivity   5
   Secondary productivity   8
   Preservation of organic matter   9
   Organic components in rocks   9
   Oxic versus anoxic deposition   10
   Sedimentation rate and grain size   13
   Lacustrine versus marine depositional settings   16
   Temporal and regional distributions of source rocks 16

2 **Organic chemistry**   18
   Alkanes: the sigma bond   18
   Alkenes: the pi bond   18
   Aromatics: benzene   19
   Structure notation   20
   Three-dimensional projections to two-dimensional space 21
   Acyclic alkanes   21
   Acyclic alkenes   23
   Monocyclic alkanes   23
   Multi-ringed cycloalkanes   24
   The isoprene rule   24
   Aromatic hydrocarbons   30
   Heteroatomic molecules   31
   Stereochemistry and nomenclature   34

   Chirality   34
   Optical activity   36
   Naming asymmetric centers (R, S, α, and β) 36
   Stereoisomerization   40
   Stereochemistry of selected biomarkers   40
   Exercise   44

3 **Biochemistry of biomarkers**   45
   Lipid membranes   45
   Membrane lipids   47
   Lipid membrane fluidity   48
   Biosynthesis of terpenoids   52
   Hopanoids and sterols in the biosphere and geosphere 58
   Porphyrins and other biomarkers of photosynthesis 64
   Carotenoids   69

4 **Geochemical screening**   72
   Source-rock screening: quality and quantity   72
   Source-rock screening: thermal maturity   88
   Geochemical logs and source potential index   96
   Reconstruction of original source-rock generative potential 97
   Tests for indigenous bitumen   100
   Detection of petroleum in prospective reservoir rocks 102
   Crude oil screening   102
   Reservoir continuity and filling history   109
   Surface geochemical exploration using piston cores 111
   Sample quality, selection, and storage   113
   Geochemical rock and oil standards   116
   Appendix: derivation of mass-balance equations   117

5   **Refinery oil assays**   119
    Basic oil assays   120
    Advanced oil assays   127
    Petroleum refining   129
    Biomarkers in refinery products   134

6   **Stable isotope ratios**   136
    Standards and notation   136
    Stable carbon isotope measurements   137
    Stable carbon isotope fractionation   138
    Converting δ values using different
    standards   139
    Applications of stable carbon isotope ratios   140
    Compound-specific isotope analysis   148
    Sulfur and hydrogen isotopes   152

7   **Ancillary geochemical methods**   157
    Diamondoids   157
    C₇ hydrocarbon analysis   162
    Compound-specific isotope analysis of
    light hydrocarbons   190
    Molecular modeling   192
    Fluid inclusions   194

8   **Biomarker separation and analysis**   198
    Organization of a biomarker laboratory   198
    Sample clean-up and separations   199
    Internal standards and preliminary
    analyses   200
    Zeolite molecular sieves   202
    Gas chromatography/mass spectrometry   208
    Mass spectra and compound
    identification   235
    Biomarker quantitation   240

9   **Origin of petroleum**   252
    Historical background   252
    Deep-earth gas hypothesis   253
    Abiogenic hydrocarbon gases   256
    Thermogenic hypothesis   259

10 **Biomarkers in the environment**   274
    Environmental markers   274
    Oil spills   276

    Processes affecting the fate of marine oil
    spills   279
    Mitigation of oil spills   282
    Modeling marine oil spills   283
    Oil spills on land   283
    Underground leakage   283
    Toxicity of petroleum   284
    Environmental chemistry field and
    laboratory procedures   287
    Chemical fingerprinting of oil spills   289
    Analysis of biomarkers and polycyclic
    aromatic hydrocarbons in oil-spill
    studies   294
    Applications of biomarkers and
    polycyclic aromatic hydrocarbons to
    oil-spill studies   298
    Biomarkers and the *Exxon Valdez*
    oil spill   305
    Gasoline and other light fuels as
    pollutants   312
    Natural gas as a pollutant   315
    Biomarkers in smoke   319

11 **Biomarkers in archeology**   322
    The ages of man   322
    Origins and transport of petroliferous
    materials in antiquity   322
    Archeological gums and resins   329
    Biomarkers in art   333
    Archeological wood tars (pitch)   334
    Paleodiets and agricultural practices   338
    Archeological beeswax   344
    Biomarkers and manuring practices   346
    Archeological DNA   347
    Ancient proteins   349
    Archeological narcotics   350
    Biomarkers and interdisciplinary studies   352

    Appendix: geologic time charts   353
    Glossary   355
    References   399
    Index   451

About the authors

Kenneth E. Peters is currently Senior Research Geologist at the US Geological Survey in Menlo Park, California, where he is involved in one-dimensional (1D), two-dimensional (2D), and three-dimensional (3D) petroleum system modeling of the North Slope of Alaska, the San Joaquin Basin, and elsewhere. He attained B.A. and M.A. degrees in geology from University of California, Santa Barbara, and a Ph.D. in geochemistry from University of California, Los Angeles, (UCLA) in 1978. His experience includes 15 years with Chevron, 6 years with Mobil, and 2 years as Senior Research Associate with ExxonMobil. Ken taught formal courses in petroleum geochemistry and thermal modeling for Chevron, Mobil, ExxonMobil, Oil and Gas Consultants International, and at various universities, including University of California, Berkeley, and Stanford University. He served as Associate Editor for *Organic Geochemistry* and the *American Association of Petroleum Geologists Bulletin*. Ken and co-authors received the Organic Geochemistry Division of the Geochemical Society Best Paper Awards for publications in 1981 and 1989. He served as Chair of the Gordon Research Conference on Organic Geochemistry (1998) and Chair of the Organic Geochemistry Division of the Geochemical Society (2001–2004).

Clifford C. Walters received Bachelor degrees in chemistry and biology from Boston University in 1976. He attended the University of Maryland, where he worked on the chemistry of Martian soil and conducted field and laboratory research on metasediments from Isua, Greenland, the oldest sedimentary rocks on Earth. After receiving a Ph.D. in geochemistry in 1981, Cliff continued with postdoctoral research on the organic geochemistry of Precambrian sediments and

meteorites. He joined Gulf Research and Development in 1982, where he implemented a program in biological marker compounds. In 1984, he moved to Sun Exploration & Production Company, where he was responsible for technical service and establishing biomarker geochemistry and thermal modeling as routine exploration tools. Mobil's Dallas Research Lab hired Cliff in 1988, where he became Supervisor of the Geochemical Laboratories in 1991. He is now Senior Research Associate with ExxonMobil Research and Engineering Company, where he conducts work on the modeling of oil generation and reservoir transformations, geomicrobiology, and processes of solids formation. Cliff published numerous papers, served as Editor of the ACS Geochemistry Division from 1990 to 1992, and is a current Associate Editor of *Organic Geochemistry*.

J. Michael Moldowan attained a B.S. in chemistry from Wayne State University, Detroit, Michigan, and a Ph.D. in chemistry from the University of Michigan in 1972. Following a postdoctoral fellowship in marine natural products with Professor Carl Djerassi at Stanford University, he joined Chevron's Biomarker Group in 1974. The Chevron biomarker team, led by the late Dr. Wolfgang K. Seifert in the mid 1970s to early 1980s, is largely credited with pioneering the application of biological marker technology to petroleum exploration. Mike joined the Department of Geological and Environmental Sciences of Stanford University as Professor (Research) in 1993. In 1986, he served as Chair of the Division of Geochemistry of the American Chemical Society, and he has twice been awarded the Organic Geochemistry Division of the Geochemical Society Best Paper Award for publications he co-authored in 1978 and 1989.

Preface

Biological markers (biomarkers) are complex molecular fossils derived from biochemicals, particularly lipids, in once-living organisms. Because biological markers can be measured in both crude oils and extracts of petroleum source rocks, they provide a method to relate the two (correlation) and can be used by geologists to interpret the characteristics of petroleum source rocks when only oil samples are available. Biomarkers are also useful because they can provide information on the organic matter in the source rock (source), environmental conditions during its deposition and burial (diagenesis), the thermal maturity experienced by rock or oil (catagenesis), the degree of biodegradation, some aspects of source rock mineralogy (lithology), and age. Because of their general resistance to weathering, biodegradation, evaporation, and other processes, biomarkers are commonly retained as indicators of petroleum contamination in the environment. They also occur with certain human artifacts, such as bitumen sealant for ancient boats, hafting material on spears and arrows, burial preservatives, and as coatings for medieval paintings.

Biomarker and non-biomarker geochemical parameters are best used together to provide the most reliable geologic interpretations to help solve exploration, development, production, and environmental or archeological problems. Prior to biomarker work, oil and rock samples are typically screened using non-biomarker analyses. The strength of biomarker parameters is that they provide more detailed information needed to answer questions about the source-rock depositional environment, thermal maturity, and the biodegradation of oils than non-biomarker analyses alone.

Distributions of biomarkers can be used to correlate oils and extracts. For example, C_{27}-C_{28}-C_{29} steranes or monoaromatic steroids distinguish oil-source families with high precision. Cutting-edge analytical techniques, such as linked-scan gas chromatography/mass spectrometry/mass spectrometry (GCMS/MS) provide sensitive measurements for correlation of light oils and condensates, where biomarkers are typically in low concentrations. Because biomarkers typically contain more than ~20 carbon atoms, they are useful for interpreting the origin of the liquid fraction of crude oil, but they do not necessarily indicate the origin of associated gases or condensates.

Different depositional environments are characterized by different assemblages of organisms and biomarkers. Commonly recognized classes of organisms include bacteria, algae, and higher plants. For example, some rocks and related oils contain botryococcane, a biomarker produced by the lacustrine, colonial alga *Botryococcus braunii. Botryococcus* is an organism that thrives only in lacustrine environments. Marine, terrigenous, deltaic, and hypersaline environments also show characteristic differences in biomarker composition.

The distribution, quantity, and quality of organic matter (organic facies) are factors that help to determine the hydrocarbon potential of a petroleum source rock. Optimal preservation of organic matter during and after sedimentation occurs in oxygen-depleted (anoxic) depositional environments, which commonly lead to organic-rich, oil-prone petroleum source rocks. Various biomarker parameters, such as the C_{35} homohopane index, can indicate the degree of oxicity under which marine sediments were deposited.

Biomarker parameters are an effective means to rank the relative maturity of petroleum throughout the entire oil-generative window. The rank of petroleum can be correlated with regions within the oil window (e.g. early, peak, or late generation). This information can provide a clue to the quantity and quality of the oil that may have been generated and, coupled with quantitative petroleum conversion measurements (e.g. thermal modeling programs), can help evaluate the timing of petroleum expulsion.

Biomarkers can be used to determine source and maturity, even for biodegraded oils. Ranking systems are based on the relative loss of *n*-alkanes, acyclic isoprenoids, steranes, terpanes, and aromatic steroids during biodegradation.

Biomarkers in oils provide information on the lithology of the source rock. For example, the absence of rearranged steranes can be used to indicate petroleum derived from clay-poor (usually carbonate) source rocks. Abundant gammacerane in some petroleum appears to be linked to a stratified water column (e.g. salinity stratification) during deposition of the source rock.

Biomarkers provide information on the age of the source rock for petroleum. Oleanane is a biomarker characteristic of angiosperms (flowering plants) found only in Tertiary and Upper Cretaceous rocks and oils. C_{26} norcholestanes originate from diatoms and can be used to distinguish Tertiary from Cretaceous and Cretaceous from older oils. Dinosterane is a marker for marine dinoflagellates, possibly distinguishing Mesozoic and Tertiary from Paleozoic source input. Unusual distributions of *n*-alkanes and cyclohexylalkanes are characteristic of *Gloeocapsomorpha prisca* found in early Paleozoic samples. 24-*n*-Propylcholestane is a marker for marine algae extending from at least the Devonian to the present.

Continued growth in the geologic, environmental, and archeological applications of biomarker technology is anticipated, particularly in the areas of age-specific biomarkers, the use of biomarkers to indicate source organic matter input and sedimentologic conditions, correlation of oils and rocks, and understanding the global cycle of carbon. New developments in analytical methods and instrumentation and the use of biomarkers to understand petroleum migration and kinetics are likely. Finally, early work suggests that biomarkers will continue to grow as tools to understand production, environmental, and archeological problems.

HOW TO USE THIS GUIDE

The Biomarker Guide is divided into two volumes. The first volume introduces some basic chemical principles and analytical techniques, concentrating on the study of biomarkers and isotopes in the environment and human history. The second volume expands on the uses of biomarkers and isotopes in the petroleum industry,

and investigates their occurrence throughout Earth history.

The Biomarker Guide was written for a diverse audience, which might include the following:

- students of geology, environmental science, and archeology who wish to gain general knowledge of what biomarkers can do;
- practicing geologists and geochemical coordinators in the petroleum industry with both specific and general questions about which biomarker and/or non-biomarker parameters might best answer regional exploration, development, or production problems;
- experienced geochemists who require detailed information on specific parameters or methodology;
- managers or research directors who require a concise explanation for terms and methodology;
- refinery process chemists requiring a more detailed knowledge of petroleum; and
- archeologists and environmental scientists interested in a technology useful for characterizing petroleum in the environment.

The text in each chapter is supplemented by many references to related sections in the book and to the literature. Various parts of the guide, such as notes, highlight detailed discussions that supplement the text.

The following is a brief overview of each chapter in the two volumes.

PART I BIOMARKERS AND ISOTOPES IN THE ENVIRONMENT AND HUMAN HISTORY

1 Origin and preservation of organic matter

This chapter introduces biomarkers, the domains of life, primary productivity, and the carbon cycle on Earth. Morphological and biochemical differences among different life forms help to determine their environmental habitats and the character of the biomarkers that they contribute to sediments, source rocks, and petroleum. The discussion summarizes processes affecting the distribution, preservation, and alteration of biomarkers in sedimentary rocks. Various factors, such as type of organic matter input, redox potential during sedimentation, bioturbation, sediment grain size, and sedimentation rate, influence the quantity and quality of organic matter preserved in rocks during Earth's history.

2 Organic chemistry

A brief overview of organic chemistry includes explanations of structural nomenclature and stereochemistry necessary to understand biomarker parameters. The discussion includes an overview of compound classes in petroleum and concludes with examples of the structures and nomenclature for several biomarkers, their precursors in living organisms, and their geologic alteration products.

3 Biochemistry of biomarkers

This chapter provides an overview of the biochemical origins of the major biomarkers, including discussions of the function, biosynthesis, and occurrence of their precursors in living organisms. Some topics include lipid membranes and their chemical compositions, the biosynthesis of isoprenoids and cyclization of squalene, and examples of hopanoids, sterols, and porphyrins in the biosphere and geosphere.

4 Geochemical screening

This chapter describes how to select sediment, rock, and crude oil samples for advanced geochemical analyses by using rapid, inexpensive geochemical tools, such as Rock–Eval pyrolysis, total organic carbon, vitrinite reflectance, scanning fluorescence, gas chromatography, and stable isotope analyses. The discussion covers sample quality, selection, storage, and geochemical rock and oil standards. Other topics include how to test rock samples for indigenous bitumen, surface geochemical exploration using piston cores, geochemical logs and their interpretation, chromatographic fingerprinting for reservoir continuity, and how to deconvolute mixtures of oils from different production zones. Mass balance equations show how to calculate the extent of fractional conversion of kerogen to petroleum, source-rock expulsion efficiency, and the original richness of highly mature source rocks.

5 Refinery oil assays

Many refinery oil assays differ substantially from geochemical analyses conducted by petroleum or environmental geochemists, although interdisciplinary use of these tools is becoming more common. Some basic oil assays include API (American Petroleum Institute) gravity, pour point, cloud point, viscosity, trace metals, total acid number, refractive index, and wax content. More advanced oil assays include chemical group-type fractionation and field ionization mass spectrometry. A brief overview of refinery processes includes the fate of biomarkers in straight-run and processed refinery products with tips on how to distinguish refined from natural petroleum products in environmental or geological samples.

6 Stable isotope ratios

This chapter describes stable isotopes and their use to characterize petroleum, including gases, crude oils, sediment and source-rock extracts, and kerogen, with emphasis on stable carbon isotope ratios. The discussion includes isotopic standards and notation, principles of isotopic fractionation, and the use of various isotopic tools, such as stable carbon isotope-type curves, for correlation or quantification of petroleum mixtures. The chapter concludes with new developments in compound-specific isotope analysis, including its application to better understand the origin of carboxylic acids and the process of thermochemical sulfate reduction in petroleum reservoirs.

7 Ancillary geochemical methods

Ancillary geochemical tools (e.g. diamondoids, C_7 hydrocarbons, compound-specific isotopes, and fluid inclusions) can be used to evaluate the origin, thermal maturity, and extent of biodegradation or mixing of petroleum, even when the geological samples lack or have few biomarkers. Molecular modeling can be used to rationalize or predict the geochemical behavior of biomarkers and other compounds in the geosphere.

8 Biomarker separation and analysis

This chapter describes the organization of a biomarker laboratory and the methods used to prepare and separate crude oils and sediment or source-rock extracts into fractions prior to mass spectrometric analysis. The concept of mass spectrometry is explained. Many of these fundamentals, such as the difference between a mass chromatogram and a mass spectrum, or between selected ion and linked-scan modes of analysis, are critical to understanding later discussions of biomarker parameters. Several key topics, including analytical procedures, internal standards, and examples of gas chromatography/mass spectrometry (GCMS) data problems, help the reader to evaluate the quality of biomarker data and interpretations.

9   Origin of petroleum

This chapter describes evidence against the deep-earth gas hypothesis, which invokes an abiogenic origin for petroleum by polymerization of methane deep in the Earth's mantle. The deep-earth gas hypothesis has little scientific support but, if correct, could have major implications for petroleum exploration and the application of biomarkers to environmental science and archeology. The discussion covers experimental, geological, and geochemical evidence supporting the thermogenic origin of petroleum.

10   Biomarkers in the environment

This chapter explains how analyses of biomarkers and other environmental markers, such as polycyclic aromatic hydrocarbons, are used to characterize, identify, and assess the environmental impact of oil spills. The discussion covers processes affecting the composition of spilled oil, such as emulsification, oxidation, and biodegradation, as well as oil-spill mitigation and modeling. Field and laboratory procedures for sampling and analyzing spills are discussed, including program design, chemical fingerprinting, and data quality control. The chapter includes sections on smoke, natural gas, and gasoline and other light fuels as pollutants, and a detailed discussion of the controversial *Exxon Valdez* oil spill.

11   Biomarkers in archeology

This chapter provides examples of the growing use of biomarker and isotopic analyses to evaluate organic materials in archeology. Some of the topics include bitumens in Egyptian mummies, such as Cleopatra, archeological gums and resins, and biomarkers in art and ancient shipwrecks. The discussion covers the use of biomarkers and isotopes in studies of paleodiet and agricultural practices, including studies of ancient wine and beeswax. Other topics include archeological DNA, proteins, and evidence for ancient narcotics.

PART II   BIOMARKERS AND ISOTOPES IN PETROLEUM AND EARTH HISTORY

12   Geochemical correlations and chemometrics

Geochemical correlation can be used to establish petroleum systems to improve exploration success, define reservoir compartments to enhance production, or identify the origin of petroleum contaminating the environment. This chapter explains how chemometrics simplifies genetic oil-oil and oil-source rock correlations and other interpretations of complex multivariate data sets.

13   Source- and age-related biomarker parameters

This chapter explains how biomarker analyses are used for oil-oil and oil-source rock correlation and how they help to identify characteristics of the source rock (e.g. lithology, geologic age, type or organic matter, redox conditions), even when samples of rock are not available. Biomarker parameters are arranged by groups of related compounds in the order: (1) alkanes and acyclic isoprenoids, (2) steranes and diasteranes, (3) terpanes and similar compounds, (4) aromatic steroids, hopanoids, and similar compounds, and (5) porphyrins. Critical information on specificity and the means for measurement are highlighted above the discussion for each parameter.

14   Maturity-related biomarker parameters

This chapter explains how biomarker analyses are used to assess thermal maturity. The parameters are arranged by groups of related compounds in the order (1) terpanes, (2) polycadinenes and related products, (3) steranes, (4) aromatic steroids, (5) aromatic hopanoids, and (6) porphyrins. Critical information on specificity and the means for measurement are highlighted in bold print above the discussion for each parameter.

15   Non-biomarker maturity parameters

This chapter explains how certain non-biomarker parameters, such as ratios involving *n*-alkanes and aromatic hydrocarbons, are used to assess thermal maturity. Critical information on specificity and the means for measurement are highlighted above the discussion for each parameter.

16   Biodegradation parameters

This chapter explains how biomarker and non-biomarker analyses are used to monitor the extent of biodegradation. Compound classes and parameters are discussed in the approximate order of increasing resistance to biodegradation. The discussion covers recent advances in our understanding of the controls and mechanisms of petroleum biodegradation and the relative significance of aerobic versus anaerobic degradation in both surface and subsurface environments. Examples show how to predict the original physical properties of crude oils prior to biodegradation.

17   Tectonic and biotic history of the Earth

The evolution of life is closely tied to biomarkers in petroleum. This chapter provides a brief tectonic history of the Earth in relation to the evolution of major life forms. Mass extinctions and their possible causes

are discussed. The end of the section for each time period includes examples of source rocks and related crude oils with emphasis on the geochemistry of the oils. These examples are linked to more detailed discussion of petroleum systems in Chapter 18.

18   Petroleum systems through time

This chapter defines petroleum systems and provides examples of the geology, stratigraphy, and geochemistry of source rocks and crude oils through geologic time. Gas chromatograms, sterane and terpane mass chromatograms, stable isotope compositions, and other geochemical data are provided for representative crude oils generated from many worldwide source rocks.

19   Problem areas and further work

This chapter describes areas requiring further research, including the application of biomarkers to migration, the kinetics of petroleum generation, geochemical correlation and age assessment, and the search for extraterrestrial life.

Purpose

The Biomarker Guide provides a comprehensive discussion of the basic principles of biomarkers, their relationships with other parameters, and their applications to studies of maturation, correlation, source input, depositional environment, and biodegradation of the organic matter in petroleum source rocks, reservoirs, and the environment. It builds upon previous books by Tissot and Welte (1984), Waples and Machihara (1991), Bordenave (1993), Peters and Moldowan (1993), Hunt (1996), and Welte *et al.* (1997). The volumes were prepared for a broad audience, including students, company exploration geologists, geochemists, and environmental scientists for several reasons:

(1) Biomarker geochemistry is a rapidly growing discipline with important worldwide applications to petroleum exploration and production and environmental monitoring.
(2) Biomarker parameters are becoming increasingly prominent in exploration, production, and environmental reports.
(3) Different parameters are used within the industry, academia, service laboratories, and the literature.
(4) The quality of biomarker data and interpretation can vary considerably, depending on their source.

The objective of this guide is to provide a single, concise source of information on the various biomarker parameters and to create general guidelines for the use of selected parameters. An important aim is to clarify the relationships between biomarker and other geochemical parameters and to show how they can be used together to solve problems. It is not intended to teach interpretation of raw biomarker data. This is a job for a biomarker specialist with years of training in instrumentation and organic chemistry. A crash-course or cookbook approach cannot provide such training without the consequence of serious interpretive errors and a tarnished view of the applicability of biomarkers in general.

A final objective of the guide is to impart in each reader a feeling for the excitement and vigor of the new field of biomarker geochemistry. Expanding research efforts at geochemical laboratories worldwide have increased the rate of change and growth in our geochemical concepts. Applications of many of the biomarker parameters presented here will undoubtedly improve with time and further research. We anticipate that more than a few readers will be directly involved in making these improvements possible.

Acknowledgments

The authors gratefully acknowledge the support of Chevron management and technical personnel (now ChevronTexaco) during preparation of the precursor to this book, which was called *The Biomarker Guide*. In particular, the authors thank G. J. Demaison, C. Y. Lee, F. Fago, R. M. K. Carlson, P. Sundararaman, J. E. Dahl, M. Schoell, E. J. Gallegos, P. C. Henshaw, S. R. Jacobson, R. J. Hwang, D. K. Baskin, and M. A. McCaffrey for discussions, technical assistance, and helpful review comments.

We acknowledge the support of Mobil and ExxonMobil management and technical personnel during preparation of much of *The Biomarker Guide* Second Edition. In particular, the authors thank Ted Bence, Paul Mankiewicz, John Guthrie, Jim Gormly, and Roger Prince for their input. We also thank Bill Clendenen, Larry Baker, Gary Isaksen, Jim Stinnett (Mobil, retired), Al Young (Exxon, retired), and Steve Koch.

We acknowledge the support of management at the US Geological Survey during preparation of the book. Special thanks are due to Les Magoon, Bob Eganhouse, Mike Lewan, Keith Kvenvolden, Fran Hostettler, Tom Lorenson, and Ron Hill.

Special thanks are due to Steve Brown and John Zumberge of GeoMark Research, Inc. for allowing us to use various oil Information sheets from their Oil Information Library System (OILS) and several cross-plots of biomarker ratios. We also thank David Zinniker of Stanford University for input on terrigenous biomarkers.

Finally, we thank the many reviewers of various drafts of the Second Edition, who are listed in the following table. Their dedication to the sometimes thankless job of peer review is to be commended.

Chapter	Title	Reviewer	Affiliation
1	Origin and preservation of organic matter	Kirsten Laarkamp	ExxonMobil Upstream Research
		Phil Meyers	University of Michigan
2	Organic chemistry	Kirsten Laarkamp	ExxonMobil Upstream Research
3	Biochemistry of biomarkers	Robert Carlson	ChevronTexaco
4	Geochemical screening	Dave Baskin	OilTracers, L. L. C.
		George Claypool	Mobil (retired)
		Jim Gormly	ExxonMobil Upstream Research
		Tom Lorenson	US Geological Survey
5	Refinery oil assays	Owen BeMent	Shell Oil Company
		Paul Mankiewicz	ExxonMobil Upstream Research
		Robert McNeil	Shell Oil Company
6	Stable isotope ratios	Mike Engel	University of Oklahoma
		Martin Schoell	ChevronTexaco (retired)
		Zhengzheng Chen	Stanford University
		John Guthrie	ExxonMobil Upstream Research
		Jeffrey Sewald	Woods Hole Oceanographic Institution

(cont.)

(*cont.*)

Chapter	Title	Reviewer	Affiliation
7	Ancillary geochemical methods	Ron Hill	US Geological Survey
		Dan Jarvie	Humble Geochemical Services, Inc.
		Yitian Xiao	ExxonMobil Upstream Research
8	Biomarker separation and analysis	John Guthrie	ExxonMobil Upstream Research
		Robert Carlson	ChevronTexaco
9	Origin of petroleum	Kevin Bohacs	ExxonMobil Upstream Research
		Barbara Sherwood Lollar	University of Toronto
10	Biomarkers in the environment	Ted Bence, Rochelle Jozwiak, Mike Smith, Bill Burns (retired)	ExxonMobil Upstream Research
		Roger Prince	ExxonMobil Strategic Research
		Bob Eganhouse, Keith Kvenvolden, Fran Hostettler	US Geological Survey
		Ian Kaplan	UCLA (retired)
11	Biomarkers in archeology	Roger Prince	ExxonMobil Strategic Research
		Max Vityk	ExxonMobil Upstream Research
12	Geochemical correlations and chemometrics	Jaap Sinninghe Damsté	Netherlands Institute for Sea Research
		Paul Mankiewicz	ExxonMobil Upstream Research
		Scott Ramos, Brian Rohrback	Infometrix, Inc.
13	Source- and age-related biomarker parameters	Jaap Sinninghe Damsté	Netherlands Institute for Sea Research
		Leroy Ellis	Terra Nova Technologies
		Kliti Grice	University of Western Australia
		Paul Mankiewicz	ExxonMobil Upstream Research
		Roger Summons	Massachusetts Institute of Technology
		David Zinniker	Stanford University
14	Maturity-related biomarker parameters	Gary Isaksen	ExxonMobil Upstream Research
		Ron Noble	BHP Billiton
15	Non-biomarker maturity parameters	Gary Isaksen	ExxonMobil Upstream Research
		Ron Noble	BHP Billiton
16	Biodegradation parameters	Dave Converse	ExxonMobil Upstream Research
		Roger Prince	ExxonMobil Strategic Research
17	Tectonic and biotic history of the Earth	Kevin Bohacs	ExxonMobil Upstream Research
		Keith Kvenvolden	US Geological Survey
18	Petroleum systems through time	Steve Creaney	ExxonMobil Exploration Company
		Les Magoon	US Geological Survey
19	Problem areas and further work	John Guthrie	ExxonMobil Upstream Research
		Mike Lewan	US Geological Survey
–	References	Jan Heagy, Marsha Harris	ExxonMobil Upstream Research
		Susie Bravos, Page Mosier, Emily Shen-Torbik	US Geological Survey

Part I
**Biomarkers and isotopes in the environment
and human history**

1 · Origin and preservation of organic matter

This chapter introduces biomarkers, the domains of life, primary productivity, and the carbon cycle on Earth. Morphological and biochemical differences among different life forms help to determine their environmental habitats and the character of the biomarkers that they contribute to sediments, source rocks, and petroleum. The discussion summarizes processes affecting the distribution, preservation, and alteration of biomarkers in sedimentary rocks. Various factors, such as type of organic matter input, redox potential during sedimentation, bioturbation, sediment grain size, and sedimentation rate, influence the quantity and quality of organic matter preserved in rocks.

INTRODUCTION TO BIOMARKERS

Biological markers or biomarkers (Eglinton *et al.*, 1964; Eglinton and Calvin, 1967) are molecular fossils, meaning that these compounds originated from formerly living organisms. Biomarkers are complex organic compounds composed of carbon, hydrogen, and other elements. They occur in sediments, rocks, and crude oils and show little or no change in structure from their parent organic molecules in living organisms. Sediments consist of unconsolidated mineral and organic detritus prior to lithification to form rock.

Biomarkers are useful because their complex structures reveal more information about their origins than other compounds. Unlike biomarkers, methane (CH_4) and graphite (nearly pure carbon) are comparatively less informative because virtually any organic compound will generate these products when sufficiently heated. However, although methane and other hydrocarbon gases are simple compounds compared with biomarkers, they still contain useful information about their origin and geologic history, as discussed later in the text.

Three characteristics distinguish biomarkers from many other organic compounds:

- Biomarkers have structures composed of repeating subunits, indicating that their precursors were components in living organisms.
- Each parent biomarker is common in certain organisms. These organisms can be abundant and widespread.
- The principal identifying structural characteristics of the biomarkers are chemically stable during sedimentation and early burial.

Before proceeding with biomarkers, we need a general discussion of the different types of organism that contribute organic matter to the carbon cycle on Earth. This is followed by a discussion of the special circumstances required to preserve organic matter during and after sedimentation.

DOMAINS OF LIFE

The three major domains of life include the archaea and eubacteria (prokaryotes) (Woese *et al.*, 1978) and the eukarya (eukaryotes or higher organisms) (Table 1.1). Unlike the prokaryotes, the eukaryotes contain a membrane-bound nucleus and complex organelles. For example, the organelles known as mitochondria and chloroplasts carry out important functions in energy generation and photosynthesis, respectively. Eukaryotic microorganisms include algae, protozoa, and fungi (molds and yeasts). All higher multicellular organisms are also eukaryotes.

The prokaryotes consist of millions of unicellular archaeal and eubacterial species, many not yet characterized. While eukaryotes are classified mainly by morphology, the prokaryotes are distinguished mainly by their diverse biochemistry and the habitats in which they grow (White, 1999). Their relatively simple shapes limit classification of prokaryotes based on morphology.

Table 1.1. *Organisms are divided into three domains: archaea and eubacteria (prokaryotes) and higher organisms (eukaryotes)*

	Prokaryotes		Eukaryotes
	Eubacteria	Archaea	Eukaryota
Single-cell microorganisms	Most	All	Algae, protozoa
Colonial with specialized cells	Cyanophyta, some others	None (?)	Algae, animals
Multicellular with differentiated cells	None	None	Plants, animals

Table 1.2. *Organisms can be classified by carbon nutrition (left column) and energy metabolism (top row)*

	Phototroph	Chemotroph
Autotroph	Photoautotroph	Chemoautotroph
Heterotroph	Photoheterotroph	Chemoheterotroph

Living prokaryotes should not be viewed as primitive. They are superbly adapted organisms that occupy all habitats. There is no universally accepted phylogeny for the prokaryotes.

The metabolism of prokaryotic species can be described as autotrophic, heterotrophic, phototrophic, chemotrophic, or by a compound name using multiple terms (Table 1.2) (Chapman and Gest, 1983). Autotrophs use carbon dioxide (CO_2) as a sole source of carbon for growth and obtain their energy from light (photosynthetic autotrophs or photoautotrophs) or from the oxidation of inorganic compounds (chemosynthetic autotrophs or chemoautotrophs). In contrast, heterotrophs (photoheterotrophs and chemoheterotrophs) obtain both their carbon and their energy for growth from organic compounds. They are saprophytes, obtaining their nutrients from dead organic matter. Chemotrophic prokaryotes obtain their energy from the oxidation of inorganic compounds, such as H_2, H_2S, and NH_3. Most chemotrophs are autotrophic, but some are heterotrophs (chemoheterotrophs), which use inorganic oxidation for energy but use organic matter for carbon as well as supplemental energy. Photosynthetic bacteria have the biochemistry for either anoxygenic photosynthesis (non O_2-producing) or oxygenic photosynthesis (O_2-producing). Most photosynthetic bacteria are autotrophs that fix CO_2 (photoautotrophs), but

some rely on organic matter for their carbon (photoheterotrophs). Adaptive prokaryotes switch their modes of metabolism depending on environmental conditions. Additional terms are commonly used to refine these basic descriptions. For example, methylotrophic bacteria are those heterotrophic organisms with the ability to utilize reduced carbon substrates with no carbon-carbon bonds (e.g. methane, methanol, methylated amines, and methylated sulfur species) as their sole source of carbon and energy. Methylotrophs that use methane are called methanotrophs. Acetogenic bacteria catalyze the reduction of two CO_2 molecules to acetate in their energy metabolism. Diazotrophs have the ability to utilize N_2 as a source of nitrogen for growth.

Oxygen tolerance, temperature, salinity, and pathology are commonly used to describe prokaryotes by habitat. Prokaryotes vary widely in their utilization and tolerance of oxygen. Obligate aerobes are prokaryotes that require oxygen. Facultative aerobes utilize oxygen, but can either tolerate its absence or switch to anaerobic metabolism, where oxygen is not required. Similarly, obligate anaerobes grow only in the absence of oxygen, while facultative anaerobes can tolerate oxygen. This tolerance usually depends on the presence of enzymes, such as superoxide dimutase and catalase, which can remove oxygen radicals. Many prokaryotes can survive wide ranges of temperature from below freezing up to boiling conditions. However, survival does not imply growth. Most prokaryotes have narrow temperature ranges for optimal growth. Prokaryotes have adapted to all temperatures at which water remains liquid. Prokaryotes that grow and thrive at very low temperatures ($\sim 0\,°C$) are called psychrophiles. For example, mats dominated by benthic filamentous cyanobacteria grow immediately below permanent ice cover at Lake Hoare in the dry valley region of southern Victoria Land, Antarctica (Hawes and Schwarz, 1995).

Prokaryotes that flourish at moderate temperatures (0–45 °C) are called mesophiles. Thermophilic organisms have specially modified lipid membranes, proteins, and genetic material that allow them to survive at temperatures above 45 °C. Prokaryotes that grow at extremely low or high temperatures are called extreme psychrophiles or extreme thermophiles, respectively. Halophiles are organisms that tolerate high concentrations of dissolved salts. Extreme halophiles not only tolerate but also require high salt concentrations for growth. Descriptions of bacteria based on their pathogenicity and/or parasitism are important to the biomedical community.

The archaea (also called archaebacteria) are common in extreme environments and include extreme halophiles (hypersaline environments), thermophiles (hot), thermoacidophiles (hot and acidic), and methanogens (methane generated as a waste product of metabolism). Their diversity is now recognized in more moderate environments, such as soils and marine waters (Torsvik *et al.*, 2002), as is their abundance in deep subsurface strata (e.g. Takai *et al.*, 2001; Chapelle *et al.*, 2002). Archaea are prokaryotes because they lack cell nuclei. While archaea share some common features with eubacteria and eukaryotes, they have genomes and biochemical pathways that are distinct from the other domains of life (Woese *et al.*, 1978).

The morphologic differences between the three domains reflect fundamental differences in biochemistry (Table 1.3). Some of these differences control the types of biomarker that originate from these groups, as discussed later in this book. For example, steranes in petroleum originate from sterols in the cell membranes of eukaryotes. Prokaryotes use hopanoids rather than steroids in their cell membranes and these precursors account for the hopanes in petroleum.

Endosymbiotic theory suggests that the eukaryotes originated by an evolutionary fusion of different prokaryotes that originally lived together as symbionts (Schenk *et al.*, 1997). According to this theory various organelles in eukaryotes, such as mitochondria, kinetosomes, hydrogenosomes, and plastids, originated as symbionts. For example, mitochondria and chloroplasts probably originated as aerobic non-photosynthetic bacteria and photosynthetic cyanobacteria, respectively. These organelles are morphologically similar to bacteria. Like bacteria, they are susceptible to antibiotics, such as streptomycin, they contain deoxyribonucleic acid (DNA) in the prokaryotic closed-circle form, and they have similar ribosomal ribonucleic acid (rRNA) phylogenetic sequences (Table 1.3).

Note: Viruses are non-cellular entities that lack the ability to reproduce without taking over the cellular apparatus of a host. For example, the T-4 bacteriophage consists of a protein coat and mechanical structure that injects the enclosed T-4 genetic material through the cell membrane of its bacterial host. Once inside, the T-4 genetic material reprograms the host cell to generate new T-4 virus. Viruses probably evolved from primitive life forms by loss of their original cellular components.

Endosymbiosis accounts for the presence of eubacterial genes within eukaryotic organisms. Some of these genes migrated to the nucleus of the host, while others remained within the organelle that was originally the symbiont (Palenik, 2002). Genomic comparisons of representatives from the three domains indicate that much of the prokaryotic DNA has been subject to horizontal gene transfer (HGT), the acquisition of new genes from other organisms typically via prokaryotic plasmids (Koonin *et al.*, 2001). Plasmids are loops of DNA that are independent of the bacterial chromosome and can be exchanged between cells during conjugation. Genes acquired by horizontal transfer may range from being non-functional to being evolutionarily advantageous. For example, archaeal genes may have allowed some eubacteria to thrive under high-temperature conditions. Photosynthesis confers such an adaptive advantage that the genome of many photosynthetic bacteria appears to have resulted from massive HGT (Raymond *et al.*, 2003). Woese (2002) advanced a theory that the three domains of life arose from early cellular evolution involving HGT from a communal pool.

PRIMARY PRODUCTIVITY

Photosynthesis is the only significant means by which new organic matter is synthesized on Earth. It is the ultimate source for the biomass in almost all living organisms and accounts for most organic matter buried in sediments and rocks, including biomarkers. Many non-photosynthetic biosystems, such as methanotrophic

Table 1.3. *Some morphologic and biochemical differences between the three domains of life*

Property	Cellular domain		
	Eukarya	Bacteria	Archaea
Cell configuration	Eukaryotic	Prokaryotic	Prokaryotic
Nuclear membrane	Present	Absent	Absent
Number of chromosomes	>1	1	1
Chromosome topology	Linear	Circular	Circular
Murein in cell wall	No	Yes	No
Cell-membrane lipids	Ester-linked glycerides; unbranched; polyunsaturated	Ester-linked glycerides; unbranched; saturated or monounsaturated	Ether-linked; branched; saturated
Cell-membrane sterols	Yes	No[1]	No[2]
Organelles (mitochondria and chloroplasts)	Yes	No[3]	No
Ribosome size	80S (cytoplasmic)	70S	70S
Cytoplasmic streaming	Yes	No	No
Meiosis and mitosis	Yes	No	No
Transcription and translation coupled	No	Yes	?
Amino acid initiating protein synthesis	Methionine	*n*-Formyl methionine	Methionine
Protein synthesis inhibited by streptomycin and chloramphenicol	No	Yes	No
Protein synthesis inhibited by diphtheria toxin	Yes	No	Yes
Histones	Yes	No	Yes

[1] Certain mycoplasmas require, but do not produce, sterols. These very small bacteria lack a cell wall and are bound by a single triple-layered membrane that is stabilized by sterols. There are reports that a few bacteria can convert partially demethylated lanosterol precursors to cholesterol, but cholesterol biosynthesis seems to be restricted to eukaryotes.

[2] Two possible exceptions include *Methylococcus capsulatus* and *Nannocystis exedens* (Tornabene, 1985). These are rare $\Delta^{8,10}$ sterols with an eight-carbon side chain and one or two methyl-groups at C-4. The sterols in these archaea may indicate a very primitive pathway for sterol synthesis (Patterson, 1994).

[3] Bacteria contain acidocalcisomes, membrane-bound organelles thought to store energy and control intercellular acidity (Seufferheld *et al.*, 2003). The evolutionary origin of these organelles is unknown.

communities (Michaelis *et al.*, 2002), rely on recycling sedimentary carbon that originally was fixed by photosynthetic organisms. In some environments, organic matter originates from non-photosynthetic chemoautotrophy (e.g. deep-sea hydrothermal vents), but their contribution to Earth's pool of sedimentary carbon is minor.

Bacterial and green-plant photosynthesis can be represented by the following generalized reaction:

$$2H_2A + CO_2 \xrightarrow{\text{light}} 2A + CH_2O + H_2O \qquad (1.1)$$

The formula CH_2O stands for organic matter in the form of a carbohydrate, such as the sugar glucose

($C_6H_{12}O_6$). Polysaccharides (i.e. polymerized sugars) are the main form in which photosynthesized organic matter is stored in living cells. Respiration is the reverse of Reaction (1.1), where energy for cell function is released by oxidizing polysaccharides.

Because energy from light is required for photosynthesis, photosynthetic organisms are restricted to the land and the photic zone in lakes and the ocean. Light intensity diminishes rapidly with water depth in the oceans and lakes, limiting the depths of the photic zone to less than $\sim$100 m. Phototrophs are organisms that derive energy for photosynthesis from light (Table 1.2). Phototrophic microorganisms are the most important photosynthesizers in aquatic environments, while higher plants dominate the land. This fundamental difference affects the types of organic matter deposited in sediments from these environments. Chemosynthesis, where chemical energy is released by oxidation of reduced substrates, can occur in deep water beyond the photic zone.

When organisms grow phototrophically, the rate of fixation of carbon from CO_2 into carbohydrate exceeds the rate of respiration. The whole cycle on Earth (Figure 1.1) is based on a net positive balance in the rate of photosynthesis over the rate of respiration.

Green-plant photosynthesis is a variation of Reaction (1.1):

$$2H_2O + CO_2 \xrightarrow{light} O_2 + CH_2O + H_2O \qquad (1.2)$$

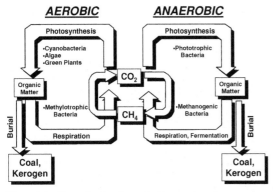

Figure 1.1. Generalized redox cycle for organic carbon. Production of new organic matter by photosynthetic fixation of carbon from CO_2 (primary productivity) can occur with (aerobic, left) or without (anaerobic, right) oxygen as a by-product. Respiration and other processes result in the nearly complete oxidation of this organic matter back to CO_2. Reprinted with permission by ChevronTexaco Exploration and Production Technology Company, a division of Chevron USA Inc.

This process is an oxidation–reduction (redox) reaction. The oxidation in Reaction (1.2) is:

$$2H_2O \rightarrow 4H + O_2 \qquad (1.2a)$$

Water is split to produce reducing power (hydrogen atoms), with O_2 as a by-product. The reduction reaction is:

$$4H + CO_2 \rightarrow CH_2O + H_2O \qquad (1.2b)$$

As shown above, the H_2O product originates by reduction of CO_2. The necessary hydrogen atoms originate from a donor molecule or reductant (in this case, H_2O) and are transferred to an acceptor molecule or oxidant, CO_2. After the reaction, the donor molecule is referred to as oxidized, while the acceptor is reduced.

In green plants and cyanobacteria (Reaction 1.2), H_2A is H_2O and 2A is O_2. In bacteria, H_2A is some oxidizable substrate and 2A is the product of its oxidation. For example, in sulfur bacteria H_2A is hydrogen sulfide (H_2S) and A is sulfur:

$$2H_2S + CO_2 \xrightarrow{light} 2S + CH_2O + H_2O \qquad (1.3)$$

Note: The equation for bacterial photosynthesis clearly shows it to be an oxidation–reduction reaction. In bacteria, water is one of the products of the reduction of CO_2 and it differs from the oxidizable component, H_2S. In green plants, the oxidant and the product of the reduction cannot be readily distinguished without isotopic labeling experiments because both consist of water (reaction 1.2).

In addition to light, the main limitation on planktonic primary productivity is the availability of nutrients, particularly nitrates and phosphates. Nitrogen fixation, or the conversion of atmospheric N_2 to NO_3^-, occurs mainly due to soil microbes, while phosphate originates from weathering of continental rocks. Thus, both nitrate and phosphate in lakes and the oceans depend on run-off from land. Because they are surrounded by land, lakes commonly contain more nutrients than most ocean areas and therefore have higher primary productivity (Table 1.4). Open-ocean areas far from land constitute $\sim$90% of the ocean surface but have few nutrients and very low productivity (Table 1.4). Vertical recycling of nutrients is negligible because of thermal

Table 1.4. *Annual rates of primary productivity in marine and freshwater ecosystems (Meyers, 1997)*

Aquatic ecosystem	Annual productivity (g carbon/m^2/year)
Open ocean	25–50
Coastal ocean	70–120
Upwellings	250–350
Oligotrophic lake	4–180
Mesotrophic lake	100–310
Eutrophic lake	370–640

and salinity stratification, which separate the shallow- and deep-water masses. In more productive coastal and upwelling areas of the oceans, these nutrients are rapidly sequestered by phytoplankton in the photic zone and rise through the various trophic levels in the food chain. The combined photosynthetic process to convert inorganic carbon and these nutrients to biomass and oxygen in diatoms is as follows (Redfield, 1942):

$$106\,CO_2 + 16\,HNO_3 + H_3PO_4 + 122\,H_2O \xrightarrow{\text{light}}$$
$$(CH_2O)_{106}(NH_3)_{16}H_3PO_4 + 138\,O_2$$

Upwelling occurs mainly in shallow ($<200\,m$) coastal waters and is the principal mechanism where deeper, nutrient-rich waters are recycled into the photic zone. Upwelling areas constitute $\sim 0.1\%$ of the ocean area, but have elevated productivities due to a combination of nutrient supply from land run-off and efficient vertical recycling of nutrients (Table 1.4). For example, upwelling and recycling of nitrates and phosphates in offshore Peru is related to wind stress. Prevailing winds push surface water northward at the same time that the Coriolis force deflects it to the west, resulting in overturn and upwelling of cold, nutrient-rich subsurface water. Many Phanerozoic phosphorites were deposited in low-latitude upwelling zones on continental shelves, such as the Permian Phosphoria Formation (e.g. Claypool *et al.*, 1978; Maughan, 1993).

Chlorophyll plays a critical role in converting the energy of light to chemical energy stored in the products of photosynthesis. This stored energy is released when the products are brought together and undergo combustion or respiration. Forest fires and warm muscles caused by exertion both utilize this type of stored energy.

All photosynthetic organisms have some type of chlorophyll. Chlorophyll a (e.g. see Figure 3.23) is the principal chlorophyll in higher plants, most algae, and cyanobacteria. Other chlorophylls absorb light in different ranges of wavelength. For example, purple and green bacteria contain bacteriochlorophylls. The various chlorophylls differ mainly in the alkyl groups attached to the tetrapyrrole ring. Chlorophylls tend to absorb light most efficiently in the red portion of the visible spectrum. Because red wavelengths do not penetrate deeply into water, aquatic plants commonly use different carotenoids as accessory pigments to increase the range of wavelengths that can be used for photosynthesis. Accessory pigments, such as carotenoids and xanthophylls, are involved at least indirectly with the capture of light and occur with chlorophylls in many phototrophic organisms. Biomarkers derived from chlorophylls, bacteriochlorophylls, and accessory pigments in living autotrophs occur in ancient rocks and petroleum and can be used to help identify past sources of organic matter.

SECONDARY PRODUCTIVITY

Much of the primary organic matter created by photosynthetic or chemosynthetic autotrophs undergoes degradation by other organisms during the secondary production of organic matter. Heterotrophic microbes in the water column, sediments, and rock continually degrade and rework primary aquatic and terrigenous organic matter. Heterotrophic microbes reduce the mass of primary organic matter preserved in sediments, while contributing some of their own characteristic secondary organic matter. For example, methane can be produced from primary organic matter in surficial anoxic sediments by methanogenic bacteria (Figure 1.1). However, these methanogens also create various characteristic compounds, such as 2,6,10,15,19-pentamethylicosane and squalane, which remain in the sediments (e.g. see Figure 2.15). Carbon dioxide is a product of various organisms that utilize fermentation, anaerobic respiration, or aerobic respiration to produce energy. As another example, when methane moves into oxic environments, methylotrophic bacteria can oxidize it to CO_2 (Figure 1.1). Details on these various organisms can be found in any modern discussion of microbial metabolism (e.g. Brock and Madigan, 1991).

Fossil fuels represent organic matter that has been temporarily removed from the carbon cycle due to burial in the lithosphere. Sedimentation and early burial normally result in the complete destruction of primary organic matter by heterotrophic degradation and other processes, such as chemical oxidation. One can view all fossil fuels, including finely disseminated organic matter, coal, and petroleum, as excess organic carbon that escaped an otherwise remarkably efficient carbon cycle due to burial (Figure 1.1). This excess organic carbon represents a small leak in the carbon cycle. It is estimated that during Earth history, 0.1% of the carbon produced as biomass by plants (primary productivity) becomes preserved in sediments and available for petroleum-related processes (Tissot and Welte, 1984). About 0.6% of the organic carbon produced in marine basins is preserved (Hunt, 1996, p. 113). Metamorphism, uplift, erosion, and combustion of fossil fuels eventually return stored organic carbon to the cycle. Organic matter, including biomarkers, is best preserved in sediments under special circumstances, as described below.

PRESERVATION OF ORGANIC MATTER

Biomarkers exist because their basic structures remain intact during sedimentation and diagenesis (Figure 1.2). In this book, the term "diagenesis" refers to the biological, physical, and chemical alteration of organic matter in sediments prior to significant changes caused by heat (typically $<50\,^{\circ}$C).

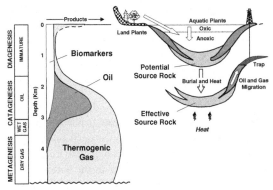

Figure 1.2. Generalized evolution of organic matter during and after sedimentation. Many biomarkers survive diagenesis and much of catagenesis prior to their complete destruction during late catagenesis and metagenesis (left). Depth scale can vary depending on different factors, including geothermal gradient and type of organic matter. High primary productivity by plants, anoxia (lack of oxygen), and other factors favor preservation of organic matter during diagenesis (right). Potential source rocks can become effective source rocks during catagenesis. Reprinted with permission by ChevronTexaco Exploration and Production Technology Company, a division of Chevron USA Inc.

Catagenesis is the process by which the organic matter in rocks is thermally altered by burial and heating at temperatures in the range of $\sim$50–150 $^{\circ}$C under typical burial conditions requiring millions of years. During catagenesis, biomarkers undergo structural changes that can be used to gauge the extent of heating of their source rocks or oils expelled from these rocks. Furthermore, because biomarkers represent a distinctive set of contributing organisms, their distribution in an effective source rock is a fingerprint that can be used to relate the rock to expelled crude oil that may have migrated many kilometers. At temperatures in the range of $\sim$150–200 $^{\circ}$C, prior to greenschist metamorphism, organic molecules are cracked to gas in the process called metagenesis. Biomarkers are severely diminished in concentration or completely destroyed because of their instability under these conditions. Deep hydrocarbon gas accumulations can result from (1) cracking of highly mature particulate organic matter, (2) decomposition of residual oil in petroleum source rocks, and (3) secondary cracking of oil in reservoir rocks (Lorant and Behar, 2002). No convincing evidence has been found for deep, commercial methane accumulations that originated by inorganic processes, such as mantle outgassing (see Chapter 9).

Biomarkers are powerful geochemical tools because many are highly resistant to biodegradation. For example, biodegraded seep oils and asphalts commonly contain unaltered biomarkers that can be used for comparisons with non-biodegraded oils.

ORGANIC COMPONENTS IN ROCKS

In sedimentary rocks that have undergone diagenesis, the organic matter consists of kerogen, bitumen, and minor amounts of hydrocarbon gases. Kerogen is non-hydrolyzable particulate organic debris that is insoluble in organic solvents and consists of mixtures of macerals and reconstituted degradation products of organic matter (Durand, 1980). Macerals are recognizable remains of different types of organic matter that can be differentiated under the microscope by their morphologies (Stach *et al.*, 1982; Taylor *et al.*, 1998). They are analogous to minerals in a rock matrix, but they differ in having less well-defined chemical compositions. Part of the kerogen represents reconstituted low-molecular-weight degradation products of biomass (e.g. Durand, 1980; Tissot and Welte, 1984). In addition, Tegelaar *et al.*

(1989) list recognized biomacromolecules (e.g. cellulose, proteins, and tannins) and rank their relative potential for preservation by incorporation into kerogen through cross-linkage during diagenesis. Transmission electron microscopy reveals that many kerogens described as amorphous using standard microscopy actually consist of bundles of thin lamellae (~10–30 nm) called ultralaminae, which represent insoluble biomacromolecules inherited from the outer cell walls of microalgae (Largeau *et al.*, 1990; Derenne *et al.*, 1993). Kerogen macerals generally reflect local (autochthonous) primary and secondary production of organic matter. However, many sediments contain appreciable transported (allochthonous) organic matter, such as land-plant detritus or pollen, soot from fires, or soil carried by rivers or wind into marine or lacustrine settings (Gogou *et al.*, 1996).

Bitumen consists of *in situ* hydrocarbons and other organic compounds that can be extracted from fine-grained sedimentary rocks using organic solvents. Biomarkers occur both free in the bitumen and chemically bound to the kerogen in petroleum source rocks. They also occur in crude oils that migrate from the fine-grained source rocks to reservoir rocks. Petroleum is a general term that refers to solid, liquid, and gaseous materials composed dominantly of chemical compounds of carbon and hydrogen, including bitumen, hydrocarbon gases, and crude oil. Petroleum accounts for ~63% of the total worldwide energy consumption (US Geological Survey, 2000).

The quantity and quality of organic matter preserved during diagenesis of sediment ultimately determine the petroleum-generative potential of the rock. Various factors play a role in the preservation of organic matter during sedimentation and burial, notably organic productivity, oxygen content of the water column and sediments, water circulation, sediment particle size, and sedimentation rate (Demaison and Moore, 1980; Emerson, 1985; Meyers, 1997). For example, anoxia, sediment size, primary productivity, organic source material, and contemporaneous diagenesis in the water column all contribute to organic matter accumulation and preservation in organic-rich shales from the Upper Devonian to Lower Carboniferous Exshaw Formation of southern Alberta (Caplan and Bustin, 1996). The relative importance of these factors remains controversial and probably differs between depositional environments.

Table 1.5. *Common terminology describing the redox conditions in sedimentary environments and the metabolism of the corresponding microbial populations (biofacies)*

Depositional environment	Microbial biofacies	Oxygen content (ml/l)	
		Rhodes and Morse (1971)	Tyson and Pearson (1991)
Oxic	Aerobic	>1	2–8
Dysoxic	Dysaerobic	0.1–1	0.2–2
Suboxic	Quasi-anaerobic	0.1–1	0–0.2
Anoxic	Anaerobic	0	0

OXIC VERSUS ANOXIC DEPOSITION

Aerobic microbes rapidly oxidize organic matter from dead plants or animals under atmospheric conditions. Aqueous sedimentation of organic matter can occur under various redox conditions, controlled mainly by the availability of molecular oxygen. The different redox conditions and their corresponding microbial biofacies (determined by metabolism) can be described using the terms in Table 1.5.

Under an oxic water column, aerobic bacteria and other organisms degrade organic matter settling from the photic zone (Figure 1.4, left). Normal seawater contains ~6–8 ml oxygen/liter. Respiratory processes create biological oxygen demand (BOD). If sufficient organic matter remains after exhausting all available oxygen, then anaerobic organisms continue to oxidize the organic matter by using other oxidants, such as nitrate or sulfate. The boundary between aerobic and anaerobic metabolism (oxic versus anoxic environments) can occur in the water column or in the bottom sediments. Benthic sediments that contain interstitial oxygen are commonly bioturbated by metazoa, including multicellular burrowing organisms, such as clams and worms. Oxic sediments that are preserved in the geologic record have massive textures without lamination.

Lakes and oceanic basins can become depleted in oxygen (stagnant) due to the combined effects of BOD and limited recharge by oxygenated water. Water recharge is controlled by various factors, including basin geometry, water temperature, and salinity gradients. For

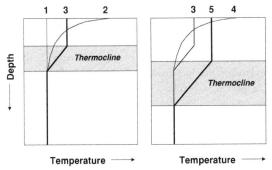

Figure 1.3. Schematic showing the formation of a thermocline in a body of water. If air temperature rises above water temperature for water having constant temperature versus depth (1), then the surface water begins to warm (2). Mixing of surface waters by wind lowers surface temperature, resulting in formation of an isothermal layer having higher temperature than the thermocline and deeper water (3, left). Further warming (4) and mixing of the surface water result in net downward transport of heat and a deeper, and commonly thicker, thermocline (5).

example, a thermocline is a layer of water where the temperature decrease with depth is greater than that in the underlying and overlying water (Figure 1.3). Thermoclines can develop in lakes and the ocean when the air temperature is high and the shallow water receives more heat than it loses by radiation and convection. The surface water warms and a shallow negative temperature gradient develops. Mixing of the surface waters by wind may lower the surface temperature, but the net effect is downward transport of heat and formation of an isothermal layer that is warmer than the underlying water. A strong thermocline thus develops between the isothermal surface layer and cooler water below. In the oceans, warm surface waters generally extend to 150–300-m depth. The underlying thermocline varies from 300 to 900 m thick. Temperature decreases more slowly below the thermocline and generally reaches 1–4 °C near the ocean bottom.

Equatorial lakes, such as Lake Tanganyika in the East African rift-lake system, are particularly prone to strong thermoclines because of humid climate without significant seasonal temperature fluctuations (Demaison and Moore, 1980). Oxygen lost from the deep water due to BOD is not readily replaced by mixing with shallower water due to the thermocline, and thus anoxia tends to develop. Lake Tanganyika is ~1500 m deep and anoxic conditions occur below ~150 m. Laminated, organic-rich marlstones of the

Eocene Green River Formation in Colorado and Utah are an ancient example of source rocks deposited in a large anoxic lake. Other examples include non-marine Lower Cretaceous source rocks from China (Songliao Basin), Brazil (Lagoa Feia Formation), and West Africa (Bucomazi Formation).

Influx of fresh water into a silled marine basin with little evaporation can result in a halocline or density-stratified water column, especially when deep saline water is isolated from the open ocean water by the sill. The Black Sea is essentially a silled marine basin. Excess fresh water from riverine input flows from the Black Sea across a 27-m deep sill at the Bosphorus into the Mediterranean Sea. This positive water balance results in low-salinity surface water compared with deeper, more saline water in the Black Sea and a permanent halocline at depth. The halocline is also a chemocline, which marks the boundary between oxic and anoxic conditions. The present-day chemocline occurs within the photic zone at 80–100-m depth, where hydrogen sulfide first appears and oxygen disappears. Distributions of isorenieratene and related compounds in Black Sea sediments indicate that photosynthetic green sulfur bacteria (*Chlorobiaceae*) were active in the Black Sea for at least 6000 years and that penetration of the photic zone by anaerobic water is not a recent phenomenon (Sinninghe Damsté *et al.*, 1993b). Ancient examples of anoxic marine source rocks that were deposited under similar conditions include the Upper Jurassic of West Siberia (Bazhenov Formation) and the North Sea (Kimmeridge hot shales) and the latest Albian in North America (Mowry Shale).

Under anoxic or suboxic water (less than ~0.2 ml oxygen/l water), aerobic degradation of organic matter is severely reduced because both metazoa (multicellular aerobic organisms) and aerobic bacteria generally require higher levels of oxygen (Figure 1.4, right). Below ~0.1 ml of oxygen/l water, bioturbation of the benthic sediments does not occur due to the absence of metazoa, leaving only anaerobic bacteria and possibly some benthic foraminifera to rework the organic matter. The general sequence of oxidants used by benthic organisms is as follows: interstitial oxygen, nitrate, Mn^{4+} oxides, Fe^{3+} oxides, and sulfate (Froelich *et al.*, 1979; Schulz *et al.*, 1994). Sulfate reduction is generally the dominant form of respiration after the onset of anaerobic conditions in marine settings because sulfate is plentiful in seawater (0.028 M) (Rice and Claypool,

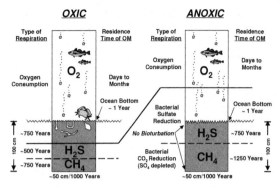

Figure 1.4. Oxic (left) and anoxic (right) depositional environments generally result in poor and good preservation of deposited organic matter, respectively (after Demaison and Moore, 1980). The solid horizontal line separates oxic (above) from anoxic (below). In oxic settings, bottom-dwelling metazoa bioturbate the sediments and oxidize most organic matter. In anoxic settings, especially where the oxic-anoxic boundary occurs in the water column, bottom-dwelling metazoa are absent and sediments are not bioturbated.

1981). Lack of bioturbation allows the development of fine laminations recording depositional cycles, which are commonly observed in effective petroleum source rocks. For example, glacial varves in fjords record yearly depositional cycles that typically include a layer of light-colored sand or silt grading upward into a thinner layer of dark-colored organic-rich clay. Laminated sediments in Saanich Inlet, a silled, fjord-like setting in British Columbia, contain up to ~9 wt.% organic matter (Nissenbaum et al., 1972). Euxinic sediments are deposited under marine, anoxic conditions with free hydrogen sulfide (H_2S) produced by sulfate-reducing bacteria (Raiswell and Berner, 1985).

Traces of animal activity provide information on water chemistry during sediment deposition (Savrda and Bottjer, 1986). For example, certain Lower to Middle Triassic mudrocks from the Barents Sea contain few ichnofossils (trace fossils) other than very shallow *Helminthoides* burrows consisting of small, monospecific horizontal mining and grazing traces (Isaksen and Bohacs, 1995). The limited evidence for burrowing suggests that these mudrocks were deposited in dysoxic to anoxic conditions, resulting in improved preservation of organic matter, high organic carbon content, and high algal/amorphous organic matter. Savrda (1995) summarized relationships between the degree of oxicity and burrow type or density in marine sediments. However, systematic relationships between redox conditions and bioturbation are poorly documented for lacustrine sediments (Harris et al., 1998).

Anaerobic degradation of organic matter is thermodynamically less efficient than aerobic degradation (Claypool and Kaplan, 1974). This observation supports the prevailing belief that anoxia is the main cause for enhanced preservation of hydrogen- and lipid-rich organic matter in petroleum source rocks (Demaison and Moore, 1980). Where oxic conditions exist, the organic matter is largely destroyed during sedimentation and diagenesis, even when organic productivity is high. For example, most polar regions in the modern oceans have high primary productivity, but the oxic bottom sediments are low in organic carbon.

Pederson and Calvert (1990) and Calvert and Pederson (1992) contend that organic productivity rather than anoxia is the major control on the accumulation of organic-rich marine sediments. They cite references, based primarily on laboratory incubation experiments, suggesting that the rates of destruction of organic matter under oxic and anoxic conditions are similar and cannot be used to invoke enhanced preservation of organic matter under anoxic conditions. In an example from the central Gulf of California, they observed no increased carbon content in alternating anoxic versus oxic sediments (Calvert, 1987). Data from anoxic sediments in the Black Sea suggest that organic carbon accumulation rates are not unusually high compared with equivalent, oxygenated environments (Calvert et al., 1991).

The worldwide distribution of modern organic-rich sediments does not correlate clearly with that of high productivity in the overlying water column (Demaison and Moore, 1980). For example, surface waters around Antarctica show high productivity, yet the underlying sediments are organic-lean because of strong circulation of cold, oxygen-rich waters that effectively satisfy all BOD imposed by settling organic matter. Modern organic-rich sediments generally occur where high productivity and bottom-water anoxia coincide.

We operate under the working assumption that anoxia is important in the development of organic-rich petroleum source rocks. Some arguments suggest that this is reasonable:

- Even if equivalent rates of aerobic and anaerobic degradation occur in laboratory incubation experiments, anaerobic reactions in nature may slow or stop prior to completion due to lack of sufficient oxidants, such as sulfate. Because of bioturbation, oxic sediments are ventilated better than anoxic sediments and depleted oxygen is replaced rapidly. Anoxic sediments approach closed systems.

• Pedersen and Calvert (1990) discuss the effects of oxic versus anoxic deposition on the quantity rather than quality of preserved organic matter. Evidence suggests that anoxic conditions favor preservation of hydrogen-rich, oil-prone organic matter. For example, like Calvert (1987), Peters and Simoneit (1982) observed similar total organic carbon content in alternating laminated (anoxic) and homogeneous (oxic) diatomaceous oozes in the Gulf of California. The laminated zones in these sediments contain more hydrogen-rich organic matter, as indicated by higher Rock-Eval pyrolysis hydrogen indices and lower oxygen indices than the homogeneous zones.

• If anoxia is unimportant in the preservation of organic matter, then it is difficult to explain the general correspondence between oil-prone, organic-rich petroleum source rocks and faunal or sedimentologic features indicating anoxia. Most source rocks are laminated and lack evidence for active infauna (Isaksen and Bohacs, 1995).

• Extracts from petroleum source rocks have biomarker and supporting parameters that indicate anoxic conditions (e.g. high vanadium/nickel porphyrin, low pristane/phytane, and high C_{35} homohopane indices).

Differences in the type of organic matter and the characteristics of the depositional environment result in lateral and vertical variations of organic facies within the same source rock. An organic facies is a mappable subdivision of a stratigraphic unit that can be distinguished on the basis of the organic components (Jones, 1987). Progress in our understanding of paleooceanographic controls on petroleum source rock deposition, as discussed above, has improved regional mapping of organic facies and our ability to predict favorable areas for further exploration (e.g. Demaison *et al.*, 1983). Biomarker compositions can be used to distinguish oils from different source rocks but are also useful in showing regional variations in organic facies within the same source rock or within oils from the same source rock. These applications are possible because the biomarker patterns in oils are inherited from their respective source rocks.

SEDIMENTATION RATE AND GRAIN SIZE

Sedimentation rate and grain size also play key roles in the preservation of organic matter. Because of similar hydraulic behavior, organic matter is preferentially deposited with fine-grained mud. Unlike sand, fine-grained mud more readily excludes oxygen-rich water below the sediment–water interface, thereby enhancing anoxia when it develops. In anoxic marine silled basins and anoxic lakes, organic matter and fine-grained sediments become concentrated in concentric or bull's-eye patterns in deep quiet water near source-rock depocenters where sediment thickness is the greatest (Huc, 1988a). Modern examples of this concentric distribution of organic carbon include the Black Sea and the Caspian Sea, while ancient examples include the Upper Jurassic in West Siberia (Figure 1.5), the lower Jurassic of the northern North Sea, and the lower Toarcian and Hettangian/Sinemurian of the Paris Basin (see Figure 18.62). Anoxic settings range from ~1 wt.% to more than 20 wt.% total organic carbon (TOC). Turbidity currents and related gravity flows may complicate the above concentric distributions because they transport organic matter or sediments into deep water along pathways not predicted by simple depositional models.

In deltaic or paralic settings dominated by terrigenous detritus, organic matter has a different pattern of increasing concentration away from the source of the organic matter at the edge of the basin toward deep water. Modern examples include the Mississippi and Mahakam deltas, while ancient examples include the Miocene Mahakam Delta and the Lower-Middle Jurassic in West Siberia (Figure 1.6). Although most deltaic settings are strongly oxic, organic matter can be preserved if it is buried rapidly, effectively removing it from attack by metazoans (Figure 1.7).

Sedimentation rate co-varies with total organic carbon content in fine-grained bottom sediments deposited under oxic conditions (Figure 1.7, bottom). Data from modern and ancient sediments suggest that rapid burial improves preservation by reducing the residence time of organic matter in the zones of bioturbation by metazoa and degradation by aerobic bacteria. TOC approximately doubles with each ten-fold increase in sedimentation rate in oxic settings (Müller and Suess, 1979; Ibach, 1982; Stein, 1986). Less than 0.01% of primary

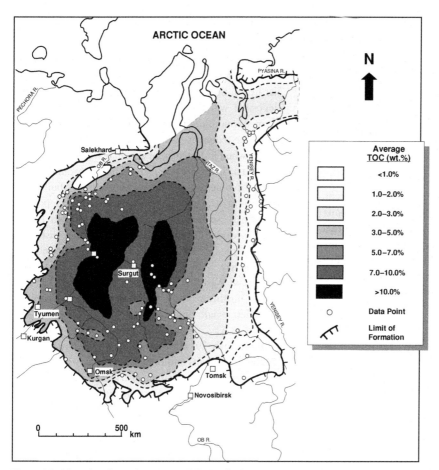

Figure 1.5. Map of total organic carbon (TOC) distribution for marine shales of the Upper Jurassic Bazhenov Formation in West Siberia, deposited in a large anoxic silled basin (modified from Kontorovich, 1984). Organic richness increases toward the basin center in a concentric pattern. Reprinted by permission of the AAPG, whose permission is required for further use.

productivity is preserved in slowly accumulating pelagic sediments of the Central Pacific (2–6 mm/1000 years). About 0.1–2% of primary productivity is preserved in more rapidly accumulating hemipelagic sediments offshore Namibia in northwest Africa or Oregon (2–13 cm/1000 years). About 11–18% of primary productivity is preserved in rapidly accumulating hemipelagic sediments from offshore Peru and the Baltic Sea (66–140 cm/1000 years) (Müller and Suess, 1979). The high percentage of organic matter preserved in sediments from offshore Peru and the Baltic Sea may be due partly to intermittent anoxic conditions. High sedimentation rates, such as those in the hypothetical examples in Figure 1.4 (50 cm/1000 years), are typical of many coastal marine and lake settings that preserve

sufficient organic matter to classify as potential source rocks.

Preservation of organic matter in anoxic settings is enhanced compared with oxic settings given the same sedimentation rate. For example, preserved TOC in anoxic Black Sea sediments is equivalent to 4–6% of primary production and is higher by a factor of five to six than that predicted for oxic sediments deposited at the same rate of 10–20 cm/1000 years (Müller and Suess, 1979).

The relationship between sedimentation rate and TOC is complicated by various other factors, such as lithology, water depth, distance from shore, and organic matter type (Ibach, 1982; Pelet, 1987; Waples, 1983). Nonetheless, the correlation can be used in

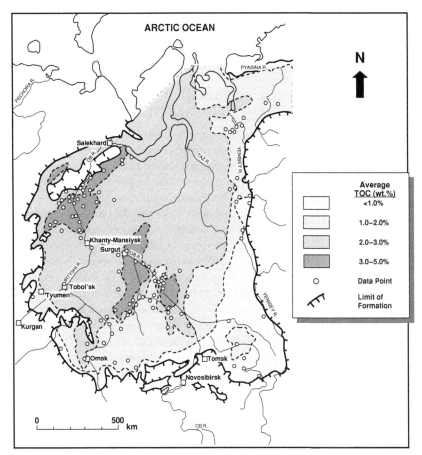

Figure 1.6. Map of total organic carbon (TOC) distribution for Lower-Middle Jurassic shales, including the Middle Jurassic Tyumen Formation in West Siberia (modified from Kontorovich, 1984). Organic richness generally decreases from the basin edge and main source of terrigenous organic matter in the west toward the east. Reprinted by permission of the AAPG, whose permission is required for further use.

frontier regions to predict source-rock richness. For example, sedimentation rates can be determined directly from seismic data by measuring the thickness of sediment between key chronostratigraphic horizons. Sedimentation-rate maps can be produced from seismic isopach maps and converted to TOC maps if the source rock lithology is known or inferred (Ibach, 1982). Some evidence suggests that the detrital mineral component may dilute organic carbon at very high sedimentation rates (>500 cm/1000 years), despite good preservation (Ibach, 1982; ten Haven, 1986; Hedges and Keil, 1995). Sedimentation rate has little influence on organic matter preservation in anoxic settings because of the lack of a shallow sediment zone of intense bioturbation and aerobic degradation (Figure 1.7, top).

Tyson (2001) used multiple regression analysis of modern marine sediment data to conclude that sedimentation rate enhances TOC preservation at only ≤5 cm/1000 years in oxic regimes. Dilution of TOC occurs at higher sedimentation rates if the carbon supply remains constant. Tyson contends that the observed positive correlation of sedimentation rate and TOC at >5 cm/1000 years is an artifact caused by including data from shallow sediment depths and/or higher productivity. Above 35 cm/1000 years, Tyson observed no appreciable difference in TOC between oxic and anoxic regimes, but they differed strongly for sedimentation rates below 10 cm/1000 years. The model indicates that oxygen is a significant variable in addition to sedimentation rate and productivity for determining TOC

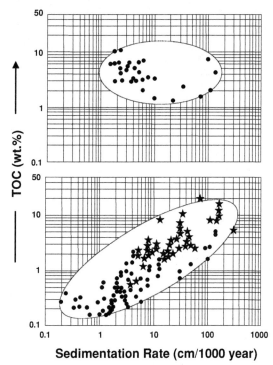

Figure 1.7. Schematic comparison of sedimentation rate versus total organic carbon in sediments (modified from Stein, 1986; Pelet, 1987; and Huc, 1988b). The degree of organic carbon preservation is independent of sedimentation rate in anoxic settings (top). Under oxic conditions, faster burial rates more effectively isolate organic carbon from degradation by burrowing metazoa, resulting in better preservation (bottom). Stars = high-productivity depostional settings.

because it predicts 2.5–4.0 times higher TOC in anoxic compared with fully oxic facies.

> **Note:** Kennedy *et al.* (2002) challenged the generally accepted mechanisms for preservation of organic matter in marine shales and mudrocks, which involve deposition of discrete organic particles with clastic mineral grains that have similar hydraulic characteristics. They present evidence that the bulk of the organic matter in certain black shales consists of dissolved organic matter from the water column, which adsorbed on or within detrital smectite and mixed-layer illite-smectite clays, thus facilitating burial and preservation. Their results imply that organic carbon preservation may be tied more closely to patterns of continental weathering and clay

mineralogy than to primary productivity or anoxia. This prediction needs to be tested for a large number of clastic source rocks worldwide.

LACUSTRINE VERSUS MARINE DEPOSITIONAL SETTINGS

The sedimentary environments of lakes and oceans differ in certain respects, resulting in different types and quantities of organic matter. Lakes are smaller than the deep ocean basins and receive proportionally more terrigenous clastic detritus. Likewise, land-derived nutrients are commonly more abundant in lakes compared with the deep oceans, enhancing primary productivity. Sedimentation rates in lakes (~1 m/1000 years) commonly exceed those in the deep oceans (~1–10 cm/ 1000 years) and organic matter is buried more rapidly, thus enhancing preservation. Coastal marine sediments are more similar to lake sediments because they receive large proportions of terrigenous organic matter and are deposited relatively rapidly (~10–100 cm/1000 years). Lacustrine rocks commonly contain up to several tens of percent TOC, while deep ocean sediments contain only a few tenths of a percent. Lacustrine benthic organisms are less diverse, and the depth of bioturbation is less than that of marine fauna (Meyers, 1997). Dissolved sulfate is a major ion in seawater, but is normally low or absent in lakes. Thus, sulfate reduction is important in the microbial reworking of marine but not lacustrine organic matter. Carroll and Bohacs (2001) proposed a three-fold lithofacies classification that accounts for the most important features of lacustrine petroleum source rocks. These include the fluvial-lacustrine, fluctuating profundal, and evaporative lithofacies, which correspond to their algal-terrestrial, algal, and hypersaline algal organic facies. For example, all three lacustrine facies occur in the Eocene Green River Formation in Wyoming and the Upper Permian non-marine facies of the southern Junggar Basin in China.

TEMPORAL AND REGIONAL DISTRIBUTIONS OF SOURCE ROCKS

Effective source rocks are not distributed evenly in the geologic record (Figure 1.8). Some depositional controls favorable for good source rocks include major marine transgressions, warm equable climate, and anoxia.

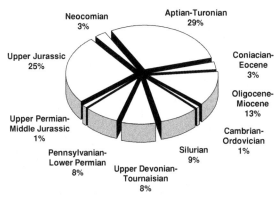

Figure 1.8. Effective source rocks are distributed unevenly through time (data from Klemme and Ulmishek, 1991). Percentages refer to the relative contribution of the periods or epochs of time to the total estimated ultimate recovery of petroleum worldwide. This figure discriminates against older source rocks (e.g. Cambrian) because they are more likely to have undergone deep burial and loss of generated petroleum. The appendix contains geological time charts.

Some of the most favorable times for source rock deposition include the Silurian, Late Devonian (i.e. Frasnian Epoch), Late Jurassic, and Cretaceous Periods (i.e. Aptian-Turonian Epochs). The temporal distribution of source rocks largely controls the present-day geography of petroleum reserves. For example, many of the world's significant petroleum source rocks occur within countries that are members of the Organization of Petroleum Exporting Countries (OPEC). The re-

gional distribution of the worlds' major petroleum systems outside of the USA is described in Table 18.2 and Figure 18.4 .

The US Geological Survey conducted a worldwide assessment to evaluate the amount of petroleum yet to be found (US Geological Survey, 2000). The study divided the world into ~1000 petroleum provinces based primarily on geologic factors and analyzed 159 petroleum systems. The assessed provinces account for ~95% of worldwide historic petroleum production. Outside of the USA, ~75% and 66% of the worldwide conventional oil and gas resources, respectively, have been discovered. For these areas, ~20% of the conventional oil and 7% of the conventional gas endowment were produced as of 1995. About 1634 billion barrels of oil equivalent (BBOE) of conventional petroleum, including crude oil, gas, and natural gas liquids (NGL) remain to be discovered outside of the USA. Of this amount, conventional oil is 649 billion barrels, natural gas is 778 BBOE, and NGL are 207 BBOE. Reserve growth, caused by advances in exploration and production technology, is expected to total an additional 612 billion barrels of oil, 551 BBOE of natural gas, and 42 BBOE of NGL. Excluding the USA, ~35.4% of the world's undiscovered conventional oil is in the Middle East and North Africa, 17.9% in the former Soviet Union, and 16.2% in Central and South America. The former Soviet Union accounts for 34.5% of undiscovered conventional natural gas, and the Middle East and North Africa account for 29.3%.

2 · Organic chemistry

A brief overview of organic chemistry includes explanations of structural nomenclature and stereochemistry necessary to understand biomarker parameters. The discussion includes an overview of compound classes in petroleum and concludes with examples of the structures and nomenclature for several biomarkers, their precursors in living organisms, and their geologic alteration products.

Carbon, hydrogen, oxygen, and nitrogen are the principal elements in living organisms. With the exception of oxygen, these elements are rare components in Earth's crust compared with silicon and the light metals. Several unusual characteristics of carbon make it the vital element around which the chemistry of life, organic chemistry, has evolved.

The atomic structure of carbon allows it to form more diverse compounds than any other element. Atomic orbital theory describes the approximate orientation of electron clouds around a single atom of carbon in the uncombined state (Figure 2.1). The outer electron shell of carbon consists of four electrons (i.e. carbon has a valence of four). Two electrons occupy the 2s orbital, which is spherical in shape. The remaining two electrons each occupy a different dumbbell-shaped 2p orbital, whose axis is at right angles to the other 2p orbitals. One of the 2p orbitals in carbon does not contain an electron. Each of the four electrons can be shared with other elements that are able to complete their electronic shells by sharing electrons to form covalent bonds.

A unique feature of carbon is that it can share electrons with other carbon atoms, resulting in large molecules dominated by carbon–carbon bonds. A few other elements contain four electrons in their outer shell and can form repetitive covalent bonds with atoms of the same element, but they are not stable under Earth conditions. Thus, although silicon can bond covalently to other silicon atoms, silicon–silicon compounds are not stable in Earth's atmosphere and readily oxidize to silica (SiO_2).

ALKANES: THE SIGMA BOND

When carbon combines with other atoms, the 2s and 2p electron orbitals hybridize into different orbital configurations. One 2s and three 2p orbitals can hybridize to form four equivalent sp3 orbitals (Figure 2.2a). Four sp3 electron orbitals that can be formed by the valence shell of carbon are directed away from the central carbon atom at angles of 109.5° from each other (Figure 2.2b). By sharing an electron with each of four hydrogen atoms, a single carbon atom can satisfy the valence requirement of eight electrons. Thus, four hydrogen atoms bound to carbon result in the highly stable compound methane, where the hydrogens are located at the corners of a tetrahedron with carbon in the center (Figure 2.2c). Methane is the simplest of the hydrocarbons, which contain only hydrogen and carbon.

In compounds like methane and ethane, where carbon atoms are bound exclusively by sp3-hybridized linkages, the carbon atoms are said to be saturated and the resulting strong bonds are known as sigma or covalent bonds. When a sigma bond is the only link between two atoms in a molecule, it is called a single bond. Stable molecules of carbon and hydrogen that contain only single bonds are called saturated hydrocarbons or alkanes.

ALKENES: THE PI BOND

The 2s and 2p electron orbitals of carbon can also hybridize into p and sp2 orbitals. In this configuration, the four valence electrons of carbon are divided between one p and three sp2 orbitals. The three sp2 orbitals, each containing one electron, are co-planar and are directed away from the central carbon atom at 120°

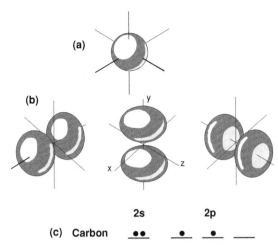

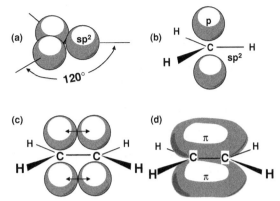

Figure 2.3. (a) Hybridized sp2 atomic orbitals of carbon lie on axes directed toward the corners of an equilateral triangle. (b) A methyl radical has a single p orbital occupied by one electron above and below the plane of the sigma bonds. (c) Adjacent p orbitals in ethylene interact to give (d) a pi bond (π cloud) above and below the plane of the sigma bonds (after Morrison and Boyd, 1987). Reprinted with permission by ChevronTexaco Exploration and Production Technology Company, a division of Chevron USA Inc.

Figure 2.1. Atomic orbitals in the outer shell of carbon showing: (a) the spherical 2s orbital, (b) the three orthogonal, dumbbell-shaped 2p orbitals, and (c) the distribution of electrons in the outer shell orbitals (after Morrison and Boyd, 1987). Reprinted with permission by ChevronTexaco Exploration and Production Technology Company, a division of Chevron USA Inc.

Such hydrocarbon (alkyl-) radicals are unstable species but can be intermediates in reactions of organic compounds.

When two p orbitals occur on adjacent atoms (e.g. C=C or C=0) a pi (π) or double bond results, as shown for ethylene in Figures 2.3c and 2.3d. Pi bonds typically are more reactive than sigma bonds. Thus, the carbon atoms in ethylene are linked by a double bond consisting of a sigma bond formed from the sp2-hybridized orbitals of each carbon atom and a pi bond formed from p orbital overlap. Hydrocarbons containing double bonds are unsaturated and include alkenes and aromatics.

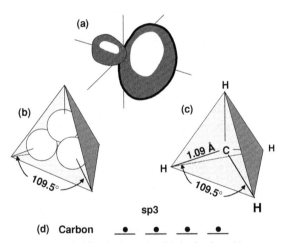

Figure 2.2. Hybridized sp3 atomic orbitals of carbon: (a) approximate shape of a single orbital, (b) four orbitals with the axes directed toward the corners of a tetrahedron, (c) shape and dimensions of a methane molecule (CH_4), an (d) distribution of electrons in the outer shell orbitals (after Morrison and Boyd, 1987). Reprinted with permission by ChevronTexaco Exploration and Production Technology Company, a division of Chevron USA Inc.

AROMATICS: BENZENE

Hydrocarbons can be divided into two broad classes: (1) aliphatic hydrocarbons, including alkanes and alkenes (discussed above), and (2) aromatic hydrocarbons. Although aromatic hydrocarbons contain pi bonds, most are highly stable because of electron de-localization, as discussed below.

Benzene is the simplest aromatic hydrocarbon (Figure 2.4), where the six carbon atoms are sp2-hybridized and connected in a flat hexagonal ring by sigma bonds. As in ethylene, each carbon atom has

from each other (Figure 2.3a). The p orbital contains the remaining electron and is oriented at 90° to the co-planar sp2 orbitals. The p orbital for the methyl radical in Figure 2.3b contains one electron.

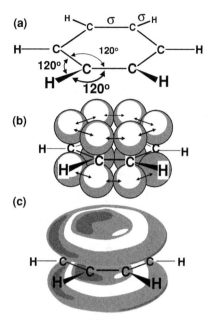

Figure 2.4. View of benzene, the simplest aromatic hydrocarbon, showing: (a) only the sigma (σ) bonds, (b) overlap of p orbitals to form pi bonds, and (c) clouds of delocalized pi electrons above and below the plane of the sigma bonds (after Morrison and Boyd, 1987). Reprinted with permission by ChevronTexaco Exploration and Production Technology Company, a division of Chevron USA Inc.

a single electron in the p orbital available for overlap in a pi bond with its neighbors. The pi bonds between carbon atoms in the benzene ring are equivalent. The six p electrons are shared equally or delocalized among the donating carbon atoms, forming a doughnut-shaped cloud of pi electrons above and below the ring. This arrangement of double bonds is more stable than an isolated double bond during diagenesis and catagenesis.

Various aromatic hydrocarbons occur in most petroleum, including benzene and toluene (one methyl group attached to a benzene ring), and ortho-, meta-, and paraxylenes (two methyl groups). Alkenes are rare or absent. Unlike the more stable aromatics, alkenes are readily hydrogenated to alkanes during deep sediment burial. In general, molecules containing functional groups, such as isolated double bonds, and hydroxyl (-OH), carboxyl (-COOH), and thiol (-SH) substituents, are more reactive than unsubstituted parent compounds and tend to be progressively eliminated during burial.

STRUCTURE NOTATION

Chemists use several different notations to describe the structures of organic compounds. The most precise, but complex, notation shows all outer shell electrons as a series of dots. In Figure 2.5a, butane is drawn in dot notation, where two electrons are shared in each sigma bond between carbon atoms or between each carbon and hydrogen atom. These two bonding electrons can be replaced by lines (Figure 2.5b) or can be omitted (Figure 2.5c). When the bonds are not shown, the proper combinations of atoms to satisfy valence requirements (stoichiometry) are assumed. For example, each carbon atom in a saturated hydrocarbon has four sigma bonds. The most abbreviated notation consists only of a zigzag pattern of lines (Figure 2.5d), which roughly depict a two-dimensional projection of the carbon skeleton of the molecule without the accompanying hydrogen atoms. Each angle and terminus of a line in this notation represents a single carbon atom with the appropriate number of hydrogens to satisfy valence requirements. Thus, the end of a line in all zigzag structures signifies a methyl group (-CH_3) unless other elements are indicated. The zigzag notation is preferred for complex organic molecules, including biomarkers, but is commonly used in combination with the other methods.

Figures 2.5e–2.5i show examples of notations for unsaturated hydrocarbons, including 1-butene and benzene. As discussed above, all carbon–carbon bonds

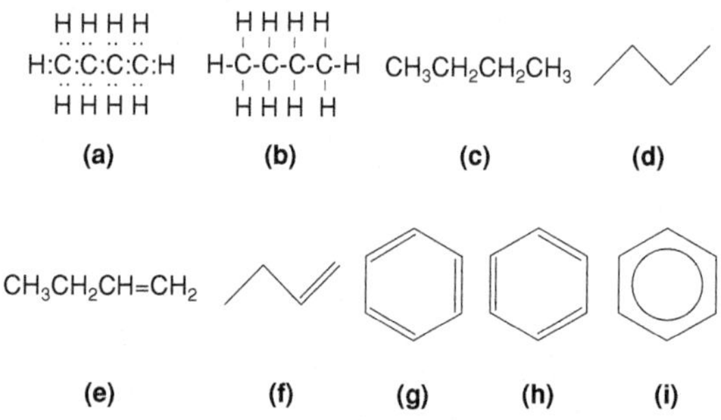

Figure 2.5. Examples of different structural notations used to describe organic compounds: (a) dot, (b) line, (c) formula, and (d) zigzag for butane; (e) formula and (f) zigzag for 1-butene; (g) and (h) two strictly non-equivalent line notations for benzene; (i) notation indicating delocalization of electrons on benzene (accounts for both (g) and (h)). Reprinted with permission by ChevronTexaco Exploration and Production Technology Company, a division of Chevron USA Inc.

in benzene are equivalent and neither Figure 2.5g nor Figure 2.5h is a true representation of the delocalized cloud of electrons in this structure. Figure 2.5i is commonly used to denote the delocalized electron cloud in benzene, although the notations in Figures 2.5g–i are all acceptable. Complex organic molecules typically require additional symbolic notation, particularly when stereochemical differences must be shown, as discussed below.

THREE-DIMENSIONAL PROJECTIONS TO TWO-DIMENSIONAL SPACE

Some attempt is usually made in structural images to convey the three-dimensional orientation of carbon atoms in organic molecules. The angles between sp^3 bonds that link carbon atoms in butane ($\sim109°$) are not readily depicted in a two-dimensional view (Figure 2.5d). Several common methods attempt to capture more accurately the three-dimensional form of molecules in two-dimensional projections. Three of these methods are shown in Figure 2.6. Stick, stick-and-ball, and space-filled projections are drawings of what one might construct using molecular model kits. These models attempt to accurately represent bond angles (stick), bond angles and radii (stick-and-ball), and bond angles, radii, and lengths (space-filled).

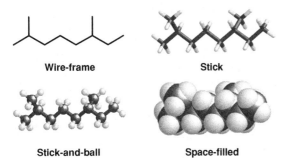

Wire-frame   **Stick**

Stick-and-ball   **Space-filled**

Figure 2.6. Structural notations for 2,6-dimethyloctane (a C_{10} isoprenoid hydrocarbon or monoterpane). The wire-frame or zigzag model shows only the carbon skeleton in two dimensions and attempts to honor bond angles. The other three representations are three-dimensional models projected as two-dimensional drawings. Carbon atoms are black, hydrogen atoms are gray. The stick model attempts to honor bond angles. The stick-and-ball model attempts to honor bond angles and radii. The space-filled model attempts to honor bond angles, radii, and lengths.

Table 2.1. *Selected acyclic alkane homologs, boiling points of the* n-*alkane isomers, and number of possible structural isomers*

Number of carbons (n)	Name	Formula	Boiling point (°C)	Possible isomers
1	Methane	CH_4	−164	1
2	Ethane	C_2H_6	−89	1
3	Propane	C_3H_8	−42	1
4	Butane	C_4H_{10}	0	2
5	Pentane	C_5H_{12}	36	3
6	Hexane	C_6H_{14}	69	5
7	Heptane	C_7H_{16}	98	9
8	Octane	C_8H_{18}	126	18
9	Nonane	C_9H_{20}	151	35
10	Decane	$C_{10}H_{22}$	174	75
11	Undecane	$C_{11}H_{24}$	195	159
12	Dodecane	$C_{12}H_{26}$	216	355
20	Eicosane	$C_{20}H_{42}$	343	366 319
30	Triacontane	$C_{30}H_{62}$	450	4×10^9

ACYCLIC ALKANES

Methane and ethane are the simplest compounds in a series of saturated hydrocarbons with the formula C_nH_{2n+2}. These compounds are called alkanes or paraffins. By increasing n, a homologous series of compounds results (Table 2.1). Compounds in the table with a linear arrangement of carbon atoms are called *normal* alkanes or *normal* paraffins (*n*-alkanes or *n*-paraffins, respectively). Compounds with a non-linear arrangement of carbon atoms are called isoalkanes, branched alkanes, or isoparaffins.

Isomers are compounds with the same molecular formula but different arrangements of their structural groups. Methane, ethane, and propane have no isomers. Butane, however, can be either of two compounds (isomers) with the molecular formula C_4H_{10} (Figure 2.7). The number of isomers increases exponentially with increasing number of carbon atoms in each compound. For example, there are >366 000 different possible positional isomers of eicosane ($C_{20}H_{42}$) (see Table 2.1). This does not include stereoisomers, e.g. enantiomers and diastereomers (see below). The number of possible

n-Butane

Isobutane (methylpropane)

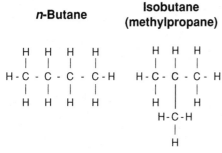

Figure 2.7. Butane (C_4H_{10}) occurs as two structural isomers. *n*-Butane (*n*C_4, left) is one homolog in the series of *n*-alkanes (*n*-paraffins), while isobutane (*i*C_4, right) is a homolog in the isoalkane series. Structures are depicted as atoms with lines for covalent bonds. Reprinted with permission by Chevron-Texaco Exploration and Production Technology Company, a division of Chevron USA Inc.

isomers would increase greatly if stereoisomers were included.

Formal rules specified by the International Union of Pure and Applied Chemistry (IUPAC) define the nomenclature for all hydrocarbons. For isoalkanes, the length of the longest continuous chain of carbon atoms determines the root name. Branching alkyl groups are termed methyl-, ethyl-, propyl- (and so on) for one, two, and three carbons, respectively. The positions of alkyl groups are numbered by counting along the carbon backbone in the direction that assigns the lowest possible carbon number to the side chains.

Figures 2.7 and 2.8 illustrate IUPAC rules for normal butane (C_4H_{10}) and heptane isomers (C_7H_{16}), respectively. As noted above, butane has two isomers, *n*-butane and 2-methylpropane. The latter compound is described using its informal name, isobutane, so frequently that this is a recognized IUPAC name. Other acyclic alkanes where informal names can be used as formal nomenclature include isopentane (2-methylbutane), neopentane (2,2-dimethylpropane), and isohexane (2-methylpentane). Heptane has nine acyclic isomers, all with the same formula of C_7H_{16}, which include *n*-heptane and eight different isoalkanes. If we remove one carbon from the linear chain and replace it with a methyl group (-CH_3), then the longest continuous carbon chain is only six carbons in length (hexane) and there are only two unique positions that the methyl group can be placed: on the second or third

carbon. These are correctly named 2-methylhexane and 3-methylhexane. Informal names for these compounds are isoheptane and anteisoheptane, respectively. Compounds named 4-methylhexane and 5-methylhexane do not exist because these are equivalent to 3-methylhexane and 2-methylhexane, respectively, if we begin numbering the carbon backbone from the other direction. If the longest chain in a C_7H_{16} hydrocarbon contains only five carbons (pentane), then five unique isomers can exist. Four of these isomers involve placing two methyl groups on various carbons; one isomer results from placing an ethyl (-C_2H_5) on the central carbon. If the ethyl group were placed at any location other than the third carbon, then the longest chain would contain six carbons, resulting in *n*-heptane or 2-methylhexane.

Note: To simplify discussions among chemists, both natural and synthetic compounds have been given names that define their chemical structures unambiguously. The IUPAC established formal rules in the *Nomenclature of Organic Compounds*, commonly known as the "*Blue Book.*" We urge the reader to become familiar with the IUPAC conventions. Advanced Chemistry Development,

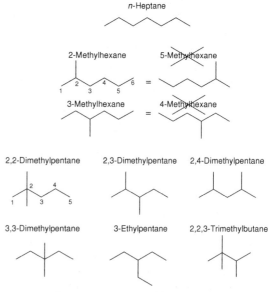

Figure 2.8. Structures and names of the nine acyclic heptane (C_7H_{16}) isomers. C-2 in *n*-heptane, 2-methylhexane, and 2,2-dimethylpentane is described as a secondary, tertiary, or quaternary carbon atom, respectively.

Inc. has published an Internet version with permission of the IUPAC at www.acdlabs.com/iupac/nomenclature, where the reader can find the conventions for naming all hydrocarbons, including biomarkers in petroleum and their biological precursors.

Chemical publications tend to honor IUPAC rules. However, many geochemical publications use informal names for organic compounds. This is particularly true for the larger biomarkers, where the IUPAC names can be ponderous. We use IUPAC nomenclature in this book, but we also use informal or incomplete naming conventions for readability.

ACYCLIC ALKENES

Alkenes are hydrocarbons that contain one or more double bonds. They are defined by the formula $C_nH_{2n+2-2z}$, where z is the number of double bonds. Hence, acyclic alkenes with one double bond have the formula C_nH_{2n}. The "z" number refers to the degree of unsaturation.

Alkenes have geometric isomers. Since the double bond is rigid, it imposes a spatial constraint that is absent in equivalent saturated hydrocarbons. Geometric isomers are compounds that can be interconverted only by rotation around a double bond or, as described below, a bond in a ring. For example, 2-butene is butane with one double bond between carbons 2 and 3. Since the double bond cannot rotate freely, the methyl groups on either side of the double bond can be either on the same sides or on opposite sides. When the methyl groups are on the same side, the compound is named *cis*-butene; when on opposite sides, the compound is named *trans*-butene (Figure 2.9). These molecules are chemically distinct and have different physical and chemical properties.

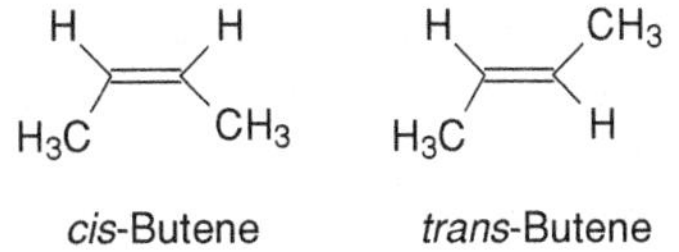

Many biomolecules contain one or more double bonds. Alkenes are stable enough to be found in low-maturity crude oils and rock extracts. Since double bonds are relatively reactive, they are absent or in trace quantities in most oils, except under unusual conditions involving radiogenic alteration (e.g. Frolov *et al.*, 1998) or entrainment of immature bitumen by migrating oil (e.g. Curiale, 1995).

MONOCYCLIC ALKANES

When two ends of an acyclic alkane join, they form a single ring. These compounds are termed monocyclic alkanes and follow the general formula C_nH_{2n}. Although cyclic alkanes containing almost any number of carbon atoms occur, only those with combinations of five (cyclopentyl) or six (cyclohexyl) carbon atoms are common in petroleum. The simplest of these hydrocarbons are cyclopentane (C_5H_{10}) and cyclohexane (C_6H_{12}), which are typically drawn as a flat pentagon and hexagon, respectively. Alkanes containing rings are sometimes called naphthenes.

Because carbon atoms in a ring cannot rotate freely, monocyclic alkanes exist as geometric isomers. Like the alkenes, alkyl side chains attached to monocyclic alkanes can extend from the same side of the plane of the ring (*cis*-configuration) or from opposite sides of the plane of the ring (*trans*-configuration) (see Figure 2.10). In wire-line illustrations, a bold wedge-shaped bond indicates that a group points out of the page. Conversely, a dashed wedge-shaped bond indicates that a group points into or behind the page.

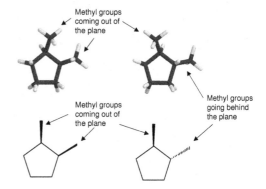

1-*cis*-2-Dimethylcyclopentane   1-*trans*-2-Dimethylcyclopentane

Figure 2.10. Stick (top) and wire-line (bottom) views of *cis*- and *trans*- geometric isomers of 1,2-dimethylcyclopentane.

H   H
 \   /
  C = C
 /   \
H₃C   CH₃

H   CH₃
 \   /
  C = C
 /   \
H₃C   H

cis-Butene   *trans*-Butene

Figure 2.9. Geometric isomers are compounds that can be interconverted only by rotation around a double bond or ring carbon.

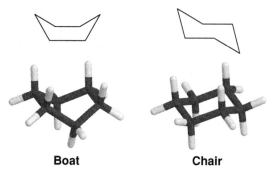

Boat   **Chair**

Figure 2.11. Three-dimensional projections (top) and stick models (bottom) of the boat and chair conformations of cyclohexane.

Cylcohexane is a puckered six-carbon ring that can assume several configurations termed chair, boat, and twisted (stretched) (Figure 2.11). Both the chair and the boat conformations are characterized by approximately tetrahedral carbon bond angles. While the chair form is quite rigid, the boat form is not and can flex into the twisted conformation without bond-angle deformation. The various conformations of cyclohexane are inseparable using techniques such as distillation because they interconvert very rapidly at room temperature.

MULTI-RINGED CYCLOALKANES

Hydrocarbons can have multiple carbon rings that are joined together. The monocyclic and multi-ringed hydrocarbons are known collectively as cycloalkanes, cycloparaffins, or naphthenes. These molecules have the same general formula as alkenes – $C_nH_{2n+2-2z}$ – where z is the number of rings and/or double bonds. Most cyclic biomarkers in crude oils are joined together in a staggered fashion (Figure 2.12). As discussed below, this is a consequence of the chemistry of their biological precursors.

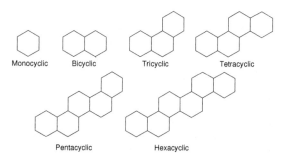

Monocyclic   Bicyclic   Tricyclic   Tetracyclic

Pentacyclic   Hexacyclic

Figure 2.12. Examples of multi-ringed cycloalkanes.

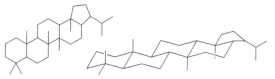

Figure 2.13. Hopane depicted as a wire-frame model (left) and as a projected three-dimensional structure (right). The six-carbon rings are all in the chair configuration.

Like cyclohexane, the rings in multi-ringed hydrocarbons can assume either chair or boat configurations. However, unlike cyclohexane, these configurations are locked together and the rings cannot interconvert easily. Ring configuration affects the shape of the molecule and is determined by enzymatic reactions that form biochemical lipids. Many petroleum biomarkers (e.g. hopanes and steranes) consist of multiple cyclohexane rings fused to a single cyclopentane ring in the most stable all-chair configuration. Some biomarkers, such as 17α-diahopanes, contain six-member rings that are stable in the boat conformation (Dasgupta *et al.*, 1995). These molecules usually are depicted using common two-dimensional wire-frame models, but they can be drawn as projections to show the three-dimensional character of the ring systems (Figure 2.13).

THE ISOPRENE RULE

Isoprene or methylbutadiene is the basic structural unit composed of five carbon atoms that is found in most biomarkers (Figures 2.14 and 2.15). Compounds composed of isoprene subunits are called terpenoids, isoprenoids, or isopentenoids (Nes and McKean, 1977). All organisms ranging from bacteria to humans require terpenoids. The isoprene rule states that biosynthesis of these compounds occurs by polymerization of appropriately functionalized C_5-isoprene subunits. Unlike other biopolymers, such as proteins and polysaccharides, terpenoids are not readily depolymerized because they are joined together by covalent carbon–carbon bonds. The biosynthesis of terpenoids is discussed in Chapter 3.

The saturated terpenoids lack double bonds and are divided into families based on the approximate number of isoprene subunits that they contain. The terpenoid families are composed of a wide variety of

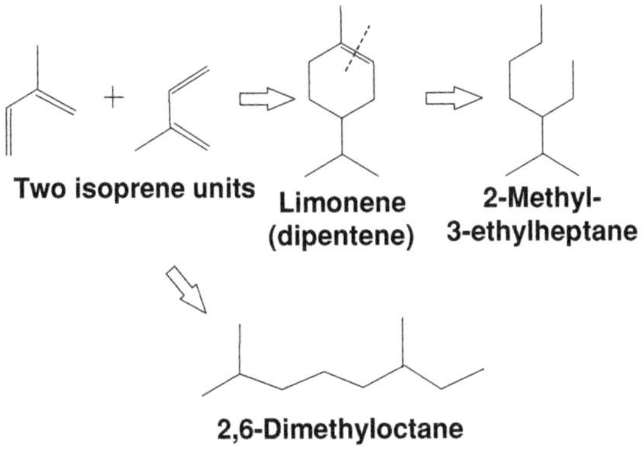

Two isoprene units   Limonene   2-Methyl-
(dipentene)   3-ethylheptane

2,6-Dimethyloctane

Figure 2.14. Proposed origin of two monoterpanes, 2,6-dimethyloctane and 2-methyl-3-ethylheptane, each by linkage of two isoprene subunits (after Mair *et al.*, 1966). These and many other low-molecular-weight hydrocarbons common in petroleum obey the isoprene rule and can be explained as thermal breakdown products of terpenoids. Note the head-to-tail linkage of isoprene subunits in 2,6-dimethyloctane. Reprinted with permission by ChevronTexaco Exploration and Production Technology Company, a division of Chevron USA Inc.

cyclic and acyclic structures (e.g. Devon and Scott, 1972; Simoneit, 1986), some of which are shown in Figure 2.15. Hemiterpanes (C_5), monoterpanes (C_{10}), sesquiterpanes (C_{15}), diterpanes (C_{20}), and sesterterpanes (C_{25}) contain one, two, three, four, and five isoprene subunits, respectively. Triterpanes and steranes (C_{30}) differ in structure but are composed of six isoprene subunits, while tetraterpanes (C_{40}) contain eight. Saturated terpenoids containing nine or more isoprene units ($C_{45}+$) are called polyterpanes.

> **Note:** Natural rubber is a polyunsaturated polymer of isoprene with a molecular weight in the thousands. Although found in various plants, most natural rubber is obtained from *Hevea brasiliensis*, a tree belonging to the *Euphorbiaceae* family from the Amazon that is now cultivated in Southeast Asia, Indonesia, and elsewhere. The Mayans from Central America were the first to collect and use rubber. They played a game whose objective was to bounce a rubber ball through a stone hoop high on a wall. Such ball courts are found at many Mayan archeological sites.

The original compounds found to obey the isoprene rule were called isoprenoids. The discovery of additional chemical structures made it clear that the linkage

between isoprene subunits can vary, with (1) head-to-tail (regular) linkage and (2) other (irregular) linkages occurring. Farnesane (Figure 2.15) is an example of a regular, acyclic (lacking rings) isoprenoid consisting of three head-to-tail linked isoprene units. Squalane and biphytane (also called bisphytane) (Figure 2.15) are examples of irregular acyclic isoprenoids. Squalane contains six isoprene units with one tail-to-tail linkage, while biphytane contains a head-to-head linkage.

Figure 2.15 shows other examples of head-to-tail or regular (Albaigés, 1980; Albaigés *et al.*, 1978), head-to-head (Moldowan and Seifert, 1979; Petrov *et al.*, 1990), tail-to-tail (Brassell *et al.*, 1981), and other irregular isoprenoids, which differ in the order of attachment of the isoprene subunits. Included in Figure 2.15 is an unusual case of three isomeric C_{40} compounds believed to originate from archaea (Albaigés *et al.*, 1985), which show head-to-head (biphytane), head-to-tail (regular C_{40}-isoprenoid), and tail-to-tail (lycopane) linkage. Volkman and Maxwell (1986) have reviewed the geochemistry of acyclic isoprenoids.

Degraded, altered, or homologous structures may still be categorized in their parent terpenoid family. The precise number of carbon atoms in a terpenoid family depends on differences in biochemistry of the source material, diagenesis, thermal maturity, and biodegradation. For example, the cyclic terpenoids cholestane (C_{27}), ergostane (C_{28}), and stigmastane (C_{29}) are three homologs in the sterane series (Figure 2.15). These compounds only approximate the isoprene rule because they do not contain an integral number of isoprene units, $(C_5)n$. However, steranes show some terpenoid character. Terpenoids that do not strictly obey the isoprene rule originate by biochemical or other reactions that cause the gain or loss of substituents.

Pristane is another example of a terpenoid that does not obey the isoprene rule. Pristane is classified as a diterpane, although it contains one less methylene group ($-CH_2-$) than phytane (C_{20}), the next largest homolog in the acyclic (linear) regular isoprenoid series. This series of regular isoprenoids extends from farnesane (C_{15}) through C_{16}, C_{17}, norpristane (C_{18}), and pristane (C_{19}) to phytane (C_{20}) (Figure 2.15). The C_{16}–C_{19} compounds can be considered degraded diterpanes derived from the C_{20} parent by consecutive loss of methylene groups. Conversely, it is equally correct to consider the C_{16}, C_{17}, and C_{18} compounds as extended homologs of farnesane by addition of methylene groups to the linear portion

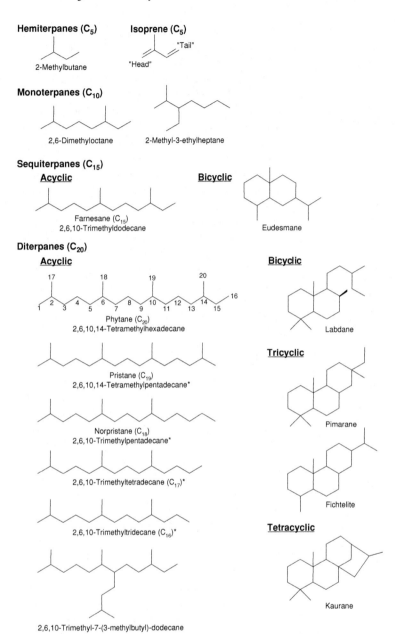

Hemiterpanes (C₅)

2-Methylbutane

Isoprene (C₅)

"Head"   "Tail"

Monoterpanes (C₁₀)

2,6-Dimethyloctane

2-Methyl-3-ethylheptane

Sequiterpanes (C₁₅)

Acyclic

Farnesane (C₁₅)
2,6,10-Trimethyldodecane

Bicyclic

Eudesmane

Diterpanes (C₂₀)

Acyclic

Phytane (C₂₀)
2,6,10,14-Tetramethylhexadecane

Pristane (C₁₉)
2,6,10,14-Tetramethylpentadecane*

Norpristane (C₁₈)
2,6,10-Trimethylpentadecane*

2,6,10-Trimethyltetradecane (C₁₇)*

2,6,10-Trimethyltridecane (C₁₆)*

2,6,10-Trimethyl-7-(3-methylbutyl)-dodecane

Bicyclic

Labdane

Tricyclic

Pimarane

Fichtelite

Tetracyclic

Kaurane

Figure 2.15. Chemical structures of terpenoid classes built from the ubiquitous isoprene (C₅) subunit. Most of the examples shown are saturated compounds (e.g. monoterpanes versus monoterpenes) because these are typical of the biomarkers in petroleum. Phytane, C₂₅ extended *ent*-isocopalane, 24-*n*-propylcholestane, and hopane are used as examples of the numbering systems for acyclic and cyclic terpenoids (see also Figure 2.23). Asterisks mark compounds that contain fewer carbon atoms than their parent class of terpanes.

Sesterterpanes (C$_{25}$)

Acyclic

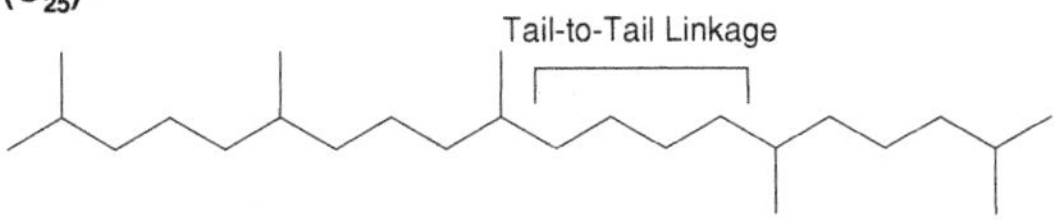

Tail-to-Tail Linkage

2,6,10,15,19-Pentamethylicosane (C$_{25}$H$_{52}$)

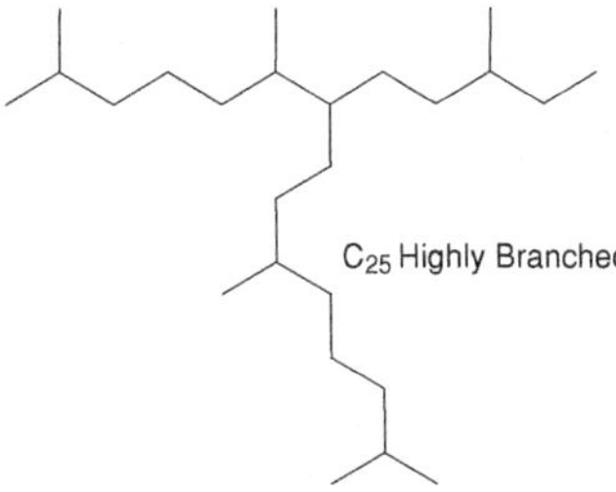

C$_{25}$ Highly Branched Isoprenoid (HBI)

Tricyclic

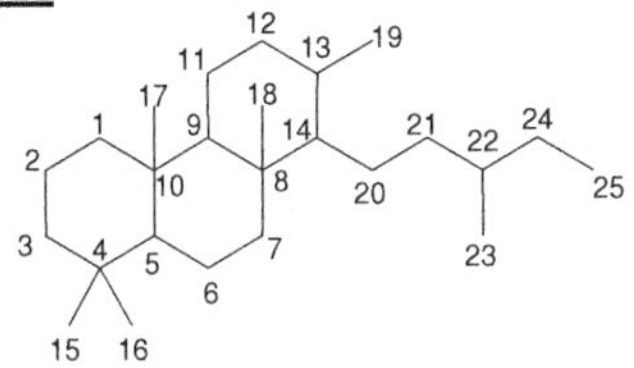

C$_{25}$ Tricyclic Terpane
(C$_{25}$ extended *ent*-isocopalane)

Tetracyclic

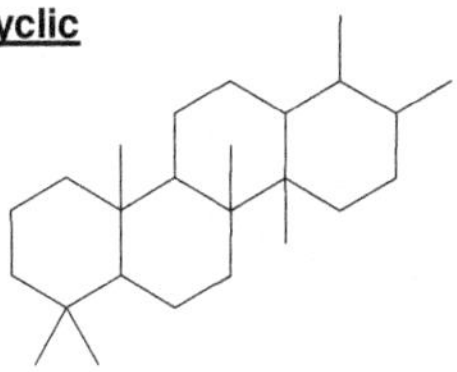

C$_{25}$ Tetracyclic Sesterterpane

Triterpanes (C$_{30}$)

Acyclic

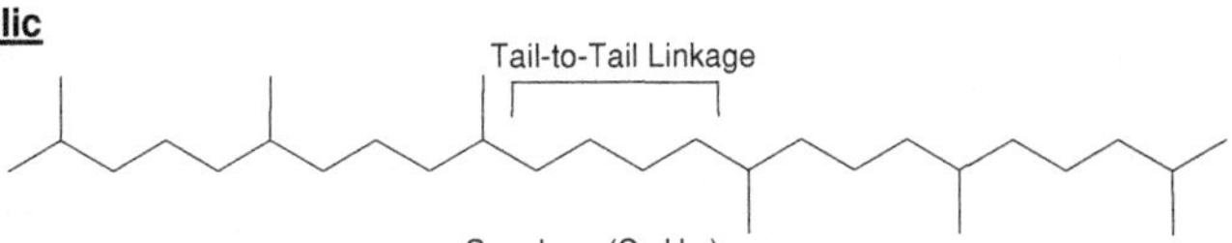

Tail-to-Tail Linkage

Squalane (C$_{30}$H$_{62}$)

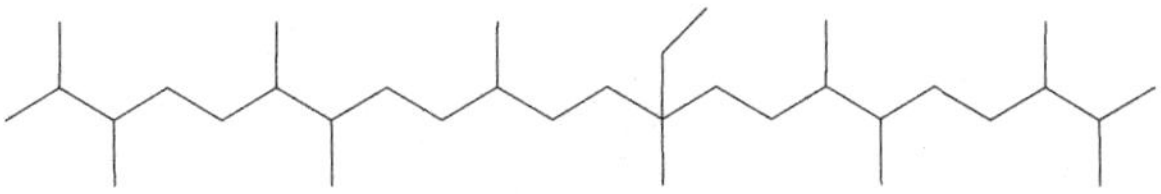

Botryococcane

Tricyclic

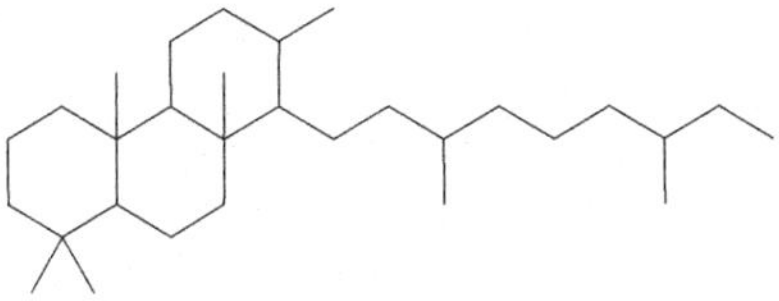

Tricyclohexaprenane
(C$_{30}$ extended *ent*-isocopalane)

Figure 2.15. (*cont.*)

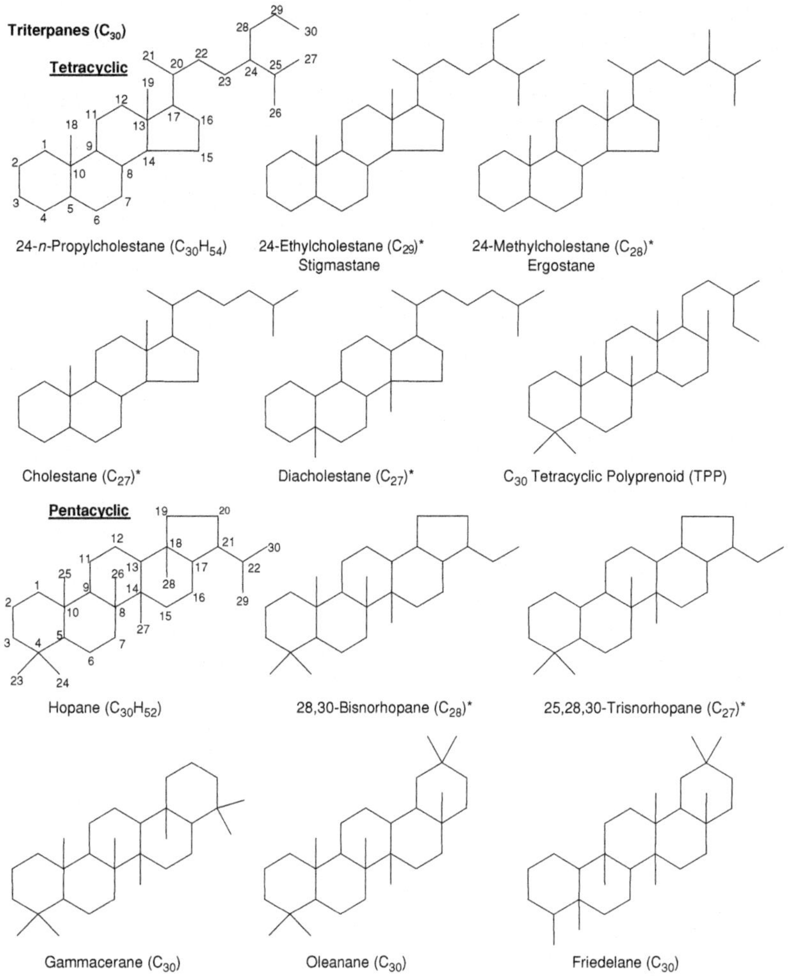

Triterpanes (C$_{30}$)

Tetracyclic

24-n-Propylcholestane (C$_{30}$H$_{54}$)

24-Ethylcholestane (C$_{29}$)*
Stigmastane

24-Methylcholestane (C$_{28}$)*
Ergostane

Cholestane (C$_{27}$)*

Diacholestane (C$_{27}$)*

C$_{30}$ Tetracyclic Polyprenoid (TPP)

Pentacyclic

Hopane (C$_{30}$H$_{52}$)

28,30-Bisnorhopane (C$_{28}$)*

25,28,30-Trisnorhopane (C$_{27}$)*

Gammacerane (C$_{30}$)

Oleanane (C$_{30}$)

Friedelane (C$_{30}$)

Figure 2.15. (*cont.*)

of the hydrocarbon chain. Pristane, however, is not an extended homolog of the C$_{18}$ acyclic isoprenoid. The nineteenth carbon atom in pristane (carbon atoms are numbered in the same manner as phytane in Figure 2.15) is a methyl branch in the fourth isoprene unit, rather than an additional carbon on the linear part of the chain.

Figure 2.15 shows several degraded compounds that contain fewer carbon atoms than their parent class of terpanes (marked by asterisks). Botryococcane has four additional carbon atoms compared with its parent class (triterpanes).

Most regular isoprenoids containing 20 or fewer carbon atoms, including pristane and phytane, originate

primarily from the phytol side chain of chlorophyll a during diagenesis (Rontani and Volkman, 2003) (see Figure 3.24). For example, pristane can originate by oxidation and decarboxylation of phytol, while phytane can originate by dehydration and reduction. Thus, pristane (C$_{19}$) and phytane (C$_{20}$) are usually the most abundant members of a whole series of regular head-to-tail-linked isoprenoids in petroleum (Figure 2.16).

Note: The regular C$_{12}$ and C$_{17}$ acyclic isoprenoids are low or absent in petroleum. Formation of these homologs is unfavorable because it requires cleavage of two carbon–carbon bonds in the C$_{19}$ or

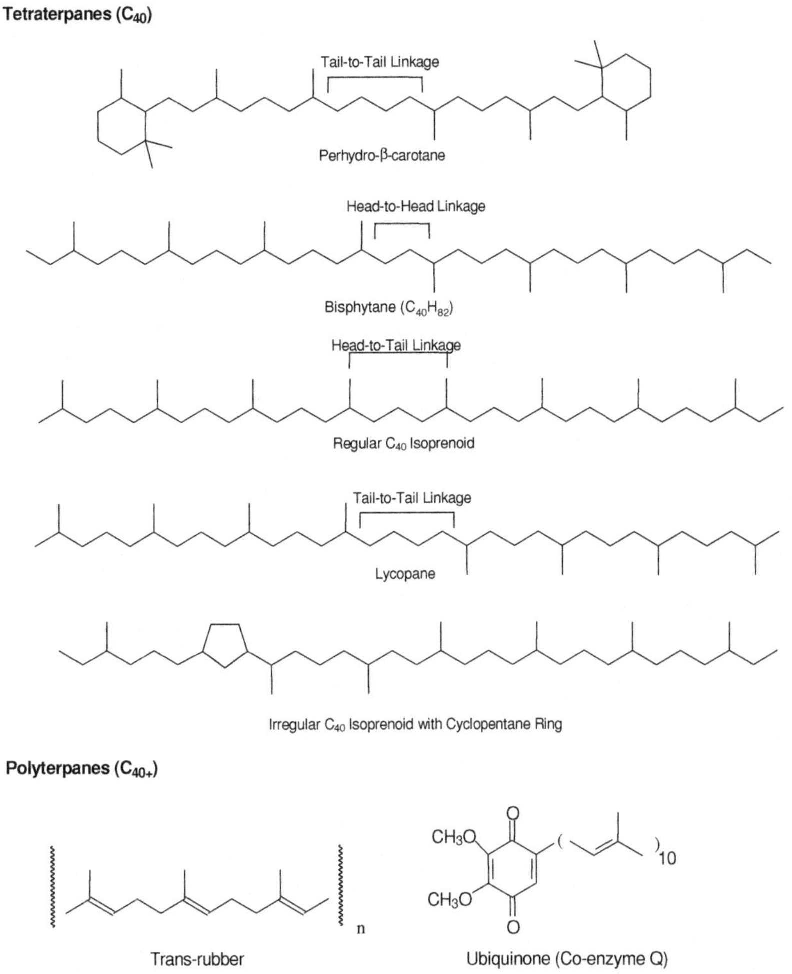

Tetraterpanes (C$_{40}$)

Tail-to-Tail Linkage

Perhydro-β-carotane

Head-to-Head Linkage

Bisphytane (C$_{40}$H$_{82}$)

Head-to-Tail Linkage

Regular C$_{40}$ Isoprenoid

Tail-to-Tail Linkage

Lycopane

Irregular C$_{40}$ Isoprenoid with Cyclopentane Ring

Polyterpanes (C$_{40+}$)

Trans-rubber

Ubiquinone (Co-enzyme Q)

Figure 2.15. (*cont.*)

C$_{20}$ isoprenoid precursor. For example, the C$_{19}$ regular isoprenoid (pristane) has two methyl groups attached to the number 14 carbon (C-14). Cracking of one of these carbon–carbon bonds yields the C$_{18}$ isoprenoid. Cracking of the carbon–carbon bond between C$_{13}$ and C$_{14}$ yields the C$_{16}$ isoprenoid. Cracking of the carbon–carbon bond between C-2 and C-3 in phytane (see Figure 2.15) results in a C$_{17}$ compound, but this compound is not a regular isoprenoid.

Other sources for acyclic isoprenoids containing up to 20 carbon atoms include chlorophyll b, bacteriochlorophyll a, α- and β-tocopherols, carotenoid pigments, and the cell membranes of archaea (Goossens *et al.*, 1984; Volkman and Maxwell, 1986). Archaea evolved from primitive organisms that existed during the Early Proterozoic time when today's extreme environments were probably more common. They include halophilic, thermoacidophilic, and methanogenic organisms with characteristics of both prokaryotes and eukaryotes (Woese *et al.*, 1978; Brock and Madigan, 1991).

Various irregular acyclic isoprenoids are important biomarkers. For example, botryoccocane is an irregular acyclic isoprenoid in crude oils and rock extracts (Figure 2.15) that is highly specific for lacustrine sedimentation. Another unusual acyclic isoprenoid,

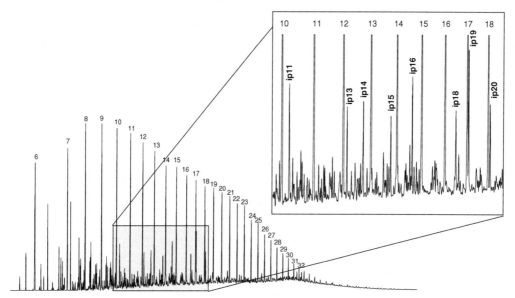

Figure 2.16. Gas chromatogram of oil generated from Cretaceous Agrio Formation marine shales in Argentina (see Figure 18.126). The regular isoprenoid hydrocarbons (labeled in the insert as ip(carbon number)) are the most abundant resolved C_{10+} compounds after the normal alkanes (labeled by their carbon number). The minor, unlabeled peaks eluting between the *n*-alkanes are mostly other isoalkanes, monocyclic alkanes, and alkylated aromatic hydrocarbons. Pristane = ip19, phytane = ip20.

2,6,10-trimethyl-7-(3-methylbutyl)-dodecane (Figure 2.15), is the second most abundant alkane in Rozel Point crude oil from Utah (Yon *et al.*, 1982). This compound and related highly branched isoprenoids originate from both marine and lacustrine phytoplanktonic algae and bacteria (Volkman and Maxwell, 1986) and may indicate hypersaline depositional conditions (see Table 13.1).

Although not biomarkers, many of the more volatile and abundant hydrocarbons in petroleum show evidence for an origin from biomarker precursors in living organisms. Many highly branched hydrocarbons and those containing aromatic or cyclohexane rings can be explained as originating by fragmentation of terpenoid precursors. Mair *et al.* (1966) show that 2,6-dimethyloctane (0.50 vol.%) and 2-methyl-3-ethylheptane (0.64%) are more abundant than all other 49 possible structural isomers (0.44%) in Ponca City crude oil because they originate from terpenoid precursors (two isoprene units) (see Figure 2.14). The high abundance of these and other compounds with terpenoid structures in petroleum is strong evidence against its origin by random polymerization of methane in the Earth's mantle, as proposed by Gold and Soter (1980).

Note: Smith (1968) explains the abundant normal alkanes in most crude oils as originating from lipids in living organisms, including naturally occurring *n*-alkanes and fatty acids. Likewise, high concentrations of iso- and anteisoalkanes (2-methyl- and 3-methylalkanes, respectively) in crude oils appear to be due to their biological origin. For example, 2-methyloctadecane is biosynthesized by archaea (Brassell *et al.*, 1981). Normal alkanes, isoalkanes, and anteisoalkanes are examples of biomarkers that are not terpenoids.

AROMATIC HYDROCARBONS

Aromatic hydrocarbons have one or more rings with delocalized π bonds and obey the formula C_nH_{2n-6y}, where y is the number of aromatic rings. Benzene, a single, six-carbon aromatic ring, is the simplest aromatic hydrocarbon (see Figure 2.5). Note that aromatic rings may be drawn with alternating double bonds or with an inner circle. Double-bond drawings are used widely and conform to IUPAC conventions. They are, however, technically inaccurate because the electrons are shared

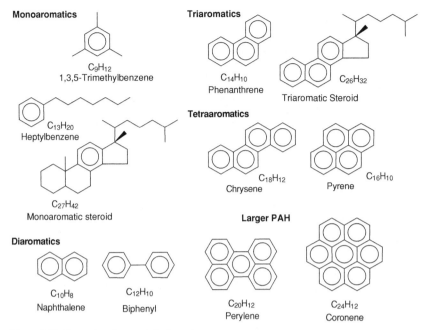

Figure 2.17. Examples of aromatic hydrocarbons.

equally by all carbon atoms within the ring. The inner circle better represents this delocalization.

Aromatic hydrocarbons can be divided into classes based on the number of aromatic rings. Monoaromatics have one ring, diaromatics have two, triaromatics have three, and so on. Collectively, any aromatic hydrocarbon that has two or more rings joined together can be called a polynuclear aromatic hydrocarbon (PAH). Various aromatic hydrocarbons with the same general formula can arise, depending on how the rings are fused (Figure 2.17). Alkylated benzenes, naphthalenes, and phenanthrenes are commonly the most abundant aromatic species in crude oils. Aromatic hydrocarbons need not be limited to structures with only six-carbon rings. As long as all adjacent atoms in the ring of a cyclic molecule have p orbitals and the total p electrons equals $4n + 2$ (Where n = any integer), then the electrons can delocalize over the entire structure.

Naphthenoaromatic hydrocarbons contain mixtures of aromatic and saturated rings. The naphthene or aromatic groups may be part of the fused ring or may be linked as pendants to the ring. In petroleum, the naphtheno portion of these compounds usually consists of five- or six-carbon rings.

HETEROATOMIC MOLECULES

Many compounds in petroleum contain elements in compounds in addition to carbon and hydrogen. Compounds that contain sulfur, nitrogen, and oxygen (NSO compounds) typically constitute ~10–15% of non-biodegraded, thermally mature oil. Sulfur-rich (type IIS) kerogens can yield low-maturity oil that is >40% NSO compounds.

Sulfur compounds

Sulfur compounds are usually the most abundant NSO compounds in petroleum. They can be subdivided into several chemical groups (Figure 2.18).

Thiols (mercaptans, R–SH) can be abundant when high-maturity condensates react with H_2S. Many alkyl- and cycloalkylthiols have been identified. Chemists often use "R" to signify a hydrocarbon group.

Disulfides (R–S–S–R) also occur primarily in high-maturity condensates and are generally alkylation products of thiols.

Thiaalkanes (sulfides) are normal branched alkyl compounds that contain a single sulfur atom within the structure (non-terminal). The sulfur is commonly incorporated in these molecules during early diagenesis.

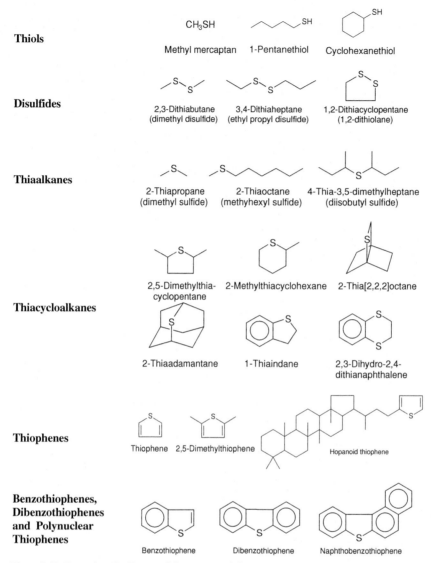

Figure 2.18. Examples of sulfur-containing compounds in petroleum.

Thiacycloalkanes (sulfides) contain a single sulfur atom incorporated in a cycloalkyl structure. These compounds generally form during early diagenesis.

Thiophenes consist of a five-member unsaturated ring containing one sulfur and four carbon atoms. Thiophenes behave like aromatic hydrocarbons. Thiophenic biomarkers form when functionalized unsaturated species react with microbial hydrogen sulfide (H_2S).

Benzothiophenes, dibenzothiophenes, and polynuclear thiophenes (sulfur aromatics) have a thiophene unit condensed with one or more aromatic rings. They behave chemically like aromatic hydrocarbons. Aromatic sulfur compounds are usually the most abundant class of non-hydrocarbons in crude oils.

Nitrogen compounds

Non-hydrocarbon compounds that contain nitrogen occur in petroleum as neutral (aromatic-like) or basic (reactive) species (Figure 2.19).

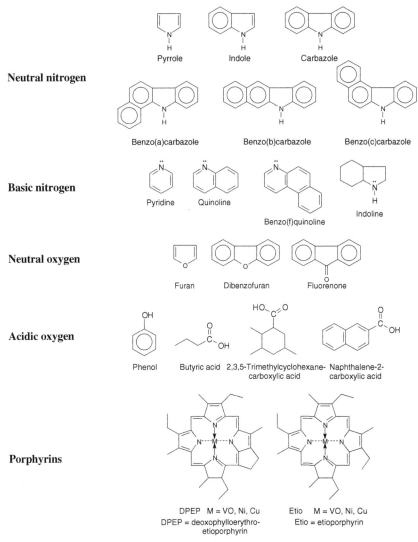

Figure 2.19. Examples of compounds containing nitrogen and oxygen that occur in petroleum.

Neutral (non-basic) nitrogen compounds chemically resemble the thiophenes. The nitrogen's electrons are shared in the π cloud and are delocalized.

Basic nitrogen compounds differ from the non-basic compounds in that nitrogen has a lone electron pair that is not part of the delocalized π cloud and can react readily with acids.

Porphyrins are a special class of nitrogen compounds that originate mainly from chlorophylls. In living organisms, porphyrin precursors contain either a magnesium atom (cholorphylls), copper, iron, or cobalt.

In petroleum, vanadium and nickel commonly replace these metals, and the side chains differ from those in the precursors.

Oxygen compounds

Oxygen compounds occur in petroleum as neutral and acidic species (Figure 2.19).

Neutral oxygen compounds are mainly furans and fluorenones where the two oxygen bonds become part of the delocalized π cloud.

Acidic oxygen compounds include phenols and carboxylic acids. Although some acids in petroleum may be inherited directly from the source rock, significant amounts can form during microbial biodegradation.

Resins, asphaltenes, and metal complexes

Resins and asphaltenes are not discrete compounds but high-molecular-weight heteroatomic molecules that are poorly defined. The difference between resins and asphaltenes is defined by their relative solubility in hydrocarbons, which roughly corresponds to size. Asphaltenes precipitate from crude oil when a large quantity of low-molecular-weight alkanes is added (e.g. *n*-pentane), while resins (polars) remain in solution. Metal complexes, such as porphyrins, are heteroatomic species with chelated metals. Most of these compounds have not been characterized.

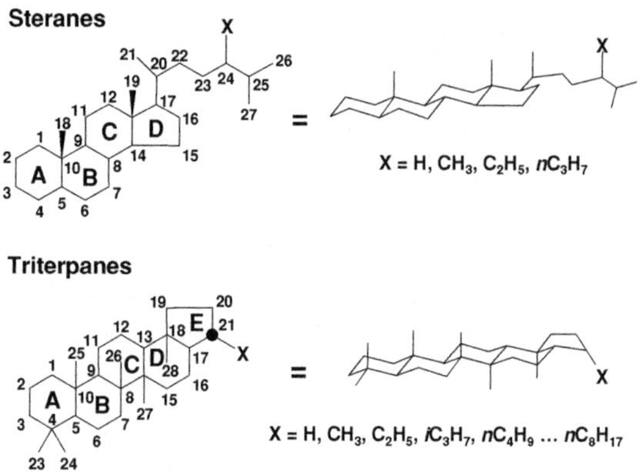

STEREOCHEMISTRY AND NOMENCLATURE

The spatial or three-dimensional relationship between atoms in molecules, called stereochemistry, is essential to the understanding of the structures of biomarkers and how they are used in geochemical studies. Each carbon atom and the rings in biomarker molecules are labeled systematically. Figure 2.15 shows the labeling for carbon atoms in the acyclic isoprenoid phytane. The labeling system for steranes and triterpanes is shown in Figure 2.20. In the text, a subscript number following a capital C refers to the number of carbon atoms in a particular compound. For example, the C_{27} sterane called cholestane contains 27 carbon atoms. A capital C followed by a dash and number refers to a particular position of a carbon atom within the compound. For example, C-20 in the steranes is the carbon atom at position 20 (Figure 2.20). The nomenclature *n*C or *i*C followed by a number indicates the normal alkane or the isoalkane, respectively, that contains the specified number of carbon atoms. For example, nC_{19} and iC_{19} refer to the *n*-alkane and the isoalkane that contain 19 carbon atoms, respectively. Rings are also specified in succession from left to right as the A-ring, B-ring, C-ring, and so forth.

The structural characteristics of biomarkers can be described using various common modifiers and other nomenclature (Table 2.2). For example, if a compound

Figure 2.20. Labeling system for carbon atoms in steranes and triterpanes (cyclic terpenoids) with examples of projected three-dimensional structures. Reprinted with permission by ChevronTexaco Exploration and Production Technology Company, a division of Chevron USA Inc.

lacks a carbon atom that occurs in the parent compound, then the prefix nor- is used, preceded by the number of the missing atom. Thus, 25-norhopanes are identical to hopanes (parent compounds), except that a methyl group (C-25) has been removed from its point of attachment at C-10 (Figure 2.20). If two or three atoms are missing, then the prefix bisnor- or trisnor-, respectively, is used. Thus, 28,30-bisnorhopanes lack the C-28 and C-30 methyl groups found in hopanes (Figure 2.15). The 8,14-secohopanes consist of hopanes where the bond between the number 8 and 14 carbon atoms in the C-ring has been broken (see Figure 16.40).

CHIRALITY

Saturated carbon atoms are linked to their substituents by means of four covalent bonds that radiate outward toward the corners of an imaginary tetrahedron (Figure 2.21). If all four substituents differ, then the carbon atom at the center of the tetrahedron is asymmetric or chiral. One or the other of two stereoisomers of the compound (mirror images of each other) can be formed by transposing any two of the substituents attached to the asymmetric carbon. These stereoisomers have the same molecular formula, but they are mirror images or enantiomers, i.e. they differ in the same manner as right and left hands (Figure 2.21). In a molecule containing more than one asymmetric center,

Table 2.2. *Common modifiers and nomenclature related to biomarker names*

Modifier	Effect on biomarker
homo-	One additional carbon (C) on structure
bis-, tris-, tetrakis- (di-, tri-, tetra-)	Two, three, four additional C, respectively
pentakis-, hexakis- (penta-, hexa-)	Five, six additional C, respectively
seco-	Cleaved C–C bond (specified)
benzo-	Fused benzene ring
nor-	One less C on structure
des-A (de-A)	Loss of A-ring from structure
iso-	Methyl shifted on structure
neo-	Methyl shifted from C-18 to C-17 on hopanes
spiro-	Two rings joined by one C
α	Asymmetric C in ring with functional group (usually H) down
β	Asymmetric C in ring with functional group (usually H) up
R	Asymmetric C that obeys clockwise convention of Cahn *et al.* (1966)
S	Asymmetric C that obeys counterclockwise convention of Cahn *et al.* (1966)

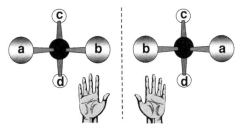

Figure 2.21. Asymmetric carbon atoms (black) are linked to four different substituents (symbolized as a, b, c, and d) by means of single covalent bonds that radiate outward toward the corners of a tetrahedron. Two mirror-image structures (similar to your right and left hands) are possible for each asymmetric center in a molecule. The vertical dashed line represents a mirror plane. The two structures shown are enantiomers because they are mirror images and cannot be superimposed without breaking bonds. Reprinted with permission by ChevronTexaco Exploration and Production Technology Company, a division of Chevron USA Inc.

inversion of all of the centers usually leads to the enantiomer. Inversion of any less than all of the asymmetric centers yields a diastereomer or epimer, as discussed below.

Note: In 1848, Louis Pasteur observed that some crystals of lactic acid spiraled to the left while others spiraled to the right. Using a hand lens and tweezers, he sorted these crystal types. The two piles of crystals were physically and chemically identical, except for one measurement: they rotated plane-polarized light in opposite directions. Pasteur had discovered stereochemistry. The two molecules of lactic acid shown in Figure 2.22 are identical,

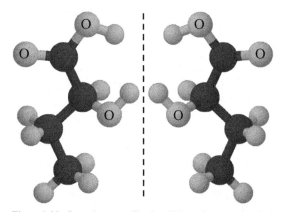

Figure 2.22. Stereoisomers of lactic acid have the same chemical formula but are mirror images of each other. When dissolved in a suitable solvent, one stereoisomer rotates plane-polarized light to the right (dextrorotatory), while the other rotates light to the left (levorotatory). O, oxygen; black, carbon; small gray, hydrogen.

except that they are mirror images of each other. They are optically active because they rotate polarized light in different directions.

OPTICAL ACTIVITY

Enantiomers have very similar chemical properties and cannot be separated using the achiral stationary phases in most gas chromatographic columns. They can be separated using gas chromatography that employs chiral stationary phases (e.g. Coleman and Lawrence, 2000), but this technique is only beginning to be applied in petroleum geochemistry (e.g. Alexander *et al.*, 1992; Bastow, 1998; Beesley and Scott, 1998).

One property of enantiomers in solution is that the direction in which they rotate plane-polarized light differs. The direction and extent of rotation of plane-polarized light induced by a chiral compound is equal but opposite of that induced by its enantiomer. Abiotic synthesis of molecules containing asymmetric centers generally results in a racemic (50 : 50) mixture of right- and left-handed molecules. These mixtures are not optically active because the rotation of each enantiomer cancels that of the other. However, many biologically formed compounds are optically active, i.e. they rotate plane-polarized light clockwise (to the right, or dextrorotatory) or counterclockwise (to the left, or levorotatory). A unique feature of living organisms is that the enzymes responsible for biosynthesis of cellular materials are chiral and generate biomolecules with only one configuration at certain asymmetric centers.

Note: The above may indicate either that life originated only once and the original organisms contained only one configuration at each asymmetric center in their enzymes, or that the alternative mirror-image organisms became extinct. For example, enzymatic generation of lactic acid by an organism results in only the levorotatory compound, while abiotic synthesis in the absence of chiral reagents or catalysts results in a racemic mixture of the levorotatory and dextrorotatory lactic acid (Figure 2.22). The configurations of amino acids and sugars can be assigned according to experimental chemical correlation with that of D- or L-glyceraldehyde as the reference compound, although the terms R- or S- are preferred. More commonly, amino acids and sugars that rotate

plane-polarized light to the right or left are described by (+) and (−). However, for compounds other than glyceraldehyde, the R- or S-configurations do not necessarily correspond to rotation of the plane of monochromatic plane-polarized light to the right or left, respectively.

Crude oil is optically active, evidence that it originated at least in part from products of once-living organisms. Silverman (1971) provided classic evidence that biological lipids, dominantly steroids and triterpenoids, represent a major source for crude oil. Compared with other fractions, the 425–450°C (atmospheric-equivalent boiling points) vacuum distillation cut in crude oil is depleted in the heavy isotope of carbon (^{13}C) and has high optical rotation, implying high concentrations of unaltered biogenic parent molecules. This distillation range corresponds to the molecular-weight range of many steroids and triterpenoids. These compounds are solids under conditions of deposition and are resistant to biodegradation in sediments during diagenesis. Whitehead (1974) also attributed the optical rotation of crude oil to steranes and triterpanes based on mass spectrometry and magnetic resonance techniques. Furthermore, biomarkers isolated from crude oils (e.g. gammacerane) have high optical activity, suggesting that they account for most of the optical activity (Hills *et al.*, 1966). Optical rotation of petroleum generally decreases with increasing thermal maturity (Williams, 1974) and increases with microbial degradation due to selective degradation (Winters and Williams, 1969).

NAMING ASYMMETRIC CENTERS (R, S, α, AND β)

Stereochemical information is included in the names of biomarkers. For example, both 5α- and 5β-steranes occur in petroleum, but they differ in physical characteristics. The name C_{27} 5α,14α,17α(H)-cholestane 20R describes a single compound among millions in petroleum. A clear understanding of these stereochemical designations is needed to apply various biomarker parameters, especially those for thermal maturity assessment (see Figure 14.3).

Figure 2.23 (top) shows the stereochemical designations for hydrogen atoms at C-3 and C-5 in the A-ring of cholestane, the C_{27} sterane, in a conformational

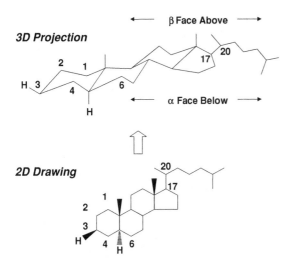

3D Projection

β Face Above

α Face Below

2D Drawing

Figure 2.23. Procedure for designating α (below the plane) and β (above the plane) hydrogens or other groups for a polycyclic biomarker. The example is the C_{27} sterane cholestane. Reprinted with permission by ChevronTexaco Exploration and Production Technology Company, a division of Chevron USA Inc.

projection as it might appear in three dimensions. Normally hydrogen atoms are not shown, but this Figure shows how α and β hydrogens lie below and above the plane of the molecule, respectively. The 3β hydrogen in the figure is equatorial because the bond between the hydrogen and C-3 is oriented in the plane defined by the ring system. The 5α hydrogen is axial because the bond between the hydrogen and the C-5 carbon is oriented along an axis perpendicular to the plane defined by the ring system.

The geochemical literature is inconsistent in the use of modifiers, such as (H) or (CH$_3$), after the symbols α or β in biomarker names. These modifiers are not needed for ternary carbon atoms that are shared by two rings because there is only one possible configuration. For example, the term 5α(H)-cholestane is redundant because it adds no additional information compared with 5α-cholestane. Where two stereochemical designations are possible, such as the 17-position in cholestane (H versus alkyl side chain), the lack of a modifier generally indicates that the symbol refers to the larger group (alkyl side chain), as is common in natural-product nomenclature. However, geochemical names usually refer to the orientation of the hydrogen at the 17-position. Therefore, the term 17β might be confusing, but the term 17β(H) clearly specifies the stereochemistry

at the 17-position. The molecule in Figure 2.23 might be named $5\alpha,14\alpha,17\alpha$(H)-cholestane 20R, but this name also is redundant. The root cholestane has $14\alpha,17\alpha$(H),20R stereochemistry by definition, because this is the stereochemistry of the natural sterol cholesterol and only stereochemical changes from the root compound need to be specified. Thus, 5α-cholestane has equivalent meaning to $5\alpha,14\alpha,17\alpha$(H)-cholestane 20R, but the latter description can be used in order to contrast this stereochemistry with other cholestane epimers when necessary.

The two possible configurations of an asymmetric carbon atom are called R and S, depending on a simple convention (Cahn *et al.* 1966). Figure 2.24 summarizes the three-step procedure for R and S stereochemical assignment using C-20 in cholestane as an example. The reader is encouraged to construct molecular models of cholestane and other biomarkers for a better understanding of stereochemistry:

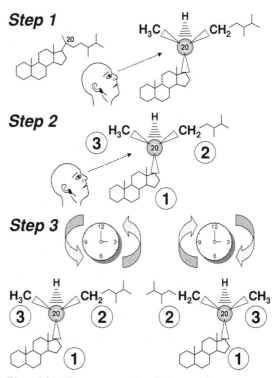

Figure 2.24. Three-step procedure for R versus S stereochemical assignments in acyclic portions of biomarkers as described in the text. Reprinted with permission by ChevronTexaco Exploration and Production Technology Company, a division of Chevron USA Inc.

Step 1. Orient the molecule so the smallest substituent attached to the asymmetric carbon atom, for example a hydrogen atom (Figure 2.25), points away from the viewer and the asymmetric carbon is closer to the viewer. Imagine that the bond between the asymmetric carbon atom and the smallest substituent represents the steering column in your car while the remaining three substituents are the steering wheel.

Step 2. Rank the three remaining substituents forming the steering wheel from largest to smallest, based on atomic number. For different carbon substituents, the larger group generally has higher priority. For example, an *n*-butyl group has higher priority than an ethyl group. However, if two or more carbon atoms are attached to the atom bound to the asymmetric carbon, then the ranking is higher than if there is only one carbon atom. An isopropyl has a higher ranking than an *n*-propyl or even an *n*-butyl group (Figure 2.25). For isotopes, priority is given to higher atomic weight.

Step 3. Draw an imaginary curved arrow from the highest to the next highest ranked group on the steering wheel. The asymmetric carbon is in the R configuration (Latin *dextral*, meaning right) if the arrow points clockwise on the wheel, or in the S

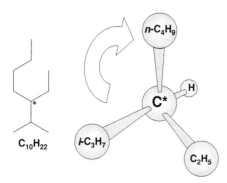

Figure 2.25. Application of the convention (see text and Figure 2.27) for describing stereochemistry of the asymmetric center in 2-methyl-3-ethylheptane. The two-dimensional stick notation on the left shows the location of the asymmetric center with an asterisk. The curved arrow shows the clockwise decrease in ranking priority for the three substituents nearest the observer. The three-dimensional example shows an R configuration at the asymmetric carbon atom (asterisk). Reprinted with permission by ChevronTexaco Exploration and Production Technology Company, a division of Chevron USA Inc.

Table 2.3. *Common symbols used to depict the stereochemistry of atoms in biomarkers*

Symbol	Description
or	Bond stereochemistry not specified
	Bond directed out of page (β)
or	Bond directed into page (α)
●	Hydrogen directed out of page (β)
○	Hydrogen directed into page (α)
‖	Double bond

configuration (*sinister*, meaning left) if it points counterclockwise. The example of 2-methyl-3-ethylheptane in Figure 2.25 shows the C-3 asymmetric carbon in the R configuration. This monoterpane was discussed earlier (see Figure 2.14).

In the text, we use the R and S nomenclature for carbon atoms that are not part of a ring and adopt the α (alpha = down) and β (beta = up) nomenclature from natural–product chemistry to describe asymmetric configurations at ring carbons. For example, cholestane can have the R or S configuration at C-20 or α or β configurations at C-14 (Figure 2.20). Table 2.3 shows symbols commonly used to indicate the stereochemistry of specific carbon atoms in two-dimensional depictions of biomarkers.

Pristane, cholestane, and most biomarkers contain more than one asymmetric center. If at least one but not all of the asymmetric centers are the same, then the resulting stereoisomers that are not mirror images are called diastereomers. Molecules that contain asymmetric carbon atoms but are superimposable on their mirror images are neither chiral nor asymmetric.

For example, pristane (Figure 2.15) contains two asymmetric carbon atoms at C-6 and C-10. When the asymmetric carbon atoms are in the 6S,10R configuration or the 6R,10S configuration, the structures are identical and are called meso-pristane (Figure 2.26). The other pair of 6R,10R- and 6S,10S-pristane isomers are related as mirror images and therefore are chiral

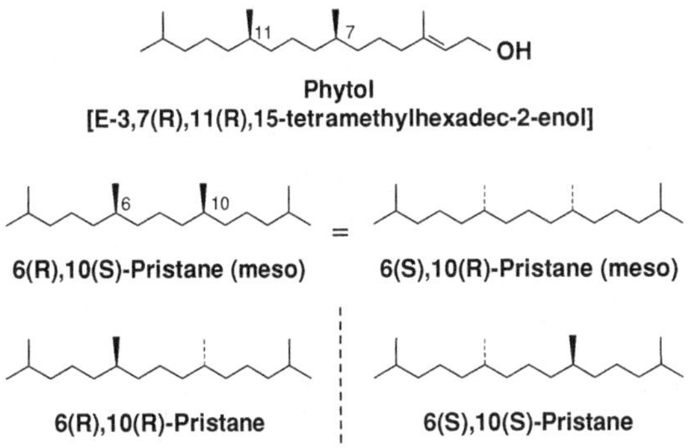

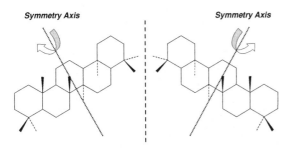

Phytol
[E-3,7(R),11(R),15-tetramethylhexadec-2-enol]

6(R),10(S)-Pristane (meso)   **6(S),10(R)-Pristane (meso)**

6(R),10(R)-Pristane   **6(S),10(S)-Pristane**

Figure 2.26. Phytol (top) is a precursor of pristane and has the same stereochemistry as the biological (meso) configuration of pristane in immature sediments (middle). Pristane contains two asymmetric carbon atoms at C-6 and C-10. The 6(R),10(S) and 6(S),10(R) configurations are identical and are called meso-pristane. The 6(R),10(R) and 6(S),10(S) configurations are mirror images (enantiomers), as shown by the vertical dashed line representing a mirror plane. Pristane has two diastereomeric pairs: meso-pristane and 6R,10R-pristane, and meso-pristane and 6S,10S-pristane. Reprinted with permission by ChevronTexaco Exploration and Production Technology Company, a division of Chevron USA Inc.

Symmetry Axis   *Symmetry Axis*

Gammacerane

Figure 2.27. Gammacerane is an example of a chiral molecule that is not asymmetric. If rotated 180° about the symmetry axis, then the resulting configuration is the same as the non-rotated molecule. However, gammacerane is not superimposable on its mirror image and these two compounds thus represent enantiomers. Reprinted with permission by ChevronTexaco Exploration and Production Technology Company, a division of Chevron USA Inc.

and are enantiomers of each other. Like gammacerane (Figure 2.27), these enantiomers are chiral but not asymmetric. As indicated in Figure 2.26, phytol is a precursor of pristane and has the same stereochemistry as the biological (meso) configuration of pristane in immature sediments. Because of the hydroxyl

group and double bond in phytol, application of the stereochemical rules of nomenclature result in a different numbering and stereochemical description for the asymmetric centers in the molecule (E-3,7(R),11(R),15-tetramethylhexadec-2-enol) compared with meso-pristane, and yet the stereochemistry at the asymmetric centers in these molecules is identical.

The relationship between 6R,10R- or 6S,10S-pristane and meso-pristane is neither enantiomeric nor equivalent, but is diastereomeric. Pristane has two diastereomeric pairs of compounds: (1) meso-pristane and 6R,10R pristane and (2) meso-pristane and 6S,10S pristane (Figure 2.26). Unlike enantiomers, diastereomers typically have different chemical and physical properties (e.g. different melting and boiling points) and are commonly separable by gas chromatography using normal, achiral stationary phases. Different diastereomers can show different mass spectra, as discussed later.

Note: Stereoisomers are a special form of structural isomer. Most structural isomers, such as *n*-pentane and 2-methylbutane, have the same molecular formula but differ in the linkage between atoms. Stereoisomers have the same formula and the same linkage between atoms; only the spatial arrangement of atoms differs.

The R versus S and α versus β designations explained above describe the relative configurations of compounds. On the other hand, the direction of rotation of plane-polarized light by a particular enantiomer is a physical property. There is no simple relationship between the configuration of an enantiomer and its direction of optical rotation.

Chirality is not strictly equivalent to asymmetry. While all asymmetric molecules are chiral, certain chiral molecules are not asymmetric. For example, gammacerane (Figure 2.27) is a biomarker that contains an axis of symmetry. If rotated 180° about this axis, then the resulting configuration is the same as the non-rotated molecule. However, the mirror image of gammacerane is not superimposable and thus represents an enantiomer. The 6R,10R- and 6S,10S-pristane isomers (Figure 2.26) have the same level of symmetry as gammacerane (i.e. 180° axis of symmetry) and are therefore chiral but not asymmetric. Texts on stereochemistry contain more detailed discussions (e.g. Mislow, 1965; Cahn *et al.* 1966).

STEREOISOMERIZATION

The configurations at asymmetric centers in biomarkers that are imposed by enzymes in living organisms are not necessarily stable at higher temperatures in buried sediments. Although the mechanism is unclear, configurational isomerization or stereoisomerization in saturated biomarkers occurs only at asymmetric carbon atoms where one of the four substituents is hydrogen. Stereoisomerization could involve removal of a hydride ion (Ensminger, 1977) or a hydrogen radical (Seifert and Moldowan, 1980). The nearly planar carbocation (carbonium ion) or radical (see sp2 hybridization, Figure 2.3a) can regain hydrogen from the same side, resulting in the same configuration, or from the opposite side, resulting in the inverted configuration (Figure 2.28). Depending on the kinetics of the reaction and the stability of the products, the isomerized compound may show any proportion of the two possible configurations (R versus S when the asymmetric

center is not part of a ring, or α versus β when part of a ring). Configurational isomerization occurs only when cleavage and renewed formation of the bonds result in an inverted configuration compared with the starting asymmetric center. A complete discussion of organic chemical mechanisms is beyond the scope of this work, but the reader can find such information in most organic chemistry texts.

The two isomeric configurations of asymmetric carbon atoms in acyclic chains generally have similar thermal stabilities. For example, the isomerization at C-20 in the side chain of steranes proceeds from nearly 100% 20R in shallow sediments to a nearly equal mixture of 20R and 20S diastereomers in deeply buried source rocks and petroleum. The endpoint or "equilibrium" ratio for 20S/(20S + 20R) in C_{29} steranes is $\sim$0.50–0.55, slightly favoring the 20S diastereomer (see Figure 14.3). Once the endpoint is reached for this and other isomerization reactions, incremental heating does not increase the corresponding isomerization ratios.

Due to steric forces imposed by a rigid cyclic structure, asymmetric centers that are part of a saturated ring system usually show two configurations with quite different thermal stabilities. Thus, C-14 and C-17 in the C_{29} steranes each isomerize to the more stable β-configuration during burial heating. The endpoint ratio for $14\beta,17\beta(H)/[14\alpha,17\alpha(H) + 14\beta,17\beta(H)]$ is $\sim$0.7 (see Figure 14.3). Intramolecular rearrangements (between different locations within the same molecule) of hydrogen or alkyl groups can also occur and are important, for example in the formation of diasterenes (rearranged sterenes) from sterenes and sterols during burial diagenesis (see Figure 13.49).

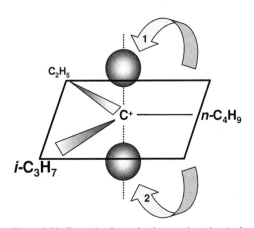

Figure 2.28. Example of a nearly planar carbocation (carbonium ion) formed by removal of a hydride ion from the asymmetric carbon atom in 2-methyl-3-ethylheptane. The p orbital, indicated by the shaded electron clouds above and below the plane of the sigma bonds, contains only one electron. Reattachment of a hydride ion from above (arrow 1) or below (arrow 2) the plane of the sigma bonds results in the R or S configuration at the asymmetric carbon, respectively. The isopropyl group has higher priority than n-butyl according to the nomenclature for describing stereochemistry (see Figures 2.24 and 2.25). Reprinted with permission by ChevronTexaco Exploration and Production Technology Company, a division of Chevron USA Inc.

STEREOCHEMISTRY OF SELECTED BIOMARKERS

The stereochemistry of hydrogen and methyl groups at asymmetric centers in compounds determines the various isomers and their chemical properties. For example, key asymmetric centers include C-6 and C-10 in pristane and phytane (Figure 2.26), C-17 and C-21 in the pentacyclic triterpanes (see Figure 2.29), and C-5, C-14, C-17, and C-20 in the steranes (see Figure 2.31). The reader may wish to construct molecular models of some of the compounds described below to help visualize the differences between the various stereoisomers.

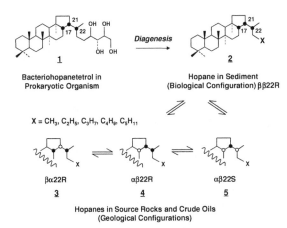

Figure 2.29. Some hopanes in petroleum originate from bacteriohopanetetrol (1) found in the lipid membranes of prokaryotic organisms (Ourisson *et al.*, 1984). Stereochemistry is indicated by open (α) and solid (β) dots, where the hydrogen is directed into and out of the page, respectively. The biological configuration [17β,21β(H)-22R] imposed on bacteriohopanetetrol and its immediate saturated product (2) by enzymes in the living organism is unstable during catagenesis and undergoes isomerization to geological configurations (e.g. 3, 4, 5). The 17β,21α(H)-hopanes (e.g. 3) are called moretanes, while all others are hopanes (e.g. 2, 4, 5). Reprinted with permission by ChevronTexaco Exploration and Production Technology Company, a division of Chevron USA Inc.

Acyclic isoprenoids

Because at least two hydrogens are bound to each carbon atom in *n*-alkanes, no asymmetric centers are possible. However, acyclic isoprenoids can have asymmetric centers where methyl groups are attached to the chain.

Phytol is a major precursor of pristane and phytane in petroleum. Phytol has a single stereoisomer (Figure 2.26) as synthesized by enzymes in living organisms, where the asymmetric carbon atoms at C-7 and C-11 are both in the R configuration (Maxwell *et al.* 1973). Because phytol is an alcohol, carbon atoms are numbered consecutively from C-1 nearest the hydroxyl group. When the hydroxyl group and double bond in phytol are removed to produce pristane and phytane, the numbering of carbon atoms changes. C-7 and C-11 in phytol become C-6 and C-10 in pristane and phytane. Loss of the hydroxyl group from phytol also changes the priority ranking of the substituents around the asymmetric carbon at C-11 (C-10 in pristane and phytane). Without changing the stereochemistry of the groups

around C-11, the nomenclature rules require that C-10 in pristane or phytane be described as being in the S configuration.

Pristane in thermally immature sediments has a configuration dominated by 6R,10S stereochemistry (equivalent to 6S,10R), (see Figure 2.26). This stereochemistry is directly comparable to that at C-7 and C-11 in phytol, as described above. During thermal maturation, isomerization at these positions results in an endpoint mixture of 6R,10S, 6S,10S, and 6R,10R in the ratio 2 : 1 : 1 (Patience *et al.*, 1980) (see Figure 2.26). Direct measurement of the eight possible isomers resulting from the three asymmetric centers in phytane has not been reported using currently available gas chromatographic columns.

Terpanes

Hopanes with 30 or fewer carbon atoms show asymmetric centers at C-21 and at all ring junctures (C-5, C-8, C-9, C-10, C-13, C-14, C-17, and C-18). Hopanes with more than 30 carbon atoms are commonly called homohopanes, where the prefix homo- refers to additional methylene groups attached to C_{30} hopane (see Table 2.2). The homohopanes show an extended side chain with an additional asymmetric center at C-22 (Figure 2.29), which results in two peaks for each homolog (22R and 22S) on the mass chromatograms for these compounds (e.g. see Figure 8.21, peaks 22–35).

The hopanes are composed of three stereoisomeric series, namely 17α,21β-, 17β,21β-, and 17β,21α(H)-hopanes. Compounds in the $\beta\alpha$ series are also called moretanes (Figure 2.29). The notations α and β indicate whether the hydrogen atoms are below or above the plane of the rings, respectively. Hopanes with the 17α,21β(H) configuration ($\alpha\beta$) in the range C_{27}–C_{35} are characteristic of petroleum because of their greater thermodynamic stability compared with the other epimeric series ($\beta\beta$ and $\beta\alpha$). The $\beta\beta$ series is generally not found in petroleum because it is thermally unstable during early catagenesis. Hopanes of the $\alpha\alpha$ series are not natural products and it is unlikely that they occur above trace levels in petroleum (Bauer *et al.*, 1983).

Note: $C_{28}\alpha\beta$-hopanes are rare or absent in petroleum for reasons similar to those for the scarcity of regular C_{17} acyclic isoprenoids. Formation of C_{28} hopanes requires cleavage of not one but two carbon–carbon bonds attached to C-22 in the C_{35} hopanoid precursor (Figure 2.15). Cleavage of the carbon–carbon bond between C-21 and C-22 (resulting in C_{27} hopane) or cleavage of either of the other two carbon–carbon bonds attached to C-22 (resulting in C_{29} hopane) is far more likely than sequential cleavage of two carbon–carbon bonds. A C_{28} compound formed by loss of a methyl group from the hopanoid ring system would represent another compound class. For example, loss of the C-25 methyl group from hopane yields 25-norhopane (see Figure 16.24).

The major precursors for the hopanes in source rocks and crude oils include bacteriohopanetetrol and related bacteriohopanes (Figure 2.29), which show the biological $17\beta,21\beta(H)$-stereochemistry. The biological configuration is nearly flat, although puckering of the carbon–carbon bonds in the rings results in a three-dimensional shape. Like the sterols discussed below, bacteriohopanetetrol is amphipathic because it contains both polar and non-polar ends (Figure 2.30). The flat configuration and amphipathic character are necessary for bacteriohopanetetrol to fit into the lipid membrane structure (Rohmer, 1987). Because this stereochemical arrangement is thermodynamically unstable, diagenesis and catagenesis of bacteriohopanetetrol result in transformation of the $17\beta,21\beta(H)$-precursors to the $17\alpha,21\beta(H)$-hopanes

Bacteriohopanetetrol

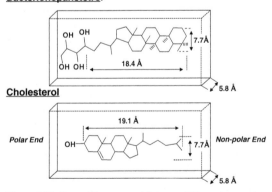

Figure 2.30. Bacteriohopanetetrol (a hopanoid in prokaryotes) and cholesterol (a steroid in eukaryotes) are similar in size and amphipathic character; both are essential components in the lipid membranes of living organisms. Reprinted with permission by ChevronTexaco Exploration and Production Technology Company, a division of Chevron USA Inc.

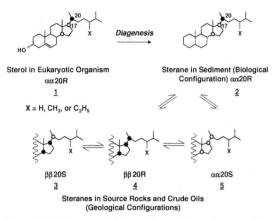

Figure 2.31. Most steranes in petroleum originate from sterols (1) in the lipid membranes of eukaryotic organisms. Stereochemistry is indicated by open (α) or solid (β) dots, which indicate that hydrogen is directed into or out of the page, respectively. The biological configuration [$14\alpha,17\alpha(H)$-20R] imposed on the sterol precursor and its immediate saturated product (2) by enzymes in living organisms is unstable during catagenesis and undergoes isomerization to geological configurations (e.g. 3, 4, 5) discussed in the text. Reprinted with permission by ChevronTexaco Exploration and Production Technology Company, a division of Chevron USA Inc.

and $17\beta,21\alpha(H)$-moretanes. Likewise, the biological 22R configuration found in bacteriohopanetetrol converts to an endpoint mixture of 22S and 22R $\alpha\beta$-homohopanes, showing 22S/(22S + 22R) for the C_{31} homolog of $\sim$0.58–0.62 in most crude oils (see Figure 14.3) (Seifert and Moldowan, 1980).

Steranes

The sterols in eukaryotic organisms are precursors to the steranes in source rocks and petroleum (Figures 2.31 and 2.32 (Mackenzie et al., 1982a; de Leeuw et al., 1989). Because of the large number of asymmetric centers in sterols, very complex mixtures of stereoisomers are possible. For example, cholesterol has eight asymmetric centers and thus might be expected to show as many as 2^8 or 256 stereoisomers. However, because of the highly specific enzymatic biosynthesis of cholesterol, only one stereoisomer of cholesterol exists in significant amounts in living organisms.

Sterols are amphipathic and have dimensions similar to bacteriohopanetetrol. Like bacteriohopanetetrol in prokaryotes, the flat configuration allows sterols to

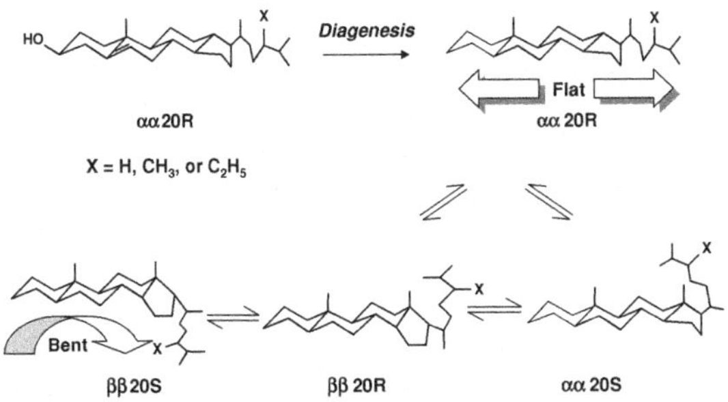

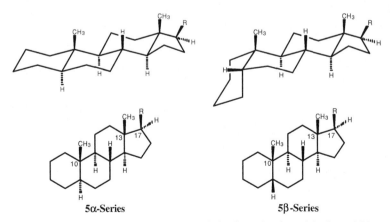

Figure 2.32. Origin of steranes from sterols as in Figure 2.31, but projecting structures from three dimensions. Reprinted with permission by ChevronTexaco Exploration and Production Technology Company, a division of Chevron USA Inc.

fit into and increase the rigidity of cell membranes in eukaryotes (Figure 2.30). The flat sterols show three-dimensional character (Figure 2.32) and are unlike benzene, where all 12 carbon and hydrogen atoms lie in one plane (Figure 2.4). Sterols in living organisms show the following configuration: $8\beta,9\alpha,10\beta(CH_3)$, $13\beta(CH_3),14\alpha,17\alpha(H)$ 20R. Many of these asymmetric carbon atoms are not particularly useful for geochemical applications and can be eliminated from further consideration, as explained below.

During diagenesis and catagenesis, the configurations at C-10 and C-13 cannot be changed because stereoisomerization mechanisms require that a hydrogen atom be attached to the asymmetric carbon atom. Furthermore, stereoisomerization at C-8 and C-9 does not occur because the biological configuration at these positions is energetically highly favorable. Thus,

the biological configurations at these positions are not changed during diagenesis and catagenesis, and they are of little use in characterizing these processes.

Both C-5 and C-24 in sterols from organisms exist as mixtures of α- and β-configurations. However, all sterols in living organisms appear to show only the 20R configuration. Tsuda *et al.* (1958) found the 20S epimer of fucosterol (sargasterol) in living organisms. Verification of this compound by additional work has been unsuccessful, and there are theoretical reasons for it not existing (Nes and McKean, 1977). C-24 in the sterols can be in the R or S configuration, but it is usually either completely R or completely S for a given organism. However, sterols in source rocks and petroleum consist of mixed 24R and 24S ergostane (C_{28}) and stigmastane (C_{29}). The 24R and 24S mixture may also originate in part by diagenetic hydrogenation of a double bond at C-24 found in some sterols.

Because most sterols carry a double bond at C-5, the stereochemistry at C-5 is determined largely by the reduction (hydrogenation) of this double bond during early diagenesis. This hydrogenation gives 5α- and 5β-mixtures favoring the 5α-epimer by $2:1$ to $10:1$ (Figure 2.33). Stanols and some sterols with saturated C-5 occur in minor concentrations in some organisms. These usually carry 5α-stereochemistry and may become 5α-steranes during early diagenesis. Bile acids are steroids secreted by the liver that carry 5β-stereochemistry, but they probably contribute little to sediments. Equilibration of C-5 in steranes greatly favors 5α- over 5β-stereochemistry. The 5β-compounds are seldom used because in thermally mature petroleum

5α-Series   **5β-Series**

Figure 2.33. Three-dimensional projections and wire-frame drawings of the 5α- and 5β-steranes.

they are generally trace components that require special analytical techniques for detection. However, when abundant, they can be useful to indicate low thermal maturity.

The important asymmetric centers during catagenesis of steranes are at C-14, C-17, and C-20 (Figure 2.31). Partly because C-20 is in the sterol side chain, relatively free from steric effects imposed by the cyclic system, the biologically derived 20R isomer converts to a near-equal mixture of 20R and 20S during thermal maturation. At equilibrium $20S/(20S + 20R) = 0.50$–0.55 for the C_{29} sterane homologs. Furthermore, the flat configuration imposed by the $14\alpha,17\alpha(H)$-stereochemistry in the sterol is lost in favor of the thermodynamically more stable $14\beta,17\beta(H)$ form (Figure 2.32).

Isomerization of the $5\alpha,14\alpha,17\alpha(H)$ 20R configuration ($\alpha\alpha\alpha R$) inherited from living organisms results in increasing amounts of the other possible stereoisomers until the equilibrium ratio for $\alpha\alpha\alpha R$, $\alpha\alpha\alpha S$, $\alpha\beta\beta R$, and $\alpha\beta\beta S$ is ~$1:1:3:3$ (Figure 2.31). These distributions were duplicated in (1) laboratory experiments, where $\alpha\alpha\alpha R$ steranes were heated using platinum–carbon catalyst (Petrov et al., 1976; Seifert and Moldowan, 1979) and (2) theoretical calculations for several cholestane (C_{27}) isomers (van Graas et al., 1982; Pustil'nikova et al., 1980). Chapter 14 describes the use of sterane isomerization ratios as thermal maturity indicators for petroleum.

EXERCISE

Using Fieser sp3 carbon models, prepare a model of the C_{30} isoprenoid squalane, as follows.

- Construct six isoprene (five carbon) units. For this exercise, disregard double bonds, i.e. make your isoprene units so that they resemble fully saturated 2-methylbutane.
- Using the isoprene units and head-to-tail linkage, construct two C_{15} isoprenoids (sesquiterpenoids). What are the formal (IUPAC) and informal names for these C_{15} isoprenoids (Figure 2.15)? Are the two C_{15} isoprenoids that you constructed identical in shape? How would you make pristane and phytane from these compounds? Identify the asymmetric carbon atoms in your C_{15} isoprenoid.

- Attach the two C_{15} isoprenoids tail to tail to make the triterpane called squalane. What is the formal name for this compound (see Figure 13.11)? Compare your squalane model with those prepared by others. Do they show different shapes? Why?

Squalane is the fully hydrogenated derivative of squalene, which is the biochemical precursor of the triterpenoid and steroid biomarkers. Without breaking or forming bonds, arrange your squalane model in a conformation like that of the steroid tetracyclic system (shown below). This part of the exercise may be difficult, but certain enzymes accomplish it readily in living systems.

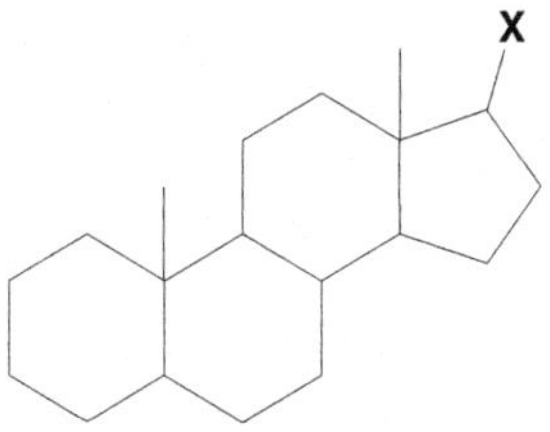

X = alkyl side chain

Steranes are an important class of steroidal biomarkers. The two-dimensional structure of the C_{27} sterane, cholestane, is shown below:

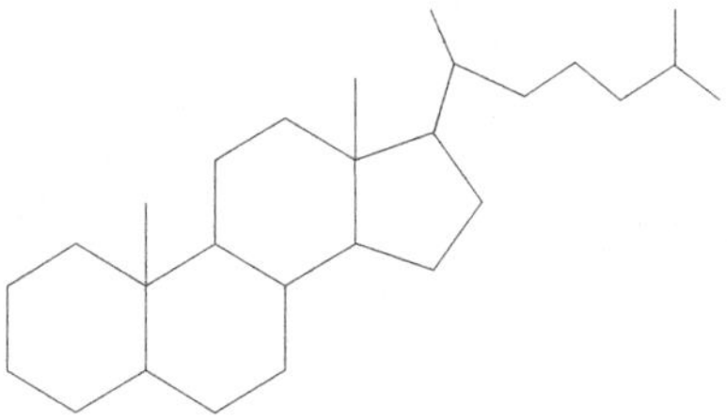

Using your squalane model, construct a three-dimensional model of cholestane by forming and breaking bonds as necessary.

- How many different three-dimensional structures (stereoisomers) of cholestane are possible? To answer this question, you need to count the number of asymmetric carbon atoms in the structure.
- What factors do you think control the cholestane stereoisomers found in thermally immature sediments compared with mature source rocks and petroleum?

3 · Biochemistry of biomarkers

This chapter provides a brief overview of the biochemical origins of the major biomarkers, including discussions of the function, biosynthesis, and occurrence of their precursors in living organisms. Topics include lipid membranes and their chemical compositions, the biosynthesis of isoprenoids and cyclization of squalene, and examples of hopanoids, sterols, and porphyrins in the biosphere and geosphere.

LIPID MEMBRANES

All living organisms have lipid membranes that are the interfaces between intercellular and extracellular environments (Figure 3.1). Lipid membranes define the boundary between life and non-life. Eukaryotes contain internal lipid membranes that further segregate the nucleus and various organelles, many of which evolved from symbiotic prokaryotes (e.g. mitochondria and chloroplasts). Lipid membranes serve various roles, but primarily they regulate the passage of water or solutes into or out of cells or their internal organelles. Solutes include inorganic ions, organic compounds for consumption, and a host of biosynthesized compounds that are excreted. Some of these solutes include metabolic waste products, biochemicals and biopolymers that protect the cell or allow motility or adhesion to surfaces, extracellular enzymes, and compounds for intracellular recognition and communication.

Lipids are the principal source for many compounds in petroleum, including the common biomarkers. Smith (1968) attributed the abundant normal alkanes in most crude oils as originating from lipids in living organisms, including naturally occurring n-alkanes and fatty acids. Likewise, high concentrations of isoalkanes and anteisoalkanes (2-methyl- and 3-methylalkanes, respectively) in crude oils appear to be due to their biological origin. For example, 2-methyloctadecane is biosynthesized by archaea (Brassell *et al.*, 1981). While Smith's

conclusion is generally true, we now recognize that some microalgae produce a highly aliphatic biopolymer called algaenan, which also can be a major source of petroleum hydrocarbons (Derenne *et al.*, 1994; Gelin *et al.*, 1994; Volkman *et al.*, 1998).

Note: The term "lipid" is poorly defined. One imprecise definition is that lipids are biomolecules that readily dissolve in organic solvents (e.g. petroleum ether, chloroform–methanol, hexane, or benzene). However, many biomolecules that meet this criterion clearly are not lipids. A preferable definition is that lipids are biomolecules composed of fixed oils, fats, and waxes. However, this excludes lipopolysaccharides, lipoproteins, sphingolipids, glycerophospholipids, glycoglycerolipids, and related compounds. We use the term "lipid" to refer to any fatty acid, fatty-acid derivative, or substance that is related by biosynthesis or function to these compounds.

Lipid membranes are typically composed of a bilayer of amphipathic compounds, molecules that contain both hydrophilic and hydrophobic surfaces, usually at opposite ends. In bilayers, the polar ends of one set of lipids face the aqueous interior of the cell, while the polar ends of another set of lipids face the external environment (Figure 3.1). The non-polar portion of both sets of lipids forms a hydrophobic zone within the membrane bilayer. Hydrocarbon-like lipids typically constitute most of the total mass of cellular membranes. The remainder is mainly proteins that are tightly integrated within the bilayer matrix or are associated more loosely (Figure 3.2). Those that are integrated within the lipid membrane always have a preferred orientation. Carbohydrates attached to enzymes or the polar heads of lipids (glycoproteins and glycolipids, respectively) always face the exterior of the membrane.

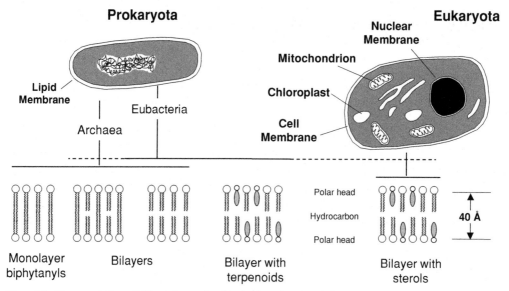

Figure 3.1. Mono– and bilayer lipid membranes found in the three domains of life.

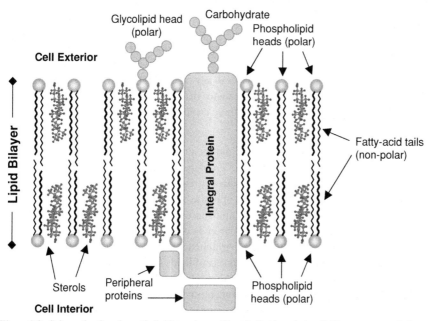

Figure 3.2. Schematic of a eukaryotic lipid membrane. Phospholipids and glycolipids are composed of two fatty acids bound to a polar head. These arrange to form a bilayer with a hydrophobic interior and a hydrophilic exterior. Sterols are used by eukaryotes to provide stability to the lipid membrane over a wide range of temperatures.

Most organisms have cell walls composed of polymers of sugars and peptides to provide structural support and to protect their lipid membranes. Eubacteria can be grouped phylogenetically based on the chemical response of the cell wall to the Gram stain, a purple-colored dye. The cell walls of bacteria consist of polymers of disaccharides cross-linked by short chains of peptides called peptidoglycan or murein. In Gram-negative bacteria, the cell wall is relatively thin, and an outer membrane composed of protein, phospholipid, and a unique lipopolysacchride surrounds the murein layer. In Gram-positive bacteria, the cell wall contains a thick layer of murein, which retains the purple color of the dye when treated (Figure 3.3). Ribosomal RNA indicates that the Gram-positive bacteria are closely related while the Gram-negative bacteria are diverse. Nevertheless, the Gram test is one of the major criteria for bacterial phylogeny. Most methanogenic archaea have cell walls similar to bacteria, but with some compositional differences in the peptide cross-linkages between the murein sheets. Other archaea have a complex cell wall composed of inorganic salts, polysaccharides, and glycoprotein or proteins. Many have a paracrystalline surface layer of proteins or glycoproteins (S-layer) that probably regulates solute flux (Claus *et al.*, 2002). Eukaryotic protists display a wide assortment of cell wall structures, including inorganic shells of carbonate and silica, and organic polymers, such as cellulose.

Note: Mycoplasmas are non-motile anaerobes or facultative aerobes that occur free-living in soil and sewage and as parasitic pathogens in higher organisms. These microorganisms lack many of the biosynthetic pathways common in other eubacteria. Some of these pathways are normally essential for life. Mycoplasmas are extremely small, typically 200–300 nm in diameter, and contain the smallest known genome, ~650 genes, which is about a fifth of that for common bacteria. Mycoplasmas should not be viewed as primitive organisms but as a highly evolved phylum that minimized its biochemistry, relying on the products of others.

Mycoplasmas are the only eubacteria that lack a cell wall. Consequently, they live in environments with minimal risk of osmotic shock. Instead of a cell wall, they have a single triple-layered membrane that, in most species, is stabilized by sterols. These sterols are not produced by mycoplasmas but are obtained from their host organisms or environment. Acholeplasma is the only mycoplasma that can survive without externally supplied sterols. This species produces polar carotenoids (see Figure 3.8).

MEMBRANE LIPIDS

The main constituents of lipid membranes in eubacteria and eukarya are glycolipids and phospholipids, which are composed of fatty acids connected to a polar head (Figure 3.4). The fatty-acid chains are typically 12–24 carbons long and may be fully saturated or contain one or more double bonds. Long-chain *n*-alkanoic acids, such as C_{24}, C_{26}, and C_{28}, are major components in the epicuticular waxes of land plants (Rieley *et al.*, 1991). Shorter-chain C_{12}, C_{14}, and C_{16} *n*-alkanoic acids are produced by all plants but dominate the lipids in algae (Cranwell *et al.*, 1987). Various moieties can be attached to the fatty acids to form the polar head. The most prevalent moieties are a family of phosphates with nitrogen bases and sugars.

Triacylglycerols (TAGs) are fatty-acid triesters of glycerol that are widespread among the eukarya. TAGs in eubacteria are reported rarely, but some species belonging to the actinomycetes (e.g. *Mycobacterium*, *Streptomyces*, *Rhodoccoccus*, and *Nocardia*) may contain up to 87% of their dry cell weight as TAGs (Alvarez and Steinbüchel, 2002). The principal use of TAGs by both eukarya and eubacteria seems to be as reserve

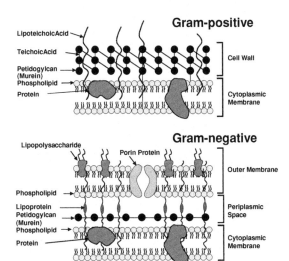

Figure 3.3. Bacterial cell wall structures of the Firmicutes (Gram-stain positive) and the Gracilicutes (Gram-stain negative).

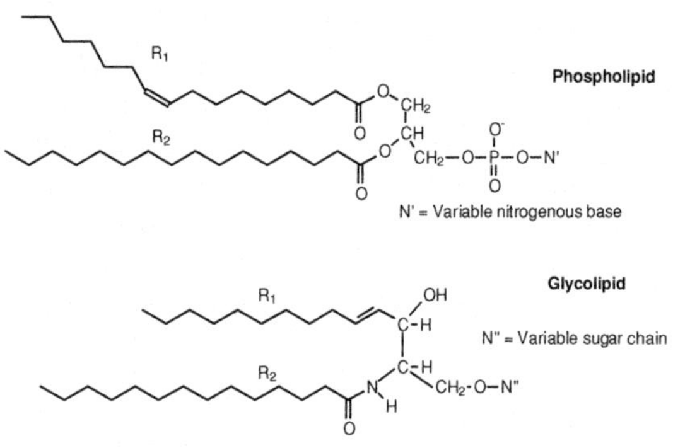

Figure 3.4. Examples of phospholipids and glycolipids that are major constituents in the lipid membranes of eubacteria and eukarya. The non-polar side chains (R1 and R2) originate from fatty acids that typically have 12–24 carbon atoms and may be saturated or contain one or more double bonds. Common additions to phospholipids include $N' = H$, ethanoamine, choline, serine, and glycerol. Sugars are added to glycolipids (N'') or to phospholipids (N'), forming phosphoglycolipids.

compounds for later consumption, although they may play a role in regulating membrane fluidity by keeping unusual fatty acids away from membrane lipids.

Fatty-acid biosynthesis is characterized by an initiation step followed by chain elongation (Figure 3.5). The initiation step between bicarbonate and acetyl-coenzyme A (acetyl-CoA) that forms malonyl-CoA is catalyzed by acetyl-CoA carboxylase (ACC). Malonyl-CoA reacts with another acetyl-CoA attached to the acyl carrier protein (ACP), forming the acetoacetyl-ACP complex. The latter can enter into the chain elongation cycle, which successively adds two carbons to form fatty acids with various chain lengths. Thus, the most common fatty acids and related membrane lipids have an even number of carbon atoms (Table 3.1).

The membrane lipids of the archaea differ from those in eubacteria and eukarya (Table 3.2). Archaeal lipids are characterized by isoprenoid hydrocarbons bound together by ether linkages (Figure 3.6). They may have a single polar head with isoprenoid side chains ranging from 15 to 25 carbons, or they may have two polar heads joined by C_{40} biphytanyl isoprenoids.

LIPID MEMBRANE FLUIDITY

Lipid membranes function as liquid crystals, an intermediate state between solid gel and molten liquid that occurs only within a narrow temperature transition.

Within this transition, the fatty-acid portion of the membrane melts and undergoes random movement, while the polar heads maintain cohesion. At lower temperatures, the bilayer lipids "freeze" into a highly ordered, densely packed configuration. At higher temperatures, the bilayer lipids become disordered.

The three domains of life evolved in separate ways to extend the transition temperature range at which their lipid membranes remain in the liquid-crystal state. Archaea use diether- and tetraether-bound isoprenoids as the main structural components of their membranes (Figure 3.6; see also Figure 13.12). Little is known about the biosynthetic pathways that give rise to these lipids (Bullock, 2000). Typical diether archaeal lipids contain C_{20}, C_{25}, and C_{40} isoprenoids and are counterparts of the diglycerides in the bacteria and eukaryotes. Diether lipids are the dominant membrane lipids for halophilic archaea and occur in other archaeal lineages, including the methanogens. Biphytanyl tetraethers, a subset of

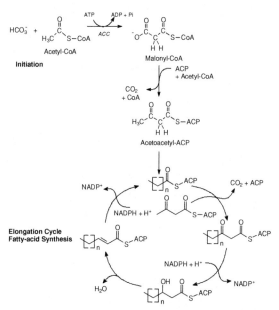

Figure 3.5. Fatty-acid synthesis begins with the formation of acetoacetyl-ACP (acyl carrier protein) from bicarbonate, acetyl-CoA (co-enzyme A), and ACP. Acetyl-CoA carboxylase (ACC) catalyzes the initial reaction. Once acetoacetyl-ACP is formed, it can enter the elongation cycle for fatty-acid synthesis, which successively adds two carbons to the chain.
ADP, adenosine diphosphate; ATP, adenosine triphosphate; $NADP^+$, nicotinamide adenine dinucleotide phosphate; NADPH, reduced form of NADP.

Table 3.1. *Names, double-bond locations, and structures of common fatty acids*

IUPAC name	Common name	Double bond	Structure
Saturated			
n-Decanoic	Capric	10:0	$CH_3(CH_2)_8-COOH$
n-Dodecanoic	Lauric	12:0	$CH_3(CH_2)_{10}-COOH$
n-Tetradecanoic	Myristic	14:0	$CH_3(CH_2)_{12}-COOH$
n-Hexadecanoic	Palmitic	16:0	$CH_3(CH_2)_{14}-COOH$
n-Octadecanoic	Stearic	18:0	$CH_3(CH_2)_{16}-COOH$
n-Eicosanoic	Arachidic	20:0	$CH_3(CH_2)_{18}-COOH$
n-Docosanoic	Behenic	22:0	$CH_3(CH_2)_{20}-COOH$
n-Tetracosanoic	Lignoceric	24:0	$CH_3(CH_2)_{22}-COOH$
Unsaturated			
cis-9-Hexadecanoic	Palmitoleic	16:1	$CH_3(CH_2)_5CH=CH(CH_2)_7-COOH$
cis-9-Octadecanoic	Oleic	18:1	$CH_3(CH_2)_7CH=CH(CH_2)_7-COOH$
cis,cis-9, 12-Octadecanoic	Linoleic	18:2	$CH_3(CH_2)_4CH=CHCH_2CH=CH(CH_2)_7-COOH$
cis,cis,cis-9,12, 15-Octadecanoic	Linolenic	18:3	$CH_3CH_2CH=CHCH_2CH=CHCH_2CH=CH(CH_2)_7-COOH$
cis,cis,cis,cis-5,8,11, 14-Eicosatetraenoic	Arachidonic	20:4	$CH_3(CH_2)_4(CH=CHCH_2)_3CH=CH(CH_2)_3-COOH$

Table 3.2. *Comparison of lipids from the three domains of life (modified from* Itoh *et al., 2001)*

Lipid class	Archaea	Eubacteria	Eukarya
Glycerolipids	+	+	+
Hydrocarbon chain	Isoprenoid	Fatty acid	Fatty acid
Carbons in compound	C_{15-25}/C_{40}	C_{12-24}	C_{12-24}
Hydrocarbon bonding type	Ether	Ester	Ester
Position in glycerol	*sn*-2,3	*sn*-1,2	*sn*-1,2
Phospholipids	+	+	+
Glycolipids	+	+	+
Phosphoglycolipids	+	+	+
Sulfoglycolipids	+	−	+
Phosphosulfogylcolipids	+	−	−
Sulfolipids	−	−/(+)	+
Sphingolipids	−	−/(+)	+
Hopanoids	−	+	−/(+)
Steroids	−	−/(+)	+

+, present in all or some species; −, absent; −/(+), absent in most species.

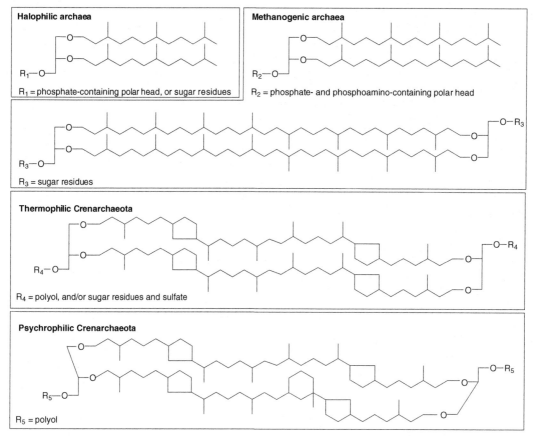

Figure 3.6. Examples of lipids in archaea.

glycerol dialkyl glycerol tetraethers (GDGTs), are the dominant membrane lipids in other archaea. These lipids form a monolayer membrane that is stabilized by their bipolar nature. Both polar ends of the tetraethers are kept taut by hydrophilic attraction to the aqueous phase, which allows little freedom of movement within the hydrophobic interior of the membrane. To grow at temperatures >80°C, hyperthermophilic Crenarchaeota incorporate one to four cyclopentyl rings within the biphytanyl ester lipids (Figure 3.6), resulting in a more densely packed, stiffer, and thermally more stable membrane. For a given hyperthermophile, the average number of cyclopentyl rings increases when grown at increasing temperature (De Rosa *et al.*, 1980, 1991; Sugai *et al.*, 2000; Uda *et al.*, 2001). The average number of cyclopentyl rings, however, does not necessarily correlate with optimal or maximum growth temperatures across archaeal species (Itoh *et al.*, 2001).

Some hyperthermophilic archaea form a bilayer of biphytanyl tetraethers, resulting in a double-thickness lipid membrane (Luzzati *et al.*, 1987).

Marine pelagic Crenarchaeota further modified the biphytanyl lipids by adding a cyclohexyl ring along with the more common three-cyclopentyl rings. This modification disrupts the dense packing structures in the related thermophilic archaea and allows the lipid membrane to remain fluid at low temperatures (Sinninghe Damsté *et al.*, 2002a). By evolving this modification, the Crenarchaeota successfully adapted to psychrophilic (<20°C) marine conditions, where they flourished to become probably the most abundant group of archaea on Earth (Karner *et al.*, 2001).

The eubacteria and eukarya evolved two strategies to moderate membrane fluidity: (1) regulation of fatty-acid length or the placement of unsaturated bonds and (2) the use of cyclic lipids. During growth, the length and

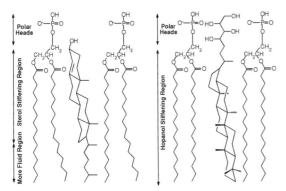

Figure 3.7. Insertion of sterols (eukarya) and hopanols (eubacteria) into their lipid membranes to provide increased ridigity and stability.

degree of unsaturation of the fatty-acid chains can be adjusted in response to temperature. Lipid membranes with shorter side chains or double bonds "melt" at lower temperatures. A membrane composed of phospholipids containing only saturated hydrocarbons has a transition temperature of ~40–50°C.

Eubacteria and eukarya use similar compounds to modify their lipid membranes, the hopanoids and the sterols, respectively. These lipids are the source of the major saturated biomarkers in petroleum, the hopanes and steranes. These amphipathic biomolecules are roughly the same shape (see Figure 2.30) and can be inserted between the glycolipids of the bilayer membranes with their polar ends in contact with the aqueous phase. The ring structures of these compounds are rigid, which restricts the random movement of the fatty-acid chains closest to the polar head groups. Depending on size and structure, the aliphatic tail of sterols may interact with the distal portion of the fatty-acid chain, causing the inner regions of the bilayer to become more fluid (Figure 3.7).

Note: Neurons in your brain communicate with each other through synapses, which require cholesterol to function properly. Mauch *et al.* (2001) found that brain neurons produce enough cholesterol to grow; however, without additional cholesterol provided by other types of brain cells called glial cells, production of synapses is limited. The brain cannot tap the cholesterol supply in the blood because the lipoproteins that carry cholesterol are too big to pass into the brain. A cholesterol-carrying brain lipoprotein is suspected

of playing a role in the loss of synaptic plasticity that occurs in Alzheimer's disease.

Your body manufactures cholesterol, but it can also be ingested in food. The average person contains 300–600 mg of sterol/100 g of wet weight. A 150-lb (68-kg) person contains ~306 g of cholesterol (Nes and McKean, 1977). A diet of fatty foods contributes to high levels of blood cholesterol, commonly found in people suffering from arteriosclerosis. Arteriosclerosis occurs when cholesterol and other fatty substances become embedded in the walls of arteries, forming plaques that gradually restrict blood flow and increase blood pressure.

Cholesterol can be transported in the blood by two classes of lipoproteins. Low-density lipoproteins (LDL) carry cholesterol to the organs and tissues, where it is used by cells. Excess LDL is responsible for deposits of cholesterol that result in plaque. For this reason, LDL is sometimes called "bad cholesterol." High-density lipoproteins (HDL; sometimes called "good cholesterol") are believed to carry excess cholesterol from cells to the liver, where it is removed from the system.

Hopanoids have been detected in only ~30% of the examined eubacteria. Their apparent absence in some cases may be due to analytical difficulties or to lack of expression of biosynthetic pathways under specific culture conditions. The number and diversity of bacterial species that contain these compounds will certainly grow. However, no hopanoids have been detected in obligate anaerobes, suggesting that these eubacteria may not utilize these compounds to rigidify lipid membranes.

Rohmer *et al.* (1979) suggested that some eubacteria might use polar carotenoids to strengthen their lipid membranes (Figure 3.8). Like the archaeal biphytanyl tetraethers, these molecules could span the width of a lipid membrane, thus enhancing rigidity. There is evidence for use of α,ω-dipolar carotenoids in the lipid membranes of some bacteria, but the use of these

Figure 3.8. α,ω-Dipolar carotenoid synthesized by *Acholeplasma* in the absence of an external source of sterols.

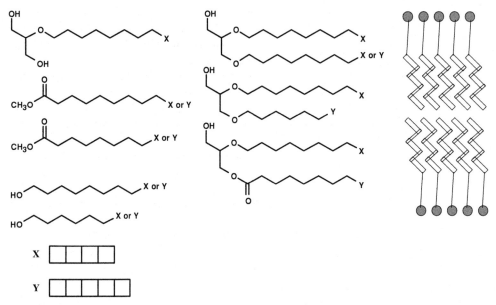

Figure 3.9. Structures of *Anammox* ladderane membrane lipids and model of the anammoxosome membrane (after Sinninghe Damsté *et al.*, 2002b). Ladderanes are linearly fused cyclobutanes (X, Y) with all *cis*-ring junctures.

compounds other than as photosensitive pigments is largely unknown (Ourisson and Nakatani, 1994).

Note: *Anammox* is a group of recently discovered anaerobic bacteria that live by oxidizing ammonia with nitrate (Strous *et al.*, 1999; Jetten *et al.*, 2001). They are autotrophs that rely on an electrochemical ion gradient to generate ATP. This unusual metabolism occurs within an organelle-like structure, the anammoxosome, where highly toxic intermediate products, such as hydrazine (N_2H_4) and hydroxylamine (NH_2OH), are retained within a lipid bilayer of unique composition. Sinninghe Damsté *et al.* (2002b) discovered that the anammoxosome core lipids are composed of alcohols, fatty acids, and glycerol diethers that contain ladderanes, which consist of up to five linearly fused cyclobutanes arranged with cis-ring junctions (Figure 3.9). Glycerol diethers are more commonly found in archaea, but they are known to occur in other, mostly thermophilic eubacteria. However, ladderanes are unprecedented in biota, although they have been synthesized and used in optoelectronics. Sinninghe Damsté *et al.* (2002b) considered this an evolutionary adaptation that allows *Anammox* to retain the reactive and highly toxic intermediates within the anammoxosome. Hydrazine and hydroxylamine tend to diffuse rapidly through typical lipid membranes. However, the ladderane lipids appear to form a dense, relatively impermeable membrane that retains these molecules. *Anammox* grow very slowly, dividing once every 2–3 weeks. As such, they must maintain an exceptionally retentive membrane that limits diffusion and protects the rest of the cell from the highly toxic intermediates.

BIOSYNTHESIS OF TERPENOIDS

Terpenoids occur in all living organisms. They include all compounds composed of two or more isoprene units and range from the C_{10} monoterpenoids to the C_{2500+} polyisoprenes of rubber (*cis*-units) and gutta-percha (*trans*-units). Terpenoids perform a broad range of biological functions. They serve as stabilizers in lipid membranes and light-gathering pigments. Terpenoids also serve as growth and sex hormones, pest repellents, and agents for external communication.

In general, the classes of terpenoids used as lipid-membrane stabilizers differ between the three domains of life. Archaea and eubacteria synthesize C_{40} biphytanyl isoprenoids and hopanoids, respectively, using pathways that do not require free oxygen. Eukarya synthesize sterols along pathways that require free oxygen. Nevertheless, these biosynthetic pathways are similar, indicating that they must have arisen among the earliest forms of life.

> **Note:** Woese (2002) proposed a new theory for the evolution of life. Instead of three domains of life diverging from a common ancestor, the earliest cells hosted many genes, which freely interchanged through horizontal gene transfer (HGT). Cellular evolution was communal, with biochemical advances being shared by the entire ecosystem. Eventually, some organisms achieved sufficient complexity to leave the communal HGT gene pool and evolve independently. Woese terms this event the "Darwinian threshold" and envisions that there were many types of cellular form. The three domains that exist today are simply those that survived extinction. The similar pathways for terpenoid synthesis among the remaining life forms

suggest that many of the major synthetic pathways may date to the time of the universal HGT ecosystem.

Biosynthesis of isopentyl pyrophosphate

As discussed above, many acyclic and cyclic biomarkers consist of repeating units of isoprene (C_5H_{10}). In biosynthesis, the basic building blocks of all terpenoids include isopentyl pyrophosphate (IPP) and dimethylalkyl pyrophosphate (DMAPP), which can be interconverted by enzymes (Figure 3.10). The condensation of three molecules of acetic acid via mevalonic acid was long thought to be the universal pathway for all organisms. An alternative pathway for the formation of isoprene units was found in eubacteria, which involves condensation of pyruvate with glyceraldehyde-3-phosphate (Rohmer, 1993; Rohmer and Bisseret, 1994; Rohmer, 1999). This then undergoes condensation with a C_2 thiamine into 2C-methyl-D-erythritol 4-phosphate, which then converts to IPP. The exact

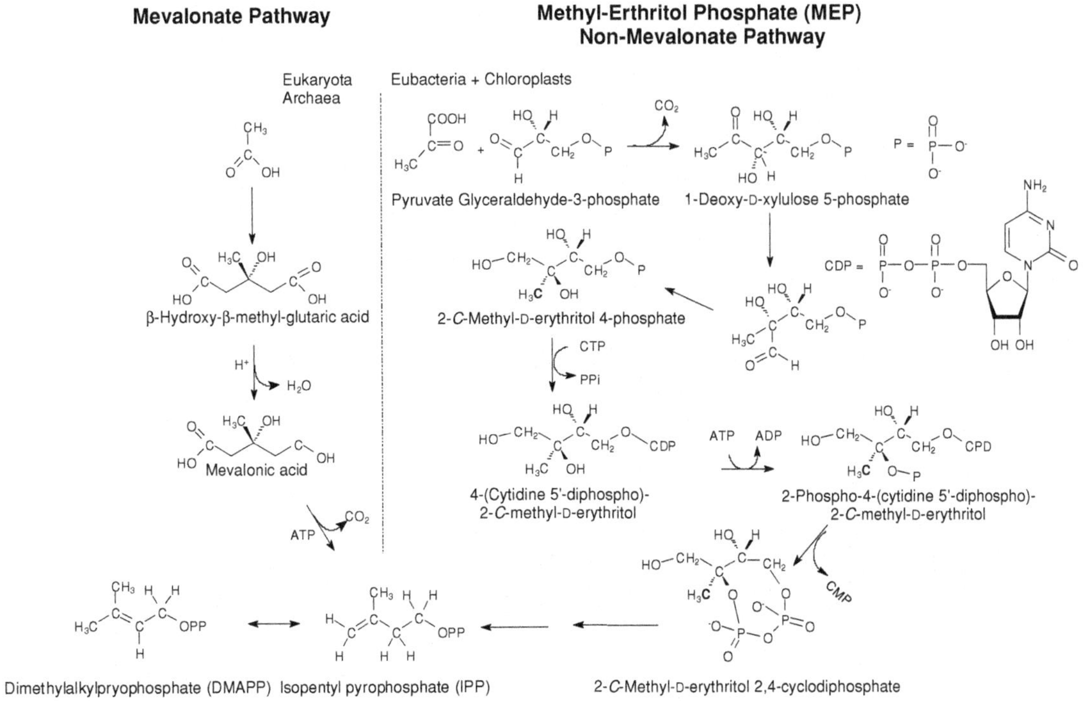

Figure 3.10. Biochemical pathways for the synthesis of isopentyl pyrophosphate. The enzymes and steps involved in final conversion to IPP are not yet fully known for the MEP pathway.

enzymes and reactions for the final step remain to be discovered.

The distribution of the mevalonate (MEV) and methyl-erthritol phosphate (MEP) pathways (Figure 3.10) in the tree of life is now being determined (Rohdich *et al.*, 2001). All animals, fungi, and archaea use the MEV pathway. Most eubacteria and some algae, protists, and land plants use the MEP pathway (Knöss and Reuter, 1998; Lichtenthaler, 2000; Meganathan, 2001; Paseshnichenko, 1998; Rohdich *et al.*, 2001). In eukaryotes with chloroplasts, IPP is synthesized by the MEV pathway in the nucleus and fluid within the cytoplasm, and by the MEP pathway in the chloroplasts. Some organisms (e.g. certain Gram-positive eubacteria) have genes for both pathways; either both genes can be activated or the genes for only one pathway may be turned on. For example, the archaea *Pyrococcus horikoshii* has genes for the MEP pathway, but only the MEV pathway is active for the production of IPP.

Note: Because all animals use only the MEV pathway, chemicals that block the MEP pathway could be ideal herbicides and bacteriocides. Drugs based on blocking the MEP pathway are being developed and investigated. Such compounds are effective against malaria (Jomaa *et al.*, 1999).

Biosynthesis of isoprenoids

All isoprenoids are constructed by the condensation of IPP or higher homologs (Figure 3.11). Table 3.3 lists various terpenoid classes and their prenol precursors. C_{10} geranyl pyrophosphate (GPP) forms by condensation of IPP and DMAPP. C_{15} farnesyl pyrophosphate (FPP) forms by condensation of GPP with IPP. The process is repeated with the condensation of FPP and IPP to yield C_{20} geranylgeranyl pyrophosphate (GGPP). Two units of FPP can condense tail to tail to form squalene, which is the precursor for all tetra- and pentacyclic triterpenoids. Condensation of two GGPP units yields the C_{40} carotenoids and archaeal lipids. Bacteria generally synthesize carotenoids as hydrocarbons or open-chain alcohols, while higher protists and plants tend to synthesize a variety of oxygenated carotenoids.

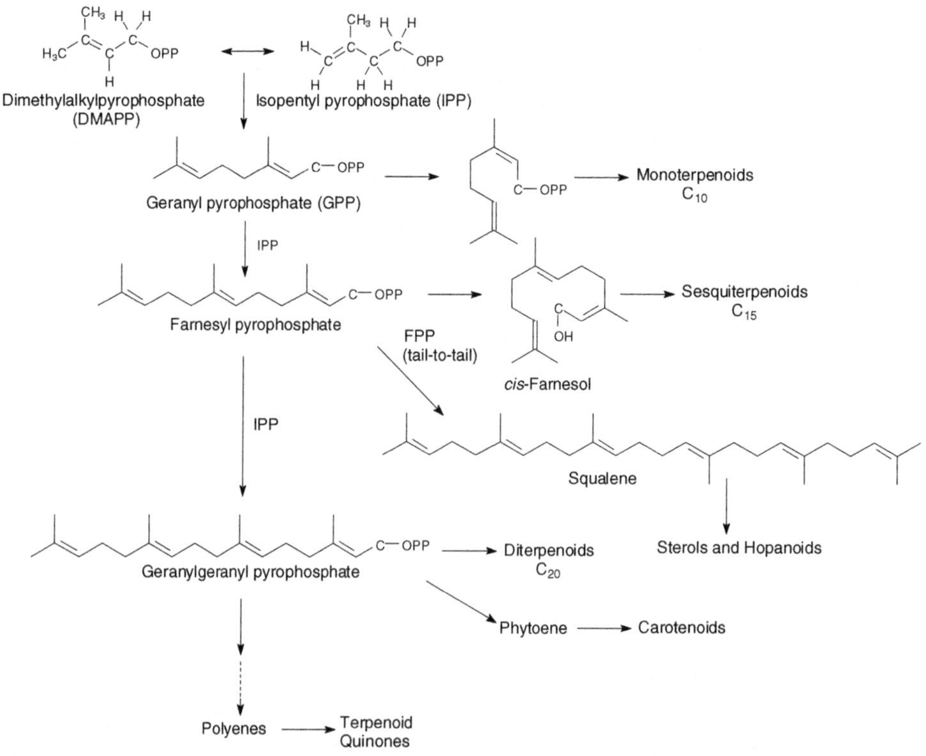

Figure 3.11. Biochemical pathways for the synthesis of terpenoids.

Table 3.3. *Names of isoprene polymers and their terpenoid classes*

Number of residues	Prenol precursor (as diphosphate, etc.)	Terpenoid class
1	Dimethylallyl alcohol	Hemiterpenoid
2	Geraniol or nerol	Monoterpenoid
3	Farnesol	Sesquiterpenoid
4	Geranylgeraniol	Diterpenoid
5	Geranylfarnesol	Sesterterpenoid
6	Farnesol[a]	Triterpenoid[a]
8	Geranylgeraniol[b]	Tetraterpenoid or carotenoid[b]
Many	–	Rubber (*all-cis*), gutta-percha (*all-trans*)

[a] Triterpenoids form from squalene, which originates by condensation of two farnesyl diphosphate precursors.

[b] Carotenoids form from phytoene, which originates from two geranylgeranyl diphosphate (GGPP) precursors. Similarly, the biphytanyl lipids in archaea form from the head-to-head condensation of two GGPP molecules.

The larger polyisoprenes (e.g. ubiquinones, rubber; see Figure 2.15) are synthesized by the head-to-tail condensation of DMAPP with successive IPP units.

Biosynthesis of mono-, sequi-, and diterpenoids

In prokaryotes, GPP, FPP, and GGPP are used to create larger terpenoids and in the synthesis of bacteriochlorophylls. Eukaryotic organisms take advantage of this pathway and divert the smaller isoprenoids along new biosynthetic pathways. For example, higher land plants generate a diverse array of acyclic, monocyclic, and bicylic monoterpenoids (Figure 3.12), sequiterpenoids, and diterpanoids (Figure 3.13). Eukaryotic marine organisms, higher plants, and fungi use these precursors to make various free diterpenoids (acids, alcohols, hydrocarbons, and phenols) and as monomers in the synthesis of resins and tissues. For example, cadinenes condense into polycadinenes in dammar resins (structures shown in Figure 13.64). The ability to synthesize these smaller terpenoids provided the eukaryotes with an evo-

lutionary advantage. These compounds disrupt further synthesis of tri- and tetra-terpenoids or detrimentally change the permeability of membranes, and thus represent useful defensive chemicals against predators or competitors (Swain and Copper-Driver, 1979). Since the mono-, sesqui-, and diterpenoids are generally not found in prokaryotes, their hydrocarbon derivatives in petroleum are excellent biomarkers for eukaryotic input. Higher plants synthesize some of these smaller terpenoids, thus providing additional source specificity.

Hemiterpanes and monoterpanes have limited use as biomarkers because their precursors are chemically reactive and highly volatile. Many of these precursors are key fragrance molecules that attract both pollinators and humans to flowers (Figure 3.14). Rosy-floral scents, such as citronellol, are among the most common. They are released from centifolia roses on exposure to sunlight. Certain moth-pollinated flowers, such as jasmine, release their white-floral scents only at night. Freesias emit ionone-floral scents with the rich odor of β-ionone, which originates from carotenes. The structures of all of these compounds contain repeating isoprene subunits.

Plants use volatile terpenoids to defend against infection and predation. Although most defenses involve terpenoids as toxins or secretions to trap predators, more complex relationships exist. For example, when injured by herbivore insects, some plants emit terpenes and other volatile compounds that attract predatory insects (Kessler and Baldwin, 2001).

Cyclization of squalene

Squalene is a critical intermediate isoprenoid for all domains of life. This unsaturated hydrocarbon is used directly by some archaea, but its principal importance is as the biosynthetic precursor for both hopanoids and steroids (Figure 3.15). The pathways for the synthesis of hopanoids do not require molecular oxygen. All oxygen in hopanoids originates exclusively from water. In contrast, several steps in the biosynthetic pathways for sterols require free oxygen. Nevertheless, considerable similarity in the enzymes and reactions suggests that the biosynthetic route to steroids evolved from that for the synthesis of hopanoid membrane lipids (Rohmer *et al.*, 1979).

In the absence of free oxygen, the enzyme squalene-hopene cyclase folds squalene into pentacyclic triterpenes. This cyclization is one of the most complex

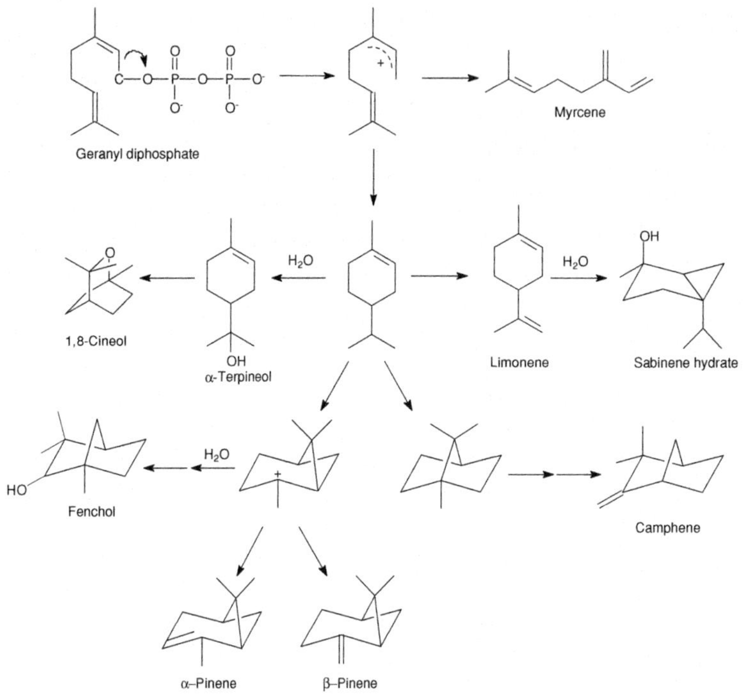

Figure 3.12. Synthesis of some monoterpenoids from geranyl pyrophosphate, also called geranyl diphosphate.

single-step reactions in biochemistry because it requires the simultaneous formation of 5 rings, alteration of 13 covalent bonds, and formation of 9 stereoisomeric centers (Kannenberg and Poralla, 1999). The initial products of the reaction are diploptene, diplopterol, and tetrahymanol. The latter compound was first isolated from the eukaryotic ciliate *Tetrahymena pyriformis*, which activates squalene-hopene cyclase in the absense of exogenous sterols. Green sulfur bacteria (Grice *et al.*, 1998) and marine ciliates (Harvey and McManus, 1991) are the major sources of sedimentary tetrahymanol, which can be converted to gammacerane during diagenesis.

Diplopterol and diploptene can be further transformed by methylation, dehydration, and the addition of various polar moieties. D-Pentose is added to form C_{35} bacteriohopanetetrol, the first functionalized hopanoid to be discovered. Further modification can occur by alteration of the side chain or by condensation of additional polar moieties attached at the C-35 carbon (Sahm *et al.*, 1993). These hopanoids with large polar groups are difficult to separate from membrane phospholipids. Analytical procedures for their detection and

identification in recent sediments were developed only recently (e.g. Fox *et al.*, 1998; Watson and Farrimond, 2000; Talbot *et al.*, 2001), and the diversity of these compounds in bacteria is only now beginning to be discovered.

Squalene can be converted to 2,3-oxidosqualene in the presence of oxygen. Enzymes convert this precursor into hydroxylated tetra- and pentacyclic terpenoids (Figure 3.15) (Ran *et al.*, 2004). The stereochemical constraints are critical and provide insight into the evolution of sterols (Nes and Venkatramesh, 1994). The common sterols originate from cycloartenol, which is synthesized by higher plants and algae (Figure 3.16) and lanosterol, which is synthesized by animals and fungi (Figure 3.17). Oxygen is required to remove the 14α-methyl group from these sterol precursors (Chapman and Schopf, 1983). Like the hopanoids, sterols can be modified by rearrangement of side chains, addition of functional groups or polar moieties to side chains or rings, and changes in the number or position of double bonds. Eukaryotes have many enzymes to catalyze the same reactions involved in the biosynthesis of sterols and other membrane lipids (Vance, 1998).

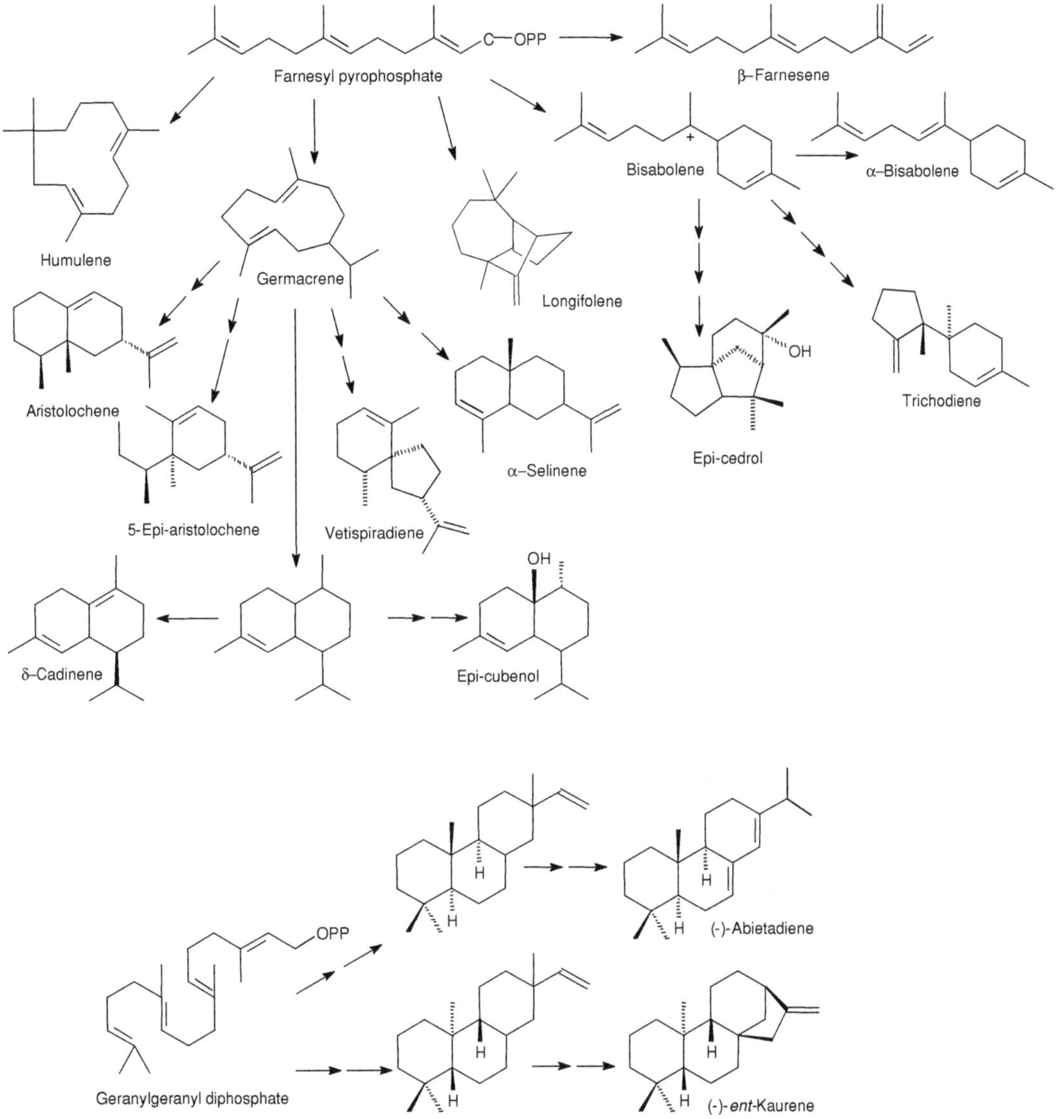

Figure 3.13. Synthesis of some common sesquiterpenoids and diterpenoids from farnesyl pyrophosphate (FPP) and geranylgeranyl diphosphate (GGPP), respectively.

It was once thought that the use of cycloartenol by higher plants and algae was the result of a modification of the lanosterol pathway. However, cycloartenol can substitute for cholesterol in lipid membranes, while lanosterol cannot. This is likely due to cycloartenol having a bent structure that partially shields the 14–methyl group, while the 14-methyl extends outward in lanosterol (Bloch, 1983). Ourisson and Nakatani (1994)

speculated that the formation of cycloartenol was not simply a late evolutionary adaptation of photosynthetic eukaryotes, but was a vestige of the earliest biosynthetic pathway for sterols. Primitive eukaryotes living in dysoxic conditions may have synthesized a mixture of terpenoids, switching between the anoxic and oxic pathways. Increasing concentrations of free oxygen in the atmosphere might drive the degradation of cycloartenol

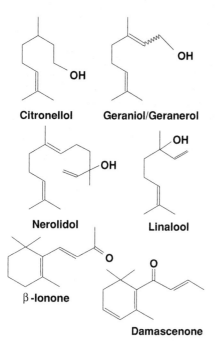

Citronellol   **Geraniol/Geranerol**

Nerolidol   **Linalool**

β-Ionone

Damascenone

Figure 3.14. Examples of floral scents. Rosy-floral scents include geraniol or geranerol and citronellol (monoterpenoids, top). White-floral scents include nerolidol (sesquiterpenoid) and linalool (monoterpenoid, middle). Ionone floral scents include β-ionone and damascenone, which are degradation products of carotenes.

toward the modern sterols. Once these pathways were established, lanosterol could substitute as a starting precursor.

One evolutionary advantage of sterol biosynthesis is that eukaryotic oxidosqualene cyclases involved with the formation of lanosterol and cycloartenol are more efficient than the squalene-hopene cyclases (Abe *et al.*, 1993). Under certain growth conditions, the bacterium *Zymomonas mobilis* produces many triterpenoids that indicate failure of squalene cyclase to control the cyclization process or to exert a high degree of substrate specificity (Douka *et al.*, 2001). Many of the hydrocarbons produced by this strictly fermentative Gram-negative bacterium have not been observed previously in surveys of bacterial neutral lipids (Figure 3.18).

HOPANOIDS AND STEROLS IN THE BIOSPHERE AND GEOSPHERE

Hopanoid lipids

One of the earliest successes of petroleum biomarker research was the prediction of bacteriohopanoid lipids

in the biosphere based on bacteriohopane in the geosphere. In many cases, the biological precursors of biomarkers were known before their discovery in the geosphere. Phytane could originate from the phytol side chain of chlorphyll (Dean and Whitehead, 1961), steranes were from saturated sterols (Burlingame *et al.*, 1965), and perhydro-β-carotane was from β-carotene (Murphy *et al.*, 1967). However, some compounds (e.g. C_{30+} hopanes, the C_{20+} tricyclic terpanes, and the highly branched isoprenoids) had no known biological precursor at the time of their identification in samples from the geosphere. Diploptene and diplopterol (Figure 3.15) were long known to be major lipids in some cyanobacteria (Gelpi *et al.*, 1970) and eubacteria (Bird *et al.*, 1971) and could account for the C_{27}–C_{30} hopanes. The biological origin of the extended hopanes was problematic (van Dorsselaer *et al.*, 1974) because they are ubiquitous in the geosphere, abundant in extracts of immature strata, and are clearly biogenic based on their optical activity. Based on the occurrence of the geohopanes, Rohmer and Ourisson (1976a, 1976b) and Langworthy and Mayberry (1976) searched for, and found, C_{35} bacteriohopanetetrol in eubacteria. Since this success, biological precursors have been found for many biomarkers. The origins of some prominent geolipids (e.g. the C_{20+} tricyclic terpanes, polyprenoids, and 3-substitituted steranes) remain controversial (Ourisson and Nakatani, 1994).

In addition to bacteriohopanetetrol, other polyfunctional bacteriohopanoids (C_{35}) with additional functional groups on the side chain are known (Figure 3.15). For example, an amino group (–NH₂) might replace the hydroxyl group (–OH) attached to C-35 (Rohmer, 1987). Further modifications of hopanoid lipids occur in eubacteria. These modifications include methylation at C-2 or C-3, double bonds at C-6 and/or C-11, and the addition of polar groups at C-34 (Sahm *et al.*, 1993). Bacteriohopanepolyols, the presumed major precursors for extended hopanes, vary in composition between sediments and may preserve specific bacterial source information. For example, Farrimond *et al.* (2000) analyzed hopanols cleaved from organic matter by treatment with periodic acid and sodium borohydride. Bacteriohopanetetrol occurred in high concentrations (up to 1500 μg/g TOC) but accounted for only a small proportion (0–26%) of the total hopanoids in all but one sample. Tetra-functionalized hopanoids dominated seven of the sediments and are probably the major source of extended hopanes in source

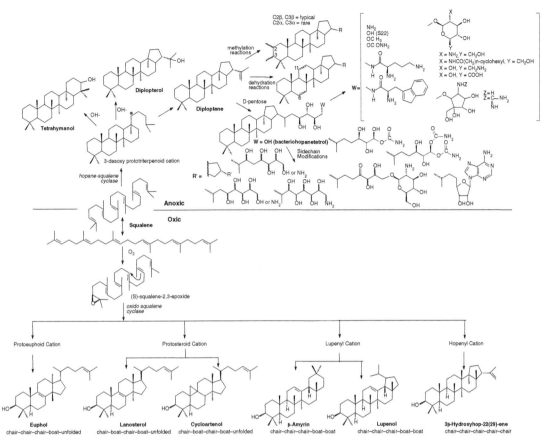

Figure 3.15. Biosynthesis of major terpenoids and terpenoid precursors.

rocks and crude oils. Hexa-functionalized hopanoids were most abundant in lacustrine sediments, particularly those from two highly productive stratified lakes (20–40% of total hopanoids), possibly due to greater contribution to the organic matter by methanotrophic bacteria. In marine sulfate-reducing sediments, hexa-functionalized hopanoids are minor constituents (2–6% of total hopanoids).

Hopanoid lipids are largely confined to eubacteria, although our knowledge of their phylogenetic distribution is incomplete. No clear pattern has emerged relating the presence of hopanoids to taxonomic classification. For example, both Gram-positive and Gram-negative eubacteria produce hopanoids. Hopanoids are relatively abundant in cyanobacteria, methanotrophic bacteria, and members of the α-proteobacteria, particularly the nitrogen-fixing eubacteria (Kannenberg and Poralla, 1999). Hopanoid contents in most of these microbes are ~0.1–5 mg/g dry cell weight, similar to the average

sterol content in eukaryotic cells. A few bacteria have exceptionally abundant hopanoid lipids (>30 mg/g dry cell weight) (Sahm et al., 1993). Hopanoids commonly occur in bacterial strains with high G + C (guanine + cytosine), characteristic of species that occupy stressful ecological niches (Kannenberg and Poralla, 1999). Their abundance may vary in response to osmotic stress (e.g. high salinity, sugar, or ethanol concentrations), increasing to provide greater membrane stability (Hermans et al., 1991). Extracellular hopanoids also may be produced in response to desiccation or as part of barriers with selective permeability to water and/or oxygen (Berry et al., 1993).

Curiously, although hopanoid biosynthesis does not require molecular oxygen, these compounds have not yet been discovered in obligate anaerobes. Geochemical analyses of sediments associated with anaerobic methane oxidation, a process involving synthropic consortia of sulfate-reducing eubacteria and archaea (Boetius et al.,

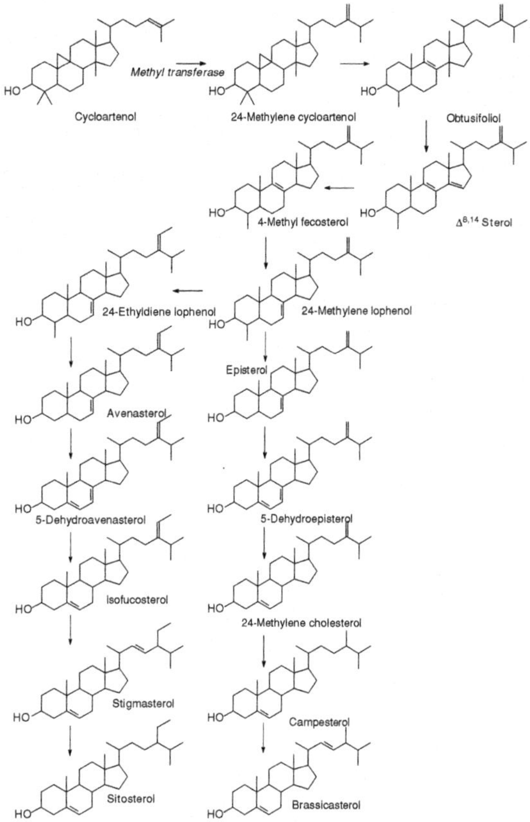

Figure 3.16. Biosynthesis of sterols from cycloartenol. This pathway is commonly used by higher plants.

2000; Orphan *et al.*, 2001b), have found ^{13}C-depleted diplopterol and diploptene (δ^{13}C from –60.5 to –74.4‰) (Elvert *et al.*, 2000) and free C_{32} homohopanoic acids (δ^{13}C as low as –78.4‰) (Thiel *et al.*, 2003). The isotopic ratios indicate *in situ* generation of these compounds by the anaerobic synthrophic consortium. Similarly, Thiel *et al.* (2001) reported ^{13}C-depleted C_{27} and C_{28} hopanes and C_{30} hop-17,21-ene in extracts from a fossilized bacterial mat in Oligocene carbonate rock (Lincoln Creek Formation, USA). These compounds (δ^{13}C from –41.5 to –52.3‰) were not as depleted in ^{13}C as the associated crocetane (δ^{13}C = –111.9‰) or PMI (δ^{13}C = −120.2‰) but were appreciably lighter than an extracted sterane (δ^{13}C = –26.8‰). Hopanoid synthesis by obligate anaerobic microorganisms may be inferred from these geochemical observations and awaits biochemical confirmation.

Hopanoids have been detected in only ~30% of the bacterial species analyzed, but they are presumed

to be widespread for several reasons. First, microorganisms grown in pure cultures may not express their full biochemistry. Laboratory-culturing procedures, which normally are optimized for growth, may not provide the necessary conditions to trigger hopanoid production. Second, the amphipathic hopanoids are difficult to isolate using standard mixtures of polar and non-polar solvents and their presence may be missed (Herrmann *et al.*, 1996). When isolated from the biological matrix, the large polar hopanoids are difficult to separate and identify. DNA sequencing offers a way to detect whether a eubacterial species has the genes necessary for hopanoid synthesis. For example, *Streptomyces coelicolor* A3(2) does not synthesize hopanoids when grown in liquid culture. However, the genome of this organism contains a cluster of genes apparently related to isoprenoid and hopanoid biosynthesis. The bacterium produces hopanoids only during the generation of aerial hyphae and spores, probably as a response to lower the water permeability across membranes (Poralla *et al.*, 2000).

Hopanoids have not been detected in archaea or animals, but small amounts occur in some ferns, mosses,

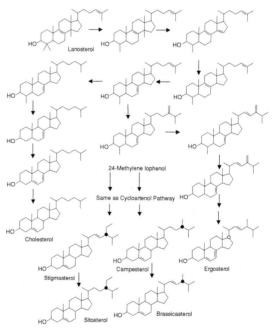

Figure 3.17. Biosynthesis of sterols from lanosterol. This pathway is commonly used by animals, protists, and many algal species.

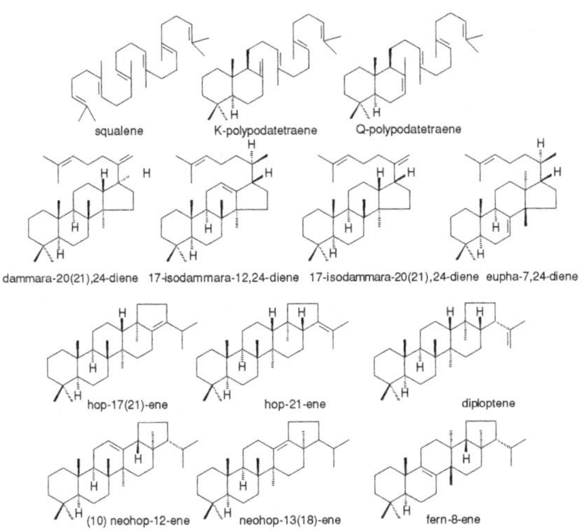

Figure 3.18. Triterpenic hydrocarbons from *Zymomonas mobilis*. All but diploptene are considered to be formed by variations in the cyclization process or lack of subtrate specificity of the squalene cyclase enzyme.

and fungi (Ourisson *et al.*, 1987; Mahato and Sen, 1997). Most of these compounds contain an oxygen group at C-3, indicating that they originated from oxidosqualene, not squalene.

Note: Hopanoids can be a major component of plant waxes and resins. They were first identified in dammar resins exuded from *Hopea*, a genus of tropical trees named after an English botanist, John Hope.

Sterols

Sterols occur in all eukarya, where they may constitute as much as half of the lipids in lipid membranes. Typical concentrations of sterols in eukaryotic cells are similar to those of hopanoid lipids in eubacteria ($\sim$0.1–5 mg/g dry cell weight or $\sim$30–3000 fg/cell, depending on the size of the cell). Some organisms and specialized differentiated cells contain much higher quantities of sterols (>300 mg/g dry cell weight). About 40–60% of the total mass of biogenic sterols may occur as a few common structures (Table 3.4), and a dozen or so other structures account for about another 20–30%. Animals tend to make cholesterol, higher land plants typically produce sitosterol and stigmasterol, and fungi preferentially synthesize ergosterol. These general distributions gave rise to the idea that sterol distributions

might be used to distinguish biotic input to sediments (Huang and Meinschein, 1978; Huang and Meinschein, 1979). Marine algae, however, produce all of the regular C_{27}, C_{28}, C_{29} sterols and some synthesize abundant C_{30} sterols (see reviews by Volkman (1988) and Volkman *et al.* (1998)). This diversity of sterols in marine biota detracts from their utility in sterol ternary diagrams to distinguish paleo-environments, although they are still useful for correlating genetically similar samples (see Figure 13.37).

Sterols and their derivatives are widespread and diverse. Modifications occur in the number and position of alkyl side chains, double bonds, and functional groups. Part of this diversity is due to the fact that sterols evolved to participate in numerous aspects of eukaryotic cell physiology. In addition to serving as membrane lipids, sterols serve as hormones that regulate growth, reproduction, and other processes. Sterol diversity is fairly well known in animals and higher plants, but it is documented more poorly in fungi, phytoplankton, and protists. Sponges are masters of sterol diversity, yielding very complex mixtures of highly functionalized compounds, many of which have no terrigenous analogs. These modifications include the addition of oxygen and double bonds to the nucleus and side chains, extensive rearrangement of the alkyl side chain, formation of sulfate esters, and unconventional nuclei (Aiello *et al.*, 1999) (Figure 3.19). These unusual steroids may

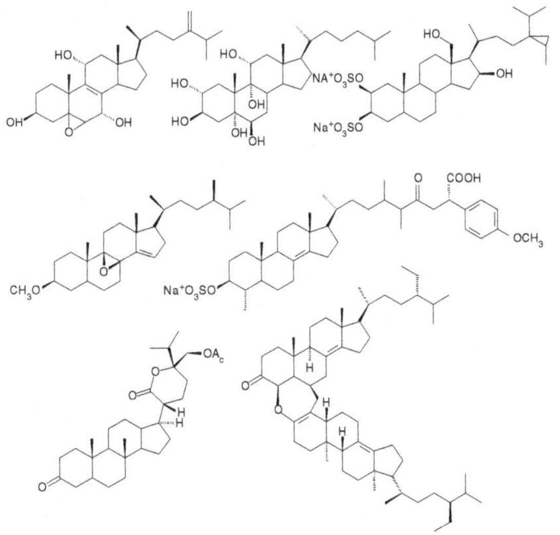

Figure 3.19. Some unusual steroids reported in sponges (Aiello *et al.*, 1999).

Table 3.4. *Chemical structures of common sterols named as the corresponding saturated hydrocarbon. All side chains except the cholane analogs are common in lipid membranes. (Allocholane and coprostane are not proper IUPAC terms.)*

Side chain	Configuration	5α-Series	5β-Series
	20R	5α-Cholane (not allocholane)	5β-Cholane
	20R	5α-Cholestane	5β-Cholestane (not coprostane)
	20R, 24S	5α-Ergostane	5β-Ergostane
	20R, 24R	5α-Campestane	5β-Campestane
	20R, 24S	5α-Poriferastane	5β-Poriferastane
	20R, 24R	5α-Stigmastane	5β-Stigmastane
	20S, 22R, 23R, 24R	5α-Gorgostane	5β-Gorgostane

occur as trace components along with the conventional cholesterols or 3β-hydroxy sterols, or in some species in large quantities, suggesting that they play a role in membrane stabilization.

Sterol biosynthesis apparently evolved from the squalene–hopene pathway, making use of free atmospheric oxygen to form oxidosqualene. One might expect that aerobic prokaryotes might benefit from adapting the sterol biosynthetic pathways. However, most prokaryotes lack sterols (<0.001% of dry cell weight) (Schubert et al., 1968). There are a few reports that suggest sterol synthesis in some prokaryotes. Bird et al. (1971) and Bouvier et al. (1976) identified abundant 4,4-dimethyl- and 4α-methyl-5α-cholest-8(14)-en-3β-ol and their 8(14),24-diene analogs in the methylotrophic archaebacterium *Methylococcus capsulatus*. Kohl et al. (1983) show substantial concentrations of cholest-8(9)-en-3β-ol in *Nannocystis exedens* (several percent). Similarly, Schouten et al. (2000a) found relatively high concentrations of 4,4-dimethylcholestenes in *Methylosphaera hansonii*, a psychrophilic methanotroph. *Thioploca* and several other bacteria also contain 4,4-dimethylsterols (McCaffrey et al., 1989). Summons et al. (2002a) found small amounts of sterols in initial cultures of the cyanobacteria *Phormidium* and *Chlorogloeopsis* in addition to the expected hopanoids and 2-methylhopanoids. However, repeated subculturing resulted in an absence of sterols, which probably represented fungal contaminants in the initial cultures.

Sterol synthesis may occur in several eubacterial lineages. However, very low concentrations of these sterols likely serve as hormones that regulate growth or other biochemical processes, not as lipid-membrane stabilizers. The genome for enzymes used in sterol biosynthesis was detected in several eubacteria, including *Mycobacterium tuberculosis* (Bellamine et al., 1999). However, the necessary genes are absent in other eubacteria, such as *Escherichia coli*.

Abundance and diagenetic products of hopanoids and sterols

Although individual biomarkers are not abundant in petroleum, they commonly represent the most abundant compounds with defined structures in the soluble portion of modern sediments and thermally immature

rocks. Excluding kerogen, which is insoluble in organic solvents and represents ~90% of the organic carbon in source rocks, hopanoids account for 5–10% of the remaining soluble organic carbon (Ourisson et al., 1984). For example, Yallourn lignite in Australia contains several hundred parts per million of a single C_{32} hopanoid acid (Ourisson et al., 1984). Although this amount may seem insignificant, a 1-m cube of the rock, weighing ~2 metric tons contains ~1 kg of the acid, thus, making it easily the most abundant structurally defined organic compound in the coal. Hopanoids in the geosphere amount to ~10^{11-12} metric tons, which is as much or more than the total mass of organic carbon in living organisms (Ourisson, 1987). Depending on the depositional setting, the sterol concentration may be substantially less or greater than that of the hopanoids.

Much of the preserved organic matter in source rocks represents the intact remains of biomass (Ourisson et al., 1984). Dehydration and reduction of bacteriohopanepolyols and sterols during diagenesis result in free bacteriohopanes and steranes. Most of these compounds, however, are first incorporated into kerogen and later released as hopanes and steranes during catagenesis (Mycke et al., 1987; Sinninghe Damsté and de Leeuw, 1990).

Note: Ten low-maturity marine crude oils from evaporitic source rocks in the Sergipe-Alagoas Basin show different distributions of *n*-alkanoic, isoprenoid, and hopanoic acids (Rodrigues et al., 2000). The functionality of the diagenetic precursor (alcohol, ether, or acid) was inferred based on comparison of the relative abundance of the neutral and acidic biomarkers (hopanoids, isoprenoids, alkyl-steranes, monoaromatic alkylsteroids). Three series of steroid-alkanoic acids and monoaromatic steroid-alkanoic acids (steroid-methanoic, ethanoic, and propanoic acids, and monoaromatic steroid-methanoic, -ethanoic, and -propanoic acids) were detected, while the neutral fraction contained only two series of each corresponding class (methyl- and ethyl-steranes and monoaromatic methyl- and ethyl-steroids). These carbon shifts suggest that decarboxylation is the main process in the formation of the alkylsteranes and monoaromatic alkylsteroids, and that carboxylic acids are the main diagenetic precursors of these classes of compound. However, no significant differences in

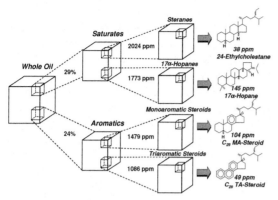

Figure 3.20. Biomarkers occur in parts per million (ppm) in most petroleum (e.g. 2024 ppm = 0.2024% steranes in oil). The figure shows the relative abundance of various compound fractions and biomarkers in biodegraded oil (17°API, 3.2 wt.% sulfur) generated from the Permian Phosphoria Formation and produced from Hamilton Dome, Wyoming. Biomarkers used as examples in the figure (far right) include C_{29} 5α,14α,17α(H) 24S- and 24R-ethylcholestane, C_{30} 17α,21β(H)-hopane, C_{29} 5α 20R, 24R- and 24S-monoaromatic steroid, and C_{28} 20S, 24R and 24S-triaromatic steroid. Reprinted with permission by ChevronTexaco Exploration and Production Technology Company, a division of Chevron USA Inc.

the molecular distributions between neutral and acidic fractions were observed for isoprenoids or hopanoids, suggesting that alcohols or ethers are the main diagenetic precursors of these compounds.

Figure 3.20 shows the quantitative distributions of various compound fractions and biomarker classes in crude oil generated from the Permian Phosphoria Formation in Wyoming. Like many others, the Wyoming oil contains individual biomarkers in concentrations ranging from tens to hundreds of parts per million. The concentrations of steroid and hopanoid hydrocarbons in petroleum vary as a function of thermal maturity of the source rock during expulsion. The precursors of these compounds may be weakly bound to the kerogen matrix by carbon–sulfur or carbon–oxygen bonds. These bonds tend to break before the main-stage oil generation and cleavage of carbon–carbon bonds. Consequently, oils expelled under low thermal stress are rich in biomarker compounds. Biomarker concentrations are substantially lower in oils expelled during main-stage generation because of this preferential early release and dilution by non-biomarker hydrocarbons. Cyclic

saturated biomarkers generally have lower thermal stabilities compared with acyclic hydrocarbons and thermally degrade at a faster rate during late-stage fluid expulsion or reservoir cracking. Other reservoir-alteration processes tend to enrich oils in biomarkers. For example, because they are commonly more resistant to microbial alteration than other compounds, biomarkers can increase by a factor of ten or more during biodegradation of oil.

PORPHYRINS AND OTHER BIOMARKERS OF PHOTOSYNTHESIS

Photosynthetic organisms occur in all three domains of life. Halobacteria have a primitive photosynthetic system unlike that in other organisms. Halobacteria use bacteriorhodopsin, a retinal-containing protein that is directly bound to cellular membranes and is not organized into an antenna/reaction center. The bacteriorhodopsin creates a proton gradient that allows photophosphorylation (Blankenship, 1992). Halobacterial photochemistry is fundamentally different from that in the other domains. It involves no oxidation/reduction electron transport, does not use chlorophylls or bacteriochlorophylls, and does not use CO_2 as a carbon source. Consequently, some biologists do not consider halobacteria truly photosynthetic (Gest, 1993).

Photosynthetic organisms are widespread among the eubacteria and eukarya. All of these organisms share features that suggest a common origin quite different from that of the halobacteria. In these organisms, the photosynthetic apparatus is organized into pigment–protein complexes that consist of a light-gathering antenna unit containing several hundred molecules of chlorophylls or bacteriochlorophylls and other accessory pigments (carotenoids and bilins) and an electron-transport system of proteins in a reaction center. The light energy absorbed by the pigments is used to fix CO_2 as a carbon source. Anoxic photosynthetic bacteria include members of the purple sulfur, purple non-sulfur, green sulfur, green non-sulfur, and Gram-positive groups. All anoxic photosynthetic bacteria rely on photosystem I with bacteriochlorophylls and carotenoids. Cyanobacteria have coupled photosystem I with photosystem II, an antenna/reaction center complex with chlorophylls, carotenoids, and phycobilins. The photosystem complexes are in membranes called thylakoids, commonly located near the

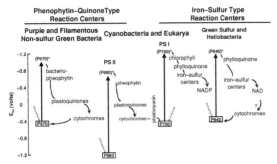

Figure 3.21. Anaerobic photosynthetic bacteria rely on two different types of reaction centers for electron transport. Cyanobacteria and photosynthetic eukaryotes combined the two types into a merged photosystem I and II (PS I and PS II) reaction center.

cellular membrane for maximum light absorption. The efficiency of the coupled photosystems allows cyanobacteria to use water as a hydrogen donor and to produce oxygen as a by-product (Figure 3.21).

Chlorophylls and bacteriochlorophylls are part of the magnesium branch of the tetrapyrrole biosynthetic pathway starting with δ-aminolevulinate (Figure 3.22). Two units of δ-aminolevulinate, which is formed by the addition of the amino acid glycine and succinyl co-enzyme A, condense to form the pyrrole porpho-

bilinogen. Four units of porphobilinogen condense into a linear tetrapyrrole, which is then cyclized into uroporphyrinogen III. Various biomolecules originate from uroporphyrinogen III by the addition of copper, nickel, and cobalt.

Decarboxylation of the acetate to methyl groups in uroporphyrinogen III produces coporphyrinogen III (Figure 3.23). Decarboxylation of two of the proprionate groups in coporphyrinogen III to vinyl groups forms protoporphyrin IX. The addition of iron results in hemes and bilins. Addition of magnesium and closure of one of the proprionate side chains to form the E-ring results in protochlorophyllide. The completely unsaturated chlorophylls c_1 and c_2 originate from protochlorophyllide. They have no esterified long-chain alcohol attached to the D-ring and occur in diatoms, dinoflagellates, macrophytic brown algae, and cryptophytes. The dihydroporphyrin, chlorophyll a, originates from protochlorophyllide and has an esterified phytol side chain. Chlorophyll b originates from chlorophyll a. Chlorophylls a and b occur in the photosystems I and II of cyanobacteria, algae, and higher plants.

All bacteriochlorophylls originate from chlorophyll a. The dihydro-bacteriochlorophylls c, d, and e have esterified farnesyl side chains, many homologs with alkyl substitutions at the C-4 and C-5 positions,

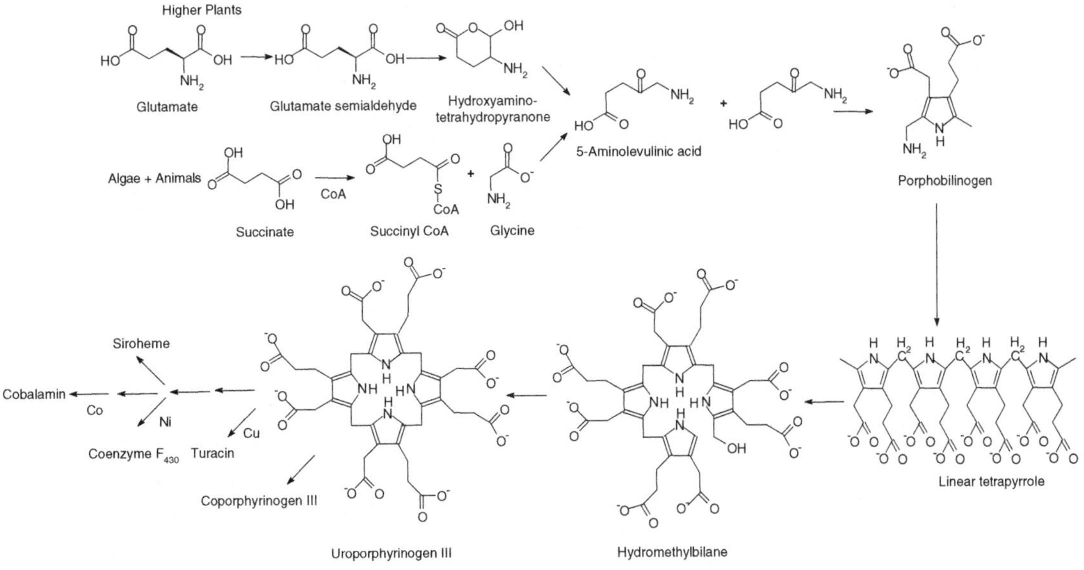

Figure 3.22. Tetrapyrrole biosynthetic pathway from δ-aminolevulinate to uroporphyrinogen III.

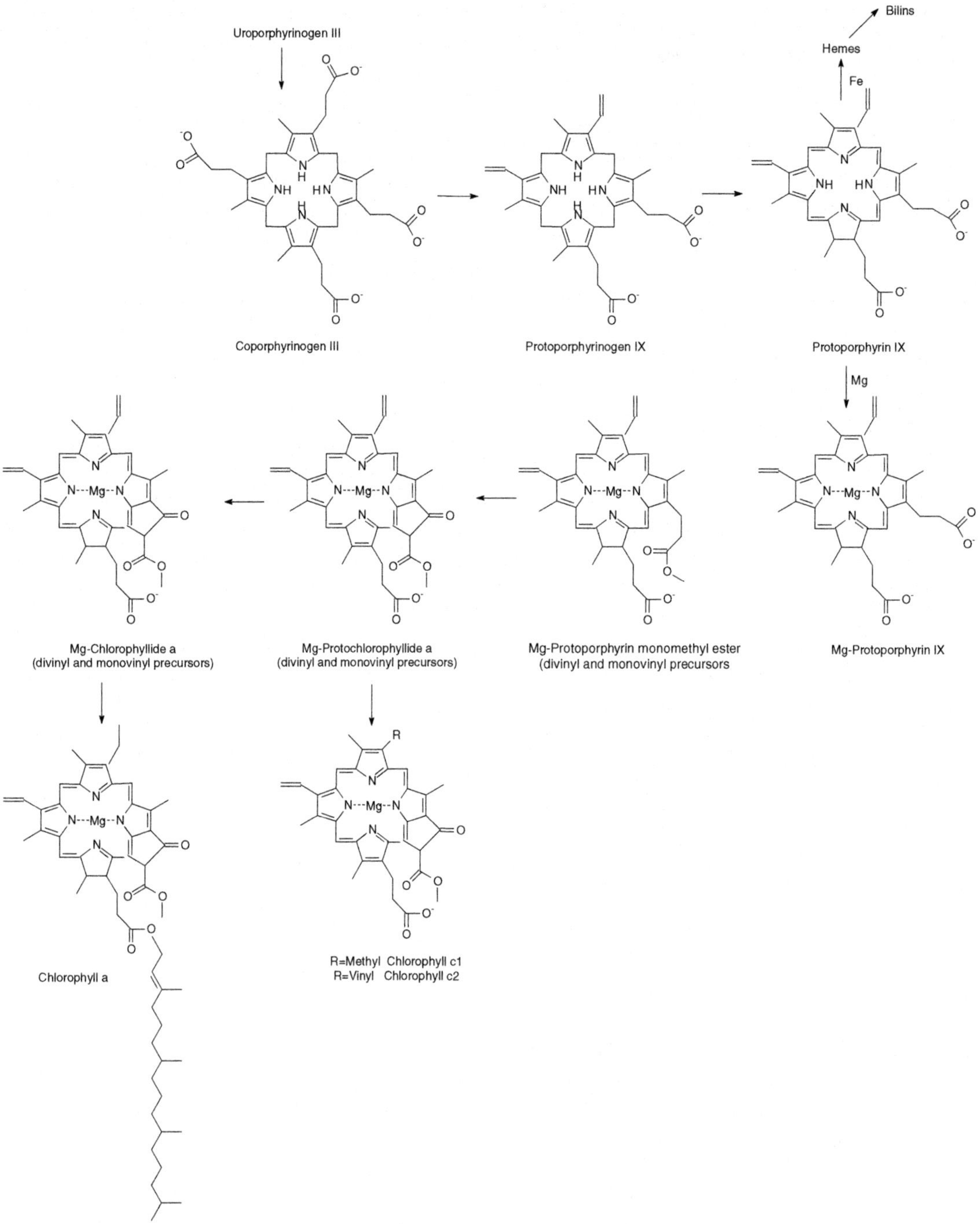

Figure 3.23. Tetrapyrrole biosynthetic pathway from uroporphyrinogen III to chlorophylls.

and differing chirality at the C2a position. The dihydro-bacteriochlorophylls reside in the light-harvesting complexes of filamentous green non-sulfur and green sulfur bacteria. The tetrahydro-bacteriochlorophylls a and b have an esterified phytyl side chain and occur in the purple sulfur and purple non-sulfur bacteria. The tetrahydro-bacteriochlorophyll g has a farnesyl side chain and occurs only in Gram-positive heliobacteria. The corresponding tetrapyrrole pigments, which lack coordinated metals, are called pheophytins or bacterio-pheophytins.

The evolutionary forces driving the synthesis of these pigments and their incorporation into complex photosystems seem too complex to give rise spontaneously to photosynthetic organisms. Nisbet *et al.* (1995) observed that the absorption spectra of bacteriochlorophyll a and b are similar to the thermal emissions of a hot body submerged in water. They proposed that ancient thermophilic chemolithotrophs might gain an evolutionary advantage if they developed thermotaxis that directed them toward heat and nutrients. The ancient thermotactic system might have evolved into bacteriochlorophyll photosynthesis based on infrared radiation. Later, chlorophyll photosynthesis could take advantage of the higher-energy light wavelengths.

Blankenship (1992) and Blankenship and Hartman (1998) proposed that the origin of the coupled photosystems in cyanobacteria is revealed by the different reaction centers in extant bacteria. Green sulfur bacteria and heliobacteria have an iron–sulfur-based reaction system similar to photosystem I in cyanobacteria. Purple and non-sulfur filamentous green bacteria use a pheophytin–quinone-based reaction system similar to photosystem II in cyanobacteria. They propose that cyanobacteria were early symbionts with bacteria, allowing each of these types of reaction center to occur in close proximity. Initially, their separate photosystems were decoupled and used inorganic sulfur species or organic substrates as hydrogen acceptors. Coupling of the systems resulted in increased efficiency, and the use of chlorophylls as light-gathering molecules allowed cyanobacteria to use water as a hydrogen acceptor (Niklas, 1996).

Chlorophylls are by far the most abundant tetrapyrroles in the biosphere. Other tetrapyrroles include heme in the blood of many animals. The ratio of chlorophyll- to heme-type tetrapyrroles in living

X = H or Alkyl Group
M = Vanadyl [V = O(II)] or Nickel [Ni(II)] Ion

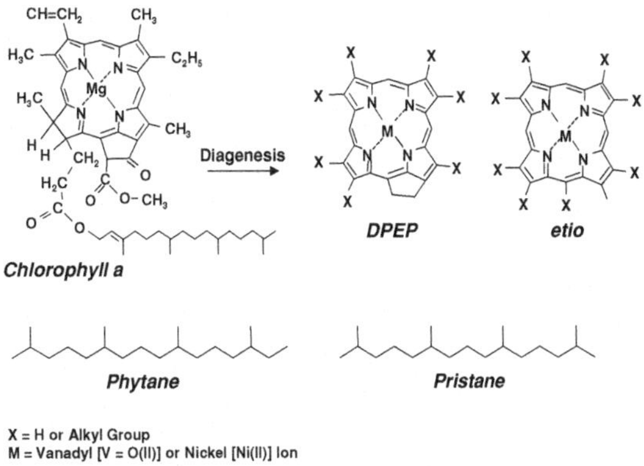

Figure 3.24. Diagenesis and catagenesis can convert chlorophylls to several biomarkers that are common in petroleum, including deoxophylloerythroetioporphyrins (DPEP), etioporphyrins (etio), pristane, and phytane. Note the tetrapyrrole nucleus and the phytyl side chain in chlorophyll a. M = nickel (Ni^{2+}), vanadyl (VO^{2+}), or other anions that are less common in petroleum. Reprinted with permission by ChevronTexaco Exploration and Production Technology Company, a division of Chevron USA Inc.

organisms is estimated at $\sim$100 000 : 1 (Baker and Louda, 1986). This biological abundance is reflected in the geosphere. During diagenesis and catagenesis, various products originate from chlorophyll pigments (e.g. Bidigare *et al.*, 1990; Keely *et al.*, 1990), including three common constituents in petroleum: porphyrins and the acyclic diterpanes pristane and phytane (Figure 3.24). Sedimentary porphyrins are complex metallated tetrapyrrolic compounds. They have been linked to chlorophylls in eukaryotes, which yield a C_{32} carbon skeleton, and to chlorophylls in prokaryotic bacteria having extended side chains. These transformations involve a complex series of reactions that are similar, but not identical, to those proposed by Treibs (1936). Keely *et al.* (1990) characterized the minimum number of intermediates between chlorophyll a and deoxophylloerythroetioporphyrin (DPEP), the major porphyrins in the biosphere and geosphere, respectively. In simple terms, the reactions summarized by Baker and Louda (1983), Filby and Berkel (1987), and Callot *et al.* (1990) proceed as follows. During early diagenesis, magnesium is removed from the chlorophylls (demetallation) and they are defunctionalized to phorbides, chlorins, and purpurins. During later diagenesis, these intermediates are aromatized in non-coaly sediments to free-base porphyrins, which chelate with metal ions (Lewan, 1984) to form immature metalloporphyrins (geoporphyrins) during late diagenesis. Subsequently, the immature metalloporphyrins can undergo alteration during catagenesis and destruction during metagenesis.

Most porphyrins originate from various chlorophylls. For example, stable carbon isotope compositions of the C_{31} and C_{32} DPEP porphyrins in Tertiary lacustrine oil shales from the Jianghan Basin indicate an origin mainly from chlorophyll (Yu $et\ al.$, 2000a). However, stable carbon isotope ratios of individual porphyrins suggest that at least one C_{32}-etioporphyrin originates from hemes (Boreham $et\ al.$, 1989; Ocampo $et\ al.$, 1989). Some porphyrins provide highly specific paleoenvironmental information. For example, high-molecular-weight cycloalkanoporphyrins ($>C_{32}$) with extended ($>C_2$) alkyl substitution in crude oil or rock extracts indicate an origin from bacteriochlorophyll d in green sulfur bacteria, which thrive under conditions of photic zone anoxia during source-rock deposition (Gibbison $et\ al.$, 1995; Rosell-Melé $et\ al.$, 1999).

Aromatization of chlorins need not always occur before metal incorporation. The major chlorin in Messel oil shale is a nickel complex of mesopyrophaeophorbide-a (Prowse $et\ al.$, 1990). The structure of Ni-mesopyrophaeophorbide-a clearly indicates an origin from chlorophyll. However, this chlorin has been metallated without prior aromatization to form a porphyrin.

The structures of many porphyrins were fully or partly characterized (Sundararaman, 1985; Chicarelli $et\ al.$, 1987; Callot $et\ al.$, 1990). Deoxophylloerythroetioporphyrins (DPEP) and etioporphyrins (etio) are the two major series of porphyrins in petroleum (Figure 3.24). DPEP and etio are commonly used as generic terms based on the number of exocyclic rings in the porphyrin structure (one or none, respectively). Precise structures for various DPEP-type porphyrins having five- to seven-member rings at different exocyclic positions were determined. Likewise, etioporphyrins may have acyclic alkyl substituents at any position on the porphyrin nucleus. They are most commonly complexed to vanadyl (VO^{2+}) or nickel (Ni^{2+}) ions in source-rock extracts and crude oil, although higher-plant coals are characterized by iron (Fe^{3+}), gallium (Ga^{3+}), and manganese (Mn^{3+}) porphyrins (Filby and Berkel, 1987). Two minor series include rhodo-DPEP and rhodo-etio porphyrins. Rhodoporphyrins have an exocyclic fused benzene ring. Barwise and Whitehead (1980) also describe di-DPEP porphyrins.

Note: Abelsonite is solid bitumen composed of a crystalline nickel (II) DPEP porphyrin derived from one of the chlorophylls. It is found only in the Parachute Creek Member of the Green River Formation, Utah (Mason $et\ al.$, 1990).

Routine biomarker studies incorporate only a few porphyrin parameters because, unlike many other biomarkers, porphyrins are especially difficult to separate and analyze reliably. Most attempts to study porphyrins rely on probe introduction of the sample and electron-impact ionization in a mass spectrometer (e.g. Quirke $et\ al.$, 1989; Beato $et\ al.$, 1991). This approach is insufficient to reveal the complexity of natural porphyrin compositions. Recent advances in high-performance liquid chromatography (HPLC) and tandem mass spectrometry allow rapid, detailed analyses of free-base porphyrins (Rosell-Melé $et\ al.$, 1999). Reviews by Baker and Louda (1986), Louda and Baker (1986), Filby and Branthaven (1987), and Callot $et\ al.$ (1990) describe porphyrin geochemistry in more detail.

Treibs (1936) showed the link between chlorophyll a in living photosynthetic organisms and porphyrins in petroleum, thus providing the first strong evidence for an organic origin of petroleum. This event marked the birth of organic geochemistry, a hybrid of natural-product chemistry, analytical chemistry, synthetic organic chemistry, physical organic chemistry, and geology (Kvenvolden, 2002). Little progress was made in our science for more than 30 years, until the development of computer-assisted gas chromatography/mass spectrometry. Today, organic geochemistry and its repertoire of analytical tools are vital components in the exploration departments of the major oil companies and at many universities and government organizations.

Organic geochemistry includes various disciplines, such as biogeochemistry, environmental geochemistry, archeological geochemistry, and petroleum exploration and production geochemistry. Some closely allied fields include analytical chemistry, biochemistry, cosmochemistry, geology, evolutionary biology, paleontology, isotope geochemistry, microbiology, paleoclimate, remote sensing, three-dimensional petroleum system modeling, molecular modeling, and chemometrics. Organic geochemists conduct studies on topics ranging from gas hydrates in shallow marine sediments to fluid

Table 3.5. *1979–2004 Alfred E. Treibs medalists*

1979	George Philippi
1980	Bernard Tissot
1981	Geoff Eglinton
1982	John Hunt
1983	Dietrich Welte
1984	Wolfgang Seifert
1985	Pierre Albrecht
1987	Tom Hoering
1989	James Maxwell
1991	Jan de Leeuw
1993	Ian Kaplan
1995	Keith Kvenvolden
1996	Patrick Parker
1997	John Hayes
2000	John Hedges
2001	John Smith
2002	Archie Douglas
2003	Roger Summons
2004	Eric Galimov

inclusions in Precambrian rocks. They study geochemical evidence for climatic change, anthropogenic pollutants in sediments or water, recent sediment diagenesis, biodegradation of petroleum in reservoirs, and many other diverse topics.

The prestigious Alfred E. Triebs Award is presented on behalf of the Organic Geochemistry Division of the Geochemical Society to scientists who have had a major impact on the field of organic geochemistry through long-standing contributions (Table 3.5).

CAROTENOIDS

Carotenoids originate from lycopene, which is produced by the condensation of two units of geranylgeranyl diphosphate followed by a series of dehydrogenations (Figure 3.25). Lycopene can be modified by hydrogenation, dehydrogenation, cyclization, and oxidation to produce an array of bioactive compounds. The carotenoids can be divided into two broad classes, hydrocarbons called carotenes and those containing oxygen functional groups, called xanthophylls. IUPAC nomenclature for carotenoids employs a semisystematic approach, where trivial names are used for the most common compounds. Specific names are

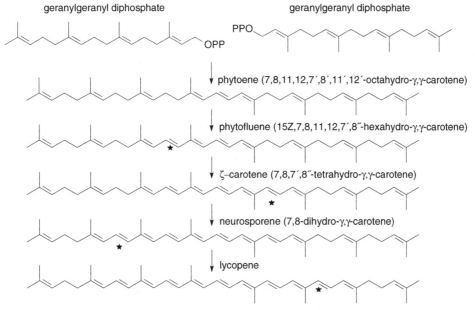

Figure 3.25. Biosynthesis of lycopene, the precursor of all carotenoids. Stars indicate new positions of unsaturation from previous structure in reaction scheme.

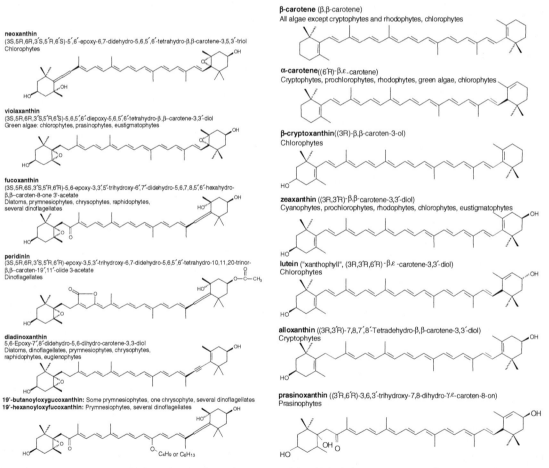

Figure 3.26. Numbering of carotenoid carbon atoms and terminal structures. The functional group R (rows 2 and 3) may consist of one or more isoprenoid subunits (bottom row).

developed from the carotene stem name using the numbering scheme shown in Figure 3.26 and two Greek letters as prefixes to determine the terminal configurations. Hence, the specific name for lycopene is ψ,ψ-carotene. Structural modifications that involve hydrogenation, ring openings, and the addition of functional oxygen groups follow the standard rules for organic-compound nomenclature. Removal of a carbon atom from one or both ends of carotenoids yields apo- or di-apocarotenoids, respectively.

Over 600 natural carotenoids are known (Britton, 1998), with new structures continuing to be identified (Mercadante, 1999). Figure 3.27 shows lycopene and the most common carotenoids in recent marine environments. Because some of these carotenoids originate from specific groups of phytoplankton, they can be used as proxies to monitor specific biotic productivity. The aryl

neoxanthin
(3S,5R,6R,3'S,5'R,6'S)-5',6'-epoxy-6,7-didehydro-5,6,5',6'-tetrahydro-β,β-carotene-3,5,3'-triol
Chlorophytes

violaxanthin
(3S,5R,6R,3'S,5'R,6'S)-5,6,5',6'-diepoxy-5,6,5',6'-tetrahydro-β,β-carotene-3,3'-diol
Green algae: chlorophytes, prasinophytes, eustigmatophytes

fucoxanthin
(3S,5R,6S,3'S,5'R,6'R)-5,6-epoxy-3,3',5'-trihydroxy-6',7'-didehydro-5,6,7,8,5',6'-hexahydro-β,β'-caroten-8-one 3'-acetate
Diatoms, prymnesiophytes, chrysophytes, raphidophytes, several dinoflagellates

peridinin
(3S,5R,6R,3'S,5'R,6'R)-epoxy-3,5,3'-trihydroxy-6,7-didehydro-5,6,5',6'-tetrahydro-10,11,20-trinor-β,β-caroten-19',11'-olide 3-acetate
Dinoflagellates

diadinoxanthin
5,6-Epoxy-7',8'-didehydro-5,6-dihydro-carotene-3,3-diol
Diatoms, dinoflagellates, prymnesiophytes, chrysophytes, raphidophytes, euglenophytes

19'-butanoyloxygucoxanthin: Some prymnesiophytes, one chrysophyte, several dinoflagellates
19'-hexanoyloxyfucoxanthin: Prymnesiophytes, several dinoflagellates

β-carotene (β,β-carotene)
All algae except cryptophytes and rhodophytes, chlorophytes

α-carotene((6'R)-β,ε-carotene)
Cryptophytes, prochlorophytes, rhodophytes, green algae, chlorophytes

β-cryptoxanthin((3R)-β,β-caroten-3-ol)
Chlorophytes

zeaxanthin ((3R,3'R)-β,β-carotene-3,3'-diol)
Cyanophytes, prochlorophytes, rhodophytes, chlorophytes, eustigmatophytes

lutein ("xanthophyll", (3R,3'R,6'R)-β,ε-carotene-3,3'-diol)
Chlorophytes

alloxanthin ((3R,3'R)-7,8,7',8'-Tetradehydro-β,β-carotene-3,3'-diol)
Cryptophytes

prasinoxanthin ((3'R,6'R)-3,6,3'-trihydroxy-7,8-dihydro-γ,ε-caroten-8-on)
Prasinophytes

Figure 3.27. Common carotenoids in modern marine environments and their biological sources. Note that fucoxanthin and peridinin contain two adjacent double bonds, while diadinoxanthin contains a triple bond.

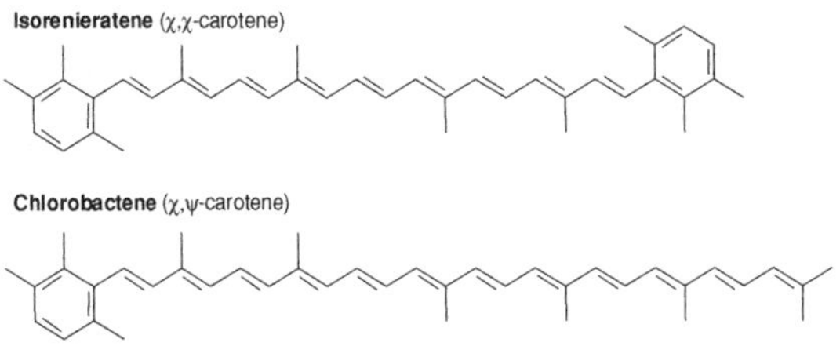

Isorenieratene (χ,χ-carotene)

Chlorobactene (χ,ψ-carotene)

Figure 3.28. Aryl carotenoids common in green sulfur bacteria.

carotenoids isorenieratene and chlorobactene are less common in marine settings but are of great significance to the geochemist. They occur only in photosynthetic bacteria and a few actinomyces (Figure 3.28), which are photosynthetic anaerobes. Thus, aryl carotenoids are proxies for euxinic conditions in the photic zone of the water column.

Plants, algae, and photosynthetic bacteria synthesize carotenoids, which serve as accessory pigments in photosynthesis (Frank *et al.*, 2000). The alternating pattern of conjugated double bonds in carotenoids absorbs energy, while the terminal groups regulate polarity and properties within lipid membranes. Chlorophyll is a potential hazard to surrounding biomolecules. To prevent cellular damage, chlorophyll is confined to photosynthetic pigment–protein complexes in thylakoid membranes. There, the carotenoid pigments serve two functions: as light harvesters that transfer energy to chlorophyll, and as molecules that quench triplet-state chlorophylls before they cause damage. Some non-photosynthetic bacteria, yeasts, and molds also produce carotenoids, which function as antioxidants and possibly in the regulation of biochemical pathways (Britton, 1998). Carotenoids are responsible for most coloration in animals, which are generally incapable of direct synthesis of these pigments and obtain them from their diet.

Note: Pink-colored salt lakes and playas and the bright red evaporation ponds of salt-recovery

plants occur in arid regions throughout the world. The red coloration of the brines results from abundant halophilic (salt-loving) archaea, which produce red carotenoid pigments for protection from ultraviolet (UV) radiation. The most abundant of these acyclic C_{40} and C_{50} carotenoids is bacterioruberin. In such hypersaline environments, archaea are dominant and halotolerant eukaryotes, such as the red algae *Dunaliella*, contribute to only a small amount of the coloration.

In the arid regions of central and east Africa, flamingos consume vast quantities of *Spirulina*. The pink color of flamingo feathers originates from carotene pigments in filaments of the *Spirulina*. Dietary carotenoids also serve as natural colorants in other organisms that lack carotenoid synthesis and are responsible for the typical colors of salmon flesh and lobster shells. The color of carotenoids in green plant tissues is masked by chlorophyll and becomes evident only after degradation of the green pigment in temperate latitudes during fall/winter months. Carotenoid pigments are also the source of β-carotene, an important antioxidant and the precursor of vitamin A. In some parts of the world, β-carotene is extracted from salt ponds containing red halophilic bacteria and algae.

The evolution of the diversity in carotenoid biosynthesis may reflect the ease of horizontal gene transfer for specific enzymes. For example, key enzymes in the carotenoid synthesis pathway of the photosynthetic bacterium *Rhodobacter sphaeroides* can be replaced with the corresponding enzyme from the non-photosynthetic bacterium *Erwinia herbicola*, resulting in carotenoids new to *Rhodobacter* (Garcia-Asua *et al.*, 1998). The ready response to the new enzyme demonstrates how a wide range of carotenoids could have evolved with only minor changes to the basic biosynthetic pathways.

4 · Geochemical screening

This chapter describes how to select sediment, rock, and crude-oil samples for advanced geochemical analyses by using rapid, inexpensive geochemical tools, such as Rock-Eval pyrolysis, total organic carbon, vitrinite reflectance, scanning fluorescence, gas chromatography, and stable isotope analyses. The discussion covers sample quality, selection, storage, and geochemical rock and oil standards. Other topics include how to test rock samples for indigenous bitumen, surface geochemical exploration using piston cores, geochemical logs and their interpretation, chromatographic fingerprinting for reservoir continuity, and how to deconvolute mixtures of oils from different production zones. Mass-balance equations show how to calculate the extent of fractional conversion of kerogen to petroleum, source-rock expulsion efficiency, and the original richness of highly mature source rocks.

Many geochemical exploration, development, and production projects are complex because of large numbers of samples. Biomarker work should not be initiated on these projects without a preliminary appraisal of samples using less costly and faster geochemical analyses. Large numbers of oils, rocks, and/or sediments can be screened using other geochemical tools, such as total organic carbon (TOC), Rock-Eval pyrolysis, vitrinite reflectance (R_o), petrographic analysis of maceral composition, gas chromatography (GC), kerogen atomic hydrogen/carbon (H/C), and stable carbon isotope ratios. Use of these practical and less expensive methods allows non-source rocks to be identified and oils and source rocks to be provisionally grouped into genetic families. Selected samples from these families can be submitted for further work using the more powerful biomarker parameters. Advanced analyses on the selected samples can be combined with the screening data to establish the most reliable interpretations.

SOURCE-ROCK SCREENING: QUALITY AND QUANTITY

Petroleum source rocks are fine-grained organic-rich rocks that could generate (potential source rock) or have already generated (effective or active source rock) significant amounts of petroleum. We prefer this definition to that which specifies that source rocks must have generated and expelled enough hydrocarbons to form commercial accumulations of oil and gas (Hunt, 1996), because of difficulties in defining the term "commercial."

Prior to any oil–oil or oil–source rock correlation, prospective oils and source rocks must be selected or screened. An effective petroleum source rock must satisfy requirements as to the quantity, quality, and thermal maturity of the organic matter. Table 4.1 shows the generally accepted criteria for describing the quantity, quality, and thermal maturity of organic matter in source rocks. Prospective source-rock samples must also satisfy criteria to ensure that they are not contaminated by organic drilling additives or migrated oil, as discussed later in the chapter.

Total organic carbon

Total organic carbon (TOC), also called organic carbon (C_{org}), measures the quantity but not the quality of organic carbon in rock or sediment samples (Figure 4.1). Total organic matter (TOM) can be calculated by multiplying TOC by 1.2, assuming that the organic matter is 83 wt.% carbon.

A common misconception is that the minimum TOC required for carbonate source rocks is less than that for shale source rocks. For example, Tissot and Welte (1984) indicate miminum TOC for effective carbonate and shale source rocks of 0.3 and 0.5 wt.%, respectively. Bordovskiy and Takh (1978) showed an inverse correlation between TOC and bitumen/TOC for both carbonate and non-carbonate sediments from the Caspian Sea, Russia (Figure 4.2). Jones (1987) used

Table 4.1. *(a) Generative potential (quantity) of immature source rock, (b) kerogen type and expelled products (quality), and (c) thermal maturity (Peters and Cassa, 1994)*

Potential (quantity)	TOC (wt.%)	Rock-Eval (mg/g rock)		Bitumen (ppm)	Hydrocarbons (ppm)
		S1	S2		
Poor	<0.5	<0.5	<2.5	<500	<300
Fair	0.5–1	0.5–1	2.5–5	500–1000	300–600
Good	1–2	1–2	5–10	1000–2000	600–1200
Very good	2–4	2–4	10–20	2000–4000	1200–2400
Excellent	>4	>4	>20	>4000	>2400

(a)

Kerogen (quality)	Hydrogen index (mg hydrocarbon/g TOC)	S2/S3	Atomic H/C	Main product at peak maturity
I	>600	>15	>1.5	Oil
II	300–600	10–15	1.2–1.5	Oil
II/III	200–300	5–10	1.0–1.2	Oil/gas
III	50–200	1–5	0.7–1.0	Gas
IV	<50	<1	<0.7	None

(b)

Maturity	Maturation			Generation		
	R_o (%)	T_{max} (°C)	TAI	Bitumen/TOC*	Bitumen (mg/g rock)	Production index (S1/(S1+S2))
Immature	0.20–0.60	<435	1.5–2.6	<0.05	<50	<0.10
Mature						
Early	0.60–0.65	435–445	2.6–2.7	0.05–0.10	50–100	0.10–0.15
Peak	0.65–0.90	445–450	2.6–2.7	0.15–0.25	150–250	0.25–0.40
Late	0.90–1.35	450–470	2.9–3.3	–	–	>0.40
Postmature	>1.35	>470	>3.3	–	–	–

(c)

* Many gas-prone coals have high bitumen yields like those of oil-prone samples, but extract yields normalized to TOC are low (<30 mg HC/g TOC). Bitumen/TOC ratios >0.25 indicate contamination, migrated oil, or artifacts caused by ratios of small, inaccurate numbers.

H/C, hydrogen/carbon ratio; R_o, vitrinite reflectance; TAI, thermal alteration index; T_{max}, maximum temperature at top of S2 peak; TOC, total organic carbon.

this plot and other evidence to conclude that no significant difference in minimum TOC exists between carbonate and non-carbonate sediments. He noted that although bitumen/TOC is statistically higher for carbonate versus non-carbonate lithologies in organic-lean samples (<0.5 wt.%), catagenesis of these sediments cannot yield significant petroleum accumulations. Bitumen/TOC for sediments with TOC >0.5 wt.% in Figure 4.2 are statistically identical, and these organic-rich sediments could generate significant petroleum during burial. Because the bitumen/TOC for organic-rich carbonate and non-carbonate lithologies

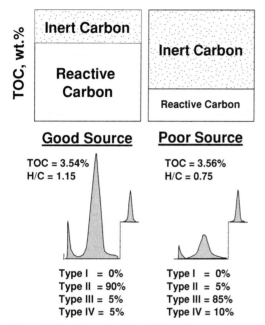

Figure 4.1. Total organic carbon (TOC) indirectly measures quantity but not quality of organic matter. Two rocks from a well in Montana show nearly identical TOC (3.5 wt.%) but differ in their oil-generative potential based on Rock-Eval pyrograms (see Figure 4.4), elemental analysis of atomic H/C, and organic petrography for proportions of the four kerogen types (data from Peters, 1986).

are essentially identical, the minimum TOC requirements are also identical. The reason for the different bitumen/TOC among organic-lean carbonates and non-carbonates is unclear, but it may relate to differences in organic matter quality, adsorption characteristics of clays versus carbonates, or generally more extensive oxidation of organic matter in carbonates (Jones, 1987).

TOC can be measured by various methods, each with limitations and potentially different results, as discussed below.

Direct combustion is the most common method, which requires acidification of the ground rock sample with 6 N HCl in a filtering crucible to remove carbonate, removal of the filtrate by washing/aspiration, and drying at ~55°C. Using a typical Leco Carbon Analyzer, the dried sample is combusted with metallic oxide accelerator at ~1000° C. The CO_2 generated during combustion is analyzed using either infrared (IR) or thermal conductivity detectors (TCDs). IR detectors are specific for CO_2, while TCDs respond to other compounds, such as sulfur dioxide and water, if they are not removed

(Jarvie, 1991). Although the direct method is rapid, it is not accurate for either organic-poor, carbonate-rich rocks or for many thermally immature sediment and rock samples. For example, immature organic matter is susceptible to acid hydrolysis and loss during filtering. Peters and Simoneit (1982) analyzed immature Deep Sea Drilling Project sediments by direct TOC and a modified method that used non-filtering crucibles so that hydrolyzate was retained. The results show that an average of more than 10% of the TOC was lost as hydrolyzate using the direct method.

The indirect TOC method is usually applied to organic-poor, carbonate-rich rocks. Total carbon (including carbonate carbon) is determined on one aliquot of the sample, while carbonate carbon is determined on another aliquot by coulometric measurement of the CO_2 generated by acid treatment. Organic carbon is the difference between total carbon and carbonate carbon. This method is more time-consuming than the direct method and requires two separate analyses of the sample.

The Rock-Eval II plus TOC (Delsi, Inc.) determines TOC by summing the carbon in the pyrolyzate with that obtained by oxidizing the residual organic matter at 600°C. For small samples (100 mg), this method provides more reliable TOC data than the methods discussed above, which require ~1–2 g of ground rock. However, mature samples, where vitrinite reflectance is more than ~1%, yield poor TOC data when determined by this method because the temperature is insufficient

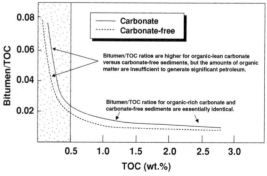

Figure 4.2. Total organic carbon (TOC) requirements for carbonate and shale source rocks are virtually identical (modified from Jones, 1987). Statistical data from Bordovskiy and Takh (1978) indicate no significant difference in bitumen/TOC for organic-rich carbonates and shales (>0.5 wt.% TOC). Bitumen/TOC differences between organic-lean carbonates and shales (stippled) are unimportant because of insufficient organic matter for generating significant amounts of petroleum.

for complete combustion. The Rock-Eval VI pyrolysis and oxidation reaches 850°C, which results in more reliable T_{max} and TOC data, especially for highly mature rock samples (Lafargue *et al.*, 1998).

Maceral composition

Because multiple lines of evidence increase the reliability of interpretations, we recommend the use of optical microscopy to substantiate biomarker and other geochemical analyses. Microscopy of kerogen preparations provides an independent assessment of the quantity, quality, and thermal maturity of organic matter. Common analyses include the relative percentages of oil-prone and other macerals, thermal alteration index, vitrinite reflectance, and qualitative or quantitative fluorescence.

The three most common preparations and techniques applied to the microscopy of coals and petroleum source rocks are as follows:

(1) Strew-mount slides of isolated kerogen for transmitted light microscopy.
(2) Slides of polished kerogen for reflected light microscopy and/or fluorescence.
(3) Polished surfaces of whole rock or coal for reflected light microscopy and/or fluorescence.

Kerogen is the organic matter in rocks and coals that is insoluble in organic solvents and survives acid digestion of the mineral matrix (Durand, 1980). Figure 4.3

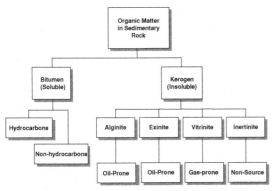

Figure 4.3. Simplified classification of organic matter in sedimentary rock. Bitumen extracted from ground rock using organic solvents consists of hydrocarbons and non-hydrocarbons. Treatment of the extracted residue with hydrochloric (HCl) and hydrofluoric (HF) acids removes most carbonate and silicate minerals, respectively. The kerogen residue consists of various macerals that can be identified by organic petrography.

shows a simplified classification of organic matter in sedimentary rock, including the various maceral groups. Macerals are individual components in the kerogen that have petrographically and geochemically distinct properties. Macerals consist of individual organic particles or phytoclasts. For techniques 1 and 2, kerogen is concentrated by means of acid maceration and/or density flotation (Bostick and Alpern, 1977) and prepared as slides (Baskin, 1979). Technique 3 is typically only for very organic-rich source rocks and coals.

Unfortunately, no universally accepted maceral classification currently exists for the finely dispersed kerogen in source rocks. Tissot and Welte (1984, p. 498) attempt to show the relationships between some of the many classifications used to describe kerogens. The principal macerals in coals and sedimentary rocks can be categorized into three groups (Stach *et al.*, 1982; Taylor *et al.*, 1998) defined below. Table 4.2 summarizes the probable origins and most commonly used nomenclature for maceral types within each maceral group.

- *Liptinites:* oil-prone macerals that have low reflectance, high transmittance, and intense fluorescence at low levels of maturity. Many liptinite phytoclasts have characteristic shapes and textures, e.g. algae (such as *Tasmanites*), resin (impregnating voids), and spores. Liptinites are divided broadly into alginites and exinites in Figure 4.3.
- *Vitrinites:* gas-prone macerals that have angular shapes, usually gelified, but sometimes with cellular structure. The reflectance of vitrinite phytoclasts is used as an indicator of the thermal maturity for rock samples. Vitrinite macerals show intermediate reflectance (low-gray) (Bostick, 1979) and transmittance, usually with no fluorescence unless impregnated by liptinites.
- *Inertinites:* inert macerals that have angular shapes, typically with cellular structure. Inertinite phytoclasts show high reflectance, show no fluorescence, and are opaque in transmitted light.

Much of the dispersed organic matter observed under the microscope is described as amorphous. Transmission electron microscopy of ultra-thin sections of amorphous organic matter from various oil shales and source rocks reveals ultralaminae, nanoscale lamellar structures thought to originate by selective preservation of non-hydrolyzable biomacromolecules from the thin outer cell walls of algae (Largeau *et al.*, 1990; Derenne *et al.*, 1993). These resistant biomacromolecules, the

Table 4.2. *Generalized maceral nomenclature and origins*

Maceral group	Kerogen type*	Maceral type	Probable origin
Liptinite	II	Resinite	Plant resins/degraded macerals
	II	Sporinite	Spores/pollens
	II	Cutinite	Plant cuticles
	II	Bituminite	Degraded algae
	I	Alginite	Algae
	II	Liptodetrinite	Mixed origin/aliphatic biological materials
Vitrinite	III	Vitrinite	Woody tissues
Inertinite		Semifusinite	Partially heated woody tissues
		Fusinite	Carbonized woody tissues
		Micrinite	Woody tissues/uncertain
	IV	Macrinite	Woody tissues/uncertain
		Inertodetrinite	Carbonized woody tissue fragments
		Sclerotinite	Fungal hyphae

* Assumes thermally immature organic matter.

so-called algaenan, are selectively preserved during diagenesis (Tegelaar *et al.*, 1989). Most, but not all, amorphous organic matter is oil-prone. Immature to mature oil-prone amorphous organic matter typically fluoresces under ultraviolet light, while other types of organic matter do not.

Kerogen types

Elemental analysis

Crude oils and kerogen are sometimes analyzed for elemental composition. For example, CHN- and sulfur analyzers can be used to provide carbon, hydrogen, nitrogen, and sulfur content of kerogen (Durand and Monin, 1980). Analysis for organic oxygen in kerogen is much more complex and is commonly not attempted. Sulfur content is the principal elemental measurement for oils.

ATOMIC C/N RATIOS

Atomic C/N ratios are used widely to distinguish between algal and terrigenous plant origins of organic matter in shallow sediments and thermally immature sedimentary rocks (e.g. Silliman *et al.*, 1996). Algae and vascular plants generally show atomic C/N ratios of 4–10 and >20, respectively (Meyers, 1994). However, C/N ratios can be affected by partial degradation of organic matter during diagenesis and the grain size of

the associated sediments (Meyers, 1997). Meyers (1994) used a plot of atomic C/N versus the stable carbon isotope ratio of bulk organic matter to distinguish input from marine or lacustrine algae and C3 or C4 land plants in sediments.

ATOMIC H/C VERSUS O/C

The most familiar method used to classify organic matter type in sedimentary rocks is the atomic H/C versus O/C or van Krevelen diagram (Figure 4.4, left). This diagram was originally developed to characterize coals (van Krevelen, 1961; Stach *et al.*, 1982; Taylor *et al.*, 1998) during their thermal maturation or coalification, but Tissot *et al.* (1974) extended its use to include the kerogen dispersed in sedimentary rocks. The plot shows different types of kerogens as type I (very oil-prone), type II (oil-prone), and type III (gas-prone). Type IV (inert) kerogens have very little hydrogen and plot near the bottom of the figure. Because kerogens represent mixtures, they contain different proportions of the three principal maceral groups in coals and sedimentary rocks (liptinites, vitrinites, and inertinites). However, type I and II kerogens are dominated by liptinite macerals, while types III and IV are dominated by vitrinites and inertinites, respectively.

Each kerogen type follows a different pathway on the van Krevelen diagram during thermal maturation as

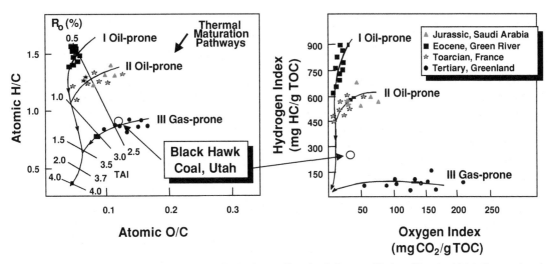

Figure 4.4. Source-rock organic matter can be described using van Krevelen (left) or modified van Krevelen (right) diagrams based on elemental analysis of kerogen or Rock-Eval pyrolysis of whole rock samples, respectively (Peters, 1986). Thermal maturity increases along converging maturation pathways, with the most mature samples in the lower left of each figure. R_o, vitrinite reflectance; TAI, thermal alteration index of pollen and spores (Jones and Edison, 1978). Rock-Eval pyrolysis can overestimate the hydrocarbon generative potential of certain coals (see also Figures 18.100 and 18.101), such as the Cretaceous Black Hawk coal from Utah.

it becomes more depleted in hydrogen and oxygen relative to carbon. At high levels of catagenesis, all kerogens approach the composition of graphite (pure carbon) near the lower left portion of the diagram. Van Krevelen diagrams provide only a crude estimate of relative maturity among kerogens from different depositional settings and thus require support by other methods (Peters, 1986).

Because it is difficult to measure organic oxygen accurately, kerogens can be plotted on the H/C versus O/C diagram using the method of Jones and Edison (1978). The measured atomic H/C defines a horizontal line on the plot (e.g. H/C = 1.5). Using independent data on maturation or organic matter type allows definition of a point on the diagram that characterizes each kerogen and infers an atomic O/C. Elemental compositions of kerogens in Figure 4.4 were calibrated to thermal maturity measurements based on microscopy (vitrinite reflectance and thermal alteration index).

Rock-Eval pyrolysis

Peters (1986) describes guidelines for evaluating or screening rock samples using Rock-Eval pyrolysis, where rock chips are heated at 25°C/min in an inert atmosphere. Figure 4.5 shows a typical Rock-Eval

pyrogram and related parameters. A flame ionization detector (FID) senses organic compounds generated during pyrolysis. The first peak in the figure (S1) represents hydrocarbons that can be thermally distilled from rock. The second peak (S2) represents hydrocarbons

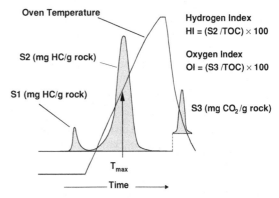

Figure 4.5. Schematic pyrogram showing evolution of organic compounds from a rock sample during pyrolysis (increasing time from left to right). Important measurements include S1, S2, and T_{max}. T_{max} intersects the oven temperature curve where the S2 peak maximizes (arrow). Hydrogen and oxygen indices are calculated as shown (Peters, 1986). For reliable Rock-Eval interpretations, we recommend pyrograms for samples collected every 30–60 feet (9–18 m) in each well.

generated by pyrolytic degradation of the kerogen in the rock. Although the literature expresses S1 and S2 in milligrams of hydrocarbons per gram of rock, the FID also detects non-hydrocarbons, providing carbon atoms are present. The third peak (S3) represents milligrams of carbon dioxide generated from a gram of rock during temperature programming up to 390°C, and is analyzed using a thermal conductivity detector (TCD). A thermocouple monitors temperature during pyrolysis. T_{max} is the oven temperature that corresponds to the maximum rate of S2 hydrocarbon generation. The hydrogen index (HI) corresponds to the quantity of pyrolyzable organic compounds from S2 relative to the TOC in the sample (mg HC/g TOC). The oxygen index (OI) corresponds to the quantity of carbon dioxide from S3 relative to the TOC (mg CO_2/g TOC). The production index (PI) is defined as $S1/(S1 + S2)$.

Modified van Krevelen diagrams characterize source-rock organic matter on an HI versus OI rather than atomic H/C versus O/C diagram (Figure 4.4, right). Both diagrams provide an assessment of hydrocarbon generative potential (Espitalié et al., 1977; Peters et al., 1983; Peters, 1986). Pyrolysis generally yields similar results to elemental analysis, but it is faster and less expensive and requires less sample, thus making it a key method for most geochemical source rock studies. However, Rock-Eval HI can underestimate kerogen quality or generative potential because highly oil-prone kerogens yield high atomic H/C but do not always show high pyrolytic yields. For this reason, the atomic H/C of selected kerogens should be used to support Rock-Eval HI measurements of the remaining generative potential of rock samples (Baskin, 2001).

Note: Coal is defined as any rock containing more than 50 wt.% organic matter. Both coals and sedimentary rocks can contain any combination of macerals and the term "coal" implies nothing about maceral composition. No universally accepted classification for dispersed organic matter and kerogen types exists in the literature at this time. In this book, we will use the type I, II, III (Tissot et al., 1974) and IV (Demaison et al., 1983) nomenclature to describe kerogens. Synthetic fuels, including synthetic oil, are commonly manufactured from coal or other hydrocarbon-containing materials.

The four principal types of kerogen in sedimentary rocks include types I (very oil-prone), II (oil-prone), III (gas-prone), and IV (inert), as shown in Table 4.1. Some discussions modify these definitions to include transitional kerogen compositions, such as type II/III, or sulfur-rich kerogens, such as type IIS, as discussed below.

Type I

Immature type I kerogen has high atomic H/C ($\sim$ 1.5), high Rock-Eval HI (>600 mg HC/g TOC), and low atomic O/C (<0.1). Type I kerogens are dominated by lipinite macerals, although vitrinites and inertinites can occur in lesser amounts. The kerogen consists mainly of aliphatic structures, suggesting major contributions from lipids during diagenesis. Sulfur is generally low in type I kerogen, although some unusual gypsiferous and anoxic lacustrine settings result in sulfur-rich type IS kerogen (e.g. Sinninghe Damsté et al., 1993a; Peters et al., 1996a; Carroll and Bohacs, 2001). Laboratory pyrolysis or burial maturation of type I kerogen results in higher yields of hydrocarbons than the other kerogen types. Alkanes dominate the pyrolysis products. Most type I kerogens are dominated by lipid-rich algal debris that has undergone extensive bacterial reworking, particularly in lacustrine settings. *Botryococcus* and similar lacustrine algae and their marine equivalents, such as *Tasmanites*, commonly are major contributors to type I kerogens.

Although type I kerogens are less common than the others, they account for many important petroleum source rocks and oil shales (Hutton et al., 1980). Examples include the organic-rich Green River Shale (actually a marl) from Utah, Colorado, and Wyoming, Chinese oil shales, boghead coals, torbanites from Scotland, and coorongite from South Australia (e.g. Cane, 1969).

Type II

Immature type II kerogen has high atomic H/C (1.2–1.5), high Rock-Eval HI (300–600 mg HC/g TOC), and low atomic O/C compared with types III and IV. The kerogen is dominated by liptinite macerals but, like type I kerogens, vitrinites and inertinites can occur in lesser amounts. Sulfur is typically higher in type II compared with other kerogen types. Unusually high sulfur in certain type II kerogens, such as the

Permian Phosphoria Formation (Lewan, 1985) and the Phosphatic Member of the Miocene Monterey Formation, may explain the tendency of these kerogens to generate petroleum at lower levels of maturity than others (Orr, 1986; Peters *et al.*, 1990; Baskin and Peters, 1992), although high atomic O/C has also been implicated (Jarvie and Lundell, 2001). Orr (1986) describes type IIS kerogens and procedures for their isolation from the Monterey Formation in the Santa Maria Basin and Santa Barbara coastal areas, California. According to Orr (1986), type IIS kerogens contain unusually high organic sulfur (8–14 wt.%, atomic S/C ≥ 0.04) and begin to generate oil at lower thermal exposure than typical type II kerogens with <6 wt.% sulfur. Pyrolysis or burial maturation of type II kerogen results in higher yields of hydrocarbons than the other kerogens, except type I.

Type II kerogen originates from mixed phytoplankton, zooplankton, and bacterial debris, usually in marine sediments. Type II kerogens account for most petroleum source rocks, including Jurassic rocks in Saudi Arabia, the North Sea, and West Siberia, Cretaceous rocks in Venezuela, and Miocene rocks in California.

Type II/III kerogen describes a transitional composition between types II and III that commonly represents a mixture of marine and terrigenous organic matter deposited in a paralic marine setting. Immature type II/III kerogen has atomic H/C and Rock-Eval HI in the range 1.0–1.2 and 200–300 mg HC/g TOC, respectively (Table 4.1).

Type III

Immature type III kerogen has low atomic H/C (0.7–1.0), low Rock-Eval HI (50–200 mg HC/g TOC) and high O/C (up to ∼0.3). Type III organic matter yields less hydrocarbons than types I or II during pyrolysis or burial maturation. This type of organic matter is common in Devonian through Tertiary rocks, usually originates from terrigenous plants, and is dominated by vitrinite and lesser amounts of inertinite macerals. For example, the Cretaceous Black Hawk coal from Carbon County in Utah is a typical humic or gas-prone coal (Figure 4.4) that contains 75% vitrinite, 20% inertinite, and 5% liptinite. More than 80% of coals worldwide are humic (Hunt, 1996). The Black Hawk sample is a low rank coal based on low vitrinite reflectance ($R_o = 0.43\%$), thermal alteration index (TAI = 2), T_{max} (429°C), and production index (0.06).

Type IV

Type IV kerogen is "dead" carbon showing low atomic H/C (<0.7), low Rock-Eval HI (<50 mg HC/g TOC), and low to high O/C (up to ∼0.3). The type IV maturation pathway is not shown in Figure 4.4 because the kerogen does not yield significant amounts of hydrocarbons. This type of kerogen is dominated by inertinite macerals. Type IV kerogen can originate from other kerogen types that were reworked and oxidized.

Sulfur content

Sulfur content is a bulk parameter commonly used to evaluate the economics of refinery operations or to support inferred genetic relationships among crude oils. For example, plots of American Petroleum Institute (API) gravity or stable carbon isotope ratio versus wt.% sulfur can be used to show tentative relationships between oils as part of sample screening.

Understanding the origin of sulfur in petroleum and kerogen is necessary in order to make reliable interpretations of source input and depositional environment. Some sulfur originates from amino acids in organic matter deposited with the sediments. However, most primary sulfur in petroleum originates from early diagenetic reactions between the deposited organic matter and aqueous sulfide species (S^{2-}), such as hydrogen sulfide (H_2S) and polysulfides (e.g. Francois, 1987). Claypool and Kaplan (1974) describe the pertinent pore water and microbial chemistry of shallow sediments.

Sulfides are produced by sulfate-reducing bacteria, such as *Desulfovibrio*, primarily in anoxic marine sediments (Figure 4.6). If H_2S migrates upward from

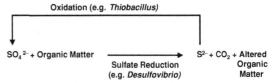

Figure 4.6. Generalized reactions of sulfur species during diagenesis of sediments. Under highly reducing to anoxic conditions, excess sulfides are generated by sulfate-reducing bacteria. Organic matter and metals, such as iron, compete for the sulfides. Because clay-poor carbonate sediments contain few metals, excess sulfides become incorporated into the immature kerogen. Reprinted with permission by ChevronTexaco Exploration and Production Technology Company, a division of Chevron USA Inc.

an anoxic into an oxic environment in sediment or the water column, then it is oxidized rapidly back to sulfate by aerobic bacteria, such as *Chlorobium* or *Thiobacillus* (Figure 4.6) (Orr and Gaines, 1974). Even in normally anoxic settings, sulfide ions produced by sulfate-reducing bacteria can be oxidized back to sulfate during periodic mixing with oxic water layers. However, because of the poor water circulation typical of anoxic basins (Demaison and Moore, 1980), H_2S may increase. H_2S is toxic to aerobic organisms.

Two sinks compete for excess sulfide in anoxic sediments: metals and organic matter. Although some H_2S may be oxidized under anoxic conditions by photosynthetic purple bacteria, the major sink is the reaction of sulfide with iron to form hydrotroilite, troilite, and eventually pyrite. Plots of organic carbon versus pyrite sulfur for rocks and sediments can be used to distinguish deposition under normal oxygenated versus euxinic (anoxic, H_2S-rich, usually deep water) marine conditions (Raiswell and Berner, 1985).

High- and low-sulfur crude oils originate from high- and low-sulfur kerogens, respectively (Gransch and Posthuma, 1974). Clay-poor carbonate muds contain insufficient iron and other metals to scavenge all available sulfide (Tissot and Welte, 1984). Under these conditions, much of the excess sulfide becomes incorporated into the kerogen. Thus, many high-sulfur kerogens and oils originate from clay-poor marine carbonates or anhydrites deposited under anoxic conditions. Type IIS kerogens are the most common sulfur-rich kerogens. They occur mainly in euxinic marine source rocks, such as the Miocene Monterey Formation in California, the Upper Jurassic Bazehnov Formation in Russia, and the Kimmeridge Clay Formation in the North Sea. Some lacustrine type IIS kerogens exist. Type IS, as found in the lacustrine Catalan oil shales in Spain (Sinninghe Damsté *et al.*, 1993a), and type IIIS are rare.

Metals associated with clays in marine siliciclastic rocks (e.g. most shales) may outcompete organic matter for reduced sulfur, leading to low-sulfur kerogens and oils. Most lacustrine kerogens and oils are also low in sulfur, but for a different reason. Lacustrine sediments usually do not contain sufficient sulfate for strong enrichment of sulfur to occur in the organic matter. However, certain salt-lake source rocks in China contain elevated sulfur (Fu *et al.*, 1986), as do the lacustrine oils from Rozel Point, Utah (up to 14 wt.%) (Meissner *et al.*, 1984; Sinninghe Damsté *et al.*, 1987).

High-sulfur kerogens generate petroleum at lower thermal exposure than other kerogens (Lewan, 1985; Orr, 1986; Baskin and Peters, 1992). Because high-sulfur kerogens form in anoxic settings, a correlation exists between high-sulfur oils derived from these kerogens and biomarker ratios indicating anoxic depositional conditions, such as high C_{35} homohopane indices or low pristane/phytane. Many sulfur-rich oils and rock extracts also have low diasterane/sterane ratios, typical of clay-poor source rocks. The same factors that control the distribution of sulfur in dispersed kerogens from marine and nonmarine source rocks also apply to the macerals in coals (Casagrande, 1987).

At temperatures above 150°C, sulfur can be incorporated into organic matter in contact with sedimentary anhydrite or gypsum by a process called thermochemical sulfate reduction (Orr, 1974). Biodegradation can result in increased sulfur in oils due to preferential removal of saturated hydrocarbons.

Sinninghe Damsté and de Leeuw (1990), Kohnen *et al.* (1991), and Guadalupe *et al.* (1991) show that biomarker distributions differ between free and sulfur-bound (resin) fractions of petroleum, apparently due to selective preservation of certain compounds by sulfur linkage during diagenesis. These sulfur-bound biomarkers can be released by treatment with Raney nickel. For example, aromatic carotenoids derived from photosynthetic bacteria and/or marine sponges are absent in the free hydrocarbon fraction but are major compounds in the desulfurized resin fraction of some samples. Thus, analysis of desulfurization products from resins may allow more accurate assessment of the original biomarkers in the depositional environment. Further, distributions of biomarkers released by Raney nickel treatment can be used as fingerprints in correlation studies (Sinninghe Damsté and de Leeuw, 1990). Distributions of sulfur-containing compounds generated by flash pyrolysis of kerogens was used to distinguish organic matter from different Ordovician rocks, including the Guttenberg oil rock from the Decorah Formation, which is composed predominantly of remains of the alga *Gloeocapsomorpha prisca* (Douglas *et al.*, 1991).

Kohnen *et al.* (1991) used MeLi/MeI for selective chemical degradation of di- and polysulfide linkages in polar and asphaltene fractions of extract from immature Italian bituminous shale. They showed that many bound biomarkers, including acyclic, branched, isoprenoid, steroid, hopanoid, and carotenoid moieties, are linked to the kerogen by di- or polysulfide bonds.

X-ray absorption near-edge spectroscopy (XANES) has been applied to identify and quantify classes of sulfur-containing compounds in oils and rock extracts (Waldo et al., 1991). This information is useful in order to describe source-rock depositional environment and thermal maturity. For example, high-sulfur oils show distinct sulfide-rich (>30% sulfide, <65% thiophene) or thiophene-rich (<15% sulfide, >75% thiophene) XANES profiles, which appear to be related to clastic versus carbonate source rocks, respectively.

Schmid et al. (1987) and Sinninghe Damsté et al. (1987) describe long-chain dialkylthiacyclopentanes in oils and suggest a mechanism for diagenetic or thermal incorporation of sulfur based on the structures of these compounds. Payzant et al. (1986) described terpenoid sulfides in petroleum. Sinninghe Damsté et al. (1987) and Valisolalao et al. (1984) tentatively identified steroid and hopanoid thiophenes, respectively, in petroleum. Abundant benzothiophenes and alkyldibenzothiophenes in petroleum were proposed as indicators of carbonate-evaporite source environments (Hughes, 1984). Sinninghe Damsté et al. (1989) described various highly branched isoprenoid thiophenes in sediments and low-maturity oils. These compounds appear to result from selective incorporation of sulfur into isoprenoid alkenes during diagenesis.

Oil shales

Oil shales are fine-grained sedimentary rocks that contain more than 10 wt.% hydrogen-rich (liptinite) kerogen that can be burned directly or pyrolyzed to yield combustible fuel (Cook and Sherwood, 1991). They can be effective petroleum source rocks if sufficiently mature. Oil shales were deposited throughout geological history, from the Proterozoic (e.g. Karalian Shungite) until more recent (e.g. Green River Shale) times. They were deposited under a wide range of geologic settings that share common traits: high productivity, anoxic bottom waters, and low rates of clastic or evaporitic sedimentation.

The terminology used to describe oil shales is based on inconsistent kerogen nomenclature. Several schemes attempt to provide consistent terminology for oil-shale classification based on petrography of the kerogen macerals within a depositional framework (Hutton, 1987; Cook and Sherwood, 1991).

Rocks and sediments with >10 wt.% TOC fall into three broad catagories: coals and coaly shales, bitumen-rich rocks and tar sands, and oil shales. Coals and coaly shales can be subdivided into humic and sapropelic types. The former contain mostly vitrinitic (humic) macerals that have low oil-generative potential. The sapropelic coals contain sufficient liptinite macerals to be oil-prone source rocks. Of these, only the cannel coals have a preponderance of lipitintic and resinitic constituents derived from higher plants. Cannel coals may be considered oil shales.

Note: Gold (1999, p. 97) asserts that many coals are abiogenic. He describes one locality where albertite fills a nearly vertical crack across sedimentary layers. He states that "biogenic theory can offer no remotely plausible causal explanations for these and other anomalous coal environments." Gold fails to recognize that albertite is solid bitumen, not coal (Curiale, 1986).

Oil shales derived from algal and microbial organic matter can be divided according to kerogen morphology under fluorescence microscopy. For example, Hutton (1987) distinguished three major maceral types: lamalginite, telalginite, and bituminite. Lamalginite is composed of thin-walled colonial or unicellular algae that occur as distinct laminae but display little or no recognizable biological structure at magnifications typical of transmitted light and fluorescence microscopy. Telaginites are composed of structured organic matter from large colonial or thick-walled unicellular organisms, such as *Botryococcus*, *Tasmanites*, and *Gloeocapsomorpha*. Bituminite is essentially amorphous kerogen that cannot be classified as either lamalginite or telalginite. Most marine oil shales contain some bituminite mixed with other kerogen macerals.

Cook and Sherwood (1991) divide the lamosite (lamalginite-rich) oil shales into two classes: marosite and lacosite (Figure 4.7). Marosite oil shales occur only in saline marine settings and contain kerogen derived mostly from dinoflagellates and acritarchs as revealed by their cysts. Few oil shales are pure marosite (e.g. the Australian Triassic Kockatea Shale and Jurassic Dingo Claystone). Most oil shales contain abundant bituminite and marosite and are called marobitosites, bitomarosites, or mixed oil shale. These mixed kerogens generally plot

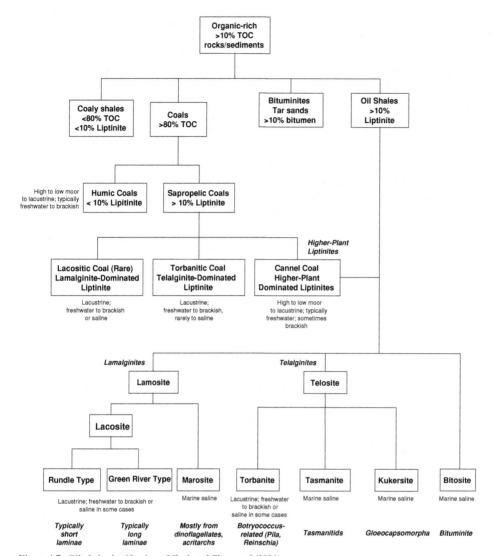

Figure 4.7. Oil-shale classification of Cook and Sherwood (1991).

along the type II van Krevelen trend. Lacosites are lacustrine deposits that contain primarily algal remains. Lacosites can be divided further into Rundle-type (i.e. similar to the Tertiary Rundle lacustrine oil shales of Queensland, Australia) and Green-River type (i.e. similar to the Mahogany Ledge facies of the Green River Formation). Various Chlorophyta (e.g. *Pediastrum*) and dinoflagellates (e.g. *Cleistosphaeridium*) contribute organic matter to Rundle-type deposits, while cyanobacteria are major contributors to Green River-type lacosites.

Cook and Sherlock (1991) divide telosite (telalginite-rich) oil shales into three classes according to

their primary algal input (Figure 4.7). Torbanite originates from *Botyrococcus* and related green algal species. These oil shales are restricted to fresh-brackish waters in lacustrine settings. Tasmanite (named after Permian oil shale in Tasmania) originates from remains of *Tasmanites*, very large (up to ~0.5 mm diameter) marine algae. The algal remains are mostly tasmanitid cysts that have high atomic H/C ratios (Figure 4.8). Facies of the Antrim/Chattanooga Shales are well known examples of Tasmanite. Kukersite (named after Estonian kukersite) originates from *Gloeocapsomorpha prisca*, an extinct organism of uncertain affinity that dominated

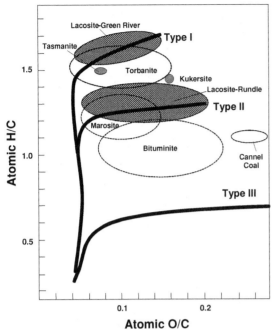

Figure 4.8. Range of different oil shale classes on the van Krevelen diagram (data from Cook and Sherwood, 1991).

calcareous marine facies in the Ordovician. *G. prisca* resembles modern colonies of *Botryococcus braunii*.

Examples of petroleum systems by kerogen type

Table 4.3 gives examples of petroleum systems according to the type of kerogen in the source rock. The table shows some of the complexities related to naming petroleum systems and quantifying their size, which are discussed further in Chapter 18. For example, the source rocks for the Mandal-Ekofisk(!), Kimmeridge-Brent(!), and Kimmeridge-Hibernia(!) petroleum systems are of about the same age, but their reservoir rocks and locations differ. The main source rock in Viking and Central grabens of the North Sea is called the Mandal or Farsund Formation in Norwegian and Danish waters, the Draupne Formation offshore northern Norway, and the Kimmeridge Clay Formation offshore the UK. On the Norwegian side of the Central Graben, the Mandal Formation is the source rock and the Ekofisk Formation within the Chalk Group is the major reservoir rock, while Brent sandstones are the main reservoir in other areas. Seafloor spreading displaced the Egret Member of the Rankin Formation on the Grand Banks of Newfoundland from its original position adjacent to

Kimmeridge source rocks of the same age in the North Sea. The sizes of petroleum systems are difficult to quantify, and the values in Table 4.3 are only estimates. These estimates are always based on incomplete data and require many assumptions to determine ultimate recovery, including conversion of gas volumes to barrels of oil equivalent or kilograms of hydrocarbons. We make no attempt to interconvert the latter in the table because the required assumptions may differ from the intent of the quoted authors.

Indirect estimates of source-rock richness

As discussed above, geochemical parameters, such as Rock-Eval HI and OI or atomic H/C and O/C, readily identify kerogen type when rock samples are available. When only geologic information is available, estimates of kerogen type can be made using Table 4.4. However, this approach is dangerous because of a weak relationship between generalized descriptions of depositional environments and kerogen types. Other methods can be used to estimate source-rock richness and quality when rock samples are not available, including conventional well-logging tools (e.g. gamma ray, resistivity) and inorganic (e.g. trace metals in rock) and organic geochemical (seep) measurements, as discussed below. Organic geochemical measurements on seeps are discussed later.

Delta log R

If the source rock interval was penetrated and analyzed by porosity and resistivity tools, then organic richness can be estimated using the delta log R method ($\Delta logR$). When correctly scaled transit-time and resistivity curves from a well are overlain for the fine-grained, non-source intervals, they show a separation or $\Delta logR$ in the nearby source-rock intervals (Passey *et al.*, 1990). The empirical relationship between TOC and $\Delta logR$ must be calibrated for different rock intervals or localities. Because $\Delta logR$ provides continuous data with depth, it allows recognition of details in the stacking patterns of source-rock richness that may not be resolved by discrete samples analyzed for TOC.

Three recurrent patterns in the vertical distribution of TOC in source rocks evident from worldwide $\Delta logR$ analyses can be explained by sequence stratigraphic concepts. Most marine organic-rich shales are composed of discrete sedimentary units with TOC that decreases upward from maxima near their bases

Table 4.3. *Examples of petroleum systems divided according to the type of kerogen in the source rock*

Petroleum system	Size*	Source-rock age	Lithology	Fields	References
Type I					
Green River(!) Uinta Basin, Utah	Significant (~500 MBOE, 86 bkg HC)	Late Paleocene–Early Eocene	Lacustrine carbonate mudstones	Altamont–Bluebell, Redwash	Fouch *et al.* (1994), Magoon and Valin (1994), Ruble *et al.* (2001)
Lagoa Feia-Carapebus(!) Campos Basin, Brazil	Large (?)	Early Cretaceous	Lacustrine saline calcareous shale	Marlim	Mello *et al.* (1994)
Bucomazi-Vermelha(!), Lower Congo Basin, Cabinda	Large (?)	Early Cretaceous	Lacustrine shale		Demaison and Huizinga (1994)
Pematang-Sihapas(!) Central Sumatra, Indonesia	Large (~12.8 BBOE)	Eocene–Oligocene	Lacustrine shale	Minas, Duri	Demaison and Huizinga (1994), Katz and Dawson (1997)
Types II and IIS					
Hanifa-Arab(!) Arab–Iranian Basin	Super Giant (~498 billion BOE)	Late Jurassic	Marine shelf shale	Ghawar	Klemme (1994)
Mandal-Ekofisk(!) Central Graben, North Sea	Large (~8.2 BBOE)	Late Jurassic	Marine mudstone/ carbonate	Ekofisk, Eldfisk, Valhal, Forties	Cornford (1994)
Kimmeridge-Brent(!) Northwest European Shelf	Super Giant (~125 BBOE, 1119 bkg HC)	Late Jurassic	Marine mudstone	Statfjord, Oseberg	Klemme (1994), Magoon and Valin (1994)
Kimmeridge–Hibernia(!) Jeanne d'Arc B., Grand Banks	Significant (?)	Late Jurassic	Marine mudstone	Hibernia, Hebron, Terra Nova	Klemme (1994)
La Luna-Misoa(!) Maracaibo Basin, Venezuela	Giant (~60 BBOE, 8160 bkg HC)	Late Cretaceous	Marine limestone/ shale	Tia Juana, Lama, Boscan, Lamar	Talukdar and Marcano (1994), Magoon and Valin (1994)

(*cont.*)

Table 4.3 (*cont.*)

Petroleum system	Size*	Source-rock age	Lithology	Fields	References
Smackover-Tamman(!) Gulf of Mexico	Giant (~98 BBOE)	Late Jurassic	Marine carbonate	Hatter's Pond, Chatom, Mary Ann	Klemme (1994)
Bazhenov-Neocomian(!) West Siberia Basin, Russia	Super Giant (~200 BBOE)	Late Jurassic	Marine shale	Fedorov, Van-Egan, Salym	Klemme (1994)
Simpson-Ellenburger(.) West Texas, New Mexico	Significant (~105 bkg HC)	Middle–Late Ordovician	Marine shale	Andector, TXL, Headlee	Katz et al. (1994), Magoon and Valin (1994)
Type II/III					
Akata-Agbada(!) Niger Delta, Nigeria	Significant (~224 bkg HC)	Middle–Late Eocene	Marine deltaic shale	Oguta, Oben, Okpai, Egemba	Ekweozor and Daukoru (1994), Magoon and Valin (1994)
Tuxedni-Hemlock(!) Cook Inlet, Alaska	Significant (~215 bkg HC)	Middle Jurassic	Marine shale	McArthur River, Swanson River	Magoon (1994), Magoon and Valin (1994)
Type III					
Bampo-Peutu(!) North Sumatra Basin, Indonesia	Significant (~531 bkg HC)	Late Oligocene	Marine deltaic shale	Arun	Buck and McCulloh (1994)
Cambay-Hazad(!) South Cambay Basin, India	Significant (<395 bkg in-place HC)	Early–Middle Eocene	Marine deltaic shale	Gandhar	Biswas et al. (1994)

* Petroleum system size is based on estimated recoverable hydrocarbons using classifications of Klemme (1994) or Magoon and Valin (1994). Super giant (>100 × 10⁹ BOE), Giant (20–100 × 10⁹ BOE), Large (5–20 × 10⁹ BOE), Significant (0.2–5 × 10⁹ BOE) (Klemme, 1994); or giant (5000–10000 bkg HC), large (500–5000 bkg HC), significant (50–500 bkg HC) (Magoon and Valin, 1994).
BBOE, billion BOE; bkg HC, billion kilograms of hydrocarbons; BOE, billion barrels of oil equivalent; MBOE, million BOE.

Table 4.4. *Relationships among common depositional environments, lithofacies, and kerogen types for petroleum source rocks*

Tectonic setting	Depositional environment	Lithofacies	Kerogen type
Marine source rocks			
Restricted basin	Anoxic with clay	Marine shale	II
Restricted basin	Anoxic, saline with or without clay	Carbonate	IIS
Epicontinental seaway	Anoxic	Shale, carbonate	II
Upwelling shelfal area	Anoxic or suboxic	Shale, carbonate, chert, phosphorite	II, IIS
Open ocean	Oxic or suboxic	Prodelta shale	III
Non-marine source rocks			
Coastal swamp	Anoxic	Coal	III
Paralic basin	Oxic or suboxic	Prodelta shale	III
Open lacustrine	Anoxic, freshwater	Oil shale	I
Restricted lacustrine	Anoxic, saline	Oil shale	IIS
Marginal lacustrine, fluvial	Oxic or suboxic	Siltstone, shale	II/III

(Creaney and Passey, 1993), as exemplified by early observations of the Kimmeridge Clay (Cox and Gallois, 1981). These "high TOC at the base, decreasing upward" units, or HTB, occur (1) alone or, more typically, (2) in stacked sequences. A third common pattern consists of TOC that increases from the base to a central TOC maximum (e.g. 1510 m in Figure 4.19), and then decreases toward the top of the unit.

The sequence stratigraphic model to explain these recurrent HTB profiles assumes two main controls on organic carbon accumulation: low oxygen conditions during deposition and variable sedimentation rates. Under anoxic conditions, sediment starvation during transgressions of the sea results in organic-rich, condensed intervals at the base of each HTB, when the shoreline is at its most landward position (e.g. Posamentier *et al.*, 1988; Peters *et al.*, 1997). These thin, deepwater marine sediments, commonly shales that may contain phosphorite, are characterized by slow depositional rates (<1–10 mm/1000 years) (Posamentier *et al.*, 1988). Under oxic conditions, non-deposition or submarine hardgrounds occur in this setting. Because the source of terrigenous organic matter is far away, high-TOC condensed sections also generally have abundant marine oil-prone organic matter and high gamma-ray response. Decreases in TOC and organic matter quality upward from the condensed section within each HTB

result from clastic dilution during progradation. Terrigenous clastics contain proportionally more oxidized, transported organic detritus, and are less oil-prone than marine organic matter in the condensed section.

Shelfal marine source rocks tend to contain more isolated and fewer stacked HTB units with generally lower TOC compared with time-equivalent basinal units, again due to greater dilution by clastic sediments. In distal basinal settings, shifts in depocenters have little effect on sedimentation rates under anoxic conditions and produce only subtle increases or decreases in TOC. The result is superposition of multiple stacked HTB units, which yield comparatively smooth TOC profiles with central maxima (e.g. see Figure 4.19). Condensed sections occur in the distal portions of all systems tracts and are best defined using high-resolution biostratigraphy, although they can be recognized in well logs using gamma ray/sonic or gamma ray/resistivity pairs (Loutit *et al.*, 1988). Lacustrine oil-prone source rocks show HTB profiles similar to marine source rocks, suggesting the same stratigraphic controls on organic richness.

Thorium/uranium

TOC can be estimated from gamma-ray spectral logs, which detect radioactive elements. Again, the empirical relationship must be calibrated for different rocks and localities. Thorium (Th) and uranium (U) are little affected by outcrop weathering and can also be used as indirect and inexpensive measures of source-rock quality (Schmoker, 1981). Samples from proximal, oxic source-rock depositional environments show lower Th/U and TOC than those from more distal dysoxic or anoxic settings. Th/U has been used to indicate the depositional environment and geochemical-lithologic facies of paralic marine shales (Adams and Weaver, 1958) and redox conditions during sedimentation (Zelt, 1985).

Sediments deposited under anoxic conditions can have high authigenic uranium (Myers and Wignall, 1987), which is estimated by assuming that the contents of detrital uranium and thorium are proportional and that both are immobile and entirely detrital, as follows:

$$U_{authigenic} = U_{measured} - (Th_{measured}/3)$$

Source-rock richness and quality generally increase with increasing authigenic uranium content. For example, Isaksen and Bohacs (1995) predicted the lower limits of TOC and hydrogen index in Lower-Middle Triassic mudrocks from the Barents Sea based on authigenic uranium content (Figure 4.9). The mudrocks with the best petroleum-generative potential were deposited under reducing conditions in distal open-marine shelf environments during late transgressive and early highstand system tracts (Figure 4.10). Relative stratigraphic position rather than measured depth on the vertical axis in the figure somewhat obscures this relationship. However, high TOC and HI clearly correspond

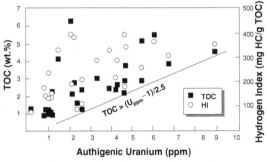

Figure 4.9. Calibration of authigenic uranium versus total organic carbon (TOC) and hydrogen index (HI) for Lower-Middle Triassic mudrocks from the Barents Sea, Norway (modified from Isaksen and Bohacs, 1995). The diagonal line represents the minimum TOC or HI predicted by authigenic uranium content, where TOC (wt.%) >[U(ppm) − 1]/2.5. Because the indicated relationship is not universal, different source rocks or organic facies require their own calibrations.

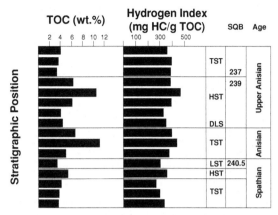

Figure 4.10. Total organic carbon (TOC) and hydrogen index (HI) for potential source rocks of Spathian through Anisian age in the Barents Sea, Norway, show systematic variations with the interpreted sequence stratigraphy (from Isaksen and Bohacs, 1995). DLS, downlap surface; HST, highstand systems tract; LST, Lowstand systems tract; SQB, sequence boundary; TST, transgressive systems tract.

to a late transgressive sample from the Anisian, and early highstand samples from the Spathian and Upper Anisian.

Note: The highest concentrations of steranes and triterpanes correspond to the maximum transgressions (HST in Figure 4.10) and C_{30} desmethylsteranes increase with the second-order rise in sea level. Lowstands show a fairly sharp base overlain by aggradationally stacked parasequences with low gamma ray and generally low TOC (~2 wt.%). The tops of the lowstands are generally sharp transgressive surfaces, commonly marked by phosphatic lag deposits. The transgressive systems tracts contain stacked retrogradational parasequences with increasing gamma ray and TOC (up to 10%) toward the mid-sequence downlap surface. Highstands contain stacked progradational parasequences with generally decreasing TOC.

Iron/sulfur

Pyrite sulfur and TOC co-vary (Berner, 1984). The iron-to-sulfur ratio (Fe/S) indicates the proportion of sulfur that is combined as pyrite versus the excess sulfur available for combination with kerogen. Pyrite formation is limited by the amount and reactivity of detrital iron in clays and organic matter in normal marine sediments deposited under oxic conditions.

Alumina/TOC

Because Fe/S is difficult to measure, Isaksen and Bohacs (1995) used the related detrital clay/TOC (Al_2O_3/TOC) as an indirect measure of organic matter quality. They assumed that (1) the proportion of total organic and inorganic sulfur to carbon is relatively constant (3 : 1) in marine mudrocks (e.g. Berner and Raisell, 1983) and (2) the iron in fine-grained sediments is associated strongly with detrital clay (Al_2O_3) (Curtis, 1987). Al_2O_3/TOC can be calculated based solely on well-log response, where detrital clay content is derived from the potassium/uranium (K/U) ratio or alumina-activation clay logging tools and TOC is estimated using the ΔlogR method (Passey et al., 1990).

Nickel/vanadium

Ni/(Ni + V) or V/(Ni + V) of source-rock extracts can be used to measure organic matter quality indirectly. Nickel and vanadium in tetrapyrrole complexes are preferentially preserved, and vanadium is higher relative to nickel in organic matter deposited under anoxic conditions (Lewan, 1984). Under oxic sedimentary conditions, Ni^{2+} is available for metallation of tetrapyrroles and vanadium exists primarily as anionic V, leading to a dominance of Ni(II) porphyrins. Figure 4.11 supports

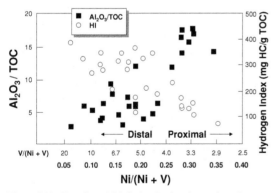

Figure 4.11. Vanadium/nickel, alumina/total organic carbon (TOC), and hydrogen index (HI) increase with source-rock quality for Lower-Middle Triassic mudrocks from the Barents Sea, Norway (modified from Isaksen and Bohacs, 1995). Al_2O_3 content is related to the amount of detrital clay in the samples. For these samples, V/(Ni + V) <3.8 (or Ni/(Ni + V) >0.26) corresponds to proximal, oxic depositional environments.

the commonly observed correlation between TOC and V/(Ni + V). Source-rock quality in Figure 4.11 is related directly to V/(V + Ni), Al_2O_3/TOC, and HI.

SOURCE-ROCK SCREENING: THERMAL MATURITY

Rock-Eval pyrolysis

Rock-Eval pyrolysis T_{max} and production index [PI = S1/(S1 + S2)] can be used to estimate thermal maturity (Table 4.1). Like vitrinite reflectance (R_o) and TAI, T_{max} is a thermal stress parameter. Changes in thermal stress parameters depend primarily on time/temperature conditions and can be related only approximately to the stage of petroleum generation for different rock types. However, PI and bitumen/TOC are generation parameters related to how much petroleum formed from the organic matter.

Rock-Eval T_{max} and PI less than ~435°C and 0.1, respectively, indicate immature organic matter that generated little or no petroleum. T_{max} greater than 470°C coincides with the wet-gas zone. PI reaches ~0.4 at the bottom of the oil window (beginning of the wet-gas zone) and can increase to as high as 1.0 when the hydrocarbon-generative capacity of the kerogen has been exhausted. Usually, some S1 will remain as adsorbed dry gas, even in highly postmature rocks. The level of thermal maturation of organic matter can be estimated roughly from HI versus OI plots (e.g. see Figure 4.4) if the kerogen quality is known. T_{max} and PI are indices of thermal maturity but depend partly on other factors, such as the type of organic matter. Conclusions on thermal maturity based on T_{max} or PI should be supported by other geochemical measurements. These might include vitrinite reflectance, TAI, or biomarker parameters. For example, Peters et al. (1983) studied the insulating effect of Cretaceous black shale on the Cape Verde Rise in the Eastern Atlantic Ocean, which was intruded by a 15-m-thick diabase sill during Miocene time. A shale sample from 11.5 m below the sill is thermally immature based on low T_{max} (424°C), PI (0.03), vitrinite reflectance (0.5%), and TAI (2.5). A nearby sample from 10.6 m below the sill contains immature 17β,21β(H)-hopanes and has low 17α,21β(H)-hopane 22S/(22S + 22R) near 0.4 (Simoneit et al. 1981).

Converting T_{max} to reflectance

The following formula can be used with caution to convert Rock-Eval pyrolysis T_{max} to vitrinite reflectance (R_o):

$$R_o(calculated) = (0.0180)(T_{max}) - 7.16$$

The formula was derived from a collection of shales containing low-sulfur type II and type III kerogen (Jarvie *et al.*, 2001b). It works reasonably well for many type II and type III kerogens, but not for type I kerogens. Use of the formula is not recommended for very low- or high-maturity samples (where T_{max} is <420°C or >500°C) or when S2 is less than 0.50 mg HC/g rock. Of course, caving of material from higher sections in a well can invalidate calculations based on analyses of cuttings samples. As with most Rock-Eval data, it is best to interpret a T_{max} trend rather than to use this formula on single samples. The above correlation is linear but corresponds reasonably well with empirical observations of T_{max} versus R_o for type III kerogens and humic coals from the Douala Basin, Cameroon (Teichmüller and Durand, 1983).

Vitrinite reflectance

Bostick (1979) showed that the optical properties of vitrinite from finely dispersed organic matter in sedimentary rocks could be used to assess thermal maturity (or rank). Vitrinite reflectance is now widely accepted by exploration geologists as a key measure of the thermal maturity. Kerogen is organic matter that is insoluble in organic solvents, and vitrinite is one of the macerals common in kerogen. Although vitrinite reflectance is related more to thermal stress than to petroleum generation, approximate R_o values have been assigned to the beginning and end of oil generation (Table 4.1; see also Figure 14.3). Interlaboratory comparison suggests that a differential of 0.1% vitrinite reflectance is significant for indicating different thermal maturities with a moderate confidence (68%), while a differential of 0.2% is required for a high confidence (95%) (Lin, 1995).

Vitrinites are a maceral group derived from terrigenous higher plants. Because land-plant communities were well developed by Devonian time, vitrinites can be important constituents in Devonian or younger sedimentary rocks. Organic petrographers subdivide the vitrinite group into telinite (retains plant structure) and collinite (unstructured). Collinite originates from the non-cellulosic parts of plants and may occur as amorphous particles or thin stringers. Although vitrinite group macerals are generally gas-prone, collinite sometimes contains hydrogen-rich resins or waxes that contribute to oil generation.

Two methods are used to prepare samples for petrographic measurement of vitrinite reflectance (R_o): (1) whole rock is polished for some very organic-rich rocks, such as coals, and (2) for less organic-rich rocks, including most source rocks, kerogen is isolated from the rock matrix. Kerogens isolated from rock samples are embedded in epoxy and polished using fine alumina grit (Bostick and Alpern, 1977; Baskin, 1979). A petrographic microscope equipped with a photometer measures the percentage of incident white light (546 nm) reflected from vitrinite phytoclasts in the kerogen preparation (Stach *et al.*, 1982; Taylor *et al.*, 1998). Phytoclasts are small particles of organic matter (Bostick, 1979). Reported R_o (%) typically represents mean or average values for ~50 phytoclasts in each polished kerogen preparation. R_o for all of the phytoclasts in a sample is commonly displayed as a histogram of R_o versus frequency (Figure 4.12).

The operator normally edits polymodal histograms to leave only those R_o values that correspond to indigenous rather than recycled or contaminant vitrinite (Figure 4.13). The mean R_o for each edited histogram is plotted versus depth to generate a reflectance profile (see Figures 4.15 and Figure 4.16).

Vitrinite contains more carbon-ring structures than the oil-prone macerals of the liptinite group. The liptinite group (also called exinite) consists of hydrogen-rich organic matter that generates significant quantities of oil during maturation. These constituents include alginite, resinite, sporinite, and cutinite (from algae, terrigenous plant resins, spores, and cuticle, respectively). Current classifications of macerals are somewhat inconsistent. In our simplified classification (Figure 4.3), we divide the liptinite group into alginite and exinite, where exinite contains all of these other maceral types.

Thermal maturation causes vitrinite to become more aromatized and reflective. The increase in reflectance of vitrinite continues throughout all thermal oil generation and appears to be caused by complex, irreversible aromatization reactions that are largely independent of rock composition. The multiple carbon-ring structures in vitrinite become *aromatized* during

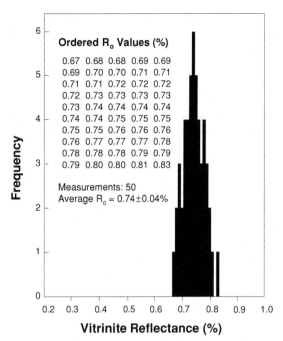

Figure 4.12. Example of a vitrinite reflectance histogram for kerogen isolated from a rock sample. Fifty reflectance values (ordered from lowest to highest by the computer, inset) were determined where each value is the percentage of incident light (546 nm) reflected from one vitrinite phytoclast. The quoted reflectance value for the sample typically is the mean of all the measurements (usually ~50) in a histogram. Reliability of reflectance as a maturity indicator increases when many histograms are obtained at different depths in each well. We recommend sampling intervals of 300 feet (~90 m) or less for vitrinite reflectance analyses. Reprinted with permission by Chevron Texaco Exploration and Production Technology Company, a division of Chevron USA Inc.

exposure to heat, similar to the aromatization of cyclohexane to benzene (Figure 4.14). Aromatic rings are more planar than their precursors and tend to align with each other, causing vitrinite to reflect more light as its internal structure becomes more ordered during maturation. The endpoint of aromatization is graphite, where essentially all carbon is found in aligned, polyaromatic sheets. Aromatization reactions in vitrinite are not necessarily associated with oil generation. Instead, they represent parallel reactions whose progress can be correlated with the stages of oil generation for different types of kerogens.

Coal petrologists were the first to use R_o to determine coal rank. Petroleum geochemists adopted R_o as a means to determine the maturity of shales and other

sedimentary rocks. Instead of analyzing coal, where vitrinite macerals are commonly abundant, petroleum geologists measure R_o in many samples where vitrinite constitutes only a small fraction of the kerogen. One extreme example might be R_o measurements on a few vitrinite phytoclasts disseminated in organic-rich type I source rock.

Measured R_o profiles with depth generally closely match predictions from mathematical basin models where the thermal history is simple, the kerogen is type III, and the rock is organic-lean. However, aromatization is inhibited in organic-rich rocks that are hydrogen-rich (types I and II), and measured R_o values are commonly depressed relative to those in type III source rock and coal of the same maturity. Some collinites that contain hydrogen-rich resins or waxes exhibit slow rates of aromatization. Telinite and collinite generally show different rates of R_o increase, especially at lower levels of thermal maturity ($R_o < 1.0\%$).

Pitfalls to interpretation

Vitrinite reflectance is one of the most common tools used by explorationists to evaluate thermal maturity. For example, R_o data are used to generate regional maturity maps on source-rock horizons, calculate erosional loss at unconformities, and calibrate basin models. For

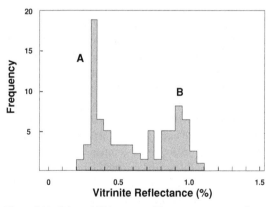

Figure 4.13. Polymodal histogram of R_o measurements made on presumed vitrinite phytoclasts in kerogen isolated from sedimentary rock. Polymodal distributions are common, but only one mode is likely to represent indigenous vitrinite. For example, if mode A is indigenous vitrinite, then mode B might represent recycled vitrinite that originated by weathering of older, more mature rocks. Alternatively, if mode B is indigenous vitrinite, then mode A might represent cavings from a shallow, low-maturity interval or liptinite that was identified incorrectly as vitrinite.

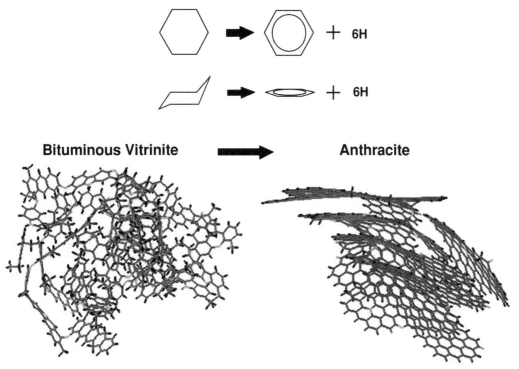

Figure 4.14. Simple aromatization reaction from cyclohexane to benzene with loss of hydrogen (top). In side view (middle), aromatization converts a puckered ring to a flat ring with sp3 and sp2 hybridized carbon atoms, respectively. Vitrinite contains polycyclic ring structures that gradually approach the polyaromatic structure of graphite during maturation (bottom). Molecular models for bituminous vitrinite and anthracite generated by J. P Mathews and P. Pappano, respectively, using the SIGNATURE program (Faulon *et al.*, 1993). Figures used with permission of J. P. Mathews, Pennsylvania State University. Light gray bonds indicate heteroatoms (e.g. N, S, O).

these reasons, interpreters must be aware of the common pitfalls associated with R_o interpretation:

- variations in kerogen or maceral type;
- contamination and sampling problems;
- misidentification of vitrinite;
- poor sample preparation.

VARIATIONS IN MACERAL TYPE

Kerogen components are categorized into maceral groups based on their morphology and genesis, but these groups actually show a range of chemical and physical properties. Thus, some vitrinites are hydrogen-rich and can yield small amounts of oil upon heating. Hydrogen content is a critical factor controlling the rate at which reflectivity changes during burial. Type I and many type II source rocks inhibit the aromatization reactions that cause R_o to increase (Figure 4.15). As another example, organic-rich (2.2–9.4 wt.% TOC) Miocene

Nodular Shale samples from the Los Angeles Basin contain abundant alginite and have lower vitrinite reflectance (0.13–0.41%) than predicted from maturation estimates based on spore coloration (0.30–0.65%) (Walker *et al.*, 1983). These results are consistent with Hutton *et al.* (1980), who observed a linear decrease in reflectance with increasing alginite content for various coals and shales. Compared with adjacent low-hydrogen, vitrinite-rich coals, hydrogen-rich, vitrinite-poor oil shales can show R_o suppression of up to 0.55% (Price and Barker, 1985).

Diverse hydrogen contents in macerals can affect R_o measurements. For example, each operator measures the reflectance of all phytoclasts that look like vitrinite. If the sample contains vitrinites with different chemical compositions, or if the operator misidentifies vitrinite, then the R_o measurements may show a large standard deviation or might even show a polymodal distribution that can affect the accuracy of the mean R_o.

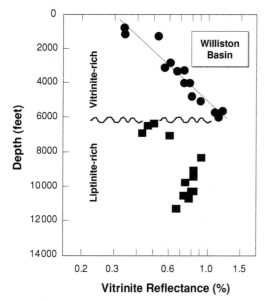

Figure 4.15. Field example of vitrinite reflectance suppression in a well from the Williston Basin, USA. (after Price and Barker, 1985). The break in the R_o trend below ~6000 feet represents suppression of R_o due to a change from terrigenous, vitrinite-rich strata overlying marine, liptinite-rich strata (Mississippian Bakken Formation source rock).

CONTAMINATION AND SAMPLING PROBLEMS

Caving of cuttings from shallow intervals during drilling, addition of lignite-based muds, and various drilling operations can affect both histograms and reflectance profiles (Figures 4.13 and Figure 4.16). When cavings or lignite additives that contain vitrinite dilute the indigenous vitrinite in a sample, then the R_o profile can be displaced toward lower values. The R_o histograms for such samples commonly show large standard deviations or polymodal distributions. The contrast in reflectance between vitrinite in caved materials (lower-reflecting) and that in the target horizon helps to distinguish these materials. Various drilling operations, such as circulating drilling fluids in the hole in zones of interest, running drill stem tests or logs, and changes in pump strokes or the mud program, can affect the quantity of cavings incorporated into the samples. Interpreters can limit these difficulties by comparing the R_o profile with notes on the mud log that was prepared during drilling. In addition, R_o data from conventional or sidewall core samples should be obtained whenever possible to corroborate the data from cuttings.

Unlike cuttings, particulate organic matter from other depth intervals does not readily contaminate cores, and their recorded depths are generally more reliable than cuttings. Where caving from coal seams or shales is suspected based on wireline log or other data, hand picking of lithologically distinct particles to exclude the caved lithology can help to reduce such contamination (Othman and Ward, 2002).

Example 1 The bimodal distribution of R_o on the histogram for a single cuttings sample in Figure 4.13 could result from several different problems:

- cavings from shallower intervals (population A);
- recycled vitrinite (population B);
- vitrinite compositional variations (e.g. collinite, telinite);
- measuring liptinite or solid bitumen rather than vitrinite.

Example 2 Figure 4.16 shows an anomalous offset at 11 500 feet (3506 m) in the reflectance profile for a well in the Western Desert, Egypt. This offset could be the result of:

(1) an erosional unconformity;
(2) a large normal fault with missing section;
(3) cuttings above 11 500 feet (3506 m) diluted by cavings. Casing was set or some other circulation/mud change occurred to minimize caving below 11 500 feet (3506 m)

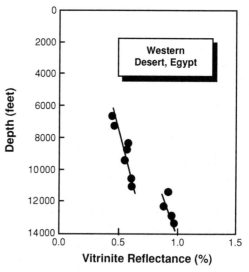

Figure 4.16. Discontinuous reflectance profile from a well in the Western Desert, Egypt (after Feazel and Aram, 1990).

(4) R_0 above 11 500 feet (3506 m) depressed due to an organic-rich, oil-prone section.

Well log data indicate that the offset in Figure 4.16 is the result of item (3), where casing was set at 11 500 feet (3506 m) and R_0 values above that point are low due to contamination by cavings from shallower intervals. Each point in Figure 4.16 is actually a mean value with a standard deviation that could be large (e.g. histogram in Figure 4.13).

If an offset between shallower and deeper R_0 profiles can be attributed reliably to an unconformity, then the minimum thickness of eroded section can be estimated graphically (Dow, 1977). The estimate is a minimum because the difference in vitrinite reflectance across an unconformity decreases with time and increasing burial depth (Katz et al., 1988). For this reason, vitrinite reflectance profiles across unconformities need not exhibit an offset, even when a substantial stratigraphic section has been removed by erosion.

MISIDENTIFICATION OF VITRINITE

Probably the most significant source of disagreement between R_0 measurements on the same samples at different laboratories is the subjectivity of the operator in selecting vitrinite phytoclasts. When R_0 profiles on the same samples differ between laboratories, they commonly reveal a consistent error throughout the well (e.g. 0.5%), or they agree at low reflectance but diverge at higher reflectance.

Geologists should insist on at least 50 measurements of different vitrinite phytoclasts per sample, when practical, and request that histograms of the raw data be provided. In return, the geologist should provide the operator with as much information as possible to assist in selection of the correct vitrinite, including sample depth, and other maturity information, such as Rock-Eval pyrolysis T_{max} and TAI. For example, indigenous vitrinite can sometimes be identified in a difficult sample that has a polymodal histogram when clear trends in R_0 are available for less difficult samples from above and below.

POOR SAMPLES OR SAMPLE PREPARATION

Low-temperature (<150°C) oxidation of vitrinite does not significantly alter R_0 in coals (Stach et al., 1982). Nonetheless, outcrop samples commonly yield unreliable R_0 measurements, possibly due mainly to lack of a depth trend that can be used to assess the data. Rock-Eval pyrolysis T_{max} is affected by weathering, thus making it difficult to corroborate outcrop R_0 data. Sample quality generally improves within a few or tens of centimeters from the recently exposed surfaces, particularly where low-dip beds outcrop along a roadcut or stream. At higher temperatures (>150°C) that might result from an ancient fire or movement of hydrothermal fluids through rocks, oxidation rims with elevated R_0 may form around vitrinite particles. High-temperature oxidation generally does not preclude useful R_0 measurements, unless the sample contains abundant charcoal deposited from an ancient fire.

Incomplete or poor polishing can reduce measured R_0 due to scattering of incident light on rough surfaces. Some polished vitrinite phytoclasts show evidence of oxidation after a few days. Stored slides should be repolished before further R_0 measurements.

Despite the potential pitfalls, reflectance profiles like those in Figures 4.15 and 4.16 have become the standard in order to evaluate subsurface maturity. Because modern basin modeling programs rely so heavily on these data, it is important that reflectance interpretations be of the highest possible quality. In general, this requires the use of all available geologic and geochemical data to evaluate measured R_0. Some general guidelines include:

- Provide the service company or operator with as much information on the samples as possible, including sample depth and other maturity measurements.
- Evaluate mud logs carefully for information on potential caving, facies changes, well additives, and thermal and drilling conditions.
- Analyze sidewall and conventional core where available to support the results from cuttings.
- Examine kerogen macerals to identify changes in organic facies.
- Compare R_0 with maturity based on T_{max}, TAI, and other maturation indices.

Maturation lines defined in terms of both TAI and vitrinite reflectance (Figure 4.4, left) were placed on the van Krevelen atomic H/C versus O/C diagram (Jones and Edison, 1978). These workers measured atomic H/C for end-member macerals (e.g. sporinite, vitrinite, exinite, resinite) isolated from organic-matter concentrates to construct the plot. Based on this plot, maceral analysis can be used to calculate the atomic H/C of kerogen for comparison with measured atomic H/C,

Table 4.5. *Examples of potential problems affecting interpretation of vitrinite reflectance*

Problem	Effect on R_o
Caving of uphole cuttings	Lower
Poorly polished vitrinite	Lower
Mud contamination (e.g. lignite additives)	Lower (usually)
Oxidation (>150°C) or recycled vitrinite	Higher (usually)
Natural reflectivity variations in vitrinite subgroups	Higher or lower
Statistical errors (insufficient number of measurements)	Higher or lower
Incorrect maceral identification (e.g. solid bitumen)	Higher or lower

thus providing independent supporting evidence for interpretations.

Probably the major limitation of vitrinite reflectance is that vitrinite group macerals do not contribute significantly to oil generation compared with the liptinite macerals. Many oil-prone source rocks, such as the Hanifa-Hadriya interval in Saudi Arabia (Ayres *et al.*, 1982), contain little or no vitrinite. Reflectance measurements determined from fewer than ~50 phytoclasts may be unreliable. Furthermore, evidence suggests that large amounts of oil-prone macerals (Hutton and Cook, 1980; Price and Barker, 1985) or bitumen (Hutton *et al.*, 1980) retard the normal progression of vitrinite reflectance with maturity. Vitrinite reflectance is subject to several other problems (Table 4.5). For these reasons, other geochemical evidence, such as TAI or biomarker maturity parameters, should always support R_o measurements of maturity.

Compared with most biomarker thermal maturity measurements on rock extracts, vitrinite reflectance has lower sensitivity and accuracy up to maturity equivalent to the onset of petroleum generation ($\sim R_o = 0.6\%$) (Table 4.1). In this range of low maturity, the biomarker parameters described below provide a more accurate assessment of rock thermal maturity than vitrinite reflectance (Mackenzie et al., 1988). Compared with the TAI, vitrinite reflectance is more precise and less subjective.

Within and beyond the zone of petroleum generation (R_o >0.6%), the usefulness of vitrinite reflectance improves compared with biomarkers. At these thermal maturities, biomarker maturity parameters are still valuable for supporting vitrinite reflectance results, and vice versa. Although vitrinite reflectance can assist in ranking the thermal maturity of organic matter in rocks, it cannot be applied to oils.

Many workers illustrate depth–reflectance plots using a log scale for reflectance versus a linear scale for depth (e.g. see Figure 4.15), which readily allows trends to be projected to greater depth. However, this introduces bias into the interpretation by minimizing nonlinear variations caused by changing chemistry of the vitrinite. For wells that penetrate rocks now at their maximum temperature, the relationship between depth and vitrinite reflectance generally has three segments: an upper segment with a linear gradient from 0.20–0.25% at the surface to ~0.6–0.7% R_o, a middle segment where reflectance increases rapidly to ~1.0% R_o, and a deeper segment where the gradient is linear but reflectance increases more rapidly than in the upper segment (Suggate, 1998). The inflection point at ~0.6–0.7% R_o is commonly obscured on semi-log plots of depth versus reflectance. Suggate (1998) prepared a generalized diagram based on many wells that relates depth, reflectance, and geothermal gradient (Figure 4.17). The diagram can be used to estimate the maximum thickness of eroded deposits or paleogeothermal gradient by making a compromise between the slope of the low-rank gradient (which projects to ~0.25% R_o at the surface), the slope of the high-rank gradient, and the shape and position of the bend between the two gradients.

Thermal alteration index and similar maturity scales

The TAI is a numerical scale based on maturity-induced color changes in organic matter under the microscope in transmitted light (Staplin, 1969). Palynomorphs generally change from yellow to brown to black with increasing burial (Bostick, 1979). Frequent comparison of samples with TAI standard microslides is necessary to describe color accurately. Measurements are best accomplished using a split-stage comparison microscope where the sample and standard can be viewed simultaneously.

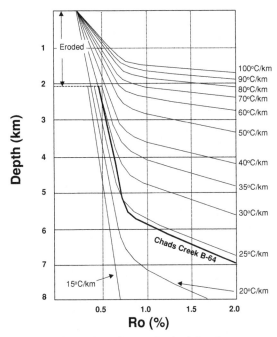

Figure 4.17. Generalized diagram showing inferred relationships between reflectance, maximum depth of burial, and paleogeothermal gradient. Fit for data from the Panarctic Chads Creek B-64 well, Arctic Canada (heavy curve) indicates a paleogeothermal gradient of ∼24° C/km and at least 2 km of eroded section (after Suggate, 1998).

Some other optical maturity scales include the conodont alteration index (CAI) (Epstein *et al.*, 1977), transmittance color index (TCI) (Robison *et al.*, 2000), and acritarch fluorescence (Obermajer *et al.*, 1997). These scales are used less commonly in geochemical studies than vitrinite reflectance or TAI. TCI of amorphous kerogen is obtained by analysis of white light from a 100-W, 6-V tungsten lamp attached to a photometric microscope. TCI is based on increasing curvature of spectra with increasing maturity. TCI curves shift from an average wavelength near 580 nm for immature, amorphous kerogen (i.e. R_o ∼0.20%) to ∼660 nm for samples containing very dark brown to black postmature kerogen (i.e. R_o ∼2.15%). The range of TCI covers all zones of petroleum generation and preservation. TCI complements R_o, TAI, and CAI, and is most useful for samples that have not yet reached the semi-anthracite stage (R_o ∼2.0%).

Acritarchs are common in Ordovician to Devonian rocks that contain marine type I or II organic matter.

The fluorescence properties of acritarchs in the 400–700-nm range parallel those of *Leiosphaeridia* alginite (Obermajer *et al.*, 1997). However, at the same maturity level, the maximum wavelength of the fluorescence intensity (I_{max}) and the red/green quotient (Q) are usually lower for acritarchs. T_{max} and Q of acritarchs vary little until the onset of oil generation for type II kerogen. At maturity corresponding to I_{max} <435°C, I_{max} and Q are commonly below 460 nm and 0.5, respectively. I_{max} shifts rapidly to 500 nm during early oil generation, and then increases throughout the oil window (500–600 nm), accompanied by slight increases in Q. For type I kerogen, no significant variations in I_{max} and Q occur up to maturity corresponding to $T_{max} = 450°C$.

TAI, TCI, CAI, and acritarch fluorescence are advantageous because they are rapid and inexpensive and do not require sophisticated instrumentation. However, the most significant advantage offered by these methods is that they provide information on the age of the organic matter, which should match that of the penetrated interval in the well. Unlike vitrinite reflectance, where identification of vitrinite can be problematic, identification of spores, pollen, or other palynomorphs with distinct morphologies is more reliable. The organisms that produce these palynomorphs commonly have distinct geologic ranges. Because many palynomorphs span known ages, the operator can use this information to identify and exclude contaminants from the maturity assessment. For example, angiosperm pollen from a species that evolved during Miocene time should not occur in cuttings from a Mesozoic well interval and probably represents particulate contamination from caving or drilling additives. TAI determined on these palynomorphs would underestimate the maturity of the Mesozoic interval. Visual examination of samples for TAI (or CAI) is also useful in order to detect reworked organic matter that overestimates the maturity of the sampled interval.

The TAI scale in this book differs somewhat from that of Staplin (1969). It ranges from 0 (very pale yellow) to 4 (black) and correlates with vitrinite reflectance (Table 4.6) and hydrocarbon generation zones (Figure 4.4) (Jones and Edison, 1978). The easiest-to-measure and most important color changes between 2.4 and 3.1 occur during beginning to peak oil generation. TAI is presumably accurate to 0.1 units, although variations can be caused by other factors, including measurements on different taxa, palynomorph thickness

Table 4.6. *Approximate relationship between thermal alteration index (TAI), spore color, and vitrinite reflectance (Jones and Edison, 1978)*

TAI

1.5	2.3	2.5	2.8	3.0	3.5	3.6	3.7	3.8	3.9	4.0

Spore color

Pale yellow		Yellow-orange	Orange-brown	Reddish-brown		Dark brown		Dark brownish-black		Black

R_o (%)

0.2	0.4	0.5	0.8	1.0	1.5	1.7	2.0	2.7	3.4	4.0

variations, weathering, and subjective errors in color assessment.

Disadvantages of the TAI method include potential errors resulting from the subjective determination of color by an operator, the need for long-ranging microfossils of the same taxon for best results, and a limited range of applicability. The method is not as accurate for measuring maturity of organic matter below or above TAI of ~2.4 and 3.1, respectively. TAI has low sensitivity to changes in maturity near the onset of oil generation compared with certain biomarker ratios, such as C_{29} sterane 20S/(20S + 20R) (Mackenzie *et al.*, 1983b).

GEOCHEMICAL LOGS AND SOURCE POTENTIAL INDEX

Geochemical logs include Rock-Eval pyrolysis and TOC data, and can be used to select appropriate source- and reservoir-rock intervals in wells for additional sampling and analysis (Figure 4.18). Extensive and closely spaced samples (every ~10–20 m) are recommended because significant changes in organic facies can occur laterally or vertically within the same source rock (e.g. Grantham *et al.*, 1980; Espitalié *et al.*, 1987; Burwood *et al.*, 1990). Additional examples of geochemical logs can be found in Peters (1986), Espitalié *et al.* (1987), and Peters and Cassa (1994). The Rock-Evaluation Expert System Advisor (REESA) is a computer module in BasinMod® 1-D (Platte River Associates) (e.g. Version 7.06) that can be used to screen large quantities of Rock-Eval and TOC data from geochemical logs (Peters and Nelson, 1992).

Geochemical logs provide some of the most reliable geochemical assessments because of depth trends established by large numbers of closely spaced samples. For example, the R_o, T_{max}, TAI, and PI trends in a geochemical log for the Pando X-1 well in Bolivia support low maturity for the Upper Devonian source rock (Figure 4.19). Thick Upper Devonian Tomachi Formation organic-rich shales in the range 1350–1590 m contain up to 16 wt.% TOC with HI up to 600 mg hydrocarbon/g TOC, indicating type I/II organic matter.

The Tomachi Formation shales are among the richest source rocks in the world based on the calculated source potential index (SPI) (Demaison and Huizinga, 1994) of 18 metric tons of hydrocarbon/m^2 (HC/m^2). This SPI surpasses those for prolific Upper Jurassic source rocks in the North Sea and Central Saudi Arabia and tertiary source rocks in the Niger Delta (15, 14, and 14 HC/m^2, respectively). Despite type III organic matter in the Akata Formation shale, the Niger Delta has a high SPI because this source rock is thick. Thus, SPI allows comparisons of the potential volumes of hydrocarbons generated from source rocks by accounting for both the quality of the organic matter and the thickness of the source-rock interval, as follows:

$$SPI = \frac{h\overline{(S1 + S2)}\rho}{1000}$$

SPI is measured in metric tons of hydrocarbons per square meter at the surface. Input is as follows: h = source-rock thickness (m), (S1 + S2) = average genetic

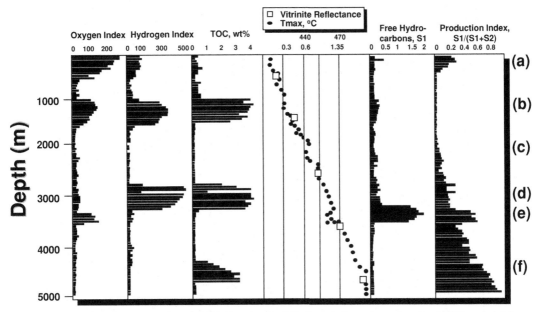

Figure 4.18. Idealized geochemical log (Peters, 1999a) based on Rock-Eval pyrolysis, total organic carbon (TOC), and vitrinite reflectance of samples from a well shows evidence for (a) oxidized organic matter that could be caused by a well additive, (b) thermally immature (potential) source rock containing type II/III kerogen, (c) non-source interval, (d) mature source rock containing type I or II kerogen, (e) reservoir rock containing petroleum possibly derived from and trapped by the overlying effective source rock, and (f) spent (postmature) source rock.

potential from Rock-Eval pyrolysis throughout source interval (kilograms hydrocarbons/metric ton), and $\rho =$ source-rock density (metric tons/m^3). In practice, only fine-grained rocks showing (S1 + S2) > 2 kg HC/metric ton of rock are used for the average genetic potential, and all source-rock densities are assumed to be 2.5 metric tons/m^3.

Lateral or vertical drainage in basins affects the significance of SPI. For example, although the SPI for Central Saudi Arabia and the Niger Delta are identical (14 kg HC/m^2) (Demaison and Huizinga, 1994), the oil potential in Saudi Arabia is much greater because oil accumulates laterally from large drainage areas compared with nearly vertical drainage from smaller areas in the Niger Delta. Because thermal maturation reduces SPI, it is calculated using Rock-Eval data from wells that penetrate immature sections of the source rock.

Geochemical logs are especially useful when they contain lithology, gamma-ray, palynology, and other geochemical data. For example, tricyclic terpane/17α-hopane for the oil at 1266 m in Figure 4.19 suggests that it is related most closely to the source rock near 1510 m, consistent with sterane and other source-related data.

Lithofacies analyses indicate that laminated shales near 1510 m in the well correspond to a maximum flooding surface associated with a major marine transgression. This interval also has the most pronounced gamma-ray response and the most oil-prone kerogen (HI $\sim$ 600 mg HC/g TOC). Prasinophyte algae are abundant, especially just below the maximum flooding surface near 1510 m. Prasinophyte algae, including *Tasmanites*, are commonly associated with high-latitude, nutrient-rich, marginal marine settings. Figure 4.19 shows an increase in tricyclic terpanes relative to hopanes in rock samples that contained elevated prasinophyte remains, consistent with published evidence that these compounds are markers of *Tasmanites* (Simoneit *et al.*, 1993).

RECONSTRUCTION OF ORIGINAL SOURCE-ROCK GENERATIVE POTENTIAL

Petroleum generative capacity depends on the original quantity (TOC°) and quality (HI°) of organic matter in

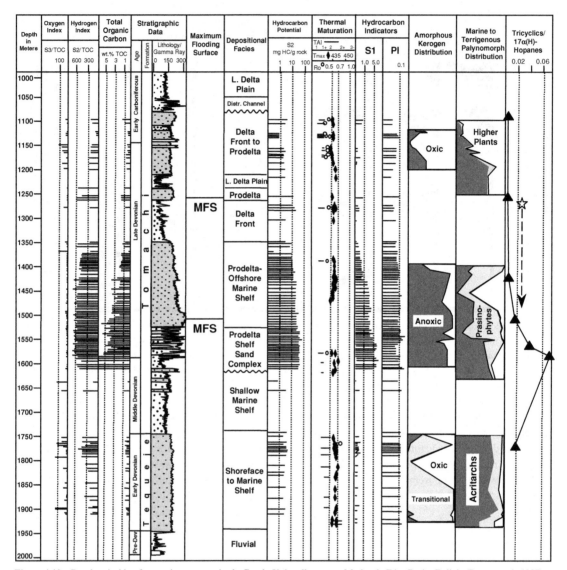

Figure 4.19. Geochemical log for continuous core in the Pando X-1 well, eastern Madre de Dios Basin, Bolivia (Peters *et al.*, 1997). The star in the far right column indicates tricyclic terpane/17α-hopane for low-sulfur (0.14 wt.%), 32° API gravity oil produced from Upper Devonian sandstones at 1266 m. Used with permission of PennWell. MFS, maximum flooding surface.

thermally immature source rock (Table 4.1). However, petroleum generation decreases the remaining generative potential as measured by TOC^x and HI^x. How does one estimate the extent of the petroleum-generation process and the volume and expulsion efficiency of petroleum generated from currently mature or post-mature source rock? By making some assumptions, the

fractional conversion (f) of source-rock organic matter to petroleum, original TOC (TOC^o), amount of expelled petroleum ($S1_{expelled}$), and expulsion efficiency (ExEf) can be calculated using the following equations. The appendix to this chapter shows derivations of these equations courtesy of G. E. Claypool (personal communication, 2002).

The extent of the petroleum-generation process, or the fractional conversion to petroleum, is calculated as follows:

$$f = 1 - \frac{HI^x\{1200 - [HI^o/(1 - PI^o)]\}}{HI^o\{1200 - [HI^x/(1 - PI^x)]\}} \quad (4.1)$$

PI^o and PI^x are the original and the measured Rock-Eval production indices $[PI = S1/(S1 + S2)]$, respectively. PI^o can be assumed to be 0.02 for most thermally immature source rocks. HI^o can be determined or estimated by various methods. For example, if thermally immature equivalents of a mature source rock are available and spatial variations in organic facies are minimal, then measured HI (HI^x) can be assumed to equal HI^o. If no immature rock equivalents are available, then petrographic and paleogeographic information can be used to estimate HI^o. Baskin (1997) concluded that estimates of original atomic H/C might be preferable to estimates of HI^o due to variations in pyrolysis results for kerogens with the same atomic H/C. However, Rock-Eval data are generally more common than elemental analyses.

The original TOC in source rock before maturation is constrained by mass balance considerations as follows:

$$TOC^o = 83.33(HI^x)(TOC^x)/[HI^o(1 - f) \\ \times (83.33 - TOC^x) + HI^x(TOC^x)] \quad (4.2)$$

where 83.33 is the percentage of carbon in generated petroleum.

$$S1_{expelled} = 1000(TOC^o - TOC^x)/(83.33 - TOC^x) \quad (4.3)$$

$$ExEf = 1 - \frac{(1 - f)[PI^x/(1 - PI^x)]}{f + [PI^o/(1 - PI^o)]} \times 100 \quad (4.4)$$

To illustrate the use of Equations 4.1–4.4, we will use data for late mature source rock ($T_{max} \sim 452^\circ C$) (Table 4.1) from the Upper Jurassic Bazhenov Formation in the Salym Field, Western Siberia (Galimov *et al.*, 1988) (2926.3 m in Borehole 312) to calculate the extent of fractional conversion, original TOC, petroleum expelled, and petroleum generation efficiency. The available data are based on 14 replicate analyses of the rock: $TOC^x = 3.65$ wt.%, Rock-Eval pyrolysis $S1^x = 7.00$ mg HC/g rock, $S2^x = 3.28$ mg HC/g rock, $HI^x = 90$ mg HC/g TOC, $PI^x = 0.68$.

The HI^o for immature equivalents of the Bazhenov source rock near the Salym Field is $\sim$550 mg HC/g TOC (Klemme, 1994). If we assume that $HI^o = 500$ mg HC/g TOC and $PI^o = 0.02$, then:

(1) What is the extent of fractional conversion of HI^o to petroleum?

$$f = 1 - \frac{90[1200 - 500/(1 - 0.02)]}{500[1200 - \{90/(1 - 0.68)\}]} = 0.86 \quad (4.1)$$

i.e. 86% of the petroleum-generation process has been completed.

(2) What was the original TOC before expulsion?

$$TOC^o = \frac{90(3.65)(83.33)}{500(1 - 0.865)(83.33 - 3.65) + 90(3.65)} \\ = 4.79 \text{ wt.%}, \quad (4.5)$$

Note that although the mature rock contains very good quantities of organic matter ($TOX^x = 3.65$ wt.%), the immature rock contained excellent quantities (Table 4.1).

(3) What was the amount of petroleum expelled from the source rock?

$$S1_{expelled} = 1000(5.42 - 3.65)/(83.33 - 3.65) \\ = 14.3 \text{ mg HC/g rock} \quad (4.3)$$

Assuming oil and shale densities of 850 mg/cm^3 and 2.4 g/cm^3, with a thickness of 10 m and area of one acre, the expelled S1 volume is $\sim$3100 barrels:

$$S1 = [14.3(2.4)/850](7758)10 = 3132 \text{ barrels/acre}$$

(4) What is the expulsion efficiency?

$$ExEf = 1 - (1 - 0.865)[0.68/(1 - 0.68)]/ \\ [0.865 + 0.02/(1 - 0.02)] = 68\% \quad (4.4)$$

We can conduct sensitivity analysis by varying the assumed HI^o to determine the effects on the calculated parameters (Table 4.7). For example, if HI^o were 600 instead of 500 mg HC/g TOC, then TOC^o would be 5.57 rather than 4.79 wt.%. These mass-balance calculations can be used to constrain the quantity and quality of organic matter in the source rock before maturation. For example, if TOC^x for the above mature Bazhenov rock was only 0.25 wt.% and the maximum HI^o was assumed to have been 500 mg HC/g TOC, then TOC^o could not have been more than 0.33 wt.%, i.e. the original

Table 4.7. *Variations in assumed initial hydrogen index (HI°) control the calculated fractional conversion (f), original total organic carbon (TOC°), total generative potential (Rock-Eval S1° + S2°), expelled petroleum (S1$_{Ex}$), and expulsion efficiency (ExEf). PI° is assumed to be 0.02*

Input assumptions		Calculated output				
HI (mg HC/g TOC)	PI°	f(%)	TOC° (wt.%)	S1° + S2° (mg HC/g rock)	S1$_{Ex}$ (mg HC/g rock)	ExEf (%)
900	0.02	97	10.8	99.3	90.2	93
800	0.02	95	8.2	67.1	57.6	90
700	0.02	93	6.6	47.4	37.6	85
600	0.02	90	5.6	34.0	24.1	78
500	0.02	86	4.8	24.4	14.3	68
400	0.02	81	4.2	17.1	7.0	50
300	0.02	71	3.8	11.5	1.2	15

rock had insufficient organic matter to be a significant source for petroleum. If we increased the assumed HI° to 700 mg HC/g TOC, then TOC° would still be poor, i.e. 0.47 wt.% (Table 4.1).

The data in Table 4.7 suggest that regardless of HI°, expulsion efficiencies for rocks containing less than 1–2 wt.% TOC° will be low. This is consistent with petrographic observations of Woodford Shale and related units, which suggest that rocks with <2.5 wt.% TOC may be incapable of establishing a continuous bitumen network to facilitate primary migration and expulsion (Lewan, 1987).

Mass-balance calculations can be used to exclude certain postmature samples from consideration as source rocks in the geologic past. For example, the above equations were used to estimate TOC° for several postmature Upper Sinian-Permian rocks from the Jianghan Basin, China (T_{max} = 464–540°C) (Peters *et al.*, 1996a). These rocks could not be correlated to seeps or produced oils from the area using biomarkers because of high maturities and low extract yields. However, seep oils from the Permian Qixia and Ordovician Baota formations may have originated from source rocks in the Qixia or Upper Sinian Doushantuo formations because their stable carbon isotope type curves project toward the isotope compositions of kerogens from Qixia and Doushantuo rock samples (−29.6‰ and −29.3‰, respectively). Assuming HI° of 500 mg HC/g TOC and PI° of 0.02 before maturation, then the calculated TOC° for Qixia and Doushantuo rock samples (3.00 and 5.96 wt.%,

respectively) are very good to excellent (Table 4.1). The isotopic composition of kerogen from the Ordovician Baota Formation (−29.7‰) is also consistent with it as a source for the seep oils. However, the low TOC^x for this rock (0.50 wt.%) argues against it as a significant source rock for oil in the geologic past. The same assumptions yield only fair quantities of calculated TOC° for the Baota rock (0.65 wt.%) (Table 4.1).

Demirel *et al.* (2001) used the same mass-balance approach to calculate the mass of petroleum expelled from Cretaceous carbonate source rocks in the Derdere, Karababa, and Karabogaz formations from southeastern Turkey (4.0–10.3 mg HC/g rock). The calculated fractional conversion and expulsion efficiency for these source rocks were 60% and 80%, respectively.

TESTS FOR INDIGENOUS BITUMEN

A critical question for oil-source rock correlation is whether the bitumen extracted from fine-grained organic-rich source rock is indigenous or whether it represents drilling contamination or migrated oil. Although migrated oil seldom invades fine-grained source rocks at depth due to the necessary entry pressures, oil-based drilling additives commonly coat or invade drilling cuttings and other well samples. If unrecognized, contamination or migrated oil in a candidate source rock could result in a spurious oil-source rock correlation. One or more of four principal tests can help to establish whether bitumen is indigenous to the rock from which it was extracted:

- Bitumen/TOC and/or PI must be consistent with the level of thermal maturity of the kerogen (e.g. T_{max} or vitrinite reflectance) (Table 4.1).
- Thermal maturity of the bitumen (e.g. CPI or biomarker maturity ratios) must match that of the kerogen.
- Isotope ratios of the bitumen and kerogen must be similar within defined limits. As discussed in Chapter 6, source-rock extracts are generally ~0.5–1.5‰ depleted in ^{13}C compared with kerogen.
- Biomarker distributions in the bitumen must be similar to those released from the kerogen by mild thermal or chemical degradation.

Bitumen/total organic carbon or transformation ratios

The ratio of extractable bitumen to total organic carbon (Bit/TOC) in fine-grained non-reservoir rocks, sometimes called the transformation ratio, ranges from near-zero in shallow sediments to ~0.25 (i.e. up to 250 mg/g TOC) at peak oil generation. At greater depths, Bit/TOC decreases because of conversion of bitumen to gas. Both Bit/TOC and hydrocarbon/TOC can be used to estimate the level of maturity of organic matter with depth in wells, because they measure generation directly, especially the threshold of oil generation (Tissot and Welte, 1984, p. 180).

Lithology and the type of solvent used for extraction of bitumen play critical roles in measured Bit/TOC. Coarse-grained rocks, such as siltstones and sandstones, commonly contain migrated hydrocarbons and show higher Bit/TOC than fine-grained rocks from similar depths. Contamination from migrated oils and drilling fluids can also affect this ratio. However, careful avoidance of coarse-grained reservoir-type rocks and analysis of many closely spaced samples can be used to establish Bit/TOC versus depth curves.

Thermal and chemical degradation

Seifert (1978) established relationships between source rocks and oils by pyrolyzing bitumen-free kerogen from the rocks and comparing the biomarkers in the pyrolyzates with those in oils. Early work was performed in an open pyrolysis system generating alkenes, while more recent studies use closed systems, such as hydrous pyrolysis, to produce more petroleum-like products (Seifert and Moldowan, 1980; Seifert et al., 1980; Lewan, 1985).

Trapped and bound biomarkers in kerogen can sometimes be differentiated. For example, Summons et al. (1988) used BBr$_3$ to release ether-linked compounds from kerogens as alkyl bromides. Using thin-layer chromatography, these compounds were separated from trapped hydrocarbons and reduced to deuterated alkanes using LiAlD$_4$ (Chappe et al., 1980). The deuterated compounds are identified readily by gas chromatography/mass spectrometry (GCMS) because their parent ions (M+) increase by one atomic mass unit (one Dalton).

Organic-lean rocks

Source rocks contain abundant extractable bitumen and kerogen unless they are immature or highly mature. Analysis of biomarkers in organic-lean samples is not recommended because of the greater potential for interference by contaminants (e.g. Rowland and Maxwell, 1984). However, in some cases, such as analysis of very old rocks for traces of life, only highly mature organic-lean samples may be available (e.g. Brocks et al., 1999; Summons et al., 1999). Special measures can be taken to identify contamination of very old, organic-lean rocks, as discussed below:

- Contamination during drilling, storage, and analysis can be identified by multiple solvent rinses of whole rock broken into smaller and smaller fragments. After each rinse is analyzed, the rock is broken into smaller fragments to expose fresh fractures and bedding surfaces. Chromatographic analysis of each rinse can be used to ensure that the extracted bitumen was associated intimately with the rock and not coating the rock surface or lining cracks. When the last rinse is clean, the sample can be ground finely and exhaustively extracted before final analysis.
- Extract yield for the sample should far exceed that of laboratory blanks prepared using the same procedures.
- Brocks et al. (1999) compared extract yields and biomarker and isotopic compositions between lithologies in the same well. They interpreted higher yields of bitumen and different biomarker distributions in shales compared with adjacent kerogen-poor,

permeable lithologies as inconsistent with contamination by migrated oil. Covariance of the quantities and isotopic compositions of bitumen and associated kerogen supported lack of contamination.

- Mild burial history, lack of significant permeability, and remoteness from younger oil-prone rocks can be used to support indigenous bitumen (e.g. Summons *et al.*, 1999).

DETECTION OF PETROLEUM IN PROSPECTIVE RESERVOIR ROCKS

Simple, inexpensive techniques, such as thermal extraction gas chromatography (Figure 7.34) (Jarvie *et al.*, 2001a) and Iatroscan chromatography (Karlsen and Larter, 1990) can be used to assess oil properties through direct analyses of reservoir rock samples. Thermal extraction is achieved by vaporization of petroleum from rock directly into a gas chromatograph. The resulting chromatograms can be used for various purposes, such as to identify bypassed pay zones or the presence of high-molecular-weight waxes that might clog production equipment. Iatroscan (thin-layer chromatography/flame ionization detection, TLC/FID) is discussed below.

Microtechniques offer inexpensive and accurate prediction of oil quality in potential reservoir zones (e.g. Baskin and Jones, 1993; BeMent *et al.*, 1996; Guthrie *et al.* 1998). For example, Guthrie *et al.* (1998) determined saturated and aromatic hydrocarbons, aromatics, resins, and asphaltenes by HPLC for crude oils from Venezuela and used them to generate a calibration to predict API gravity, sulfur, and viscosity from sidewall core extracts. Multivariate linear regression showed that the HPLC calibration works as well as more expensive and time-consuming analyses based on combined HPLC, biomarker, and pyrolysis data. The biomarker parameters included the same demethylated hopane ratios used by McCaffrey *et al.* (1996). The pyrolysis data included digitized S1 and S2 peaks, where S1 consists of volatile petroleum (<400°C) and S2 consists of a mixture of the volatilized high-molecular-weight compounds and cracked components (>400°C). A similar pyrolysis approach was used to estimate the API gravity of oil in ditch cuttings and core samples (Mommessin *et al.*, 1981). Guthrie *et al.* (1998) used their data to generate oil-quality profiles for wells that penetrated stacked reservoirs in the Cerro Negro area. The

downhole profiles show significant vertical variations in oil quality, which can be correlated laterally using similar data in adjacent wells. Bypassed zones of higher oil quality can be identified and targeted for exploitation, which might include horizontal drilling.

CRUDE OIL SCREENING

Oils can be screened rapidly and inexpensively using various measurements, such as API gravity, TLC/FID, and gas chromatography. Vanadium/nickel, stable carbon isotopes, and biomarkers are also used to evaluate oils, but these measurements are generally reserved for more detailed studies, and they are discussed later.

API gravity

API gravity is a bulk physical property of oils that can be used as a crude indicator of thermal maturity. API gravity is related inversely to specific gravity:

$$°API \text{ gravity} = (141.5/\text{specific gravity at} 15.6°C) - 131.5$$

Generalized relationships between API and various parameters, such as gas–oil ratio (GOR), reservoir depth, percentage sulfur, and trace metal contents are described in Tissot and Welte (1984) and Hunt (1996). These relationships are only approximate, and they should be used with caution because of many exceptions.

During thermal maturation, the heavy components in oil, NSO compounds, asphaltenes, and heavy saturated and aromatic compounds undergo increased cracking, resulting in increased API gravity. However, API gravity is also affected by other factors, including original organic matter input, biodegradation, water washing, migration (phase separation, molecular partitioning), and inspissation (evaporation). For example, Walters and Cassa (1985) observed no correlation between API gravity and reservoir depth for a suite of offshore Gulf of Mexico oils. Some of the shallow oils were biodegraded, while other oils migrated into reservoirs from deeper source rocks of varying maturities.

Thin-layer chromatography/flame ionization detection

Iatroscan TLC/FID is a rapid screening tool that separates petroleum in small rock samples into compound-class fractions of various polarities. Iatroscan can be used to distinguish petroleum populations in reservoirs before more detailed geochemical work and to predict barriers to reservoir continuity, such as carbonate-cemented horizons and asphaltic-rich zones. The method can be used to define lateral or vertical changes in the gross composition of petroleum in reservoirs and to assist in the selection of samples for more detailed geochemical work. Karlsen and Larter (1990) used TLC/FID to separate up to 70 samples per day into saturated, monoaromatic, diaromatic, polyaromatic, and polar fractions. Differences in composition were used to distinguish petroleum populations and to predict barriers to reservoir continuity, such as carbonate-cemented horizons and asphaltic-rich zones.

Nickel and vanadium

Concentrations and ratios of nickel/vanadium can be used to classify and correlate crude oils. These metals exist in petroleum largely as porphyrin complexes. Oils from marine carbonates or siliciclastics show low wax content, moderate to high sulfur, high concentrations of nickel and vanadium, and low nickel/vanadium (≤ 1) ratios (Barwise, 1990). Dominance of vanadium over nickel in these oils is caused by the greater relative stability of vanadyl porphyrins under low redox potential (Eh) conditions associated with sulfate reduction during diagenesis of marine source rocks (Lewan, 1984). Oils from lacustrine source rocks show high wax, low sulfur, moderate quantities of metals, and high nickel/vanadium (>2). Non-marine oils derived from higher-plant organic matter show high wax, low sulfur, and very low metals.

Comparisons of oils using bulk parameters, such as total nickel and vanadium content, sulfur, or API gravity, require that the thermal maturity of the samples be taken into account. Metal content and sulfur decrease with increasing maturity (e.g. API) for related crude oils.

Vanadium/nickel ratios can be useful in order to genetically classify crude oils, even when they differ in thermal maturity and degree of biodegradation. For example, rapid and inexpensive measurements of vanadium and nickel separate crude oils from eastern Venezuela into genetic groups (Figure 4.20) (Alberdi et al., 1996). Two groups of oils with high V/Ni ratios originated from different facies of the same source rock in the Upper Cretaceous Guayuta Group. Biomarker

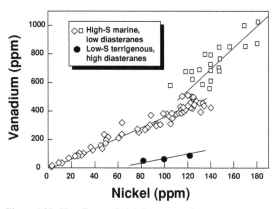

Figure 4.20. Vanadium and nickel concentrations separate genetically distinct crude oils from the Eastern Venezuelan Basin (Alberdi et al., 1996).

parameters indicate that the group with the higher V/Ni ratio (mean = 5.2 versus 3.6) originated from a more reducing, carbonate-rich source-rock facies that received more marine organic matter. The three oils with low V/Ni ratios (mean = 0.7) occur in a frontier exploration area east of the Orinoco Oil Belt. They contain oleanane, high diasteranes, and high C_{29} Ts and 17α-diahopane, consistent with an origin from Upper Cretaceous or Tertiary deltaic marine shales.

Gas chromatographic fingerprints

Gas chromatography using high-resolution capillary columns is used widely to screen and correlate crude oils and source-rock extracts because:

- it is less expensive and more versatile than many analytical methods, and it requires little sample;
- little sample preparation is necessary;
- high-resolution capillary gas chromatography columns can be used to generate a reproducible fingerprint of petroleum consisting of hundreds of peaks in the range C_1–C_{40};
- a large database of peak ratios for each oil or bitumen can be recorded digitally and compared statistically by computer.

Oil composition as measured by gas chromatography is sensitive to organic matter input (Figure 4.21), biodegradation (Figure 4.22), thermal maturation (Figure 4.23), and evaporative loss or weathering (Figure 4.24). Many compounds in petroleum

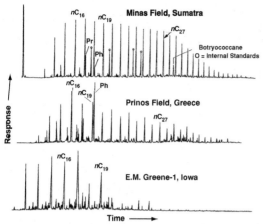

Figure 4.21. (Top) Bimodal *n*-alkane distribution in gas chromatogram for waxy Minas oil from Sumatra. The presence of botryococcane indicates that the source rock contains remains of the alga *Botryococcus braunii*. Although most high-molecular-weight *n*-alkanes (near *n*C$_{27}$ on the figure) originate from terrigenous higher plants, the *n*-alkanes in this oil appear to be from algal lipids (Gelpi *et al.*, 1970; Moldowan *et al.*, 1985). Pyrolyzates of the non-marine alga *Tetraedron* and of laminated Messel Shale from Germany containing densely packed *Tetraedron* remains have similar distributions of *n*-alkanes and *n*-alkenes (Goth *et al.*, 1988).

(Middle) Even-numbered *n*-alkane predominance in the gas chromatogram of oil from Prinos, Greece, which originated from carbonate source rock (Moldowan *et al.*, 1985).

(Bottom) Odd-numbered *n*-alkane predominance in the gas chromatogram of bitumen from Middle Ordovician rock in the Greene No. 1 well, Iowa, dominated by input from *Gloeocapsomorpha prisca* (Jacobson *et al.*, 1988). Reprinted with permission by ChevronTexaco Exploration and Production Technology Company, a division of Chevron USA Inc.

co-elute when using low-resolution, packed-column chromatography. These types of columns are used rarely in modern geochemical studies. Because the relative contributions of different compounds to each gas chromatography peak are unknown, comparison of peak-height or peak-area ratios for correlation of different samples becomes problematic. However, grouping of oils by high-resolution gas chromatography is a useful step during preliminary screening of samples before detailed biomarker analysis.

Gas chromatography fingerprints can indicate certain types of source organic matter input. Bimodal *n*-alkane distributions, and those skewed toward the range *n*C$_{23}$–*n*C$_{31}$, are usually associated with terrigenous higher-plant waxes. The C$_{27}$, C$_{29}$, and C$_{31}$

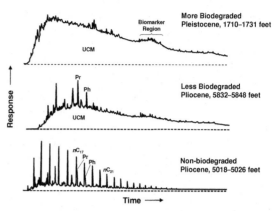

Figure 4.22. Gas chromatograms for three related Gulf Coast oils that differ in the extent of biodegradation. Compared with the non-biodegraded oil (bottom), the moderately biodegraded oil (middle) lost *n*-alkanes while the more biodegraded sample (top) lost both *n*-alkanes and acyclic isoprenoids. Reprinted with permission by ChevronTexaco Exploration and Production Technology Company, a division of Chevron USA Inc.
Ph, phytane; Pr, pristane; UCM, unresolved complex mixture. Dashed line, baseline.

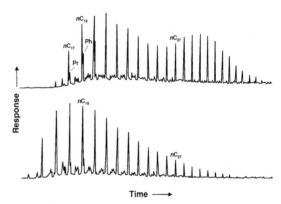

Figure 4.23. Gas chromatograms of two related oils generated from the Permian Phosphoria Formation in Wyoming. The least mature oil based on biomarker analyses (Dillinger Ranch Field, top) has a bimodal *n*-alkane distribution maximizing at *n*C$_{20}$ and *n*C$_{30}$. The most mature oil (Dry Piney Field, bottom) has a unimodal *n*-alkane distribution. The higher-molecular-weight *n*-alkanes were cracked to lighter products during maturation. Reprinted with permission by ChevronTexaco Exploration and Production Technology Company, a division of Chevron USA Inc.
Ph, phytane; Pr, pristane.

n-alkanes in crude oils and source-rock extracts originate mainly from higher-plant epicuticular waxes (Eglinton and Hamilton, 1967). However, certain algae (e.g. *Botryococcus braunii*) that also contain

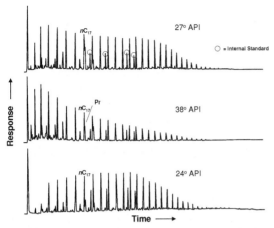

Figure 4.24. Gas chromatograms for three oils from Trinidad are dissimilar, despite biomarker and isotopic evidence indicating that they are related. The least mature oil (top, 27° API) has reached the early to middle oil window based on thermal maturation-dependent biomarkers. This oil has a bimodal n-alkane distribution that is absent in a more mature oil (middle, 38° API). The heavier n-alkanes in the top oil originated from terrigenous higher plants (however, see Figure 4.21). The most mature oil (bottom) has reached the late stages of the oil window based on biomarkers. Lack of a bimodal n-alkane distribution in the two more mature oils is due largely to thermal maturation. Despite high thermal maturity, the oil is depleted in compounds below $\sim nC_{15}$ due to evaporative loss or weathering after sampling. This may account for the low gravity of the bottom oil (24° API) compared with the others. Reprinted with permission by ChevronTexaco Exploration and Production Technology Company, a division of Chevron USA Inc.

higher-molecular-weight n-alkanes (Figure 4.21, top) complicate this interpretation. Algal contributions to source rocks and oils are commonly indicated by abundant shorter-chain n-alkanes, particularly nC_{17} (Blumer et al., 1971). Bitumens and oils related to carbonate source rocks commonly show even carbon-number n-alkane predominance (Figure 4.21, middle). An odd predominance in n-alkanes is common in many lacustrine and marine oils derived from shaly source rocks. Many Lower Paleozoic bitumens and oils (e.g. Middle Ordovician) show an unusual predominance of odd-numbered n-alkanes below nC_{20} (Figure 4.21, bottom), typical of the alga Gloeocapsomorpha prisca (Reed et al., 1986; Rullkötter et al., 1986; Jacobson et al., 1988).

Note: Ruthenium tetroxide (RuO_4) oxidation, Fourier transform infrared analysis, and flash

pyrolysis GCMS of Ordovician rocks containing abundant Gloeocapsomorpha prisca indicate the polymeric structure of these microfossils (Blokker et al., 2001). For example, Estonian Kukersites contain a characteristic set of C_5–C_{20} mono-, di-, and tricarboxylic acids. These compounds suggest that the Estonian Kukersites are composed of a polymer consisting of mainly C_{21} and C_{23} n-alkenyl resorcinol building blocks. Although tricarboxylic acids are absent, RuO_4 degradation product mixtures of rocks from the Guttenberg Member of the Decorah Formation also suggest a poly(n-alkyl resorcinol) structure. Higher thermal maturity is most likely responsible for the different chemistry and morphology of the G. prisca microfossils in these samples. The poly(n-alkyl resorcinol) structure could represent selectively preserved cell-wall or sheath components or materials that polymerized during diagenesis of the organism.

Figures 4.22–4.23 show examples of chromatograms where genetic relationships among crude oils are obscured by secondary processes. The three Gulf Coast crude oils in Figure 4.22 are genetically related based on biomarker and isotopic compositions, yet their chromatograms are dissimilar because of differing degrees of biodegradation. Compared with the non-biodegraded oil in Figure 4.22, the mildly biodegraded oil lost n-alkanes, while the more biodegraded oil lost both n-alkanes and isoprenoids. Figure 4.23 shows gas chromatograms for two related oils derived from the Permian Phosphoria Formation in Wyoming. The gas chromatograms of the related Wyoming oils differ due to different maturation histories. Figure 4.24 shows different gas chromatograms for three related oils from Trinidad. In this case, some of the differences are caused by thermal maturity and others appear due to weathering of the highly mature 24° API oil after production.

Gas chromatographic correlations

Slentz (1981) and Kaufman et al. (1990) used peak-height or peak-area ratios from gas chromatography to correlate crude oils and bitumens. The method has been used effectively to determine reservoir continuity and mixing relationships between different reservoirs in the same field. Thompson (1983) found various light hydrocarbon ratios from gas chromatography analyses useful for assessment of correlation, maturity, biodegradation, and water washing. These methods are described in greater detail below.

Several limitations affect correlation of oils and bitumens using gas chromatography fingerprints. Because of their high concentrations in most oils compared with other compounds, n-alkanes and acyclic isoprenoids dominate many gas chromatograms. These compounds are readily altered by secondary processes, including biodegradation, maturation, and migration. For example, biomodal n-alkane distributions and even- or odd-carbon number predominance are lost with increasing thermal maturity. Thus, oils may show different n-alkane distributions by gas chromatography because they are from different source rocks or because they have experienced different secondary processes. For this reason, n-alkanes are generally not used for reservoir correlations.

Biodegraded or low-maturity petroleum is particularly difficult to correlate using gas chromatography alone because a substantial portion of the hydrocarbons in these samples cannot be resolved. This unresolved complex mixture (UCM) of compounds or "hump" rises significantly above the baseline, and is especially pronounced in biodegraded (Figure 4.22) and low-maturity (Figure 4.25) oils (Milner *et al.*, 1977; Rubinstein *et al.*, 1977; Killops and Al-Juboori, 1990). Because the gas chromatograms of many biodegraded and low-maturity non-biodegraded oils have similar characteristics (e.g. low n-alkanes relative to pristane and phytane), other methods must be used to distinguish them. For example, maturity-related biomarker parameters and the geologic history of the source rock indicate that a sample from the Baghewala-1 well in northwestern India in Figure 4.25 represents low-maturity, non-biodegraded oil.

Pristane/phytane

Pristane/phytane (Pr/Ph) is discussed in detail in Chapter 13, but it is also noted at this point because the abundance of these two compounds in most crude oils allows direct measurement from gas chromatography traces without GCMS. Pr/Ph is commonly used in correlations. For example, Powell and McKirdy (1973) showed that high-wax Australian oils and condensates from non-marine source rocks had Pr/Ph in the range 5–11, while low-wax oils from marine source rocks had Pr/Ph in the range 1–3. Although Pr/Ph of petroleum reflects the nature of the contributing organic matter, this ratio should be used with caution. Pr/Ph typically increases with thermal maturation (e.g. Alexander *et al.*, 1981; ten Haven *et al.*, 1987).

Isoprenoids/n-alkanes

Pristane/nC_{17} and phytane/nC_{18} are sometimes used in petroleum correlation studies (Figure 4.26). For example, Lijmbach (1975) noted that oils from rocks deposited under open-water conditions showed Pr/nC_{17} <0.5, while those from inland peat swamps had ratios greater than 1. These ratios should be used with caution for several reasons. Both Pr/nC_{17} and Ph/nC_{18} decrease with thermal maturity of petroleum. Alexander *et al.* (1981) suggested use of the ratio (Pr + nC_{17})/(Ph + nC_{18}) because it is less affected by variations in thermal maturity than Pr/nC_{17} or Ph/nC_{18}. Biodegradation increases these ratios because aerobic bacteria generally attack the n-alkanes before the isoprenoids.

Allocation of commingled fluids

For many production applications, gas chromatography is more appropriate than biomarker analysis because it

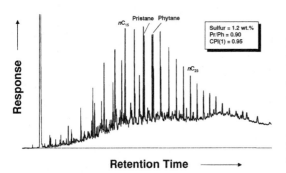

Figure 4.25. Gas chromatogram of low-maturity, non-biodegraded crude oil from the Baghewala-1 well, India, showing high pristane/nC_{17} and phytane/nC_{18} and a rising baseline (Peters *et al.*, 1995). The rising baseline (above dotted line), high sulfur (1.2 wt.%), low pristane/phytane (0.90), low even-to-odd n-alkane predominance (CPI(1) = 0.95), and low API gravity (18°) are consistent with biomarker data (e.g. homohopane distribution as shown in Figure 13.86) indicating anoxic source-rock deposition. Reprinted by permission of the American Association of Petroleum Geologists (AAPG), whose permission is required for further use.

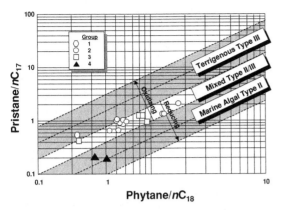

Figure 4.26. Pristane/nC_{17} versus phytane/nC_{18} for crude oils from eastern Indonesia can be used to infer oxicity and organic matter type in the source-rock depositional environment (Peters *et al.*, 1999a). Most Tertiary (groups 2 and 3) and Triassic-Jurassic (groups 1 and 4) oils occur north and south of $2°$S latitude, respectively. Increasing thermal maturation and biodegradation displace points toward the lower left and upper right, respectively. Reprinted by permission of the AAPG, whose permission is required for further use.

is rapid and inexpensive. Light hydrocarbons are more volatile, structurally less complex, elute faster during chromatography, and are generally more sensitive to reservoir alteration than biomarkers. Oils in different reservoirs within the same field commonly originate from the same source rock and show essentially identical biomarker distributions. However, these oils may have experienced slightly different histories that affect hydrocarbon compositions. For example, sealing faults may separate an accumulation into two reservoirs that subsequently undergo slightly different secondary processes or filling histories.

Note: Although generally less sensitive to reservoir processes than light hydrocarbons, biomarkers still show statistically significant heterogeneities in the Gullfaks Field, Norwegian North Sea, that are attributed to slight differences in biodegradation, maturity, and source (Horstad *et al.*, 1990).

England (1990) provides evidence that both lateral and vertical differences in petroleum composition exist *within* individual reservoirs because of slow intra-reservoir mixing. Lateral variations are attributed to how the field filled from its source rock(s) and vertical differences are explained by gravitational segregation.

Some examples of the use of GC to solve production problems include the following:

- Reservoir continuity, e.g. do compositions of oils on opposite sides of a mapped fault support the existence of a seal resulting in two distinct reservoirs in a field?
- Production allocation, e.g. what is the relative production from each interval in a well where the total production is commingled? Are certain intervals within a reservoir unit non-productive?
- Detection of drilling fluid contamination or leaks, e.g. is a sample composed of petroleum from the formation or a drilling additive?

Kaufman *et al.* (1990) described a useful gas chromatography approach for correlation production allocation, as explained below. For a typical production problem, samples might be obtained from the wellhead, drill stem tests (DST), repeat formation tests (RFT), swab runs, or reversed circulation. Large-volume samples from the flowing wellhead are less susceptible to contamination than others. Because performance varies between different chromatographic columns and for the same column over long periods, analyses for each study are completed promptly using the same column and operating conditions. Adjacent peaks and peaks with similar retention times that can be identified on every chromatogram are selected in order to calculate the ratios. Accurate peak heights and areas are obtained using short, nearly horizontal baselines that are drawn in identical fashion among all chromatograms.

The specific ratios used in the above approach may differ from one study to the next. The ratios are chosen to maximize differences among the samples that are consistent with geologic and engineering information. For reservoir applications, this gas chromatographic approach is always used with other information, such as structural geology, wire-line logs, and pressure tests. Typically, no more than a dozen ratios are used. The ratios can be plotted on polar coordinate paper (star diagrams), which facilitate visual comparisons. Cluster analysis and other multivariate techniques can be used to generate dendrogram plots for showing relationships among large numbers of samples.

Chromatography can be used to determine the relative proportions of end-member oils from different zones contributing to commingled production (e.g. Kaufman *et al.*, 1987). In the Gulf of Mexico

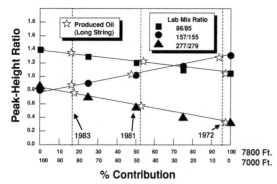

Figure 4.27. Laboratory mixtures of end-member oils from different production zones can be used as calibration to reveal the relative contributions of these zones to mixed oils (modified from Kaufman *et al.*, 1990). % Contribution refers to the relative proportion of oil contributed from the 7000-feet (2134-m) versus 7800-feet (2378-m) reservoir zones to the combined production of the short- and long-string tubing.

and elsewhere, historical production data are used to plan further development of oil fields. However, dual-completed wells complicate this task. Over time, the plumbing in these wells can develop leaks that may not be detected by field measurements. Gas chromatography can also deconvolute mixed production due to string tubing leaks and indicate the correct amount of oil produced from different zones, as described below.

In the Main Pass 299 Field, the OCS-G 1315 No. A-2 well was completed in 1967 as a dual producer with short- and long-string tubing terminating in the 7000-feet (2134-m) and 7800-feet (2378-m) sands, respectively (Kaufman *et al.*, 1990). Leakage between the short- and long-string tubings was suspected in 1986. Stored samples of produced oil from both string tubings were collected in 1967, 1972, 1981, 1983, and 1986. Capillary gas chromatograms of these samples show distinct differences among 16 selected peak ratios in the C_9–C_{19} range. Distinct chromatograms of the samples from 7000 feet (2134 m) and 7800 feet (2378 m) collected in 1967 are end members representative of the original oil from these reservoirs (Figure 4.27).

For the above well, laboratory mixtures of the two 1967 end-member oils were made in 75 : 25, 50 : 50, and 25 : 75 ratios. Gas chromatography of the mixtures and the end-member oils was used to generate the same 16 peak-height ratios described above. Figure 4.27 is a simplified binary mixing diagram that illustrates how the mathematical model works if only three peak-height

ratios are used. In reality, the mathematical model calculates the mixing curves using the end-member oils, and the final number of peak-height ratios used is determined by finding the optimum number of ratios that calculate most closely the proportions measured from the laboratory mixtures.

The three selected peak-height ratios for the 1967 end-member oils from the short-string tubing (left, 100% 7000 feet (2134 m)) and long-string tubing (right, 100% 7800 feet (2378 m)) are indicated by solid symbols in Figure 4.27. The data for the laboratory mixtures (solid symbols) show that each ratio changes along a mixing line between each end member. The 1972-produced oil from the long-string tubing (open stars, right) is within experimental error (~5%) of the composition of the 1967 long-string tubing end member, indicating that no leakage occurred between the long- and short-string tubing from 1967 to 1972. By 1981, the long-string tubing was producing a 50 : 50 mixture of oil. By 1983, only ~13% of the long-string tubing oil originated from the 7800-feet (2378-m) reservoir. Analysis of the 1986 deep-string oil (not shown) showed that the 7800-feet (2378-m) sand no longer contributed any of the produced oil and that the long string produced 100% 7000-feet oil. Kaufman *et al.* (1990) combined these calculated proportions with the historical production data (e.g. Figure 4.28) for the well to estimate that ~500 000 barrels of oil thought to originate from 7800 feet (2378 m) actually came from the 7000-feet (2134-m) reservoir. This information affected the location of new development wells in the field. A later

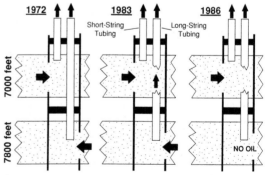

Figure 4.28. Schematic indicates the interpreted flow of oils in 1972, 1981, and 1983 from reservoirs at 7000 feet (2134 m) and 7800 feet (2378 m) based on gas chromatographic data for short- and long-string tubing samples in Main Pass 299 Field, well OCS-G 1315 No. A-2 (modified from Kaufman *et al.*, 1990).

work-over of the well confirmed broken-string tubing in the 7000-feet (2134-m) reservoir.

RESERVOIR CONTINUITY AND FILLING HISTORY

The amount of petroleum that can be recovered from reservoirs ranges from 10 to 80%, but the global average may be as low as 20% (Miller, 1995). Petroleum geochemistry offers rapid, low-cost assessment of reservoir-related issues that can increase recovery of oil previously abandoned in reservoirs.

Different fluid compositions in different parts of a field imply compartmentalization. Identifying these compartments and their distribution can help to guide development of the reservoir. Identification of distinct reservoir compartments is important because it allows:

- more accurate estimates of reserves;
- better recovery and placement of new wells;
- corroboration of geologic maps;
- a baseline for evaluation of future production problems.

The composition of expelled petroleum changes during thermal maturation of source rocks. If mixing of reservoir hydrocarbons is hindered during filling due to compartmentalization, then the least mature hydrocarbons are assumed to be the furthest migrated. At high maturity, gas charge to reservoirs can displace oil toward up-dip traps. Geochemical analyses of petroleum from different portions of reservoirs can be used to identify bypassed reserves near previously discovered accumulations. Use of existing platforms, pipelines, and/or refineries in the area might then make such new reserves economic. Enhanced understanding of migration directions can improve interpretations of petroleum composition along migration pathways and to rank exploration or development targets.

Walters *et al.* (1999) used petroleum geochemistry of light and heavy hydrocarbons and the petroleum system concept to improve understanding of reservoir continuity and filling history of the Beryl Complex in the North Sea. These highly faulted fields are located on the west flank of the South Viking Graben, with production from Upper Triassic–Upper Jurassic sandstones. A multiple source model explains the chemistry of the Beryl Complex oils and gases. Differences in organic richness, kerogen kinetics, and composition of expelled

products from Kimmeridge Clay and Heather Formation source systems explain the observed chemical variations. The organic-rich Kimmeridge Clay Formation contains type II and some type IIS kerogen. Early expulsion of oil may occur at temperatures as low as $\sim$90°C, with peak expulsion at $\sim$120°C. The Heather Formation contains mixed type II/III kerogen with lower generative potential and requiring higher temperatures for expulsion than the Kimmeridge Clay. Initial and peak expulsion occur at $\sim$120°C and $\sim$140–150°C, respectively. Middle Jurassic coals also may contribute gas and condensate.

Biomarkers suggest that Beryl Complex oils originate primarily from the Kimmeridge Clay, while light hydrocarbons indicate significant contribution from older strata. This apparent conflict between the light and heavy hydrocarbon data results from the disproportional contributions of C_{15+} biomarker compounds from low to moderately mature Kimmeridge Clay sources and light hydrocarbons from high-maturity Heather Formation sources. While some oils represent near-end-member fluids, most of the Beryl Complex oils are mixtures from multiple sources.

The Beryl Complex can be subdivided into three overlapping petroleum systems. Oils produced from the eastern flank of Beryl Field originated by sequential expulsion of fluids from multiple sources in both the Frigg and Beryl kitchens and are largely segregated by stratigraphic and structural controls. In the western satellite fields of the Beryl Complex, reservoirs filled episodically with petroleum from the Frigg kitchen. The produced oils are mixtures of low- to moderate-maturity oils and high-maturity condensates. In the eastern flank and in the central trough, oils originated from local sources in the Kimmeridge Clay Formation.

Strontium isotopes in residual salts

Ratios of the stable isotopes of strontium (^{87}Sr/^{86}Sr) in residual salts from core samples can be used to evaluate reservoir compartmentalization (Smalley and England, 1994; Mearns and McBride, 1999). Strontium isotope data are best used to complement interpretations of reservoir compartmentalization based on pressure trends and organic geochemical data. For example, core descriptions, petrophysical behavior, and log correlations were used to divide the Tilje Formation into three reservoir zones in the Smørbukk North

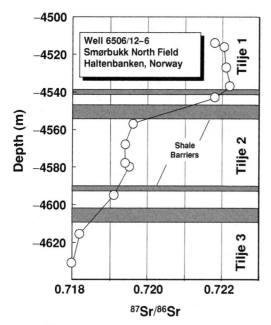

Figure 4.29. Strontium isotope ratios of residual salts in the oil leg show that certain shales impede vertical flow of petroleum and thus define reservoir compartments in the Smørbukk North Field well 6506/12–6, Haltenbanken area, Norway (after Smalley and England, 1994). The isotope data suggest that some shales are more effective barriers to mixing of petroleum (e.g. ~4550 and ~4605 m) than others (e.g. ~4592 m).

Field, Norway. The Tilje Formation consists of sandy and shaly units deposited in a marginal marine fan delta environment. Residual salts analyses for well 6506/12–6 in the field show a stepped trend of decreasing $^{87}Sr/^{86}Sr$ with depth (Figure 4.29). Significant shifts in $^{87}Sr/^{86}Sr$ occur across shales at ~4550 and ~4605 m, corresponding to the established boundaries between the Tilje 1 and 2 and Tilje 2 and 3 reservoir zones, respectively. Variations in $^{87}Sr/^{86}Sr$ within these oil legs represent fossilized water compositions that existed when oil filled these reservoir zones. The strontium isotope evidence for the shales as barriers to vertical flow is supported by pressure information from RFT and DST in this and other wells in the field (Smalley and England, 1994).

Faults: seals or migration conduits?

Direct geochemical evidence for cross-stratal migration along faults or migration immediately beneath the seal rock within the carrier bed is difficult to obtain, partly because of the limited samples available from well bores. However, various geochemical observations for cores that intersect a large growth fault support episodic fluid injection into Eugene Island Block 330 in the US Gulf Coast (Losh et al., 1999). Losh et al. compared core and cuttings samples from two wells that penetrated the same normal fault ~300 m apart. The US Department of Energy Pennzoil Pathfinder well showed little paleothermal or geochemical evidence for fluid flow through the fault zone. However, faulted rocks in the nearby A6ST well showed elevated vitrinite reflectance compared with those outside the fault zone, supporting a paleothermal anomaly. The faulted sediments contain carbonates that are depleted in ^{13}C due to incorporation of carbon related to kerogen maturation or organic acid decarboxylation, supporting a deep source for the diagenetic fluid. The mudstone gouge zone at the top of the fault cut in the well is depleted in sodium and enriched in calcium, and this effect diminishes with distance from the gouge zone. The sealing capacity of faults depends on the pressure difference across the fault zone. Fluid-pressure analysis in the A6ST well shows that the fault is a strong permeability barrier, with up to 1800 psi (12 411 kPa) of water pressure differential across it. A charge of high-pressure fluid moving upward along the fault lowers the effective stress in the fault zone. If the fluid enters an area of the fault adjacent to the downthrown, relatively low pressure reservoir sands, then the fluid discharges into them. The resulting decrease in permeability does not allow fluid to re-enter the fault zone to escape from the reservoir.

Recognition of migration along faults in outcrop can also be difficult when the original migrating hydrocarbons are low or absent due to extensive weathering. Inorganic geochemistry can provide evidence for cross-stratal migration of petroleum and/or associated brines in weathered outcrops. Spectacular exposures of Jurassic sandstones in southeastern Utah are red-colored due to hematite (Fe_2O_3) coatings formed by the oxidation of iron minerals. The immobile iron in these rocks occurs as Fe^{3+}. However, two of these sandstones, the Navajo Sandstone and the Moab Tongue, were locally bleached white by reduction and removal of hematite adjacent to the Moab fault. Brines formed by solution of deep Pennsylvanian salt are interpreted to have moved up the fault and bleached the adjacent sandstones (Chan et al., 2000). These brines were reducing due to interaction with petroleum, organic acids, or hydrogen sulfide, and thus removed iron as Fe^{2+} and bleached the sandstones near the fault. For example, for methane:

$$CH_4 + 4Fe_2O_3 + 16H^+ = 10H_2O + CO_2 + 8Fe^{2+}$$

In some locations tar sands and bitumen veins remain in the sandstone, but in many places hydrocarbons are low or absent and bleaching is the only evidence that they affected the migrating fluids. After bleaching, much of the mobilized, reduced iron was oxidized when the migrating saline fluids mixed with more oxic, meteoric waters. This resulted in diagenetic hematite and manganese oxide deposits.

SURFACE GEOCHEMICAL EXPLORATION USING PISTON CORES

While trap geometry and petroleum reservoir zones can be predicted from seismic data with some confidence, methods for predicting the composition of reservoir fluids are only marginally reliable. The amplitude variation with offset (AVO) method (Castagna and Backus, 1997) provides some control but is not always reliable, as evidenced by dry exploration holes and incorrect predictions of fluid type. Fortunately, nearly all petroleum accumulations leak, and surface geochemical exploration can be used to recover and identify the escaped oil or gas before the purchase of blocks and exploratory drilling.

The goals of surface geochemical exploration are to identify low-risk areas within basins and to rank prospects by using geochemical screening of piston-cored seep samples. Piston-core seep surveys and related technologies are rapidly growing research topics because they provide information on the geographic extent of petroleum systems before drilling (e.g. Brooks et al., 1986). Seeps provide information on the quality, thermal maturity, age, and distribution of the underlying source rock. Surface geochemical exploration is the only method available for investigating the character and extent of petroleum systems in frontier basins before drilling. The method takes advantage of the fact that biomarkers in seeps and produced oils can be used to identify and predict the regional distributions of source-rock facies, as discussed below. If seep oils are available from deepwater exploration areas, then they can be used to provide information on the quality, thermal maturity, age, and distribution of the source rock before costly drilling decisions. To reduce sampling costs, other pro-

grams, such as heat-flow measurements and additional seismic acquisition, usually accompany piston coring.

Selecting and sampling piston core targets

In marine settings, core sites are chosen based on seismic evidence for leakage and subsurface disturbance of unconsolidated sediments by rising petroleum (e.g. Haskell et al., 1999). Target core sites are best located where faults link the source rock to the seabed, as occurs commonly in geologically active areas, such as the Gulf of Mexico and the Niger Delta. Ideal faults are those associated with: (1) seismic-amplitude anomalies and/or bottom-simulating reflectors associated with gas hydrates (Kvenvolden and Lorenson, 2001); (2) seabed leakage features, such as authigenic carbonate accumulations and mud-gas mounds or pits; and (3) thermogenic gas chimneys (MacDonald, 1998).

Core sites can be positioned using differential global positioning satellite (GPS) technology, commonly with precision of $\pm 5\,m$ and within $\pm 30\,m$ of the preselected location (Cameron et al., 1999). Precision bathymetric and sub-bottom 3.5-kHz or Chirp sonar can be used to refine core positions in the field. Heavy-duty piston corers (e.g. 2000 pounds (1905 kg)) offer several advantages over less expensive conventional gravity coring, including (1) greater penetration, (2) better core recovery, and (3) less disturbed samples. A sliding piston inside the core barrel reduces wall friction with the sediment and assists evacuation of displaced water from the top of the corer. A typical 6-m core length allows three sections for each piston core to be analyzed. The deeper sections are commonly less influenced by bioturbation, anthropogenic pollution, and surface diffusion of gases within the top meter of each core. Piston cores can be taken at water depths of up to 4500 m. After retrieval, the cores are processed shipboard, logged, subsampled, and frozen ($-20^\circ C$).

Screening procedures

Interpretation of piston-core data is complex because of variable biodegradation of the seep and mixing with recent organic matter near the sediment–water interface. Analyses focus on sediment below the top meter of core, thus minimizing the effects of bioturbation, anthropogenic pollution, and diffusion of gases from the water column. Screening methods identify samples

Table 4.8. *Examples of criteria used to classify piston core samples. (Peters and Fowler, 2002)*

Parameter	No seep	Oil seep	Gas seep
UCM (ppm)	<30	>30	<30
n-Alkanes (ppm)	<1500	>1500	<1500
TSF (fluorescence units)	<4000	>4000	<4000
C_{2+} (ppm)	<1	>1	>1

C_{2+}, ethane, propane, butanes, and pentanes; TSF, fluorescence response; UCM, unresolved complex mixture (hump) on whole-oil gas chromatogram.

that provide the least ambiguous data for interpretation. For example, Table 4.8 shows four criteria used to classify the quality of piston-cored sediment samples. The unresolved complex mixture, *n*-alkanes, and C_{2+} gases are measured using gas chromatography. Total scanning fluorescence (TSF) provides a rapid, semi-quantitative measure of petroleum aromatic hydrocarbons that is insensitive to all but severe biodegradation (Brooks *et al.*, 1986). These methods are described below.

Assignment of seep samples to a particular petroleum system requires geochemical oil–oil or oil–source rock correlation, as discussed below. For seep samples, special care must be taken to avoid the use of correlation parameters affected by interfering materials from the associated sediment. Wenger *et al.* (1994) used maps of oil or seep types to delineate the complex regional distributions of petroleum systems in the Gulf of Mexico by age of the source rock and by chemical composition of the generated products. These maps can be used to predict the geochemical character of petroleum that might be discovered by drilling in selected areas. These geochemical correlations are particularly useful in settings like the Gulf of Mexico, where faults, salt domes, and highly variable lithologies contribute to complex migration plumbing from potential source rocks to reservoirs.

Total scanning fluorescence

TSF provides a rapid, semi-quantitative measure of petroleum aromatic hydrocarbons (Brooks *et al.*, 1986). Increasing TSF intensity (arbitrary units) corresponds

to higher aromatic hydrocarbons in piston-core sediment extracts. Comparisons of TSF intensity for data obtained using different instruments require calibration. Migrated oil has higher concentrations of larger aromatic compounds containing three or more benzene rings and fluoresces at longer wavelengths. Extracts containing gas or condensate fluoresce at shorter wavelengths. TSF patterns are insensitive to all but severe biodegradation. A three-dimensional spectrum is acquired using Perkin-Elmer Model LS 50B fluorometers. A sample extract is placed in the precalibrated fluorometer and scanned over a range of excitation wavelengths while measuring the fluorescence emission intensities over a range of emission wavelengths (Figure 4.30, courtesy of J. Brooks).

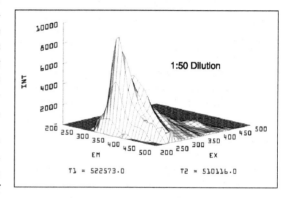

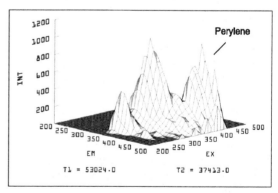

Figure 4.30. Spectral features of background material (bottom) differ markedly from those for a seep dominated by crude oil (top, courtesy of J. Brooks at www.tdi-bi.com/analytical_services/geochem/sge-tsf.htm). Increasing total scanning fluorescence (TSF) intensity (expressed in arbitrary intensity units) generally corresponds to increasing aromatic hydrocarbon concentrations in sediment extracts.

Gas chromatography of piston-core sediment samples

Gas chromatography of piston-core sediment extracts can be used to distinguish immature lipids in uncontaminated sediment (background), migrated crude oil, and biodegraded oil. For example, a strong chromatographic response for non-biodegraded oil consists of a large unresolved complex mixture (UCM) (Figure 4.22) and an *n*-alkane distribution similar to that of thermally mature crude oil that covers a broad range of carbon numbers with little preference for odd- or even-numbered homologs. Background samples yield a weak chromatographic response, consisting of a flat or nearly flat baseline with scattered *n*-alkane peaks unlike those in mature oil. The *n*-alkanes generally show a preference for odd- or, less commonly, even-numbered *n*-alkanes, typical of thermally immature organic matter.

Most piston-core sediment samples with crude oil are modified by biodegradation before analysis. In some cases, sufficient concentrations of alkanes and other compounds remain to indicate an origin from petroleum. More commonly, all *n*-alkanes are biodegraded, resulting in a pronounced "hump" or unresolved complex mixture of compounds that rises significantly above the baseline. Qualitative comparison of chromatogram character and the amounts of extractable hydrocarbons can be used to identify samples dominated by petroleum for further analysis. Figure 4.31 shows examples of gas chromatograms for (1) background, (2) fresh crude oil, and (3) biodegraded crude oil. Interpretations must account for the fact that virtually all piston-core samples contain some contamination by recent sediment organic matter. When contamination is severe, it is best to reject the sample.

Symbols designating the different classes of piston-core samples can be plotted on bathymetric maps to assist geologic interpretation.

Headspace gas analysis

Headspace gas analysis is used to determine the composition of interstitial hydrocarbon gases. Various parameters, such as gas wetness, total alkanes, and ethane/ethene, can be used to distinguish thermogenic from microbial gas seeps. If sufficient methane is available, then stable carbon isotope ratios of the methane help to determine origin.

SAMPLE QUALITY, SELECTION, AND STORAGE

Sample availability and quality are complicating factors that can determine the success or failure of geochemical studies. The geochemist and regional geologists or environmental engineers typically play critical roles in acquiring the proper samples with adequate and accurate sample information. Samples of prospective source rocks or suspected sources of contamination may not be readily available. Source rocks may lie far beneath the reservoir and may not have been drilled or sampled. Produced oils may have migrated great distances from their source rocks. Critical samples may be owned by competitors, contaminated by drilling additives, lost due to improper storage, or depleted because of other work.

Information that should be included with rock and oil samples includes well name, operator, location, date of sampling, formation, age (rock or reservoir), depth (or production interval), and type of sample (e.g. conventional or sidewall core, outcrop, cuttings, separator oil, DST oil, reservoir rock or piston-core sediment extract, seep oil). Maps and lithology logs assist the investigators in the evaluation of sample quality and geologic relationships. If manmade additives or migrated oil could represent contaminants, then samples of these materials are helpful.

Rock samples

Rock sample quality generally decreases in this order: whole core, sidewall core, cuttings, outcrop samples. Outcrop samples are commonly weathered, resulting in alteration of organic matter. Biomarker analyses of cuttings are not recommended because they are most readily affected by flushing with drilling fluids. These fluids can alter or contaminate the indigenous biomarker distributions, thus nullifying or limiting useful geochemical interpretations. When only outcrop or cuttings samples are available, critical geochemical questions can still be answered using biomarkers. However, these samples should be used only as a last resort, and they should be discussed on an individual basis with the biomarker

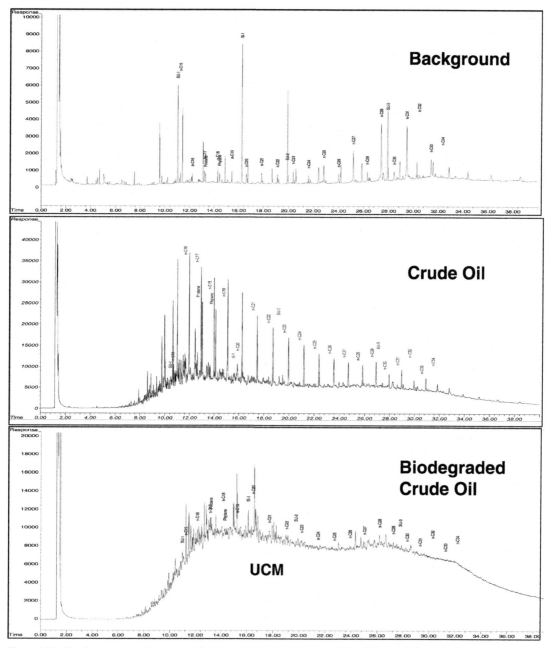

Figure 4.31. Gas chromatograms for extracts of sediment piston cores (courtesy of Bernie Bernard). Extracts dominated by non–biodegraded oil (middle) and biodegraded oil (bottom) differ from that of an uncontaminated sediment with terrigenous background (top).

UCM, unresolved complex mixture.

Figure 4.32. The Lokbatan mud volcano, located ~18 km southwest of Baku in the South Caspian region, erupted on October 24, 2001, and continued for six months. Hydrocarbon gases flared to a height of ~30–45 m during the initial eruption, which was followed by a large flow of mud that carried fragments of deeply buried Miocene source rocks. The Lokbatan mud volcano is one of the largest and most active on the Apsheron Peninsula, having erupted 23 times since 1828. Photos from the website of K. Scholte, Delft University of Technology at www.ta.tudelft.nl/aw/local/section/mudvolcano/.

personnel. Screening analyses, such as Rock-Eval pyrolysis and gas chromatography, can be used to eliminate samples showing unfavorable characteristics, indicating weathering and contamination.

Note: There are >800 mud volcanoes that occur in areas of recent and rapid sedimentation. Over half of these occur in the South Caspian region of Azerbaijan, with ~200 of them onshore and many currently active (Figure 4.32). They are formed when unconsolidated mud and sand are squeezed upward by compressional forces that may be generated by rapid sedimentation or seismic activity. In Azerbaijan, the mud volcanoes are associated with the emission of oil and gas, which can ignite spontaneously. Such events have been occurring for centuries and are tied closely to the development of the Zoroastrian religion of fire worship. In addition to petroleum, mud volcanoes can transport pieces of deeply buried strata to the surface. When tied to specific potential source units through geochemical or fossil evidence, such samples may be the only ones available for study (e.g. Isaksen *et al.*, 1999).

We do not recommend evaluating biomarkers from organic-lean rocks (<1 wt.% TOC) because these rocks are unlikely sources for commercial accumulations of petroleum. Information on the quantity, quality, and thermal maturity of organic-lean rocks is best obtained using rapid, inexpensive screening techniques, such as Rock-Eval pyrolysis. Furthermore, biomarker analyses of bitumens from organic-lean rocks can be misleading. In some organic-lean rocks, biomarkers that are characteristic of the organisms contributing to the depositional environment are overwhelmed by those associated with recycled organic matter (Rowland and Maxwell, 1984). For example, Farrimond *et al.* (1989) noted that unlike organic-rich rocks, adjacent organic-lean Toarcian rocks contained bitumens showing anomalously high maturities based on hopane and sterane isomerization ratios. They concluded that recycled organic matter in the organic-lean samples had experienced high thermal maturity before deposition of the sediments.

Rock-sample size for biomarker analysis varies, depending on organic richness. Typically ~50 g of rock is advised, although successful analyses can be completed on some samples as small as 1–5 g. Core, sidewall, and outcrop samples can be brushed or scraped to remove mudcake and residues from marking pens. Remove weathered surfaces before sampling for geochemistry. Because of contamination, avoid cores enclosed in wax-based sealing agents and core plugs acquired using oil-based lubricants or oil-based drilling mud.

Cuttings are normally washed with fresh or salt water to remove mudcake and are air-dried at low temperature (below ~50°C) on paper towels at the drill site. Cuttings washed with organic solvents must be avoided because these solvents extract biomarkers.

Before analysis, cuttings should be examined using a binocular microscope, and obvious contaminants, such as walnut hulls and woodchips, should be removed. We call this "negative picking." We do not normally recommend positive picking, where one lithology is selected for analysis from a mixture in each cuttings sample. Although informative about isolated lithologies, positive picking may result in non-representative samples.

The compositions of extracts and kerogen do not change appreciably during storage for years, provided that the rock samples are dry and remain in an air-conditioned facility. However, major changes in the composition of saturated hydrocarbons occurred when aliquots of freeze-dried recent sediments were reanalyzed after 1 month of wet storage at $25°C$ (Grimalt et al., 1988). These experiments resulted in transformation of a strong odd-to-even C_{25}–C_{33} n-alkane distribution to C_{22}–C_{30} n-alkanes with no carbon-number predominance, generation of an unresolved complex mixture of hydrocarbons between nC_{16} and nC_{22}, and degradation of hopanes, apparently due to aerobic microbial reworking.

Oil samples

Oils are best collected and stored in tightly sealed glass containers. Metal containers can react gradually with emulsified water and leak. Oil can leach contaminants, such as phthalates, from plastic containers or rubber-lined caps. Teflon^{TM} liners are recommended for bottled oils. Usually 100–250 cm^3 of oil is sufficient for all geochemical analyses, with some remaining for storage. However, certain samples as small as a few milligrams have been analyzed successfully for biomarkers because of high concentrations of these compounds. Each sample container should be labeled carefully using waterproof ink.

Care must be taken in the collection of produced oils because they may represent mixtures from various zones or fields. The date of collection is important because of changes that may occur during production history. For example, workover (well maintainence) may alter the depth of the completion interval for production.

Bottled oils do not appear to be affected by biodegradation over periods of years under typical storage conditions. However, oils heavily biodegraded in the reservoir present special problems in analysis and interpretation of residual biomarkers compared with unaltered oils. Condensates and light oils (>35° API) typically have low biomarker concentrations due to extensive thermal degradation.

GEOCHEMICAL ROCK AND OIL STANDARDS

Most geochemical laboratories generate internally consistent data based on in-house calibration rock and oil standards. However, interlaboratory data consistency can vary (e.g. Claypool and Magoon, 1985) because of a general lack of suitable standards with certified analytical reference values (e.g. Lin, 1995; Isaacs, 2001). The Norwegian Geochemical Standards (NGS) project is an example of an attempt to solve this problem (http://eaog.ncl.ac.uk/newsletters/nl9page4.html). This project was established to provide the petroleum industry and research institutions with rock and oil standards. Reference values for the NGS standard samples are based on analyses conducted according to the *Norwegian Industry Guide to Organic Geochemical Analyses* (Patience et al., 1993). The standards include the following:

- Svalbard rock (NGS SR-1), Anisian (Middle Triassic) Botneheia Formation, Eastern Spitsbergen, 2.17 wt.% TOC, 0.41% mean vitrinite reflectance, 4800 mg extractable organic matter/kg rock.
- Jet rock (NGS JR-1), Toarcian Whitby Mudstone Formation, Port Mulgrave area, Yorkshire, England, 12.4 wt.% TOC, 0.47% mean vitrinite reflectance, 16 000 mg extractable organic matter/kg rock.
- North Sea oil (NGS NSO-1), Oseberg Field Block 30/9, Norwegian North Sea, 32° API, mean C_{15+} fraction 77% of whole crude, 1.9 wt.% asphaltenes.

The NGS samples are available mainly to the petroleum industry and research institutions performing geochemical projects on exploration or reservoir exploitation in the Norwegian Continental Shelf. However, access to the samples may also be granted based on evaluation of geochemical aspects of the application. Application forms and newsletters with sample documentation, including all procedures and results of the interlaboratory calibration, are available from the Norwegian Petroleum Directorate at Box 600, 4003 Stavanger,

Norway or on the Internet at http://npd.no.webdesk/netblast/pages/standard.html.

APPENDIX: DERIVATION OF MASS-BALANCE EQUATIONS

We can use the principle of mass balance to derive equations needed to calculate the original TOC and HI (TOC^o and HI^o) of source rocks that are now thermally mature or postmature and have reduced TOC and HI (TOC^x and HI^x) due to petroleum generation (G.E. Claypool, personal communication, 2002). The extent of fractional conversion, f, of the original petroleum-generative potential of a source rock (G^o) to petroleum (P) is defined as:

$$f = P/G^o = (G^o - G^x)/G^o = 1 - G^x/G^o \qquad (4.6)$$

where G^x is the remaining petroleum generative capacity.

If we assume that Rock-Eval pyrolysis S2 (mg HC/g rock) is a measure of G, then:

$$f = (S2^o - S2^{x'})/S2^o \qquad (4.7)$$

where $S2^o$ is the initial S2 of the source rock and $S2^{x'}$ is the S2 of the equivalent amount of the initial source rock after generation and expulsion of petroleum. The measured S2 of the mature source rock is $S2^x$ and the relation between $S2^x$ and $S2^{x'}$ is:

$$S2^{x'} = S2^x(CF) \qquad (4.8)$$

where CF is the weight-loss correction factor.

The distinction between $S2^{x'}$ and $S2^x$ is needed in order to properly compare the S2 between the immature and mature stages of the source rock. If 1 g of immature source rock were heated to generate and expel petroleum, then the remaining mass of the rock would be <1 g by the amount of petroleum expelled. Rock-Eval S2 is always normalized to 1 g of rock. Therefore, the measured $S2^x$ exceeds that which should be subtracted from $S2^o$ in the numerator of Equation (4.7) and $S2^x$ must be corrected using the appropriate correction factor (CF).

CF has a value <1.0 that can be calculated from TOC and Rock-Eval pyrolysis data using several different equations:

$$CF = 1 - (S1 \text{ expelled}/1000) = [1000 - (S1^o + S2^o)]/[1000 - (S1^x + S2^x)] \qquad (4.9)$$

$$CF = RC^o/RC^x = [TOC^o - 0.0833(S2^o + S2^o)]/[TOC^x - 0.0833(S1^x + S2^x)] \qquad (4.10)$$

where RC is residual or inert carbon, and the generated petroleum is assumed to contain 83.33 wt.% carbon (Espitalié et al., 1987).

$$CF = 1 - (TOC^o - TOC^x)/(83.33 - TOC^x) = (83.33 - TOC^o)/(83.33 - TOC^x) \qquad (4.11)$$

Basic mass-balance expression

Combining equations (4.7) and (4.8) gives the basic mass-balance expression:

$$f = \frac{S2^o - S2^x(CF)}{S2^o} \qquad (4.12)$$

Rearranging and substituting for CF gives:

$$1 - f = \frac{S2^x(RC^o/RC^x)}{S2^x} = \frac{S2^x/RC^x}{S2^o/RC^o} \qquad (4.13)$$

Equation (4.13) is similar to the formulation of Cooles et al. (1986), but it does not require residual or inert carbon to remain constant during petroleum generation.

Substituting for RC in Equation (4.13) gives:

$$1 - f = \frac{S2^x/[TOC^x - 0.0833(S1^x + S2^x)]}{S2^o/[TOC^o - 0.0833(S1^o + S2^o)]} \qquad (4.14)$$

Because HI is defined as S2/TOC and PI is defined as S1/(S1 + S2):

$$S2 = (HI)(TOC)/100$$

and

$$S1 + S2 = (HI)(TOC)/100(1 - PI)$$

Equation (4.14) becomes:

$$1 - f = \frac{\dfrac{(HI^x)(TOC^x)/100}{TOC^x - 0.0833[(HI^x)(TOC^x)/(100(1 - PI^x))]}}{\dfrac{(HI^o)(TOC^o)/100}{TOC^o - 0.0833[(HI^o)(TOC^o)/(100(1 - PI^o))]}} \qquad (4.15)$$

Canceling TOC/100 gives:

$$1-f = \frac{\dfrac{HI^x}{100 - 0.0833[(HI^x)/(1 - PI^x)]}}{\dfrac{HI^o}{100 - 0.0833[(HI^o)/(1 - PI^o)]}} \quad (4.16)$$

Fractional conversion

Rearranging and multiplying the numerator and denominator in Equation (4.16) by 12/12 gives a convenient expression for calculating the extent of fractional conversion. The calculation requires measured HI and PI for the mature rock and appropriate assumptions about the original HI and PI before maturation:

$$f = 1 - \frac{HI^x[1200 - HI^o/(1 - PI^o)]}{HI^o[1200 - HI^x/(1 - PI^x)]} \quad (4.1)$$

Note: If one assumes that PI^o and PI^x are equal to zero, then Equation (4.1) becomes:

$$f = \frac{HI^x(1200 - HI^o)}{HI^o(1200 - HI^x)}$$

which is the expression used by Cooles *et al.* (1986). However, this calculation overestimates f when PI^x is appreciable.

The same basic mass-balance expression (Equation 4.12) is used as the starting point to derive the equation for reconstruction of original TOC (TOC^o):

$$f = \frac{S2^o - S2^x(CF)}{S2^o} \quad (4.17)$$

Substituting for S2 and CF gives:

$$f = \frac{\dfrac{HI^o(TOC^o)}{100} - \dfrac{HI^x(TOC^x)}{100} \times \dfrac{(83.33 - TOC^o)}{(83.33 - TOC^x)}}{HI^o(TOC^o)/100} \quad (4.2)$$

Rearranging gives:

$$HI^o(TOC^o)(1 - f)$$
$$= HI^x(TOC^x) \times \frac{(83.33 - TOC^o)}{(83.33 - TOC^x)} \quad (4.3)$$

$$= \frac{HI^x(TOC^x)(83.33) - HI^x(TOC^x)(TOC^o)}{83.33 - TOC^x} \quad (4.4)$$

$$HI^o(TOC^o)(1 - f)$$
$$= \frac{HI^x(TOC^x)(83.33)}{83.33 - TOC^x} - \frac{HI^x(TOC^x)(TOC^o)}{83.33 - TOC^x} \quad (4.5)$$

Original total organic carbon

Combining terms and rearranging terms in Equation (4.5) gives:

$$TOC^o = \frac{HI^x(TOC^x)(83.33)}{HI^o(1 - f)(83.33 - TOC^x) - HI^x(TOC^x)} \quad (4.5)$$

5 · Refinery oil assays

Many refinery oil assays differ substantially from geochemical analyses conducted by petroleum or environmental geochemists, although interdisciplinary use of these tools is becoming more common. Some basic oil assays include API gravity, pour point, cloud point, viscosity, trace metals, total acid number, refractive index, and wax content. More advanced oil assays include chemical group-type fractionation and field ionization mass spectrometry. A brief overview of refinery processes includes the fate of biomarkers in straight-run and processed refinery products, with tips on how to distinguish refined from natural petroleum products in environmental and geological samples.

Environmental and petroleum geochemists approach description of the chemical composition of crude oil from a different perspective compared with refinery chemical engineers. Geochemists commonly want to evaluate the geohistory of crude oil, including its genetic origin and extent of catagenesis or other subsurface alteration. This type of information can be used to improve upstream (exploration and production) success. Refinery chemists are interested mainly in crude oil as feedstock that can be processed into marketable products. Process models predict and optimize refinery output and evaluate the costs associated with refining. Downstream (refining, chemicals, and marketing) chemical assays are conducted to provide input parameters for these models and to determine the value of each crude oil. Biomarkers and other tools are likely to play increasingly important roles in solving technical problems related to downstream operations.

The analytical methods used by geochemists and refinery chemists to characterize crude oils reflect their different needs. Geochemists commonly rely on subtle variations in isomer distributions and isotopic ratios of individual compounds for their interpretations about

source, thermal maturity, and extent of secondary alteration. Refinery chemists emphasize analytical techniques that can be used to group compounds according to their behavior during refinery processes. Biomarker hydrocarbons are grouped with non-biomarker hydrocarbons having similar boiling points and gross chemical compositions, while isomeric distributions and stable isotopic ratios are rarely used.

One major difference between upstream and downstream analytical approaches is standardized procedure. In petroleum geochemistry, there is little uniformity in methods between laboratories. To optimize resolution and sensitivity, the latest advances in chromatographic and spectrometric technologies tend to be adapted quickly without much concern for compatibility with older techniques. Data from individual laboratories usually are internally consistent but do not compare well with each other (Lin, 1995; Isaacs, 2001b). In contrast, refinery chemists generally employ strict standardized procedures. The American Society of Testing Materials (ASTM) and the Gas Processors Association (GPA) established analytical methods and procedures that are universal for refinery oil assays. These methods specifically describe necessary instrumentation, procedures, calculations, data processing, and report formats. New methods must pass formal review, verification of reproducibility, comparison with established methods, and committee approvals before use. Refinery chemists commonly analyze oil properties or compositions using both ASTM and superior, but uncertified, methods.

Ideally, crude oil samples analyzed by environmental or petroleum geochemists are from single-spill events or preproduction tests of specific reservoirs penetrated by individual wells, respectively. Refinery crude oils are usually composites of fluids from multiple production zones and wells or even multiple fields. Many oil assays are designed to monitor the consistent quality of the feedstock so as to meet the specifications of purchase contracts. For these and other reasons discussed above, geochemists and refinery chemists rarely use or consider the methods of oil characterization used by their counterparts.

Applied petroleum biomarker research focuses almost exclusively on upstream business (exploration and production). However, there are downstream problems that could be addressed using an interdisciplinary approach. Identification and remediation of spilled crude oils and refined products is one area where such interdisciplinary studies already have proven successful. We believe that an increased awareness of downstream operations (refining, chemicals, and marketing) can broaden the applications of biomarkers and other trace organic compounds. For example, petroleum geochemists can employ their expertise to improve their understanding of the molecular transformations (e.g. biodegradation) responsible for altering oil quality (e.g. viscosity, sulfur, trace metals, and acidity) that impact economic value. Conversely, refinery crude assays can provide the petroleum geochemist with a more complete picture of oil composition, particularly for the higher-molecular-weight heteroatomic species. Sophisticated refinery models that use equation of state (EOS) properties for chemically "lumped" oil components are being incorporated into basin models that predict oil and gas compositional changes during expulsion and migration.

BASIC OIL ASSAYS

Crude oil assays are used to determine the economic value of refinery feedstock and how best to refine it. Physical properties are needed to determine oil flow characteristics, combustibility, and volatility. Chemical properties and compositions are needed to determine the optimal refinery conditions for processing the oil into profitable products. Although refinery oil assays have been standardized and documented in detail, there is no consistent set of analyses that defines a standard refinery assay. The measurements described below include the most common standardized analyses as well as several advanced, non-standard techniques that may be particularly interesting to environmental and petroleum geochemists.

Specific gravity: API gravity

Density is a fundamental property used to evaluate oil quality. The density (mass per unit volume) of any substance depends on temperature. To normalize

temperature effects, oil density is compared with that of pure water at a fixed temperature. This relative density or *specific gravity* is usually measured at 60°F (15.6°C) and is reported in units of grams/milliliter (g/ml, g/cc, g/cm^3) or as kg/m^3. Petroleum-fluid specific gravities are ~0.74–0.80 for condensates and light oils, 0.80–0.93 for normal oils, and 0.93–1.0 for biodegraded heavy oils. Very heavy biodegraded oils have specific gravities >1 and sink in water.

The American Petroleum Institute (API) introduced an arbitrary scale, where the density of water is 10 (specific gravity = 1.0; 6.30 barrels/metric ton). On this API gravity scale:

$$°API = \frac{141.5}{\text{specific gravity (at 60°F (15.6°C))}} - 131.5$$

°API gravity ranges over an expanded scale (~0–60°), where heavy or dense oils have low values and less dense oils or condensates have high values. Examples of very heavy oil in tar sands include Athabasca (6–10°API), Wabasca (10–13°API), and Cold Lake (10–12°API) in Western Canada, Cerro Negro-Morichal-Jobo (8–12°API) in Eastern Venezuela, and Asphalt Ridge and Sunnyside (8–12°API) in Utah (Tissot and Welte, 1984). Condensates have API gravities > 45°.

Density is usually measured following ASTM method D4052 (Standard Test Method for Density and Relative Density of Liquids by Digital Density Meter) or D5002 (Standard Test Method for Density and Relative Density of Crude Oils by Digital Density Analyzer), which rely on an Anton Parr DMA 48 digital density meter. The precision of these measurements is ± 0.0001 g/ml, with an accuracy of ± 0.0005 g/ml (~± 0.2°API). Highly viscous oils can be measured with the same accuracy using a pycnometer.

Pour point

Pour point is defined as the lowest temperature that oil will flow without disturbance (e.g. agitation or shaking) under specified test conditions. Solidification below this temperature is usually attributed to precipitation of waxes from the parent oil, but it can also be due to increased viscosity in heavy oils.

In ASTM method D97 (Standard Test Method for Pour Point of Petroleum Products), the pour point is defined as 3°C (5°F) above the temperature at which the oil in a test vessel does not move when the container is held horizontally for five seconds. This rather sub-jective test is still performed but has been superseded by mechanical test equipment as specified by ASTM methods D5949 (Automatic Pressure Pulsing), D5950 (Automatic Tilt), and D5985 (Rotational).

Cloud point

As oil or fuel is cooled, waxes begin to precipitate, thus imparting a cloudy appearance. The cloud point is the temperature at which this occurs. Cloud point depends on the rate of cooling and the temperature program. ASTM method D2500 is used routinely but may be su-perseded by methods D5771 (Stepped Cooling), D5772 (Linear Cooling Rate), or D5773 (Constant Cooling Rate). In non-opaque samples, the cloud point is re-producible to $\sim \pm 3\,^\circ$C.

Flash point

Flash point is the lowest temperature to which the oil or fuel must be heated to produce a vapor/air mixture above the liquid that is ignitable when exposed to an open flame. Flash point is used as a measure of the po-tential fire hazard when shipping oils and refined fuels.

Flash point is extremely sensitive to the ASTM method used, such as the Cleveland Open Cup (ASTM D92), the Pensky-Martens Closed Cup (ASTM D93), and the Tag Closed Tester (D56) methods. Flash points may be reported with designations O.C or C.C. to iden-tify the method. The closed cup tester is needed for volatile fuels or condensates (flash point $<10\,^\circ$C). Care-fully measured flash points are reproducible to $\pm 4\,^\circ$C.

Fire point

The fire point is the lowest temperature at which an oil or fuel will ignite and burn for at least five seconds when a test flame is applied to its surface using ASTM method D92. The fire point is typically $\sim 30\,^\circ$C above the flash point.

Viscosity

Viscosity measures the internal resistance to flow of a fluid and is therefore useful in order to help determine the amount of recoverable reserves, producibility, and economic value of oil accumulations. Fluid viscosity is determined by its shear stress and shear rate. Shear

stress is the frictional force that results from sliding one layer of fluid along another, while the shear rate is the velocity of flow. For Newtonian fluids, such as oil, the shear stress varies in proportion to shear rate. Ab-solute or dynamic viscosity is the ratio of shear stress to shear rate and is constant for a fixed temperature. In non-Newtonian fluids, such as some polymers and lu-bricants, the shear stress is not proportional to the shear rate. For these fluids, only an apparent or relative vis-cosity can be determined at a specified shear rate and temperature.

Absolute viscosities are measured by a standard method. The SI unit of absolute viscosity is the millipascal-second (mPa-s), which is equivalent to the commonly used unit of centipoise (cP). Poise is the force required to move a plane surface with a velocity of 1 cm/s when the surfaces are separated by a layer of fluid 1 cm thick (shear force/shear rate = dynes-s/cm^2). Modern rotational viscometers are capable of making absolute viscosity measurements for both Newtonian and non-Newtonian fluids at various shear rates. The viscosities of non-Newtonian fluids are reported at a specific shear rate. Unfortunately, no ASTM stan-dard method exists that makes use of these viscome-ters. Nonetheless, these instruments are in widespread use. With computer-controlled shear rates, the preci-sion of the dynamic viscosity for both Newtonian and non-Newtonian fluids can be measured to $\pm 5\%$ of the mean. In theory, absolute viscosity measurements do not depend on the instrumentation, and results can be compared between laboratories.

Kinematic viscosity is the absolute viscosity divided by the density of the fluid measured at the same temper-ature and is usually expressed in stokes or centistokes (cSt). Unfortunately, relative viscosity, where the vis-cous drag of a fluid is measured without known or uni-formly applied shear rates, is also termed kinematic vis-cosity. There arc several ASTM methods for measuring the viscosity of oil. However, only methods D445 (Stan-dard Test Method for Kinematic Viscosity of Transpar-ent and Opaque Liquids (the Calculation of Dynamic Viscosity)) and D4486 (Standard Test Method for Kine-matic Viscosity of Volatile and Reactive Liquids) yield absolute viscosity measurements for Newtonian fluids. In these methods, kinematic viscosity is measured as the time required for a fixed amount of oil to flow through a glass capillary tube under the force of gravity.

There are several other obsolete methods that were used to measure relative velocity, such as Saybolt

Universal Viscosity (SUV), Saybolt Furol Viscosity, Engier Viscosity, and Redwood Viscosity. The Saybolt viscometer was the most widely used. Results were expressed as Saybolt Universal Seconds (SUS) and defined as the time in seconds for 60 ml of oil to flow through a Saybolt standard orifice at a given temperature (ASTM D88). ASTM D2161 (Standard Practice for Conversion of Kinematic Viscosity to Saybolt Universal Viscosity or to Saybolt Furol Viscosity) established the official equations using SUS kinematic viscosity units, mm^2/s.

Viscosity typically increases with decreasing temperature. All viscosity values must also report the temperature at which they were measured. Crude oil viscosities typically are measured at three or more temperatures. Intermediate values can be estimated by a logarithmic fit between temperature and viscosity.

Water and sediment

The amount of water and sediment suspended or emulsified in oil affects its economic value and refinery treatment procedures. The water and sediment may be added to oil during production, transportation, or storage. Methods that separate oil from water and sediment using a centrifuge are rapid but routinely underestimate the amount of water (D96 (Standard Test Methods for Water and Sediment in Crude Oil by Centrifuge Method (Field Procedure)) or D4007 (Standard Test Method for Water and Sediment in Crude Oil by the Centrifuge Method (Laboratory Procedure))). ASTM method D473 (Standard Test Method for Sediment in Crude Oils and Fuel Oils by the Extraction Method) is a more accurate procedure, where crude oil is dissolved in hot toluene and filtered to remove sediments. Water content in crude oils can be determined more accurately using several ASTM methods. Method D4006 (Standard Test Method for Water in Crude Oil by Distillation) involves refluxing the oil with xylene. The xylene–water vapors condense, and the water settles into a graduated trap. The accuracy of this procedure is only marginally better than the centrifuge methods. Methods D4377 (Standard Test Method for Water in Crude Oils by Potentiometric Karl Fischer Titration) and D4928 (Standard Test Methods for Water in Crude Oils by Coulometric Karl Fischer Titration) provide improved accuracy but may be subject to interference by mercaptans and sulfides. The precision and accuracy of D4928

are $0.056x^{2/3}$ and $0.112x^{2/3}$, respectively, where x is the measured mean value from 0.02 to 5% water.

Salt content

Most of the salts in crude oil are dissolved in colloidal water and can be removed in desalting units before distillation. Small amounts of salts are dissolved in the oil. If not treated, these salts can accumulate or react to form corrosive acids. The salt content of oil is determined using ASTM method D3230 (D3230-99 Standard Test Method for Salts in Crude Oil (Electrometric Method)), which compares the conductivity of oil dissolved in water with standard salt solutions.

Petroleum distillation

Composition by boiling point ranges is a fundamental assay that is critical in order to determine the value of crude oil and the processing needed to maximize the yield of desirable products. There are standard procedures for the distillation, but no universal conventions for the ranges or names of the distillate cuts or for the refined products exist. Table 5.1 is an example of such a system.

Traditional oil assays involve atmospheric and vacuum distillation. ASTM method D2892 (Standard Test Method for Distillation of Crude Petroleum (15-Theoretical Plate Column)) is the standard method for crude oil distillation. The distillation starts at ambient pressure but switches to vacuum to extend volatilization to the atmospheric-equivalent boiling point of ~400°C. The residuum is then analyzed using ASTM method D5236 (Standard Test Method for Distillation of Heavy Hydrocarbon Mixtures (Vacuum Potstill Method)) that continues the distillation at 0.5 torr to obtain the atmospheric equivalent boiling point limit of 560°C. Distillation of large volumes (5–50 l) of crude oil is performed to obtain petroleum cuts within specific boiling point ranges for additional analyses.

Traditional distillations have largely been replaced by gas chromatography, which simulates distillation (Figure 5.1). These chromatographic methods rely on short capillary columns with thin films that provide a rapid distribution of petroleum components based on boiling point but generally do not separate hydrocarbon isomers. Elution temperatures can be converted to equivalent true boiling points. ASTM methods D2887 (Standard Test Method for Boiling Range Distribution

Table 5.1. *Approximate true boiling point (TBP) distributions for straight-run and processed refinery products*

Distillation cut	Liquid natural gas	Gasoline	White spirits	Kerosene JP-4	Kerosene/jet A1	Diesel	Lubricant stock	Heating oils	Bunker C	Residual/asphalt	Hydrocarbon range	TBP (°C)
Natural gas											C_1–C_4	<0
Light straight-run gasoline											C_4–C_6	0–70
Light naphtha											C_6–C_7	70–100
Medium naphtha											C_7–C_9	100–150
Heavy naphtha											C_9–C_{11}	150–190
Light kerosene											C_{11}–C_{13}	190–235
Heavy kerosene											C_{13}–C_{15}	235–265
Atmospheric gas oil											C_{15}–C_{20}	265–343
Vacuum gas oil											C_{20}–C_{30}	343–565
Atmospheric residue											>C_{30}	>450
Vacuum residue											>C_{60}	>615

■ (grey) Refinery product partially spans distillation cut   ■ (black) Refinery product includes entire distillation cut

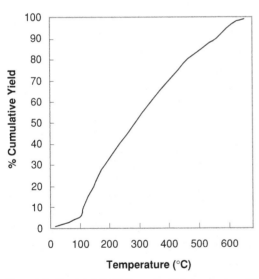

Figure 5.1. Simulated distillation using high-temperature gas chromatography. Cumulative detector response is plotted versus extrapolated boiling temperature (calculated from elution temperature).

of Petroleum Fractions by Gas Chromatography) and D3710 (Standard Test Method for Boiling Range Distribution of Gasoline and Gasoline Fractions by Gas Chromatography) use *n*-alkanes as external standards.

ASTM method D5307 (Standard Test Method for Determination of Boiling Range Distribution of Crude Petroleum by Gas Chromatography) is similar to D2887 but requires two runs for each sample, one of which uses an internal standard. The amount of material boiling above 560°C (reported as residue) is calculated from the differences between the two runs. Modern high-temperature gas chromatographic (HTGC) analyses are capable of equivalent boiling points in excess of 750°C and have recently been approved as a standard method (D6352 (Standard Test Method for Boiling Range Distribution of Petroleum Distillates in Boiling Range from 174–700°C by Gas Chromatography)). The cut-point yields obtained by HTGC and distillation methods (D2892/D5236) are very similar, varying by less than 2 wt.%, except for the range ∼400–480°C (750–900°F), which corresponds to the cross-over between the D2892 reflux column method and the D5236 vacuum potstill method.

Elemental analysis (CHNOS)

Elemental analyses are used to characterize whole petroleum or distillate fractions. Two methods can be used for CHN analyses (ASTM D5291 (Instrumental Determination of Carbon, Hydrogen, and Nitrogen in

Table 5.2. *Precision and accuracy of CHNOS elemental analyses*

Analysis	Range (%)	± Precision	± Accuracy
Carbon (D5291)	75–87	$(x + 48.48)0.0072$	$(x + 48.48)0.018$
Hydrogen (D5291)	9–16	$(x^{0.5})0.1162$	$(x^{0.5})0.2314$
Hydrogen (D4808)	Light distillates	$0.23(x^{0.25})$	$0.72(x^{0.25})$
Hydrogen (D4808)	Middle distillates	$0.0015(x^2)$	$0.0031(x^2)$
Hydrogen (D4808)	Residua	$33.3(x^{-2})$	$70.2(x^{-2})$
Nitrogen (D5291)	0.75–2.5	0.1670	0.4456
Nitrogen (D5762)	40–10 000 ppm	$0.099x$	$0.291x$
Oxygen (D5622)	1–5	0.06	0.26
Sulfur (D1552)	0.06–2	<0.09	<0.27
Sulfur (D1552)	2–5	<0.16	<0.49
Sulfur (D2622)	0.001–0.0049	$0.60x$	$0.60x$
Sulfur (D2622)	0.005–0.0149	$0.20x$	$0.40x$
Sulfur (D2622)	0.015–5	$0.05x$	$0.16x$
Sulfur (D4294)	0.015–5	$0.02894(x + 0.16691)$	$0.1215(x + 0.05555)$
Sulfur (D5452)	0–400 ppm	$0.1788\,x^{0.75}$	$0.5797\,x^{0.75}$
Sulfur (D5452)	>400 ppm	$0.02902x$	$0.1267x$

x, mean measured value.

Petroleum Products and Lubricants); D5622-95 (Standard Test Methods for Determination of Total Oxygen in Gasoline and Methanol Fuels by Reductive Pyrolysis)). Hydrogen can be also be determined using nuclear magnetic resonance (NMR) (ASTM D4808 (Standard Test Methods for Hydrogen Content of Light Distillates, Middle Distillates, Gas Oils, and Residua by Low-Resolution Nuclear Magnetic Resonance Spectroscopy)). Low concentrations of nitrogen are determined using a chemiluminescence detector following ASTM method D5762 (Standard Test Method for Nitrogen in Petroleum and Petroleum Products by Boat-Inlet Chemiluminescence).

There are many ASTM methods available to determine sulfur in crude oil and refined products. In method D1552 (Standard Test Method for Sulfur in Petroleum Products (High-Temperature Method)), sulfur is combusted to SO_2 and measured by an infrared (IR) detector. This method commonly relies on a LECO SC32 or SC132 sulfur analyzer and is limited to samples with boiling points >177°C ($\sim C_{10+}$) and at least 0.06% sulfur. Method D2622 (Standard Test Method for Sulfur in Petroleum Products by Wavelength Dispersive X-ray Fluorescence Spectrometry) uses soft X-ray spectroscopy. It is suitable over a broad range of

concentrations and can be used for crude oils, liquid products, and solids that can be dissolved. Method D4294 (Standard Test Method for Sulfur in Petroleum Products by Energy-Dispersive X-Ray Fluorescence Spectroscopy) is rapid (2–4 min/sample) and is suitable for oils, distillate fractions, and refined products. Method D6481 (Standard Test Method for Determination of Phosphorus, Sulfur, Calcium, and Zinc in Lubrication Oils by Energy Dispersive X-ray Fluorescence Spectroscopy) is a similar procedure. Method D5453 (Standard Test Method for Determination of Total Sulfur in Light Hydrocarbons, Motor Fuels and Oils by Ultraviolet Fluorescence) is based on conversion of sulfur to SO_2 in an Antek 711 pyroreactor, which is quantified by ultraviolet (UV) fluorescence using an Antek 714 sulfur detector. The preparative combustion requires sample fractions with boiling points in the range 25–400°C. The method is sensitive and the detector has a narrow, linear range. High-sulfur oils can be measured if the samples are diluted.

ASTM methods establish precision and accuracy by comparing intra- and interlaboratory results. Most methods report the precision and accuracy of elemental analyses as a function of the mean measured value for a given range (Table 5.2).

Note: One requirement for adaptation of an analytical procedure as an industry standard is that its accuracy and precision be defined for given ranges and detection limits. Because these terms are frequently used incorrectly, definitions are given below:

Accuracy: the closeness of agreement between the value that is accepted either as a conventional true value or as an accepted reference value and the measured value. This is sometimes called trueness.

Precision: the closeness of agreement between a series of measurements obtained from multiple sampling of the same homogeneous sample or standard solution using the defined analytical procedure. Analytical precision may be measured as repeatability (precision of the same analytical procedure conducted sequentially or over a short time period) or reproducibility (precision between laboratories). High precision is not equivalent to high accuracy, nor does high accuracy imply high precision. The ideal method should yield both high accuracy and high precision.

Detection limit: the lowest amount of detectable analyte that may not necessarily be quantified as an exact value.

Quantitation limit: the lowest amount of analyte that can be quantified within acceptable precision and accuracy.

Range: the interval between the upper and lower amounts of analyte for which the method yields values with acceptable precision and accuracy.

Linearity: the ability of a method to yield test values that are directly proportional to the concentration of an analyte within a given range.

Robustness: the ability of an analytical procedure to provide measurements with acceptable precision and accuracy, while accommodating small but deliberate variations in method parameters. Robustness also may be a subjective term indicating the methods reliability during normal operations.

Carbon residue

Carbon residue is the material remaining after evaporation (distillation) and pyrolysis of oil. However, it is not composed entirely of carbon. Carbon residue provides an indication of the tendency of a sample to form coke. In ASTM method D189 (Condrason Carbon Residue of Petroleum Products), a sample is placed in a crucible, heated by a flame burner for a fixed period of time, cooled, and weighed. The residue is the Conradson carbon residue and is reported as weight percent. ASTM D4530 (Determination of Carbon Residue (Micro Method)) is a modification of D189 that fixes the heating temperature at 500°C, performs the residuum pyrolysis under nitrogen, and uses a small amount of sample. Results from this method, reported as carbon residue (micro), are equivalent to the Conradson carbon residue but more precise.

ASTM D524 (Ramsbottom Carbon Residue of Petroleum Products) also measures residue carbon. Samples are heated at 550°C in glass bulbs. Evaporated or pyrolyzed volatiles emanate from a capillary opening in each bulb, where the residuum undergoes additional cracking, coking, and oxidation. Ramsbottom and Conradson carbon residue values differ from each other but correlate in a non-linear fashion.

Trace metals

Trace metal analysis is used to evaluate refinery processing and monitor internal engine wear. For example, metals in feedstock can interfere with catalysts used in refinery catalytic reactors. Early ASTM methods based on atomic absorption and flame-emission spectroscopy [ASTM 3605 and 5056] have been largely replaced by those using inductively coupled plasma atomic emission spectrometry (ICP-AES) (ASTM 5185 and 5600). ICP-AES quantifies many elements (Al, Ba, B, Ca, Cr, Cu, Ir, Pb, Mg, Mn, Ni, P, K, Si, Ag, Na, S, Sn, Ti, V, and Zn) with sub-ppm precision and accuracy. Nickel and vanadium concentrations are of particular interest to petroleum geochemists because their ratio is indicative of redox conditions during source-rock deposition (see Figure 4.11).

Total acid number

Crude oils contain organic acids that collectively (although improperly) are called naphthenic acids after the first group of such compounds characterized in petroleum. Naphthenic acids consist only of the saturated cyclic carboxylic acids, but there are many other chemical classes of organic acids in crude oils. Oils with high acid contents are commonly biodegraded. High-acid oils are corrosive and may require costly installation of corrosion-resistent metals for processing. Only a few refineries worldwide are capable of handling high-acid crude without prior blending with better-quality oil. Refinery chemists also must consider acid content in finished products and performance specifications. Organic acids can form by oxidation of lubricating oil in engines, and their increased concentration with time is a measure of oil breakdown.

The standard methods to characterize organic acids are ASTM D664 (Standard Test Method for Acid Number of Petroleum Products by Potentiometric Titration) and D974 (Standard Test Method for Acid and Base Number by Color-Indicator Titration). Both methods define total acid number (TAN) as milligrams of potassium hydroxide (KOH) required to titrate a gram of oil to a specified endpoint. In method D974, the endpoint is defined as the color change of p-naptholbenzein in a toluene–water–isopropanol solvent. In method D664, the endpoint is defined by a potentiometric reading in millivolts that compares the initial reading with that of freshly prepared non-aqueous basic buffer solution or with a well-defined inflection point. The advantage of method D974 is that it requires only inexpensive glassware and reagents. However, the p-naptholbenzein color change can be obscured by the color of the oil sample. Method D664 lacks this limitation but requires a potentiometric titrator. Both methods can also be used to determine the total base number (TBN) by substituting HCl as the titrating solution and methyl orange as the color-change indicator.

Crude oils can have TAN ranging from <0.1 to 8 mg KOH/g. Samples with >0.5 mg KOH/g are high-acid oils. However, TAN may not reflect the true acid content or corrosive nature of oil. The ASTM method dissolves oil in a toluene–water–isopropanol solution (500 : 5 : 495 ml). Isopropyl alcohol is miscible with both toluene (needed to dissolve the hydrophobic oil) and water (needed to ionize the acid for titration). Without the two organic solvents, the water would not dissolve the carboxylic acids in the crude oil. The ASTM methods measure all compounds with dissociation constants in water greater than 10^{-9} and salts with hydrolysis constants greater than 10^{-9}. Hence, some neutral species in crude oils, such as calcium naphthenates, are reported in the TAN. In refined products, many of the additives

of lube oils yield high TAN (2–5), even though they are not corrosive. TAN measurements do not necessarily reflect the electrochemical or corrosive behavior of a sample.

Naphthenic acids may occur in all distillate fractions. Low-molecular-weight fatty acids are particularly problematic because even low-TAN oils with volatile fatty acids may be rejected for refining because of their offensive odor. Higher-molecular-weight naphthenic acids include acyclic, cyclic, aromatic, and heteroatomic species.

Refractive index

The refractive index (RI) is a ratio of the velocity of light at a specific wavelength in air to its velocity in the sample at a specific temperature. RI is frequently measured for oils, oil fractions, and refined products. The method (D1218 Standard Test Method for Refractive Index and Refractive Dispersion of Hydrocarbon Liquids) uses a Bausch and Lomb precision refractometer. Readings are taken between 20 and 30°C and are accurate to the fifth decimal place. RI is used in other methods to calculate average molecular and structural properties (e.g. ASTM method D3238).

Asphaltene content

ASTM method D6560 (Determination of Asphaltenes (Heptane Insolubles) in Crude Petroleum and Petroleum Products) is used routinely to precipitate and quantify asphaltenes. The oil sample is mixed with heptane and refluxed. The precipitated asphaltenes, waxes, and inorganic material collect on a filter, and the waxes can be removed by washing with hot heptane in an extractor. Asphaltenes can be separated from the inorganic material by dissolution in hot toluene, filtering, and evaporation of the solvent.

Wax content

Waxes are high-molecular-weight alkanes. Many petroleum geochemists use the term "wax" to indicate alkanes above C_{35}–C_{40}, the upper limit of conventional gas chromatography. However, refinery chemists consider waxes to include saturated hydrocarbons as low as C_{16} that are not readily distilled into transportable fuels. The methods that refinery chemists use to determine wax content reflect this distribution. Oils are typically dissolved in hexane and treated with sulfuric acid to precipitate asphaltenes. Dissolving the deasphalted oil in methylene chloride cooled to $-30°C$ precipitates waxes. ASTM method D5442 (Standard Test Method for Analysis of Petroleum Waxes by Gas Chromatography) describes a routine method for characterizing alkanes from C_{17} to C_{44}. Advances in high-temperature gas chromatography can extend the analysis to C_{100+} (del Río and Philp, 1992).

Aniline point

The aniline point is the minimum temperature for the complete solution of equal volumes of aniline and oil sample. A mixed aniline point is the minimum equilibrium solution temperature for a mixture of two volumes of aniline, one volume of n-heptane, and one volume of oil. Aromatic hydrocarbons have the lowest aniline temperatures, normal and isoalkanes have the highest, and cycloalkanes and olefins have intermediate values. In hydrocarbon mixtures, the aniline or mixed aniline point is used as a rough estimate of aromatic content. Aniline points are reproducible to $\sim \pm 0.6$–$1.0°C$ using the procedure described in ASTM D611 (Standard Test Methods for Aniline Point and Mixed Aniline Point of Petroleum Products and Hydrocarbon Solvents).

ADVANCED OIL ASSAYS

In recent years, chemical models of refinery processes have become increasingly sophisticated, necessitating advanced molecular assays of crude oil. These assays begin with chromatographic analysis of C_1–C_5 hydrocarbons, followed by distillation of the oil into multiple fractions or cuts. The number of distillation cuts may exceed 50, with each cut point determined by the true boiling point of individual n-alkanes. Each cut can then be analyzed using gas chromatography, liquid chromatography, mass spectrometry, or combined methods (e.g. GCMS) to determine distributions of normal and isoalkanes, cycloalkanes, aromatic hydrocarbons by ring number, and various sulfur-containing species. Physical properties, such as viscosity and gravity, and chemical properties, such as sulfur and nitrogen content and average molecular weight, can also be measured directly or calculated from the compositional data. The results of advanced oil assays are used as input data for refinery simulators that predict and optimize refinery performance for different crude oil feedstocks. While the basic oil assays described above are standardized, few of the advanced oil assay methods have been approved by the ASTM.

Molecular weight

ASTM methods determine the average molecular weight of hydrocarbon fractions. Method D2502 (Standard Test Method for Estimation of Molecular Weight (Relative Molecular Mass) of Petroleum Oils From Viscosity Measurements) is based on a relationship between kinematic viscosity (measured at $100°F$ $(37.8°C)$ and $210°F$ $(98.9°C)$) and average molecular weight. The method is applicable to average molecular weights in the range 250–700 Daltons. Method D2503 (Standard Test Method for Relative Molecular Mass (Molecular Weight) of Hydrocarbons by Thermoelectric Measurement of Vapor Pressure) measures the relative change in temperature when a drop of sample is added to a closed chamber saturated with solvent. Since the vapor pressure of the solution is less than that of pure solvent, solvent condenses on the sample drop, causing a decrease in temperature. This method has been calibrated for liquid samples with average molecular weight of 800 Daltons but can be used for samples with average molecular weight of up to 3000 Daltons.

Two instrumental methods provide a distribution of average molecular weight. Size-exclusion or gel-permeation chromatography relies on the physical interaction between components in the sample and the variable pore space of the substrate, typically cross-linked polystyrene. Size-exclusion chromatography, however, measures hydrodynamic volume rather than molecular weight. The correlation between hydrodynamic volume and molecular weight can be established for pure compounds and polymers but is only approximate for oil and oil fractions. Various mass spectrometric methods can provide molecular-weight distributions. Of these, matrix-assisted laser desorption/ionization (MALDI) mass spectroscopy has the widest application. The technique mixes the sample with a UV-absorbing compound or matrix of compounds. A UV laser excites the absorbing matrix and ionizes the sample molecules, which desorb from the matrix. The chemistry of these reactions is poorly understood. However, MALDI mass spectroscopy can yield highly accurate mass distributions for both polar and non-polar compounds. The technique is used widely in biochemical and polymer applications. Hyphenated techniques that couple size-exclusion chromatography with MALDI mass spectroscopy have been developed recently.

Chemical group-type fractionation

Environmental and petroleum geochemists and refinery chemists fractionate oils into four chemical classes. The analysis is called SARA or PARA, where the first letter refers to s̲aturates or p̲araffins and the next three letters refer to a̲romatic hydrocarbon, r̲esins (polars), and a̲sphaltenes.

The distribution of compounds in each SARA compound class depends on the method. Asphaltenes are non-hydrocarbons that precipitate in a large volume of a low-molecular-weight n-alkane. The amount and chemical nature of the asphaltenes differs depending on the hydrocarbon solvent (nC_5, nC_6, or nC_7) and the time and temperature used for precipitation. While the strict chemical distinction between aromatic hydrocarbons and NSO heterocompounds is clear, group-type separations of these classes rely primarily on differences in solubility and polarity. Non-polar sulfur-containing compounds, such as benzothiophenes, and non-basic nitrogen compounds, such as benzocarbazoles, may elute in the aromatic fraction. Highly condensed aromatic hydrocarbons with more than four or five rings have low solubility in hydrocarbon solvents and typically elute in the polar fraction. Saturated normal, branched, and cyclic hydrocarbons are usually well defined and separated. In some methods, however, alkylbenzenes may elute in the saturate fraction (Chunqing *et al.*, 2000a). There are no widely accepted standard procedures for separating these chemical classes. However, the literature describes many alternative separation procedures based on thin-layer chromatography (Iatroscan), liquid chromatography (open-column, medium-pressure, and high-performance), and supercritical fluid chromatography.

Each of the SARA group-types can be fractionated further. Saturated hydrocarbons can be separated into normal and branched hydrocarbons (isoalkanes) and naphthenes (cycloalkanes). Aromatic hydrocarbons can be separated according to ring number. Heteroatomic compounds can be separated based on polarity. For example, Willsch *et al.* (1997) separated five heterocompound fractions (acids, bases, and three

NSO fractions of differing polarity) in addition to the conventional saturated and aromatic hydrocarbon fractions. Other methods separate sulfur aromatic compounds from polynuclear aromatic hydrocarbons (Mansfield *et al.*, 1999).

Refinery chemists classify oils and products into four hydrocarbon classes: paraffins (normal and branched alkanes), olefins, naphthenes (saturated cycloalkanes), and aromatics. Many standard and non-standard methods are used to measure the hydrocarbon classes in whole crude oils, individual distillate cuts, and refined products.

For samples with boiling points up to 200°C, ASTM method D5443 (Standard Test Method for Paraffin, Naphthene, and Aromatic Hydrocarbon Type Analysis in Petroleum Distillates Through 200°C by Multi-Dimensional Gas Chromatography) is routinely used. The sample is injected into a gas chromatograph, where it passes through a polar (OV-275) column that retains all aromatics, binaphthenes, and compounds with boiling points >200°C. The components that are not retained on the column enter a platinum reactor that hydrogenates olefins, which are trapped on a molecular sieve (13X and 5 Å) column. When temperature-programmed, the molecular sieve separates the C_6–C_{11} hydrocarbons by carbon number and structure, where naphthenes elute before paraffins. Hydrocarbons retained on the polar column elute on to a Tenax trap in three distinct temperature pulses. These pulses are desorbed sequentially on to a non-polar (OV-101) column, where they can be partially separated into C_6–C_{10} aromatic and naphthenic hydrocarbons. Saturated hydrocarbons with boiling points >200°C that are transferred to the Tenax or non-polar column are backflushed at the end of the analysis. Additional traps and columns allow separation of oxygenates from gasolines and separation of olefins by carbon number. The instrumentation and method are described in ASTM D6293-98 (Standard Test Method for Oxygenates and Paraffin, Olefin, Naphthene, Aromatic (O-PONA) Hydrocarbon Types in Low-Olefin Spark Ignition Engine Fuels by Gas Chromatography).

Various methods can be used to obtain chemical group-type data from refinery oil assays. One ASTM method (D3238-95 (Standard Test Method for Calculation of Carbon Distribution and Structural Group Analysis of Petroleum Oils by the n-d-M Method)) does not rely on direct measurement but is based on correlations of refractive index, viscosity, and density. Results are reported in terms of proportions of aromatic rings (R_A), naphthene rings (R_N), and paraffin chains (R_P). Alternatively, the composition may be expressed in terms of carbon distribution, i.e. the percentage of the total number of carbon atoms ($\%C_P$).

Molecular analysis

Detailed molecular refinery assays rely heavily on gas chromatographic separation of saturated and aromatic hydrocarbons. Other refinery assays measure non-hydrocarbon gases or volatiles (H_2, He, CO_2, CO, H_2S, mercaptans). Older ASTM methods for analyzing the light hydrocarbons based on packed-column gas chromatography and mass spectrometry were largely replaced by capillary-column gas chromatographic methods (e.g. ASTM D5623 (Standard Test Method for Sulfur Compounds in Light Petroleum Liquids by Gas Chromatography and Sulfur Selective Detection)). Light hydrocarbon analyses can be merged with simulated distillation gas chromatographic results that effectively begin at C_5 (Figure 5.2).

Distillate cuts and their chemical group fractions can be analyzed individually. With the exception of the low-molecular-weight alkanes, the refinery methods are designed to provide molecular-weight distributions of individual compound classes, but they do not differentiate between isomers. Gas chromatography, HPLC, mass spectrometry, Fourier transform infrared (FTIR), NMR, UV, other forms of spectrometry,

and hyphenated methods can be used to measure the molecular composition of these fractions (Mansfield et al., 1999).

Field ionization mass spectrometry (FIMS) has become an important tool for classifying the molecular distribution of chemical classes in crude oil. The method, which uses soft ionization to characterize individual components by their mass, was applied by Lijmbach et al. (1983) to describe the composition of crude oils and source-rock extracts. FIMS can be applied to high-molecular-weight alkanes that do not yield molecular ions by electron impact ionization (del Río and Philp, 1999).

Advances in molecular process models allow downstream chemists to optimize refinery operations. These predictive models depend on detailed chemical descriptions of crude oil and product composition, methods to represent the molecular structures and complex reactions for large numbers of components, and molecular structure–properties relationships to calculate kinetic, thermodynamic, physical, and quality parameters (Quann and Jaffe, 1992; Quann, 1998; Jacob et al., 1998). The accuracy of these process models is limited by incomplete knowledge of molecular composition. Recent advances in separation and detection techniques promise unprecedented crude oil characterization. For example, Hughey et al. (2002a) detected over 14 000 NSO acidic compounds using negative ion electrospray (ESI) and Fourier transform/ion cyclotron resonance/mass spectrometry (FT/ICR/MS).

As refinery crude oil assays become increasingly directed toward molecular characterization, their value to petroleum geochemists increases. While the isomeric separations that are so vital to biomarker studies are not part of these assays, the FIMS technique offers a broad overview of total composition. We believe that significant insight will emerge by combining geochemical and refinery separation and characterization techniques. This seems to be particularly true in examining the relationship of heteroatomic species to conventional biomarkers using ultra-high-resolution mass spectrometry techniques (Tomczyk et al., 2001; Hughey et al., 2002a; Hughey et al., 2002b).

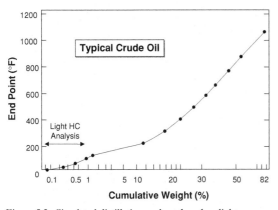

Figure 5.2. Simulated distillation analyses based on light hydrocarbon gas chromatography and high-temperature gas chromatography can be merged to provide a complete description of crude oil composition from methane to residuum.

PETROLEUM REFINING

Petroleum refining is a complex, interdependent series of processes that convert crude oil into various products (Table 5.3), where the most economically important are

Table 5.3. *Refinery processes require specific feedstocks to produce various products* (OSHA Technical Manual TED 1-0.15A)

Action	Method	Purpose	Feedstock(s)	Product(s)
Fractionation processes				
Atmospheric distillation	Thermal separation	Separate fractions	Desalted crude oil	Gas, gas oil, distillate, residual
Vacuum distillation	Thermal separation	Separate fractions without cracking	Atmospheric tower residual	Gas oil, lube stock, residual
Conversion processes: decomposition				
Catalytic cracking	Catalytic alteration	Upgrade gasoline	Gas oil, coke distillate	Gasoline, petrochemical feedstock
Coking polymerizing	Thermal alteration	Convert vacuum residuals	Gas oil, coke distillate	Gasoline, petrochemical feedstock
Hydrocracking	Catalytic hydrogenation	Convert to lighter hydrocarbons	Gas oil, cracked oil, residual	Lighter, higher-quality products
*Hydrogen steam reforming	Thermal/catalytic decomposition	Produce hydrogen	Desulfurized gas, O_2, steam	Hydrogen, CO, CO_2
*Steam cracking	Thermal decomposition	Crack large molecules	Atmospheric tower heavy fuel/distillate	Cracked naphtha, coke, residual
Visbreaking	Thermal decomposition	Reduce viscosity	Atmospheric tower residual	Distillate, tar
Conversion processes: unification				
Alkylation combining	Catalytic	Unite olefins and isoparaffins	Tower isobutane/cracker olefin	Isooctane (alkylate)
Grease compounding	Thermal combining	Combine soaps and oils	Lube oil, fatty acid, alkyl metal	Lubricating grease
Polymerizing	Catalytic polymerization	Unite two or more olefins	Cracker olefins	High-octane naphtha, petrochemical stocks

(*cont.*)

Table 5.3 (*cont.*)

Action	Method	Purpose	Feedstock(s)	Product(s)
Conversion Processes – Alteration or Rearrangement				
Catalytic reforming	Catalytic alteration and dehydration	Upgrade low-octane naphtha	Coker/hydrocracker naphtha	High-octane reformate/ aromatic
Isomerization	Catalytic rearrangement	Convert straight chain to branch	Butane, pentane, hexane	Isobutane, isopentanes, isohexanes
Treatment Processes				
*Amine treating	Absorption	Remove acidic contaminants	Sour gas, hydrocarbons with CO_2 and H_2S	Acid-free gases and liquid hydrocarbons
Desalting	Dehydration, absorption	Remove contaminants	Crude oil	Desalted crude oil
Drying and sweetening	Absorption/thermal	Remove water and sulfur compounds	Liquid hydrocarbons, LPG feedstock	Sweet and dry hydrocarbons
*Furfural extraction	Solvent extraction, absorption	Upgrade mid–distillate and lubes	Cycle oils and lube feedstocks	High-quality diesel and lube oil
Hydrodesulfurization	Catalytic treatment	Remove sulfur, contaminants	High-sulfur residual/gas oil	Desulfurized olefins
Hydrotreating	Catalytic hydrogenation	Remove impurities, saturate hydrocarbons	Residuals, cracked hydrocarbons	Cracker feed, distillate, lube
*Phenol extraction	Solvent extraction, absorption/thermal	Improve viscosity, color	Lube oil base stocks	High-quality lube oils
Solvent deasphalting	Absorption	Remove asphalt	Vacuum tower residual, propane	Heavy lube oil, asphalt
Solvent dewaxing	Cool/filter	Remove wax from lube	Vacuum tower lube	Dewaxed lube basestock
Solvent extraction	Solvent extraction, absorption/precipitation	Separate unsaturated oils	Gas oil, reformate, distillate	High-octane gasoline
Sweetening	Catalytic treatment	Remove H_2S, convert mercaptans	Untreated distillate/ gasoline	High-quality distillate/ gasoline

* These processes are not depicted in the refinery process flow chart (Figure 5.4).

LPG, liquefied petroleum gas.

transportation fuels: gasoline, jet fuel, and diesel. Other products are used as feedstock for chemicals, heating fuels, lubricating oils, waxes, and asphalt. High API gravity crude oils contain more of the lighter hydrocarbons that make up the transportation fuels. They are considered easy to refine and generally have lower sulfur, nitrogen, and trace metal contents. Low API gravity oils contain a proportionally smaller amount of the hydrocarbons directly used to make transportation fuels. Modern refinery methods, however, are capable of converting a significant amount of low API gravity oil into the more valuable products. These conversion processes are complex and expensive. Oils containing high sulfur, nitrogen, and trace metals must be treated to eliminate these elements during refining because they can poison catalysts used in the upgrading process. Further processing may be needed to meet sulfur-content specifications for refined fuels.

The first petroleum refineries used distillation towers to separate crude oil into boiling-point fractions (Figure 5.3). This separation process still remains the first step in most modern refineries. After the removal of entrained salt water, crude oil is fed continually into a heated distillation column. The lightest C_3–C_4 hydrocarbons make it to the top of the column and condense. Gasoline-range hydrocarbons are less volatile and can be drawn off the side of the column. Below the gasoline-range hydrocarbons is kerosene, and below that

are diesel fuels. The crude-oil fraction that is not volatile under atmospheric conditions (atmospheric bottoms) is removed from the first distillation column and sent to a second distillation tower that is under reduced pressure. The lower pressure allows for some of the heavier components to be vaporized and condensed (vacuum gas oil). The non-volatile remains are called the vacuum residuum. Other methods of crude oil fractionation include narrow-cut distillation (superfractionation), gas-phase selective absorption, solvent extraction, and wax crystallization.

For a typical crude oil, distillation alone does not produce sufficient volumes of transportation fuels with desirable qualities to be economic. The fractions obtained by distillation are termed "straight-run products" and typically are routed for further refining, blending, and mixing with additives before being sent to market (Figure 5.4). The octane rating of gasoline can be upgraded using metal catalytic reforming processes that break down long-chain hydrocarbons into smaller saturated isoparaffins. Other processes designed to increase the octane rating of gasoline are alkylation and isomerization. The akylation process is an acid-catalyzed polymerization of undesirable C_3 and C_4 olefins to produce high-octane isoparaffins. Isomerization reactions catalytically convert the low-octane normal alkanes into higher-octane isoparaffins. Kerosene and diesel fuel, in particular, must be upgraded by hydroprocessing. Hydotreating is a catalytic reaction in the presence of hydrogen to saturate aromatic rings and remove nearly all NSO compounds.

Note: Octane number ranks the engine knock (pre-ignition) characteristics of fuels on a 0–100 scale based on tests using pure hydrocarbons in gasoline engines. In these tests, *n*-heptane and isooctane (2,2,4-trimethylpentane) caused the most and least engine knock and were assigned octane ratings of 0 and 100, respectively. Fuels with low octane ratings combust unevenly, commonly resulting in engine knock. High-octane gasolines reduce engine knock because they contain abundant structural groups that retard precombustion. Aromatics, olefins, naphthenes, and highly branched alkanes tend not to oxidize in the engine cylinder until temperatures are high enough for complete combustion. However, *n*-alkanes begin to oxidize at low temperatures and combust early, causing knock. Alkylated lead additives were once

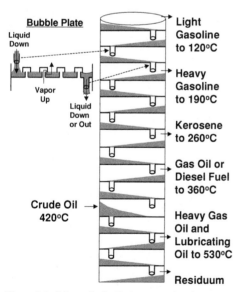

Figure 5.3. Schematic distillation tower showing an expanded view of a bubble plate (modified from Hunt, 1996).

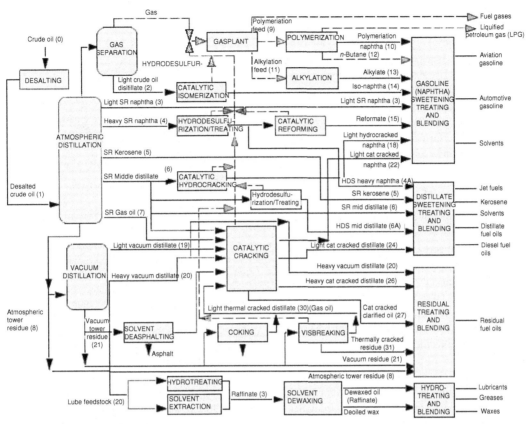

Figure 5.4. Refineries process crude oil into marketable products. Each refinery is uniquely designed to process specific types of crude oil and generate products within precise specifications. Basic processing units in refineries perform one of three functions: separate crude oil into fractions of more closely related properties, chemically convert the separated fractions into more desirable reaction products, and purify the products of unwanted elements and compounds (*OSHA Technical Manual TED 1-0.15A*, Section IV: Safety Hazards, Chapter 2: Petroleum Refining Processes).

added to fuels (see Figure 10.29) to increase octane ratings and prevent engine knock. In April 1994, the UN Commission on Sustainable Development called upon governments worldwide to eliminate lead from gasoline. To date, more than 50 countries have complied, representing more than 80% of the gasoline sold worldwide.

Heavy crude oil fractions can be cracked to yield lighter, more valuable products. Thermal cracking was the earliest method used by the refining industry. High-molecular-weight hydrocarbons left in the distillation bottoms are thermally broken into smaller hydrocarbons. This process is inefficient and gives low yields. Thermal cracking has been replaced by catalytic cracking. There are many refining techniques based on

catalysts, but fluid catalytic cracking is the most common. Fluid catalytic cracking can convert a large percentage of the feedstock to gasoline or light cycle oil, which is sometimes blended into diesel. The atmospheric and vacuum gas oil fraction may require hydrocracking, i.e. catalytic cracking in the presence of hydrogen. The process saturates aromatic rings, destroys most heteroatomic molecules, and adds hydrogen when carbon–carbon bonds are broken. Hydrocracking typically produces kerosene- and diesel-range fuels. The vacuum distillation residuum is used as asphalt or feed for the coker unit, where it is subjected to high-temperature ($\sim$500°C) pyrolysis. Volatile pyrolyzates are drawn off, treated, and added to the transportation fuels. Coke, the carbon residue, can be sold as a solid fuel.

Removal of trace impurities and fuel blending are the final refinery processes before the fuels are ready for market. The removal of hydrogen sulfide, mercaptans, and some olefins is achieved by oxidation (sweetening), extraction, selective absorption by clay or molecular sieves, or hydrofinishing (a catalytic reaction that is less severe than hydrotreating). Various product streams can be blended and mixed with additives to formulate products that meet performance specifications, environmental regulations, and economic needs.

BIOMARKERS IN REFINERY PRODUCTS

Geochemists are frequently asked to distinguish refined from natural products. For example, an oil slick might be caused by natural seepage or by a refined product that escaped from a pipeline. Drilling additives, such as diesel in oil-based mud or pipe dope used to lubricate the drill bit, might be mistaken for an oil show, resulting in needless drill stem tests or exploration wells. Biomarkers are potentially useful in order to distinguish natural petroleum from some synthetic products, but little work has been published.

Peters *et al.* (1992) compared gas chromatographic and biomarker distributions for straight-run versus processed product refinery streams generated from San Joaquin Valley heavy crude oil feedstock (Figure 5.5). Straight-run streams yield jet, diesel, gas oil feed for the hydrocracker (GOF), vacuum gas oil (VGO), and residuum. Processed streams yield hydrocracker, hydrofiner, fluid catalytic cracker (FCC), and coker products, as discussed below.

Compositions of straight-run products are controlled mainly by compound volatilities and sharp temperature gradients in the distillation tower, which complicate interpretation of source- and maturation-dependent biomarker parameters. Straight-run products, except residuum, show narrow ranges of peaks and large unresolved chromatographic humps with clear maxima, consistent with a narrow range of temperature and volatility for each distillation cut. Such narrow ranges of peaks are not typical of natural products. Maxima and ranges differ for jet fuel ($\sim nC_{12}$, nC_5-nC_{18}), diesel ($\sim nC_{17}$, nC_5-nC_{20}), GOF feed ($\sim nC_{22}$, $nC_{18}-nC_{26}$), and VGO ($\sim nC_{28}$, $nC_{22}-nC_{32}$). These ranges also differ somewhat, depending on the study (e.g. Table 5.1). Residuum has a broader range of peaks than the other straight-run products and a distinct

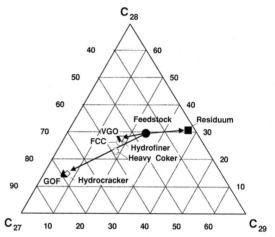

Figure 5.5. Ternary diagram showing the relative abundances of C_{27}, C_{28}, and C_{29} regular steranes in saturated fractions of San Joaquin Valley heavy crude oil feedstock, straight-run refinery distillation products (solid symbols), and processed materials (open symbols) (Peters *et al.*, 1992). Arrows emphasize enrichment of C_{27} steranes in the volatile gas oil feed (GOF) and vacuum gas oil (VGO) fractions and C_{29} steranes in the residuum relative to the feedstock.

FFC, fluid catalytic cracker products.

unresolved hump that increases above $\sim nC_{28}$. Jet fuel and diesel lack most biomarkers (<1–2 ppm), except for highly volatile C_{19} and C_{20} tricyclic terpanes. Only 31.5% of the C_{19} tricyclic terpane from the feedstock survives in the corresponding diesel. The distillation range of GOF overlaps only the most volatile biomarkers. For example, GOF is enriched in C_{27} versus C_{29} steranes, diasteranes, and monoaromatic steroids compared with the feedstock, obscuring their genetic relationship on $C_{27}-C_{28}-C_{29}$ ternary diagrams (Figure 5.5). The volatility range of VGO corresponds approximately to that of most biomarkers. VGO is slightly enriched in C_{27} versus C_{29} steranes compared with the feedstock on the $C_{27}-C_{28}-C_{29}$ ternary diagram, but their $C_{27}-C_{28}-C_{29}$ diasterane and monoaromatic steroid distributions are nearly identical. The volatility range of the residuum overlaps only the least volatile biomarkers. For example, residuum is enriched in C_{29} versus C_{27} steranes, diasteranes, and monoaromatic steroids compared with the feedstock on $C_{27}-C_{28}-C_{29}$ ternary diagrams.

Partitioning of biomarkers by volatility among distillation cuts affects those thermal maturity ratios where the numerator and denominator consist of more and less volatile (early- and late-eluting) biomarkers, respectively. For example, the residuum is enriched in the least volatile biomarkers and thus has diasterane/sterane, tricyclic terpane/17α-hopane, and triaromatic steroid TA(I)/(TAI + II) ratios, indicating lower maturity than the feedstock. The residuum also has lower sterane isomerization maturity ratios than the feedstock, possibly due to release epimers having the immature stereochemical configuration (e.g. 20R) from heavy precursors in the oil during high-temperature atmospheric distillation. This result is consistent with laboratory experiments. Many bound biomarkers released during laboratory maturation have stereochemistry indicating low thermal maturity, probably because of steric hindrance of isomerization in the bound state (Peters *et al.*, 1990; Peters and Moldowan, 1991).

Factors controlling biomarkers in processed materials are more complex than straight-run products and include volatility, thermal stability, generation from heavier precursors, and the effects of catalysts and hydrogen pressure. For example, hydrocracked products lack mono- and triaromatic steroids due to their destruction at high temperatures and hydrogen pressures, but little change occurs in the distributions of terpanes or steranes from the GOF feed. Source- and maturity-related biomarker parameters for the hydrocracked product and GOF are nearly identical. Hydrofining of VGO feed does not alter the distribution of most biomarker ratios significantly, except Ts/(Ts + Tm). For example, the C_{27}–C_{28}–C_{29} sterane and diasterane distributions for hydrofining product and VGO are nearly identical. Tm appears to be less stable to hydrofining than other terpanes. FCC contains reduced amounts of biomarkers with similar compositions to the feedstock, but it lacks monoaromatic steroids. Coking of residuum severely reduces and alters distributions of biomarkers.

Although Peters *et al.* (1992) tracked biomarkers through various refinery processes, only naturally occurring compounds were analyzed; new biomarkers generated from natural compounds in the feedstock by these processes were not evaluated. Carlson *et al.* (1995) simulated refinery transformations of pure cholestane, which occurs naturally in all feedstocks. They isolated and characterized the major mono-, di-, and triaromatic products of these transformations using HPLC, proton NMR, and GCMS. Based on product distributions, Carlson *et al.* (1995) proposed a series of reaction pathways from cholestane to these products that offers a method for probing molecular transformations in refinery process streams.

6 · Stable isotope ratios

This chapter describes stable isotopes and their use in characterizing petroleum, including gases, crude oils, sediment and source-rock extracts, and kerogen, with emphasis on stable carbon isotope ratios. The discussion includes isotopic standards and notation, principles of isotopic fractionation, and the use of various isotopic tools, such as stable carbon isotope type curves, for correlation or quantification of petroleum mixtures. The chapter concludes with new developments in compound-specific isotope analysis, including its application to better understand the origin of carboxylic acids and the process of thermochemical sulfate reduction in petroleum reservoirs.

Stable isotope compositions of carbon, sulfur, nitrogen, and hydrogen (Kaplan, 1975; Fuex, 1977; Hoefs, 1997; Schoell, 1984; Macko and Quick, 1986) are used with biomarkers to determine genetic relationships among oils and bitumens. Isotopes are atoms whose nuclei contain the same number of protons but different numbers of neutrons. The discussion focuses on carbon because it is the dominant element in petroleum and because more is known about it due to the relative ease of its analysis by bulk and compound-specific methods.

Figure 6.1 shows the subatomic composition of the two stable isotopes of carbon. Carbon-12 (^{12}C) and carbon-13 (^{13}C) are called the light and heavy stable isotopes and account for ~98.89 and 1.11 wt.% of all carbon, respectively. The unstable ^{14}C atom (not shown) contains six protons and eight neutrons in the nucleus. ^{14}C is radioactive and accounts for only a trace of naturally occurring carbon. Because it has a half-life of ~5730 years, ^{14}C cannot be measured reliably in samples more than ~50 000 years old. Petroleum that has not been contaminated by recent carbon lacks ^{14}C.

Note: Microorganisms facilitate the oxidation of sedimentary organic matter to inorganic carbon when sedimentary rocks are exposed to erosion.

Petsch *et al.* (2001) cultured prokaryotes from weathered Upper Devonian New Albany Shale. Based on isotopic analysis of phospholipid fatty acids isolated from these prokaryotes, they showed that 74–94% of the lipid carbon originated from assimilation of ^{14}C-free kerogen in the shale. Phospholipids are degraded rapidly after cell death and thus do not occur in ancient sedimentary rocks.

STANDARDS AND NOTATION

Stable isotope data are presented as delta (δ) values representing the deviation in parts per thousand (‰, per mil, or ppt) from an accepted standard:

$$\delta(‰) = [(R_{sample} - R_{standard})/R_{standard}] \times 1000 \tag{6.1}$$

R represents the isotope abundance ratio, such as ^{13}C/^{12}C, ^{18}O/^{16}O, ^{34}S/^{32}S, ^{15}N/^{14}N, or D/H (^{2}H/^{1}H). The δ value for carbon, for example, is a convenient means to describe small variations in the relative abundance of the ^{13}C in organic matter. A negative δ value implies that the sample is depleted in the heavy isotope relative to the standard. A positive value means that the sample is isotopically enriched in the heavy isotope relative to the standard. For carbon, we avoid the terms "light" and "heavy" and use instead "^{13}C-depleted" and "^{13}C-enriched", respectively, to describe relative isotope composition.

Table 6.1 shows reference standards and guidelines for reporting stable hydrogen, carbon, nitrogen, oxygen, and sulfur isotope ratios (Coplen, 1996; Hoefs, 1997). Isotopic reference standards can be obtained from two organizations: National Institute of Standards and Technology, Standard Reference Materials Program, Room 204, Building 202, Gaithersburg, Maryland 20899-0001, USA; email: srminfo@enh.nist.gov; and International Atomic Energy Agency, Section of Isotope Hydrology, Wagramerstasse 5, PO Box 100, A-1400 Vienna, Austria; email iaea@iaea1.iaea.or.at

Table 6.1. *Isotopic reference standards and guidelines for reporting hydrogen ($^2H/^1H$), carbon ($^{13}C/^{12}C$), nitrogen ($^{15}N/^{14}N$), oxygen ($^{18}O/^{16}O$), and sulfur ($^{34}S/^{32}S$) isotopic compositions (Coplen, 1996; Hoefs, 1997)*

Element	Standard abbreviation	Guidelines for reporting
Hydrogen in water	VSMOW	δ^2H on scale where VSMOW = 0‰, SLAP water = −428‰
Other hydrogen		Same as above, but also report δ^2H of NBS-22 oil or other appropriate reference standard
Carbon in organic matter (old method)	PDB	$\delta^{13}C$ on scale where PDB = 0‰ (PDB primary standard is now depleted)
Carbon in carbonate	VPDB	$\delta^{13}C$ on scale where NBS-19 calcite (secondary standard) = +1.95‰
Other carbon		Same as above, but also report $\delta^{13}C$ of NBS-22 oil or other appropriate reference standard
Nitrogen	−	$\delta^{15}N$ on scale where atmospheric nitrogen = 0‰
Oxygen in carbonate	VPDB or VSMOW	$\delta^{18}O$ on scale where SLAP water is −55.5‰ relative to VSMOW, NBS-19 = −2.2‰, and stating the oxygen isotopic fractionation factor used to calculate the $\delta^{18}O$ or the carbonate and NBS-19 if they are not identical $\delta^{18}O$ on scale where SLAP water is −55.5‰ relative to VSMOW, VSMOW = 0‰, stating all isotopic fractionation factors upon which $\delta^{18}O$ measurement depends
Sulfur	CD	$\delta^{34}S$ on scale where CD = 0‰

CD, Canyon Diablo Troilite; NBS, National Bureau of Standards; PDB, *Belemnitella americana* from the Cretaceous Peedee Formation (this standard is depleted); SLAP, Standard Light Antarctic Precipitation; VPDB, Vienna PDB; VSMOW, Vienna Standard Mean Ocean Water.

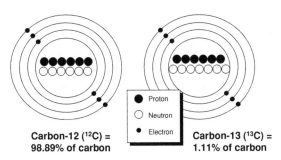

Figure 6.1. Comparison of proton, neutron, and electron configurations in the stable isotopes of carbon. The ^{12}C atom contains six protons and six neutrons in the nucleus (indicated by the inner circle) and accounts for ∼98.89% of all carbon. The ^{13}C atom contains an additional neutron compared with ^{12}C. Reprinted with permission by ChevronTexaco Exploration and Production Technology Company, a division of Chevron USA Inc.

STABLE CARBON ISOTOPE MEASUREMENTS

Sealed-tube combustion was the most popular method to convert organic matter to carbon dioxide for isotope analysis until ∼1990 because it yields reproducible results but is generally faster and less expensive than dynamic combustion using a vacuum line (e.g. Stuermer *et al.*, 1978; Sofer, 1980). However, prolonged storage of combustion tubes for weeks or more can result in $\delta^{13}C$ depleted by up to 1–3‰ (Engel and Maynard, 1989), probably due to reaction of the CO_2 with the copper oxide (CuO) wire used for the oxidation, yielding copper carbonate (e.g. Cu_2CO_3). Isolating the CO_2 gas from the copper oxide wire or simply reheating the tubes immediately before analysis avoids this problem. Carbonate contribution to CO_2 can be removed before loading of samples into the combustion tubes by reaction with H_3PO_4 in the range 25–100°C (Wachter and Hayes, 1985; Hoefs, 1997).

Today, most analyses for stable carbon isotope composition are completed using online combustion systems with a coupled elemental analyzer and isotope ratio mass spectrometer (combustion/IRMS) (Hoefs, 1997). This approach allows the use of much smaller sample sizes than the previous methods described above. In addition, a gas chromatograph can be coupled to the combustion system, thus allowing carbon isotope ratios to be determined on individual organic compounds (GC/combustion/IRMS, also called compound-specific isotope analysis (CSIA); see Figure 6.12).

STABLE CARBON ISOTOPE FRACTIONATION

Stable carbon isotope ratios are used to describe small variations of ^{13}C abundance in organic and inorganic materials (Figure 6.2). The expanded portion of Figure 6.2 shows the range of isotopic values for oils from different sources. Bacterial decomposition of plants produces methane (marsh gas) containing considerably less ^{13}C than the decaying plants. Laws et al. (1995) and

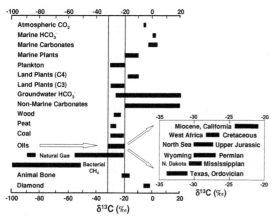

Figure 6.2. Variations in stable carbon isotope ratios (VPDB standard) for different organic and inorganic compounds (modified from Mook, 2001). C3 and C4 plants are discussed in the text. The expanded scale shows ranges of isotopic values for various crude oils from some petroleum source rocks, arranged from oldest at the bottom to youngest at the top. Stable isotope ratios can be used with biomarkers to show relationships between crude oils and their source rocks. Reprinted with permission by ChevronTexaco Exploration and Production Technology Company, a division of Chevron USA Inc.

Riebesell et al. (2000) discuss many of the factors that control stable carbon isotope fractionation during primary production, including growth rate and carbon dioxide concentration.

The difference in mass between isotopes of the same element results in measurable isotopic fractionation during physical and chemical processes. These fractionations are more pronounced for light elements because their isotopes (e.g. hydrogen, ^1H, versus deuterium, ^2H) show proportionally larger differences in mass than isotopes of heavier elements (e.g. ^{12}C versus ^{13}C). The main mechanisms for fractionations include:

- equilibrium (or thermodynamic) exchange reactions;
- kinetic effects associated with irreversible physical or chemical reactions.

Equilibrium exchange reactions depend on temperature, where increasing temperature decreases the magnitude of isotopic fractionation (e.g. O'Neil, 1986). For example, evaporation/condensation of water is a reversible chemical equilibrium, where

$$H_2{}^{16}O_{vapor} + H_2{}^{18}O_{liquid} = H_2{}^{18}O_{vapor} + H_2{}^{16}O_{liquid}$$

Kinetic fractionation occurs when evaporating water is removed without allowing condensation. In nature, so-called non-equilibrium isotope fractionations are common. For example, evaporation of water from oceans and lakes is neither a one-way kinetic process (some water vapor condenses) nor a reversible equilibrium (net evaporation occurs).

Carbon isotope exchange reactions occur among the various carbon species in the atmosphere, hydrosphere, and carbonate minerals: $CO_2(gaseous)$, $CO_2(aqueous)$, $HCO_3{}^-$, $CO_3{}^{2-}$, and $CaCO_3$ (aragonite or calcite). Zhang et al. (1995) provide experimental calibrations of these fractionations. The order of relative enrichment in ^{13}C among these species at surface temperatures on Earth is as follows:

$$CO_2(aq) < CO_2(g) < CO_3{}^{2-} < HCO_3{}^- < calcite$$
$$< aragonite$$

The carbon isotope fractionation between dissolved bicarbonate ($HCO_3{}^-$) in natural waters and the carbonate minerals is very small, but significant equilibrium fractionation leads to temperature-dependent enrichment of ^{13}C in dissolved bicarbonate compared with atmospheric carbon dioxide across the air–water interface (Figure 6.2):

$$H^{13}CO_3^- + {}^{12}CO_2 = H^{12}CO_3^- + {}^{13}CO_2$$

Kinetic fractionations result from irreversible or one-way physicochemical processes, such as the evaporation of water with immediate removal of the vapor, adsorption and diffusion of gases, photosynthesis, and bacterial decay of plant remains. These fractionation effects are mainly controlled by the binding energies of the compounds (Mook, 2001). During physical processes, molecules containing the lighter isotope (e.g. ^{12}C) have higher velocities and lower binding energies, resulting in faster reactions than molecules with the heavier isotope (e.g. ^{13}C).

The carbon isotope abundances in organic matter are controlled by an important kinetic isotope effect associated with photosynthesis, which leads to selective incorporation of ^{12}C into organic matter. (However, Galimov (1973) suggests a different explanation for isotopic compositions of organic matter based on intramolecular equilibrium isotope effects.) The most important isotope effects during carbon fixation appear to occur during assimilation of CO_2 by autotrophic organisms, which can be described as a two-step process:

$$CO_2(external) \leftrightarrow CO_2(internal) \rightarrow organic\ product$$

The first step is diffusion of CO_2 into the photosynthetic reaction centers, which has a relatively small isotope fractionation near $-4.4‰$ in air and even less for plants in water (Schildowski and Aharon, 1992). In the second step, the CO_2 is fixed enzymatically as a carboxyl group in an organic acid. The magnitude of this fractionation depends on various factors, including the specific photosynthetic pathway used for carbon fixation, other species-specific differences in isotope fractionation by marine photosynthetic organisms (Popp et al., 1998), the intensity of irradiance, growth rate or availability of growth-rate-limiting nutrient, and physical isolation of primary producers, for example in sea ice (e.g. Riebesell et al., 2000). Fractionations resulting from kinetic isotope effects generally exceed those from equilibrium processes.

Most green plants, including eukaryotic algae and many autotrophic bacteria, fix carbon using the C3 or Calvin pathway. Two additional photosynthetic pathways include the C4 or Hatch–Slack pathway and the crassulacean acid metabolism (CAM) pathway. C4 plants are mainly tropical grasses, desert plants, and salt marsh (marine) plants that originated during the Paleocene Epoch or later. Plants using the C4 photosynthetic pathway are enriched in ^{13}C compared with C3 plants (Figure 6.2). For example, the carbon isotopic compositions of n-alkanes measured by compound-specific isotope analysis (CSIA) in Miocene soils and sediments from the Indian subcontinent record a shift from low $\delta^{13}C$ ($\sim-30‰$) to higher values ($\sim-22‰$) prior to 6 Ma, consistent with a rapid transition from dominantly C3 vegetation to an ecosystem dominated by C4 plants typical of semi-arid grasslands (Freeman and Colarusso, 2001). Meyers (1994; 1997) used plots of atomic C/N versus $\delta^{13}C$ of bulk organic matter to distinguish marine or lacustrine algae and C3 or C4 plant input in thermally immature sedimentary rocks. CAM plants are mainly succulents that use either C3 or C4 pathways and consequently show $\delta^{13}C$ spanning the range of both C3 and C4 plants.

CONVERTING δ VALUES USING DIFFERENT STANDARDS

Unfortunately, not all δ values are given relative to a single standard. Because of the peculiar form of Equation (6.1), conversion of δ values between different standards is not straightforward. The following equation can be used to convert δ values from one standard to another (after Hoefs, 1997):

$$\delta_{X(a)} = \{[(\delta_{b(a)}/1000) + 1][(\delta_{X(b)}/1000) + 1] - 1\} \times 1000 \tag{6.2}$$

where $\delta_{X(a)}$ and $\delta_{X(b)}$ are the isotopic compositions of the sample with respect to oil standards a and b, respectively, and $\delta_{b(a)}$ is the isotope composition of standard b relative to standard a.

For example, a belemnite from the Cretaceous Peedee Formation in South Carolina (PDB) is the primary standard for carbon, although the original sample no longer exists. Isotopic measurements are reported relative to PDB by calibrations relative to secondary standards. Most workers calibrate the PDB scale by adopting a $\delta^{13}C$ of $+1.95‰$ for NBS-19 carbonate relative to PDB. When the PDB scale is calibrated in this manner, the measurements can be quoted relative to Vienna PDB (VPDB).

Some older petroleum industry publications report stable carbon isotope measurements relative to NBS-22, currently used by some as a secondary standard. The PDB and NBS-22 standards are both defined as zero for their respective δ value scales. On the PDB scale, NBS-22 oil measures $\sim -29.81\%o$ (Schoell, 1983). The δ value for a sample relative to NBS-22 cannot generally be converted to the PDB scale simply by adding 29.81‰. The substituted version of equation (6.2) becomes:

$$\delta_{X(PDB)} = \{[(0.9702)(\delta_{X(NBS)}/1000) + 1] - 1\} \times 1000 \tag{6.3}$$

From equation (6.3), NBS-22 values of 10, 0, -10, -20, -30, and $-40\%o$ correspond to calculated PDB values of -24.96, -29.81, -39.51, -49.21, -58.92, and $-68.62\%o$, respectively. These PDB values differ from those obtained simply by subtracting 29.81‰ from the NBS value by -0.30, 0, 0.30, 0.60, 0.89, and 1.19‰, respectively. Thus, proper conversion of NBS to PDB values becomes more important for ^{12}C-rich materials, such as methane, where $\delta^{13}C_{PDB}$ of $-50\%o$ or less are common.

APPLICATIONS OF STABLE CARBON ISOTOPE RATIOS

Hydrocarbon gases

The isotopic composition of light hydrocarbon gases is controlled kinetically because $^{12}C-^{12}C$ bond strengths are slightly less stable than $^{12}C-^{13}C$ bonds (Tang et al., 2000). Hydrocarbon gases generated by cracking in nature and in the laboratory show progressive depletion in ^{13}C from n-butane to propane to ethane to methane, consistent with mathematical models (e.g. James, 1983; Clayton, 1991). Likewise, thermal maturation results in loss of ^{12}C-enriched methane and concentration of ^{13}C in the residual kerogen, as modeled in the laboratory (Peters et al., 1981). On the other hand, polymerization of methane to higher homologs by spark discharge or other abiogenic mechanisms results in the reverse distribution, where δ^{13}methane $> \delta^{13}$ethane $> \delta^{13}$propane $> \delta^{13}n$-butane (Des Marais et al., 1981). Nearly all analyzed gases from significant petroleum accumulations worldwide, regardless of age, depth, or other factors, show isotopic distributions indicative of thermal cracking rather than abiogenic polymerization (Figure 6.3). Published quantitative results for large numbers of gas samples are shown in Chapter 9.

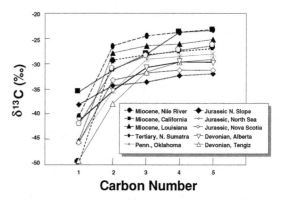

Figure 6.3. Examples of stable carbon isotope compositions for C_1–C_5 hydrocarbon gases in worldwide petroleum accumulations of various ages (data from Chung et al. (1988) and Jenden et al. (1993a)). The relative amount of ^{13}C decreases from n-pentane through n-butane, propane, and ethane to methane, consistent with kinetic isotope fraction imposed by cracking from heavier components.

Most hydrothermal and geothermal methane originates from biogenic rather than abiogenic sources. For example, Des Marais et al. (1981) showed that geothermal methane becomes enriched in ^{13}C with increasing temperature at Yellowstone National Park due to natural pyrolysis of sedimentary organic matter. Des Marais et al. (1988) simulated the production of hydrocarbon gases from the Cerro Prieto geothermal field in Baja California by pyrolyzing lignite from rock samples obtained from this field. In both the pyrolysis and the field gases, higher temperatures favored higher relative concentrations of methane, ethane, and benzene and more positive $\delta^{13}C$ for individual hydrocarbons. The best correlations between laboratory and well data were obtained when gases from lower- and higher-temperature pyrolysis experiments were mixed, suggesting that the geothermal gases are also mixtures of gases that originated by cracking of organic matter at various depths.

Isotopic differences between gaseous hydrocarbons of the same origin can be used to estimate the thermal maturity of the gas because higher temperatures reduce kinetically controlled isotopic fractionation (e.g. James, 1983). Thus, smaller isotopic differences between methane and ethane of the same origin can be interpreted to indicate higher maturity.

Figure 6.4 can be used to determine gas maturity, correlate gas in the reservoir to its source, correlate different reservoir gases, and recognize gas mixtures. For

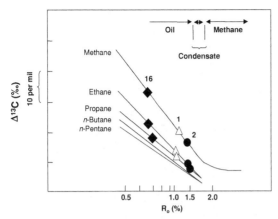

Figure 6.4. Plot relates calculated separations of stable carbon isotope values (10‰ scale) for $C_1–C_5$ hydrocarbons to source-rock maturity measured by vitrinite reflectance (modified from James, 1983). Gas samples 1, 2, and 16 were obtained from the Central Appalachian Basin, Pennsylvania and Ohio (data from Table 2 in Laughrey and Baldassare, 1998). Inset at upper right indicates approximate range of maturity for oil, condensate, or methane. These three gases are included in Figure 6.15.

example, although gas 1 lacks butane and pentane, the $C_1–C_3$ hydrocarbons indicate an origin from source rock in the late oil window ($\sim$1.1% R_o equivalent). Sample 2 has a composition that indicates maturity in the principal zone of gas-condensate formation near 1.5% R_o equivalent. If gas 2 were found in a low-maturity reservoir rock (e.g. measured $R_o = 0.5\%$), then one could conclude that it migrated updip from the mature source and that intervening downdip reservoirs might contain condensate. The isotopic composition of sample 16 indicates that it is an oil-associated gas that originated from early-mature source rock ($\sim$0.7% R_o equivalent). Geologic information suggests that the hydrocarbons in samples 1 and 2 originated from the Ordovician Utica Shale, while sample 16 is from the Upper Devonian Huron Shale (Laughrey and Baldassare, 1998).

Correlation of whole oils, bitumens, and kerogens

Several general rules apply to correlation based on stable carbon isotope ratios of whole oils, bitumens, and kerogens.

- A positive correlation is supported, but not proven, when oils of similar maturity differ by no more than

$\sim$1‰. Based on our experience, maturity differences among related oils account for isotopic variations of up to $\sim$2–3‰ (e.g. Monterey oils showing a large maturity range show δ^{13}C differences of no more than 1.5‰; see Figure 19.18).
- Oils that differ by more than $\sim$2–3‰ are usually from different sources, although there are exceptions. The 3.6‰ range from least to most mature oils in the Big Horn Basin, Wyoming, appears to be due solely to differences in thermal maturity (Chung *et al.*, 1981). Large variations can occur for oils from a widespread source rock characterized by major changes in organic facies (e.g. Hwang *et al.*, 1989). For oils from source rocks deposited in certain restricted depositional environments, large isotopic variations occur, apparently without major differences in the type of contributing organic matter. An extreme example is the 8.1‰ variation for Middle Ordovician organic matter (Hatch *et al.*, 1987), which is reflected in the isotopic composition of related oils. This variation was caused by limited water circulation and variable organic productivity and their effects on carbon cycling by living organisms rather than changes in organic matter type.
- Bitumens are generally $\sim$0.5–1.5‰ depleted in ^{13}C compared with kerogen in the source rock. Likewise, crude oil is depleted in ^{13}C by $\sim$0–1.5‰ compared with the source-rock bitumen. These general relationships assume approximately equivalent levels of maturity for the oil, bitumen, and kerogen.

Note: Crude oils from Eocene Kreyenhagen and Miocene Monterey and Antelope Shale source rocks from California can be distinguished reliably by their respective ^{13}C-poor versus ^{13}C-rich carbon isotope compositions (Jones, 1987; Peters *et al.*, 1994). These differences are reflected in the isotope compositions of kerogens above and below the basal Neogene boundary.

Although many Italian crude oils have biomarker characteristics of carbonate source rocks, stable carbon isotope ratios identify two genetic families derived from Triassic-Toarcian and Cretaceous-Oligo-Miocene intervals (Katz *et al.*, 2000a). Triassic-Toarcian oils are isotopically depleted in ^{13}C (δ^{13}C < −26‰) and include the Rospo Mare and Malossa oils, and the degraded Maiella oil. Cretaceous-Oligo-Miocene oils are isotopically enriched in ^{13}C (δ^{13}C > −26‰) and include the Monte Alpi and Bagnolo oils and the Tramutola seep.

Isotopic analyses indicate an unexpected origin for numerous flattened tar balls collected from shorelines along the northern and western parts of Prince William Sound, Alaska (Kvenvolden *et al.*, 1995). Although the sound still contains traces of the 1989 Exxon Valdez oil spill, the tar balls are clearly not related to this North Slope oil. The 61 tar-ball samples show stable carbon isotope ratios $(-23.7 \pm 0.2\%o)$ similar to crude oils from the Monterey Formation source rocks of California but distinctly different from 28 samples of residues from the Exxon Valdez spill $(-29.4 \pm 0.1\%o)$. Unlike the Exxon Valdez oil and its weathered residues, the tar balls contain 28,30-bisnorhopane, 25,28,30-trisnorhopane, and oleanane (see Figure 10.23) and have low C_{29} sterane 20S/(20S + 20R) and C_{31} hopane 22S/(22S + 22R) ratios (0.34 $\pm$ 0.02 and 0.54 $\pm$ 0.01, respectively), typical of Monterey crude oils. The authors speculate that the ubiquitous tar balls represent Monterey asphalt and fuel oil imported to Alaska that spilled into the sound during the 1964 Alaska earthquake.

Plots of pristane/phytane versus whole-oil stable carbon isotope ratio are commonly used to support tentative relationships among oils and source-rock extracts (Figure 6.5). The plot also provides some information on the source-rock depositional environment. The isotopic and gas chromatographic analyses required for this plot are rapid and inexpensive. However, interpretations based on the plot are best supported by additional geochemical analyses, such as biomarkers. Biodegradation can alter pristane/phytane or completely remove pristane and phytane from petroleum.

The bonds formed by the heavy isotope of an element require more energy for cleavage than bonds of the light isotope. This is the basis for the kinetic isotope effect. Thus, early thermal products including methane and other light gases become enriched in the light isotope relative to the reactants, such as crude oil and kerogen. Oils can become isotopically enriched in ^{13}C with maturation (e.g. Sofer, 1984) or isotopically depleted (e.g. Hughes *et al.*, 1985), depending on which fraction of the evolving oil is sampled. A series of oils might consist of: (1) the volatile ^{12}C-rich products derived from progressive maturation of an original oil, or

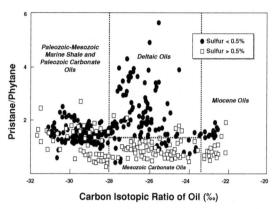

Figure 6.5. Pristane/phytane versus stable carbon isotope ratio of whole oil can be used to support genetic relationships among oils and infer depositional environments (Chung *et al.*, 1992). Samples with similar plot locations are not necessarily related genetically. Caution must be applied when inferring depositional environments from this plot because many exceptions occur. Reprinted by permission of the AAPG, whose permission is required for further use.

(2) the residual ^{13}C-rich oil remaining after removal of more volatile products.

Secondary processes, such as migration and biodegradation, can cause variations in the compound class distributions of related oils (e.g. percentage saturated versus percentage aromatic hydrocarbons) and can, therefore, result in different isotopic compositions for related oils. Care must be used when comparing the stable carbon isotope compositions of whole oils showing large differences in API gravity, or when comparing oils with bitumens. Gasoline-range hydrocarbons are enriched in ^{13}C compared with the whole oil (Silverman, 1971) and are prominent in high API gravity oils and condensates. If high API gravity petroleum is not topped by distillation to remove these gasoline-range components, then comparisons with bitumen may be misleading. Most gasoline-range hydrocarbons in bitumens are lost during sample preparation because bitumens are extracted from ground rock using organic solvents. Removal of the solvent by rotoevaporation results in loss of gasoline-range and other low-molecular-weight compounds below $\sim nC_{15}$.

Kerogens may contain large amounts of inert or gas-prone macerals that do not contribute to oil generation but overwhelm the isotopic signature of the oil-prone macerals. Likewise, migrated oil or contaminants may obscure the isotopic composition of indigenous bitumen. When either of these problems occurs, oil-source rock studies based on stable carbon isotope ratios of bitumens and kerogens may be of limited value. Bailey *et al.* (1990) avoided this problem by measuring isotopic compositions of the liquid pyrolyzates obtained from pre-extracted rocks. By using rapid anhydrous pyrolysis, they were able to generate pyrolyzates from many closely spaced rock samples (pyrolyzate isotopic profiling) for isotopic and biomarker comparison with oils in the North Sea. An added benefit of this approach is that geochemically heterogeneous intervals within thick source rock units are less likely to be overlooked due to inadequate sampling. The same approach was used by Burwood *et al.* (1990) to distinguish organic facies within subsalt source rocks from Angola.

Stable carbon isotope type curves

The shapes and trends of stable carbon isotope type curves (Galimov, 1973; Stahl, 1978) are used to identify relationships between crude oils, bitumens, and kerogens. Petroleum has increased ^{13}C for fractions of increasing polarity and boiling point (e.g. Chung *et al.*, 1981). Similar isotope type curves support, but do not prove, genetic correlation. Figure 6.6 shows examples of type curves for oils from the Middle Ob region, West

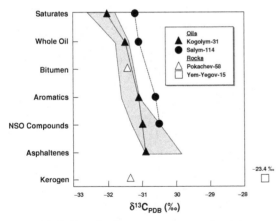

Figure 6.6. Stable carbon isotope type curves can be used to show relationships between oils or between oils and source-rock organic matter in the Middle Ob region of West Siberia (Peters *et al.*, 1993). The shaded area includes Kogolym-31 oil and four other representative oils generated from nearby Upper Jurassic Bazhenov Formation source rock. Reprinted with permission by ChevronTexaco Exploration and Production Technology Company, a division of Chevron USA Inc.

Siberian Basin, Russia. The Kogolym-31 oil has a type curve similar to other oils from the same area (stippled trend), supporting biomarker results, indicating that they are related (Peters *et al.*, 1993). Biomarkers for the Salym-114 oil indicate a relationship with Kogolym-31 and the other West Siberian oils. Because of its higher thermal maturity (late oil window) compared with the other oils (early to peak oil generation), the Salym-114 oil has ^{13}C-rich compound class fractions. The trend indicated by all of the type curves suggests that the kerogen in the source rock for the oils has a stable carbon isotope value in the range of ~−29 to −31‰. Kerogen isolated from prospective Upper Jurassic Bazhenov Formation source rock in the Pokachev-58 well has an isotope value (−31.4‰) that is consistent with an oil–source rock relationship. Likewise, the bitumen from this rock (−31.5‰) falls within the stippled trend established by the related oils, again supporting a relationship between the oils and rock. The bitumen also correlates with the oils based on biomarkers. The carbon isotope composition of kerogen isolated from siltstone clasts in Middle Jurassic Tyumen Formation sandstone from the Yem-Yegov-15 well (−23.4‰) confirms the biomarker results, indicating that it is unrelated to the oils.

The reason for increased ^{13}C for fractions of increasing polarity and boiling point is unclear, but it could be explained as follows. If the fractions are generated during disproportionation reactions by thermal cracking of kerogen, then kinetic fractionation is expected to enrich the kerogen in ^{13}C (heavy), while the fractions become relatively depleted (light) in the order of decreasing polarity: asphaltenes > NSO compounds > aromatics > saturates. Isotope type curves in the literature show various components on the *y*-axis, but they are always in order of decreasing polarity with increasing distance from the origin of the figure. Whole oil or bitumen generally has an isotopic composition between the saturated and aromatic hydrocarbons because these account for the bulk of most petroleum samples.

Extrapolation of stable carbon isotope type curves for crude oils and their fractions can be used to predict the approximate isotopic composition of kerogen in the source rock. Proposed source rocks that contain bitumen or kerogen and that have isotopic compositions inconsistent with this extrapolation can be eliminated. When bitumen has an isotopic composition inconsistent with the associated kerogen, it is unlikely to be indigenous and may represent migrated oil or contaminant.

Isotope type curves should be used cautiously in predicting source rock because some kerogens contain abundant inert or gas-prone macerals with isotopic compositions that differ from the associated oil-prone macerals (e.g. Bailey *et al.*, 1990). Thus, the isotopic relationship between the oil-prone macerals and expelled oil might be obscured. Organic petrography can be used to help identify kerogens with maceral compositions that may present this type of problem.

The trends of many stable carbon isotope type-curves are irregular (Chung *et al.*, 1981) because secondary processes, including thermal maturation, migration, and deasphalting, can influence the isotopic composition of each fraction. For example, biodegradation can increase the $\delta^{13}C$ of the residual saturate fraction. Irregular stable carbon isotope type curves characterize three asphaltene-poor oils from the Timan-Pechora Basin, Russia (Figure 6.7). In this example, the typecurves do not include bitumen or kerogen on the y-axis. Two of the Timan-Pechora oils have ^{13}C-depleted asphaltene fractions, resulting in irregular type curves that are similar in shape. The anomalously ^{13}C-depleted asphaltene fractions in oils A and C could be caused by coprecipitation of ^{13}C-depleted saturates with the asphaltene fraction during sample preparation. HPLC separation of oil B yielded insufficient asphaltenes for isotopic analysis. The three oils each contain more than 54 wt.% saturates and less than 0.4 wt.% asphaltenes.

Biomarker results indicate that oil C originated from a different organic facies of the same source rock that generated the other oils. Based on extrapolation of the type curves without including the anomalous asphaltenes, the predicted isotopic composition of kerogen in the source rock for these oils is $\sim-28\permil$. We recommend that anomalous asphaltene isotopic values, particularly for waxy or saturate-rich crude oils, be disregarded when extrapolating type curves to predict a carbon isotopic composition of the source-rock kerogen.

Note: Based on our analyses, coprecipitation of ^{13}C-depleted saturated hydrocarbons with asphaltenes also occurs during preparation of waxy oils from Sudan. Both the Sudan and the Timan-Pechora oils are rich in saturated hydrocarbons (more than 45% and 54%, respectively) and poor in asphaltenes (less than 3% and 0.4%, respectively).

Quantitative estimate of oil co-sources

Because stable carbon isotope ratios are bulk properties of petroleum, they can be used in quantitative estimates of source inputs to mixed oils once the co-sources have been established. Peters *et al.* (1989) show how widely differing carbon isotopic compositions of whole oil from the Beatrice Field and bitumens from two co-sources for the oil (Middle Jurassic and Devonian) were used to calculate the approximate contributions from each to the oil (however, see Bailey *et al.* 1990). Because the mixture involved only two oils, they used the following simple equation:

$$(\delta^{13}C_{oil})\,(100) = (\delta^{13}C_{source\ A})(X) \\ + (\delta^{13}C_{source\ B})(100 - X) \quad (6.4)$$

The contribution of source A measured by X can be estimated only if the $\delta^{13}C$ of the mixture and the contributing sources are known. Such estimates require that the samples of mixed oil and contributing oils or bitumens be non-biodegraded and have approximately the same level of thermal maturity.

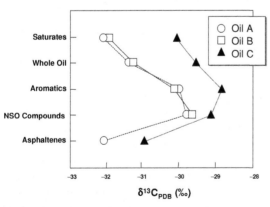

Figure 6.7. Irregular stable carbon isotope type curves for three related oils from the Timan-Pechora Basin, Russia. In addition to trends, the shapes of type curves are considered in evaluating oil–oil and oil–source rock correlations. Reprinted with permission by ChevronTexaco Exploration and Production Technology Company, a division of Chevron USA Inc.

Age and depositional environment

Isotopic analyses of crude oils indicate a general trend of ^{13}C enrichment with decreasing age of their source rocks (e.g. Figure 6.2). Andrusevich *et al.* (1998) compared

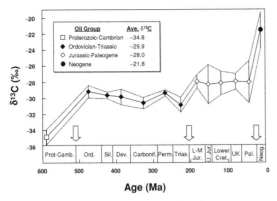

Figure 6.8. Average stable carbon isotopic ratios for C_{15+} saturate fractions of crude oils versus age. Proterozoic-Cambrian (17 oils), Ordovician (5), Silurian (25), Devonian (17), Carboniferous (11), Permian (7), Triassic (12), Lower-Middle Jurassic (12), Upper Jurassic (52), Lower Cretaceous (51), Upper Cretaceous (UK) (200), Paleogene (71), and Neogene (34). Vertical bars on sample points are standard deviations, which generally increase with decreasing age. Arrows indicate Cambrian-Ordovician, Triassic-Jurassic, and Paleogene-Neogene boundaries, where episodic shifts toward increased ^{13}C may occur. Inset shows four groups of oils defined by similar age and average isotope compositions of the saturate fractions. Reprinted from Andrusevich et al. (1998). Copyright 1998, with permission from Elsevier.

stable carbon isotope compositions of the saturated and aromatic hydrocarbon fractions from 514 crude oils representing 13 age divisions from Proterozoic to Neogene (Figure 6.8). The ^{13}C content of both saturated and aromatic hydrocarbon fractions increases episodically with decreasing age and appears to be independent of source rock type. Figure 6.8 shows only the saturated hydrocarbon data. Three episodic shifts toward increased ^{13}C occur at the Cambrian-Ordovician, Triassic-Jurassic, and Paleogene-Neogene boundaries. The range of isotopic values also tends to increase with decreasing age, although this could be due to the limited number of samples and their limited geographic distribution. For example, Permian marine organic matter from a large database, including the USA, Germany, the North Sea, Peru, and Russia, has a considerably broader range of $\delta^{13}C$ (-26 to $-31‰$) (Simoneit et al., 1993) than the range for saturate fractions from seven Permian crude oils indicated in Figure 6.8. Nonetheless, others obtained similar results to those in the figure for the isotopic compositions of whole oils (e.g. Stahl, 1977; Botneva et al., 1984; Chung et al., 1992).

The reasons for this general trend are unclear. Andrusevich et al. (1998) suggest that episodic enrichment in ^{13}C of oils and isotopic variations in ^{13}C of carbonates are linked. The three largest ^{13}C enrichments in the saturate fractions of oils at the Cambrian-Ordovician, Triassic-Jurassic, and Paleogene-Neogene boundaries are preceded by large depletions in the ^{13}C content of carbonates at the base of the Cambrian, Triassic, and Paleogene periods, respectively. As discussed earlier, photosynthesis results in preferential fixation of ^{12}C. Burial of ^{12}C-rich organic carbon occurs when strong phytoplankton productivity is followed by mass mortality. The remaining biomass tends to become enriched in ^{13}C. At the same time, removal of ^{12}C-rich carbon dioxide and bicarbonate from the water due to increased productivity may increase the carbonate ion concentration, until ^{13}C-rich calcium carbonate precipitates. Thus, times of major ^{13}C enrichment of carbonate commonly correlate with deposition of organic-rich source rocks. Each sharp increase in ^{13}C of oils occurs after a maximum depletion of ^{13}C in carbonate and the onset of new phytoplankton skeletal types. These authors believe that in the Precambrian and Paleozoic time intervals, kerogen in sediments (-31 to $-35‰$) was enriched in more resistant, ^{13}C-poor lipids because ^{13}C-rich carbohydrates and proteins were readily degraded during diagenesis of organic-walled phytoplankton. However, during the Mesozoic and Cenozoic eras, more phytoplankton had calcareous or siliceous tests, which protected carbohydrates from degradation during diagenesis, resulting in ^{13}C-enriched kerogen (-25 to $-21‰$).

Paleolatitude during source-rock deposition affects crude oil isotopic compositions. Andrusevich et al. (2000) focused on Upper Jurassic crude oils to better constrain the controls on isotopic variation in a specific time interval. Crude oils from high-latitude Upper Jurassic source rocks (e.g. Neuquén, North Sea, West Siberia basins; Figure 6.9) are depleted in ^{13}C compared with those from equatorial areas (e.g. Sureste and Gotnia basins). Upper Jurassic source rocks containing type II kerogens account for ~25% of the worldwide petroleum reserves (Klemme and Ulmishek, 1991). Most Upper Jurassic petroleum occurs in the Arabian-Iranian (46%), West Siberian (22%), Gulf of Mexico (13%), and North Sea (11%) basins.

Crude oils from the Grand Banks and the Interior Salt Basin are more enriched in ^{13}C than might be predicted by the paleolatitude of their source rocks

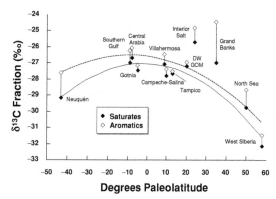

Figure 6.9. Average stable carbon isotope compositions of C_{15+} saturated and aromatic hydrocarbon fractions of 482 Upper Jurassic oils vary systematically with paleolatitude (Andrusevich *et al.*, 2000). Solid and dashed lines are second-order polynomial best-fit curves for the saturated and aromatic fractions, respectively. Data for oils from the Interior Salt Basin in the Gulf of Mexico and the Grand Banks in Nova Scotia were not included in the curve fitting.

DW GOM, deepwater Gulf of Mexico.

(Figure 6.9). Oils from these basins were not included in the curve fit shown in Figure 6.9 because their Upper Jurassic source rocks originated in highly restricted basins characterized by enhanced assimilation of ^{13}C. For example, the Grand Banks oils have high gammacerane, suggesting a stratified water column that was hypersaline during deposition of the Egret Member of the Rankin Formation source rock (Fowler and McAlpine, 1995). These oils also have abundant 4-methylsteranes and high C_{27} steranes, indicating high-productivity dinoflagellate blooms that might cause enhanced assimilation of ^{13}C. Like the Grand Banks oils, oils from the Smackover source rock in the Interior Salt Basin contain high 4-methylsteranes and other evidence of restricted source-rock deposition, including high sulfur and low pristane/phytane.

Various factors might explain the effect of paleolatitude on crude oil composition, including surface water temperature and differences in biomass at different latitudes. Surface water temperature, pH, and dissolved carbon dioxide are primary factors influencing the stable carbon isotopic composition of marine biomass. Dissolution of atmospheric carbon dioxide in water and the carbon isotope fractionation during photosynthesis increase with decreasing temperature. As carbon dioxide concentration decreases due to increased water tem-

perature and/or phytoplankton growth, proportionally more ^{13}C is incorporated from bicarbonate during photosynthesis due to thermodynamic and kinetic isotope effects.

Figure 6.10 suggests that species diversity supplements the thermodynamic and kinetic isotope effects of temperature as a control on crude oil isotopic composition. The equatorial Arabian and Gulf of Mexico basins contain oils from carbonate source rocks with different organic matter compared with those from high-latitude siliciclastic source rocks of the same age. The equatorial carbonate sources contain relatively greater bacterial input than the high-latitude siliciclastic source rocks based on lower steranes/hopanes. In addition, different C_{27}/C_{29} sterane ratios between the high- and low-latitude oils suggest differences in precursor algal input.

The isotopic-age trend shown in Figure 6.8 has many exceptions that make it difficult to use as a tool to accurately predict source-rock ages for oils. The range of isotopic values in specific time intervals generally does not allow reliable age dating. For example, Precambrian (Hayes *et al.*, 1983) and Ordovician (Hatch *et al.*, 1987) organic matter is commonly depleted in ^{13}C compared with younger organic matter, but these differences are *not* diagnostic. Schoell and Wellmer (1981) show that some Precambrian organic matter is unusually depleted in ^{13}C, but most Precambrian organic matter has values in the same range as that from the Phanerozoic. Marine organic matter at several worldwide locations is enriched in ^{13}C by $\sim$3–4‰ at the Cretaceous-Tertiary

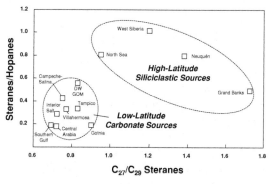

Figure 6.10. Average C_{27}/C_{29} sterane and steranes/hopanes ratios are lower for 482 Upper Jurassic crude oils from low-latitude carbonate source rocks than for those from high-latitude siliciclastic source rocks (Andrusevich *et al.*, 2000). DW GOM, deepwater Gulf of Mexico.

boundary (Arthur *et al.*, 1988). Similarly, hypersaline environments commonly contain ^{13}C-rich organic matter (Schildowski *et al.*, 1984). However, again these differences are not diagnostic.

For the above reasons, isotopic data cannot be used to indicate the age or environment of deposition of organic matter. Stable carbon isotope ratios must be used with caution in indicating source input because of fractionations during and after formation of the organic matter (Fuex, 1977; Deines, 1980a). However, combined use of stable carbon isotopes and biomarkers allows more confident assessment of organic matter input, as exemplified below.

Note: Oils from Pre-Ordovician carbonate source rocks commonly show low carbon isotope values, pristane/phytane, and Ts/Tm (<1), and a predominance of C_{28}- or C_{29}-steranes (McKirdy *et al.*, 1983). Oil from north Oman has a very negative δ^{13}C of $-33.14\%o$, Pr/Ph of 0.85, and Ts/Tm of 0.89, but it is not particularly enriched in C_{28}- or C_{29}-steranes. Compared with other oils in the area, this oil has a higher gammacerane index and a lower C_{30}-sterane index. These data suggest that the source-rock depositional environment was isolated from marine conditions, possibly in a starved euxinic basin or lagoonal carbonate evaporite. Gammacerane is a marker for a stratified water column during source-rock deposition, commonly associated with elevated salinity, as in alkaline lakes and lagoonal carbonate evaporites. The combined biomarker, isotopic, and geologic information for the Oman oil indicates that it was generated from nearby Infra-Cambrian Huqf Formation source rocks, consistent with published interpretations of similar oils from North Oman (Terken and Frewin, 2000).

Marine versus terrigenous organic input

Despite early attempts (Silverman and Epstein, 1958), crude oils that originate from marine and terrigenous organic matter cannot be distinguished clearly using stable carbon isotope ratios of the whole oil. However, many marine and terrigenous crude oils can be distinguished using the stable carbon isotopic ratios of the C_{15+} saturated and aromatic hydrocarbons (Figure 6.11). This distinction refers to provenance of the organic matter, not depositional environment. For example, many

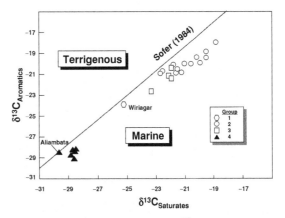

Figure 6.11. Stable carbon isotope ratios ($\%o$ relative to PDB) for saturated versus aromatic hydrocarbons differ between oil groups from eastern Indonesia (Peters *et al.*, 1999), where the inferred source rock ages for each group are as follows: 1, Lower-Middle Jurassic; 2, Tertiary (higher oleanane); 3, Tertiary (lower oleanane); 4, Triassic-Jurassic. The dotted best-separation line is based on statistical analysis of 339 terrigenous (above) and marine (below) oils (Sofer, 1984). The plot locations suggest that the source rocks for these four oil groups contained mainly marine organic matter. See also Figure 13.61. Reprinted by permission of the AAPG, whose permission is required for further use.

source rocks deposited in deltaic marine environments are dominated by allochthonous, terrigenous organic matter and yield crude oils that plot in the terrigenous field in Figure 6.11.

The isotopic relationship for terrigenous oils ($\delta^{13}C_{aro} = 1.12\delta^{13}C_{sat} + 5.45$) differs from that for marine oils ($\delta^{13}C_{aro} = 1.10\delta^{13}C_{sat} + 3.75$) and can be formalized by a statistical parameter called the canonical variable (Sofer, 1984):

$$CV = -2.53\delta^{13}C_{sat} + 2.22\delta^{13}C_{aro} - 11.65$$

where CV is the canonical variable.

Based on stepwise discriminant analysis of 339 nonbiodegraded oils, CV above 0.47 indicates mainly waxy terrigenous oils, while values under 0.47 indicate mainly nonwaxy marine oils (Sofer, 1984). The CV represents the perpendicular distance of a sample from the best separating line between terrigenous and marine oils on plots of $\delta^{13}C_{saturates}$ versus $\delta^{13}C_{aromatics}$. More negative CV indicates more marine input, and vice versa.

To better quantify the subjective chromatographic classification of oils as waxy or non-waxy, Sofer (1984) fitted the *n*-alkane chromatographic envelope in the range $nC_{19}-nC_{33}$ to a quadratic equation:

$$\%n\text{-alkane} = \text{A(carbon number)}^2 + \text{B(carbon number)} + \text{C}$$

Waxy and non-waxy distributions show negative and positive quadratic coefficients (QC), respectively.

Caution must be exercised when interpreting waxiness and CV. For example, Peters *et al.* (1986) used two-group discriminant analysis to show that only 66% of the oils in a suite of marine and non-marine crude oils were separated correctly using stable carbon isotope ratios of the saturated and aromatic hydrocarbon fractions. However, they successfully distinguished marine and non-marine oils by statistical analysis of combined isotope and biomarker data. Biodegraded oils are not readily interpreted because microbes preferentially attack waxes and alter CV. Many highly mature terrigenous oils (high CV) are non-waxy because cracking destroys waxes. For example, highly mature Red Desert oils from Wyoming originated from terrigenous coaly source rock. Despite non-waxy *n*-alkane distributions typical of marine oils, the Red Desert oils have CV values that still indicate a terrigenous origin (Sofer, 1984). Some problematic oils show marine CV values but are waxy (e.g. Udang from Borneo, Latrobe from Australia, and Minas from Sumatra) (Sofer, 1984). Where conflicts occur, stable isotopes are generally more reliable than wax content in order to indicate marine versus terrigenous input, but the final interpretation is best supported by additional source-related data.

COMPOUND-SPECIFIC ISOTOPE ANALYSIS

CSIA (Figure 6.12) or isotope ratio monitoring/gas chromatography/mass spectrometry (IRM/GCMS) allows stable carbon isotopic analysis of biomarkers and other compounds as they elute from a gas or liquid chromatograph (Matthews and Hayes, 1978; Hayes *et al.*, 1987; Hayes *et al.*, 1990; Freedman *et al.*, 1998, Jasper, 1999). Random errors and variance associated with CSIA are described by Jasper (2001).

Gas and liquid chromatography can affect the measured stable carbon isotope compositions of individual organic compounds due to partitioning that occurs during their repeated sorption and desorption between the column stationary and mobile phases (Liberti *et al.*, 1965). For example, capillary gas chromatographic

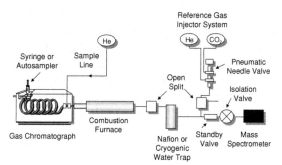

Figure 6.12. Diagram of an isotope ratio monitoring/gas chromatograph/mass spectrometer (IRM/GCMS) for compound-specific isotope analysis (CSIA) (modified from Jasper, 2001). The three major components include: (1) a gas chromatograph, where organic compounds are separated into peaks; (2) a high-temperature ($\sim$900°C) combustion furnace, where individual peaks are combusted to CO_2, H_2O, and other products; and (3) an isotope ratio mass spectrometer, where the $^{13}CO_2$ and $^{12}CO_2$ in each peak are measured to give isotope ratios for the corresponding compounds. The sample line can be used to introduce purified samples that do not require chromatographic separation.

columns yield peaks for individual compounds where the ^{13}C-rich components elute at the front of the peak, while their ^{13}C-poor counterparts with the same chemical structure elute toward the tail (Hayes *et al.*, 1990). Accurate carbon isotope analysis using CSIA requires analysis of the entire peak without co-elution of other compounds. High-performance liquid chromatography (HPLC) is commonly used for preparative separation of organic compounds before further analysis. However, HPLC can also fractionate the isotope composition of individual compounds. For example, fractions collected across a chlorophyll a peak obtained by C_{18} reverse-phase HPLC showed significant stable carbon isotope variations, indicating that quantitative recovery is necessary for reliable CSIA results (Bidigare *et al.*, 1991). Preparative HPLC on ultra-stable-Y zeolite shape-selective molecular sieve separates steroids and hopanoids without isotopic fractionation (Kenig *et al.*, 2000). This shape-selective approach does not involve the same type of sorption–desorption phase partitioning described above.

Hayes *et al.* (1987) show that the stable carbon isotopic compositions of individual biomarkers are preserved during diagenesis. Biomarkers in the Eocene Messel Shale show a broad range in carbon isotopic compositions, from -20.9 to $-73.4\%o$. These biologically controlled isotopic compositions can be used to identify specific sources for some compounds and to help reconstruct the carbon cycling within the lake and sediments that formed the Messel Shale. For example, structures and isotopic compositions for several porphyrins allowed their differing origins from algal or bacterial sources to be determined.

Bridging the gap between isotope and biomarker geochemistry is important in order to understand paleoenvironments. By assuming that compounds with common biological origins show similar isotopic compositions, CSIA becomes a powerful tool for reconstructing biogeochemical pathways for organic carbon. Using this hypothesis, Freeman et al. (1990) reconstructed various primary and secondary pathways for carbon flow for ancient Lake Messel. The term "primary" refers to all products of photosynthesis, while "secondary" indicates derivation by subsequent processes, both biological and thermal. As another example, isotopic results (Hayes et al., 1990) show contrasting sources for acyclic isoprenoids and n-alkyl hydrocarbons in samples from the marine Cretaceous Greenhorn Formation. Pristane, phytane, and porphyrins comprise a group of compounds showing co-variant isotopic compositions, while n-alkanes and total organic carbon form another group. The results are consistent with derivation of the pristane, phytane, and porphyrins from primary material (i.e. chlorophyll), while the n-alkanes originate from secondary inputs.

CSIA for reconstruction of paleoenvironment

Summons and Powell (1986; 1987) identified a series of aryl isoprenoids (1-alkyl-2,3,6-trimethylbenzenes) (see Figure 13.118) in crude oils from Silurian and Devonian reef reservoirs in the Michigan and Western Canada basins, respectively. The structures and isotope compositions of these compounds suggest that they are diagenetic products of aromatic carotenoids in anoxygenic green sulfur bacteria (Chlorobiaceae), which utilize an unusual biosynthetic pathway to fix inorganic carbon during photosynthesis (reductive tricarboxylic acid cycle). The occurrence of aryl isoprenoids in petroleum was thus inferred to indicate photic zone anoxia in the depositional environment of the source rock. Later work showed that aryl isoprenoids could originate from at least two sources: isorenieratene or β-isorenieratene in Chlorobiaceae and β-isorenieratene from the ubiquitous

carotenoid pigment β-carotene (Koopmans et al., 1996). They found that $\delta^{13}C$ for β-carotane and β-isorenieratane in a North Sea crude oil are similar ($\sim -26\permil$), consistent with a common origin, while isorenieratane is $\sim 15\%$ enriched in ^{13}C, consistent with an origin from Chlorobiaceae. When ^{13}C- rich aryl isoprenoids and/or isorenieratane occur in petroleum, Chlorobiaceae contributed to organic matter in the depositional environment. Thus, a new class of biomarkers was identified and its significance understood by combined use of biomarker and isotopic analysis

Riebesell et al. (2000) recommend caution when assigning different sources to lipids that might have $\delta^{13}C$ that differ by only a few parts per thousand (‰). They measured stable carbon isotope compositions of bulk organic matter, alkenones, sterols, fatty acids, and phytol in the coccolithophorid Emiliania huxleyi over a wide range of CO_2 concentrations and found large differences in $\delta^{13}C$ for individual lipids in the same alga. Relative to the bulk organic matter, individual fatty acids were depleted in ^{13}C by 2.3–4.1‰, phytol was depleted by 1.9‰, and the major sterol, 24-methylcholesta-5,22E-dien-3β-ol, was depleted by 8.5‰.

Schoell et al. (1994) used CSIA to measure carbon isotope variations among biomarkers in gilsonites from the Uinta Basin, Utah. Individual biomarkers show isotopic compositions indicative of their paleoecological origin within the Eocene Uinta and Greater Green River paleolake system. For example, stable carbon isotope ratios for the C_{28} and C_{29} steranes (−25 to −32‰), pristane and phytane (−33 to −34‰), and β-carotane (−33.2‰) suggest an origin from photic zone organisms. Many of these algal biomarkers in the gilsonites are isotopically uniform throughout the study area (~ 1500 km^2) and match isotopic ratios of the same compounds in the proposed Mahogany Ledge oil shale source rocks. The isotope measurements distinguish two groups of hopanes in the gilsonites. C_{29}, C_{31}, and C_{32} hopanes and moretanes (−40.9 to −44.3‰) probably originated from midwater bacteria, while C_{30} hopanes and moretanes (−51.9 and −60.5‰) show strong ^{13}C depletion, suggesting at least partial derivation from methylotrophs. Gammacerane was abundant, but coeluting C_{31} methylhopane precluded accurate isotope measurements. The isotopic data suggest that the source rocks for the gilsonites were deposited in a stratified paleolake system under conditions similar to those that formed the Mahogany Ledge oil shales. Schoell et al.

(1994) conclude that isotopic compositions of the algal biomarkers, in particular the steranes, are the best tool for genetic correlation of expelled products from the Mahogany Ledge, while hopanes allow recognition of subfacies. Similar differences between sterane and hopane isotope patterns occur in the Monterey Formation, where steranes are isotopically uniform and hopanes vary laterally and vertically within the formation (Schoell et al., 1992).

Yu et al. (2000a) examined the $\delta^{13}C$ of individual biomarkers from the Maoming oil shale. The $\delta^{13}C$ ($\sim-24\%o$) for pristane and phytane were consistent with their origin from the phytol side chain of chlorophyll a. The $\delta^{13}C$ values of the 4-methylsteranes, however, were isotopically very negative ($\sim-29\%o$) and more similar to the associated hopanes than to the steranes. Results from the Maoming Shale contrasted with analysis of biomarkers from the Jianghan oil shale, where the $\delta^{13}C$ of the 4-methylsteranes were similar to pristane, phytane, and the normal steranes. Yu et al. (2002a) inferred that $\delta^{13}C$ affinity of the 4-methylsteranes to the hopanes indicated that these compounds originated from bacteria.

We believe that the Yu et al. (2000a) study illustrates the limitations of relying on $\delta^{13}C$ to assign specific biological origins for individual biomarkers. The origin of 4-methylsteranes appears to be firmly established by a correlation between abundant dinoflagellate cysts in Maoming strata that are enriched in these compounds (Brassell et al., 1985). A bacterial source for the 4-methylsteranes is unlikely. Only a few bacteria are known to synthesize sterols (Methylococcus capsulatus, Nannocystis exedens, Polyangium sp., Methylobacterium organophilum, and Methylosphaera hansonii) (Bird et al., 1971; Bouvier et al., 1976; Patt and Hanson, 1978; Schouten et al., 2000a). The sterol content in most of these bacteria is low and varies under different temperature and growth conditions. Thus, the microbial diversity and potential biomass for bacterial contributors is limited. Furthermore, the $\delta^{13}C$ for sterols in methanotrophs reflects the isotopic signature of the metabolized methane, which, if biogenic, yields sterols very rich in ^{12}C (Schouten et al., 2000b). The isotopically negative values for the Maoming 4-methylsteranes probably reflect an unusual fractionation by dinoflagellates under specific environmental conditions.

Yu et al. (2000a; 2000b) also developed a novel derivatization method to measure the $\delta^{13}C$ of individual geoporphyrins. In order to make the porphyrins volatile for gas chromatographic separation, the compounds were demetallated with methanesulfonic acid. Silicon was then inserted by reacting the demetallated porphyrin with hexachlorodisilane. Finally, the SiOH groups resulting from the silicon insertion were silylated using N-methyl-N-tert-butyldimethylsilyltrifluoroacetamide (MTBSTFA). The process does not alter the isotopic composition of the porphyrin carbon. The $\delta^{13}C$ of C_{31}DPEP and two C_{32}DPEP porphyrins from a Jianghan Basin oil shale extract were nearly identical (−23.1, −23.5 and −23.9‰, respectively), suggesting a common origin from chlorophyll. In contrast, while the C_{31}DPEP and C_{32}ETIO porphyrins from the Maoming oil shale were similar (−26.7 and −26.9‰, respectively), the C_{32}DPEP was over 4‰ more negative (−22.1‰), suggesting a different origin.

The recent development of CSIA for hydrogen isotopes (Scrimgeour et al., 1999; Sessions et al., 1999) opened new avenues of research to study source-rock paleoenvironments and diagenesis. Andersen et al. (2001) presented the first hydrogen isotope data for biomarkers, n-alkanes, and isoprenoids in the sulfur-bound fractions of Messinian source rocks in the eastern Mediterranean Sea. The $\delta^{13}C$ and δD co-vary and show ranges of 14 and 160‰, respectively. The δD of the original source water used by photosynthetic organisms in the paleoenvironment was estimated to range from −31 to +66‰. The heavier isotope compositions are consistent with episodes of extreme evaporation during the Messinian salinity crisis. The offset between δD for n-alkanes and isoprenoids for Messinian samples is similar to that in modern biological cultures, suggesting that diagenesis did not significantly affect the original deuterium isotope distributions. These results are consistent with other data, indicating that carbon-bound hydrogen in lipids is stable and does not readily exchange with the surrounding water during diagenesis (Schimmelmann et al., 1999).

CSIA for correlation

CSIA is a valuable method in oil–oil and oil–source rock correlation (Figure 6.13; see also Figure 10.19). The method can also be used to identify commingled oils from different sources. Guthrie et al. (1996) used CSIA to characterize a suite of organic-rich, immature shales

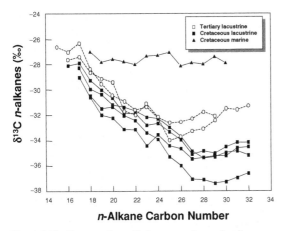

Figure 6.13. Compound-specific isotope analyses of n-alkanes differentiate extracts from Tertiary lacustrine, Tertiary normal marine, and Cretaceous normal marine organic-rich source rocks in Brazil (from Guthrie *et al.*, 1996). Similar analyses can be used to differentiate crude oils derived from these and other source rocks.

from various basins in Brazil. Stable carbon isotope compositions of the n-alkanes (e.g. n-heptadecane), pristane, phytane, 28,30-bisnorhopane, hopane, and gammacerane distinguish both the rock samples and crude oils generated from them. These compounds are consistently more depleted in ^{13}C among the lacustrine shales compared with marine shales. The correlation of high hopane/sterane ratios with depleted ^{13}C compositions for these compounds supports more abundant input of methanogenic bacteria to the lacustrine shales. CSIA of light hydrocarbons is especially useful in order to correlate light oils and condensates (see Figure 7.39).

Distributed source-rock sampling

Curiale and Sperry (1998) observed that molecular distributions in extractable organic matter from discrete rock samples in the Cretaceous Morro do Barro Formation, Camamu-Almada Basin, offshore Brazil, correlate with oils from the basin, but δ^{13}C of individual n-alkanes do not (Figure 6.14). Explanations for different CSIA δ^{13}C values over short depth intervals range from secular variations in the isotopic composition of atmospheric carbon dioxide (Santos Neto *et al.*, 1998) to photic zone carbon dioxide limitation (Hollander and McKenzie, 1991). Curiale and Sperry (1998) demonstrated the importance of distributed sampling, rather than random, discrete sampling, for correlation of these lacustrine

extracts with the crude oils. Random sampling of Morro do Barro rocks having a hydrogen index (HI) greater than 400 mg HC/g TOC indicates that individual n-alkanes in the source unit can vary in δ^{13}C by up to 5–6‰. They contend that the Camamu-Almada oils formed by mixing of hydrocarbons from numerous discrete source intervals within the Morro do Barro Formation. In the distributed sampling approach, they computed δ^{13}C for n-alkanes in a composite oil by weighting the δ^{13}C input of each individual n-alkane from nC_{20}–nC_{28} according to the HI of the source rock sample.

CSIA of carboxylic acids

Because carboxylic acids are weakly acidic, they may influence the development of secondary porosity in petroleum reservoirs and carrier beds or the transport of ore-forming metals. Concentrations of carboxylic acids from fluids in petroliferous basins and sediment-covered hydrothermal systems vary over many orders of magnitude, but the relative abundances of individual acids decrease systematically with increasing carbon chain length (Shock, 1988; Seewald, 2001).

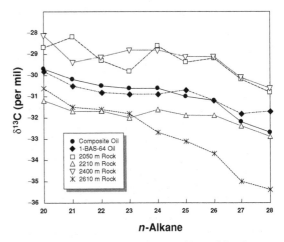

Figure 6.14. Stable carbon isotope compositions of C_{20}–C_{28} n-alkanes in crude oil and rock extracts from the Camamu-Almada Basin, Brazil (Curiale and Sperry, 1998). The isotopic distribution of n-alkanes in the 1-BAS-64 well oil does not match any of the four rock extracts from discrete depths in the Cretaceous Morro do Barro Formation source rock in the 1-BAS-71 well. However, the δ^{13}C distribution of the oil closely matches that of a calculated composite extract, defined by weighting the extract results for each n-alkane according to the hydrogen index of each rock sample.

Carboxylic acids in the subsurface can originate by different processes. For example, acetate and other volatile carboxylic acids are important intermediates in the anaerobic degradation of organic matter, but their concentrations in sediments are usually low ($<15\ \mu M$) due to rapid consumption by microbes (Wellsbury et al., 1997). The close association of carboxylic acids with organic matter that has been heated during burial or hydrothermal activity suggests that they originate partly from this source, as supported by laboratory experiments (e.g. Knauss et al., 1997). Up to 2.5 wt.% of the organic matter in source rocks may be converted to short-chain C_2–C_5 carboxylic acids during thermal maturation, and possibly half of this amount may be expelled with petroleum as dissolved organic acids (Lewan and Fisher, 1994). Acetate concentration increased by more than three orders of magnitude during heating of surface coastal marine sediments to simulate increased burial (Wellsbury et al., 1997). They observed that porewater acetate concentration at two sites in the Atlantic Ocean increased at depths below about 150 m and was also associated with increased bacterial activity. Others propose that reactions involving mineral oxidants are another source of acids (e.g. Eglinton et al., 1987; Borgund and Barth, 1994). In addition, the relative abundances of some carboxylic acids may be regulated by redox-dependent metastable thermodynamic equilibrium involving water, carbon dioxide, and carbonate minerals (Helgeson et al., 1993; Shock, 1994).

Laboratory heating experiments support the idea that n-alkanes generated during thermal maturation of source rocks undergo stepwise aqueous oxidation to produce n-alkenes, alcohols, ketones, and, finally, carboxylic acids (Seewald, 2001). Likely oxidation agents include ferric iron in aluminosilicates, oxides and hydroxides, pyrite, sulfate-bearing minerals, and water. A model based on these laboratory results accurately predicts the relative distributions of carboxylic acids in oilfield brines. Because the oxygen required to produce carboxylic acids originates from minerals and/or water, more carboxylic acids may be generated and available for creation of secondary porosity than expected from models where the kerogen is the only source of oxygen for these acids.

Franks et al. (2001) and Dias et al. (2002a) showed that short-chain C_2–C_5 organic acids in oilfield waters from the San Joaquin Basin in California generally become isotopically depleted in ^{13}C with increasing carbon number but are consistently more ^{13}C-rich than co-produced Miocene Antelope (Monterey equivalent) or Eocene Kreyenhagen crude oils. They used plots of δ^{13}C for each acid versus $1/n$, where n is the carbon number, to support their hypothesis that organic acids with higher carbon numbers have more ^{13}C-poor aliphatic carbon atoms to dilute the isotopic effect of ^{13}C-rich carboxyl carbon. Based on this relationship, carboxyl carbon atoms were inferred to be enriched in ^{13}C by 10–38‰ compared with aliphatic carbon atoms in these acids. Carbon isotope ratios of the aliphatic carbon atoms in the organic acids were inferred to fall in the ranges −22.5 to −25.6 and −27.8 to −29.8‰, reflecting the two major sources of oils in the basin (Antelope oils ~−22 to −25‰ and Kreyenhagen oils ~−28 to −30‰). Thus, waters associated with these two sources can be readily identified based on the carbon isotope composition of the aliphatic carbon atoms in the organic acids. The authors concluded that the carboxyl carbons are enriched in ^{13}C due either to inherited isotopic compositions from biological precursors or to exchange with carbon in dissolved bicarbonate.

SULFUR AND HYDROGEN ISOTOPES

Plots of the stable isotope ratios for carbon versus sulfur (e.g. Orr, 1974) or carbon versus hydrogen (e.g. Schoell, 1984; Peters et al., 1986) can be used to distinguish petroleum groups. For example, samples 1 and 2 in Figure 6.15 are hydrocarbon gases from Upper Cambrian reservoir rocks in wells from eastern Ohio and that originated from Ordovician Utica Shale source rock (Laughrey and Baldassare, 1998). The plots of δ^{13}C versus δD of methane and δ^{13}C methane versus gas wetness for samples 1 and 2 indicate that these samples are thermogenic, condensate-associated gases generated by oil and kerogen cracking. Carbon isotope separations among the methane, ethane, and propane in samples 1 and 2 indicate TAI maturities near 3− and 3, respectively, corresponding to the late oil window (Figure 6.4), which is consistent with the estimated maturity based on the vitrinite reflectance (R_o) scale in Figure 6.15.

Gas sample 3 from the Grugan Field in western Pennsylvania has methane with δ^{13}C of −27.2‰, which is the most ^{13}C-rich methane ever reported in the Appalachian Basin (Laughrey and Baldassare, 1998). This unusual gas also has a reversed isotopic trend, where δ^{13}C ethane and propane are −35.8 and −37.4‰.

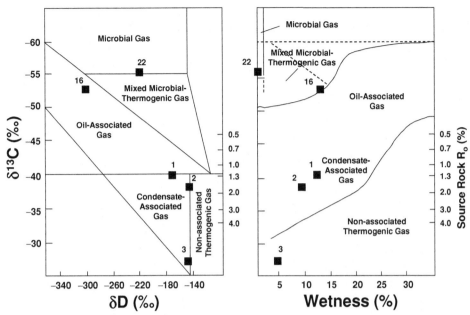

Figure 6.15. Plots of the stable carbon isotope ratio of methane versus hydrogen (or deuterium) isotope ratio of methane (left) and gas wetness ($(C_2–C_5)/(C_1–C_5)$; right) for selected gases from eastern Ohio and western Pennsylvania (modified from Schoell, 1983; data and sample numbers from Laughrey and Baldassare, 1998). Vitrinite reflectance scale (R_o) (Jenden *et al.*, 1993b) is approximate and does not apply to all gas samples (e.g. sample 16). Gases 1, 2, and 16 are also shown in Figure 6.4. Reprinted by permission of the AAPG, whose permission is required for further use.

Although sample 3 might be interpreted to be abiogenic gas, the reversed isotopic trend for methane, ethane, and propane probably results from heterogeneity in the source organic matter (probably mostly Ordovician Utica Shale), mixing of gases from different sources, oxidation of thermogenic gas, or partial diffusive leakage of the gas reservoir (Jenden *et al.*, 1983a; Laughrey and Baldassare, 1998).

Sample 16 in Figure 6.15 consists of gas from Upper Devonian Huron Shale Member reservoir rock in a well from western Pennsylvania and that originated from Upper Devonian shale source rock (Laughrey and Baldassare, 1998). Figure 6.4 indicates that sample 16 has a TAI maturity near 2+ and is an early-mature, oil-associated gas.

Sample 22 is coalbed gas produced from high-volatile bituminous coal with vitrinite reflectance (R_o) of 0.9% (Laughrey and Baldassare, 1998). Methane in coalbed gas is typically depleted in ^{13}C at low ranks (to $-60‰$) and more enriched in ^{13}C (to $-40‰$) at higher ranks (Rice *et al.*, 1993). However, the composition of coalbed gas is also influenced by mixing of thermogenic

methane with late-stage microbial methane generated by anaerobic microbes introduced into the coal by ground-water (Rice *et al.*, 1993; Scott *et al.*, 1994). Sample 22 gas has an isotopic composition indicating lower thermal maturity than that measured in the surrounding coal (0.9% R_o), probably due to admixture of this ^{13}C-poor microbial methane. Figure 10.31 is another example of how stable isotopes of carbon and hydrogen can be used to distinguish methane from different origins.

Burwood *et al.* (1990) distinguished two families of Angolan oils using carbon versus hydrogen isotopic plots. Accretionary Wedge oils from Tertiary reservoirs are depleted in deuterium and contain abundant 18α-oleanane (e.g. Riva *et al.*, 1988). These oils are believed to originate from Upper Cretaceous-Paleogene marine Iabe-Landana source rock. Carbonate Platform oils produced from the Pinda Formation and some Pre-Salt reservoirs are rich in deuterium, suggesting deposition under aquatic conditions with high evaporative loss. These oils show many features of the lacustrine oils from Brazil, including prominent tricyclic terpanes, 28,30-bisnorhopane, and gammacerane. Many of these Angolan oils also have 25,30-bisnorhopane. The Carbonate Platform oils and their Brazilian analogs originate from the Pre-Salt Lower Cretaceous lacustrine Bucomazi Formation, deposited before rifting of South America from Africa. Three oil-prone organic facies were recognized within the Bucomazi Formation based primarily on stable carbon isotopes and the type of organic matter. Interestingly, the 25,30-bisnorhopane observed in many of the Carbonate Platform oils occurs exclusively in one organic facies that is enriched in ^{13}C and contains type II organic matter.

The Potiguar Basin in Brazil has two source-rock intervals consisting of shales and marls deposited during the rift-to-marine transition (Aptian Alagamar Formation) and lacustrine shales of the rift section (Neocomian Pendência Formation). Carbon and hydrogen isotopic mixing models were used together to determine the relative contributions of these marine-evaporitic and lacustrine source rocks to mixed oils from the basin (Santos Neto and Hayes, 1999). The hydrogen isotopic compositions of the end-member oils were related mainly to the δD of water in the source rock depositional environment.

Thermochemical sulfate reduction

Because sulfur can be incorporated into oils during secondary processes (Orr, 1974), sulfur isotope compositions may not be related directly to the original organic matter input. Nonetheless, sulfur isotope ratios show potential for oil–oil correlations (Gaffney et al., 1980). Premuzic et al. (1986) used sulfur isotopes to: (1) support source groups for oils from Prudhoe Bay, Alaska, previously established by biomarkers (Seifert et al., 1980; 1983) and (2) distinguish the Prudhoe Bay oils from those in similar formations in the Point Barrow area (Magoon and Claypool, 1981; 1983; 1984).

Thermochemical sulfate reduction (TSR) is the abiological reduction of sulfate by hydrocarbons in reservoirs close to anhydrite at high temperatures (e.g. Worden et al., 1995). TSR occurs in the Smackover Trend in the Gulf of Mexico (Claypool and Mancini, 1989), the Western Canada Basin (Krouse et al., 1989), and the Big Horn Basin in Wyoming (Orr, 1974). Other examples of TSR occur in the Permian Zechstein Formation in northwestern Germany (Orr, 1977), the Aquitaine Basin in France (Connan and Lacrampe-

Couloume, 1993), and Abu Dhabi (Worden et al., 1995). The following is a simplified TSR reaction scheme (Orr, 1974).

$$SO_4{}^{2-} + 3H_2S \rightarrow 4S_0 + 2H_2O + 2OH^- \qquad (6.5)$$

$$4S_0 + 1.33(CH_2) + 2.66H_2O \rightarrow 4H_2S + 1.33CO_2 \qquad (6.6)$$

$$SO_4{}^{2-} + 1.33(CH_2) + 0.66\,H_2O \rightarrow$$
$$H_2S + 1.33CO_2 + 2OH^- \quad \text{(net reaction)}$$

In the above equations, (CH_2) represents reactive organic matter. Some types of organic matter are more susceptible to TSR than others. For example, the C_2–C_5 hydrocarbon gases are more reactive than methane. Many sour gas reservoirs lack C_2–C_5 hydrocarbons but contain methane, H_2S, and other non-hydrocarbon gases, such as carbon dioxide. As discussed later, the relative rates of reaction of higher-molecular-weight hydrocarbons also vary. Because the oxidation state of sulfur ranges from $+6$ to -2 during the reduction of sulfate to sulfide, S_0 in reactions (6.5) and (6.6) includes elemental sulfur and other sulfur intermediates, such as polysulfides and thiosulfates (Steinfatt and Hoffmann, 1993; Goldstein and Aizenshtat, 1994). Anhydrite is an effective seal rock that is generally the source of the sulfate. Because anhydrite is not particularly soluble, reaction (6.5) is considered to be the rate-determining step.

The minimum temperature to initiate TSR was controversial for many years. Some authors proposed temperatures as low as 80°C (Orr, 1977), while others argued that it was unlikely to occur below 200°C (e.g. Trudinger et al., 1985), partly because of disputes over the meaning of laboratory data (Goldhaber and Orr, 1995). Recent work suggests that TSR begins in the range 127–140°C, depending on the hydrocarbons in the reservoir, and that higher temperatures are required to initiate TSR for methane than for heavier hydrocarbons (Rooney, 1995; Worden et al., 1995; Machel et al., 1995a).

While TSR is commonly associated with gas accumulations, the lower temperature range for TSR corresponds to that for generation of light oils and condensates. TSR can modify the compositions of these fluids. Various parameters distinguish the effects of TSR from those of thermal maturity on the composition of liquid hydrocarbons (Table 6.2).

Table 6.2. *Comparison of geochemical changes in liquid hydrocarbons due to increasing thermochemical sulfate reduction (TSR) or increasing thermal maturation (modified from Peters and Fowler, 2002). Stable carbon isotope ratios are presented as delta values ($\delta^{13}C$), representing the deviation in parts per thousand (‰ or per mil) from a standard*

Parameter	Increasing TSR	Increasing maturity
Saturate/aromatic	Decrease	Increase
Organosulfur compounds	Increase	Decrease
API gravity	Slight increase or decrease; decrease at high levels of TSR (diamondoids)	Increase
$\delta^{34}S$ of sulfur compounds	Approaches $CaSO_4$ (Figure 6.16)	Little change
$\delta^{13}C$ of saturates	Increase	Increase
$\delta^{13}C$ (CSIA) of gasoline range	Normal/branched alkanes increase up to 20‰; cyclics and aromatics less so	2–3‰ for all compounds

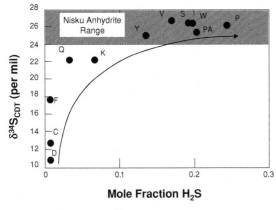

Figure 6.16. Variation of $\delta^{34}S$ in crude oils versus hydrogen sulfide concentration in associated gas for Nisku Formation reservoirs in the Brazeau River area of west central Alberta. The $\delta^{34}S$ of the samples increased from ~10.8 to 26.3‰ relative to Canyon Diablo Troilite (CDT) standard, approaching the $\delta^{34}S$ of Upper Devonian anhydrite (24–28‰) with increasing H_2S. Reprinted from Peters and Fowler (2002); modified from Manzano *et al.* (1997). Copyright 2002, with permission from Elsevier.

affected by TSR had more positive values. CSIA of the gasoline-range hydrocarbons in oils (~C_6 and C_7 hydrocarbons) shows greater isotopic shifts for *n*-alkanes and branched alkanes than for monoaromatic compounds, such as benzene and toluene, indicating that the saturated compounds are more reactive (Rooney, 1995).

Without TSR, the main controls on aromatic sulfur compounds in oils are source-rock depositional environment and thermal maturity (Ho *et al.*, 1974;

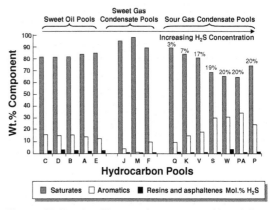

Figure 6.17. Bar chart illustrates the gross composition of oils and condensates from Nisku Formation reservoirs in the Brazeau River area of west central Alberta. Pools with higher H_2S (more affected by thermochemical sulfate reduction (TSR)) have less saturated and more aromatic hydrocarbons than the sweet oil and gas pools. Most Nisku pools have only a letter designation; however, PA is the Peco A pool. Reprinted from Peters and Fowler (2002); modified from Manzano *et al.* (1997). Copyright 2002, with permission from Elsevier.

With increasing thermal maturation of crude oil, saturated hydrocarbons increase relative to aromatics (e.g. Tissot and Welte, 1984, p. 187). The opposite trend occurs during TSR due to the greater reactivity of saturated compared with aromatic hydrocarbons (Figure 6.16). While both TSR and thermal maturation increase saturated hydrocarbon $\delta^{13}C$ (Table 6.2), Claypool and Mancini (1989) noted that condensates

Hughes, 1984). However, the concentration of aromatic sulfur compounds increases with H_2S content during TSR because these compounds are formed as by-products (Orr, 1974). This results in a slight increase or even a decrease in API gravity with increasing maturity (Claypool and Mancini, 1989; Manzano *et al.*, 1997). Because anhydrite is the source of sulfur in the neo-formed sulfur compounds, the $\delta^{34}S$ of whole oils increase toward the $\delta^{34}S$ of the anhydrite with increasing TSR (Figure 6.16) (Orr, 1974; Manzano *et al.*, 1997).

CSIA of gasoline-range hydrocarbons is a sensitive method for detecting TSR in condensates (Rooney, 1995; Whiticar and Snowdon, 1999). The change in $\delta^{13}C$ due to TSR appears to correlate with both molecular structure and reservoir temperature. Rooney (1995) showed substantial isotopic shifts in the $\delta^{13}C$ of some gasoline-range hydrocarbons in TSR-affected oils relative to the maximum shifts caused by thermal maturation alone. Variations in $\delta^{13}C$ also depended on the types of hydrocarbon. The *n*-alkane and branched hydrocarbons in TSR-affected oils vary in $\delta^{13}C$ by up to 22‰, whereas monoaromatics, such as toluene, show much smaller shifts in the range 3–6‰. Oils not affected by TSR show a maximum increase of 2–3‰ for each molecular species with increasing maturation, whereas much larger shifts occur with increasing reservoir temperature among TSR-affected oils. TSR may accelerate the destruction of some hydrocarbons compared with thermal cracking, with the remaining hydrocarbons becoming enriched in ^{13}C due to higher fractional conversion for each compound. This was supported by much lower concentrations of branched and normal alkanes with increasing reservoir temperature in TSR-affected oils relative to other oils (Rooney, 1995).

Changes in the composition of gasoline-range hydrocarbons caused by TSR can complicate correlation of condensates. For example, TSR affects the Mango parameters (ten Haven, 1996) as exemplified for crude oils generated from the Upper Devonian Duvernay Formation in Western Canada (Figure 6.18). Mango (1987; 1990) hypothesized that steady-state catalytic isomerization involving metal catalysts controls preferential

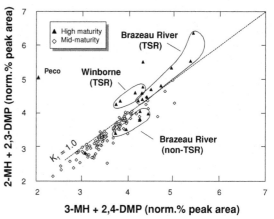

Figure 6.18. Relative amounts of 3-methylhexane + 2,4-dimethylpentane versus 2-methylhexane + 2,3-dimethylpentane (Mango parameters) for petroleum generated from the Upper Devonian Duvernay Formation in Alberta. High-maturity, non-thermochemical sulfate reduction (TSR) light oils and condensates from the Brazeau River area plot near mid-mature oils below the $K_1 = 1$ line. Condensates affected by TSR and light oils from the Winborne Field that are associated with high H_2S plot above the $K_1 = 1$ line. Peco condensate was altered severely by TSR and plots far from the other samples. Reprinted from Peters and Fowler (2002). Copyright 2002, with permission from Elsevier.

ring opening of cyclopropane (three-ring) intermediates to form the isoheptanes. Based on this kinetic model (van Duin and Larter, 1997), 2-methylhexane + 2,3-dimethylpentane should co-vary with 3-methylhexane + 2,4-dimethylpentane depending on temperature, as in Figure 6.18. Mid-mature Duvernay oils plot in a narrow band just below the $K_1 = 1$ line in the figure. Higher-maturity Duvernay oils from the Brazeau River Field that are unaffected by TSR plot along a trend similar to the mid-mature oils. Oils that are associated with H_2S plot above the $K_1 = 1$ line. The Peco sample has the highest concentration of H_2S, appears to be most affected by TSR based on CSIA data, and plots farthest from the $K_1 = 1$ line. Data from other oil families suggest a tendency for all TSR-affected oils to plot above the $K_1 = 1$ line (Fowler, unpublished results).

7 · Ancillary geochemical methods

Ancillary geochemical tools (e.g. diamondoids, C_7 hydrocarbons, compound-specific isotopes, and fluid inclusions) can be used to evaluate the origin, thermal maturity, and extent of biodegradation or mixing of petroleum, even when the geological samples lack or have few biomarkers. Molecular modeling can be used to rationalize or predict the geochemical behavior of biomarkers and other compounds in the geosphere.

DIAMONDOIDS

Diamondoids are small, thermally stable, cage-like hydrocarbons in petroleum, where carbon–carbon bonds are arranged according to the structure of diamond (Figure 7.1). They consist of pseudo-homologous series with the general formula, $C_{4n+6}H_{4n+12}$, including adamantane, dia-, tri-, tetra-, and pentamantane ($n = 1$–5, respectively) and higher polymantanes, plus various alkylated series (Wingert, 1992; Lin and Wilk, 1995). However, there are non-isomeric polymantanes that do not obey this formula. Diamondoids occur in the saturated hydrocarbon fraction of petroleum and can be characterized using their M+ molecular ion.

Dahl *et al.* (2002) isolated and determined the crystal structure for a series of higher diamondoids ranging from 4 to 11 adamantane units. These higher polymantanes were isolated from Paar bomb pyrolysis products of the 345–550°C vacuum distillation fraction of condensate from the Norphlet Formation, Gulf of Mexico. This condensate contains abundant diamondoids, which were further concentrated during the pyrolysis due to their exceptional stability. They were separated from the char by column chromatography and collection of the saturated fraction. Dahl *et al.* (2002) found that the higher diamondoids have many three-dimensional shapes, including rods and screws with resolvable chirality (Figure 7.2). Structural complexity of the diamondoids increases rapidly above tetramantane,

which has four possible isomers. The pentamantanes have nine isomers with the formula $C_{26}H_{32}$ and one $C_{25}H_{30}$ compound. There are 39 hexamantanes, 28 of which are $C_{30}H_{36}$ isomers, 10 are $C_{29}H_{34}$ isomers, and one, the pericondensed cyclohexamantane, is a $C_{26}H_{30}$ compound (Dahl *et al.*, 2003). A cyclohexamantane molecule can be thought of as a nanometer-sized diamond of $\sim 10^{-21}$ carats (1 carat $= 0.2$ g).

Note: Cyclohexamantane may form by rearrangement of suitable C_{26} precursors during oil generation. Alternately, lower diamondoids may be converted to more complex adamantologs by clay-catalyzed reactions involving existing alkyl and naphthenic substitutions and/or by alkylation with methane at high pressures and temperatures, followed by cyclization (Lin and Wilk, 1995). If correct, this mechanism might account for the formation of diamondoids substantially larger than cyclohexamantane, and possibly may even explain the origin of carbonados. Carbonados are microcrystalline diamonds that are unlike other diamonds because they appear to originate in the Earth's crust rather than deep within the mantle (Kamioka *et al.*, 1996). Carbonados have $\delta^{13}C$ in the range -23 to -30% (Shelkov *et al.*, 1997), which is similar to that of petroleum and unlike mantle-derived carbon ($\sim -5\%$).

Diamondoids do not occur in living organisms but have been synthesized from a wide variety of organic precursors via carbonium ion intermediates on superacids (Schleyer, 1957, 1990). This mode of formation combined with their ubiquitous occurrence, even in oils of low thermal maturity, suggests that diamondoids form by hydrocarbon rearrangement reactions on acidic clay minerals in petroleum source rocks. Unlike biomarkers, diamondoids in crude oils and source rocks are structurally very different from their probable precursors in living organisms.

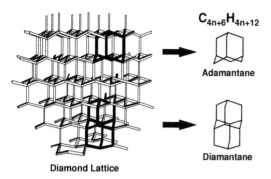

$$C_{4n+6}H_{4n+12}$$

Adamantane

Diamantane

Diamond Lattice

Figure 7.1. Diamondoids are hydrocarbons with carbon structures composed of small subunits of the diamond lattice that are exceptionally resistant to thermal cracking.

Production problems caused by diamondoids

Diamondoid concentrations increase in oils that have undergone cracking in the reservoir due to their greater thermal stability compared with other compounds, including biomarkers. Sublimates of diamondoids can cause problems during production of deep natural gas (Figure 7.3). For example, within two weeks after initial production, solid deposits of nearly pure adamantane and diamantane (also called congressane) precipitated due to temperature and pressure reduction as deep, hydrogen sulfide-rich natural gas was brought to the surface in the Mary Ann Field, lower Mobile Bay, Alabama. These diamondoid deposits clogged the production string and required remedial treatment to maintain gas flow. Mary Ann gas formed by cracking of crude oil that migrated from the Upper Jurassic Smackover Formation source rock to the Norphlet Formation reservoir. The gas consists of methane with subordinate hydrogen sulfide, carbon dioxide, and ethane produced from ~6300-m depth at reservoir temperatures near 196°C (Wingert, 1992). Similar production problems caused by diamondoids occurred at the Swan Hills Field, Alberta, Canada (King, 1988).

Extent of oil-to-gas cracking

Thermal maturity is one factor that limits worldwide exploration success for crude oil. At high temperatures, crude oil cracks to light oil, condensate, and finally gas plus pyrobitumen in reservoirs buried below the oil deadline, which generally occurs at ~5-km depth at temperatures of 150–175°C (Hunt, 1996). Some controversy exists as to whether significant oil cracking occurs at depths penetrated by commercial drilling. The thermal stability of some hydrocarbons (Mango, 1991) and the occurrence of oils at reservoir temperatures in the range 150–175°C (Horsfield et al., 1992; Price, 1992; Pepper and Dodd, 1995) suggest that oil cracking may be less important than thought previously. However, other evidence indicates that oil cracking in reservoirs can be extensive (Claypool and Mancini, 1989).

Various computer models simulate oil generation from the source rock and predict the extent of natural oil cracking in the reservoir in order to reduce the risk associated with petroleum exploration (Welte et al., 1997). Oil-to-gas conversion can be estimated using assumptions about carbon mass balance and measured gas-to-oil ratios (GOR) (standard ft^3/barrel) as follows:

$$f(\%) = GOR/[GOR + (7.98 \times 10^5) \, density/ \\ average \, molecular \, weight] \times 100$$

where f is the fractional conversion of oil to gas (Claypool and Mancini, 1989). The calculation assumes that the reservoir is a closed system and that gas and pyrobitumen originate solely from oil in a simple reaction. Typical values for the density and average molecular weight of the original oil are 0.80 g/cm^3 and 210 g/mol, respectively.

Unfortunately, GORs are readily influenced by secondary processes, such as retrograde condensation during migration, addition of gas from deeper reservoirs,

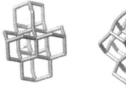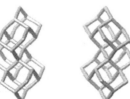

Figure 7.2. Some molecular structures of hexamantanes. Reprinted with permission from Dahl et al. (2002), © Copyright 2002, American Association for the Advancement of Science.

Table 7.1. *Oil and condensate properties are related to reservoir maturity and the extent of oil cracking*

Reservoir temperature °C	R_o (%)	Fraction oil cracked*	GOR (standard cubic feet/barrel)	CGR (barrels/ million cubic feet)	API gravity	$C_1/(C_1-C_6)$ wetness
140	0.86	0.012	1000	–	–	–
150	1.00	0.057	1000	–	–	–
160	1.15	0.200	1000	–	–	–
170	1.33	0.470	2750	–	40	0.84
175	1.43	0.630	5300	190	44	0.91
180	1.52	0.770	10 400	96	48	0.93
190	1.76	0.880	23 000	44	50	0.95
200	2.01	0.996	300 000	3	45	1.00

*Extent of oil cracking estimated for reservoirs currently at maximum burial depth and temperature for a typical Gulf Coast burial history (3°C/my).
CGR, condensate/gas ratio; GOR, gas/oil ratio.

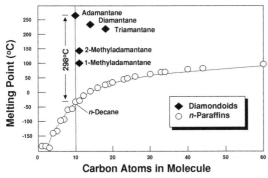

Figure 7.3. Diamondoids have remarkable physical properties, including high melting points and the tendency to sublimate directly as solids from the gas phase during production. The example in the figure shows that the melting point of adamantane at atmospheric pressure is 298°C higher than that of *n*-decane, although both compounds have ten carbon atoms. Thus, cooling of a mixture of light hydrocarbons might result in early precipitation of solid adamantane, but *n*-decane and other hydrocarbons of similar molecular weight would not precipitate from solution until much later.

leakage through the seal rock, biodegradation, and dissolution by moving groundwater. Direct measurement of the thermal maturity of light oils and condensates is difficult because biomarkers are commonly low or absent.

Diamondoids can be used to determine the extent of oil-to-gas cracking, which can take place in the source rock or reservoir (Dahl *et al.*, 1999). This approach is

the first documented method suitable for calibrating the extent of cracking predicted by computer simulations. It can be used together with calibrated simulations to determine the oil deadline, i.e. the depth at which most liquids are cracked to gas. The approach also allows direct identification of mixed oils. The method is based on mass spectrometric measurement of diamondoids and biomarkers (i.e. 24-ethylcholestane 20R was used) in produced oils of various maturities. Diamondoids occur at different concentrations in >1000 analyzed oils worldwide (Dahl *et al.*, 1999).

The carbonate-rich Smackover Formation of Mississippi, Alabama, and Florida consists of source and interbedded reservoir rock plus underlying Norphlet sandstone reservoir rock. These closed or nearly closed units are effectively sealed above and below by the Buckner anhydrite and the Pine Hill anhydrite-Louanne salt, respectively. Because of these effective seal rocks, vertical migration of generated oil is minimal. Furthermore, these units now reside at their maximum burial depths and temperatures (Table 7.1). The Smackover oil trend at depths from 3000 to 4300 m changes to a gas-condensate trend extending to 5500 m and finally to a deep dry gas trend below 6000 m (Mancini *et al.*, 1989). Hydrogen sulfide associated with mature Smackover condensates and gases results from thermochemical sulfate reduction (TSR), where hydrocarbons contact anhydrites at high temperatures (Wade *et al.*, 1989; Claypool and Mancini, 1989), as described below.

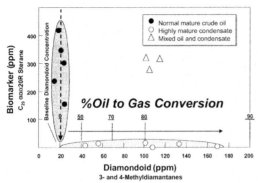

Figure 7.4. Schematic relationship between concentrations of 3- and 4-methyldiamantanes (diamondoids) and stigmastane [5α,14α,17α(H)-24-ethylcholestane 20R] for a series of hypothetical oils and condensates (modified from Dahl et al., 1999). Maturation of Smackover oils causes dilution of biomarkers by generation of other components, eventually yielding light oils or condensates (from top toward bottom in vertical stippled area). The baseline concentrations of diamondoids among related oils can be established by this vertical trend. Intense oil cracking causes destruction of the major oil components and concentration of diamondoids (from left toward right in horizontal stippled area). Methyldiamantane and stigmastane concentrations were measured by metastable reaction monitoring/gas chromatography/mass spectrometry (MRM/GCMS) using m/z 202 → 187 (Figure 7.5) and m/z 400 → 217 transitions, respectively. Samples were spiked with d_3-1-methyldiamantane (labeled with three deuterium atoms on the methyl group) and 5β-cholane internal standards before a series of concentration procedures and analysis.

Figure 7.4 compares the relative concentrations of two diamondoids (3- and 4-methyldiamantane) with a common biomarker found in most petroleum (stigmastane) for a series of oils generated from the Smackover Formation. The systematic relationship between methyldiamantanes (as measured in Figure 7.5) and stigmastane concentrations represented by the curve in the figure is controlled by the relative thermal maturity of the samples, as measured by various parameters, including methylphenanthrene and dimethylpentane ratios. Criteria used to select these samples were that they (1) originated from similar facies of the same source rock and (2) did not migrate far from the source rock. All samples were produced from either fractured source rock or stratigraphically deeper reservoir rock. Production depths of these samples correlate with their relative thermal maturities (Dahl et al., 1999). API gravities of

the Smackover samples range from 32 to 56°. Because little vertical migration occurred, reservoir depth and temperature also correlate closely with the maturity parameters. We observed similar systematic relationships among various suites of related oils worldwide. The concentrations of diamondoids are similar for related low- to middle-maturity oils (diamondoid baseline), but highly mature equivalents show elevated diamondoids. The baseline biomarker and diamondoid concentrations may differ substantially for oils generated from different source rocks.

The decrease in stigmastane concentration during thermal maturation of crude oil (Figure 7.4) is caused by (1) cracking of stigmastane at high temperatures and (2) dilution with other compounds as more mature oil is generated from kerogen in the source rock (Dahl et al., 1999). Thermal maturation of the source rock induces structural rearrangements among steroid and terpenoid precursors for diamondoids by a carbonium ion mechanism involving acidic sites on clays (Dahl et al., 1999). During maturation within the oil window, kerogen generates oil that is progressively depleted in biomarkers (Requejo, 1992). However, diamondoids and many light hydrocarbons are stable in this range of maturity, and the "baseline" diamonoidoid concentration remains nearly constant for related crude oils. This baseline concentration commonly differs for oils from different source rocks. Diamondoid concentrations begin to increase near the end of the oil window only because of their relative stability. Thus, as the stigmastane concentration approaches zero, diamondoid concentrations increase due

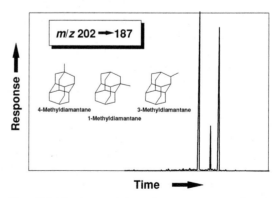

Figure 7.5. The three methylated diamantanes in crude oils and condensates are readily detected using metastable reaction monitoring/gas chromatography/mass spectrometry (MRM/GCMS) m/z 202 → 187 transition.

to their extreme thermal stability compared with other components in the oil. Diamondoid concentrations increase at higher maturity due primarily to depletion of less stable hydrocarbons by cracking. This is supported by data showing that (1) diamondoids originate by carbonium ion reactions on superacid sites in source rocks rather than by free-radical cracking reactions in reservoirs, and (2) laboratory heating of oils concentrates diamondoids (below). Mixed oils commonly show high concentrations of both diamondoids and biomarkers (Figure 7.4, upper right).

Because diamondoids are concentrated rather than created during cracking, the oil-to-gas cracking or diamondoid cracking ratio for a highly mature oil in the reservoir can be estimated from the increase in diamondoid concentration above the baseline concentrations in low-maturity oils. We use the simple formula:

$$OTG_d = (1 - U/C) \times 100$$

where OTG_d is the percentage oil-to-gas conversion based on diamondoids, U is the average diamondoid concentration in oils that have not undergone reservoir cracking (baseline), and C is the diamondoid concentration in cracked oil. For example, assume that Smackover oil contains ~20 ppm of 3- and 4-methyldiamantanes before significant reservoir cracking (diamondoid baseline, Figure 7.4). Highly mature Smackover condensate that contains 40 ppm of these compounds represents residual oil that lost ~50% of its liquids by cracking to gas or pyrobitumen.

The above method for estimating oil-to-gas conversion is validated by both laboratory and field observations. Laboratory cracking experiments were performed on waxy lacustrine oil from Indonesia and aromatic marine oil from the North Slope of Alaska (Dahl et al., 1999). Both experiments show excellent linear correlations between the extents of oil cracking determined using the diamondoid method and the final versus initial weight of oil. For both experiments, the estimated oil-to-gas conversion based on diamondoids is higher than that determined directly by weight difference. Gas chromatographic analysis of the heated samples suggests that evaporative loss of light ends occurred in the severely cracked samples before diamondoid analysis. Cracking ratios based on GOR for deep Tuscaloosa Formation oils agree with those calculated from diamondoids. Similar agreement between GOR and diamondoid estimates of oil-to-gas conversion occurs in

Smackover oils (Table 7.1) and other data sets where little gas has leaked from the reservoir.

Application of the above method to various basins suggests that oil-to-gas cracking is common. In the absence of TSR, which can drastically lower the temperature required for oil destruction (Orr, 1974), cracking appears to occur when reservoir temperatures exceed 140–160°C. This temperature range is consistent with previous observations on the depth of oil destruction, various laboratory experiments, and thermodynamic calculations (Quigley and Mackenzie, 1988; Ungerer et al., 1988). Cracked liquids occur in some reservoirs that have not exceeded 150°C, suggesting that oil-to-gas cracking occurred at greater depths before charging the reservoir, possibly in the source rock.

Deconvoluting mixtures using diamondoids

The above approach also offers a direct means to recognize mixtures of high- and low-maturity oils. Detection of such mixed oils is important because it can result in new petroleum exploration play concepts and better understanding of migration paths. Some oils show both high diamondoid and high biomarker concentrations (e.g. upper right in Figure 7.4). These oils represent mixtures derived from low-maturity source rock rich in biomarkers (e.g. stigmastane) and high-maturity source rock rich in diamondoids. The diamondoid approach to assessing mixtures can be applied to many parts of the world where oils might originate from multiple sources, including the Gulf Coast, Venezuela, Colombia, West Africa, Brazil, the North Sea, and Australia.

Diamondoid source parameters

Although published diamondoid maturity parameters for their samples did not vary throughout the oil window, Schulz et al. (2001) found several dimethyldiamantane facies ratios that distinguished extracts from terrigenous, marine carbonate, and marine siliciclastic facies. The ethyladamantane ratio was also useful as a source tool but was affected by maturity in the late gas window.

The stable carbon isotope compositions of alkyldiamantanes that occur in certain dry gas accumulations can be used to identify the source of the accumulations and correlate the gases with related oils and condensates. Alkyldiamantanes in gases from Norphlet (Gulf of Mexico) and Swan Hills reservoirs (Alberta, Canada) have $\delta^{13}C$ similar to the C_{15+} saturated hydrocarbons in oils or condensates generated from the Smackover ($\sim-24‰$) and Duvernay ($\sim-30‰$) source rocks, respectively (Schoell *et al.*, 1997). The adamantanes, however, show ^{13}C enrichment with lower molecular weight, possibly due to isotope fractionation associated with progressive dealkylation of the alkyladamantanes during cracking.

Diamondoid maturity parameters

Various diamondoid maturity parameters have been proposed to measure highly mature samples, but their usefulness remains unclear. Chen *et al.* (1996) used two diamondoid indices to evaluate the thermal maturity of crude oils and condensates from the Tarim, Yinggehai, Qiongdongnan, and other Chinese basins:

Methyladamantane (MA) index,

$$\% = 1\text{-MA}(1\text{-MA} + 2\text{-MA})$$

Methyldiamantane (MD) index,

$$\% = 4\text{-MD}/(1\text{-MD} + 3\text{-MD} + 4\text{-MD})$$

The methyladamantane and methyldiamantane indices show initial values of $\sim50\%$ and 30%, respectively, at an equivalent vitrinite reflectance of $\sim0.9\%$. However, the methyldiamantane index (MDI) ranges from 40 to 65% for source-rock extracts from the Lower Ordovician Majiagou Formation of the central gas field, Shanganning Basin, China (Li *et al.*, 2000). Changes in the MDI index in the highly mature section ($R_o > 2.0\%$) are small, and no linear correlation between MDI and R_o or between MDI and depth occurs, suggesting that, contrary to Chen *et al.* (1996), MDI may have limited applicability as a maturity indicator.

Biodegradation of diamondoids

The methyladamantane/adamantane ratio rises with increasing biodegradation of diamondoids in Australian crude oils from the Carnarvon and Gippsland basins (Grice *et al.*, 2000). Significant changes in the ratio occur at extreme levels of biodegradation, indicating that diamondoids are useful indicators of biodegradation when most other hydrocarbons have been removed. The methyladamantane/adamantane ratio can be used

1-Methyl-2-thiaadamantane   1,5-Dimethyl-2-thiaadamantane

Figure 7.6. Bridgehead alkylated thiaadamantanes identified in thermochemical sulfate reduction (TSR)-altered oils by Hanin *et al.* (2002).

to identify mixtures of severely biodegraded and non-biodegraded oils and to assess the extent of biodegradation of crude oils from the Gippsland Basin, where 25-norhopanes are generally absent.

Thiaadamantanes: indicators of thermochemical sulfate reduction

Thiaadamantanes are thought to be molecular markers for petroleum that has undergone TSR. Hanin *et al.* (2002) observed a series of these homologous sulfides, characterized by an intense molecular ion at 154 + (n•14) in crude oils from the Smackover and Nisku formations that had been altered by TSR. The mass spectra of thiaadamantanes are similar to those of adamantanes with minor ion fragments m/z 93 and 107. The most abundant compounds are thiaadamantanes with bridgehead alkylation (Figure 7.6). Adamantanes may form by acid-catalyzed rearrangement of cycloalkanes during early diagenesis. Their high concentrations in oils that are highly mature or altered by TSR may result from enrichment as less stable hydrocarbons are destroyed. In contrast, bridgehead alkylated thiaadamantanes, the most thermally stable of the thiaadamantane isomers, were absent in Smackover oils not affected by TSR. Hanin *et al.* (2002) found that the sulfur isotopic ratios of the thiaadamantanes ($\delta^{34}S = +21$ to $+22‰$) are similar to those of reservoir sulfates ($\delta^{34}S$ +18 to +24‰), indicating that these compounds form by a sulfurization process that occurs during TSR.

C_7 HYDROCARBON ANALYSIS

Light hydrocarbons (C_4–C_9) are not biomarkers because their carbon skeletons are too small to preserve evidence of a unique, biological origin. Most light hydrocarbons form by catagenetic breakdown of larger molecular precursors. Some isomers may have direct biological origins or may be cleaved from biomarkers, but there are so many possible precursors that they cannot be tied to a specific origin. The C_{10} monoterpenoids are the smallest hydrocarbons that may be considered biomarkers. However, assignment of even these compounds to a distinct biological origin is uncertain. Nevertheless, light hydrocarbons contain considerable information on their source, thermal maturity, and post-expulsion history. Thus, the geochemistry of light hydrocarbons complements biomarkers. Molecular and isotopic distributions of both light hydrocarbons and biomarkers are critical in order to derive complete and accurate models of petroleum systems, particularly for mixtures where light and heavy ends may have different source rocks.

Light hydrocarbons are a significant portion of most crude oils. In unaltered oils, the amounts of light hydrocarbons correlate with thermal maturity. Low-maturity, early-expelled oils may have <15%, typical mid-oil window marine oils have ~25–40%, and high-maturity condensates may be nearly 100% light hydrocarbons (Figure 7.7).

Various secondary processes can alter the abundance of light hydrocarbons in crude oils. Biodegradation, water washing, TSR, and evaporation can remove light hydrocarbons. Late-stage generation, reservoir cracking, and various migration processes, including phase separation, evaporative fractionation, and admixture of condensate, can increase light hydrocarbons.

Note: The C_6 and C_7 hydrocarbons are commonly called gasoline-range or light hydrocarbons. Technically, these designations are imprecise and

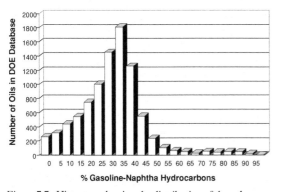

Figure 7.7. Histogram showing the distribution of the volume percentage of gasoline-naphtha (~C_4–C_{11}) hydrocarbons in 9078 oils in the Department of Energy (DOE) Bartlesville Crude Oil Analysis Database.

do not define clearly the range of hydrocarbons by molecular weight or carbon number. "Gasoline" is a refinery term and may include hydrocarbons ranging from C_4 to C_{10+}, depending on formulation. Light hydrocarbons may refer to hydrocarbons ranging from C_1 to ~C_{15}, depending on context.

Origins of light hydrocarbons

Many early studies focused on light hydrocarbons in recent sediments and downhole profiles to determine the temperatures of oil generation (Philippi, 1975; Johathan *et al.*, 1975; Leythaeuser *et al.*, 1978; Leythaeuser *et al.*, 1979; Hunt, 1984). Recent marine sediments lack light hydrocarbons, except for trace amounts that may be microbial or migrated (Hunt *et al.*, 1980a). Systematic changes in the abundance and distribution of light hydrocarbons occur with increasing temperature and depth. For example, low-temperature (<90°C) diagenetic rearrangement reactions of carbocations favor tertiary over secondary carbon atoms (i.e. carbons bound to three versus two other carbon atoms; see Figure 2.8), resulting in an increase in the isopentane/n-pentane ratio with depth (Hunt *et al.*, 1980b). At the top of the oil window, light hydrocarbons with tertiary carbon structures become more abundant, while at the bottom of the oil window (>150°C), hydrocarbons with quaternary carbon structures increase due to free radical reactions (Hunt, 1984). However, within the oil window, generation rather than rearrangement reactions dominate light hydrocarbon distributions. For example, n-alkanes are generated in high relative abundance, resulting in an overall decrease in the isopentane/n-pentane ratio at the top of the oil window that continues with depth.

Thermal cracking was long assumed to be the major mechanism accounting for light hydrocarbons in petroleum. The C_7 hydrocarbons in 18 oils having a wide range of geological ages were out of thermodynamic equilibrium (Martin *et al.*, 1963). Because acid clay catalysis reactions typically lead to equilibrium, the non-equilibrium distribution suggested that thermal rather than catalytic cracking is the dominant process in light hydrocarbon formation.

Subsequent studies of light hydrocarbons in oil focused less on mechanism and more on their use in correlation. Smith (1968) demonstrated the regularity in

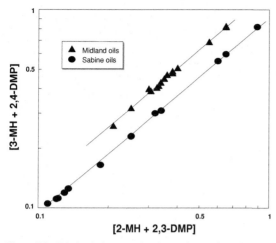

Figure 7.8. Relative isoheptane abundances for two homologous oil suites. The Sabine and Midland oil suites bracket the range of K_1 observed for most other primary oils. Mango attributed the difference in K_1 between the homologous suites to structural controls imposed by different kerogens on the kinetic reaction rates of solid-state catalysis. Reprinted from Mango (1987). © Copyright 1987, with permission from Elsevier. $K_1 = $ [2-methylhexane + 2,3-dimethylpentane]/ [3-methylhexane + 2,4-dimethylhexane]; ▲ = Permian Spraberry Formation, Midland Basin, West Texas. $K_1 = 0.786$; ● Upper Cretaceous Saratoga Chalk Formation, Sabine Pass, Louisiana. $K_1 = 1.09$.

the distribution of C_4–C_7 hydrocarbons based on the structural or functional group similarities among normal alkanes, branched alkanes, cyclic alkanes, and aromatic hydrocarbons. Erdman and Morris (1974) and Williams (1974) used light hydrocarbons for early oil correlation studies. In all of these studies, the C_7 hydrocarbons received special consideration, partly because their isomers can be resolved completely by standard gas chromatographic methods. Furthermore, compared with the C_6 and lighter hydrocarbons, the C_7 species are less susceptible to alteration by well-site sampling procedures and evaporation that can occur during storage and sample preparation.

During the late 1970s and early 1980s, oil–oil and oil–source rock correlations dominated petroleum geochemistry with increasing use and sophistication of biomarkers. Few studies focused on light hydrocarbons because their accepted origin by thermal cracking from larger molecules reduces geochemical information content. Thus, the light hydrocarbons were recognized as useful for describing secondary effects but were viewed

as inferior to biomarkers for correlation. Stable isotope studies focused on either C_1–C_4 gases or the C_{15+} fractions, due partly to limitations in analytical procedures. The gasoline-range hydrocarbons are difficult to isolate and prepare without causing isotopic fractionations.

Mango (1987) revitalized interest in light hydrocarbons in a series of papers that ultimately advocate their origin via steady-state metal catalysis. He found that while the relative abundance of the isoheptanes varied by orders of magnitude, the ratio of certain compounds, specifically $K_1 = $ (2-methylhexane + 2,3-dimethylpentane)/(3-methylhexane + 2,4-dimethylpentane), was constant (1.06, 0.336 standard deviation). Furthermore, oils from the same source exhibited remarkably constant K_1 values (Figure 7.8). Homologous oils consist of oils derived from the same source that show this high degree of correlation.

Mango (1990) proposed that the invariance in the isoheptane ratio results from a steady-state kinetic scheme where the methylhexanes originated from a common n-heptane precursor and the dimethylpentanes were daughter products of the methylhexanes (Figure 7.9). These reactions occurred via a transition metal catalyst forming a three-ring (cyclopropyl) intermediate. Under constant temperature and pressure, 2-methylhexane/3-methylhexane and 2,3-dimethylpentane/2,4-dimethylpentane ratios would be constant. Under changing temperature and pressure conditions, these ratios change, but in opposite directions. Hence, the isoheptane invariance among homologous oil suites is due to offsetting changes. Mango (1987; 1990) proposed that the variance in K_1 between oil suites was due to controls on the competing kinetic pathways imposed by different kerogen compositions.

Figure 7.9. Mango's steady-state metal catalyzed reaction scheme for the formation of methylhexanes from an n-heptyl moiety bound to kerogen. Mango (1990) proposed that a cyclopropyl ring forms by reaction of an olefinic precursor. Breakage of the "a" bond yields 2-methylhexane, while breakage of the "b" bond yields 3-methylhexane. According to this scheme, if $k_1 = k_2$, then the ratio of 2- to 3-methylhexane remains constant and independent of temperature and pressure. If $k_1 \neq k_2$, then the ratio changes with pressure and temperature.

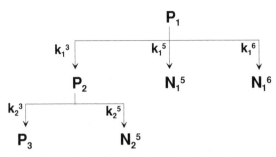

Figure 7.10. Metal-catalyzed steady-state kinetic reaction scheme for the formation of C$_7$ hydrocarbons (Mango, 1994). Mango modified the nomenclature of his reaction scheme several times leading to some confusion (see ten Haven, 1996). P$_1$ = n-heptane (nC$_7$); P$_2$ = 2-methylhexane (2-MH) + 3-methylhexane (3-MH); P$_3$ = 3-ethylpentane (3-EP) + 3,3-dimethylpentane (3,3-DMP) + 2,2-dimethylpentane (2,2-DMP) + 2,3-dimethylpentane (2,3-DMP) + 2,4-dimethylpentane (2,4-DMP) + 2,2,3-trimethylbutane (2,2,3-TMB); N$_1^5$ = ethylcyclopentane (ECP) + 1,2-(cis + $trans$)-dimethylcyclopentanes (1,2-DMCP); N$_1^6$ = toluene + methylcyclohexane (MCH); N$_2^5$ = 1,1-dimethylcyclopentane (1,1-DMCP) + 1,3-(cis + $trans$)-dimethylcyclopentanes (1,3-DMCP). 1,2-(cis + $trans$)-Dimethylcyclopentanes (1,2-DMCP) can also be formed by the five-ring closure of 3-MH and could be included in N$_2^5$.

Mango (1990) developed his reaction scheme further to include five- and six-member ring closures (Figure 7.10). The six-carbon ring compounds methylcyclohexane and toluene can originate by cyclization of the n-heptyl precursor. Ethylcyclopentane and 1,2-(cis + $trans$)-dimethylcyclopentanes also can form from an n-heptyl precursor via a five-ring closure. 1,1-Dimethylcyclopentane and 1,3-(cis + $trans$)-dimethylcyclopentanes, however, can be formed only by ring closure of methylhexanes and are thus daughter products in much the same manner as the dimethylpentanes.

Mango's reaction scheme predicts several invariant ratios among the C$_7$ hydrocarbons (Figure 7.11). K$_2$ is the ratio of the three-ring closure products (P$_3$) to the sum of the methylhexane parents (P$_2$) and the five-ring closure products (N$_2$). The invariance of K$_2$ within homologous oil sets couples the C$_7$ isoalkanes and dimethylcyclopentanes in a manner consistent with a kinetic steady-state model. The Mango reaction scheme predicts several other invariant ratios. 3,3-Dimethylpentane and 3-ethylpentane form only from 3-methylhexane, and 2,4-dimethylpentane forms only

from 2-methylhexane, each through three-ring intermediates. Ratio K$_3$, defined as (2-MH + 3,3-DMP + 3EP)/(3-MH + 2,4-DMP), should be as invariant as K$_1$. Similarly, the cis- and $trans$-1,3-dimethylpentane isomers form only from the five-ring closure of 3-methylhexane, while 1,1-dimethylpentane forms only from the five-ring closure of 2-methylhexane. Ratio K$_4$, defined as (2-MH + 1,3(cis + $trans$)-DMCP)/(3-MH + 11-DMCP), also should be invariant. These ratios are invariant within homologous oils suites from the North Viking Graben (Chung et $al.$, 1998).

Even though Mango's reaction scheme appears to explain the observed distributions of C$_7$ hydrocarbons, many do not accept the metal-catalysis mechanism for several reasons. Metal catalysis requires that organic substrates be associated intimately with metals in a solid kerogen matrix. Furthermore, the kerogen environment associated with oil generation is believed to be poisonous to metal catalysts. Mango's reaction mechanism also requires the formation of olefinic bonds and hydrogen by kerogen precursor compounds to form the metal-catalyzed ring closures. There is

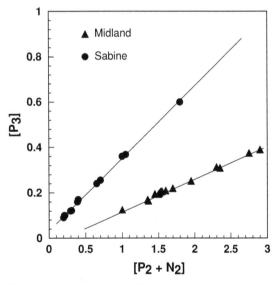

Figure 7.11. Concentrations of P$_3$ versus (P$_2$ + N$_2$) fractions, expressed as wt.% of the total oil, for two homologous oil suites. The invariance of these compounds, expressed as K$_2$ = P$_3$/ (P$_2$ + N$_2$), is predicted from Mango's (1990) steady-state kinetic scheme involving three-ring and five-ring closure reactions. Reprinted from Mango (1990). © Copyright 1990, with permission from Elsevier.

compelling evidence that initial oil formation results from the thermal decomposition of kerogen. A special mechanism for the formation of light hydrocarbons, therefore, seemed unnecessary. Molecular models, however, were developed that supported Mango's ring-closure reactions (van Duin and Larter, 1997; Xiao and James, 1997; Xiao, 2001).

Much of the criticism of Mango's hypothesis is directed toward the formation of natural gases. Mango (1992; 1996) and Mango et al. (1994) proposed that the C_1–C_4 gases also are generated via transition-metal-catalyzed reactions that are controlled by steady-state kinetics. Such reactions lead to a comparatively dry gas (60–95+ wt.% methane). In contrast, laboratory simulations of the thermal cracking of oil and kerogen produce wet gas (10–60 wt.% methane). Mango argued that the drier composition of gases produced from the catalyzed reactions is more consistent with natural observations. Price and Schoell (1995) countered that production data showing methane enrichment are not a true reflection of generation but reflect a combination of generation and fractionation during migration. They noted that in Bakken Shale, which serves as a closed system of both source and reservoir, associated methane is ~45 wt.%, consistent with laboratory simulations of thermal cleavage. Mango (1997) countered that if the generative gas compositions were so wet and drier gas reservoirs are the result of fractionation during migration, then there should be corresponding reservoirs with associated gases highly depleted in methane (~7 wt.%). Since associated gases rarely are observed to have such low amounts of methane, Mango reasoned that methane enrichment due to fractionation during migration is not relevant. This argument, however, is valid only if the system is closed and if there are no secondary contributors of methane to the petroleum system. Such simple geologic systems are rare, if they occur at all. Secondary methane is known to arise from coals, organic-lean source rocks, and microbial sources. McNeil and BeMent (1996) proposed that methane might be generated from the side-chain cleavage of alkylaromatics. Mango (1997) dismissed this hypothesis, arguing that thermal decomposition of model compounds yields the entire alkyl side chain and does not yield appreciable amounts of methane (e.g. 1-dodecylpyrene → pyrene + dodecane). He does not, however, address the fact that high-maturity kerogens are essentially polynuclear aromatics with methyl side chains and that the thermal decomposition of such sources yields primarily methane.

Although Mango presented additional evidence for natural gas formation via metal and metal oxide catalysis in laboratory simulations (Mango, 1998; Mango and Elrod, 1998; Mango and Hightower, 1997), the theory still did not address some major criticisms. For example, how do metals become catalytically active? How would they be able to react with organic matter in a bound, solid-phase environment? Why do natural samples lack high concentrations of the required olefinic precursors? Mango (2000) revised his hypothesis, addressing several of these factors. He now recognized that initial oil (C_{9+}) generation from kerogen results from thermal decomposition, light hydrocarbons result from metal-catalytic decomposition of saturated hydrocarbons, and natural gases result from the metal-catalytic decomposition of the light hydrocarbons. The revised model eliminates olefins as reactants, shifts light hydrocarbons as intermediates rather than final products, and allows the metal catalyst to be in the kerogen, rock matrix, or liquid petroleum interacting with liquids and gases:

$$\text{Kerogen} \xrightarrow{\Delta} \text{oil} \xrightarrow{[M*]} \text{light hydrocarbons}$$
$$+ \text{ wet gas} \xrightarrow{[M*]} \text{dry gas}$$

Most of the elements involving three-ring, five-ring, and six-ring closure reactions that Mango previously advocated remain in the new model, with the addition of four-ring species as possible intermediates. In the model, saturated hydrocarbon precursors form a triad relationship with daughter products by reaction through a metal- or metal-oxide-induced cyclopropyl intermediate:

$$S \longrightarrow \underset{MO_x}{\overset{C_2}{\underset{\downarrow}{C_1+C_3}}} \longrightarrow C_1-C_2-C_3 \;+\; C_1-C_3^{\,C_2} \;+\; C_1-C_3^{\,C_2}$$

In the above reaction, S is a saturated hydrocarbon precursor and C_1 and C_3 can be any alkyl groups. Because the rates of the reactions for homologous series should be similar, there should be proportional corresponding compound ratios. For example, the reactions that give rise to n-hexane (x) and the methylpentanes (x_i, x_{ii}) should be similar to those that give rise to n-heptane (y) and the methylhexanes (y_i, y_{ii}) (Figure 7.12).

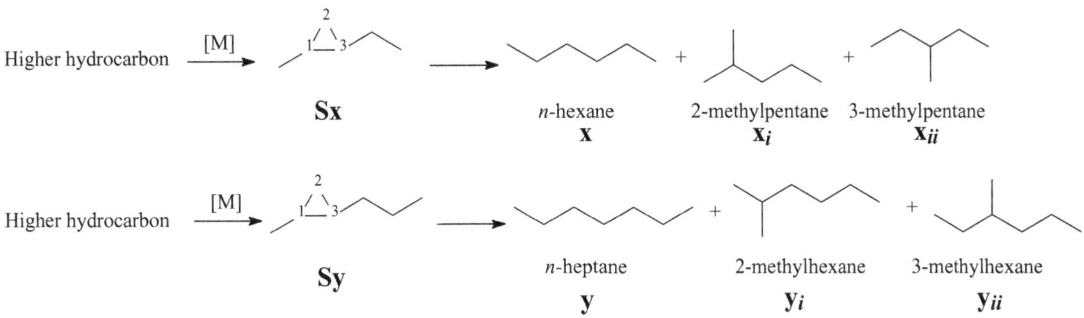

Figure 7.12. Mango reactions that give rise to *n*-hexane should be similar to those yielding *n*-heptane. M, metal catalyst. Other symbols are explained in the text.

If so, then the ratio of $x/(x_i + x_{ii})$ and $y/(y_i + y_{ii})$ should be proportional, regardless of the concentrations of the precursors S$_x$ and S$_y$. This proportionality is expressed as $x/(x_i + x_{ii}) = \alpha y/(y_i + y_{ii})$ or $\alpha = (x)(y_i + y_{ii})/(x_i + x_{ii})(y)$. In the above example, $\alpha = [(nC_6)(2\text{-MH} + 3\text{-MH})]/[(2\text{-MP} + 3\text{-MP})(nC_7)]$. Mango argued that such product relationships are characteristic only of catalyzed reactions and would not occur if the light hydrocarbons originated by thermal breakdown of higher hydrocarbons. Reactions involving cycloalkanes follow similar reactions with a cyclopropyl–metal oxide intermediate allowing for the catalyzed isomerization of 1,3-dimethylcyclopentanes to 1,2-dimethylcyclopentanes.

Theoretical models (Xiao, 2001) show that the above reactions are feasible in the presence of transition metals or mineral catalysts. Whether metal oxides or complexed transition metals can become active in the subsurface remains a critical issue for Mango's theories of light-hydrocarbon generation. He now suggests that the increase in light hydrocarbons and wet gas that occurs at reservoir temperatures >150 °C could indicate the conditions needed for metal catalysts to become active (higher temperatures, higher hydrogen partial pressure, and absence of suppressing agents). The decomposition of hydrocarbons would be due not to thermal cracking but to the activation of metal catalysts. This position, however, returns us to the initial problem that Mango tried to address by evoking metal-catalyzed reactions. What are the mechanisms that are responsible for the generation of light hydrocarbons within the oil window between temperatures of 90 and 150 °C? Mango calls upon his older model, where metals in the source rocks become active catalysts at temperatures

lower than those in reservoir rocks. He also believes that clay-mediated acid catalysis may play a role in these low-temperature reactions.

The chemical mechanisms responsible for the generation of light hydrocarbons under thermal conditions associated with oil generation are unknown. The higher hydrocarbons do not thermally decompose at these temperatures, so the light hydrocarbons must form by decomposition of kerogen and/or polar complexes. In our opinion, Mango's unproven ring-closure reactions offer an attractive explanation for light-hydrocarbon generation within the oil window. Such reactions can describe the parent–daughter relationships and invariance observed among the C$_6$–C$_7$ hydrocarbons, eliminate the need for exotic precursor compounds, and explain the observed non-equilibrium distributions. Acid or metal catalysis of these reactions remains unproven. Metal-catalyzed reactions may not be necessary in order to generate dry gas. Laboratory simulations show that hydrocarbons can thermally decompose at high temperatures (~190 °C) at realistic geologic heating rates (Horsfield *et al.*, 1992). Furthermore, high-maturity coals and kerogens produce substantial amounts of methane (Boreham and Powell, 1993; Law and Rice, 1993).

C$_6$–C$_7$ chromatographic separations

C$_7$ is the highest carbon number where all (17) hydrocarbon isomers can be resolved fully. However, baseline separation is difficult to achieve because the C$_7$ isomers must be resolved from each other and from coeluting C$_6$ and C$_8$ hydrocarbons. Baseline separation of all 17 C$_7$ isomers cannot be achieved using standard

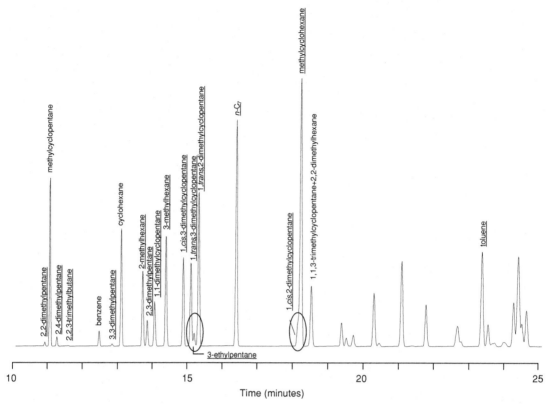

Figure 7.13. Optimized gas chromatographic separation of C_7 hydrocarbons on a single-column gas chromatographic system, a Hewlett Packard 5890 Series II Plus gas chromatograph equipped with a 100-m 100% methylsiloxane capillary column (J&W DB-1; 0.25 mm internal diameter (ID) × 0.5 μm film thickness). Hydrogen was used as the carrier gas and set with a linear velocity of ~18 cm/s. Injector temperature was 270°C and run in split mode (~50 : 1). Flame-ionization detector temperature was 350°C. The optimized program begins at 30°C and ramps immediately at 1.0°C/min to a final temperature of 107°C. The oven is then heated rapidly to 325°C and held for 10 minutes to flush heavier hydrocarbons. This method achieves baseline resolution for all C_7 isomers, except for 3-ethylpentane (partially separated from 1-*trans*-3-dimethylcyclopentane) and 1-*cis*-2-dimethylcyclopentane (shoulder on methylcyclohexane). C_7 hydrocarbons are underlined. Reprinted from Walters and Hellyer (1998). © Copyright 1998, with permission from Elsevier.

gas chromatographic methods, capillary columns, and chemically bound stationary phases. For example, using a 100% methylsilicone column, 3-ethylpentane is separated only partially from the 1-*trans*-3-dimethylcyclopentane, and 1-*cis*-2-dimethylcyclopentane is a shoulder on methylcyclohexane (Figure 7.13). Resolution of the latter compounds can be improved using conditions optimized for their separation, but this results in the loss of resolution of other compounds. Standard methods for the analysis of petroleum light hydrocarbons used by the American Society for Testing Materials (1992) (i.e.

ASTM D 5134–92) and the Gas Processors Association (1995) (i.e.) GPA 2186–95 resolve even fewer of the C_7 isomers.

Separation of all C_7 isomers can be achieved using highly apolar stationary phases, such as squalane or hexadecane–hexadecene, but these phases are thermally labile because they are not chemical bonded to the capillary column (maximum temperature <100°C). The low temperature limits demand that only the volatile hydrocarbon fraction be introduced on to the column. With care, the gasoline-range fraction can be distilled and

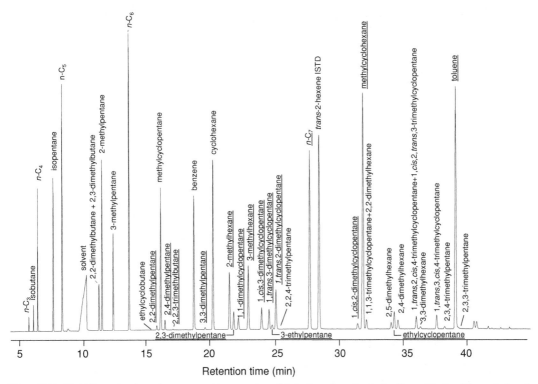

Figure 7.14. Multidimensional gas chromatographic separation of C$_7$ hydrocarbons (underlined) provides baseline resolution. A Siemens Sichromat 2-8 dual-oven GC with dual flame ionization detector (FID) is fitted with a 100-m 100% methylsiloxane (J&W DB-1, 0.25 mm internal diameter (ID) × 0.5 μm film thickness) precolumn and a 60-m 50% methyl–50% octylsiloxane (Supelco SPB Octyl 25 mm ID × 0.25 μm film thickness) column. The injector temperature is 250°C. Hydrogen is used as a carrier gas, with a linear velocity of ~17 cm/s in the precolumn and ~43 cm/s in the analytical column. The temperature program for the precolumn begins with an 18-minute hold at 29°C, ramps at 3°C/min to 53°C, and is held for 15 minutes. The precolumn is then backflushed and heated rapidly to 230°C to remove heavier hydrocarbons. The temperature program for the analytical column begins with a 35-minute hold at 28°C, ramped at 0.5°C/min to 31°C, then heated quickly to 130°C to remove heavier hydrocarbons. FIDs are set at 330°C for both ovens. Oil samples are prepared by dilution in CS$_2$ spiked with *trans*-2-heptene, which serves as an internal standard. Reprinted from Walters and Hellyer (1998). © Copyright 1998, with permission from Elsevier.

collected without alteration of the C$_7$ hydrocarbons, as in the procedure used for the C$_7$ database of Mango. An alternative approach uses a precolumn or temperature-programmed injector so that only light hydrocarbons are introduced on to the temperature-sensitive columns.

Walters and Hellyer (1998) developed a multidimensional, dual-oven gas chromatographic method for C$_7$ hydrocarbon analysis that achieves baseline resolution of all isomers (Figure 7.14). The method requires little sample preparation beyond an internal standard, uses commercially available capillary columns and

minimal amounts of cryogenic cooling liquids, and produces rapid and reproducible quantitative results.

C$_6$–C$_7$ light hydrocarbon parameters

Many C$_6$–C$_7$ light hydrocarbon parameters can be used to correlate oils and condensates, determine thermal maturity, and indicate various reservoir alteration processes. The proposed ratios are based on one of three principles: (1) models of generation, (2) physicochemical and biological processes, and (3) insensitivity to physicochemical and biological processes.

In our presentation of light-hydrocarbon parameters, we reference original published data and apply the parameter to an independent oil suite. The purpose of this test is not to validate the parameter but to provide the reader with a better understanding of the variance and difficulties that can be experienced when using light-hydrocarbon parameters. The data sets include:

- *Smackover Formation (Alabama):* Oil samples from onshore and offshore wells range from low to high maturity, including several oils that underwent TSR (Rooney, 1995).
- *Deep Tulscaloosa Formation (Louisiana):* Samples include oils and condensates produced from hot reservoirs ranging from ~135 to 190°C.
- *Scotian Shelf (Canada):* Samples include offshore oils and condensates generated from gas-prone terrigenous organic matter deposited in pro-deltaic shales. The Middle-Upper Jurassic Verrill Canyon pro-deltaic shales are the most likely source rock. There is good evidence for a second, marly source in some fields. The suite includes multiple DST tests from individual wells. Reservoir temperatures range from <50°C where biodegradation has occurred to >160°C.
- *Middle East:* Samples include oils from Iraq, Iran, and Kuwait that originated from Upper Jurassic and Lower Cretaceous carbonates, Cretaceous shales/marls, and Eocene shales.
- *Viking Graben (North Sea):* Samples include oils from the Beryl Field and satellite fields generated from low- to moderate-maturity Kimmeridge Clay Formation, moderate- to high-maturity Heather Formation, and high-maturity Jurassic coals. Many oils are mixtures of two or more end-member fluids.

Hunt parameters

Thermal maturation parameters based on empirical observations of the depth of maximum yields for light hydrocarbons with tertiary and quaternary carbon atoms. Rarely used for determining oil maturity.

Hunt *et al.* (1980b) observed that the light-hydrocarbon ratios of quaternary to tertiary carbon species (e.g. 2,2-dimethylbutane/2,3-dimethylbutane) increase with depth in mature sediments. For the fine-grained

rocks in the Gulf of Mexico, the maximum yield of the quaternary isoheptanes (2,2-DMP + 3,3-DMP) occurs ~3500 feet (1067 m) deeper than for the tertiary isoheptanes (3-EP + 2,3-DMP + 2,4-DMP), presumably due to greater stability of the intermediate tertiary carbonium ion or free radical (Hunt *et al.*, 1980; Hunt, 1984b).

These parameters have seen limited use to indicate oil maturity. Chung *et al.* (1998) found that the ratio of quaternary to tertiary isoheptanes correlated with 2,4-DMP/2,3-DMP for North Sea oils. Ratios of quaternary/(quaternary + tertiary) C_7 isoheptanes and cyclopentanes (1,1-dimethylcyclopentane/Σ dimethylcyclopentanes) correlate with reservoir depth on the Scotian Shelf (Figure 7.15). The correlations are weak, although some of the scatter could be due to vertical migration. For oils, the Hunt ratios only approximate thermal maturity and we recommend that these ratios be used only as supporting evidence for other, more reliable parameters.

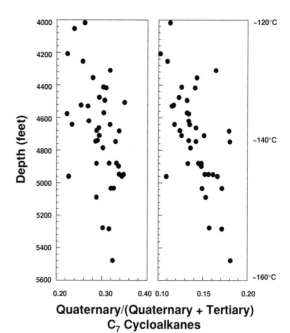

Figure 7.15. Quaternary/(quaternary + tertiary) C_7 isoheptanes (left) and cyclopentanes (right) for oils from the Scotian Shelf believed to originate from type II/III source rocks where reservoir depth approximates thermal maturity. The Hunt ratios exhibit only an approximate correlation with depth. Some of the scatter could be due to vertical migration but, in general, the ratios are unreliable indictors of thermal maturity.

Schaefer parameters

Thermal maturation parameters based on empirical observations of the light hydrocarbon ratios and vitrinite reflectance. Correlations are basin-specific and are rarely used to determine oil maturity.

Schaefer and Littke (1988) and Schaefer (1992) analyzed light hydrocarbons in Posidonia Shale, Lias ϵ (Lower Toarcian age, Saxony Basin) of varying thermal maturity (R_m, mean vitrinite reflectance ranging from 0.48 to 1.45%). They found strong correlations between R_m and several C$_7$ ratios. In particular, the ratios of C$_7$ paraffins/naphthenes (V) and (2-methylhexane + 3-methylhexane)/(1,2-(cis + $trans$)- + 1,3 (cis + $trans$) dimethylpentanes) (J) gave linear fits with R_m (Figure 7.16). The latter ratio approximates the $P_2/N_2{}^5$ (parent/daughter) ratio of the Mango reaction scheme.

We found that the Schaefer parameters V and J indicate thermal maturation. However, the relationship between these C$_7$ ratios and vitrinite reflectance also depends on source-rock organofacies (Chung et $al.$, 1998). For example, oils from the Scotian Shelf show consistent relationships between reservoir depth and parameters V and J. Inferred mean vitrinite reflectance (R_m)

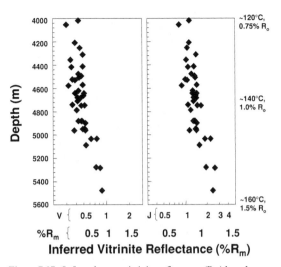

Figure 7.17. Inferred mean vitrinite reflectance (R_m) based on Schaefer parameters V and J (left and right panels, respectively) for a suite of oils from the Scotian Shelf. The Scotian Shelf oils are believed to originate from type II/III source rocks, where reservoir depth approximates thermal maturity. Temperature and measured vitrinite reflectance (R_o, right vertical axis) are averages from numerous wells.

for the Scotian Shelf oils, however, are shifted too low for parameter V and too high for parameter J, relative to measured R_o (Figure 7.17).

Thompson parameters

C$_6$–C$_7$ parameters that reflect kerogen type, maturity, and fractionation phenomena are based largely on empirical observations.

Thompson developed various C$_6$–C$_7$ ratios that describe light hydrocarbon distributions by compound class (Table 7.2). Several of these ratios correlate empirically with kerogen type and maturity, while others correlate with fractionation that occurs during migration experiments. Most of the ratios are influenced by secondary reservoir alteration events.

Thompson (1983) used plots of the heptane versus isoheptane ratios as indicators of source, thermal maturity, and biodegradation (Figure 7.18). By examining the light hydrocarbons extracted from cuttings and outcrop samples, he established that the relationships between the heptane and isoheptane ratios follow different

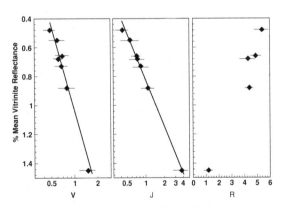

Figure 7.16. Averaged maturity-sensitive C$_7$ hydrocarbon ratios for Lias ϵ samples versus vitrinite reflectance. Reprinted from Schaefer (1992). © Copyright 1992, with permission from Elsevier.
V = C$_7$ paraffins/naphthenes, where "paraffins" are the acyclic alkanes. J = (2-methylhexane + 3-methylhexane)/(1,2-(cis + $trans$)- + 1,3 (cis + $trans$) dimethylpentanes). R = 1-$trans$-2-dimethylpentane/1-cis-2-dimethylpentane). Linear regression yields: R_m = 1.0 + 1.8 • log V and R_m = 0/84 + 1.1 • log J.

Table 7.2. *C_6–C_7 Thompson ratios describe processes affecting light hydrocarbons*

Name	Ratio	Property	Process
A	Benzene/n-hexane	Aromaticity	Fractionation, water washing, TSR
B	Toluene/n-heptane	Aromaticity	Fractionation, water washing, TSR
X	(m-Xylene + p-xylene)/n-octane	Aromaticity	Fractionation, water washing, TSR
C	$\dfrac{n\text{-hexane} + n\text{-heptane}}{\text{cyclohexane} + \text{methycyclohexane}}$	Paraffinicity	Maturity, biodegradation
I	$\dfrac{2\text{-} + 3\text{-methylhexane}}{1c3\text{-} + 1t3 + 1t2\text{-DMCPs}}$	Paraffinicity	Maturity, source, biodegradation
F	n-Heptane/methylcyclohexane	Paraffinicity	Maturity, biodegradation
H	$\dfrac{100 \times n\text{-heptane}}{(\Sigma \text{ cyclohexane} + C_7 \text{ HCs})}$	Paraffinicity	Maturity, source, biodegradation
S	n-Hexane/2,2-dimethylbutane	Paraffin branching	Maturity, source, biodegradation
R	n-Heptane/2-methylhexane	Paraffin branching	Maturity, source, biodegradation
U	Cyclohexane/methylcyclohexane	Naphthene branching	Maturity, source

DMCP, dimethylcyclopentane; H, heptane ratio. The heptane ratio was defined by Thompson (1983) as the percentage n-heptane relative to the sum of [cyclochexane + 2-methylhexane + 1,1-DMCP + 3-methylhexane + 1-*cis*-3-DMCP + 1-*trans*-3-DMCP + 1-*trans*-2-DMCP + n-heptane + methylcyclohexane]. Thompson noted that gas chromatographic separations then in use failed to resolve all isomers. The denominator of the heptane ratio as calculated in Thompson (1983) also contains 2,3-dimethylpentane, 3-ethylpentane, and 1-*cis*-2-DMCP; I, isoheptane ratio.

maturation curves for aliphatic and aromatic kerogens. The relationships established from source rocks can be applied to oils to indicate relative maturity and source kerogen. Thompson used oils largely from domestic marine shales to test this hypothesis and, except for three Paleozoic oils, the samples generally followed the kerogen curves. Thompson found that a significant number of the test oils were within 10% of a heptane ratio of 19 and an isoheptane ratio of 0.9. Compared with the light hydrocarbon ratios for a calibrated suite of source rock extracts, this range corresponded to a vitrinite reflectance of ~0.9%. Thompson concluded that this maturity represents normal oil generation and expulsion. Oils that had lower heptane and isoheptane ratios were biodegraded, while those with significantly higher values were thermally cracked. Oils with heptane ratios from 18 to 22, 22 to 30, and >30 were called normal, mature, and supermature, respectively.

Several different data sets show that the variance in oil compositions is greater than that described by Thompson (Figure 7.18). Oils from carbonate/marl (type IIS) source rocks generally follow a trend above the

aliphatic curve. Highly aliphatic source rocks from lacustrine and *G. prisca* organic facies yield high heptane ratios. Marine (type II) and pro-deltaic (type II/III) source rocks generally fall between the aliphatic and aromatic curves. Thompson stated that oils with heptane ratios <18 were biodegraded. While it is true that biodegradation can reduce the heptane ratio to zero, low-maturity, non-biodegraded oils can yield heptane ratios as low as ~12.

There are several notable exceptions to the generalized trends. Normal and isoalkanes are exceptionally abundant compared with cycloalkanes in Precambrian oils (Riphean source, Russia), resulting in curiously high isoheptane ratios (Figure 7.19). The depletion in cycloalkanes may be due to the lower thermal stability of these hydrocarbons relative to branched and normal alkanes that manifest only over a billion years. More likely, the light-hydrocarbon distribution reflects limited biodiversity of organic matter input to Precambrian sources. TSR also can alter light-hydrocarbon distributions. While the relative proportion of light to heavy hydrocarbons increases with TSR, there is a preferential

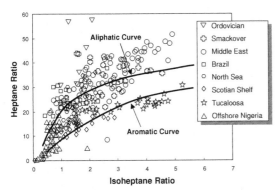

Figure 7.18. Heptane versus isoheptane ratios for various oils and the aliphatic and aromatic kerogen type curves determined by Thompson (1983). Ordovician oils are from the Michigan Basin and originated from kerogens rich in *G. prisca*. Smackover oils are from the Jurassic Smackover Formation carbonates in Alabama that contain type IIS kerogen. Middle East oils are from Iran, Iraq, and Kuwait and originate mainly from Jurassic and Cretaceous carbonates and marls containing type IIS kerogen. Brazil oils are from various facies of pre-salt lacustrine source rocks. North Sea oils are from the Viking Graben and Møre Basin and originated mainly from the Kimmeridge Clay (type II kerogen), with secondary contribution from the underlying Heather Formation (type II/III kerogen). Oils and condensates from the Scotian Shelf originated from the Middle-Upper Jurassic Verrill Canyon pro-deltaic shales (type II/III kerogen), although a few have a different carbonate source. Deep Tuscaloosa condensates are high-maturity fluids generated from the Tuscaloosa Shale (type II/III). Offshore Niger Delta oils originated from low-maturity Oligocene-Miocene pro–delta shale with a high terrigenous input (type II/III).

decrease in cycloalkanes relative to the normal and isoalkanes. This alteration results in a decrease in the isoheptane ratio and an increase in the heptane ratio (Figure 7.20).

The heptane and isoheptane ratios can be used to define broad oil classifications. Because the parameters have such a strong source influence, their use in defining thermal maturity is limited to homogeneous suites of oils that have not experienced significant reservoir fractionation or alteration.

Thompson (1987; 1988) and Thompson and Kennicutt (1990) described evaporative fractionation to account for certain light-hydrocarbon distributions. The evaporative fractionation concept is a modification of the phase-separation process described by Silverman (1965). In phase separation, a two-phase reservoir fluid fractionates into a gas phase and volatile-depleted

residual oil. The volatile-rich gas phase may migrate preferentially into a shallower reservoir, where decreases in temperature and pressure allow further fractionation into a gas phase and a retrograde condensate. The process can be repeated with the preferential migration of the secondary gas phase into still shallower reservoirs. Evaporative fractionation requires repeated recharging of the residual oil with methane. With each recharge, hydrocarbons in the residual oil re-equilibrate into gas and liquid phases. The gas phase then preferentially migrates to form retrograde condensate, as described by phase separation. Sequential recharging of the residual fluid with additional methane distinguishes evaporative fractionation from simple phase separation.

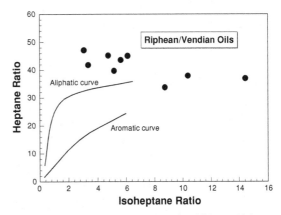

Figure 7.19. Precambrian oils from Russia exhibit very high isoheptane ratios. Thompson's aliphatic and aromatic curve lines show the typical limit for younger oils.

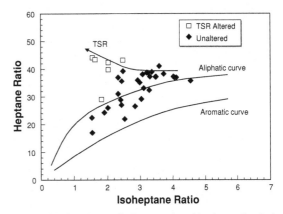

Figure 7.20. Smackover oils that were altered by thermochemical sulfate reduction (TSR) have decreased isoheptane and increased heptane ratios relative to normal unaltered oils.

Thompson (1987) conducted a series of experiments in which normal, non-fractionated oil was charged repeatedly with methane and the resulting equilibrated gas phase was removed and condensed. With each methane recharge, the residual oil becomes progressively depleted in heavy hydrocarbons. Because of different gas solubilities, light aromatic hydrocarbons preferentially partition into the residual oil relative to saturated species. Because normal and branched alkanes partition into the gas phase more readily than aromatic and cyclic hydrocarbons, the initial condensates had higher concentrations of normal and branched alkanes than the parent oil. The residual oil is enriched in aromatic and cyclic hydrocarbons relative to the parent oil, and addition of methane results in a gas-phase condensate that is also enriched in aromatic and cyclic hydrocarbons relative to the initial condensate. Subsequent methane recharges yielded both condensates and residual oils that become progressively depleted in normal and branched alkanes and enriched in aromatic and cycloalkanes relative to the previous residual oil.

Thompson (1987) used toluene/n-heptane (aromaticity ratio, B) and n-heptane/methylcyclohexane (paraffinicity ratio, F) to describe evaporative fractionation. The compounds used in these ratios show the largest difference in gas–liquid solubility among the C_7 hydrocarbons (toluene/n-heptane $= 0.22_g$, 0.34_l; methylcyclohexane/n-heptane $= 0.21_g$, 0.29_l) (Carpentier $et\ al.$, 1996) and toluene/n-heptane have a significant difference in volatility ($\Delta°C_{boiling\ point} = 12.3$). Plots of these ratios from various data sets resulted in the B-F diagram that Thompson used to illustrate several alteration processes (Figure 7.21).

Using data from three basins (Cook Inlet, Denver, and Powder River), Thompson (1987) found that normal, unaltered oils occupy a common region defined by (nC_7/MCH $\sim$0.4–0.8, toluene/nC_7 $\sim$0.2–0.6). We found that nC_7/MCH depends on both source and maturity and that the range for normal oils is broader than originally defined by Thompson. Oils from marine shales have nC_7/MCH $\sim$0.4–1.5, while oils from marine carbonates have nC_7/MCH $\sim$0.4–5.0+. In contrast, non-biodegraded oils from Tertiary coaly-source rocks yield nC_7/MCH $\sim$0.1–0.5. Because of this source-dependency, the normal oil area in the B-F diagram is not universal but should be defined for each petroleum system.

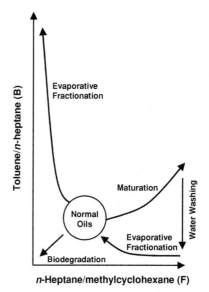

Figure 7.21. Thompson (1987) B-F diagram summarizing several reservoir alteration vectors in a plot of aromaticity ratio (toluene/n-heptane, B) versus his paraffinity ratio (n-heptane/methylcyclohexane, F). The area occupied by normal, unaltered oils is source-dependent and must be calibrated from individual petroleum systems. Based on laboratory simulations and empirical observations, Thompson proposed that evaporative fractionation initially results in a decrease in F and an increase in B. Subsequent evaporative fractionation will result in a rapid increase in B. Thompson proposed further that other alteration processes can also be illustrated with the B-F diagram. Vectors for maturation, water washing, and biodegradation are indicated.

Based on experiments and empirical observations, Thompson (1987) proposed that oils that experienced fractional evaporation will deviate from the normal oil region initially with decreasing nC_7/MCH and toluene/nC_7, followed by a rapid increase in toluene/nC_7 with little change in nC_7/MCH. The B-F diagram can be used to illustrate these deviations and provide indications that evaporative fractionation has occurred (e.g. Dzou and Hughes, 1993; Masterson $et\ al.$, 2001). We find that many sets of genetically related oils from single-source petroleum systems behave in this manner. Furthermore, the largest fractionation effects occur in deltaic basins, where conditions are commonly favorable for continual methane recharge. Thompson's B-F plot, however, must be used with caution because factors other than evaporative fractionation, such as biodegradation and water washing, can influence these ratios.

Several studies question whether evaporative fractionation accounts for toluene-rich fluids. Walters (1990) described condensates from High Island 511A in the Gulf of Mexico that are highly enriched in light aromatic hydrocarbons and appear to follow Thompson's evaporative fractionation model (Figure 7.22). Condensate from the A-6D well exceeds the maximum toluene/n-heptane ratio Thompson (1987) observed in

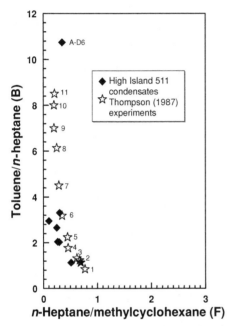

Figure 7.22. B–F diagram showing data from High Island 511A (♦) compared with Thompson's (1987) laboratory simulated evaporative fractionation. A parent oil (★) that was charged with methane was allowed to equilibrate. When the gas phase was removed, the initial condensate (☆1) was enriched in n-heptane relative to the parent oil. The residual oil was recharged with methane, equilibrated, and the condensed gas-phase hydrocarbons removed again. This process was repeated 11 times (☆2–11). Because the residual oil is always depleted in n-heptane relative to the previous experiment, the resulting condensate is also depleted, but not to the degree of the residual oil. Many crude oils from single-source petroleum systems follow the evaporative fraction alteration pathway toward the upper left in the figure. For example, condensates from High Island 511A in the Gulf of Mexico appear to be highly fractionated, although their compositions may not be due exclusively to evaporative fractionation (Walters, 1990).

simulations after 11 sequential methane-recharge experiments. This condensate would have to be depleted in C_7 hydrocarbons to reach such a composition by evaporative fractionation, and one might expect the n-alkane envelope to shift to a maximum near C_{12}–C_{15}. However, the High Island 511A condensates show no correlation between aromaticity, light-hydrocarbon depletion, and n-alkane maxima. There was, however, a correlation between aromaticity, GOR, and the amount of microbial methane as extrapolated from $\delta^{13}C$ of co-produced gases. Walters (1990) concluded that immature hydrocarbons (primarily toluene and methylcyclohexane from immature type III kerogens) were extracted by microbial methane as it migrated to the reservoir. Thus, the reservoirs with the highest GOR ($\sim$9500 MCF/BBL) and most microbial gas ($\sim$30%) exhibited the highest aromaticity.

Mango (1990) examined the Shell database of 2258 oils for evidence of evaporative fractionation. Based on his steady-state catalysis model, the rate of formation of methylcyclopentane (MCP) from a straight-chain precursor should be equal to the rate of formation of 1,3-dimethylcyclopentane (1,3-DMCP) from an isoalkane precursor, and the MCP/1,3-DMCP ratio should be invariant. Such invariance appears to occur in the data. However, the difference in boiling points for these compounds is significant ($\Delta^\circ C = 19.7$) and is greater than toluene/n-heptane ($\Delta^\circ C = 12.2$) or n-heptane/methylcyclohexane ($\Delta^\circ C = 2.3$). Mango argued that if evaporative fractionation occurs, then the invariance in the MCP/1,3-DMCP ratio would be perturbed to a greater extent than compound pairs with a smaller $\Delta^\circ C$. In the Shell database, 145 oil samples had toluene/$nC_7 >1$ and nC_7/MCH <0.5. Assuming that this subset underwent evaporative fractionation, Mango reasoned that the oils should exhibit the greatest variance in the MCP/1,3-DMCP ratio, with MCP being depleted. Instead, he found the opposite trend, with the mean MCP/1,3-DMCP for the subset being higher than for the complete data set. Mango concluded that evaporative fractionation does not occur. Mango's arguments must be viewed with caution because evaporative fractionation is driven as much by hydrocarbon solubility in methane as by volatility. The difference in solubility of individual cyclopentanes will be less than that between n-heptane and toluene or methylcyclohexane; therefore, the MCP/1,3-DMCP ratio may not be particularly diagnostic. Jarvie (2001)

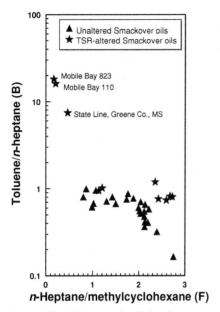

Figure 7.23. B–F diagram for oils/condensates generated from Smackover Formation (Jurassic) carbonate mudstone source rock (Rooney, 1995). Mild to moderate thermochemical sulfate reduction (TSR) appears to have little effect, perhaps elevating the toluene/nC_7 ratio slightly. Severely TSR-altered fluids, however, are influenced strongly with very high toluene/nC_7 and lower nC_7/MCH ratios. Condensates from Mobile Bay are so much altered by TSR that adamantanes (diamondoids) constitute most of the surviving liquid hydrocarbons.

found that the MCP/1,3-DMCP ratio was source-dependent.

TSR is another factor that can strongly influence the B–F ratios. Moderate TSR appears to slightly elevate toluene/nC_7 while having little influence on nC_7/MCH. However, in cases of severely altered TSR oils, toluene/nC_7 can be elevated drastically while the nC_7/MCH is reduced. This is illustrated on a B–F diagram for a set of oils derived from the Smackover carbonates (Figure 7.23). Rooney (1995) used compound-specific isotopic analysis to classify Smackover oils that were altered only slightly or moderately by TSR from those altered by thermal cracking. Such a distinction is not readily apparent simply from compositional data. The advanced degree of TSR in the State Line and Mobile Bay fluids is evident in both CSIA and compositional variations. Mobile Bay (offshore Alabama) condensates are altered so much that most of

the conventional hydrocarbons were destroyed, leaving a fluid highly enriched in adamantanes (diamondoids).

Several other alteration vectors specified by Thompson (1987) on the B–F diagram must be examined critically. Thompson considered oil and condensates with nC_7/MCH > 1.5 to be supermature, i.e. at maturity levels associated with late-stage generation or thermal cracking (R_o > 1.2%). The proposed maturation vector, indicated on the B–F diagram by an increase in both ratios from the normal oil region, is based on empirical observation of high-maturity condensates from the Zechstein Basin (Figure 7.24). We believe that the Thompson maturation vector is incorrect. The nC_7/MCH is highly dependent on the source kerogen type. Algal-rich kerogens yield high nC_7/MCH (1.5–>5) during the course of normal oil generation/expulsion. Thompson's B–F alteration vectors would

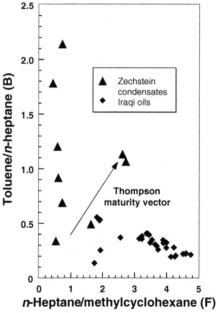

Figure 7.24. B–F diagram showing the relationship of Zechstein high-maturity oils/condensates and the derived Thompson (1987) maturity vector, and oils from Iraq. The latter were generated from Lower Cretaceous carbonates in the Sulaiy and Ratawai formations at low to moderate levels of thermal maturity (oil window). nC_7/MCH ratios >1.5, therefore, are not indicative of high maturity but are characteristic of algal-rich source rocks. Furthermore, there is no correlation between the 2,4-DMP/2,3-DMP ratio and the B or F ratios.

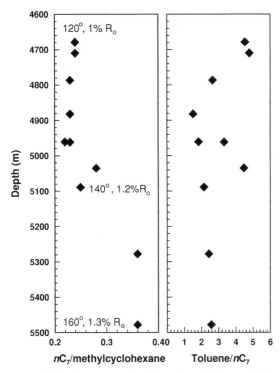

Figure 7.25. Drill stem test (DST) data from the Venture D-23 well, Scotian Shelf, showing an increase in nC_7/MCH with increasing thermal cracking. The magnitude of the effect, however, is small and could be related to other factors. No systematic depth trends are observed for toluene/nC_7.

quantities in severely degraded oils, a ratio of these two components may produce spurious results.

To summarize, we believe that evaporative fractionation occurs, particularly in deltaic systems where multiple source units can charge stacked reservoirs connected by faults. The Thompson plot can be used to illustrate this phenomenon, but conclusions should be supported by other data. For example, Masterson *et al.* (2001) used the B–F plot to show the effects of oil fractionation on a suite of Kuparuk and West Sak oils from the North Slope, Alaska. In this case, phase separation occurred due to loss of solution gas or gas from a paleo-gas cap in the Kuparuk reservoirs, which then migrated into West Sak. Supporting evidence for this fractionation was provided by molecular and isotopic analyses of the oils and by general agreement with the geologic model.

Thompson's B–F plot should not be used without such supporting data. We demonstrated that Thompson maturation vector is incorrect because high nC_7/MCH oils result from algal-rich kerogens, not thermal cracking. The area of normal oils is much broader and much more source-dependent than Thompson originally proposed. Jarvie (2001) effectively used the B–F plot to differentiate various sources of Paleozoic oils from the Williston Basin.

Halpern parameters

C$_7$ hydrocarbon ratios are useful in oil–oil correlations and in detecting subtle variations in chemical composition due to alteration processes. These ratios may be plotted in star diagrams for comparison.

Halpern (1995) proposed eight C$_7$ process (transformation) ratios that can be used in star diagrams to assess reservoir alteration processes for related oils (Table 7.3). Seven of the eight process ratios have 1,1-dimethylcyclopentane (1,1-DMCP) as the denominator, which Halpern stated is the C$_7$ hydrocarbon most resistant to biodegradation. Because of the enhanced solubility of light aromatics in water compared with saturated hydrocarbons, the ratio of toluene to 1,1-DMCP (TR1) is a measure of water washing, which may occur during or before significant biodegradation. While there are several processes that may enrich oil in light aromatic hydrocarbons (e.g. evaporative

argue that such oils are supermature and that the toluene has been extracted by water washing. This, however, is not the case, as when the B and F ratios are compared against independent estimates of thermal maturity (including biomarkers and isotopic data), the Iraqi oils provide no indication of high-temperature generation (Figure 7.25, left). We have observed minor increases in nC_7/MCH in thermally cracked, genetically related oils; however, the corresponding increase in toluene/nC_7 has not been demonstrated (Figure 7.25, right).

The other two alteration vectors in the B-F diagram are consistent with observations. Water washing selectively solubilizes toluene, reducing toluene/nC_7. Microbes selectively remove *n*-heptane, thus reducing nC_7/MCH. Since water washing always accompanies (and possibly precedes) microbial alteration, biodegradation drives normal oils toward the origin. As both toluene and *n*-heptane may be absent or in trace

Table 7.3. *Halpern (1995) C_7 ratios for use in star diagrams to differentiate oils*

Name	Ratio	ΔBP (°C)	ΔSolubility (ppm)	Process
TR1	Toluene/X	22.8	496	Water washing
TR2	nC_7/X	10.6	−21.8	
TR3	3-Methylhexane/X	4.0	−21.4	
TR4	2-Methylhexane/X	2.2	−21.5	
TR5	P2/X	(3.2)	(−21.4)	
TR6	1-*cis*-2-Dimethylcyclopentane/X	11.7	−11.0	Evaporation
TR7	1-*trans*-3-Dimethylcyclopentane/X	3.0	−4.0	
TR8	P2/P3	(6)	(−2.4)	
C1	2,2-Dimethylpentane/P3	(−5.8)	(−0.6)	
C2	2,3-Dimethylpentane/P3	(4.8)	(0.3)	
C3	2,4-Dimethylpentane/P3	(−4.5)	(−0.6)	Correlation
C4	3,3-Dimethylpentane/P3	(1.1)	(0.9)	
C5	3-Ethylpentane/P3	(8.5)	(−2.0)	

X = 1,1-dimethylcyclopentane, boiling point 87.8°C, solubility 24 ppm. P2 = 2-methylhexane + 3-methylhexane, boiling point 91°C, solubility 2.6 ppm. P3 = 2,2-dimethylpentane + 2,3-dimethylpentane + 2,4-dimethylpentane + 3,3-dimethylpentane + 3-ethylpentane, boiling point 85°C, solubility 5 ppm.
ΔBP = boiling point numerator − boiling point denominator (°C).
ΔSolubility = solubility of numerator − solubility of denominator (ppm in distilled water).
Parentheses indicate average values for mixtures.

fractionation, thermal cracking, and TSR), water washing is the only major process that can deplete these compounds in oil. Ratios of *n*-heptane, methylhexanes, and dimethylcyclopentanes to 1,1-DMCP provide six parameters (TR2–TR7) that are affected by biodegradation to varying degrees. The last parameter (TR8) is the ratio of methylhexanes to dimethylpentanes, which is likely to be the least affected by microbial activity. TR6 can be used to measure evaporation because the compounds in the ratio have very different boiling points and are not influenced greatly by differences in solubility or susceptibility to biodegradation. The eight ratios can be plotted using polar coordinates on the C_7 oil transformation star diagram (C_7-OTSD).

Halpern selected five C_7 ratios for correlation (Table 7.3). These five correlation parameters are the proportion of individual C_7 alkylated pentanes to the sum of these compounds. As these hydrocarbons have essentially the same solubility in water and the same susceptibility to microbial alteration, subtle differences in the distribution of these compounds will mostly reflect their source. Migration and phase behavior effects could come into play, as there are small differences in volatility between these compounds. These five ratios can be plotted using polar coordinates on the C_7 oil correlation star diagram (C_7-OCSD).

Published applications of the Halpern parameters are limited. Halpern (1995) demonstrated the use of the C_7 transformation and correlation star diagrams in determining the source of casing leakage in several Arabian fields, and in oil and condensate correlations in Saudi Arabia and the Red Sea. Wever (2000) used Halpern's parameters and star diagrams to differentiate oils and condensates from Egyptian basins in the Gulf of Suez, Western Desert, and the Nile Delta. Both studies demonstrated the value of light-hydrocarbon ratios used with biomarker and isotopic data.

We used Halpern parameters and star diagrams in a study of DST samples from the Sable Island E-48 well from the Scotian Shelf, offshore Eastern Canada. Fifteen individual zones were tested from 1460 to 2285 m using the Halpern correlation parameters. Most of the DST samples cluster tightly within a narrow band on the star diagram (Figure 7.26). This group includes both degraded and non-degraded fluids. DST 1 and 9 define

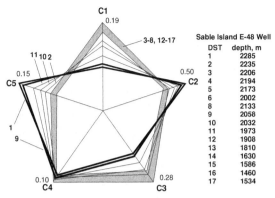

Figure 7.26. Halpern C₇ oil correlation star diagram for drill stem test (DST) samples from the Sable Island E-48 well, Scotian Shelf, Canada. The numbers at the end of each axis are endpoints for individual ratios. Two end-member oil groups are apparent. One group includes DST from depths of 1460–1908 m and 2133–2206 m (DST 12–17, 3–8). The other end member consists of DST 1 and 9. DST 2, 10, and 11 are mixtures of the two end-member fluids.

a second petroleum group. The ratios are not very different, but the variance is evident when plotted on the star diagram. DST 1 and 9 correlate very well, even though DST 1 lost appreciable volatile hydrocarbons during sampling and/or storage (Figure 7.27).

DST 2, 10, and 11 appear to be mixtures of the two end-member fluid groups based on the correlation parameters. By taking the averaged, normalized values for each Halpern correlation ratio, we can estimate the percentage contribution of each end member to the mixed samples by minimizing the combined error. Following this procedure, DST samples with intermediate compositions can be expressed as DST 11 = 67%, DST 10 = 38%, and DST 2 = 23% of the end-member composition, as defined by DST 1 and 9.

The Halpern transformation ratios reflect the source differences in the correlation ratios, plus an overprint largely due to differences in the degree of biodegradation (Figure 7.28). The relatively low 1,1-dimethylcyclopentane concentrations in the DST 1 and 9 samples cause them to show high TR1–TR7 ratios in the figure. Oils in the other end-member group either cluster tightly or show the effects of biodegradation. DST 14 and 15, which still retain small amounts of n-heptane and C₇ isoalkanes, are less altered than DST 16 and 17, which lack all but the most biodegradation-resistant C₇ cycloalkanes. TR7,

the most resistant ratio to biodegradation, is nearly identical for all samples in this group. DST samples 2, 10, and 11 have intermediate compositions, suggesting that they are mixtures. We could calculate relative end-member proportions for these samples, but transformation ratios yield substantially greater error than the correlation ratios.

The Halpern ratios and star diagrams provide a framework for investigating correlations and reservoir alteration processes using light hydrocarbons. They should not be used in isolation. As with all other approaches using light hydrocarbons, confirming evidence is required for valid conclusions. Ideally, this evidence might include supporting biomarker and isotopic data as well as a consistent geologic model. For example, Carrigan et al. (1998) used Halpern ratios and stable carbon isotopic evidence to show systematic differences among condensates from the Devonian Jauf reservoir in the giant Ghawar Field, Saudi Arabia. Although all of the condensates originated from the basal organic-rich hot shale of the Qusaiba Member in the Silurian Qalibah Formation, the data show distinct north–south trends, indicating at least six compartments within the reservoir that have not allowed mixing of the trapped hydrocarbons. The geochemical differences between condensates in these compartments indicate distinct migration pathways into the reservoir that drained different areas of the source kitchen. Early identification of these compartments assists design of efficient production strategies.

Halpern C₇ oil correlation star diagrams can be used to distinguish oils from a common source that were affected by TSR (Figure 7.29). Figure 7.29 shows four oils generated from Jurassic Smackover Formation carbonate mudstones that differ in the degree of alteration by TSR.

Mango parameters

C₇ hydrocarbon ratios based on Mango's steady-state kinetic model of light-hydrocarbon generation are useful for oil–oil and oil–condensate correlations and to determine the temperature of generation.

2,4-Dimethylpentane/2,3-dimethylpentane

Mango (1987; 1990) suggests that temperature controls preferential ring opening of cyclopropane

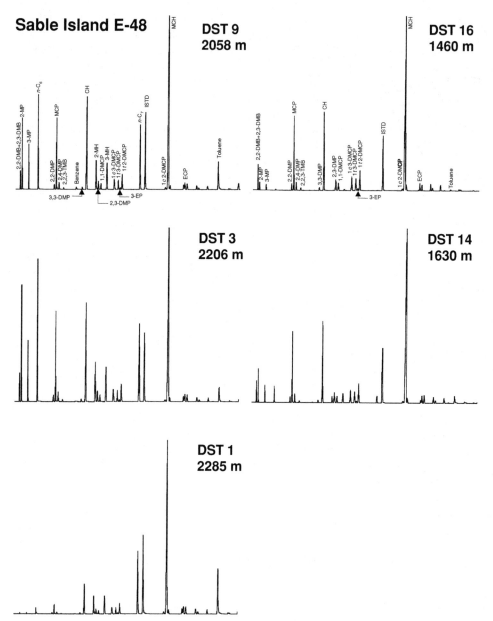

Figure 7.27. C$_7$ gas chromatograms of selected drill stem test (DST) samples from the Sable Island E-48 well. Two oil families can be distinguished based on light-hydrocarbon correlation ratios and compound-specific isotopic analyses. DST 1 and 9 define one family. The other family includes DST 3 and most remaining DST samples. DST 14 and 16 are examples of biodegraded condensates in the latter family.

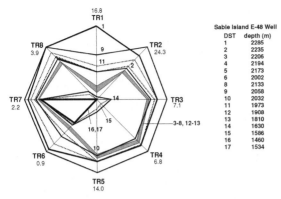

Figure 7.28. Halpern C$_7$ oil transformation ratio star diagram for drill stem test (DST) samples from the Sable Island E-48 well, Scotian Shelf, Canada. The numbers at the axis apex are the endpoints for the individual ratios. DST 14–17 were biodegraded (e.g. DST 14 in Figure 7.27).

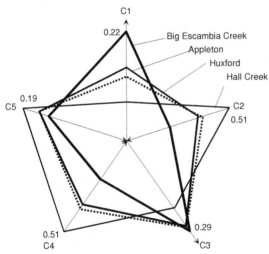

Figure 7.29. Halpern C$_7$ oil correlation star diagram for four Smackover oils from Escambia County, Alabama. The oils share a common source and thermal history, but differ in the degree of alteration by thermochemical sulfate reduction (TSR). Oil from the Hall Creek Field is unaltered, while oil from Big Escambia Creek Field has been subjected to strongest TSR. Oils from Appleton and Huxford were less altered by TSR. The arrows on the axes indicate the direction that Halpern correlation parameters change when altered by TSR. Numbers at the ends of each axis are the endpoints for individual ratios.

(three-ring) intermediates to form the isoheptanes. Based on this model, 2-methylhexane/3-methylhexane and 2,4-dimethylpentane/2,3-dimethylpentane (2,4-DMP/2,3-DMP) should depend on temperature, with

the latter being more sensitive. BeMent et al. (1995) examined source rocks of different ages and kerogen types (Santa Maria (Miocene, type IIS), Unita (Eocene, type I), Gulf Coast (Cretaceous, type II), Williston (Devonian-Mississippian, type II), and East Coast (Jurassic, type III)). Measured 2,4-DMP/2,3-DMP ratios were compared with calculated vitrinite reflectance determined from basin reconstructions of maximum burial temperature for each source rock. Based on these calibrations, BeMent et al. (1995) concluded that the 2,4-DMP/2,3-DMP ratio was independent of source and heating rate and corresponds to the temperature of generation, as discussed further in Mango (1997)

$$°C_{temp} = 140 + 15(\ln[2,4\text{-DMP}/2,3\text{-DMP}])$$

Based on a large oil database, BeMent et al. (1995) found that the calculated $°C_{temp}$ for primary oils ranged from ~95 to 140°C, with an average at ~125°C, values consistent with current kinetic models of oil generation. The $°C_{temp}$ formula constrains the value to temperatures within the window of oil generation. 2,4-DMP/2,3-DMP rarely exceeds 1.0 for oils, setting an upper limit for $°C_{temp}$ at 140°C. $°C_{temp}$ below 95°C result when 2,4-DMP/2,3-DMP <0.05. Uncertainty associated with these measurements is such that the allowable error always permits a lower $°C_{temp}$ limit of ~95°C. As additional support for $°C_{temp}$ as determined by 2,4-DMP/2,3-DMP, BeMent et al. (1995) showed several examples of genetically related oils showing good correlation between $°C_{temp}$ and API gravity. The original C$_7$ data from source rocks used to calibrate 2,4-DMP/2,3-DMP were not published.

Despite the potential significance of this work, few publications test the validity of 2,4-DMP/2,3-DMP. Chung et al. (1998) examined 2,4-DMP/2,3-DMP for a suite of oils from the North Viking Graben, where they found good correlation with other light hydrocarbon maturity ratios (e.g. C$_7$ quaternary/tertiary) but poor correlation with biomarker maturity parameters. However, they also observed good correlations between 2,4-DMP/2,3-DMP and various source parameters (e.g. pristane/phytane, δ^{13}C whole oil). They concluded that the North Viking oils are mixtures of low to middle-maturity oil generated from the Kimmeridge Clay Formation, middle to high-maturity oils from the more terrigenous Heather Formation, and high-maturity condensates from Jurassic coals. By examining

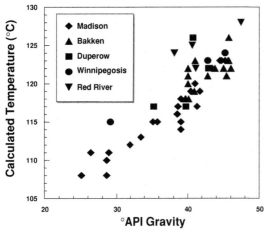

Figure 7.30. Correlation of API gravity and calculated temperature ($^\circ C_{temp}$) from 2,4-dimethylpentane/2,3-dimethylpentane (2,3-DMP/2,4-DMP) for oils from the Williston Basin. Oils generated from the Madison Group carbonates yield the lowest $^\circ C_{temp}$, consistent with a thermally labile type IIS kerogen (Jarvie, 2001).

the $\delta^{13}C$ isotopic variations between specific pairs of light hydrocarbons, Chung *et al.* (1998) isolated the influence of organofacies variations and demonstrated that 2,4-DMP/2,3-DMP indicates thermal maturity.

Jarvie (2001) published C_7 data for oils from the Williston Basin that show a correlation between API gravity and $^\circ C_{temp}$ (Figure 7.30). These oils were selected because they show no evidence of mixing in the reservoir, extensive vertical migration, or secondary alteration. Oils generated from the Madison Group carbonates yield the lowest $^\circ C_{temp}$, consistent with a thermally labile type IIS kerogen. Oils from the Bakken and older formations yield higher $^\circ C_{temp}$, with no obvious differences between sources.

Our experience indicates that relationships between 2,4-DMP/2,3-DMP and thermal maturity parameters are rarely as convincing as those in BeMent *et al.* (1995). For example, suites of Smackover oils from the Mississippi Salt Basin (BeMent *et al.*, 1995) and the Alabama Embayment (Walters and Hellyer, 1998) yield the same general trends with roughly the same variance (Figure 7.31). However, there is little correlation between 2,4-DMP/2,3-DMP and reservoir temperature. In this petroleum system, reservoir temperatures are approximately equivalent to source-rock temperature, because the lower Smackover Formation is in close

proximity to reservoir rocks in the overlying Smackover or the immediately underlying Norphlet formations.

Although some suites of oils from the same source rock exhibit correlations between $^\circ API$ gravity and 2,4-DMP/2,3-DMP, most do not. Most suites show fairly consistent $^\circ C_{temp}$ that do not correlate with gravity, other bulk properties, or biomarker maturity ratios. For example, oils from the Scotian Shelf and condensates from the deep Tuscaloosa Formation yield fairly consistent $^\circ C_{temp}$ of $\sim$132°C (Figure 7.32). Reservoir temperatures, which in the case of the deep Tuscaloosa exceed $^\circ C_{temp}$ by up to $\sim$65°C, show no correlation with $^\circ C_{temp}$.

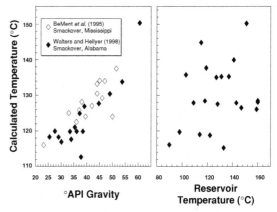

Figure 7.31. Relationship of API gravity and reservoir temperature to calculated temperature ($^\circ C_{temp}$) from 2,4-dimethylpentane/2,3-dimethylpentane (2,4-DMP/2,3-DMP) for Smackover oils.

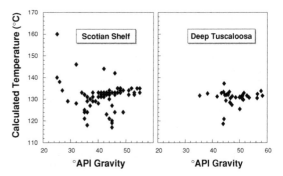

Figure 7.32. Relationship between calculated temperature ($^\circ C_{temp}$) and API gravity for oils from the Scotian Shelf and the deep Tuscaloosa Formation. Reservoir temperatures for the Scotian Shelf range from $\sim$70 to 160°C, while those from the deep Tuscoolsa range from $\sim$135 to 195°C.

We believe that 2,4-DMP/2,3-DMP and the derived $°C_{temp}$ can be used to indicate thermal maturity with certain restrictions. The reactions that drive this ratio appear to be restricted to the source rock. Subsequent thermal decomposition of expelled oil in reservoirs hotter than the source appears to be unimportant. This limits the effective temperature range to conditions associated with oil expulsion, between ~95 and 135°C. 2,4-DMP/2,3-DMP correlates with °API gravity only in relatively closed petroleum systems, where the source and reservoir are adjacent. The relationship between 2,4-DMP/2,3-DMP and °API gravity fails in basins where oil has undergone substantial vertical and/or lateral migration. In these cases, bulk oil properties can be so influenced by secondary processes that they no longer reflect the thermal maturity of the source rocks. There are no convincing studies that correlate biomarker thermal maturity parameters to 2,4-DMP/2,3-DMP. On the other hand, 2,4-DMP/2,3-DMP correlates well with other light-hydrocarbon ratios thought to depend on maturity. It is possible that differences between light-hydrocarbon and biomarker maturity parameters reflect mixing of light and heavy hydrocarbon components of differing thermal maturities.

K_1 and K_2: invariance of isoheptanes

Mango (1987) showed that all oils exhibit invariance in the proportion of isoheptanes: $K_1 = $ (2-methylhexane + 2,3-dimethylpentane)/(3-methylhexane + 2,4-dimethylpentane). K_1 for his data set of 2000 oils is ~1.06. Ten Haven (1996) verified the isoheptane invariance, finding K_1 ~1.07 for a data set of 500 oils. K_1, however, can vary between individual sets of oils. Mango (1987) showed that his data set was constrained by two such trends. Oils from Sabine Pass yield $K_1 = 1.09$, while Midland Basin oils yield $K_1 = 0.786$. Mango attributed differences in K_1 between the oil suites to different kinetic reaction rates of solid-state catalysis associated with different kerogens and source rocks.

K_1 can be used for oil–oil and oil–condensate correlations because oils derived from the same source rock yield identical K_1 values at all levels of maturity. For example, differences in K_1 exist between six oils from the northwestern basins of Argentina that were thought to originate from the Upper Cretaceous Yacoraite Formation (ten Haven, 1996). The oils could not be differentiated based on biomarker distributions or stable carbon isotopes of the C_{15+} hydrocarbon fractions. However,

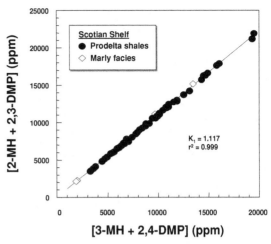

Figure 7.33. Oils and condensates from the Scotian Shelf show invariance of the Mango isoheptane ratio, K_1. For these samples, K_1 is insensitive to source differences established from biomarkers and compound-specific isotope analysis.

these oils can be divided into two groups based on their light hydrocarbon distributions ($K_1 = 1.87 \pm 0.01$ versus 1.47 ± 0.04). The two groups could represent different source rocks or two organofacies within the Yacoraite Formation. It is unknown why these K_1 values are so high compared with other oils.

Oil groups can be distinguished by statistically different K_1 values, but identical K_1 values do not assure correlation. For example, oils and condensates from the Scotian Shelf illustrate invariance of K_1 (Figure 7.33), even though biomarkers and compound-specific isotope data show them to originate from different shale versus marl facies. Subtle variations, such as those revealed by Halpern C_7 correlation plots for oils from the Sable Island E-48 well (Figure 7.26), are not clear using K_1.

Obermajer et al. (2000b) analyzed 189 crude oils from Paleozoic reservoirs in the Williston Basin for gasoline-range hydrocarbons ($iC_5H_{12}–nC_8H_{18}$) to verify the biomarker classification of oil families. Four distinct families of oils confined to specific stratigraphic intervals can be distinguished based on C_7 Mango parameters. The families are also evident based on their n-alkane and biomarker signatures. Most of the oils within each family show little to no variance in K_1, but K_1 is different for each oil family. Other Mango parameters (N2, P2, P3) support the oil families. They conclude that gasoline-range analysis complements classical biomarker correlations of crude oils, especially

TEHRGC of Rock Sample

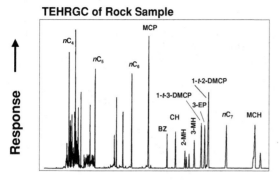

Gas Chromatography of Oil Sample

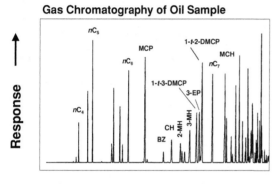

Retention Time ⟶

Figure 7.34. Thermal extraction high-resolution gas chromatogram of Bakken source rock (TEHRGC, top) and C_7 gas chromatogram of Lodgepole mound oil (bottom) supports an oil-to-source rock correlation (from Jarvie *et al.*, 2001a). For example, both the rock and oil show higher 1-*trans*-2-dimethylcyclopentane (DMCP)/nC_7 than other samples from the study area. Different columns and chromatographic conditions explain the different relative retention times for nC_7 and methylcyclohexane (MCH) near the end of each chromatogram (right). The column used for these chromatograms differs from that in Figure 7.13, giving slightly different retention times for various compounds. For example, 3-EP elutes midway between 1-*t*-3-DMCP and 1-*t*-2-DMCP in this figure, while in Figure 7.13 it elutes closer to 1-*t*-3-DMCP. Published with permission of the GCSSEPM Foundation. BZ, benzene; CH, cyclohexane; 2-MH, 2-methylhexane. 3-EP, 3-ethylpentane; MCP, methylcyclopropane.

when high thermal maturity decreases the usefulness of biomarkers. Biomarkers separate oils from Bakken, Lodgepole, Winnipegosis, and Red River reservoirs into four distinct groups. Oils from Bakken (K_1 = 0.90) and Lodgepole (K_1 = 0.86) could not be distinguished from each other but were clearly separate from

oils from Winnipegosis (K_1 = 1.23) and Red River (K_1 = 1.15). The latter two groups cannot be distinguished from each other based on K_1 alone but were differentiated by the relative amounts of light hydrocarbons. However, the Lodgepole Formation mound oils contain an unusual distribution where *trans*-1,2-dimethylcyclopentane exceeds the amount of *n*-heptane (Jarvie and Walker, 1997). Other oils in the Madison Group, and mature Bakken oils, do not exhibit this relationship. They found that Bakken Formation rocks immediately below the Lodgepole Formation mound oil discoveries have the same high 1-*trans*-2-DMCP/nC_7 as the oils (Figure 7.34). This correlation was confirmed by biomarker analysis.

$K_2 [P_3/(P_2 + N_2)]$ can also be used for correlation of genetically related oils. Plots showing correlations based on K_2 can be based on P_3 versus $(P_2 + N_2)$ or P_2 versus (N_2/P_3). Kornacki (1993) used such plots to distinguish oils generated from the phosphatic and siliceous facies of the Monterey Formation. Ten Haven (1996) used K_2 to show different sources for Northwest Basin oils of Argentina (Figure 7.35). However, such correlations are rare and considerable scatter on K_2 plots is typical, even for oils that can be shown to be similar based on other light-hydrocarbon distributions (e.g. ten Haven, 1996). Obermajer *et al.* (2000b) found K_2 plots to be only partially useful for separating oil families from the Williston Basin. No oil family produced straight-line K_2 correlations.

As with K_1, statistically different K_2 values indicate genetically different oils, but identical K_2 values do not assure correlation. For example, a suite of Middle East oils can be separated into four genetic groups based on biomarkers and isotopes. However, oils from Upper Jurassic and Lower Cretaceous carbonate source rocks within the suite cannot be distinguished using either K_2 values or parent–daughter plots (Figure 7.36). One interpretation of the parent–daughter plot might be that the scatter between the oils is due to mixing from two or more sources.

Parent–daughter ratio plots

> Plots of light-hydrocarbon distributions that follow Mango's kinetic reaction scheme, where parent compound(s) are related to daughter ratios. Rarely used for correlation.

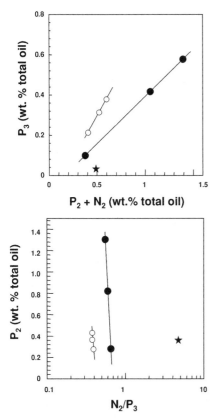

Figure 7.35. K$_2$ (top) and parent–daughter (bottom) plots for Devonian oil from Bolivia (stars) and two oil families from the northwestern basins of Argentina (dots). The oil families from Argentina may originate from different organic facies of the Cretaceous Yacoraite Formation, or one of the families could originate from the Paleocene Olmeda Formation. Reprinted from ten Haven (1996). © Copyright 1996, with permission from Elsevier.

Mango (1997) does not favor the use of K$_1$ and K$_2$ as correlation tools. Instead, he promotes classifications based on ring-preference reactions. Following his reaction scheme, P$_1$ (*n*-heptane) gives rise to P$_2$ (methylhexanes) via three-ring closures, N$_1{}^5$ (ECP + 1,(*cis*,*trans*)-2-DMCP) via five-ring closures, and N$_1{}^6$ (MCH + toluene). Parent–daughter ratio plots (e.g. *n*-heptane versus P$_2$/N$_1{}^5$, P$_2$/N$_1{}^6$, or N$_1{}^5$/N$_1{}^6$) or normalized ternary plots of the daughter products differentiate oils based on ring preference. Mango reasoned that various source rocks would differ in the conditions of metal catalysis and kerogen-free volume, thus exerting control on K$_1$, K$_2$, and ring preference.

Use of the Mango ring-preference plots as a correlation tool is limited. Mango claims that the daughter ratios are determined by source-rock conditions. Thus, similar source rocks will yield similar daughter ratios. As an example, he found that oils from Sabine Parish and Eugene Island were indistinguishable, suggesting common source conditions. In practice, we find that similar source facies yield different daughter ratios. For example, our Middle Eastern data set contains oils that were generated from both Upper Jurassic and Lower Cretaceous carbonates. The depositional settings for these source rocks were nearly identical and are difficult to distinguish even with complete biomarker and isotopic analyses. The daughter ratios for oils from these two sources span a considerable range and overlap nearly completely (Figure 7.37). The plots differentiate oils from Lower Cretaceous and Upper Jurassic carbonate, Cretaceous shales/marls, and Eocene shale source

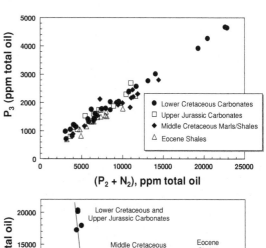

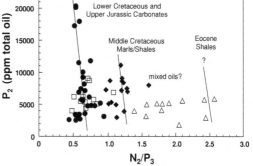

Figure 7.36. K$_2$ (top) and parent–daughter (bottom) plots for oils from Iraq, Iran, and Kuwait. The groups, determined from biomarker and isotopic analyses, are not distinguished by K$_2$. The scatter in the parent–daughter plot could be interpreted as being due to mixing of oils from multiple sources, but this must be verified independently.

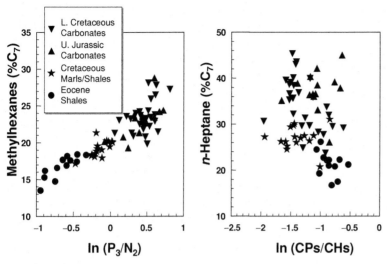

Figure 7.37. Parent–daughter ratio plots for Middle Eastern oils (Iraq, Iran, Kuwait) (after Mango, 1994). These correlation plots follow Mango's kinetic reaction scheme. In the first plot, the parent species are the methylhexanes (P_2), and the daughter ratio (P_3/N_2) is (sum of dimethylpentanes + ethylpentane + 2,2,3-trimethylbutane)/(1,1- + 1-(cis, trans)-3-dimethylcyclopentanes). In the second plot, the parent species is n-heptane (P1) and the daughter ratio (CPs/CHs or N_1^5/N_1^6) is [1-(cis, trans)-2-methylcyclopentanes + ethylcyclopentane]/ (methylcycohexane + toluene).

rocks. The spread in daughter ratios for these samples could indicate a transitional environment of source-rock deposition or mixing of crude oils from the two sources.

Primesum

> A single value derived from the Mango reaction scheme and empirical data that is used to determine whether secondary reservoir processes have altered oil. Rarely used and not recommended.

Kornacki and Mango (1996) describe a method to evaluate whether light-hydrocarbon distributions of crude oil were altered by secondary processes based on the concept that the C_7 distributions for all unaltered oils can be described as a linear combination of three end-member fluids (Mango, 1997). If correct, then the coefficients of the three end members should sum to one, the primesum:

$$\text{Primesum} = a + b + c$$

Calculation of the coefficients that make up the primesum (Table 7.4) is based on the Mango reaction scheme and semi-empirical determination of the

end-member fluid compositions defined by three C_7 ratios:

$$k^3 = (P_2 + P_3 + N_2)/P_3$$
$$k^5 = (N_2^5 + N_1^5)/2(P_3)$$
$$k^6 = N_1^6/P_3$$

End-member A is composed mainly of the six-ringed C_7 compounds (methylcyclohexane + toluene), end-member B is mainly dimethylpentanes, and end-member C is mainly methylhexanes and cyclopentanes. An oil sample can be defined as a linear mix of these theoretical end members by three equations:

$$\ln(k^3) = a(1.64) + b(1.11) + c(2.65)$$
$$\ln(k^5) = a(0.76) - b(3.62) + c(3.12)$$
$$\ln(k^6) = a(3.98) - b(1.46) + c(1.51)$$

By matrix inversion, the coefficients a, b, and c can be determined using the equations:

$$a = -0.0197\ln(k^3) - 0.1199\ln(k^5) + 0.28226\ln(k^6)$$
$$b = -0.2436\ln(k^3) - 0.1745\ln(k^5) - 0.0671\ln(k^6)$$
$$c = -0.2875\ln(k^3) - 0.1473\ln(k^5) - 0.1469\ln(k^6)$$

Mango (1997) indicated that primary oils, not altered by reservoir processes, yield primesum ~1, where

Table 7.4. *Calculated coefficients for the primesum (Mango, 1997)*

End member	k^3	k^5	k^6	$\ln(k^3)$	$\ln(k^5)$	$\ln(k^6)$
A	5.16	2.14	53.52	1.64	0.76	3.98
B	3.03	0.03	0.23	1.11	−3.62	−1.46
C	14.15	22.65	4.53	2.65	3.12	1.51

all coefficients are positive. He stated further that biodegraded oils yield primesum <0.8, with the *a* coefficient being negative (0 to −0.9). Based on our experience, this is only partially correct. Biodegraded oils yield primesum values <0.8, but the *a* coefficient is rarely negative. Rather, it is the *c* coefficient that reflects this behavior. Primesum data for the Sable Island E-48 DST fluids illustrate this effect (Table 7.5). Oils from depths >1800 m are unaltered and yield primesum values from 0.90 to 1.07 (average 0.96). Oils from shallower strata are biodegraded and yield primesum values from 0.54 to 0.65. While the primesum can indicate biodegradation, other light-hydrocarbon parameters are much more suited for this task. Because *n*-alkanes are readily biodegraded, parameters such as the heptane index, which include nC_6 or nC_7, are excellent indicators of microbial alteration. Other ratios that compare compound classes having different susceptibilities to biodegradation, such as the isoheptane or the Halpern transformation ratios, are also more diagnostic than primesum for biodegradation.

Mango (1997) adds that oils affected by TSR yield primesum >1 and are distinguished further by K_1 >1. Again, we find this generalization to be only partially correct. Many unaltered oils, particularly those from carbonate source rocks, have primesum values >1. Furthermore, oils altered by TSR based on compound-specific isotopic analysis yield primesum values indistinguishable from unaltered oils. K_1 values appear to reflect TSR, but only oils that were severely altered by TSR yield K_1 values appreciably greater than unaltered oils from the same source rocks (Table 7.6).

We conclude that primesum is not very useful in order to determine whether secondary processes altered crude oil. Unaltered oils have primesum values that span the range ∼0.8–1.2, with the average at ∼0.9. Oils that have been altered based on other methods may fall within this range. Only severely altered oils are readily distinguished from unaltered oils using primesum.

Molecular class or ring-preference plots

> Ternary diagrams showing normalized distributions of C_7 compounds by a molecular class that may or may not follow Mango's kinetic reaction scheme. These diagrams have numerous uses in illustrating correlations, maturation, and reservoir alteration processes. There are, however, no universal guidelines for interpretation.

Plots based on grouping compounds by molecular class are common in petroleum geochemistry. The approach has been used inconsistently to describe light-hydrocarbon distributions, partly because of the three-component limitation of ternary diagrams. C_7 hydrocarbons ideally require four or more components to express meaningful molecular class distributions. Researchers tend to lump those components that support a specific hypothesis, such as Mango's catalysis model (Table 7.7), or have similar ranges of values within the data set.

Dai (1992) plotted *n*-heptane (P_1), the sum of three cyclopentanes (1-*cis*-3, 1-*trans*-3, and 1-*trans*-2), and methylcyclohexane to give a ternary plot that distinguished products from source rocks with high terrigenous versus marine input. The terrigenous source was enriched in methylcyclohexane. Odden *et al.* (1998) used this plot to distinguish light hydrocarbons originating from the Spekk and Åre formations, offshore Norway.

Ten Haven (1996) used a ternary diagram to differentiate lacustrine from terrigenous oils in Vietnam. The diagrams were based roughly on Mango's scheme and showed the normalized distributions of three-ring $(P_2 + P_3)$, five-ring $(N_1^5 + N_2^5)$, and six-ring (N_1^6) compounds. Terrigenous and lacustrine oils were enriched in six-ring and three-ring compounds, respectively. Interpretation of the diagram from a restricted data set easily identified those oils thought to be mixtures. Such simple interpretations, however, are not universal. When ten Haven plotted his worldwide data set, he found considerable overlap between oil sources. The six-ring bias for terrigenous and three-ring bias for lacustrine oils was apparent but not well defined. Marine

Table 7.5. *Primesum values for Sable Island E-48 DST compared with Thompson ratios*

DST/ Test	Depth (m)	k^3	k^5	k^6	a	b	c	Primesum	Heptane ratio	Isoheptane ratio
1	2285	6.30	1.96	26.09	0.80	0.11	0.15	1.07	19.36	1.44
2	2235	5.15	1.28	12.16	0.64	0.19	0.14	0.97	20.48	1.63
3	2206	4.59	1.07	8.47	0.57	0.22	0.13	0.92	19.47	1.66
4A	2194	4.56	1.06	8.33	0.56	0.22	0.13	0.91	19.24	1.65
5	2173	4.65	1.10	8.81	0.57	0.21	0.14	0.92	19.30	1.64
6	2002	5.18	1.42	14.24	0.68	0.16	0.14	0.97	18.41	1.49
8	2133	4.53	1.05	8.05	0.55	0.22	0.14	0.91	19.30	1.67
9	2058	4.77	1.24	11.17	0.62	0.18	0.13	0.93	18.10	1.47
10	2032	6.11	1.70	20.11	0.75	0.15	0.16	1.05	21.51	1.68
11	1973	5.69	1.58	17.42	0.72	0.15	0.15	1.02	19.53	1.56
12	1908	5.01	1.21	10.07	0.60	0.20	0.15	0.95	20.74	1.67
13	1810	4.50	1.04	7.53	0.54	0.23	0.14	0.90	20.06	1.65
14	1630	2.77	1.16	9.38	0.59	0.07	−0.01	0.65	2.25	0.34
15	1586	2.64	1.19	9.75	0.60	0.05	−0.03	0.63	1.18	0.25
16	1460	2.26	1.17	8.69	0.58	0.03	−0.06	0.54	0.10	0.03
17	1534	2.25	1.14	8.50	0.57	0.03	−0.06	0.54	0.12	0.04

oils tended to show a preference for five-ring compounds but significantly overlapped the other fields. He concluded that the overlap between these groups was too great for the ternary plot to be applied universally.

Obermajer *et al.* (2000b) presented a ternary diagram that showed both the limitations and the utility of this technique. Their diagram followed Mango's classifications and plotted P_2, N_1^6, and ($P_3 + N_1^5 + N_2^5$) for oils generated from the Bakken and Madison formations. The overlap of these two groups was nearly complete, and the diagram could not differentiate the two sources. However, within the Bakken group, oils in the north were enriched in dimethylcyclopentanes compared with those from the central and southern parts of the basin, apparently due to minor mixing with Madison oils.

Jarvie (2001) also effectively used ternary diagrams to differentiate and identify mixing of oils from the Williston Basin (Figure 7.38). The diagrams plot *n*-heptane (P_1), six-ring (N_1^6) compounds, and the sum of all C_7 isoalkanes and cyclopentanes ($P_2 + P_3 + N_1^5 + N_2^5$). Pre-Devonian oils show enhanced *n*-heptane. Separating the Bakken and Madison oils is again problematic. Many of the Madison oils show high toluene and form a separate group. Other oils from Madison

reservoirs, however, plot with the Bakken oils and originate from the Bakken source rock.

There are no established guidelines describing the influence of source on the relative proportions of light hydrocarbons. Figure 7.39 shows the average C_7 values for a suite of oils separated by their source-rock depositional environment. Considerable variability exists within any source type, but the average compositions illustrate some general principles. Although there are fluctuations in the proportions of isoalkanes and dimethylcyclopentanes, most of the variance can be attributed to the proportions of *n*-heptane, methylcyclohexane, and toluene. The proportion of *n*-heptane can be traced to contributions of algal input to the source rock. Source rocks dominated by algal input (e.g. pre-Devonian) or that promote the preservation of algal organic matter (e.g. lacustrine, marine carbonates) yield oils that are enriched in *n*-heptane. In contrast, source rocks dominated by terrigenous or bacterial organic matter (e.g. shales, coals, evaporites) yield oils relatively depleted in *n*-heptane.

The proportion of methylcyclohexane appears to be related primarily to input from terrigenous organic matter. Oils derived from terrigenous coals and coaly shales are enriched in methylcyclohexane, where it is

Table 7.6. *Primesum values for oils derived from Smackover carbonate source rocks*

Field	County	Depth (m)	k^3	k^5	k^6	a	b	c	Primesum	K$_1$
South Cypress Creek	Wayne	4361	5.45	1.31	7.66	0.509	0.230	0.228	0.967	1.132
Pool Creek	Jones	3834	4.94	0.80	3.78	0.370	0.338	0.233	0.940	1.154
South Cypress Creek	Wayne	4406	5.57	1.02	5.26	0.433	0.304	0.253	0.990	1.161
Cypress Creek	Wayne	3866	5.77	1.10	6.19	0.469	0.288	0.251	1.007	1.162
Mt Carmel	Santa Rosa	4697	5.18	1.12	6.30	0.473	0.257	0.220	0.951	1.164
Blacksher	Baldwin	4773	5.39	1.20	8.12	0.536	0.238	0.204	0.979	1.167
Pool Creek	Jones	3462	4.83	0.77	2.51	0.260	0.369	0.279	0.908	1.169
Movico	Mobile	5156	4.94	0.93	6.66	0.512	0.274	0.172	0.957	1.173
Pachuta Creek	Clarke	3978	5.20	0.85	5.44	0.466	0.317	0.201	0.984	1.173
Walkers Creek	Monroe	4435	5.57	1.43	7.11	0.477	0.224	0.259	0.960	1.182
Nancy	Clarke	4075	5.17	0.70	4.49	0.434	0.361	0.200	0.995	1.183
Nancy	Clarke	4112	4.94	0.68	4.12	0.414	0.361	0.196	0.971	1.189
Hall Creek	Escambia	4578	5.57	0.93	6.22	0.491	0.308	0.216	1.014	1.198
Pachuta	Clarke	3945	4.92	0.78	4.26	0.407	0.334	0.210	0.951	1.200
Turkey Creek	Choctaw	3775	5.74	0.99	6.89	0.512	0.299	0.217	1.028	1.202
Hatters Pond	Mobile	5589	7.70	1.14	11.27	0.627	0.311	0.251	1.190	1.208
Jay	Santa Rosa	4737	5.63	1.14	7.07	0.502	0.267	0.230	0.999	1.210
Sugar Ridge	Choctaw	3527	5.74	0.73	5.35	0.478	0.369	0.209	1.056	1.215
Chunchula	Mobile	5632	7.05	1.10	9.50	0.586	0.308	0.245	1.139	1.245
Cold Creek	Mobile	5627	6.58	1.08	7.89	0.537	0.307	0.250	1.094	1.251
Brantley Jackson	Hopkins	2820	5.57	0.86	5.90	0.485	0.325	0.211	1.022	1.269
Gin Creek	Choctaw	4139	5.96	0.97	6.85	0.512	0.311	0.227	1.049	1.345
Bryan Mills	Cass	3134	6.36	1.34	6.08	0.438	0.278	0.311	1.027	1.355
Chatom	**Washington**	4921	7.48	1.92	10.59	0.548	0.218	0.329	1.095	1.424
Crosby Creek	**Washington**	4997	6.91	1.40	9.24	0.549	0.263	0.280	1.091	1.430
Huxford	**Escambia**	4451	6.65	1.53	9.00	0.532	0.240	0.285	1.057	1.455
Vocation	**Monroe**	4245	7.02	1.65	10.24	0.558	0.231	0.293	1.083	1.493
Appleton	**Escambia**	3940	6.38	1.28	11.52	0.624	0.245	0.211	1.080	1.562
Como	**Hopkins**	3848	9.87	2.85	16.40	0.619	0.188	0.402	1.209	2.024
Big Escambia Creek	**Escambia**	4645	6.73	1.46	19.17	0.750	0.200	0.171	1.122	2.250
State Line	**Greene**	5287	5.81	1.35	91.14	1.203	0.073	−0.111	1.165	4.862

Oils in **bold** were shown to be altered by thermochemical sulfate reduction (TSR) based on δ^{13}C of individual C$_7$ hydrocarbons.
All oils from Alabama, except State Line (Mississippi).

usually the largest single component. Lacustrine oils also tend to have higher proportions of methylcyclohexane than those from marine systems. The data suggest that there is an important biological precursor for methylcyclohexane in higher plants. However, this compound is ubiquitous in source rocks and oils, including those that lack higher-plant organic matter (e.g.

pre-Devonian), so methylcyclohexane must also have algal or microbial precursors.

Oils derived from type III kerogens are enriched in toluene. In these source rocks, lignin provides a direct biological source for the light aromatic hydrocarbons. Anoxic, iron-poor depositional environments, such as evaporites and marine carbonates, also are relatively

Table 7.7. *Molecular classes of C₇ hydrocarbons by structure and by Mango's kinetic reaction model*

					C₇H₁₆				C₇H₁₄			C₇H₈
n-Alkane				Branched				Naphthenic				Aromatic
		Mono		Di+				Five-ring		Six-ring		
nC_7	P_1	2-MH 3-MH	P_2	2,2-DMP 2,3-DMP 2,4-DMP 3,3,-DMP EP 2,2,3-TMB	P_3	ECP 1*t*2-DMCP 1*c*2-DMCP	N_1^5	MCH	Toluene			
						1,1-DMCP 1*t*3-DMCP 1*c*3-DMCP	N_2^5	N_1^6				

c, *cis*; DM, dimethyl; E, ethyl; H, hexane; M, methyl; P, pentane; *t*, *trans*; TM, trimethyl.

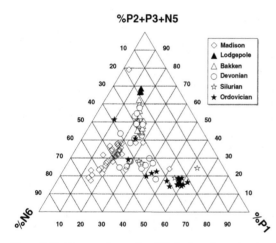

Figure 7.38. Ternary diagram for all C₇ hydrocarbons, showing differences between crude oils from various Williston Basin petroleum systems (Jarvie, 2001). The diagram plots *n*-heptane (P_1), six-ring (N_1^6) compounds and the sum of all C₇ isoalkanes and cyclopentanes ($P_2 + P_3 + N_1^5 + N_2^5$), as defined in Table 7.7.

enriched in toluene. This is attributed to the cyclization and aromatization of straight-chain alkanes during diagenesis as promoted by sulfur (Sinninghe Damsté *et al.*, 1991; 1993a). Because of the divergent pathways that give rise to toluene and methylcyclohexane, we do not recommend summing these components in ternary plots or statistical analyses.

COMPOUND-SPECIFIC ISOTOPIC ANALYSIS OF LIGHT HYDROCARBONS

Volatile hydrocarbon fractions of crude oils can have $\delta^{13}C$ that differ from the non-volatile C₁₅₊ hydrocarbons. The extent of these differences was not realized until advances in CSIA allowed routine determination of the $\delta^{13}C$ of individual hydrocarbons. Early CSIA studies of crude oils focused on the *n*-alkanes (e.g. Bjorøy *et al.*, 1991; Sofer *et al.*, 1991) and biomarkers (Freeman et al., 1990). The technique is particularly well suited for light

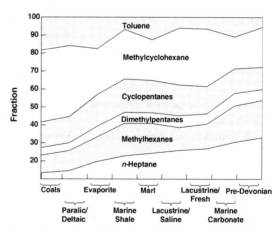

Figure 7.39. Average distribution of C₇ hydrocarbons for a suite of well-characterized oils classified by their source-rock depositional environment, as defined by biomarker analysis and oil–source rock correlations.

hydrocarbons, although it is not used widely. For this reason, CSIA of light hydrocarbons is discussed as an ancillary geochemical method, while CSIA of *n*-alkanes and biomarkers was discussed in Chapter 6. Baseline chromatographic separation, coupled with little or no column bleed, allows for highly precise measurement of C_1–C_7 hydrocarbons (typically $\pm$ 0.5‰, but as good as $\pm$ 0.1‰). While *n*-alkanes showed little or systematic changes with carbon number, wide variations in $\delta^{13}C$ between individual light hydrocarbons occur. Bjorøy *et al.* (1994) used $\delta^{13}C$ patterns of both *n*-alkanes and other light hydrocarbons for oil-condensate correlation. By coupling CSIA with conventional geochemistry, Clayton and Bjorøy (1994) and Chung *et al.* (1998) showed how the effect of thermal maturity could be discerned in complex source systems. Rooney (1995) showed that large isotopic shifts occur for selected light hydrocarbons when altered by thermochemical sulfate reduction. CSIA of light hydrocarbon continues to be developed for detailed correlation and petroleum systems analysis (e.g. Whiticar and Snowdon, 1999).

The most obvious use for CSIA of light hydrocarbons is for correlation. Unless altered by reservoir processes, oils and condensates generated from the same source rocks should have the same pattern of $\delta^{13}C$ for individual light hydrocarbons. The Sable Island 48 well DST samples discussed earlier illustrate the power of CSIA. Halpern correlation (Figure 7.26) and transformation (Figure 7.28) plots indicated two groups of end-member oils: DST 1 and 9 represent one end member, while DST 3–8 and 12–13 represent the other. DST 2, 10, and 11 were mixed oils. DST 14–17 were biodegraded, and DST 14 and 15 were less altered than 16 and 17. If these conclusions are correct, then they should be consistent with isotopic values of individual light hydrocarbons.

CSIA supports the groups of Sable Island 48 well DST samples established using the Halpern parameters (Figure 7.40). The error in these measurements is estimated to be $\pm$ 0.3‰. Biodegradation dramatically altered the $\delta^{13}C$ of the <C_7 normal and branched hydrocarbons, while leaving the C_6 cyclic and C_7 hydrocarbons unaltered. The end-member fluids from DST 1 and 9 based on Halpern parameters are isotopically enriched in ^{13}C compared with other fluids, particularly for the C_6 hydrocarbons. The other end-member fluids form a tight cluster with less internal variation. The mixed oils have a stronger affinity to the DST 1 and

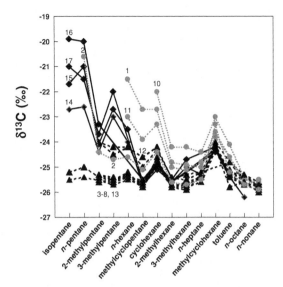

Figure 7.40. Compound-specific isotopic analysis of light hydrocarbons in oil samples from the Sable Island 48 well supports the oil groups established using Halpern parameters (e.g. Figure 7.26). Data labels refer to drill stem test (DST) numbers in the well.

9 end members, exhibiting heavier than expected $\delta^{13}C$ for some C_6 and C_7 hydrocarbons. From the CSIA data, DST 7 and 12 (end-member fluids based on C_7 ratios) show isotopic signatures suggesting some mixing.

By integrating molecular and isotopic analysis of both light hydrocarbons and biomarkers, complex petroleum systems can be revealed in far greater detail than by using only one technique. In many cases, apparent conflicts in interpretations indicate that simple explanations do not describe reality adequately. Oil reservoirs charged by a single source within a narrow range of thermal maturity without subsequent alteration are probably rare. Most accumulations probably received multiple charges from one or more sources and/or were modified by reservoir processes.

Preliminary data suggest that CSIA of light hydrocarbons may be useful in order to distinguish geochemically similar crude oils from Silurian and Devonian source rocks in Algeria (Peters and Creaney, 2003). Silurian-sourced oil samples from Hassi Messaoud and Zemlet fields differ from four Devonian oil samples in the isotopic composition and pattern of isotopic compositions among certain *n*-alkanes and gasoline-range branched and cyclic hydrocarbons (Figure 7.41). The origins of the oil samples were established by geologic

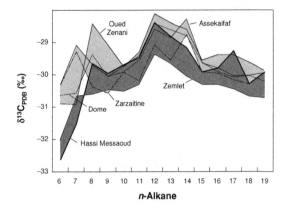

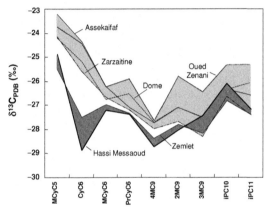

Figure 7.41. Compound-specific isotopic analyses of *n*-alkanes (top) and selected branched and cyclic light hydrocarbons (bottom) for six Algerian crude oils (Peters and Creaney, 2004). The largest isotopic difference between Hassi Messaoud and Zemlet oil samples (Silurian source rock, dark stipple) and the four Devonian samples (light stipple) occurs at *n*C$_6$. The Hassi Messaoud and Zemlet samples have more negative stable carbon isotope ratios for methylcyclopentane (McyC5), cyclohexane (CyC6), methylcyclohexane (McyC6), and several other light hydrocarbons compared with the Devonian samples. Unlike the other samples, the Hassi Messaoud and Zemlet oil samples have cyclohexane that is isotopically more negative than methylcyclopentane and methylcyclohexane.

constraints (e.g. see Figure 9.7). Except at *n*C$_7$ and *n*C$_9$, the Zemlet oil sample has more negative stable carbon isotope ratios for the C$_6$–C$_{19}$ *n*-alkanes than the other Algerian oil samples. The isotopic composition of *n*-alkanes in Hassi Messaoud oil resembles more closely the other samples than the Zemlet sample, except at *n*C$_6$ and possibly *n*C$_7$. Admixture of Devonian oil may affect the heavier components in the Hassi Messaoud

oil sample, or their isotopic compositions simply may not be diagnostic of Silurian versus Devonian oil. However, the *n*C$_6$ in Hassi Messaoud and Zemlet oil samples are substantially depleted in ^{13}C (1.1–2.3‰) compared with the other samples. Hassi Messaoud and Zemlet oil samples also have more negative CSIA stable carbon isotope ratios for several gasoline-range cyclic hydrocarbons, especially methylcyclopentane, cyclohexane, and methylcyclohexane ($\geq$0.6, 1.9, and 0.2‰, respectively) compared with the other Algerian oil samples. The pattern of isotopic ratios among these three compounds also varies systematically. Unlike the other oil samples, cyclohexane is more depleted in ^{13}C than methylcyclopentane and methylcyclohexane in the Hassi Messaoud and Zemlet oil samples.

MOLECULAR MODELING

Quantum mechanics or molecular mechanics can be used to compute the geometries and properties of molecules, including biomarkers (Figure 7.42). Both methods assume an initial set of nuclear positions. In the quantum mechanical method, the initial nuclear positions are used to compute molecular orbitals and electron density, which are used to calculate potential energies of molecules. The forces on atoms in a molecule are calculated by computing the change in potential energy as the molecular geometry is changed. As these forces change, they are used to adjust the positions of the atoms. This process is repeated until the most stable geometry with minimum energy is obtained. The minimum-energy geometry, molecular orbitals, and electron density are then used to compute molecular properties, such as molecular volume or dipole moment. Quantum mechanical techniques are slow for large molecules because molecular orbitals and electron density are interdependent and must be determined iteratively.

In molecular mechanics, the potential energy of the molecule is expressed as a function of geometric variables, such as bond lengths and angles, rather than in terms of molecular orbitals and electron density. Molecular mechanics uses empirical energy functions to describe the geometries, energies, and properties of molecules. Internal coordinates, including bond lengths, bond angles, torsion angles, and non-bonding interactions, control the geometry of a molecule. The relation between the internal coordinates and the potential

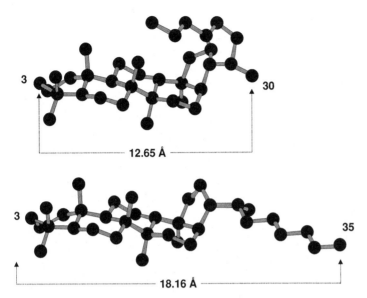

Figure 7.42. Molecular mechanics calculations explain the observed preferential demethylation at C-10 in the C_{35} 17α-hopane favouring the 22R versus the 22S epimer during in-reservoir biodegradation. Geometry optimization of the 22S (top) and 22R (bottom) epimers results in so-called scorpion and rail conformations, respectively. These preferred conformations suggest that microbial attack of the methyl group attached to C-10 will be sterically hindered more by 22S than by 22R stereochemistry. However, another mechanism must explain conservation of 18α-oleanane, which has a methyl group at C-10 but lacks a side chain like that in the extended hopanes. Reprinted from Peters *et al.* (1996b). © Copyright 1996, with permission from Elsevier.

or steric energy is described by various energy functions. The set of functions is developed using a training set of compounds and is called an empirical force field. The MM+ force field is an extension of MM2 (Berkert and Allinger, 1982). The nuclear positions can be adjusted based on the computed forces, and new forces can be computed without the need to recompute molecular orbitals or electron density. Removal of this step makes molecular mechanics computations of molecular geometry much faster than quantum mechanical methods.

Geometry optimization identifies favorable geometries with respect to the internal coordinates. Geometry optimization techniques minimize the potential energy of a structure. A conformation resulting from geometry optimization represents the closest energy minimum to the starting structural model, as described above. This closest minimum energy conformation is generally a local minimum on the potential energy surface describing the molecule. The global energy minimum is the overall lowest energy conformation on the potential energy surface.

Kolaczkowska *et al.* (1990) compared the thermodynamic stability of selected C_{27}–C_{31} alkylated, dealkylated, and rearranged 17α- and 17β-hopanes using MM2 methods. Calculated equilibrium ratios of common maturity-dependent ratios (e.g. Ts/Tm and 22R/22S) were comparable to those observed in thermally mature crude oils. Van Duin *et al.* (1996a) developed a molecular force-field method tailored specifically to address tertiary carbocation reactions that are believed to occur during biomarker isomerization. Using this approach, the diagenesis of Δ^7-5α-sterenes (van Duin *et al.*, 1996b) and homohopanes (van Duin *et al.*, 1997) and their product distributions were modeled, providing insight into the reactive pathways and intermediates. Using a different force-field approach, van Duin and Sinninghe Damsté (2003) modeled reaction pathways for the cyclization of the diaromatic carotenoid isorenieratene. They concluded that the formation of tetracyclic isorenieratene derivatives proceeds by means of an A-ring-initiated reaction mechanism, while the formation of monoaromatic derivatives involves B-ring-initiated cyclization reactions.

Molecular dynamics is another computation approach that describes liquid–liquid and liquid–solid interactions (Frenkel and Smit, 2001). Using molecular dynamics, van Duin and Larter (1998) studied the partitioning behavior of benzocarbazoles in water and oil phases. These simulations suggested that benzo[c]carbazole has a slightly higher affinity for the hydrocarbon phase than benzo[a]carbazole, supporting the use of the ratio of these compounds as an indictor of migration distance. Molecular dynamics simulations of the partitioning behavior of benzocarbazoles on water-wet and hydrocarbon-wet clay surfaces, however, showed that the benzocarbazoles have a slight preference for being absorbed on water-wet kaolinite surfaces over being desorbed into the water phase (van Duin and Larter, 2001). In this case, the isomers show no significant differences in adsorption behavior.

Computational chemistry can be applied to modeling hydrocarbon generation from kerogen. Using *ab initio* quantum mechanical calculations, the kinetics and mechanisms of the thermal alteration of kerogen can be modeled (Xiao, 2001). These computations are sufficiently complex to predict accurately the product distribution from hypothetical kerogen structures. Such calculations have provided insight into the underlying reactions behind C_7 isoheptane ratio invariances (van Duin and Larter, 1997; Xiao, 2001). They are also capable of modeling isotopic compositions of gases during generation. For example, Xiao (2001) showed how *ab initio* modeling could explain the $\delta^{13}C$ correlations seen in the "natural gas plot" of Chung *et al.* (1988), which plots the stable carbon isotope composition of individual gaseous hydrocarbons versus the inverse of the carbon number of each molecule.

FLUID INCLUSIONS

Fluid inclusions are imperfections in minerals that trap small quantities of oil, gas, and water. Primary inclusions form during initial mineral growth within intracrystalline microcavities, while secondary inclusions form during cementation and occur in intergranular pore space or in microfractures. Many minerals may contain fluid inclusions, but the most common in sedimentary basins are carbonates, silicates (quartz and feldspar), and evaporites (halite, anhydrite, and fluorite). Inclusion

diameters are typically measured in several microns but range from submicroscopic to centimeter scales. Typical mass contents are in the order of nano- to femtograms.

Many fluid inclusions preserve the chemistry and physical properties of the original parent fluids from which they formed. Consequently, they are small "time capsules" of the composition, temperature, and pressure of static and migrating subsurface fluids. For this reason, fluid inclusions are commonly used in studies of petroleum generation and migration. Fluid inclusion analysis is also used in studies of the petrogenesis of igneous, metamorphic, and sedimentary rocks, in metals transport and ore formation, in geomechanics, stress analysis, and paleo-earthquakes, geothermal energy, and mobilization of subsurface contaminants, including radionuclides (Roedder, 1984; De Vivo and Frezzotti, 1994; Goldstein and Reynolds, 1994).

When fluid inclusions form, the cavity within the mineral matrix contains a homogeneous liquid. When cooled, inclusions may remain as a single liquid phase or separate into liquid aqueous, liquid hydrocarbon, and/or a vapor phase that contains both non-hydrocarbon and hydrocarbon gases. Inclusions may also include inorganic and organic solids that precipitated after formation. Because inclusions are self-contained systems, the temperature of entrapment can be determined by heating samples and measuring the temperature at which the separated phases homogenize. Typically, microthermometric measurements are made on multiple inclusions, yielding a histogram of homogenization temperatures. These analyses can distinguish different inclusion-forming events and identify inclusions that may have been perturbed. Homogenization temperatures are used in thermal models to constrain the temperature and age sequence of mineral cementation, episodes of fluid (water, oil, and gas) migration, and fracture healing. Commonly, many different events can be discerned within a single rock sample (e.g. Burruss *et al.*, 1983). Aqueous inclusions can also be cooled, allowing salinity to be determined by freezing-point depression, provided that certain assumptions are made about chemical composition.

If the phase behavior of petroleum and coeval aqueous fluid inclusions is known, then the geopressure at the time of formation can be determined (Aplin *et al.*, 2000; Pironon *et al.*, 2000; Thiéry *et al.*, 2000; Tseng *et al.*, 2002). This calculation requires knowing the homogenization temperature, the composition of the

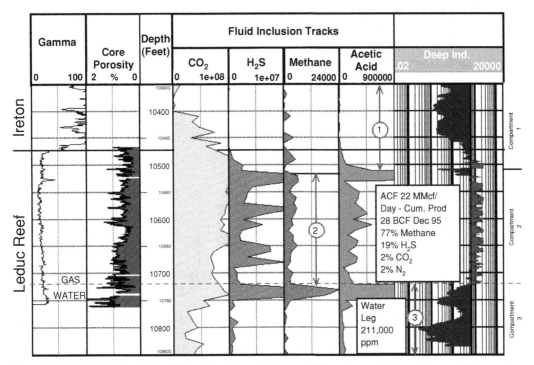

Figure 7.43. Fluid inclusion volatiles (FIV) well log helps to delineate seal (1), gas reservoir (2), and water leg (3) for a well penetrating the Devonian Leduc Reef, Canada (Hall *et al.*, 2002b). Additional columns show gamma-ray response, core porosity, well depth, and deep induction log response. Reprinted by permission of the AAPG, whose permission is required for further use.

fluids, and the volume of the vapor and liquid within the inclusion. The latter measurements are obtained using confocal scanning laser microscopy (CSLM). Paleo-pressure can then be calculated using these data as input in pressure, volume, temperature (PVT) simulators that describe the equations of state for the inclusion components. Thus, fluid inclusions are capable of constraining both temperature and pressure for paleohistory reconstruction.

Various methods are used to determine the chemical compositions of fluid inclusions. Nuclear microprobe techniques measure elemental compositions, which are needed to understand ore formation from hydrothermal fluids. These include proton-induced X-ray emission (PIXE), proton-induced gamma-ray emission (PIGE), backscattering spectrometry, elastic recoil detection analysis (ERDA), nuclear reaction analysis (NRA), channeling contrast microscopy (CCL), and ionoluminescence (Ryan and Griffin, 1993; Timofeeff *et al.*, 2000). Analysis of petroleum compositions in fluid inclusions relies on indirect spectroscopy (e.g.

laser Raman, FTIR, UV fluorescence) or direct analysis by mass spectrometry or GCMS (Munz, 2001). Spectroscopic methods have the advantage of being able to resolve individual inclusions, but they are limited to methane, light hydrocarbon gases, and bulk petroleum compositions, such as alkane/aromatic ratios (Burke, 2001). UV fluorescence spectra of petroleum have been corrected to bulk properties, such as API gravity, and provide a means in order to differentiate petroleum inclusions (McLimans, 1987; Wang and Mullins, 1994; Kihle, 1995). Interpretations of fluorescence colors of petroleum inclusions face numerous artifacts and ambiguities (e.g. George *et al.*, 1997; 2001).

Direct measurement of composition requires breaking open the inclusions, usually by crushing or thermal decrepitation. Volatile components can be introduced directly into a mass spectrometer (Barker and Smith, 1986), or the inclusions can be extracted and the released petroleum analyzed as conventional bitumen (Jones and Macleod, 2000). The disadvantage of direct measurement is that the composition is an average based on multiple rather than individual inclusions. Because there is no way to control the breakage, the measured composition may reflect inclusions formed during multiple events or may include bitumen that escaped removal by prior solvent extraction.

Mass spectral analysis of volatiles released by crushing inclusions is emerging as a routine tool in exploration. The technique, termed fluid inclusion stratigraphy (FIS) or fluid inclusion volatiles (FIV), involves running a suite of core or cuttings samples from one well to construct a geochemical log from the detector response of diagnostic mass fragments versus well depth (Figure 7.43). The full utility of this technique is still under investigation, but claims have been made for its success in defining pay zone, present and paleo-fluid contacts, petroleum quality, seal integrity, proximity to pay, and reservoir delineation (Parnell *et al.*, 2001; Fishman *et al.*, 2002; Hall *et al.*, 2002a; Hall *et al.*, 2002b).

The most abundant light hydrocarbons in fluid inclusions are commonly analyzed by direct introduction (either by a crushing injector or by thermal decrepitation) into the analyzer. However, the analysis of biomarkers in fluid inclusions is currently limited to "off-line" procedures. The techniques are relatively simple, involving crushing the rock in a mortar under a solvent (Karlson *et al.*, 1993). The rock samples, however, must be previously extracted thoroughly and treated with chemical oxidants to assure that surface bitumens are removed and only organic matter encapsulated in inclusions remains (Jones *et al.*, 1996; George *et al.*, 1997; Jones and Macleod, 2000). Extracted hydrocarbons can then be separated and analyzed by conventional procedures.

Various studies have combined fluid inclusion and biomarker analysis of free bitumens and oil to constrain thermal conditions (e.g. Cazier *et al.*, 1995; Honghan *et al.*, 1998) or to examine Mississippi Valley-type ore deposits (e.g. Henley and Hoffmann, 1987; Etminan and Hoffmann, 1989; Rowan and Goldhaber, 1995; Rowan and Goldhaber, 1996; Rowan *et al.*, 1994a; Rowan *et al.*, 1994b; Rowan *et al.*, 1995; Hulen and Collister, 1999; Guilhaumou *et al.*, 2001). Other studies analyzed biomarkers in fluid inclusions primarily to constrain migration pathways and filling histories (George *et al.*, 1997; George *et al.*, 1998a; George *et al.*, 1998b; Isaksen *et al.*, 1998; Scotchman *et al.*, 1998; Bhullar *et al.*, 1999; Jones and Macleod, 2000; Ruble *et al.*, 2000).

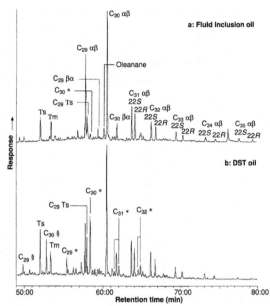

Figure 7.44. Reconstructed *m/z* 191 ion chromatograms, showing the triterpane distribution of included and free oil from Toro sandstones, Papua Fold Belt. Reprinted from George *et al.* (1996b). © Copyright 1996, with permission from Elsevier. * Diahopanes; § unidentified hopanoids.
DST, drill stem test.

For example, George *et al.* (1997) examined the biomarker distributions in free oil and fluid inclusions in the Lower Cretaceous Toro sandstones from the Papuan Fold Belt (Figure 7.44). They showed that the oils differed in both source and maturity. The free oil probably originated from Middle-Upper Jurassic shales. In contrast, oil in the inclusions contained 1,2,7-trimethylnaphthalene and oleanane, indicating angiosperm input and a Cretaceous or younger source rock. Maturity indicators suggested that the oil in the inclusions was more mature than the produced oil. George *et al.* (1997) concluded that the oil in the inclusions represented initial reservoir charge to the Toro sandstones from deeply buried, probably Cretaceous source rock during the Miocene. This oil was later diluted and/or displaced by a larger volume of oil generated from Jurassic shales. Fluid inclusions did not form from this latter charge, because the reservoir was already oil-saturated, thus inhibiting mineral diagenesis.

Isaksen *et al.* (1998) characterized fluid inclusions and associated oils and oil-shows in Mesozoic reservoirs from the Sleipner area, North Sea. They showed that the oil in the inclusions was less mature, although the

28,30-bisnorhopane contents were lower than in the free oil. This pattern is not expected if bisnorhopane were diluted progressively, as more hydrocarbons were generated from the source rocks. Consequently, they concluded that the difference in bisnorhopane concentrations between the fluid inclusions and free oils is due primarily to variations in the generative yields from different source rocks.

Biomarker analyses of free oils, adsorbed oils, and oil-bearing fluid inclusions indicate at least two oil-charging events for reservoir rocks collected from the Kuche Depression in the Tarim Basin, China (Pan *et al.*, 2000). The compositions of oil-bearing fluid inclusions differ from the free oils. Compared with free oil in the reservoirs, oils in the fluid inclusions have relatively high Pr/nC_{17} and Ph/nC_{18}, low Pr/Ph, hopanes/steranes, C_{30} diahopane/hopane, and Ts/Tm, low content of $C_{29}Ts$ terpane, and high maturities, as indicated by C_{29} sterane $20S/(20R + 20S)$. The early-migrated oils in the fluid inclusions correlate with the oils from the northern and central Tarim Basin, which originated from Cambrian-Ordovician marine source rocks. The later oils migrated from Triassic-Jurassic terrigenous source rocks and strongly diluted the earlier charge in the reservoirs. The adsorbed oils appear to be an intermediate type between free oils and oil-bearing fluid inclusions.

Pan *et al.* (2003) conducted a similar study that compared the biomarker distribution of free and included oils from the Junggar Basin, China. Samples from the northwestern and eastern edges of the basin had similar biomarker distributions and correlated with local Permian source rocks. However, included oil from the central area differed substantially from produced oils, indicating that this portion of the basin received petroleum charges from multiple sources. This conclusion was supported by regional field and seismic data.

8 · Biomarker separation and analysis

This chapter describes the organization of a biomarker laboratory and the methods used to separate and prepare crude oils and sediment or source-rock extracts into fractions before mass spectrometric analysis. The concept of mass spectrometry is explained. Many of these fundamentals, such as the difference between a mass chromatogram and a mass spectrum, and that between selected ion and linked-scan modes of analysis, are critical to understanding later discussions of biomarker parameters. Several of the notes in this chapter, including analytical procedures, internal standards, and examples of gas chromatography/ mass spectrometry (GCMS) data problems, help the reader to evaluate the quality of biomarker data and interpretations.

ORGANIZATION OF A BIOMARKER LABORATORY

A well-balanced biomarker laboratory requires input and support from various disciplines. The following description of a biomarker laboratory may not represent the only good way to organize this type of group, but our experience shows that it works well.

The most critical stage in any geochemical project is the input required: (1) define the problem and (2) designate a coordinator to function as a liaison between the generalists (e.g. regional geologists) and the biomarker specialists. The coordinator evaluates the suitability of geochemical methods for answering the problem, assembles appropriate samples based on a clear understanding of the regional geology or environmental setting, and prepares a detailed list of objectives for the biomarker specialist. The coordinator generally screens samples by requesting the appropriate routine geochemical analyses (such as Rock-Eval pyrolysis, total organic carbon (TOC), vitrinite reflectance, and gas chromatography) from an in-house laboratory or a service company.

The biomarker team consists of specialists with expertise covering several topics:

- Chromatographic separation methods, including column chromatography and high-performance liquid chromatography (HPLC).
- Natural product chemistry, required to evaluate precursor–product relationships.
- Biomarker and GCMS interpretive skills obtained by experience and knowledge of the literature.
- Geology, reservoir engineering, or other appropriate disciplines required to evaluate the condition and quality of samples and to place the biomarker results in perspective for the generalists.
- Familiarity with conventional geochemical parameters and their applications.

Supporting technologies that are essential for the smooth operation of the biomarker group include:

- Advanced mass spectral and electronics technology.
- Synthetic organic chemistry, nuclear magnetic resonance (NMR) spectroscopy, and X-ray diffraction crystallography for identification and structural elucidation of unknown compounds.
- Computer science required for data collection and processing.

Additional support includes access to relevant literature on each problem through computerized library searches, and supporting geochemical analyses conducted within the organization or elsewhere. Finally, constructive and timely peer review is essential in order to ensure a uniformly high quality of released technical material.

SAMPLE CLEAN-UP AND SEPARATIONS

Biomarkers in petroleum normally must be concentrated before analysis. For example, bitumen can be extracted from ground sedimentary rocks using ultrapure solvents to avoid contamination. Because the crude oils, bitumens, and sediment extracts that contain biomarkers are complex mixtures, glass-column and high-performance liquid chromatography are used to clean and separate them into fractions before analysis by GCMS. For example, the saturated hydrocarbon fraction represents the non-aromatic organic compounds in petroleum, including normal and branched alkanes and cycloalkanes. The aromatic hydrocarbon fraction contains organic compounds with one or more unsaturated rings, such as monoaromatics (C_nH_{2n-6}) and polycyclic aromatic hydrocarbons as well as some compounds that contain sulfur, nitrogen, and oxygen. Various sophisticated mass spectral techniques, such as GCMS/MS, multisector (i.e. triple quadrupole) mass spectrometry, and high-resolution GCMS, offer the possibility of direct analysis of whole oils for biomarkers.

Solvent extractions and separations are based on the like-dissolves-like principle. Solvents most readily dissolve solutes of approximately the same polarity. A polar molecule contains regions of opposite electrical charge. Methanol (CH_3OH), for example, has positive and negative charge regions on the methyl and hydroxyl groups, respectively. Normal hexane (*n*-hexane) is nonpolar compared with methanol. In column chromatography, the oil or bitumen sample is poured on to alumina or other suitable adsorbent in a glass column, and fractions of increasing polarity are obtained by pouring increasingly polar solvents through the column. Soxhlet extraction is a common method used to obtain soluble bitumen from rock samples. A hot organic solvent, such as chloroform or dicholormethane, or an azeotrope mixture of solvents (e.g. choroform : methanol) is refluxed through a crushed rock sample, extracting the bitumen.

Column chromatography

In traditional column chromatography, saturated and aromatic hydrocarbon fractions can be separated from samples by adding *n*-pentane or *n*-pentane/dichloromethane, respectively, to a glass column packed with silica gel and alumina. A modified column chromatography method uses neutral alumina to isolate

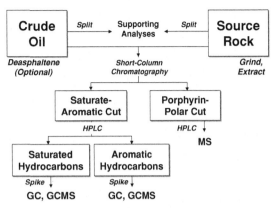

Figure 8.1. Flow chart showing procedure to separate oils and bitumens into fractions for analysis. The short-column chromatography apparatus is shown in Figure 8.2. Reprinted with permission by ChevronTexaco Exploration and Production Technology Company, a division of Chevron USA Inc. GC, gas chromatograph; GCMS, gas chromatography/mass spectrometry; HPLC, = high-performance liquid chromatography; MS, mass spectrometry; spike, internal standard.

polar nitrogen fractions for use as petroleum migration tracers (i.e. benzocarbazoles) (Li *et al.*, 1992; Larter *et al.*, 1996). The latter procedure allows deasphaltened oil and bitumen to be separated into saturated and aromatic hydrocarbon fractions and a nitrogen-enriched fraction by sequential elution with *n*-hexane, toluene, and chloroform/methanol. However, the neutral alumina method yields severely altered distributions of monoaromatic hydrocarbons in the aromatic hydrocarbon fraction compared with the traditional method (Chunqing *et al.*, 2000a).

Details of typical sample clean-up and separation procedures are described below. Figure 8.1 shows a generalized flow chart of procedures.

Adsorption of sample on alumina

The oil or bitumen sample is dissolved in methylene chloride (HPLC grade), mixed with 15 times its weight of alumina (e.g. Baker alumina oxide acid powder for chromatography, deactivated to Brockman activity II by adding water to 2.9 wt.%), and distributed evenly at the bottom of a round-bottomed flask. The solvent is removed by rotoevaporation at low speed to avoid bumping and generation of fine alumina particles. Final rotoevaporation is done at full-house vacuum (~170–200 mBar) with the flask in a 40°C waterbath.

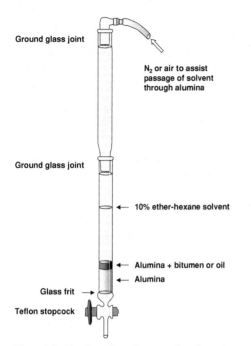

Ground glass joint

N₂ or air to assist
passage of solvent
through alumina

Ground glass joint

← 10% ether-hexane solvent

← Alumina + bitumen or oil
← Alumina

Glass frit →

Teflon stopcock

Figure 8.2. Alumina column for separation of saturate–aromatic
and porphyrin–polar fractions from petroleum samples
(short-column chromatography in Figure 8.1). In this figure, the
column containing the alumina and sample (bottom) has been
extended by another section of column (top) to accomodate the
ether : hexane mixture used to elute the fractions. Reprinted with
permission by ChevronTexaco Exploration and Production
Technology Company, a division of Chevron USA Inc.

Preparation and loading of column

The chromatography column is half filled with di-
ethyl ether : hexane (10 : 90 vol : vol) to which alumina
(50 times the weight of the sample) is added (Figure 8.2).
The alumina with the adsorbed sample is loaded into
the column and rinsed from the column walls on to the
top of the clean alumina bed using a few milliliters of
ether : hexane.

Saturates and aromatics are eluted together from
the column using 10 : 90 ether : hexane (HPLC grade,
ten times the weight of the clean alumina). The column
can be extended for this step by adding another section of glass column above the original column. Positive
pressure of nitrogen or air, cleaned using molecular sieve
traps, can be applied at the top of the solvent stream to
assist passage of the solvent through the column.

Porphyrins and polar compounds are eluted together using 100% chloroform (HPLC grade) until a
dark brown band comes off the column. The eluent can

be checked by ultraviolet (UV)–visible spectrophotom-
etry (350–600 nm) for quantitative isolation. After rotoevaporation and weighing, the saturate–aromatic and
porphyrin–polar fractions are ready for further separations using HPLC.

High-performance liquid chromatography

The cleaned saturate–aromatic fraction of the oil or bitumen can be separated using an HPLC pump equipped
with a silica guard column (Figure 8.3). The guard column protects the main column from irreversibly adsorbed compounds.

The eluent can divided into three cuts (saturates,
mono- plus di- and triaromatics, and polar compounds)
using a programmable fraction collector and 400-ml
glass bottles. Under normal flow conditions (Figure 8.3,
lower left), saturates and aromatics elute with hexane,
while methylene chloride is used for the polar compounds. Before the polars elute from the column with
methylene chloride, the backflush valve is switched, resulting in passage of the eluent from the injection port
to the main column and back through the guard column.
In this reversed flow valve position (Figure 8.3, lower
right), the direction of flow through the main column
remains the same but flow through the guard column
is reversed (backflushed). The columns are partially re-
equilibrated with hexane before the next separation.

UV (254 nm) and refractive index (RI) detectors
monitor the fractions using a multichannel chromatographic data system. The cut point between saturates
and aromatics is based on retention times of cholestane
and monoaromatic steroid standards, as discussed below. The cut between triaromatics and polars is made
immediately after elution of dimethylphenanthrene.

Fractions are rotoevaporated under house vacuum,
transferred to tared vials, and rotoevaporated again using a 45°C water bath for ten minutes and a vacuum of
35 mmHg. Unusually low concentrations of biomarkers,
as in certain condensates, may require further treatment
of the saturate fraction by urea adduction or molecular
sieves to remove normal alkanes (e.g. see Figure 13.26)
(Michalczyk, 1985).

INTERNAL STANDARDS AND
PRELIMINARY ANALYSES

Subsamples of whole oil or bitumen and separated fractions are always taken for auxiliary geochemical analyses,

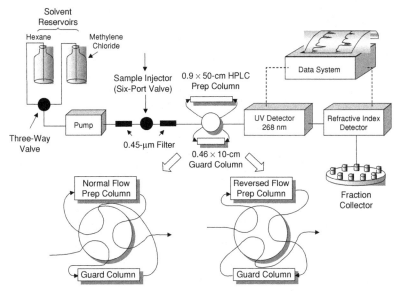

Figure 8.3. Configuration of the high-performance liquid chromatograph (HPLC) for automated separation of saturate, aromatic, and polar fractions from petroleum samples. Except for the sample inlet valve, all operations are automated using a microprocessor in the pump. Concept for figure courtesy of F. J. Fago. Reprinted with permission by ChevronTexaco Exploration and Production Technology Company, a division of Chevron USA Inc.

such as gas chromatography, stable carbon isotope ratios, sulfur content, and API gravity. Internal standards added to the separated saturated and aromatic fractions facilitate quantitation of chromatographic peaks. For example, a small aliquot of the saturated fraction is prepared for injection into a gas chromatograph for a preliminary evaluation of its behavior during GCMS. This sample is spiked with four branched alkanes (3-methylheptadecane, 3-methylnonadecane, 2-methyldocosane, and 3-methyltricosane) to facilitate peak measurements and determine whether further treatment is necessary before GCMS analysis. This gas chromatographic analysis commonly confirms that certain bitumens or heavy oils require dilution with a solvent, such as hexane. Extremely waxy oils may require removal of n-alkanes by urea adduction to allow adequate concentrations of biomarkers for GCMS analysis.

Quantities of biomarkers in oils and bitumens may be determined using steroid internal standards. These steroids do not occur in nature and offer the advantage of fragmenting to give the same principal ion as the steroids being measured. If quantities of biomarkers are not determined, then differences in detector response between the various compounds being analyzed are ignored. This

approach is not valid when comparing compounds with different mass spectral characteristics.

Many laboratories add 5β-cholane to the saturate fraction as an internal standard for sterane and terpane measurements before all GCMS analyses (Seifert and Moldowan, 1979). 5β-Cholane is not found in significant abundance in crude oils, does not interfere with the indigenous compounds, and fragments to give the same principal ion (mass/charge = m/z 217) by the same mechanism as other steranes (Figure 8.4). Figure 8.21 (upper left) shows the m/z 217 mass chromatogram for oil from Hamilton Dome, Wyoming, used as a standard. The mass chromatogram includes the internal standard peak identified as 5β-cholane (peak 1 in Figure 8.21).

Two internal standards are added to the aromatic fraction (Figure 8.4). For monoaromatic steroids, the standard is a synthetic C_{30} monoaromatic-steroid mixture of four epimers (5β(20S), 5α(20S), 5β(20R), and 5α(20R)), which fragment to give the same principal ion (m/z 253) as natural monoaromatic steroids. Figure 8.22 shows the monoaromatic-steroid mass chromatogram for oil standard from Carneros, California, including the four aromatic-steroid epimers used as the internal standards (peaks 17–20). For triaromatic steroids, the

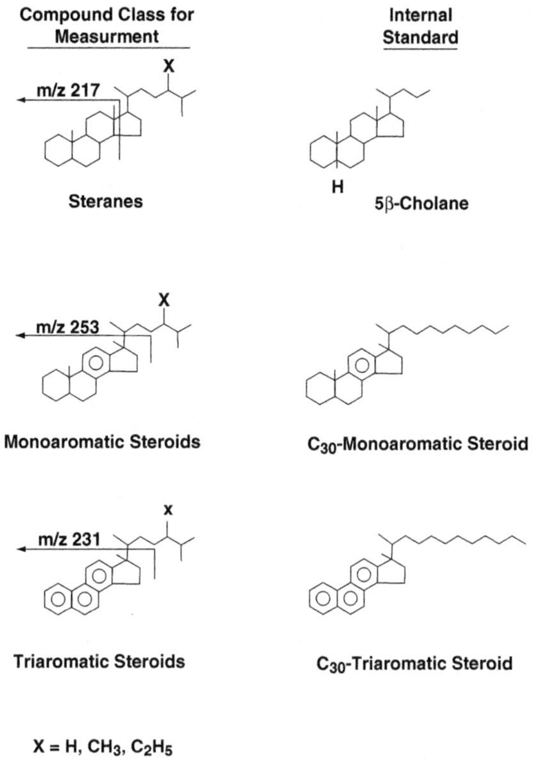

Compound Class for Measurement

m/z 217

Steranes

Internal Standard

5β-Cholane

m/z 253

Monoaromatic Steroids

C₃₀-Monoaromatic Steroid

m/z 231

Triaromatic Steroids

C₃₀-Triaromatic Steroid

X = H, CH₃, C₂H₅

Figure 8.4. The steroids used as internal standards for the saturate [5β-cholane] and aromatic (C₃₀ mono- and triaromatic steroids) fractions are not found in petroleum in significant concentrations. However, these steroids fragment to give the same principal ions in the source of the mass spectrometer as the steroids being measured. This allows the quantities of biomarkers in the petroleum fractions to be determined. Peaks corresponding to the above internal standards are identified on mass chromatograms in Figures 8.21–8.23. C₃₀ mono- and triaromatic steroid standards were prepared in collaboration with D. S. Watt and co-workers. Reprinted with permission by ChevronTexaco Exploration and Production Technology Company, a division of Chevron USA Inc.

internal standard is a synthetic C₃₀ triaromatic-steroid mixture of two epimers (20S and 20R), which fragment to give the same principal ion (m/z 231) as natural triaromatic-steroids. Figure 8.23 shows the triaromatic-steroid mass chromatograms for the standard oil from Wyoming, including peaks representing the above internal standards.

Some deuterated compounds are available commercially (e.g. Chiron Laboratories, Norway) for use as internal biomarker quantitation standards for GCMS.

In deuterated standards, one or more deuterium atoms (a heavy isotope of hydrogen, atomic weight 2) are substituted synthetically for hydrogen atoms (atomic weight 1). The result is a molecule that has very similar physical and chemical properties to the non-deuterated analog but that is one unit heavier for each hydrogen atom that has been replaced by deuterium. Thus, the advantage of these deuterated standards is that they behave in almost the same way as the compounds being analyzed under mass spectrometric conditions. A disadvantage is that the standards and compounds to be analyzed must be recorded on chromatograms with different mass/charge.

ZEOLITE MOLECULAR SIEVES

Because of their sieve-like properties, dehydrated zeolites or molecular sieves can be used to concentrate compounds from petroleum, thereby allowing more reliable analysis of otherwise trace components. Zeolites are hydrated crystalline aluminosilicates composed of three-dimensional networks of aluminum and silicon tetrahedra (AlO₄ and SiO₄) that are linked together by shared oxygen atoms (e.g. Breck, 1974). Group I or II metals or other cations between the networks balance the negative charge of the aluminosilicates. Dehydration of zeolites generates pores or channels that allow them to be used in order to separate compounds from complex mixtures based on subtle differences in size and shape.

For many years, organic geochemists have routinely separated the branched and cyclic alkanes from normal alkanes in the saturate fraction of petroleum using zeolites, such as Linde 5A molecular sieves (Murphy, 1969; Breck, 1974; Jasra and Bhat, 1987; Ruthven, 1988). An automated extension of this process, in which two silicalite molecular sieves with different retention characteristics are arranged in series, separates the normal, branched, and cyclic alkanes from crude oils (Nolte, 1991). Unfortunately, these conventional separation methods commonly fail to yield mixtures that can be resolved fully using capillary gas chromatography. Further concentration procedures offer the potential for detailed analysis of minor components in these complex mixtures. However, the literature contains few references on the use of zeolites to further concentrate selected biomarkers or other compounds from either the saturated or aromatic fractions (Kenig et al., 2000). The following is a brief summary of the various structural types of zeolite and some of the pioneering efforts to use them to selectively concentrate compounds from petroleum, as discussed below.

> **Note:** Silicalite is a term that has been used incorrectly to designate high Si/Al ZSM-5 zeolite (Flanigen *et al.*, 1978). ZSM-5 and so-called silicalite are equivalent (Fyfe *et al.*, 1982). Budiansky (1982) describes the patent dispute that arose due to the use of this term. It is now generally recognized that the Si/Al for synthetic ZSM-5 is limited only by Al impurities in the starting materials. Silicalite is not an item of commerce.

Major impetus to improve zeolite-based methods for concentrating selected compounds from petroleum has been provided by compound-specific isotope analysis (CSIA), also called isotope ratio monitoring/gas chromatography mass spectrometry (IRM/GCMS) (Schoell and Hayes, 1994). The use of CSIA to reconstruct paleoenvironments and characterize petroleum is limited mainly by co-elution of compounds and the resulting inaccuracies in measured isotope compositions. For example, Schoell *et al.* (1992) reported a precision of 0.2–0.3‰ for the stable carbon isotope ratios of *n*-alkanes with no background and 0.1–1.5‰ for complex mixtures. Co-elution is more severe for other compounds in petroleum because they are generally far less abundant than *n*-alkanes.

Ellis and Fincannon (1998) examined the stable carbon isotope ratios of whole crude oil, the saturate fraction, the molecular sieve fraction, and the urea adduct. They found no isotopic fractionation effects by molecular sieves for partially adducted *n*-alkanes. They also found that pretreatment of whole oil is critical for reliable isotopic analyses of *n*-alkanes. The mean difference between urea and molecular sieve samples was only 0.2‰. The mean difference between sieving (urea or molecular sieve) and the isolated saturate fraction was 0.5‰, while that between sieving and untreated whole oil was 1.1‰. More surprisingly, they found that the use of perdeuterated internal standards for CSIA without *n*-alkane isolation is dangerous. Isotopic internal standards apply a correction factor to all measured components. If the internal standard co-elutes with another peak, then the resulting spurious correction factor is applied to every component of interest in the sample.

Source- and age-related biomarker applications provide another impetus to develop better methods to concentrate specific compounds. Two examples of biomarkers that bear source and age information are 24-*n*-propylcholestanes and oleanane, respectively (Moldowan *et al.*, 1985; 1994a). When present in sufficient concentrations for unequivocal identification, 24-*n*-propylcholestanes in a crude oil are diagnostic of marine source rock, while oleanane indicates angiosperm (flowering plant) input to Cretaceous or younger source rock. Identification of these compounds by GCMS or GCMS/MS becomes difficult for saturate fractions from condensates or light oils where biomarker concentrations are low. In such cases, additional concentration of the biomarkers using zeolites is a viable option.

> **Note:** Trace biomarkers can be introduced into light oils or condensates during migration through carrier beds or as carry-over contaminants from improperly cleaned laboratory equipment that was used previously to prepare biomarker-rich samples. These problems can be detected if the analyst recognizes that they can occur. For example, trace oleanane with $18\alpha/18\beta$ indicating low thermal maturity in highly mature condensate is likely to have originated from low-maturity carrier beds traversed by the condensate during migration.

Zeolite structures

Zeolites are crystalline, hydrated aluminosilicates that typically contain group I or group II elements when not in hydronium form. They may be represented by the generic formula:

$$M_{2/n}O \cdot Al_2O_3 \cdot xSiO_2 \cdot yH_2O$$

These are complex crystalline structures, comprising the major group of the framework silicates. Most structural information on zeolites is obtained by X-ray diffraction (e.g. McCusker, 1994), with additional data provided by infrared and nuclear magnetic resonance (NMR) spectroscopy (e.g. Engelhardt and Michel, 1987). Details on the methods used to analyze concentrated compounds, such as CSIA, NMR spectroscopy, and X-ray crystallography, are beyond the scope of this book.

The fundamental unit of zeolites consists of the silicon tetrahedron (silicon in tetrahedral coordination with four oxygen atoms). A second component, typically aluminum, coordinates tetrahedrally as well as octahedrally with oxygen. Silicon, aluminum, and other tetrahedrally coordinated atoms in the zeolite frame-work are described as T-atoms.

The substitution of aluminum (Al^{3+}) for silicon (Si^{++}) results in an electrical charge that must be lo-cally neutralized by an additional positive ion, such as sodium, within the pores of the zeolite structure. These charge-compensating cations are mobile and can be ex-changed with other cations (Vaughan, 1988). Larger charge-compensating cations result in smaller effective pore sizes.

Synthetic zeolites are produced under carefully controlled hydrothermal conditions (Vaughan, 1988; Kerr, 1989). Much of the commercial success of syn-thetic zeolites is due to their availability in pure forms with unique structures that are not found in naturally occurring zeolites. Zeolites with high aluminum content show high polarity and hydrophilic properties, while those with high silica show low polarity and hydropho-bic properties (Olson *et al.*, 1980; Hoering and Freeman, 1984). Different types of aluminosilicates also result from differences in the way in which the tetrahedra link in one, two, or three dimensions and from the types of other ions that substitute within the interstices. If SiO_4 and/or AlO_4 tetrahedra are linked in three dimensions by a mutual sharing of oxygen alone, then a framework structure results. The kind of tetrahedral linkage has a profound effect on zeolite structure and properties. Some of these properties include cation-exchange ca-pacity and selectivity, stability of the crystal structure, density and void volume, degree of hydration, size of channels (i.e. size of sorbed molecules), and catalytic properties.

Framework density, or the number of T-atoms per 1000 $Å^3$, can be used to distinguish zeolites from denser silicate crystalline solids (Baelocher *et al.*, 2001). A sig-nificant gap in framework density exists between the ze-olites and other materials. Framework density is related inversely to pore volume but does not indicate the size of the pore openings. Framework densities range from ~12.5 T-atoms/1000 $Å^3$ for zeolites with the largest pore volume to around 20.5 T-atoms/1000 $Å^3$.

Framework structures can be classified further in terms of secondary building units (SBUs), consisting of linked tetrahedra that can lead to various polyhe-dral zeolite structures. A unit cell is the smallest re-peating crystallographic unit of the zeolite framework and consists of an integral number of SBUs. For exam-ple, faujasite (zeolites A, X, and Y) consists of frame-

Table 8.1. *International Union of Pure and Applied Chemistry (IUPAC) structural codes and notations for four common zeolites*

Code	Zeolite	Notation
FAU	Faujasite	<111> **12** 7.4***
MOR	Mordenite	[001] **12** 6.5 × 7.0*
		↔ [010] **8** 2.6 × 5.7*
MFI	ZSM-5	{[010] **10** 5.3 × 5.6
		↔ [100] **10** 5.1 × 5.5}***
LTA	Linde Type A	<100> **8** 4.1***

works of linked truncated octahedra. The nature of the void spaces and channels resulting from the dif-ferent possible arrangements of the SBUs determines the physical and chemical properties of each struc-ture. The channel types can be identified in several ways: (1) one-dimensional systems without intersect-ing channels, (2) two-dimensional systems, (3) three-dimensional intersecting channels of equal size, and (4) three-dimensional intersecting channels of differ-ing size depending on the crystallographic direction. Finally, certain arrangements of the SBUs may gener-ate large internal cavities or supercages at intersecting channels.

The channel dimensions are defined by the crystal-lographic free diameters and are generally determined by the spatial arrangement of tetrahedra making up 8- or 10- or 12-rings. These apertures can range from nearly circular to elliptical to a severely puckered shape. The nature of these differences has a profound influence on the absorption and molecular sieving effect for molecules entering the pores (Table 8.1). The crystal-lographically determined pore sizes offer only an ap-proximate correlation to the effective pore size because of variations caused by factors such as temperature, type of cation, and hydration. For example, potassium, sodium, and calcium forms of Linde type A (LTA) zeo-lite show effective pore sizes of ~3, 4, and 5 Å (0.3, 0.4, and 0.5 nm), respectively.

A shorthand notation describes the channels in var-ious silicate frameworks (Baelocher *et al.*, 2001). The notation includes: (1) the channel direction relative to the axes *x*, *y*, and *z* of the type structure, (2) the num-ber of either T- or O-atoms (bold) forming the rings and controlling diffusion through the channels, and

(3) the crystallographic free diameters of the channels in angstroms. The four types of zeolite discussed in this section are as follows, where the three capital letters represent the International Union of Pure and Applied Chemistry (IUPAC)-approved structural code.

Asterisks indicate whether the channel system is one-, two-, or three-dimensional. Double arrows ($\leftrightarrow$) separate interconnecting channel systems. For example, mordenite (MOR) has two interconnecting channel systems that consist of somewhat elliptical 12-ring apertures (6.5×7.0 Å) and strongly elliptical eight-ring apertures (2.6×5.7 Å) that limit diffusion in the [001] and [010] directions, respectively. Baelocher *et al.* (2001) provide more details on zeolite structures and nomenclature.

Zeolites to remove *n*-alkanes from petroleum

The structure of Linde 5A (LTA, above) allows selective adsorption of *n*-alkanes from petroleum saturate fractions (Murphy, 1969; Breck, 1974; Ruthven, 1988). The *n*-alkanes are commonly major components in crude oils, and their removal is essential for detailed study of other, less abundant components. An eight-ring window of oxygen atoms with aperture $\sim$3–5 Å in diameter controls access by *n*-alkanes and other compounds to the pores of Linde 5A. However, the adsorbed molecules and the size of the aperture can be deformed to some extent. *n*-Alkanes show cross-sectional diameters near 5 Å, but will diffuse into the Linde 5A molecular sieve, especially when heat is applied. Linde 5A also contains large cavities of 11.4 Å diameter at the intersections along its three-dimensional channel system of 4-Å eight-member oxygen rings. *n*-Alkanes less than *n*-decane are readily adsorbed inside these large cavities. Sorption rates of *n*-alkanes larger than *n*-decane decrease with increasing chain length, probably because they do not fit entirely within the cavity and extend into adjacent channels (Jasra and Bhat, 1987).

n-Alkanes containing more than 20 carbon atoms are completely adsorbed into high Si/Al ZSM-5 in only two minutes, while the same effect requires heating the Linde 5A molecular sieve for $\sim$24 hours (West *et al.*, 1990). Like most zeolites, ZSM-5 is a polar adsorbent due to aluminum in the tetrahedral framework and the presence of charge-compensating cations (Ruthven, 1988; Olson *et al.*, 1980). At high Si/Al, the ZSM-5 structure has low substituted aluminum and few

charge-compensating cations, making van der Waals' forces the only major factor in adsorption (Flanigen *et al.*, 1978). *n*-Alkanes are adsorbed strongly by ZSM-5 because these molecules are of similar size to the channel dimensions. Ellis and Fincannon (1998) tabulate adduction efficiencies for *n*-alkanes on high Si/Al ZSM-5.

Zeolites to concentrate compounds in petroleum

New instrumentation and several key geochemical problems are driving additional research aimed at concentrating biomarkers and other compounds from petroleum. Improved zeolite-mediated concentration of compounds is required for a breakthrough in the developing applications of compound-specific isotope analysis to source-rock paleoenvironmental reconstruction and petroleum characterization. Better concentration methods will enhance our understanding of the following: (1) the presence and significance of key source and age-related biomarkers in petroleum, (2) the structures and information content of novel compounds, and (3) the origin of condensates and light oils, where geochemically significant components are low. To better understand and predict the sorption characteristics of biomarkers on zeolites will require further development of a new research avenue that links the disciplines of zeolite mineralogy with biomarker conformational analysis by computational chemistry. Table 8.2 lists some of the early applications of zeolites to separate organic compounds, including biomarkers, from petroleum.

Pioneering work by Whitehead (1974) demonstrated the use of 8-Å 10X and NaX (FAU) (Table 8.2) molecular sieves to enrich pentacyclic terpenoids in petroleum. He treated branched and cyclic alkane fractions from Nigerian crude oil with activated 10X molecular sieve to enrich the hopanes. Extended hopanes were enriched in the occluded fraction, and the 22S diastereomers were sorbed more strongly by the sieve compared with 22R. The author suggested that large triterpanes, such as lupanes, oleananes, and taraxastanes, would be excluded from the sieve.

Zeolite NaX (13X) molecular sieves with channel dimensions near 8 Å were used by Dimmler and Strausz (1983) to enrich polycyclic hydrocarbons from the branched and cyclic alkane fraction of an Athabasca tar sand. Hopanes and tricyclic terpanes selectively sorbed into the sieve. These compounds were desorbed

Table 8.2. *Some pioneering applications of zeolite molecular sieves to separate organic compounds from petroleum (modified from Armanios, 1995)*

Zeolite	Pore aperture (nm)	Occluded (or adsorbed) hydrocarbons	References
5 Å	0.43	*n*-Alkanes	Murphy (1969), Breck (1974)
ZSM-5 "Silicalite"	0.51 × 0.56	*n*-Alkanes, methylalkanes, alkylcyclopentanes, alkylcyclohexanes, alkylbenzenes, *p*-alkyltoluenes	Hoering and Freeman (1984)
Mordenite	0.67 × 0.70	Isoprenoids, *o*, *m*-alkyltoluenes, methylalkylcyclohexanes, methylnaphthalenes, some alkylxylenes, di-/tri-/tetramethylnaphthalenes, methyl and dimethyl phenanthrenes, steranes	Curran *et al.* (1968), Ellis *et al.* (1992; 1994), Fisher *et al.*, (1996b)
10X	0.80	17α-Hopanes, 17α-diahopanes	Whitehead (1974)
13X	0.80	17α-Hopanes, tricyclic terpanes, steranes and drimanes	Dimmler and Strausz (1983)
US-Y	0.74	17α-Diahopanes, lupanes, oleanane, bicadinanes, 18α-norneohopane	Armanios *et al.* (1992; 1994; 1995a), Armanios (1995)

by exhaustive extraction with isooctane for 36 hours. Shorter desorption times resulted in selective removal of the 22S C_{31}–C_{35} 17α-hopanes, while the 22R diastereomers were sorbed more strongly by the sieve (Table 8.2).

Hoering and Freeman (1984) describe the use of both Linde 5A (calcium-exchanged) and silicalite (high Si/Al ZSM-5) molecular sieves (LTA and MFI, respectively) (Table 8.1) to isolate monomethylalkanes from petroleum. Monomethylalkanes, which show kinetic diameters between 5 and 6 Å, are selectively occluded in high Si/Al ZSM-5 and excluded by Linde 5A molecular sieves. These authors also used high Si/Al ZSM-5 column chromatography to separate 2-, 3-, and 4-methylalkane isomers. West *et al.* (1990) enriched the branched and cyclic components in the saturate fraction of petroleum by using 6-Å-high-Si/Al ZSM-5 molecular sieve. This sieve selectively removed *n*-alkanes, methylalkanes, alkylcyclohexanes, alkylbenzenes, and p-substituted alkyltoluenes from a petroleum alkane fraction.

Separation of diastereomers of 3,4-dimethylhexane was achieved using two in-series gas chromatographic columns coated with the sodium form of ZSM-5

molecular sieve (Weitkamp *et al.*, 1991). The separation was not possible using smaller (ZSM-23) or larger (13X) pore sieves.

A 7-Å mordenite molecular sieve (MOR) (Table 8.2) was used to concentrate selected alkanes from a complex mixture in extract from the Green River Formation (Curran *et al.*, 1968). This sieve has a pore size intermediate between the Linde 5A and 10X sieves. Mordenite selectively occludes branched alkanes, such as pristane and phytane, but excludes larger cyclic compounds, such as steranes and triterpanes. Because of the low polarity, high-silica mordenite sieves, where silicon/aluminum exceeds 10, are useful for shape-selective separations of aromatic components.

Ellis *et al.* (1992) used a dealuminated mordenite sieve and pentane as stationary and mobile phases, respectively, to selectively sorb and concentrate monoaromatic hydrocarbons, including *n*-alkylbenzenes, *n*-alkyltoluenes, and some *n*-alkylxylenes from the aromatic fraction in petroleum. Ellis *et al.* (1994) used the same mordenite to separate alkylnaphthalene and alkylphenanthrene isomers from a crude oil fraction containing di- and triaromatic hydrocarbons. Alkylnaphthalenes with substituents at C-1 and C-4 were excluded from the sieve, some alkylnaphthalenes with substituents at C-1, C-3, and C-7 were weakly sorbed, and other isomers were strongly sorbed. Methylphenanthrenes and dimethylphenanthrenes with substituents at C-9 or C-10 were excluded from the sieve. Methylphenanthrenes with substituents at C-4 were partially sorbed, while phenanthrenes substituted at C-2 were strongly sorbed. These techniques allowed quantitation of certain dimethylphenanthrenes that are difficult or perhaps impossible to separate using other, more conventional chromatographic techniques (Fisher et al., 1996).

Armanios et al. (1992) developed a liquid chromatographic method using ultrastable-Y zeolite (US-Y, FAU) (Table 8.2) to separate structurally similar petroleum hopanoid hydrocarbons based on their size and shape. Column chromatography with US-Y zeolite concentrated the 17α-diahopanes, 18α-norneohopane, 22S 17α-hopanes, 22R 17α-hopanes, and the moretanes. 17α-Diahopanes show large molecular cross-sections compared with most hopanoids, causing them to elute through the sieve without being occluded. 18α-Norneohopane eluted more slowly than the diahopanes because some possible conformers are less hindered and thus become occluded more readily in the sieve. The extended 22S 17α-hopane diastereomers were also retarded slightly by the column because their higher-energy conformers are occluded by the sieve while their lower-energy conformers are hindered. The 22R 17α-hopane diastereomers showed the longest retention time on the column because their most stable conformers fit readily into the sieve. 17α-Norhopane and 17α-hopane, which both lack extended side chains, and the 17β-moretanes, which have a more planar skeleton compared with the 17α-hopanes, were also retained and elute with the 22R 17α-hopanes.

The above liquid chromatography and US-Y molecular sieves method was used to demonstrate very low levels of bicadinanes in Jurassic mudstone and siltstone samples from the Eromanga Basin, Australia (Armanios, 1995; Armanios et al., 1995). Bicadinanes were originally thought to indicate Oligocene or younger organic matter due to their association with dammar resins from *Dipterocarpaceae* and other angiosperms (van Aarssen et al., 1990). However, recent evidence summarized by Armanios et al. (1995) suggests a Jurassic or older evolution for angiosperms. Bicadinanes are totally excluded from the sieve channels (Armanios et al., 1994) and thus

were selectively enriched and separated from other co-eluting components in the Eromanga Basin samples. Removal of these co-eluting components improved the detection limits for the trace amounts of bicadinanes. If not from angiosperms, then the low levels of bicadinanes in the samples might originate from other plant types that are capable of producing these compounds.

Armanios et al. (1992) observed preferential occlusion of the 22R compared with the 22S extended hopanes in the US-Y molecular sieves, which they attributed to different sorption energies of these diastereomers. These results are consistent with Dimmler and Strausz (1983), who noted incomplete recovery of 22R C_{31}-C_{35} 17α-hopanes during desorption of the 13X-occluded terpenoid fraction with isooctane. However, the results appear to conflict with a published chromatogram for the 10X-occluded terpenoid fraction, which preferentially adsorbs the 22S rather than 22R diastereomers (Whitehead, 1974).

Armanios (1995) examined computer-generated molecular models and steric energies of C_{34} 22S and 22R 17α-hopane (tetrakishomohopanes) and found significant differences in the steric energies for the most and least stable rotational conformers of each epimer. The most stable conformation of the 22S epimer (368 kJ mol^{-1}) was found to show a large effective cross-sectional diameter (>9.0 Å), which hinders the molecule from entering US-Y molecular sieve channels (aperture diameter ~7.4 Å). The least stable conformation of the same C_{34} 22S epimer (380 kJ mol^{-1}) showed a smaller diameter (7.4 Å). Compounds having molecular diameters greater than ~8.1 Å were not sorbed, while those with diameters ranging from 7.0 to 7.3 Å were sorbed within the sieve channels. The steric energies of the most and least stable conformers of the C_{34} 22R hopane (382 versus 379 kJ mol^{-1}) did not differ as greatly as those for the 22S epimer, suggesting that a greater proportion of the 22R epimer molecules would have the more strongly sorbed (smaller diameter) conformation. Thus, C_{34} 22R 17α-hopane is sorbed more strongly by the sieves than its 22S epimer. Armanios (1995) also observed an increase in the selectivity of the molecular sieves for separating hopanes with increasing carbon number (i.e. increase in length of the C-22 side chain). Increased length of the C-22 side chain results in energetically less favorable confined conformations within the sieve.

Computational organic geochemistry provides additional insight into the structures of extended hopanes that helps to explain the above observations. Peters *et al.* (1996b) show that geometry-optimized C_{31}–C_{35} hopane 22S and 22R epimers consist of distinct scorpion and rail-shaped conformations, respectively, controlled by different 21–22–29–31 and 17–21–22–30 torsion angles. Based on quantitative structure–activity relationships (QSAR) applied to the geometry-optimized conformations, the molecular volumes of the C_{31} or C_{32} 22S and 22R epimers are similar, but the volumes of the C_{33}–C_{35} 22S epimers are consistently greater than their corresponding 22R epimers. These data suggest a direct link between the shape and size of the hopane epimers and their zeolite sorption characteristics, as observed by Armanios (1995).

Armanios *et al.* (1994) used US-Y molecular sieves to concentrate non-hopanoid pentacyclic triterpanes from Indonesian crude oils, including bicadinanes, spirotriterpane, lupane, oleananes, taraxastanes, as well as other compounds, such as cadinanes and homocadinanes. Several novel triterpanes with unknown structures were concentrated sufficiently to allow more detailed structural studies. The molecular-sieve enrichment also enabled lower amounts of bicadinanes to be analyzed than was previously possible using selected ion monitoring/gas chromatography/mass spectrometry (SIM/GCMS) of the saturated hydrocarbon fractions. The approach increased signal-to-noise ratio but did not change the relative proportions among the bicadinanes compared with an unsieved aliquot of the same sample.

The above initial findings suggest a new avenue for geochemical research, where computational chemistry is applied to both zeolites and biomarkers or other compounds to predict their adsorption interactions. This computational approach could also help to provide mechanistic explanations for laboratory measurements of adsorption by zeolites. Different levels of sophistication for the computational approach range from relatively simple comparisons of the free energies of the zeolites and occluded compounds (J. E. Dahl, 1999, personal communication) to detailed molecular dynamics calculations. To the best of our knowledge, the latter approach has been used only to model the stable geometries of simple compounds, such as methanol, in zeolite frameworks. For example, Shah *et al.* (1996) used *ab initio* techniques and massively parallel computing to investigate the mechanism of adsorption of methanol on zeolites (and its implications for the catalytic generation

of dimethyl ether and gasoline). This new research avenue of computational organic geochemistry could improve our understanding and ability to predict the sorption of biomarkers and other compounds on zeolites.

GAS CHROMATOGRAPHY/MASS SPECTROMETRY

Computerized GCMS (McFadden, 1973; Watson, 1997) is the principal method used to evaluate biomarkers (Figure 8.5). A typical GCMS system performs six functions, indicated in the figure as follows:

(1) Compound separation by gas chromatography.
(2) Transfer of separated compounds to the ionizing chamber of the mass spectrometer.
(3) Ionization.
(4) Mass analysis.
(5) Detection of the ions by the electron multiplier.
(6) Acquisition, processing, and display of the data by computer.

GCMS can be used to detect and provisionally identify compounds using relative gas chromatographic retention times, elution patterns, and the mass spectral fragmentation patterns characteristic of their structures.

Stringent criteria are applied to GCMS procedures (Seifert and Moldowan, 1986) to ensure meaningful interpretations. For example, GCMS data are obtained using high-resolution capillary columns (generally 50 m or more long), high signal-to-noise output from a finely tuned mass spectrometer, and rapid scanning as discussed below.

Gas chromatography in GCMS

The theory and practice of gas chromatography, sometimes called gas/liquid chromatography, is described extensively in the literature (e.g. Poole and Schuette, 1984; Kitson *et al.*, 1996; Beesley and Scott, 1998; Grob, 2001). A syringe is used to inject a known amount (typically <0.1 μl) of the saturated or aromatic hydrocarbon fraction, which may or may not be dissolved in solvent (usually toluene), into the gas chromatograph (Figure 8.6, top). In gas chromatography, each injected sample is vaporized and mixed with an inert carrier gas, typically helium or hydrogen (e.g. David and Sandra,

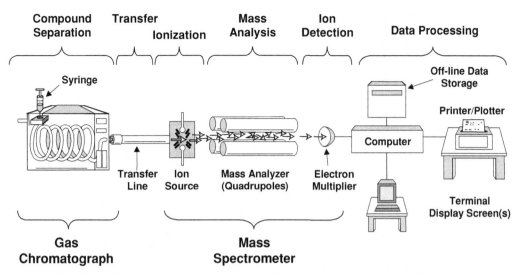

Figure 8.5. A typical gas chromatograph/mass spectrometer performs six functions (from left to right): (1) compound separation by gas chromatography; (2) transfer of separated compounds to the ionizing chamber of the mass spectrometer; (3) ionization and acceleration of the compounds down the flight tube; (4) mass analysis of the ions; (5) detection of the focused ions by the electron multiplier; and (6) acquisition, processing, and display of the data by computer. The quadrupole mass analyzer is an important component in quadrupole mass spectrometers. Mass analysis can be accomplished using four parallel quadrupole rods. By varying a combination of radiofrequency and direct currents within the rods, a beam of ions can be scanned, thus allowing only ions of a given mass to reach the detector at any moment during the scan. Reprinted with permission by Chevron-Texaco Exploration and Production Technology Company, a division of Chevron USA Inc.

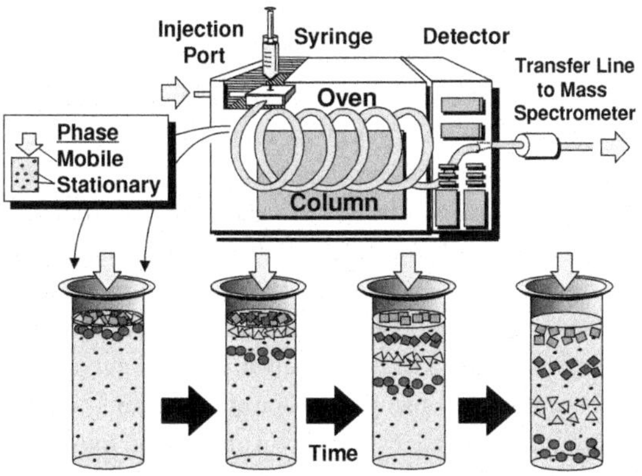

Figure 8.6. Detailed view of a typical gas chromatograph used to separate mixtures of compounds. The blow-up (bottom) shows the separation of compounds during movement down the chromatographic column, which results from their repeated partitioning between the mobile and stationary phases. Reprinted with permission by ChevronTexaco Exploration and Production Technology Company, a division of Chevron USA Inc.

1999). The vaporized mixture of sample and carrier gas then moves through a capillary column as discussed below.

Injection

Various injection techniques can be used for capillary chromatography, depending on the objectives of the analyst (Grob, 2001). Some of these techniques include classical vaporizing injection, programmed temperature vaporizing (PTV) injection, and on-column injection. In classical vaporizing injection, the sample evaporates in a hot vaporizing chamber before transfer to the column. However, PTV, in which the sample is injected into a cool chamber that is later heated to vaporize the sample, has largely replaced this method.

In PTV, larger molecules in the sample can be retained in the chamber and the stationary phase at the head of the gas chromatographic column in a process called cold trapping. The temperature of the column is raised gradually using a temperature-programmed oven, causing the cold-trapped compounds to move. PTV may involve split, splitless, solvent-split, or direct injection. Only a small portion of the vapor enters the column in split injection. This is the method of choice for concentrated samples and for gas and headspace analysis. In splitless injection, nearly the entire sample is transferred to the column. This method is commonly used to analyze biomarkers in petroleum and trace components in contaminated samples. In solvent splitting, most of the solvent vapor is vented and the solute is transferred into the column in splitless mode, allowing large-volume injections for trace analysis. The entire vapor is transferred to the column in direct injection. This method is used for trace analysis and usually involves instruments that were converted from packed-column gas chromatography to capillary gas chromatography (see below).

In on-column injection, the liquid sample is injected into the column inlet or an oven-thermostatted capillary precolumn. This method provides excellent results but is generally not suited for highly contaminated samples. Classical on-column injection requires small volumes of sample. However, larger volumes can be injected on-column when using the retention gap technique or precolumn solvent splitting. For the retention gap technique, an uncoated precolumn is used to overcome band broadening caused by sample liquid flooding the column inlet. In precolumn solvent splitting, most of the solvent vapor is released by injection into a precolumn connected to a vapor exit.

Compound separation by gas chromatography

The gas (mobile phase) and sample mixture moves through a long, thin capillary column (typically 0.20–0.25 mm ID, 30–60 m long) whose inner surface is coated with a film (~0.25 μm thick) of non-volatile liquid (stationary phase). Different components separate during movement down the column as they are repeatedly retained by the stationary phase and released into the mobile phase depending on their volatility and affinity for each phase (Figure 8.6, bottom).

Flexible fused silica capillary columns have largely replaced the older glass or stainless-steel capillary columns used in GCMS because of several advantages:

- The stationary phase may be chemically bonded to silica, increasing the thermal range of the column and minimizing column bleed.
- Lower activity reduces peak tailing and loss of sample by adsorption on the stationary phase.

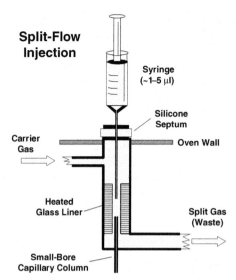

Figure 8.7. A typical split-flow injector is used to inject small samples of whole oil or saturated or aromatic hydrocarbon fractions on to a small-bore chromatographic column.

- Flexibility eliminates many difficulties related to installation.
- Ease and lower cost of column preparation with consistent performance.

One consequence of small-bore capillary columns, particularly those with thin film thickness that are favored for use in GCMS, is their low sample capacity. The amount of material injected must be less than ~100 ng/peak to prevent overloading. The most common method for adjusting the injected amount is sample dilution and the use of a split-flow injector, which diverts most of the vaporized components to a vent (Figure 8.7). Modern split-flow injectors are capable of routine operation at ~350°C and are designed to minimize mass discrimination. Nevertheless, the response for hydrocarbons over ~C_{35} is lower than expected due to decreased volatility. Mass discrimination during injection can be minimized using temperature-programmed injectors and eliminated using direct on-column injection. These injection methods, however, require more attention to minimizing peak overload by sample dilution.

Most published gas chromatographic data for petroleum were obtained using 100% methyl- or 95% methyl–5% phenylpolysiloxane stationary phases (e.g. OV-101, DB-1, DB-5) that are chemically bonded to the silica column (Figure 8.8). Hydrocarbon retention

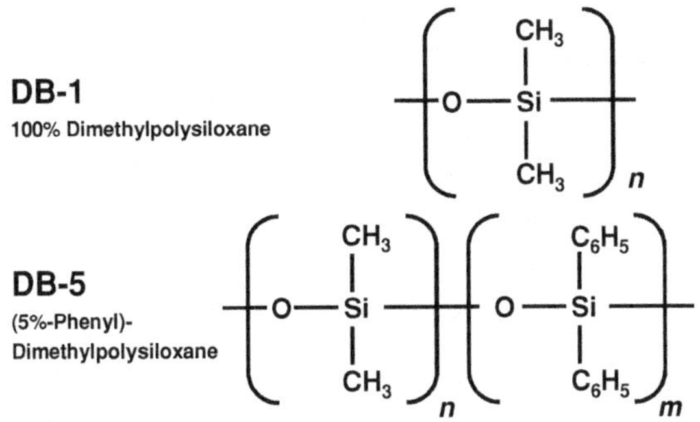

DB-1
100% Dimethylpolysiloxane

$$\left(\!-\mathrm{O}-\underset{\underset{\mathrm{CH_3}}{|}}{\overset{\overset{\mathrm{CH_3}}{|}}{\mathrm{Si}}}-\!\right)_{\!n}$$

DB-5
(5%-Phenyl)-
Dimethylpolysiloxane

$$\left(\!-\mathrm{O}-\underset{\underset{\mathrm{CH_3}}{|}}{\overset{\overset{\mathrm{CH_3}}{|}}{\mathrm{Si}}}-\!\right)_{\!n}\left(\!\mathrm{O}-\underset{\underset{\mathrm{C_6H_5}}{|}}{\overset{\overset{\mathrm{C_6H_5}}{|}}{\mathrm{Si}}}-\!\right)_{\!m}$$

Figure 8.8. Common polysiloxane stationary phases used in gas chromatographic studies of petroleum. Values of n and m for the DB-5 stationary phase are in the ratio 95:5 for dimethyl- and phenylmethylpolysiloxane.

on these stationary phases is mostly a function of relative volatility. Consequently, the elution orders of most biomarkers are similar and GCMS results are comparable. Separation of individual biomarkers may vary slightly. For example, gammacerane separates completely from $17\alpha,21\beta(\mathrm{H})$-homohopane (22R) using a 95% methyl–5% phenylpolysiloxane phase but is typically separated only partially using a 100% methylpolysiloxane phase. Substantial improvement in the separation of aromatic biomarkers is achieved using

50% methyl–50% phenylpolysiloxane or liquid crystalline stationary phases. Other non-bonded stationary phases, such as squalene, Dexsil, and cyclodextrin, have been used for the separations of specific biomarker isomers. For best standardization of analytical conditions, an oil standard (e.g. oil from Hamilton Dome Field, Wyoming) is run with each set of samples to evaluate column performance in separating various compounds and to facilitate calibrations of sensitivity (response) for peak measurement.

Using standard polysiloxane stationary phases, many biomarkers elute from the gas chromatograph in the range $n\mathrm{C_{24}}$–$n\mathrm{C_{36}}$ and are generally in much lower abundance than the n-alkanes (Figure 8.9). Exceptions include pristane, phytane, and various diterpanes and tricyclic terpanes, which elute prior to this range, and porphyrins, which do not elute under normal gas chromatographic conditions because of their high molecular weights and low volatilities.

Chiral gas chromatography

Chiral chromatographic phases using modified cyclodextrins are becoming more popular for specialized studies that require highly selective enantiomeric separations of analytes based on their shapes. Cyclodextrins

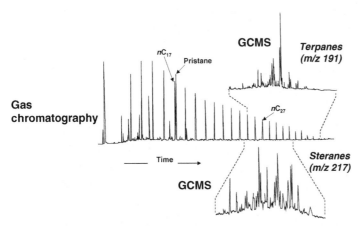

Figure 8.9. The gas chromatogram of oil obtained by routine gas chromatography using a flame ionization detector (center) is dominated by the homologous series of n-alkanes. Mass chromatograms of the steranes and terpanes (bottom and top, respectively) in the same oil are obtained by gas chromatography using a mass spectrometer as detector (GCMS). Steranes and terpanes are monitored using their principal fragment ions at m/z 217 and m/z 191, respectively. Note that the sensitivity of this analysis allows detection of sterane and terpane peaks too small to see on the gas chromatogram. Many biomarkers, including the steranes and terpanes, elute from the gas chromatograph between the n-alkanes with 24 and 36 carbon atoms ($n\mathrm{C_{24}}$–$n\mathrm{C_{36}}$). Pristane and phytane are acyclic isoprenoid biomarkers, which elute before this range of retention time. Reprinted with permission by Chevron-Texaco Exploration and Production Technology Company, a division of Chevron USA Inc.

are oligomers of 1,4-linked α-D-glucose units that contain several chiral centers in the shape of a truncated cone that results in a cavity. α-, β-, and γ-Cyclodextrins have six, seven, and eight glucose units, respectively. Cyclodextrins are readily acylated with different functional groups due to differing reactivity of the hydroxyl groups in the glucose units (König, 1992). The macrocyclic conformation of cyclodextrin changes with the degree of acylation, allowing variations in the size of the internal cavity and the extent of interaction between the modified cyclodextrin and the analyte. Capillary gas chromatography using cyclodextrins began with Juvancz et al. (1987), who coated a column with molten permethylated β-cyclodextrin. Schurig and Nowotny (1988) dissolved the permethylated β-cyclodextrin in OV-1701 silicone phase. Thermal stability can be increased by chemically binding the permethylated β-cyclodextrin to a polydimethylsiloxane phase (Fischer et al., 1990; Schurig, 1994). Permethylated β-cyclodextrin phases show chiral selectivity for a wide range of compounds, including hydrocarbons in petroleum (Bastow, 1998).

Extended hold-time gas chromatography
Separation of saturated and aromatic biomarkers can be improved substantially if the chromatographic oven is held at a low temperature ($\sim$150°C) for an extended time (Gallegos and Moldowan, 1992). The analysis proceeds using a normal temperature ramp after the initial hold period. Using hold times of up to 48 hours after injection, Gallegos and Moldowan (1992) partially resolved the C_{29} steranes into 14 components, of which 9 were pure enough to yield discrete mass spectra. These 14 steranes are isomers assigned to the 4 peaks that are resolved under normal chromatographic methods (5α,14α,17α 20S + 20R and 5α,14β,17β 20S + 20R) as well as diastereomers that are normally not resolved, such as the 24S and 24R and the 5α,14β,17α or 5β,14α,17α epimers. The technique also improved the separation of mono- and triaromatic steroidal hydrocarbons, but it is used rarely because of time requirements and restrictions on the range of compounds that can be analyzed for each run. Increased resolution from using extended hold times is believed to result from enhanced thermal diffusion and different interactions between biomarkers and the stationary phase at 150°C versus those at the more typical $\sim$300°C temperature.

Compound separation by two-dimensional gas chromatography
The above discussion focuses on gas chromatography as the means to separate compounds in petroleum. However, readers should be aware of new research in two-dimensional gas chromatography (GC × GC), where two different chromatographic separation mechanisms act together to improve component separation. Early multidimensional gas chromatographs linked together two gas chromatographs, with the columns joined by a mechanical valve or pressure-controlled Dean's splitter. These linked instruments allowed for so-called heart cutting of individual peaks but were not effective for complex mixtures. Multidimensional gas chromatography is still used but is being replaced by comprehensive two-dimensional gas chromatography (see reviews by Bertsch (1999), Ong and Marriott (2002), and Blomberg et al. (2002)).

Note: Single-column, high-resolution, whole-oil gas chromatography with a slow temperature ramp is commonly used in order to evaluate petroleum reservoir continuity. The peaks used for comparing oils from different reservoir compartments may be resolved only partially and unidentified. Consequently, these reservoir continuity studies require that all samples be run under the same chromatographic conditions (Kaufman et al., 1990). Changes in column and instrument performance require that the complete set of oils be rerun with each addition of a new sample to the suite. If the eluting compounds can be baseline separated and identified, then a database can be established that allows oil comparisons over time without rerunning the samples. For example, Walters and Hellyer (1998) used a multidimensional gas chromatography method to separate C_7 hydrocarbons in crude oils. Likewise, Nederlof et al. (1994) developed a multidimensional gas chromatography method to separate C_6–C_{12} monoaromatic hydrocarbons.

Unlike multidimensional gas chromatography, comprehensive two-dimensional gas chromatography uses a modulator between the columns that refocuses the effluent from the first column for reinjection as a discrete band on to the second column. The columns have different stationary phases that are selective for different molecular properties. For example, the first

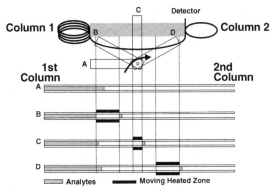

Figure 8.10. Example of a thermal modulator based on a slotted heating element that rotates clockwise from position A to position D during each modulation phase.

column usually has a non-polar phase that separates hydrocarbons mostly by boiling point; the second column may have a phase that is selective for polarity or shape. Compounds introduced on to the short second column are detected before peaks arrive from the next modulation, which typically is ~10 seconds (Figure 8.10). The result is two-dimensional separation, where the elution time from the first column is separated into discrete intervals and can be plotted relative to the elution time from the second column (Figures 8.11 and 8.12). Figure 8.13 shows a GC × GC chromatogram for a marine diesel fuel. Loss of resolution caused by condensing the effluent from the first column is minimal because the modulation time is typically a fraction of the total elution time for a single peak. Any loss of resolution is offset by improved separation from the second column and by increases in sensitivity. Phillips and Beens (1999) and Bertsch (2000) discuss the theory, methods, and applications of comprehensive two-dimensional gas chromatography. Pursch *et al.* (2002) review modulator design, advantages and limitations.

The complexity of crude-oil composition is ideal for demonstrating the power of comprehensive two-dimensional gas chromatography. Most published studies concentrate on the separation of the <C_{30} fraction of whole crude oils, distillate cuts, and refined products. Figure 8.13, for example, shows a GC × GC chromatogram for a marine diesel fuel. The first column separates the hydrocarbons by boiling point. The second column separates by polarity, which resolves aromatic from saturated components. Application of GC × GC methods to biomarker compounds is in its infancy.

Xu *et al.* (2001) demonstrated the power of this method by separating novel long-chain ketones from Holocene Black Sea sediments (Figure 8.14). Using a GC × GC system equipped with a non-polar methylsiloxane column coupled to a polar trifluoropropylmethyl column, they resolved the well-known methyl- and ethyl-alkenones into two discrete homologous series, allowing identification of a novel $C_{36:2}$ ethyl-alkenone as well as other possible alkenones that occur in trace abundance.

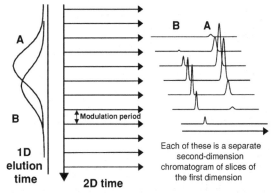

Figure 8.11. Example of multiple chromatograms obtained by second gas chromatograph (GC-2) (right) by modulation (slicing) of unresolved peaks A and B from first gas chromatograph (GC-1) (left).

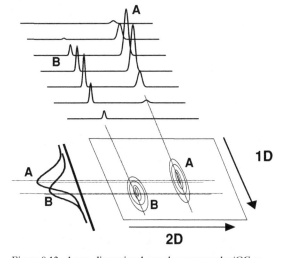

Figure 8.12. A two-dimensional gas chromatography (GC × GC) plot (lower right) is based on chromatographic results from first gas chromatograph (GC-1) (peaks A and B, lower left) and chromatographic slices of these peaks from second gas chromatograph (GC-2) (top) (from Gaines *et al.*, 1999).

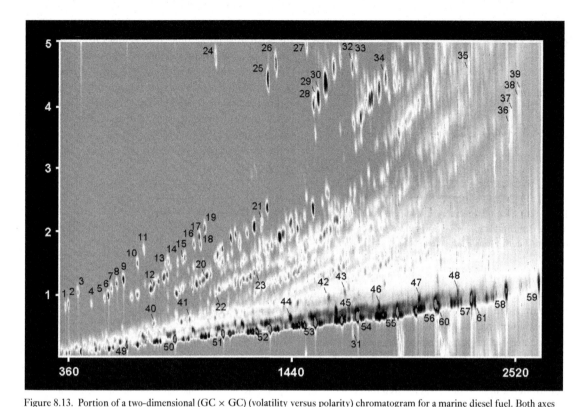

Figure 8.13. Portion of a two-dimensional (GC × GC) (volatility versus polarity) chromatogram for a marine diesel fuel. Both axes are in seconds. Peak identities are given below. Signal intensities are assigned a shade from black (high) through gray (medium) to white (low). The baseline is gray to facilitate visualization. A small range of signal intensity is plotted so that the small peaks can be seen. The tallest peaks have been truncated, resulting in the appearance of a uniform black center. Reprinted with permission from Gaines *et al.* (1999); original in color. © Copyright 1999, American Chemical Society.

1 = ethylbenzene; 2 = *m*-xylene/*p*-xylene; 3 = *o*-xylene; 4 = isopropylbenzene; 5 = propylbenzene; 6 = 3-ethyltoluene/4-ethyltoluene; 7 = 1,3,5-trimethylbenzene; 8 = 2-ethyltoluene; 9 = 1,2,4-trimethylbenzene; 10 = 1,2,3-trimethylbenzene; 11 = indan; 12 = *n*-butylbenzene; 13–14 = methylindan (a); 15 = 1,2,4,5-tetramethylbenzene; 16–17 = methylindan (a); 18 = 1,2,3,4-tetramethylbenzene (a); 19 = tetrahydronaphthalene; 20 = *n*-pentylbenzene; 21 = pentamethylbenzene; 22 = 1,4-diisopropylbenzene; 23 = *n*-hexylbenzene; 24 = naphthalene; 25 = 2-methylnaphthalene; 26 = 1-methylnaphthalene; 27 = biphenyl (a); 28 = 2-ethylnaphthalene; 29 = 1-ethylnaphthalene; 30 = 2,6-dimethylnaphthalene; 31 = 1,8-dimethylnaphthalene; 32–33 = methylbiphenyl (b); 34 = 2,3,5-trimethylnaphthalene; 35 = anthracene/phenanthrene; 36–39 = methylanthracene/phenanthrene (a); 40 = *trans*-decahydronaphthalene; 41 = *cis*-decahydronaphthalene; 42–43 = pentamethyldecahydronaphthalene (b); 44 = *n*-heptylcyclohexane (b); 45 = *n*-octylcyclohexane (b); 46 = *n*-nonylcyclohexane (b); 47 = *n*-decylcyclohexane (b); 48 = *n*-undecylcyclohexane (b); 49 = decane; 50 = undecane; 51 = dodecane; 52 = tridecane; 53 = tetradecane; 54 = pentadecane; 55 = hexadecane; 56 = heptadecane; 57 = octadecane; 58 = nonadecane; 59 = eicosane; 60 = pristane; 61 = phytane.

Biomarker analysis requires transfer of the GG × GC effluent to a mass spectrometer. Frysinger and Gaines (1999) demonstrated the feasibility of GC × GCMS detection of various hydrocarbon classes in whole oil using a quadrupole mass spectrometer (Frysinger and Gaines, 1999). GC × GCMS is capable of separating and distinguishing saturated and aromatic biomarkers, but the slow scan rates of quadrupole mass spectrometry are inadequate for GC × GC separations with rapid thermal modulation (Frysinger and Gaines, 2001). This problem was solved by coupling GC × GC and time-of-flight mass spectrometry (TOF-MS) using a high scan rate (van Deursen *et al.*, 2000). The major limitation in GC × GCMS analysis is the inability of

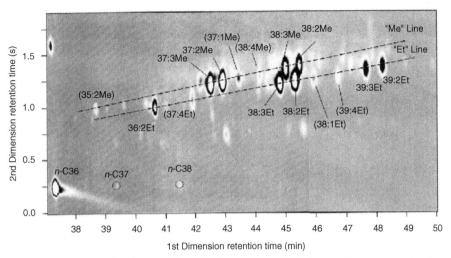

Figure 8.14. Partial two–dimensional (GC × GC) chromatogram of a Black Sea sediment extract, showing the separation of methyl-and ethyl- families (dashed lines) of alkenones. Reprinted from Xu *et al.* (2001). © Copyright 2001, with permission from Elsevier.

current processing systems to handle three-dimensional data effectively (Shellie *et al.*, 2001).

Transfer

The interface between the gas chromatograph and mass spectrometer serves to transfer the separated compounds to the ion source (Figure 8.6). In the past, various types of separator were used to concentrate the effluent from the gas chromatograph by removing much of the carrier gas. Low carrier flow rates for capillary columns and high-capacity diffusion pumps on newer GCMS systems have largely eliminated the need for these separators. Most new GCMS systems transfer all effluent directly to the ion source without a separator. Flexible capillary columns can be introduced directly into the ion source, thereby eliminating loss of chromatographic resolution, which previously arose due to the separator.

Mass spectrometry in GCMS

Detailed descriptions of mass spectrometry are available to supplement the following discussion (e.g. Burlingame *et al.*, 1980; Kitson *et al.*, 1996; Watson, 1997).

Ionization

Electron impact is the usual mode of ionization in GCMS. Other modes of ionization include chemical ionization and field ionization, as discussed below.

ELECTRON-IMPACT IONIZATION

After the separated compounds elute from the high-resolution capillary column, they are analyzed by the mass spectrometer (Figure 8.5). In electron-impact ionization, eluting compounds pass directly from the column to the source or ionizing chamber of the mass spectrometer, where they are ionized by an electron beam. An electron impact source consists of a filament, an electron trap, a repeller, and appropriate focusing plates. Passing a current (<1 mA) through a 10-μm rhenium or tungsten wire filament generates the beam. Resistance heats the filament and causes electrons to escape. The electrons are then accelerated using ~70 eV toward the trap. Typically, a mass range of 50–600 amu is scanned in three seconds or less. Low pressures are maintained within the mass spectrometer (<10^{-5} Torr).

Most mass spectrometry systems ionize the eluting compounds in the electron-impact mode using 70 eV. The choice of 70 eV ionizing voltage is based on the empirical observation that molecules are ionized most efficiently in the range 50–90 eV. Below 50 eV, the electrons do not impart sufficient energy to the target molecules to cause efficient ionization. Above 90 eV, the electrons are so energetic that they do not react with the target molecules.

Each molecule (M) eluting from the gas chromatograph forms molecular ions (M+.) when bombarded by energetic electrons in the ionizing chamber, as follows:

$$M + e- \rightarrow M + . + 2e-$$

The molecular ion can undergo fragmentation or rearrangement to form other ions (F+., F1+.), neutral molecules (N1, N2), or radical ions:

$$M + . \rightarrow F + . + N1$$
$$F + . \rightarrow F1 + . + N2$$

Fragment ions are electrically charged dissociation products from a parent ion. Fragment ions may dissociate further to form other electrically charged molecular or atomic moieties of successively lower formula weight. Fragment ions generated in the ion source chamber are accelerated toward the detector through the mass analyzer by a high differential voltage (Figure 8.5). Ions formed in the source of the mass spectrometer are analyzed according to their mass/charge ratio using a magnetic or quadrupole mass spectrometer. An electron multiplier detects positive ions. The result is a characteristic fragmentation pattern or mass spectrum of molecule M.

The mass of the molecular ion helps to identify each compound analyzed by GCMS. However, 70 eV ionizing energy commonly reduces the molecular ion to very low levels relative to fragment ions. By lowering the ionizing energy to ~20 eV or less and injecting more sample, the operator can reduce the efficiency of ionization and, thus, increase the relative abundance of the M+. ion compared with the fragment ions. This choice results in loss of measured intensity.

Electron impact is the most commonly used ionization technique because it usually provides all of the necessary spectral information required to identify organic compounds. This is particularly true for routinely analyzed compounds whose structures and retention times are already known from previous investigations. For routine work, compound identification is a necessary first step, but the principal objective is to quantify the compounds for use in biomarker parameters. For less routine work, where structural elucidation of unknown compounds is the principal objective, other ionization techniques are frequently used in conjunction with electron impact, as described below.

CHEMICAL IONIZATION

Techniques other than electron impact, such as chemical-ionization and field-ionization GCMS have been used to determine molecular weight. Chemical ionization involves ionization of components by ion–molecule reactions rather than by electron impact or other forms of ionization. A large excess of reagent gas, R, is normally ionized by electron impact followed by ion–molecule reactions involving neutral analyte molecules (compounds to be analyzed, M) and the reagent gas ions (R+). These neutral molecules (M) can undergo further reactions, forming additional fragment ions (M+, F+, F1+, etc.) and neutral species (N, N1, N2, etc.). The principal reactions are as follows:

$$R + e- \rightarrow R + +2e-$$
$$(R+) + M \rightarrow (M+) + N$$
$$(M+) \rightarrow (F1+) + N1$$
$$F+ \rightarrow (F2+) + N2$$

In chemical ionization, relatively high pressures (~1 mm Hg) of reagent gas, usually a low-molecular-weight hydrocarbon or ammonia, are introduced into the ion source of the mass spectrometer. The filament ionizes this gas to give primary and secondary ions formed by fragmentation of the primary ions and/or recombination of the fragments. The analyte passes through the gas chromatograph and reacts with the secondary ions in the ion source. The analyte occurs in low concentration, usually <1% of the reagent gas. The ions are generally formed by proton transfer or hydride abstraction, and the principal ion is usually (M + 1)+ or (M − 1)+. Little additional fragmentation occurs compared with electron-impact spectra.

FIELD IONIZATION

Like chemical-ionization mass spectrometry, field ionization is another soft ionization technique that can provide molecular weight. In field ionization, a high voltage (~10^8 V/cm) is applied to an electrode system consisting of an anode or emitter (usually a 10-μm tungsten wire on which carbon dendrites, i.e. hair-like filaments, were grown) and a cathode or extractor plate with a narrow slit. The compounds to be analyzed pass from the gas chromatograph into the field of this electrode system, where they acquire a positive charge by a quantum mechanical tunneling mechanism. Field-ionization spectra typically show only the molecular ion with few if any fragments. Field ionization or GC/FIMS of biomarkers has not been reported.

Mass analysis

Fragment ions are focused into a concentrated beam in the mass analyzer so that only positive ions of a given mass/charge (m/z) impinge on the detector at any moment. The two principal methods used to analyze the ion beam use either magnets or quadrupole rods, as described below:

MAGNETIC MASS SPECTROMETRY

Magnetic mass spectrometers consist of single- or double-focusing systems. A double-focusing instrument uses an electrostatic sector to energy-focus the ion beam before entering or after leaving the magnetic sector to achieve high mass resolution of 2000 or more. For example, an electrostatic lens can be used to energy-focus the beam into a magnetic field surrounding the curved flight tube (Figure 8.15). Heavier ions are deflected less than light ions of the same charge in the magnetic field. Scanning the magnetic field by varying the field strength focuses ions of a given m/z on to the detector. Adjustable slits are used at the source and detector to improve the mass resolution of the system. The curved-flight-tube ion-focusing arrangement typical of magnetic instruments is not shown in Figure 8.5, which instead shows a mass analyzer composed of quadrupole rods (see below).

Single-focusing magnetic systems consist only of the magnet, do not have electrostatic focusing, and, thus,

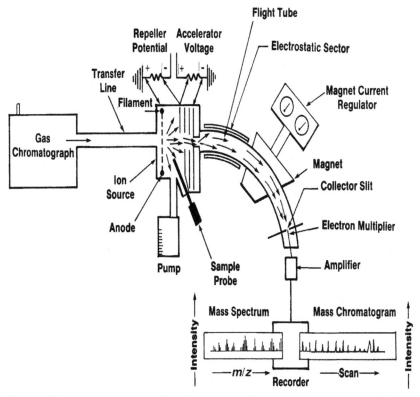

Figure 8.15. An electrostatic mass analyzer is an example of one type of mass spectrometer. Mass analysis can be accomplished using an electrostatic lens, which focuses the ion beam into a magnetic field surrounding the curved flight tube. The flight paths of ions are changed (scanned) by varying the magnetic field strength. A narrow collector slit admits only ions of a given mass to the detector at any moment during the scan. Reprinted with permission by ChevronTexaco Exploration and Production Technology Company, a division of Chevron USA Inc.

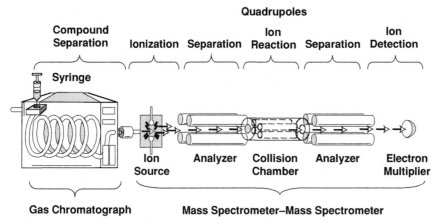

Figure 8.16. Schematic of a gas chromatograph equipped with a triple quadrupole mass spectrometer consisting of three sets of quadrupole rods.

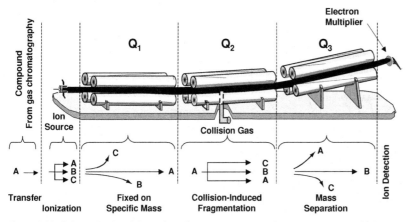

Figure 8.17. Schematic of a triple quadrupole mass spectrometer for gas chromatography/mass spectrometry/mass spectrometry (GCMS/MS) analysis. Combined use of the three quadrupoles (Q_1, Q_2, and Q_3) allows highly selective compound detection (e.g. only daughter ion B from parent compound A). Reprinted with permission by ChevronTexaco Exploration and Production Technology Company, a division of Chevron USA Inc.

achieve only low resolution (defined below) of $\sim$1000 or less. Both single- and double-focusing GCMS systems scan the magnetic field to obtain mass spectra.

Note: Resolution in mass spectrometry is defined as the ratio of mass (amu) to the difference in mass between two adjacent masses that the instrument is just able to separate completely ($M/\Delta M$). Low resolution is considered to be $\sim$1000 (no units), whereas high resolution is over $\sim$2000. Most quadrupole instruments are low-resolution mass filters. High-resolution mass spectrometers can be

used to separate ions of the same nominal mass but having different exact mass.

SINGLE/TRIPLE QUADRUPOLE AND THREE-DIMENSIONAL ION TRAPPING

Single (Figure 8.5) or triple quadrupole (Figures 8.16 and 8.17) and three-dimensional ion trap instruments all make use of quadrupole radio frequency (RF)/electrical fields to select ions of a given mass/charge ratio. Quadrupole mass filters use no focusing at all but achieve mass selection because, for a given set of conditions, only

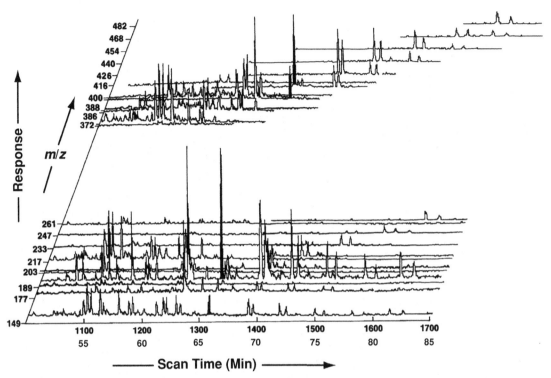

Figure 8.18. Three-dimensional plot demonstrating the principle of scan analysis in mass spectrometry. The x- (horizontal), y-(vertical), and z- (into page) axes represent the scan number or gas chromatographic retention time, detector response, and mass/charge ratio (m/z), respectively. Scans typically cover the mass range m/z 50–600 every three seconds. For example, scan 1200 is complete after 60 minutes (3600 seconds). Black peaks on the plot are the response for a single compound: C_{30} 17α,21β(H)-hopane (peak 20 in Figure 8.21). The Wyoming oil used as a standard contains 145 ppm of this compound (Figure 3.20). Reprinted with permission by ChevronTexaco Exploration and Production Technology Company, a division of Chevron USA Inc.

ions with a narrow range of mass/charge ratios are able to remain within the instrument during mass analysis. For example, a two-by-two arrangement of four parallel quadrupole rods can be used to scan the ion beam by varying a combination of RF and direct currents (DC) within the rods. This arrangement is used as the example of the mass analyzer portion of Figure 8.5 and is shown in greater detail in Figure 8.17. The three-dimensional ion trap uses no focusing. In this case, the ions are trapped in three dimensions by RF electrical fields. Thus, only those ions of interest are trapped and detected.

Detection and scan analysis

During scan analysis, the detector typically measures ions over the mass range m/z 50–600 every three seconds (i.e. >500 ions in three seconds). Each peak that

elutes from the gas chromatograph yields a distribution of fragment ion masses. Thus, a time slice of the ions generated by each peak is detected every three seconds. A three-dimensional diagram shows the principle of a scan analysis (Figure 8.18). The x-, y-, and z-axes on this diagram represent the time or scan number (i.e. total number of scans at three seconds/scan, related to the gas chromatographic retention time), the mass/charge ratio of the ions (m/z), and the detector response, respectively.

The difference between a mass chromatogram and a mass spectrum is critical for understanding the sections that follow. A mass spectrum plots m/z versus response at constant scan number or time. A mass chromatogram plots scan number versus response at constant m/z. Each mass spectrum consists of a series of fragment ions that can be used to help elucidate the structure of a single compound. A mass chromatogram can be used to monitor a series of compounds of various molecular weights that all fragment to form a particular ion (e.g. m/z 217 is a fragment ion common to the pseudohomologous series of steranes).

Detection of ions separated by the mass analyzer is achieved using a multistage electron multiplier whose output either is recorded directly on to an analog recorder or is digitized and acquired by a data system. Virtually all mass spectrometry units rely on electron multiplier (discrete dynode) detection, where the electron beam is amplified by a cascade of collisions with special metal surfaces (Watson, 1997).

Each compound eluting from the gas chromatograph yields a particular distribution of fragment ion masses called a mass spectrum (Figures 8.18 and Figure 8.27). Mass spectra of biomarkers are useful because they usually show the molecular mass (some of the molecules ionize but do not fragment further) and characteristic fragmentation patterns that can be used to infer structures. Ideally, each gas chromatographic peak represents a separate compound that gives a unique mass spectrum. In reality, many chromatographic peaks are unresolved mixtures of two or more compounds, thus complicating interpretation.

The magnitude of the total ion current for all mass spectra in each sample can be plotted versus retention time on a reconstructed ion chromatogram (RIC) (also called a total ion chromatogram, ITC) to show a series of peaks that represent the relative amounts of eluting compounds. RIC and gas chromatography traces of petroleum are potentially identical, except that an RIC requires mass spectrometric detection while gas chromatography uses the more conventional flame ionization detection (FID) (Figure 8.19). However, several factors introduce greater discrimination into RIC compared with FID data. Full-scan GCMS is commonly limited to a range of mass. For example, fragment ions with m/z <60 are not recorded in a scan run limited to m/z 60–600 and do not contribute to the RIC, resulting in sensitivity discrimination against compounds that give abundant low-mass fragments. Some molecules ionize more readily in the ion source of the mass spectrometer, thus forming relatively more ions than others. Discrimination against ions with high m/z ratios is common in quadrupole instruments, resulting in mass discrimination in the RIC. Conversely, FID discriminates against compounds that do not

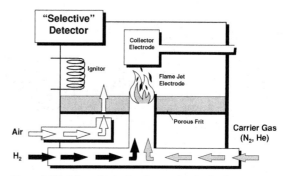

Figure 8.19. A flame ionization detector is selective for compounds that contain carbon and hydrogen but does not detect non-hydrocarbons, such as carbon dioxide.

burn completely (e.g. carbon tetrachloride, CCl_4, is not detected).

Data processing and calibration

A computer is necessary to operate the GCMS and store and process the large amounts of data generated. During a typical scan run requiring ~90 minutes, the mass spectrometer scans the mass range every three seconds and generates 1800 mass spectra for each sample [(90 minutes × 60 seconds/minute)/(3 seconds/scan)]. Most GCMS data systems consist of a central computer with one or more display terminals and various peripheral devices for data storage, printing, and plotting (Figure 8.5). The computer contains a library of thousands of electron-impact spectra for known compounds. As the computer acquires data from each GCMS analysis, unknown compounds in the injected sample can be identified provisionally by automated comparison of their spectra with the library spectra. The computer typically provides two or more best choices for the compound to be identified with accompanying purity and fit information. Purity and fit are each ranked on a scale of 0–1000, where 1000 represents a perfect correlation between the known and unknown spectra.

Computer terminals are used to manipulate data acquired and processed by the computer. Software accompanying the data system usually includes programs for instrument calibration, data acquisition, background subtraction, and monitoring the variation in intensity of one or more ions (selected ion monitoring).

Every GCMS system requires periodic calibration of the mass scale (m/z axis in Figure 8.18). This may be accomplished using a standard compound, such as

Table 8.3. *Modes of gas chromatography/mass spectrometry (GCMS) analysis for biomarkers*

Mode	Method	Result	Advantages/disadvantages
Selected ion monitoring (SIM)/GCMS or multiple ion detection (MID)/GCMS or selected ion recording (SIR)/GCMS	Scan only selected ions (e.g. m/z 217, 191, 253, etc.) using a dwell time of ~100 ms/ion. Depending on the instrument, ~1 pmol is required for a reliable response when only two or three selected ions are monitored	Chromatograms of selected ions can be used as fingerprints for the selected compound type (e.g. steranes, hopanes, monoaromatic steroids)	Longer dwell time per ion results in better sensitivity compared with full-scan data. Requires knowledge of the retention times and fragmentation characteristics of the molecules to be studied. Forfeits full spectral data that are sometimes required in order to identify unknown compounds
Full-scan GCMS	Scan over range m/z 60–600 every three seconds (i.e. over 180 ions/second). About 1 nmol of each compound is required for a reliable spectrum, comparable to that required for a discernible peak by flame ionization detection in gas chromatography	Provides both mass spectra for structural elucidation and chromatograms of all ions	Mass spectra allow tentative structural elucidation. Unlike SIM mode, no data are discarded. Uses more computer disk space and has lower sensitivity (signal/noise) due to shorter dwell times than SIM mode
Parent-mode GCMS/MS	In the parent ion mode of a triple quadrupole or other tandem instrument, the third quadrupole (Q3) monitors daughter ions (e.g. m/z 217) and the first quadrupole (Q1) scans for all possible precursor (parent) ions. Both Q1 and Q2 are normally operated in SIM mode for higher sensitivity. Selected precursor molecules that are allowed to pass through Q1 undergo collision-activated decomposition (CAD) when they collide with neutral target gas in the second quadrupole (Q2) (see Figure 8.17)		GCMS/MS offers improved selectivity and signal-to-noise ratio compared with other modes, but it is more expensive than most GCMS instrumentation

(*cont.*)

Table 8.3. (*cont.*)

Mode	Method	Result	Advantages/disadvantages
Parent-mode metastable reaction monitoring (MRM)/ GCMS	Monitor a selected daughter ion (e.g. m/z 217) formed from molecular ions that decompose in the first field-free region of a double-focusing mass spectrometer. This is a commonly used form of GCMS/MS		MRM/GCMS offers advantages in selectivity and signal-to-noise ratio that are similar to parent mode GCMS/MS using a tandem instrument. Uses a double-focusing magnetic instrument, which may be more readily available than a tandem instrument. May be quantitatively more reproducible than GCMS/MS because of fewer variables to control (e.g. no collision cell or Q3)

FC-43, which gives known fragments over the desired mass range. FC-43 consists of perfluorotributylamine and has a molecular weight of 671 Daltons.

GCMS operating modes

Depending on the available instrumentation, GCMS analyses can be made in various operating modes. Each mode provides a different type and quality of information. Biomarker parameters are sensitive to the GCMS method employed, and only one method should be used for any study (e.g. Steen, 1986; Fowler and Brooks, 1990). For example, sterane data obtained in multiple ion-detection versus GCMS/MS mode commonly differ. Table 8.3 summarizes some of the more important GCMS operating modes, as discussed in more detail below.

Selected ion monitoring GCMS

Selected ion monitoring (SIM), sometimes called multiple ion detection (MID), is the usual mode of GCMS data acquisition for biomarker analysis. For most biomarker studies, familiar classes of compounds are used, such as hopanes and steranes. One ion of a given m/z together with the gas chromatographic retention time is often diagnostic of the structure of these compounds. Computer-assisted plots of intensity

of a specific ion versus gas chromatographic retention time are called mass chromatograms (or mass fragmentograms when referring to fragment ions). Figure 8.9 shows mass chromatograms for the steranes and terpanes, and their relation in terms of retention time on a gas chromatogram of oil.

Note: Early use of mass chromatograms in organic geochemistry led to a stereochemical understanding of steroid and other carboxylic acids in petroleum (Seifert, 1975) and the first practical method of fingerprinting steranes and terpanes for correlation purposes (Seifert, 1977). Since these early applications, the use of mass chromatography aided by computers has grown rapidly.

Mass chromatograms allow identification of the carbon number, isomer, and homolog distribution of the compound type. In SIM, several diagnostic ions for various compounds of interest are selected for analysis. For each compound type, the selected ion to monitor is usually the most abundant in the mass spectrum and is called the base peak. For example, steranes, hopanes, monoaromatic steroids, and triaromatic steroids are monitored using m/z 217, 191, 253, and 231, respectively, which are the base peaks in the mass spectra of these compounds.

Figure 8.20 shows diagnostic mass spectrometric fragmentations and their mass/charge ratios, which are

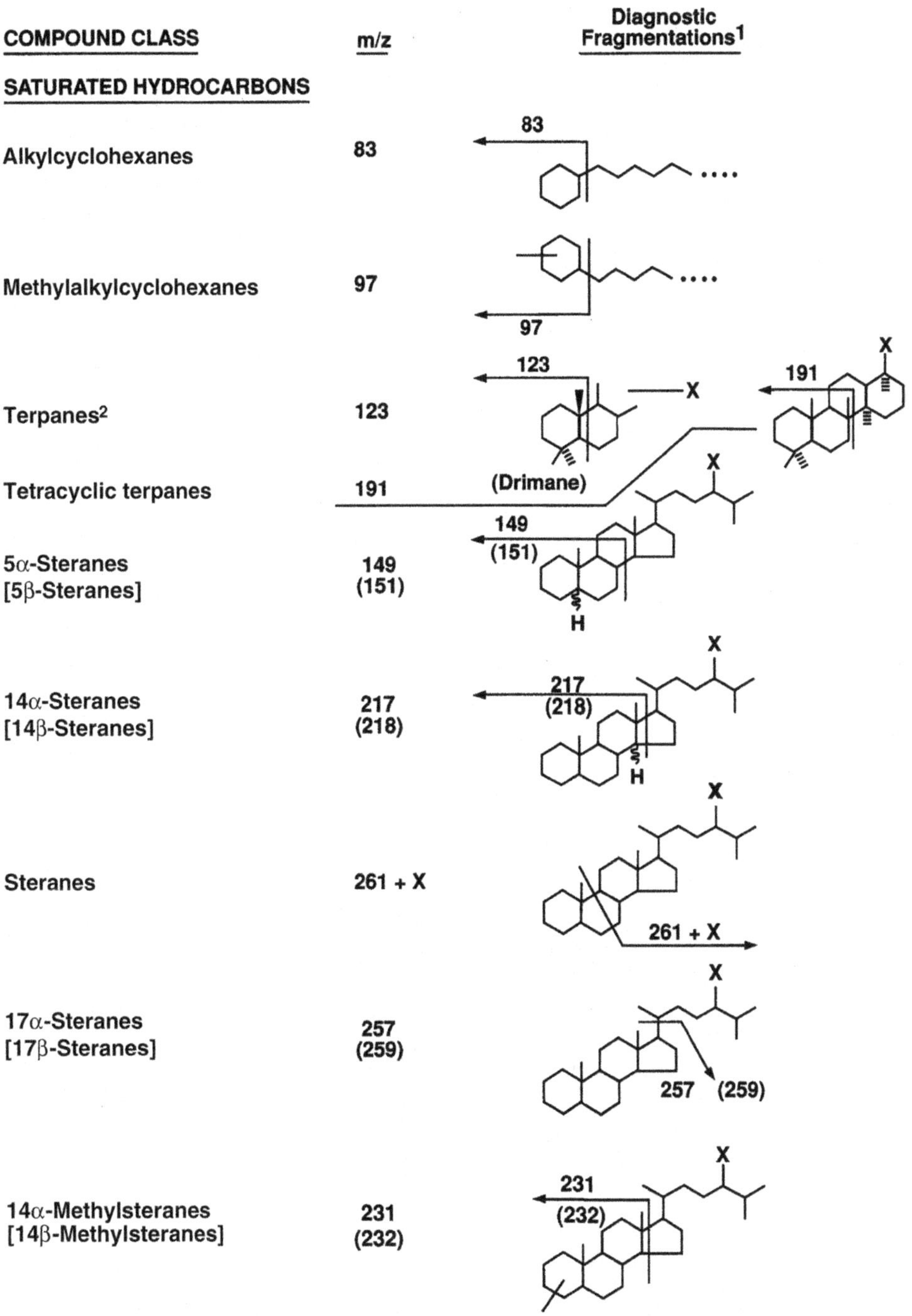

Figure 8.20. Diagnostic mass spectrometric fragmentations (and their mass/charge ratio) used to monitor various biomarkers and other compounds. Reprinted with permission by ChevronTexaco Exploration and Production Technology Company, a division of Chevron USA Inc.

[1] Fragmentation arrows indicate carbon–carbon bond cleavage. Many fragmentations also include hydrogen transfers and rearrangements, which are not shown.

[2] Includes most terpanes with a drimane structure or substructure.

Dinosteranes   98

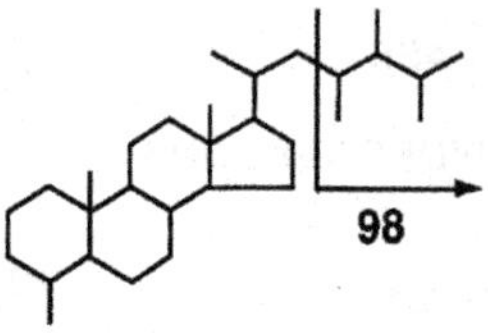

Diasteranes   259

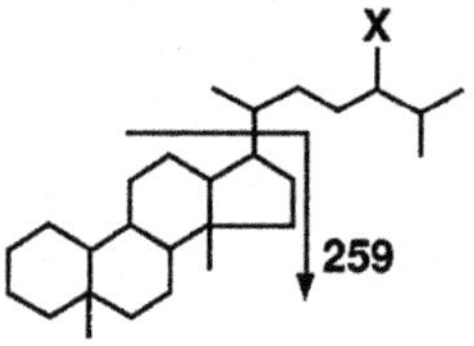

13α,17β(H)-Diasteranes   232

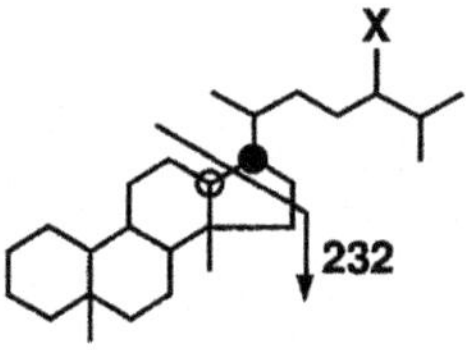

Gammacerane
$C_{30}H_{52}$ 191, 412

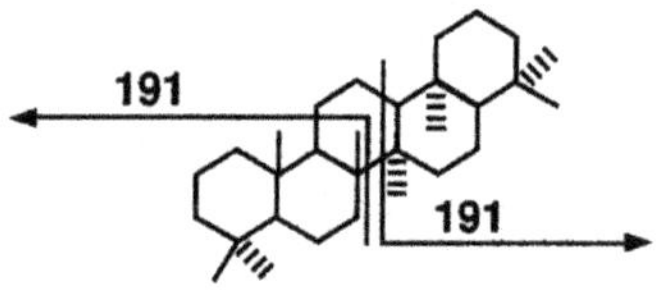

Oleanane
$C_{30}H_{52}$ 191, 412

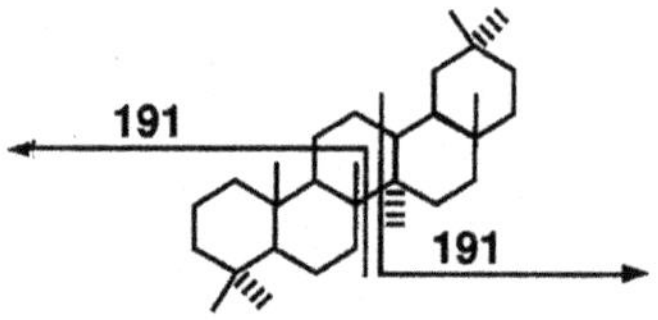

Cheilanthanes[3]   191

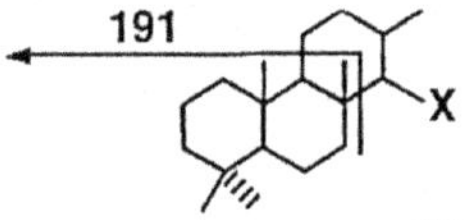

Hopanes (Ring A+B)[3]   191

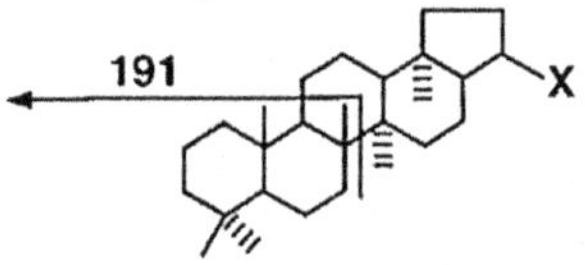

Hopanes (Ring D+E)
$(C_{27}-C_{35})$ 148+X
(149, 163, 177, 191,
205, 219, 233, 247, 261)

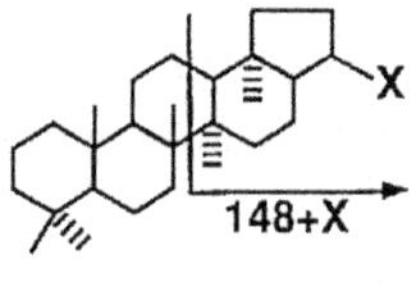

Figure 8.20. (*cont.*)

[3] Includes most terpanes with a cheilanthane structure or substructure, such as degraded and extended hopanes, gammacerane, and oleanane.

Hopanes   369

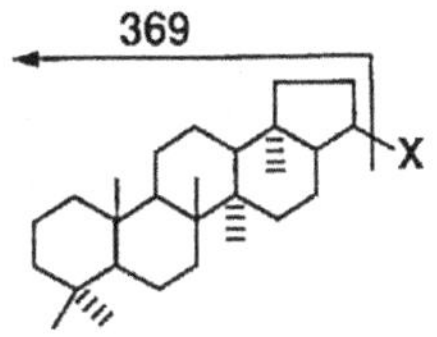

22,29,30-Trisnorneohopane[4]
(Ts or Trisnorhopane-II)   149, 191

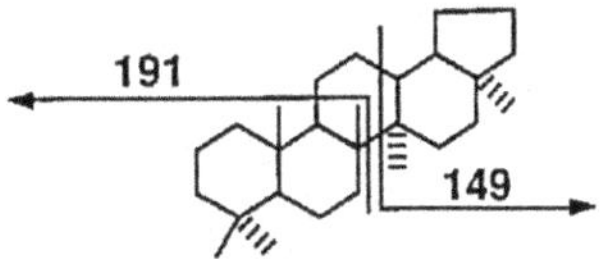

25-Norhopanes (Ring A+B)   177

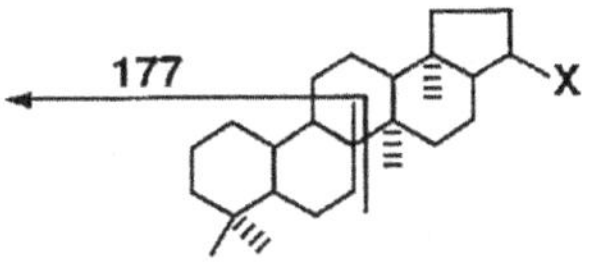

28,30-Bisnorhopanes[5]   163, 191

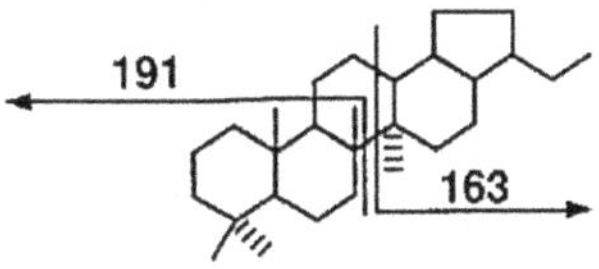

Hexahydrobenzohopanes   191, 216+X, 94+X

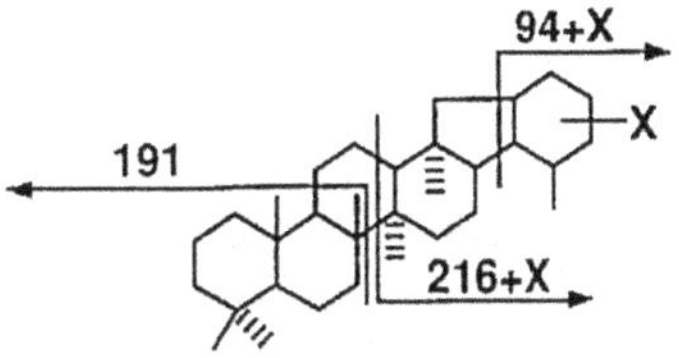

Botryococcane
C₃₄H₇₀   238, 239, 294, 295

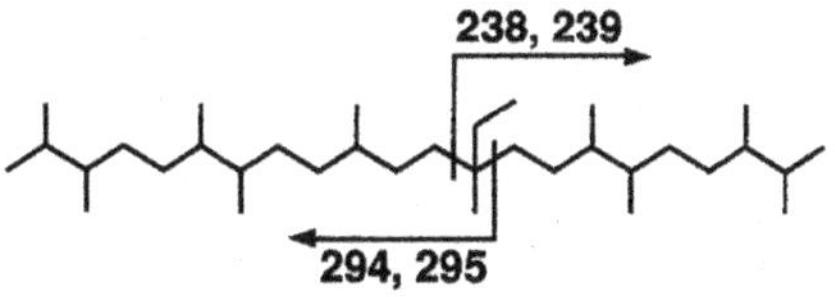

β-Carotane
C₄₀H₇₈   125, 558

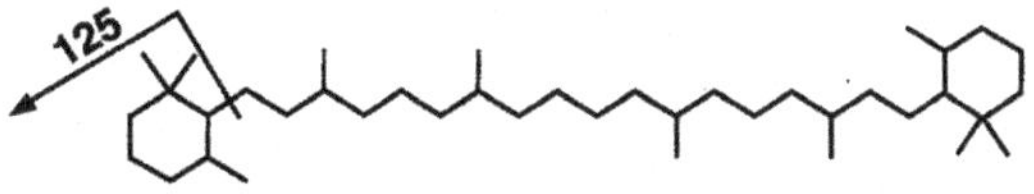

Regular
Isoprenoids   113+70n

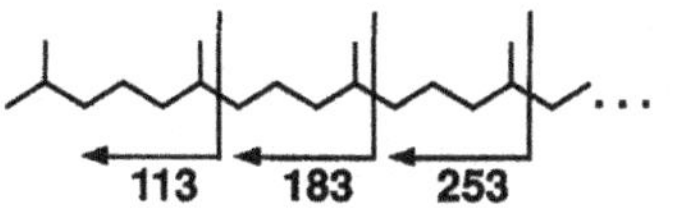

Head-to-Head
Isoprenoids
[1,1′- Bis(phytane)]   (323)

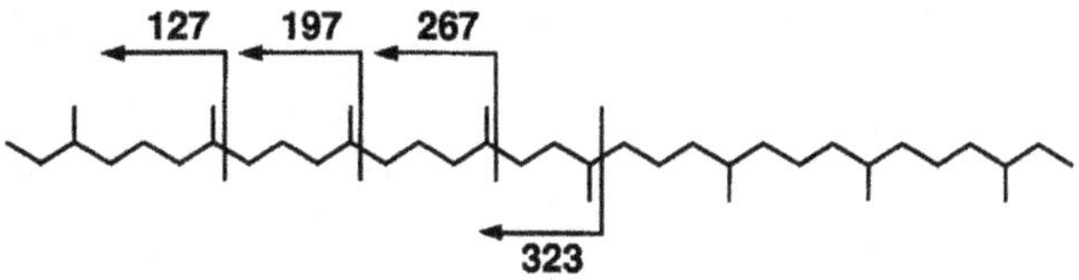

Figure 8.20. (*cont.*)

[4] The ring D + E fragment is the same as that for the hopanes.
[5] The 25,28,30–trisnorhopanes have the *m/z* 163 fragment plus an *m/z* 177 fragment for ring A + B.

183

Squalane
$C_{30}H_{62}$ 183, 422

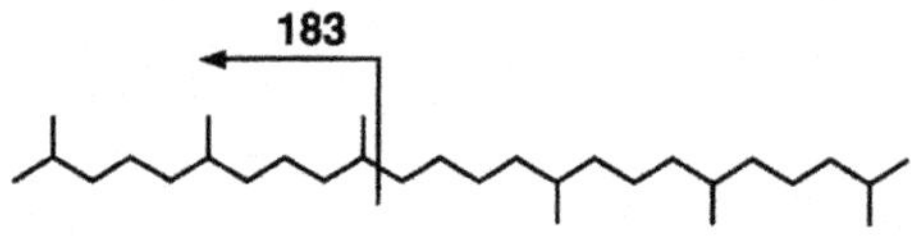

AROMATIC HYDROCARBONS

91

Alkylbenzenes   91

105

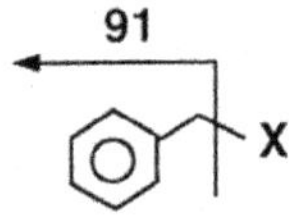

Substituted Methylbenzenes   105

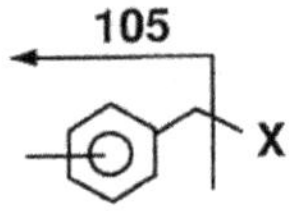

225+X
210+X

Benzohopanes   191,
225+X,
210+X

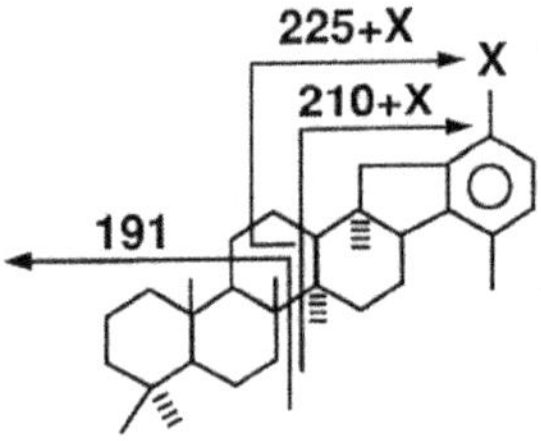

191

C-Ring Monoaromatic
Steroids   253
(267)

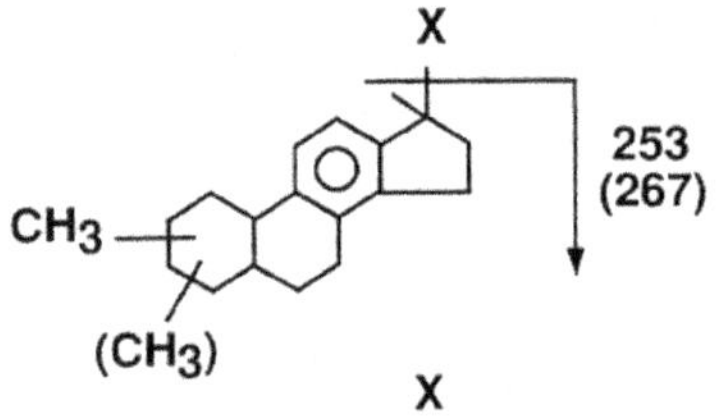

CH_3 —

(CH_3)

253
(267)

X

Triaromatic Steriods   231
(245)

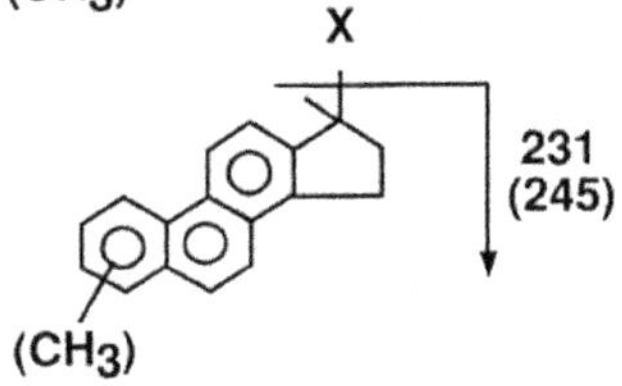

(CH_3)

X

231
(245)

172
159

Aromatic 8,14-Secohopanoids   159,
172,
186+X,
365,
366

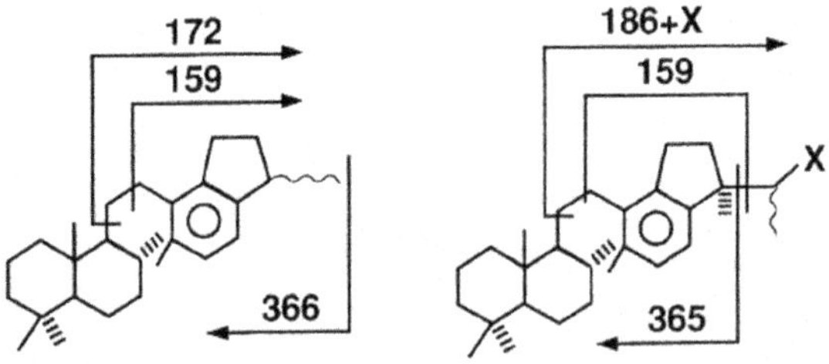

186+X
159

366   365

X

Perylene, Benzopyrenes   252

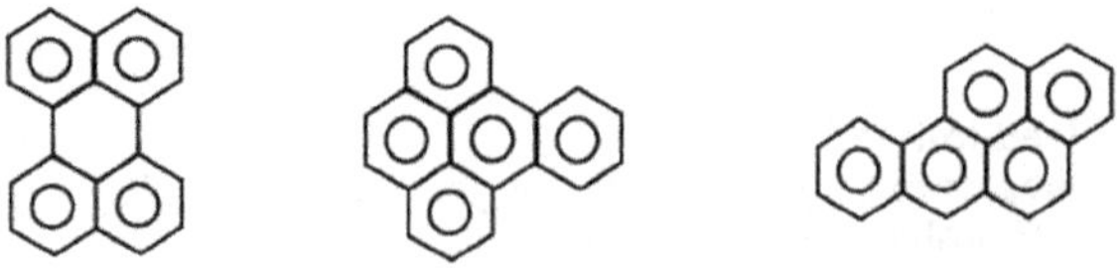

Perylene   1,2-Benzopyrene
[Benzo(e)pyrene]   3,4-Benzopyrene
[Benzo(a)pyrene]

Figure 8.20. (*cont.*)

used to monitor various biomarkers and other compounds in petroleum. Figures 8.21–8.23 show selected mass chromatograms for several common biomarker classes obtained in SIM mode for oil standards from Hamilton Dome in Wyoming and Carneros in California.

Full-scan GCMS

In full-scan GCMS, virtually all fragment ions generated with each scan of the magnet or quadrupoles (i.e. 50–600 amu) are recorded. Unlike SIM, no data are lost, but extensive computer storage is required. Complete mass spectra for identification of compounds can be generated from full-scan analysis. Full-scan mode records several hundred ions per scan (~3 seconds), resulting in <0.0075 seconds of dwell time per mass. By comparison, SIM acquisitions record at ~10 ions/second, resulting in a longer dwell time (~0.01 second/ion) and an order of magnitude better sensitivity and signal-to-noise ratio. Thus, SIM is preferred over full-scan data acquisition for quantitative biomarker analysis, even though the same kind of mass chromatograms can be displayed from full-scan analysis.

Bench-top quadrupole GCMS

Bench-top quadrupole GCMS systems offer many of the capabilities of larger, more expensive floor models and are designed primarily for routine and inexpensive biomarker analysis in SIM mode. Many bench-top systems provide mass chromatograms for a dozen or more ions in a single analysis. Hwang (1990) showed that sterane and triterpane distributions obtained using a bench-top system were in qualitative and quantitative agreement with those produced using a more versatile floor-model GCMS in SIM mode. Screening before detailed biomarker analysis can be provided by bench-top quadrupole GCMS analysis.

Bench-top systems are useful for many applications, such as screening samples for more detailed biomarker analysis and rapid correlation of groups of oils. However, bench-top SIM and GCMS/MS (see below) data are not interchangeable. Bench-top quadrupole systems are currently incapable of GCMS/MS analyses of parent–daughter ion relationships. Certain maturity ratios, such as the C_{29} 20S/(20S + 20R) are best determined by GCMS/MS because of possible interfering peaks on SIM analyses.

GCMS/MS

GCMS/MS is based on the fact that complex organic molecules (parents) ionized in the ion source of a mass spectrometer break down into smaller charged fragments (daughters). Some of these daughter ions are characteristic of their parent molecules (e.g. m/z 217 is a daughter of most steranes). GCMS/MS allows the operator to determine the parents of selected daughter ions.

GCMS/MS refers to tandem mass spectrometry, where triple quadrupole, tandem magnetic, and hybrid magnetic-quadrupole instruments can be used for parent, daughter, and neutral loss experiments (e.g. Futrell, 2000), as discussed below. These systems typically consist of three mass analyzers or sectors and use a neutral collision gas to facilitate decompositions of the compounds to be analyzed in the second sector.

Tandem mass spectrometers have a major advantage over conventional GCMS systems because they can resolve individual compounds or compound families from complex petroleum mixtures using linked-scanning techniques, generally called GCMS/MS (Table 8.3). Triple quadrupole or multisector mass spectrometers are the most common type of tandem mass spectrometer, consisting of three quadrupoles connected in series (Figure 8.17): the first or parent quadrupole (Q1), the middle or collision cell quadrupole (Q2), and the third or daughter quadrupole (Q3). In the collision cell quadrupole, all ions formed in the ion source and focused through Q1 undergo collision-activated decomposition with argon or another inert gas in Q2. The fragment ions formed in the collision process are separated or selectively monitored by the daughter quadrupole Q3 and recorded using an electron multiplier and computer.

Because of the selectivity offered by the use of three quadrupoles, triple quadrupole mass spectrometry offers the possibility of direct analysis of whole oils without preparatory separations of fractions.

Triple quadrupole mass spectrometers can be operated in three GCMS/MS modes:

- parent
- daughter
- neutral loss.

Philp *et al.* (1988) described applications of the triple quadrupole mass spectrometer for determining parent–daughter and daughter–parent ion relationships.

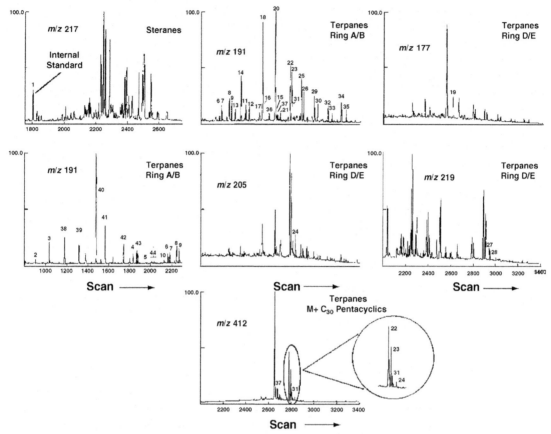

Figure 8.21. Selected ion monitoring (SIM) mass chromatograms for biomarkers in the saturate fraction of crude oil from Hamilton Dome, Wyoming, which is used as a standard. The oil was generated from the Permian Phosphoria Formation. Numbered peaks are identified in the accompanying table. The structure of the internal standard, 5β-cholane, is shown in Figure 8.4. The Wyoming oil standard was spiked with 5β-cholane internal standard, and 0.1 μl of sample was injected into a gas chromatograph equipped with a 60-m fused-silica capillary column with DB-1 stationary phase. The gas chromatograph was programmed to hold at 150°C for 10 minutes followed by heating from 150 to 325°C at 2°C/minute. Electron multiplier voltage in the VG Micromass mass spectrometer was set at 160 V, gain set at 1×10^{-6}; and range scans 1751–2750 recorded.

As indicated in Figure 8.20, m/z 217 = steranes, m/z 191 = ring A/B fragment in terpanes, m/z 177, 205, 219 = ring D/E fragment in terpanes, and m/z 412 = molecular ion of C_{30}-pentacyclic terpanes. All peaks are normalized to the fragment with the greatest mass spectrometric response (100% on y-axis). Reprinted with permission by ChevronTexaco Exploration and Production Technology Company, a division of Chevron USA Inc.

Other tandem mass spectrometers combine various series of magnetic and electrostatic sectors and collision cells. A hybrid mass spectrometer combines a magnetic sector instrument with a collision cell quadrupole and a mass filtering quadrupole. These other types of tandem mass spectrometer offer some potential advantages over triple quadrupole instruments. For example, hybrid mass spectrometers with a high-resolution magnet in place of Q1 allow the selection of parent ions at high resolution.

GCMS/MS allows determination of specific parent–daughter relationships with little interference from other reactions and their related ions. The analytical specificity of these methods eliminates most interference by co-eluting gas chromatographic peaks. This chemical specificity may boost signal-to-noise ratios to

Peak	Name	Carbon Number
1	5β-Cholane	24
2	C$_{19}$ Tricyclic (Cheilanthane)	19
3	C$_{20}$ Tricyclic (Cheilanthane)	20
4	C$_{24}$ Tetracyclic	24
5	C$_{25}$ Tricyclic (Cheilanthane)	25
6	C$_{28}$ Tricyclic (Cheilanthane)	28
7	C$_{28}$ Tricyclic (Cheilanthane)	28
8	C$_{29}$ Tricyclic (Cheilanthane)	29
9	C$_{29}$ Tricyclic (Cheilanthane)	29
10	C$_{26}$ Tetracyclic	26
11	C$_{30}$ Tricyclic (Cheilanthane)	30
12	C$_{30}$ Tricyclic (Cheilanthane)	30
13	22,29,30-Trisnorneohopane (Ts)	27
14	22,29,30-Trisnorhopane (Tm)	27
15	17α(H)-30-Nor-29-Homohopane	30
16	18α (H)-30-Norneohopane (C$_{29}$Ts)	29
17	C$_{31}$ Tricyclic (Cheilanthane)	31
18	17α,21β(H)-30-Norhopane	29
19	17β,21α(H)-30-Norhopane (Normoretane)	29
20	17α,21β(H)-Hopane	30
21	17β,21α(H)-Hopane (Moretane)	30
22	17α,21β(H)-29-Homohopane 22S	31
23	17α,21β(H)-29-Homohopane 22R	31
24	17β,21α(H)-29-Homohopane 22S + 22R	31
25	17α,21β(H)-29-Bishomohopane 22S	32
26	17α,21β(H)-29-Bishomohopane 22R	32
27	17β,21α(H)-29-Bishomohopane 22?	32
28	17β,21α(H)-29-Bishomohopane 22?	32
29	17α,21β(H)-29-Trishomohopane 22S	33
30	17α,21β(H)-29-Trishomohopane 22R	33
31	Gammacerane	30
32	17α,21β(H)-29-Tetrakishomohopane 22S	34
33	17α,21β(H)-29-Tetrakishomohopane 22R	34
34	17α,21β(H)-29-Pentakishomohopane 22S	35
35	17α,21β(H)-29-Pentakishomohopane 22R	35
36	C$_{29}$ 17β,21α(H)-30-Norhopane	29
37	18α-Neohopane (C$_{30}$Ts; putative)	30
38	C$_{21}$ Tricyclic (Cheilanthane)	21
39	C$_{22}$ Tricyclic (Cheilanthane)	22
40	C$_{23}$ Tricyclic (Cheilanthane)	23
41	C$_{24}$ Tricyclic (Cheilanthane)	24
42	C$_{25}$ Tricyclic (Cheilanthane) (22S + 22R)	25
43	C$_{26}$ Tricyclic (Cheilanthane) (22S + 22R)	26
44	C$_{27}$ Tricyclic (Cheilanthane) (22R + 22S)	27

Figure 8.21. (*cont.*)

levels that are orders of magnitude better than those obtained using conventional GCMS in SIM mode.

SIM/GCMS analysis of oil for the m/z 217 fragment ion reveals a complex mixture of structural isomers and stereoisomers of steranes and may also include some non-sterane compounds (Figure 8.24).

The top chromatogram in Figure 8.25 is a hypothetical example of a simple mixture dominated by three homologous steranes with 27, 28, and 29 carbon atoms. Each homolog yields a base peak at m/z 217 on this trace. GCMS/MS analysis of the sterane parent ion transitions corresponding to m/z 372 → 217, m/z 386 → 217, and m/z 400 → 217 allows separate mass chromatograms for the C$_{27}$, C$_{28}$, and C$_{29}$ steranes, respectively (bottom three chromatograms, Figure 8.25).

GCMS/MS analyses can be used to determine marine input to oils (C$_{30}$ steranes) and for correlations using triangular diagrams of C$_{27}$–C$_{28}$–C$_{29}$ steranes, diasteranes, triaromatic steroids, and other compounds. Figure 8.21 includes an example of m/z 217 for oil from Wyoming, which is much more complex than the simple hypothetical mixture shown in Figure 8.25. GCMS/MS of the same oil (Figure 8.26) allows differentiation of the steranes using mass chromatograms by carbon number.

Metastable reaction monitoring/GCMS

Metastable reaction monitoring (MRM)/GCMS provides many of the same features as GCMS/MS by monitoring decompositions that occur in the first field-free region of a double-focusing mass spectrometer. These methods are also called selected metastable ion monitoring (SMIM) (e.g. Steen, 1986). A triple-sector instrument uses a neutral gas in the middle sector to induce decomposition. Double-focusing magnetic instruments can be configured in two ways for MRM/GCMS. In the first method, the electrostatic voltage is delinked from the accelerating voltage (Gallegos, 1976; Warburton and Zumberge, 1982). The magnet is set to focus only the daughter fragments of interest, and the accelerating voltage is scanned to measure the metastable transitions that occur in the first field-free region of the mass spectrometer. Only the daughter ions of interest that form from metastable transitions of the selected parent ions are detected. In the second method, called linked scan, the magnet and the electrostatic sectors are linked (Haddon, 1979). For example, a linked-scan experiment could involve scanning the magnetic sector field strength (B) and the electric sector field strength (E) simultaneously, holding the accelerating voltage constant so as to maintain a constant ratio of B to E. This constant value is determined by the ratio of the two field strengths, which transmit the desired ions of a predetermined mass/charge ratio. For parent ion experiments, the sectors are linked according to the constant B^2/E. For daughter ion experiments, the sectors are linked according to the constant B/E.

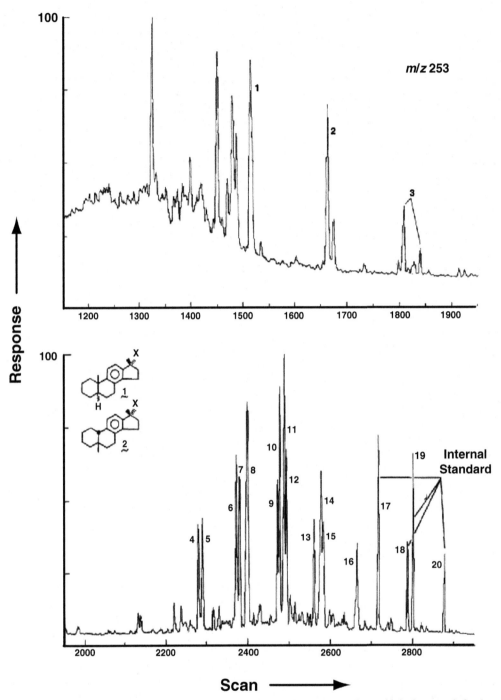

Figure 8.22. Selected ion monitoring (SIM) mass chromatograms for the monoaromatic steroids in the aromatic fraction of a standard oil from Carneros, California. This standard is more suitable than the Wyoming oil standard because it contains all compounds used in routine studies. Numbered compounds are identified in the table. Figure 8.4 shows the structures of the C_{30} monoaromatic-steroid standards. Reprinted with permission by ChevronTexaco Exploration and Production Technology Company, a division of Chevron USA Inc.

Peak	Structure	Identification*	Carbon Number
1	?	Pregnane (X = ethyl)	21
2	?	20-Methylpregnane (X = 2-propyl)	22
3	?	20-Ethylpregnane (X = 2-butyl)	23
4	1	5β-Cholestane 20S	27
5	2	Diacholestane 20S	27
6	1 / 2	5β-Cholestane 20R; diacholestane 20R	27
7	1	5α-Cholestane 20S	27
8	1 / 2	5β-Ergostane 20S; diergostane 20S	28
9	1	5α-Cholestane 20R	27
10	1	5α-Ergostane 20S	28
11	1 / 2	5β-Ergostane 20R; Diergostane 20R	28
12	1 / 2	5β-Stigmastane 20S; Diastigmastane 20S	29
13	1	5α-Stigmastane 20S	29
14	1	5α-Ergostane 20R	28
15	1 / 2	5β-Stigmastane 20R; Diastigmastane 20R	29
16	1	5α-Stigmastane 20R	29
17	1	5β-n-Nonylpregnane 20S (X = 2-undecyl)	30
18	1	5α-n-Nonylpregnane 20S (X = 2-undecyl)	30
19	1	5β-n-Nonylpregnane 20R (X = 2-undecyl)	30
20	1	5α-n-Nonylpregnane 20R (X = 2-undecyl)	30

*Name of sterane (structure 1) or diasterane (structure 2) with the same side chain (X).

Figure 8.22. (cont.)

While not truly GCMS/MS as performed on a tandem mass spectrometer, MRM/GCMS yields similar results. Thus, further references to GCMS/MS in the text also include MRM/GCMS, unless stated otherwise. Due to the greater sensitivity and reliability of these techniques, their use in biomarker applications is likely to increase gradually compared with conventional GCMS.

Parent-mode GCMS/MS
In the parent mode, ions formed in the ion source of the mass spectrometer enter the collision cell quadrupole (Q2 in Figure 8.17), collide with the inert gas, and dissociate to form various daughter ions. One or more daughter ions for each biomarker family to be monitored are selected by the third quadrupole (Q3) to focus on the electron multiplier for analysis. For example, compounds increase by one methylene group ($-CH_2-$, i.e. 14 amu) in the series $C_{26}-C_{30}$ steranes. GCMS/MS can be used to differentiate between these compounds. Molecular ions (parents) of the $C_{26}-C_{30}$ steranes consist of m/z 358, 372, 386, 400, and 414, respectively. Each of these parent ions produces a major daughter ion at m/z 217 (Table 8.4) following collision with the

Table 8.4. *Parent-daughter transitions for selected compounds of geochemical interest*

Compound	Parent ion separated by first quadrupole, Q1*	Daughter ion generated in second quadrupole, Q2*
C_{19} Tricyclic terpane (cheilanthane)	262	191
C_{20} Tricyclic terpane	276	191
C_{27} Tricyclic terpane	374	191
C_{28} Tricyclic terpane	388	191
C_{24} Tetracyclic (des-E-hopane)	330	191
Des-A-Lupane	330	191
Phyllocladane	274	123
Isopimarane	276	247
Drimane	208	123
Eudesmane	208	165
C_{25} HBI	239 (352)	238
C_{24} Sterane (5β-cholane)	330	217
C_{26} Steranes	358	217
C_{27} Steranes	372	217
C_{28} Steranes	386	217
C_{29} Steranes	400	217
C_{30} Steranes	414	217
C_{30} Methylsteranes	414	231
C_{27} Trisnorhopanes	370	191
C_{28} Bisnorhopanes	384	191
C_{30} Hopane	412	191
C_{31} Homohopanes	426	191
C_{35} Hopanes	482	191
Bicadinanes	412	369
Methylbicadinanes	426	383
Spirotriterpanes	412	342
Botryococcane	294	197
Dinor-lupanes	384	177
25-Norhopane	398	177
25,30-Bisnorhopane	384	177
Bishomohopanes	440	191
Carotanes	558	125

[a] Refer to Figure 8.17.
HBI, highly branched isoprenoid.

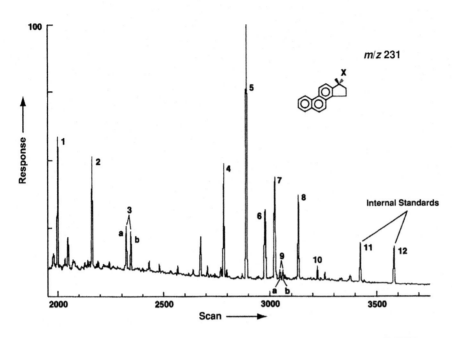

Figure 8.23. Selected ion monitoring (SIM) mass chromatogram for the triaromatic steroids in the aromatic fraction of oil from Hamilton Dome, Wyoming, which is used as a standard. Numbered compounds are identified in the table. Figure 8.4 shows the structures of the triaromatic-steroid internal standards. Reprinted with permission by ChevronTexaco Exploration and Production Technology Company, a division of Chevron USA Inc.

Peak	Identification*	Carbon Number
1	Pregnane (X = ethyl)	20
2	20-Methylpregnane (X = 2-propyl)	21
3	20-Ethylpregnanes (X = 2-butyl; a and b are epimeric at C_{20})	22
4	Cholestane 20S	26
5	Cholestane 20R; ergostane 20S	26, 27
6	Stigmastane 20S (24-ethylcholestane 20S)	26
7	Ergostane 20R (24-methylcholestane 20R)	27
8	Stigmastane 20R	28
9	24-*n*-Propylcholestane 20S (a and b are epimeric at C-24)	29
10	24-*n*-Propylcholestane 20R	29
11	20-*n*-Decylpregnane 20S (X = 2-dodecyl)	30
12	20-*n*-Decylpregnane 20R	30

*Name of sterane with the same side chain (X).

inert gas in the collision cell (Q2, Figure 8.17) that can be monitored using the daughter quadrupole (Q3). Both parents and daughters can be selectively monitored to improve signal-to-noise ratio. Thus, a mass chromatogram showing only C_{27} steranes can be obtained by setting the parent quadrupole (Q1) to pass only ions having m/z 372 and by setting the daughter quadrupole to pass only m/z 217 ions. A signal is obtained only if the selected molecular ion produces the selected daughter ion. The parent ions do not necessarily need to be molecular ions. For example, although the molecular ion for the C_{25} highly branched isoprenoid (HBI) (Table 8.4) is m/z 352, the m/z 239 → 238 transition is preferred to m/z 352 → 238 or 239 because

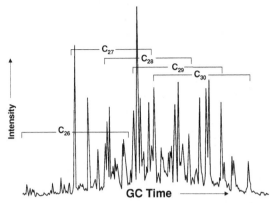

Figure 8.24. Selected ion monitoring/gas chromatography/ mass spectrometry (SIM/GCMS) analysis of steranes (m/z 217) for a typical crude oil, showing regions where different homologs overlap or interfere.

ionization of the compound yields only a small m/z 352 fragment that is more difficult to measure than either m/z 238 or m/z 239.

The GCMS/MS approach to identification of the C_{27} steranes described above is based on the relationship between the parent (m/z 372) and daughter (m/z 217) fragments. This approach is superior to direct monitoring of m/z 372 in SIM mode because SIM is affected by interference. For example, C_{28} steranes can lose a methyl group (15 amu) during ionization, resulting in a fragment at m/z 371. However, some m/z 372 will also result from the C_{28} steranes because heavy isotopes of carbon (^{13}C) or hydrogen (D) are also present. Compounds other than steranes may also fragment to yield m/z 372 ions under SIM/GCMS conditions.

GCMS/MS (including MRM/GCMS) represents a significant refinement over routine GCMS. For example, monitoring m/z 217 by routine GCMS in SIM mode provides a single mass chromatogram containing all steranes, many of which co-elute. However, GCMS/MS provides nearly complete separation of individual steranes by carbon number (Figure 8.26), and mass chromatograms for each are obtained. These GCMS/MS data are critical in constructing C_{27}–C_{28}–C_{29} sterane ternary diagrams for oil–oil and oil–source rock correlations. Routine SIM/GCMS, as offered by many service companies, cannot be used for constructing these diagrams as accurately as GCMS/MS because of interference from compounds containing different numbers of carbon atoms. During any single GCMS/MS analysis, several parent–daughter

relationships can be monitored simultaneously, allowing resolution of several carbon numbers of the same family (e.g. C_{27}–C_{30} steranes) or of several different families of biomarkers (e.g. steranes, tricyclic and pentacyclic terpanes). GCMS/MS analyses usually involve numerous parent–daughter relationships that must be deconvoluted by computer.

Collision-activated decomposition (CAD) GCMS/MS is superior to other GCMS modes for deconvoluting biomarker compositions. For example, Fowler and Brooks (1990) compared sterane distributions and maturity parameters for oils and bitumens from the Egret Member of the Rankin Formation,

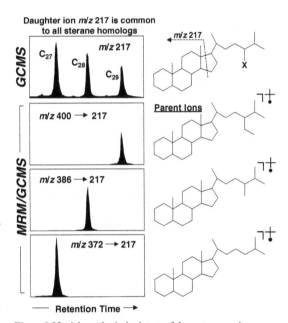

Figure 8.25. A hypothetical mixture of three sterane epimers (C_{27}, C_{28}, and C_{29}) yields three peaks when analyzed for m/z 217 by selected ion monitoring/gas-chromatography/mass spectrometry (SIM-GCMS), a fragment ion common to all steranes (top). A typical m/z 217 mass chromatogram for petroleum contains a much larger number of overlapping peaks representing various epimers and sterane homologs (e.g. see Figure 8.24). Gas chromatography/mass spectrometry/mass spectrometry (GCMS/MS) of the mixture (bottom three panels) measures specific parent/daughter ions, e.g. from m/z 400 (parent ion for C_{29} steranes) to daughter m/z 217. Parent/daughter GCMS/MS reduces interference for more complex mixtures, such as petroleum. MRM/GCMS, metastable reaction monitoring/gas chromatography/mass spectrometry.

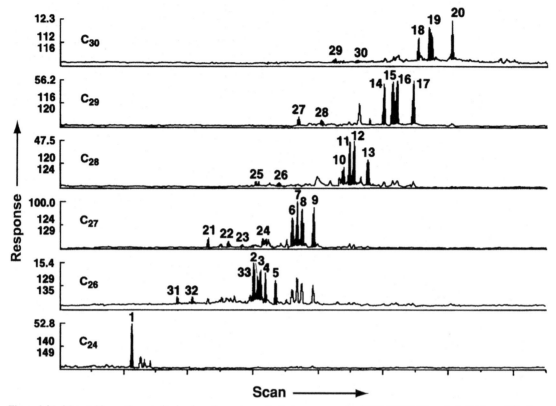

Figure 8.26. Metastable reaction monitoring/gas chromatography/mass spectrometry (MRM/GCMS) analysis (M+ → 217) of the same Wyoming oil shown in Figure 8.21 (steranes) allows differentiation of sterane epimers by carbon number. Note the complex epimer distributions at each carbon number compared with the hypothetical example shown in Figure 8.25. Peaks corresponding to 6, 7, 8, and 9 on the C_{27} chromatogram also appear on the C_{26} chromatogram. This is interference caused by the inherent low resolution (~100) of the MRM method. Similar experiments using gas chromatography/mass spectrometry/mass spectrometry (GCMS/MS) on a tandem mass spectrometer are typically performed at higher resolution (~1000) to reduce this type of interference. Reprinted with permission by ChevronTexaco Exploration and Production Technology Company, a division of Chevron USA Inc.

offshore eastern Canada using four methods. The abundant 4-methylsteranes in these samples interfere with regular (4-desmethyl) steranes on m/z 217 mass chromatograms because both groups of steranes yield m/z 217 fragments. As expected, neither low- nor high-resolution SIM/GCMS systems effectively remove the interference caused by the 4-methylsteranes because the fragment ions of both groups of steranes have the same atomic composition. MRM/GCMS results for the same samples were more reliable, because specific parent–daughter transitions could be monitored. However, even more reliable results were obtained using CAD/GCMS/MS due to greater resolution of both parent and daughter ions.

Daughter-mode GCMS/MS

In the daughter mode, a selected parent ion enters the collision cell, where it undergoes CAD, and a complete daughter mass spectrum is collected by setting the daughter quadrupole on full scan. This approach allows identification of components with specific molecular weights from their mass spectra. It also allows a mass spectrum to be obtained for compounds within complex mixtures without interference from co-eluting compounds of different mass.

Neutral-loss-mode GCMS/MS

In the neutral loss mode, the parent and daughter quadrupoles (sectors) are adjusted to scan over the

Peak	Identification	Carbon Number
1	5β-Cholane	24
2	5α,14α,17α(H)-27-Norcholestane 20S	26
3	5α,14β,17β(H)-27-Norcholestane 20R	26
4	5α,14β,17β(H)-27-Norcholestane 20S	26
5	5α,14α,17α(H)-27-Norcholestane 20R	26
6	5α,14α,17α(H)-Cholestane 20S	27
7	5α,14β,17β(H)-Cholestane 20R	27
8	5α,14β,17β(H)-Cholestane 20S	27
9	5α,14α,17α(H)-Cholestane 20R	27
10	5α,14α,17α(H)-Ergostane 20S	28
11	5α,14β,17β(H)-Ergostane 20R	28
12	5α,14β,17β(H)-Ergostane 20S	28
13	5α,14α,17α(H)-Ergostane 20R	28
14	5α,14α,17α(H)-Stigmastane 20S	29
15	5α,14β,17β(H)-Stigmastane 20R	29
16	5α,14β,17β(H)-Stigmastane 20S	29
17	5α,14α,17α(H)-Stigmastane 20R	29
18	5α,14α,17α(H)-24-n-Propylcholestane 20S	30
19	5α,14β,17β(H)-24-n-Propylcholestane 20R + 20S	30
20	5α,14α,17α(H)-24-n-Propylcholestane 20R	30
21	13β,17α(H)-Diacholestane 20S	27
22	13β,17α(H)-Diacholestane 20R	27
23	13α,17β(H)-Diacholestane 20S	27
24	13α,17β(H)-Diacholestane 20R	27
25	13β,17α(H)-Diaergostane 20S (24S + 24R)	28
26	13β,17α(H)-Diaergostane 20R (24S + 24R)	28
27	13β,17α(H)-Diastigmastane 20S	29
28	13β,17α(H)-Diastigmastane 20R	29
29	13β,17α(H)-Dia-24-n-propylcholestane 20S	30
30	13β,17α(H)-Dia-24-n-propylcholestane 20R	30
31	13β,17α(H)-Dia-27-norcholestane 20S	26
32	13β,17α(H)-Dia-27-norcholestane 20R	26
33	5α,14α,17α(H)- + 5α,14β,17β(H)-21-Norcholestanes	26

Figure 8.26. (*cont.*)

desired mass range. The daughter quadrupole is set to scan at a given mass below the parent quadrupole based on hypothetical loss of that mass from the parent compounds as a neutral fragment. Thus, to monitor the distribution of sulfur-containing compounds that undergo a neutral loss of mass 32, the daughter quadrupole is set 32 mass units below the parent quadrupole. Only compounds that lose m/z 32 will be detected.

MASS SPECTRA AND COMPOUND IDENTIFICATION

Mass spectra are an important tool for interpreting structures of unknown compounds (McLafferty, 1980). A mass spectrum commonly indicates the mass of a molecule and the masses of its fragments. One mass spectrum is generated approximately every three seconds as the detector scans through the desired mass range (usually 50–600 amu). Thus, each mass spectrum plots the mass/charge (m/z) for ions impinging on the

detector during each scan versus response (Figure 8.18). Figure 8.27 shows mass spectra for several common biomarkers. Philp (1985) compiled the mass spectra for many biomarkers.

Note: Stable isotopes in biomarkers affects their mass spectra and can be used to help in structural elucidation. Biomarkers fragment into ions in the ion source of the mass spectrometer. The probability that one of the atoms in any ion is a ^{13}C isotope increases with the number of atoms. For example, the molecular ion for cholestane (m/z 372) contains 27 carbon atoms and has a much greater probability of containing a ^{13}C atom than a molecule containing one carbon atom (e.g. $27 \times 1.1 = 29.7\%$). The factor of 1.1% per carbon atom varies slightly depending on the organic matter source (~2% relative). The mass spectrum of cholestane (Figure 8.27) shows a significant m/z 373 after the m/z 372 molecular ion resulting from the contribution of ^{13}C. If this mass spectrum were for an unknown compound, then the maximum number of carbon atoms in the compound could be estimated by comparing the two isotope peaks for the molecular ion (McLafferty, 1980).

Isotope peaks can sometimes complicate the interpretation of mass chromatograms. For example, the Beatrice oil contains abundant benzohopanes characterized by a mass spectral base peak at m/z 252. Because of their high concentrations in this oil and their isotope peak at m/z 253, the benzohopanes interfere with the monoaromatic steroid mass chromatogram (m/z 253) for this oil (Figure 8.36).

Combinations of analyses, which might include mass spectroscopy, nuclear magnetic resonance (NMR) spectroscopy, X-ray diffraction crystallography, and gas chromatography of co-injected authentic standards, are necessary to prove the chemical structures of individual components eluting from the gas chromatograph. The structure of an unknown component is proven only when these analyses are identical for both the unknown and an authentic standard compound synthesized in the laboratory or when X-ray diffraction results are conclusive. Rigorous identification of compounds using mass spectra (McLafferty, 1980) combined with other methods, such as two-dimensional NMR spectroscopy

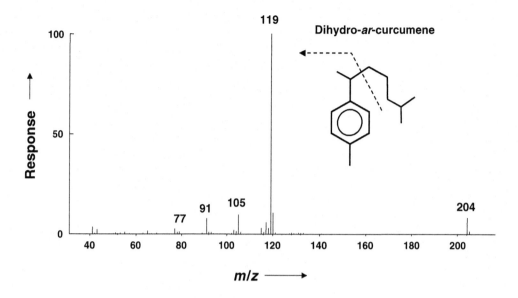

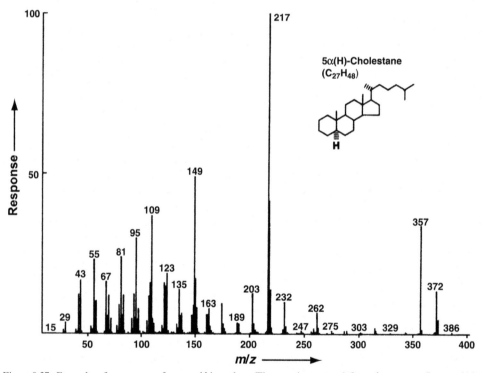

Figure 8.27. Examples of mass spectra for several biomarkers. The most intense peak for each spectrum (base peak) is defined as showing a response of 100 units on the *y*-axis. Philp (1985) shows mass spectra for many biomarkers. Reprinted with permission by ChevronTexaco Exploration and Production Technology Company, a division of Chevron USA Inc. Mass spectrum of dihydro-*ar*-curcumene from Ellis (1995).

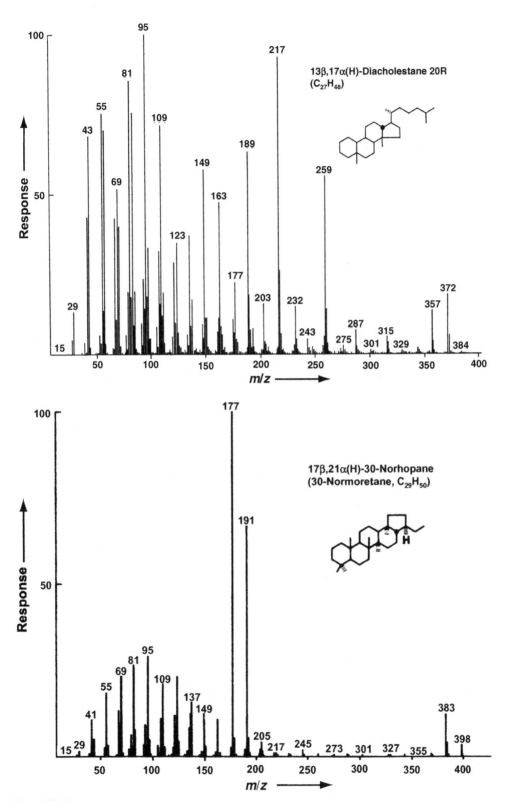

Figure 8.27. (*cont.*)

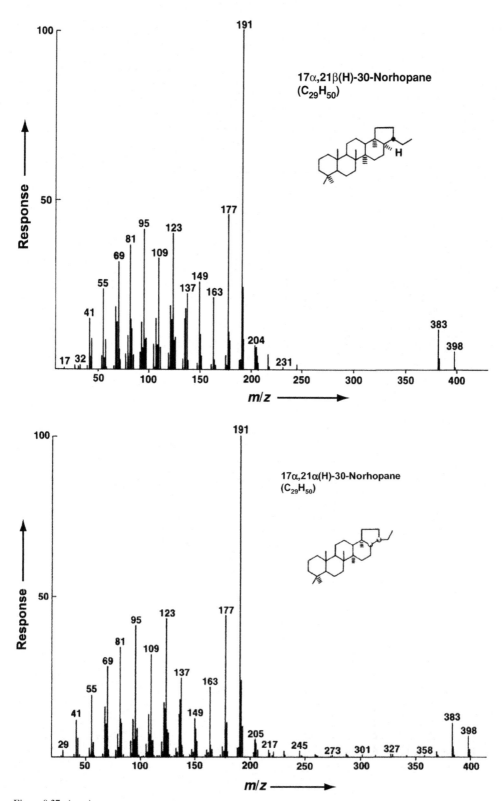

Figure 8.27. (*cont.*)

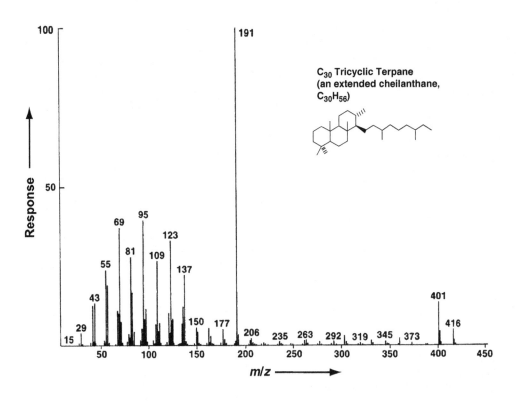

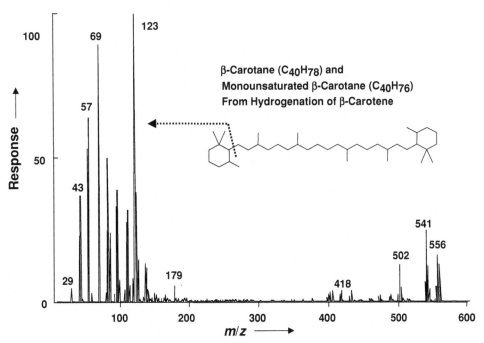

Figure 8.27. (*cont.*)

(Croasmun and Carlson, 1987), is beyond the scope of this book, but two examples follow. Smith *et al.* (1970) used X-ray crystallography to determine the structures of 18α-oleanane and a spirotriterpane in a Nigerian crude oil, while Balogh *et al.* (1973) used ^{13}C NMR to determine the structure of 17α-hopane in Green River oil shale. Fortunately, routine geochemical studies do not require such detailed proof of chemical structures because the same key compounds are used and their principal fragments and retention times are known.

Acceptable proof of compound structure is usually obtained by co-elution with a chemically synthesized standard of known structure using two high-resolution columns showing different polarities. If they represent the same structure, then the unknown and standard compounds also show identical mass spectra. Co-injection is a chromatographic technique used in co-elution experiments. A synthesized or commercial standard is mixed with the sample containing the compound to be identified in a process called spiking. If the co-injected standard and unknown compounds co-elute

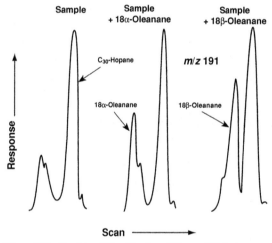

Figure 8.28. Provisional identification of 18α- and 18β-oleanane in bitumen (sample 1 in Figure 14.9) from offshore Eel River Basin, California, by co-injection of authentic standards. The terpane mass chromatogram (m/z 191) for the neat (unspiked) bitumen (left) shows two unknown peaks, which elute before the C_{30} $17\alpha,21\beta$(H)-hopane peak. Synthetic 18α-oleanane co-elutes with the first-eluting unknown peak (center). The peak height for 18α-oleanane increased relative to the C_{30} hopane from the neat sample (left) to the co-injected sample (center). Synthetic 18β-oleanane co-elutes with the second unknown peak (right). Reprinted with permission by ChevronTexaco Exploration and Production Technology Company, a division of Chevron USA Inc.

from the gas chromatograph, then the relative peak intensity of the unknown compound on chromatograms of the mixture will be higher than that for the neat (unspiked) sample. Figure 8.28 shows the provisional identification of 18α- and 18β-oleanane peaks in bitumen from the offshore Eel River area, California. Co-elution of a standard with the unknown compound suggests, but does not prove, that the compounds are identical. Commonly, co-elution experiments are repeated using another chromatographic column with a different stationary phase. Two different compounds that fortuitously co-elute are unlikely to do so on another column with different polarity.

After co-elution is established, matching of mass spectra for the unknown and the standard can be used to further support that the two are identical. However, spectra obtained using different instruments or the same instrument under different conditions may differ. Figure 8.29 compares our mass spectrum for an unknown peak in oil from the North Sea with a published mass spectrum for 25-nor-17α-hopane. (25-Norhopanes as indicators of heavy biodegradation are discussed later.) The similarity of the spectra, combined with co-elution of the standard and the unknown peak, represents provisional identification of the compound. As indicated above, rigorous proof of structure might also include co-elution of the unknown and standard on various chromatographic columns, and NMR or X-ray structural verification.

Plots of scan numbers (retention times) for precursors versus proposed products can be used to assist in structural elucidation of homologs. For example, 25-norhopanes are believed to originate from the series of hopane homologs by loss of a methyl group at C-25. Thus, scan numbers for hopane homologs Ts and Tm (C_{27}), bisnorhopane (C_{28}), norhopane (C_{29}), hopane (C_{30}), and the C_{31}–C_{35} homohopanes (m/z 191) should show a linear relationship versus scan numbers for demethylated 25-norhopanes (m/z 177).

BIOMARKER QUANTITATION

Quantitation of individual compounds using mass chromatograms can be automated by computer programs that identify peaks in GCMS analysis based on (1) gas chromatographic retention time relative to an internal standard co-injected with the sample and (2) comparison of mass spectra of unknown compounds with a library of standard spectra.

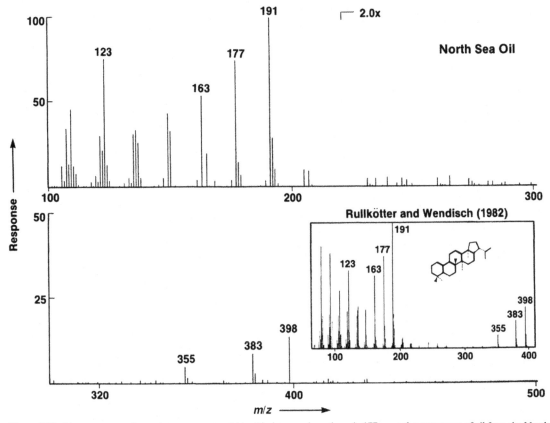

Figure 8.29. Mass spectrum of an unknown compound identified as a peak on the m/z 177 mass chromatogram of oil from the North Sea. The similarity of the mass spectrum of the unknown compound with that of the published spectrum (inset) assisted in the provisional identification of the compound as 25-nor-17α-hopane (structure shown in inset). The mass spectrum of the compound has major fragment ions at m/z 191 and m/z 177. Figure 14.2 shows m/z 191 and m/z 177 mass chromatograms for the same oil. Reprinted with permission by ChevronTexaco Exploration and Production Technology Company, a division of Chevron USA Inc.

Internal standards are described below. For saturates, 5β-cholane is commonly used as the internal standard (Figures 8.4 and Figure 8.21) because it is not present in significant amounts in crude oils, its fragmentation in the mass spectrometer is similar to other steranes (Seifert and Moldowan, 1979), and it does not co-elute with other steranes. Synthetic aromatic steroids that are not found in crude oils and do not co-elute with natural aromatic steroids are commonly used to quantify the monoaromatic and triaromatic steroids (Figures 8.22 and 8.23).

Quantitation requires the area of each identified peak on a given mass chromatogram and that of the internal standard, which is added in a known amount from a standard solution. The amount of each compound in parts per million (ppm) of the sample is calculated as follows:

Amount compound
$$= \frac{(\text{area compound peak}) (\text{amount standard}) (\text{response factor})}{(\text{area standard})}$$

where

$$\text{response factor} = \frac{(\text{area standard})/(\text{amount compound})}{(\text{area compound})/(\text{amount standard})}$$

Because response factors change with instrument conditions, standard oil containing known amounts of the key compounds is run periodically. Response factors for these compounds are adjusted in the quantitation program, based on measurement of the oil standard. It is necessary to quantify the amounts of individual compounds to use many biomarker ratios. For example, area ratios for two compounds showing similar fragmentation patterns could be substituted for ratios of amounts if the samples were run under the same operating conditions. However, different instrument conditions can change the relative responses of various compounds in a biomarker ratio, particularly when the samples are run at different times or with different instruments. Such variations in response are especially deleterious for ratios that involve compounds with different fragmentation characteristics, such as regular steranes versus 17α-hopanes and triaromatic versus monoaromatic steroids. Others used similar approaches to quantify the amounts of biomarkers from GCMS (Rullkötter *et al.*, 1984; Mackenzie *et al.*, 1985; Eglinton and Douglas, 1988). These approaches also rely on internal standards that do not co-elute with the natural biomarkers to adjust response factors for quantitation.

Examples of GCMS data problems

Some of the factors affecting the quality of GCMS data include scan rate, sampling frequency, electronic zero level, background noise, sampling thresholds, amplifier saturation, availability of disk space, and sample size (gas chromatographic column overload). Some of the more common GCMS data problems are described below.

Poor gas chromatography resolution
Chromatographic resolution of biomarker compounds is necessary for the most accurate interpretations. Older papers commonly show poorly resolved biomarkers compared with more recent work because of innovations in chromatographic column technology. Improper column maintenance or prolonged use without replacement can also result in poor resolution of peaks and peak tailing. Degrading column performance is obvious when chromatograms for a familiar standard mixture are compared through time.

Poor signal-to-noise ratio
Poor signal-to-noise ratios can result from several different causes. Mass chromatograms of mixtures containing very low concentrations of biomarkers may show low signal-to-noise ratios. Low biomarker concentrations are typical of highly mature rock extracts, light oils, and condensates, where nearly all biomarkers were destroyed (Figure 8.30). Some samples are inherently low

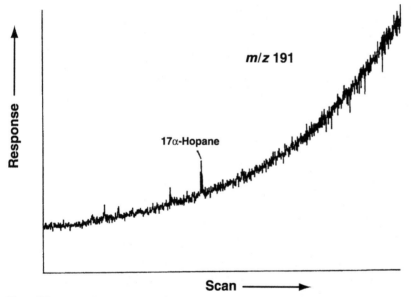

Figure 8.30. Terpane mass chromatogram or fingerprint (*m/z* 191) showing low signal-to-noise ratios for a highly mature condensate (49° API) from the Eel River Basin, California. The low concentrations of biomarkers in this condensate preclude detailed biomarker analysis. Compare this fingerprint with those for various other oils in Figure 13.72. The rising baseline is attributed to chromatographic bleed from the DB-1 column (see Figure 9.2). Reprinted with permission by Chevron-Texaco Exploration and Production Technology Company, a division of Chevron USA Inc.

in certain biomarker classes. For example, some crude oils from lacustrine or terrigenous-dominated marine source rocks are low in steranes.

Mass spectrometer problems frequently cause low signal-to-noise ratios. Dirty lenses, gain settings being too high or too low, dirty source or quadrupole rods, weak multiplier, poor calibration, and insufficient dwell time for the mass being analyzed can cause signal-to-noise ratio problems. Mass spectrometer sensitivity and stability also vary according to manufacturer and model. A sample with low biomarker concentrations can show poor signal-to-noise ratio on one instrument, while another will give acceptable results.

Ion sampling frequency

Ion sampling frequency is an important parameter affecting the precision and accuracy of mass chromatography. Sampling frequency is usually expressed as the number of seconds per scan over a given ion. Each scan is recorded by the data system as a data point having a height above the baseline or intensity proportional to the number of ions arriving at the detector of the mass spectrometer. A typical biomarker peak representing a C_{30} compound might have an elution width spanning 12–15 seconds. Figure 8.31 compares sampling rates of 3 and 1.5 seconds and how these rates affect definition of the chromatographic peak. The peak sampled at a

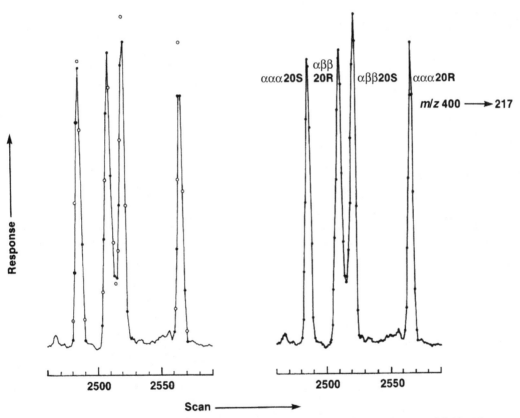

Figure 8.31. Comparison of the effects of 3-second (left) and 1.5-second (right) sampling (scan) rates on definition of chromatographic peaks. The figure shows a portion of a mass chromatogram for steranes obtained by gas chromatography/mass spectrometry/mass spectrometry (GCMS/MS) (M+, m/z 400 $\rightarrow$ m/z 217). Note that the three-second scan rate missed the top of the $\alpha\beta\beta$20S and $\alpha\alpha\alpha$20R peaks, causing a spurious increase in 20S/(20S + 20R) compared with that obtained using the 1.5-second scan rate. A three-second scan results in only half the sampled points of a 1.5-second scan rate. For comparison, the chromatogram obtained using the three-second scan rate shows points (open circles) that would have been available using a 1.5-second scan rate. Reprinted with permission by ChevronTexaco Exploration and Production Technology Company, a division of Chevron USA Inc.

1.5-second scan rate is better defined and has a smoother profile than that at 3 seconds. The 1.5-second scan rate results in more accurate and reproducible peak measurements on repetitive runs. We generally quantify peaks using a sampling rate of at least ten scans/peak or ~1.5 second/scan. A three-second scan rate is usually suitable for mass spectra.

Column bleed

The mass spectrum of DB-1 stationary-phase column bleed at 325°C (maximum program temperature for most analyses) is shown in Figure 8.32. Although programming low chromatograph temperatures can minimize column bleed, the effects of column bleed can

be background-subtracted. Excessive column-bleed results in rising chromatographic baselines and is usually corrected by replacement of the column and/or reduction of the maximum oven temperature. Even with background subtraction, column-bleed ions sometimes appear in mass spectra, complicating structural assessment.

Column overload

Injecting too much sample, or injecting sample overly enriched in a particular compound class, can result in negative peaks, tailing peaks, poor resolution, and altered retention times. Examples of column overload are shown in Figure 8.33.

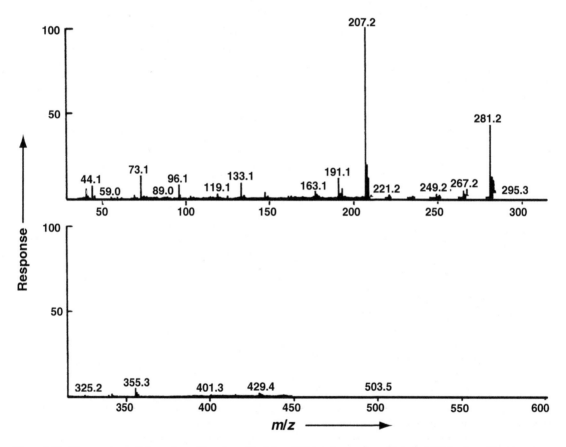

Figure 8.32. Mass spectrum of chromatographic column bleed from DB-1 stationary phase. Note that DB-1 column bleed has a significant m/z 191 peak, which can account for a rising baseline in the m/z 191 chromatogram used for terpane analysis (e.g. see Figure 8.30). Reprinted with permission by ChevronTexaco Exploration and Production Technology Company, a division of Chevron USA Inc.

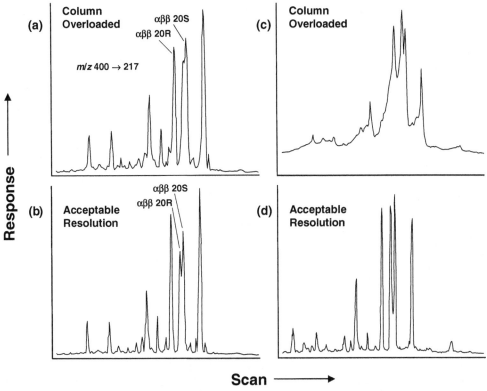

Figure 8.33. Examples of column overload on peak resolution in metastable reaction monitoring (MRM)/GCMS mass chromatograms of C_{29} steranes (m/z 400 → 217). (a) Injection of 0.2 µl of a toluene solution of a sterane-rich seep oil from Turkey results in characteristics typical of an overloaded column, including poor resolution and peak broadening compared with (b). Poor resolution is particularly evident for the $\alpha\beta\beta$20R and $\alpha\beta\beta$20S peaks; (b) injection of only 0.1 µl of the same solution as in (a) results in narrower peaks and acceptable resolution; (c) severe overloading of the column with 0.2 µl of a toluene solution of the Wyoming standard oil results in very broad, poorly resolved peaks; (d) analysis of 0.1 µl of a toluene solution of the same oil as in (c) that has acceptable peak resolution. Reprinted with permission by ChevronTexaco Exploration and Production Technology Company, a division of Chevron USA Inc.

Data system overload

All data systems are limited by the rate at which data can be acquired. When this rate is exceeded because of excessive signal from the mass spectral analyzer, the data system is overloaded or saturated. The effect on a mass chromatographic peak is a height limitation, which results in flattening of the top of the peak and spurious quantitation (Figure 8.34). However, the change in appearance of the overloaded peaks may be subtle.

Incorrect mass range monitored

Figure 8.35 (top) is an example of overlap of m/z 218 on the m/z 217 trace for standard oil from Wyoming. Note the unusually high C_{29} 14β,17β-sterane peaks

compared with the same peaks in an acceptable analysis of the same sample (bottom) where no overlap with m/z 218 occurs. The m/z 218 and m/z 217 overlap resulted from an incorrect mass calibration, which set the m/z 217 window in the mass spectrometer at a position to also collect data from m/z 218. Less extreme cases of monitoring the incorrect mass range can result in less obvious, but still incorrect, results.

Interfering peaks

Several classes of compounds commonly interfere with the m/z 217 mass fragmentogram used for sterane analysis. Some pentacyclic triterpanes (notably 28,30-bisnorhopane, which elutes from the gas chromatograph

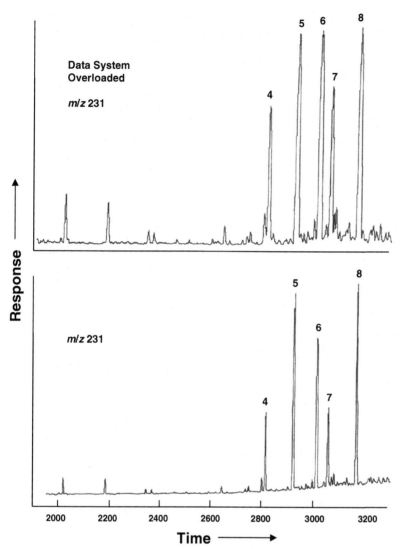

Figure 8.34. Example of the effect of a data system overload on the character of an m/z 231 (triaromatic steroids) mass chromatogram obtained by selected ion monitoring/gas chromatography/mass spectrometry (SIM/GCMS) of the aromatic fraction of a seep oil from Turkey (top). The three tallest peaks on the mass chromatogram have saturated the data system, resulting in overloading of the computer caused by too high a rate of data acquisition. Consequently, the size of these peaks is underestimated compared with the same peaks on the normal mass chromatogram (bottom). The top and bottom mass chromatograms were obtained by injecting 0.2 and 0.02 μl of neat (without solvent) sample, respectively. Peak numbers correspond to identified compounds in Figure 8.23. Reprinted with permission by ChevronTexaco Exploration and Production Technology Company, a division of Chevron USA Inc.

among the C_{29} steranes) (Moldowan *et al.*, 1984) have significant m/z 217 fragments in their spectra. Mass spectra of 4-methylsteranes have small m/z 217 fragments. In some oils or extracts, usually from lacustrine source rocks, 4-methylsteranes predominate over the 4-desmethylsteranes and interfere with their analysis using m/z 217.

Another example of interfering peaks occurs when saturated and aromatic hydrocarbons are analyzed together. Analysis of C-ring monoaromatic steroids

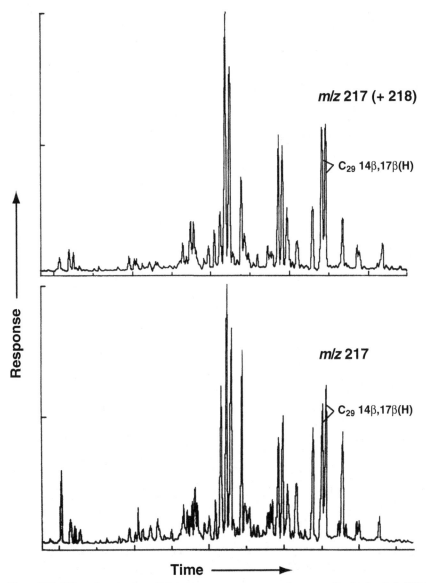

Figure 8.35. Effect of monitoring a slightly incorrect mass range on the sterane distribution (*m/z* 217) for Wyoming oil standard. The top mass chromatogram is an example of overlap of *m/z* 218 on the *m/z* 217 trace. Note the unusually high C$_{29}$ 14β,17β(H) sterane peaks on this trace compared with the normal mass chromatogram at the bottom. Reprinted with permission by ChevronTexaco Exploration and Production Technology Company, a division of Chevron USA Inc.

requires selected ion monitoring of *m/z* 253, but many acyclic saturated hydrocarbons show the same major fragment, which interferes with the monoaromatic-steroid analysis. These compounds can be separated, however, by using a higher-resolution (exact mass) mass spectrometer set at *m/z* 253.20 for monoaromatics or *m/z* 253.29 for saturates (Mackenzie *et al.*, 1983a).

An example of severe interference is where the isotope peaks for two abundant compounds with molecular weights of 252 registered as major peaks on the *m/z* 253 mass chromatograms. Several analyses of the sample showed that the *m/z* 253 trace was not reproducible, primarily because of variations in the intensity of these two peaks. Figure 8.36 shows these two peaks on the

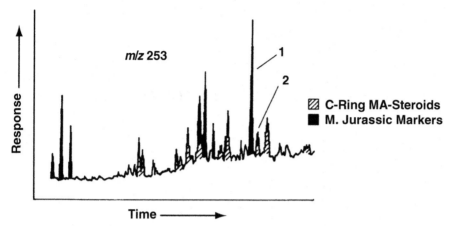

Figure 8.36. Mass chromatogram of m/z 253 for the aromatic fraction of Beatrice oil, Inner Moray Firth, UK. Note the interference of polynuclear aromatics (black peaks) with analysis of the monoaromatic steroids (hatched peaks). Reprinted with permission by ChevronTexaco Exploration and Production Technology Company, a division of Chevron USA Inc.

m/z 253 trace for Beatrice oil from Well 11/30–2 in the Inner Moray Firth, UK (see Figure 18.118). This oil is a mixture of hydrocarbons from lacustrine Devonian and marine Middle Jurassic source rocks (Peters *et al.*, 1989). The Beatrice oil and bitumen extracted from a prospective Middle Jurassic source rock nearby contain these unusual aromatic markers, which are much less abundant in other samples from the area. We detected compounds with identical retention times on m/z 253 chromatograms in petroleum from onshore Eel River, California.

Mass spectra for each of the two most prominent aromatic peaks in the Middle Jurassic and Eel River samples are essentially identical, showing an intense base peak at m/z 252 with a subordinate peak at m/z 126. Many polynuclear aromatic hydrocarbons generate prominent molecular ions at m/z 252, including 1,2- and 3,4-benzopyrene (Figure 8.20). The ion at m/z 126 appears to represent a doubly charged species of the molecular ion. The mass spectra of these peaks and co-injection of authentic standards show that the two compounds represent 1,2- and 3,4-benzopyrene. Apparently, the quantities of these compounds in the Beatrice oil are so large that an isotope peak, representing benzopyrenes containing one deuterium or ^{13}C atom, registers on the m/z 253 trace. Differences in the slit size used to monitor m/z 253 or in mass calibration between repeated analyses appear to be responsible for the variability of peak intensity for the isotope peaks of these compounds.

Cold spots in the transfer line

Occasionally, there may be a problem with heat applied to the section of tubing at the interface between the gas chromatograph and the mass spectrometer (transfer line). In modern instruments, the gas chromatography column is threaded through the transfer line so that the effluent from the column elutes directly at the ion source. A cold spot in the transfer line acts as a barrier to the transmission of compounds with boiling points above the temperature of the cold spot. The resulting chromatograms show loss of resolution for peaks representing the higher-boiling compounds, while lower-boiling compounds show normal resolution (Figure 8.37).

Dirty ion source

A dirty ion source can cause increased background and lower signal-to-noise ratios. A dirty source prevents proper instrument calibration.

Defocusing by n-*alkanes*

Elution of *n*-alkanes from the gas chromatograph into the ion source can cause negative peaks on mass chromatograms of other compounds, simply by overwhelming or defocusing the ion source (Figure 8.38). Defocusing is particularly troublesome when analyzing paraffinic saturate fractions on GCMS systems where the mass spectrometer source has a low ion volume. Low ion volumes are typical of certain older quadrupole instruments and mass-selective detectors. Analyses of

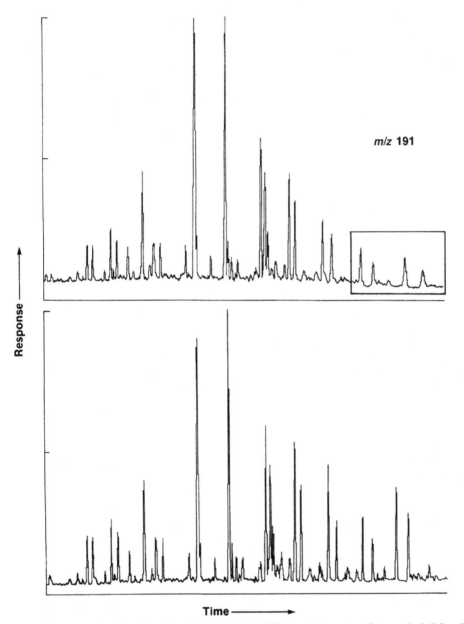

m/z **191**

Figure 8.37. Example of the effects of a cold spot on the *m/z* 191 mass chromatogram for a standard oil from Hamilton Dome, Wyoming. The *m/z* 191 mass chromatogram at the top has peak broadening (boxed area) due to a cold spot in the transfer line, while the lower chromatogram has acceptable resolution of peaks in this area. Reprinted with permission by ChevronTexaco Exploration and Production Technology Company, a division of Chevron USA Inc.

Table 8.5. *Precision calculations from ten consecutive analyses of the same standard*[1]

Steranes (M $\pm$ $\rightarrow$ m/z 217) by gas chromatography/mass spectrometry/mass spectrometry (GCMS/MS). Saturate cut of oil standard, Hamilton Dome, Wyoming, plus 5β-cholane internal standard.

Ratio or compounds	Use	Standard deviation (%)
C_{27} 20R/$(C_{27}-C_{29})$	Correlation (ternary diagram)	1.4
C_{28} 20R/$(C_{27}-C_{29})$	Correlation (ternary diagram)	1.0
C_{29} 20R/$(C_{27}-C_{29})$	Correlation (ternary diagram)	1.0
C_{30}/$(C_{27}-C_{30})$	Marine input	8.8
C_{29} 20S/(20S + 20R)	Maturity	2.0[2]
C_{29} $\beta\beta/(\beta\beta + \alpha\alpha)$	Maturity	5.6[2]
C_{27} 20S/(20S + 20R)	Maturity	2.0
C_{27} $\beta\beta/(\beta\beta + \alpha\alpha)$	Maturity	3.6
ppm C_{27} (in crude)	Concentration	9.9
ppm C_{28}	Concentration	8.0
ppm C_{29}	Concentration	10.2
ppm total regular steranes	Concentration	9.4

[1] For details of operating conditions, see Peters *et al.* (1990).

[2] Based on ten consecutive analyses of the same samples, Steen (1986) notes standard deviations of 2.5% and 1.6% for the C_{29} 20S/(20S + 20R) and C_{29} $\beta\beta/(\beta\beta + \alpha\alpha)$ ratios, respectively.

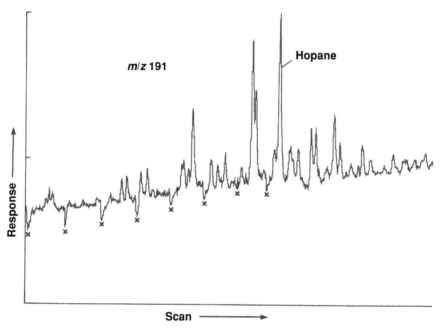

Figure 8.38. Terpane mass chromatogram (m/z 191) of a highly mature, paraffinic oil from the Nugget Sandstone, Pineview Field, Wyoming. The example shows defocusing caused by high concentrations of *n*-alkanes compared with biomarkers. The negative peaks marked by an $\times$ are at scan times corresponding to the *n*-alkanes. The character of the mass chromatogram can be improved by removal of the *n*-alkanes using molecular sieves or urea adduction. Reprinted with permission by ChevronTexaco Exploration and Production Technology Company, a division of Chevron USA Inc.

samples affected by defocusing can be improved by removal of *n*-alkanes using molecular sieves.

Maximum column temperature too low

If the maximum column temperature is set too low, then the peaks for the higher-boiling analytes will broaden progressively as the isothermal or constant-temperature portion of the program is reached. The resulting data resemble those from systems with a cold spot in the transfer line (see above), except that the onset of peak broadening is less abrupt.

Baseline threshold set too high

If the baseline threshold is set too high, then the data system sets up a default baseline above the actual baseline. Using this default baseline as the peak base for measurement causes changes in relative peak intensities, which favor the tallest peaks on the chromatogram.

The smallest peaks may be lost completely if they are lower in intensity than the threshold.

Error analysis

Table 8.5 is a compilation of error analyses for many parameters described in this book. These error analyses can be used with caution as a general guide to the significance of specific results, provided that it is understood that standard deviations about the mean value for any parameter can differ between instruments and samples. Therefore, any experimental data that the reader wishes to evaluate with respect to tabulated error data must be obtained using (1) optimum operating conditions, (2) the same instrument with identical settings, and (3) the same operating mode as specified in the table. The ideal approach to error evaluation is for the analyst to include replicate analyses of standards with every sample suite.

9 · Origin of petroleum

This chapter describes evidence against the deep-earth gas hypothesis, which invokes an abiogenic origin for petroleum by polymerization of methane deep in the Earth's mantle. The deep-earth gas hypothesis has little scientific support, but, if correct, it could have major implications for petroleum exploration and the application of biomarkers to environmental science and archeology. The discussion covers experimental, geological, and geochemical evidence supporting the thermogenic origin of petroleum.

HISTORICAL BACKGROUND

Petroleum has been used since Biblical times, yet its origin remained a mystery for much of human history. Classical literature makes little or no reference to petroleum. In his 1268 treatise *Opus Tertium*, Roger Bacon commented on the lack of discussion of the origins of oils and bitumen by Aristotle and other natural philosophers. Two hypotheses on the origin of petroleum emerged during the Renaissance. In his 1546 text *De Natura eorum quae Effluunt ex Terra*, Georgius Agricola expanded on Aristotle's concept of exhalations from deep within the Earth and proposed that bitumen condensed from sulfur. Andreas Libavius theorized in his 1597 text *Alchemia* that bitumen formed from the resins of ancient trees. These early discussions mark the beginning of a long scientific debate as to whether petroleum forms by abiogenic processes or represents once-living organic matter that was altered within the Earth.

Historic records are unclear, but Lomonosov proposed the idea as early as 1757 (Kenney, 1996) and certainly by 1763 (Wellings, 1966) that liquid oil and solid bitumen originate from coal due to subsurface heat and pressure. By this time, most scientists had accepted fossil evidence that coal originates from plant remains. Various biogenic hypotheses emerged during the early

nineteenth century suggesting that petroleum originated either directly from biological remains or by a distillation process (Dott, 1969).

Modern concepts of the origin of petroleum from ancient organic-rich sedimentary rocks emerged during the nineteenth century. Hunt (1863) concluded that organic matter in some North American Paleozoic rocks came from marine animals or plants and that its transformation to bitumen must be similar to the process of coal formation. Lesquereux (1866), the American father of paleobotany, reached similar conclusions after studying Devonian shales from Pennsylvania, as did Newberry (1873) in his study of Devonian shales from Ohio. Early twentieth-century field (Arnold and Anderson, 1907) and chemical (Clarke, 1916) studies by the US Geological Survey provided convincing evidence that crude oil in the Monterey Formation of California originated from diatoms in organic-rich shales. Studies of organic-rich shales in Europe at this time resulted in similar conclusions (Pompecki, 1901; Schuchert, 1915).

Support of the biogenic hypothesis became widespread in the mid twentieth century, with converging scientific advances in paleontology, geology, and chemistry. Alfred Treibs (1936) established a link between chlorophyll in living organisms and porphyrins in petroleum (Figure 3.24). Additional geochemical evidence followed with the discoveries that low- to moderate-maturity oils retain fractions that are optically active (Oakwood *et al.*, 1952), that stable carbon isotope compositions of petroleum preserve evidence of biological isotope fractionation (Craig, 1953), and that oils contain many "chemical fossils" (biomarkers) in addition to porphyrins that can be traced to their biological precursors (Eglinton and Calvin, 1967). Concurrent with these findings were field studies recognizing that organic-rich strata occur in all petroliferous sedimentary basins, that this organic matter (kerogen) originates from biota (Forsman and Hunt, 1958; Abelson, 1963), that it has been chemically altered from its initial state, and that it generates oil and gas as the sediments are buried and heated (Tissot, 1969).

Although there is overwhelming evidence for a biogenic origin of petroleum, some still advocate abiotic mechanisms. The modern abiogenic concept is rooted in the mid nineteenth century. Berthelot (1860) described experiments where n-alkanes formed during the acid dissolution of steel. Mendeleev (1877) proposed that surface water percolates deep into the Earth and reacts with metallic carbides to form acetylene, which then condenses into larger hydrocarbons. Mendeleev's abiotic hypothesis, further refined in 1902, was attractive at the time because it offered an explanation for the growing awareness of widespread petroleum deposits that suggested some sort of deep, global process.

However, support for the abiotic hypothesis dwindled with mounting evidence for a biogenic origin of petroleum. By the 1960s, there was little support for the abiotic hypothesis, except among a small group in the Soviet Union. First proposed by Kudryavtsev (1951) and advanced over the years in some Soviet publications (see Kenney, 1996), a modern version of Mendeleev's hypothesis has emerged. This hypothesis relies on a thermodynamic argument, which states that hydrocarbons greater than methane cannot form spontaneously except at the high temperatures and pressures of the lower-most crust (Kenney et al., 2002). The hypothesis ignores the fact that all life is in thermodynamic disequilibrium, as discussed later in this chapter.

In the West, a few astronomers have been among the most vocal advocates for abiotic petroleum. Carbonaceous chondrites and other planetary bodies, such as asteroids, comets, and the moons and atmospheres of the Jovian planets, clearly contain hydrocarbons and other "organic" compounds that were generated by abiotic processes (Cronin et al., 1988). Hoyle (1955) reasoned that since the Earth formed from similar materials, there should be vast amounts of abiogenic oil. More recently, Thomas Gold (1985; 1999) has become the principal proponent of abiotic petroleum, despite criticism by nearly all geochemists and petroleum geologists. Gold convinced the Swedish government to drill two deep wells (Gravberg-1 in 1986–90 and Stenberg-1 in 1991–92) into fractured granite under the Siljan Ring, the site of an ancient meteorite crater. The wells failed to find economic reserves, and evidence for even trace amounts of abiotic hydrocarbons is controversial (Kerr, 1990), as discussed later in this chapter.

The concept that petroleum originates from sedimentary organic matter from once-living organisms is consistent with natural observations, laboratory analyses and experiments, theoretical considerations, and basin simulations. Geochemists recognize that some abiogenic hydrocarbons occur in the geosphere. Small amounts of abiotic hydrocarbon gases are generated by rock–water interactions involving serpentinization of ultramafic rocks (Sherwood Lollar et al., 1993; McCollom and Seewald, 2001), the thermal decomposition of siderite in the presence of water (McCollom, 2003), and during magma cooling as a result of Fischer–Tropsch type reactions (Potter et al., 2001). However, commercial quantities of abiotic petroleum have never been found, and the contribution of abiogenic hydrocarbons to the global crustal carbon budget is inconsequential (Sherwood Lollar et al., 2002).

Given the preponderance of evidence for a biogenic origin of petroleum, the reader may ask why a discussion of the origin of petroleum is needed in this text. Our discussion is designed to refute a small but vocal community that continues to popularize the abiogenic hypothesis, particularly in the press. Some advocates of the abiogenic hypothesis dismiss the evidence supporting biogenic origins and accuse petroleum geochemists of being conformists to an outdated concept. In our opinion, they use many of the same tactics that the supporters of "scientific creationism" use to ridicule evolutionary theory. We have prepared a response to the arguments put forth as evidence for abiogenic petroleum and a rebuttal of criticisms of biogenic origin. Other chapters in this book provide more details on stable isotopic compositions that reflect biological fractionations (Chapter 6), the occurrence of biologically derived molecules in petroleum (Chapters 13 and 14), and source rocks and petroleum systems (Chapter 18).

DEEP-EARTH GAS HYPOTHESIS

Lithoautotrophic microbes are chemotrophs (Table 1.2) that reside deep in the Earth and feed on carbon dioxide and hydrogen generated by rock reacting with water. They are thus independent of photosynthesis and all surface life (Stevens and McKinley, 1995). Chapelle et al. (2002) discovered a subsurface hydrogen-based ecosystem where methanogens account for more than 90% of the 16S ribosomal DNA sequences. In deep hydrothermal waters (200 m below ground) from Lidy Hot

Springs in Idaho, methanogens thrive exclusively on geothermal hydrogen and carbon dioxide. No reduced form of carbon is available. The discovery of this ecosystem suggests that the contribution of archaeal lipids to the geosphere may be greater than thought previously. Furthermore, the findings expand the possibility that extraterrestrial hydrogen-based ecosystems could exist in the subsurface of Mars, Europa, and elsewhere.

Note: Archaea form microbial methane using one of two pathways: (1) by the reduction of simple acids or (2) by the direct reduction of CO_2 by H_2. These reactions produce small amounts of energy compared with biochemical pathways involving other electron acceptors (e.g. oxygen, nitrate, and sulfate). Consequently, methanogens exist only in environments that are too hostile or nutrient-deficient for other microorganisms to live. In typical sediment, they account for only ~2–3% of microbial life.

Speculating on the above observations, Gold (1999) presented his deep hot biosphere and deep-earth gas hypotheses, which incorporate many of the ideas of previous publications (e.g. Ponnamperuma and Pering, 1966; Szatmari, 1989). Gold believes that most subsurface organisms obtain energy from abiogenic hydrocarbons (rather than CO_2) upwelling from the mantle and that the mass of these organisms far exceeds that of life on the Earth's surface. In the deep-earth gas hypothesis, abiogenic methane from the mantle migrates upward and polymerizes to form vast amounts of untapped petroleum in igneous rocks. However, the evidence against the deep–earth gas hypothesis is overwhelming (e.g., Stinnett, 1982; Bromley and Larter, 1986; Philp and Brassell, 1986; Brassell, 1987; Apps and van de Kamp, 1993; Peters, 1999b). This evidence can be categorized as geological, geochemical (optical activity, biomarkers, isotopes, oil–oil and oil–source rock correlation, basin modeling), and experimental (hydrous pyrolysis). Petroleum originated from biologically produced materials, not from abiogenic methane. Furthermore, most commercial methane was generated by thermal cracking of larger molecules and not vice versa. The proportion of methane commonly increases with depth in petroliferous basins because of increased temperature and cracking of heavier hydrocarbons (e.g. Lorant and Behar, 2002), not because of closer proximity to a mantle source of methane.

The deep hot biosphere hypothesis conflicts with conventional concepts of the Earth's biosphere. Most subsurface microbial communities anaerobically oxidize portions of sedimentary organic matter to produce CO_2 and electrons (H_2). Continued oxidation of organic matter requires removal of electrons by reducing a succession of inorganic electron acceptors, such as oxygen, manganese, nitrate, ferric iron, sulfate, and CO_2 (Froelich *et al.*, 1979). Oxidized moieties of sedimentary organic matter can also serve as electron acceptors, but these fermentation processes are not well documented. Growing evidence suggests that acetate generation from organic matter during deep burial explains the presence of deep microbial populations (Wellsbury *et al.*, 1997). An abiogenic source of hydrocarbons and other metabolites is not necessary. Furthermore, Gold's estimate of the abundance of subsurface microbes is many orders of magnitude too high compared with measured subsurface microbial populations (Cragg *et al.*, 1996).

Many of the arguments in support of the deep-earth gas hypothesis are distorted or incorrect (Peters, 1999b). For example, Gold (1999, p. 32) states incorrectly that others interpreted a microbial origin for the methane in Columbia River basalt because they could not believe that basalt contained abiogenic methane. Gold ignores isotopic analyses and microbial culture experiments showing that the methane originated from archaea (Stevens and McKinley, 1995). As another example, Gold (1999, p. 84) states that biomarkers in crude oils "can all be linked to constituents of bacteria or archaea, and none is linked exclusively to macroflora or fauna." For this reason, "there is no evidence that any surface life must be invoked to explain the presence of these biological molecules in subsurface hydrocarbons." Gold fails to mention that virtually all crude oils contain at least some biomarkers from sources that preclude a deep mantle origin. For example, oleanane originates from flowering plants (Moldowan *et al.*, 1994), and desmethylsteranes originate from sterols in eukaryotes (Huang and Meinschein, 1979), including diatoms (Holba *et al.*, 1998), Chrysophyte algae (Moldowan *et al.*, 1990), and sponges (McCaffrey *et al.*, 1994). Porphyrins originate from photosynthetic organisms or respiratory pigments in animals (Baker and Louda, 1986).

One might argue that biomarkers are contaminants solubilized from low-maturity carrier beds during the last stages of upward migration of abiotic petroleum from the mantle. However, most reservoir and carrier rocks (e.g. sandstones) are organic-lean, and the concentrations of biomarkers solubilized from these sources are low compared with those in migrating oil of low to medium API gravity. However, solubilization can be important in somes cases, as, for example, when high API gravity, biomarker-poor condensate migrates through organic-rich coal seams. Solubilization has been observed in some areas, including the Mahakam Delta (Durand, 1983; Hoffmann *et al.*, 1984; Jaffé *et al.*, 1988a; Jaffé *et al.*, 1988b), Australia (Philp and Gilbert, 1982; 1986), offshore Brunei (Curiale *et al.*, 2000), and Angola. Solubilization can be recognized because the solubilized materials are commonly of lower thermal maturity than those comprising the migrating oil (see Figure 19.1). The maturity of the biomarkers in most crude oils is inconsistent with a shallow origin and matches that of the other components in the crude oil. Furthermore, the crude oils can be linked to thermally mature source rocks, as discussed later in this chapter.

Remarkably, many of the arguments for the deep-earth gas hypothesis have not been modified significantly based on previous critiques of the idea. For example, Gold (1999, p. 73) states that the "association of helium with hydrocarbons is probably the most striking fact that the biogenic theory fails to account for . . . " Gold's model requires that mantle methane flushes primordial helium (^{3}He) from established migration pathways, thereby lowering ^{3}He/^{4}He. Radiogenic decay of uranium and thorium accounts for nearly all ^{4}He. Gold fails to mention that these radioactive elements can be concentrated in petroleum source-rock organic matter (Hunt, 1996). For this reason, many organic-rich petroleum source rocks, such as the Kimmeridge "hot" shale in the North Sea, show strong response on gamma-ray well logs. Gold's suggestion that low ^{3}He/^{4}He does not rule out mantle-derived hydrocarbons conflicts with evidence that volcanic or geothermal activity is always associated with elevated ^{3}He/^{4}He (Jenden *et al.*, 1993a; Ballentine *et al.*, 2001). Gold (1999, p. 27) himself argues that extensive flushing by upwelling hydrocarbons occurs near volcanoes and causes the dominance of CO_2 over methane at these sites. Thus, if Gold were correct on this point, then volcanoes would be sites where one would expect low rather than high ^{3}He/^{4}He.

Most view the results of deep drilling at the Siljan Ring impact crater in Sweden as strong evidence against the deep-gas theory. The two Siljan Ring wells were failures because no commercial petroleum was

discovered and no credible evidence was gained for a dominant mantle source for the small amounts of gaseous hydrocarbons (Castaño, 1993). Furthermore, the small amounts of liquids that were recovered from the area either correlate geochemically with conventional source rocks (Vlierboom *et al.*, 1986; Hedberg, 1988) or represent contamination, as discussed below.

In his book, Gold (1999, p. 111) states that the Gravberg-1 well at Siljan was drilled with water-based drilling mud, "so as not to contaminate the well with introduced oils." Gold assigns major significance to 60 kg of oily black paste from the well that consisted mainly of magnetite but was mostly lost at the well site. He also states (p. 121) that a downhole pump in the second Siljan well (Stenberg-1) retrieved "eighty-four barrels of oil" and that "the theory of the abiogenic origin of petroleum had thus been confirmed." No analyses of the oil are reported. Gold fails to mention that organic additives to the Gravberg-1 well interfered with the geochemical analyses and included various lubricants (Idlube, Torque Trim), organic polymer, diesel, and oil-based drilling mud (Castaño, 1993). Gold also fails to mention that the magnetite paste formed in the drill pipe when Torque Trim was added, and that analyses indicate that the paste contained C_{11}–C_{21} alkanes, biomarkers, and other petroleum-like components that could be mistaken as originating from the rock being drilled (Jeffrey and Kaplan, 1989). Furthermore, when Idlube and calcium carbide were used, drill-bit metamorphism resulted in the production of C_2–C_6 hydrocarbon gases.

Kenney *et al.* (2002) propose that the hydrocarbons in petroleum are abiogenic and form at depths of $\sim$100 km, pressures over $\sim$30 kbar, and temperatures of $\sim$900°C. Their thermodynamic calculations indicate that only methane and elemental carbon are stable at the shallower depths and lower temperatures and pressures of petroleum reservoirs. Their model requires that to avoid reverting to methane and elemental carbon during upward migration, complex hydrocarbons formed at great depths must be "quenched" by a rapid decrease in temperature while maintaining pressure. They support their model with experiments where iron oxide, calcium carbonate, and water were heated in an apparatus that achieves up to 50 000 bars and 1500°C. The hydrocarbon products, including *n*-alkanes up to $C_{10}H_{22}$, were cooled rapidly to near room temperature at 700°C/s, while maintaining pressure. Stable carbon and hydrogen isotope analyses of these hydrocarbons were not completed.

The Kenney *et al.* (2002) model is fatally flawed in several respects:

- It requires that complex hydrocarbons be quenched rapidly in order to remain stable as they rise to drilling depths. Although rapid vertical migration of subsurface fluids can occur, most evidence indicates that petroleum migration occurs over time intervals that are many orders of magnitude longer than a few seconds.
- The model assumes that hydrocarbons heavier than methane cannot form spontaneously, except at high temperatures and pressures, such as those in the lower-most crust of the Earth. According to this model, transformation of complex biological compounds to hydrocarbons heavier than methane requires decomposition of these compounds to methane followed by polymerization. This ignores the fact that living organisms routinely biosynthesize functionalized biomarker precursors that require only minor modification during diagenesis and catagenesis to yield structurally similar biomarkers and other compounds in petroleum. All life is in thermodynamic disequilibrium with its environment. The formation of complex biomarker precursors is not thermodynamically favorable, yet organisms biosynthesize these compounds using energy from photosynthesis and other sources. These transformations are an important topic of this book.
- The model indicates that highly oxidized biomolecules, specifically glucose ($C_6H_{12}O_6$), cannot be converted readily to petroleum hydrocarbons. However, glucose or other carbohydrate precursors are not part of the modern view of the organic origin of petroleum. Carbohydrates and proteins, which comprise up to ~40 and 50 wt.% of marine plankton, respectively, are degraded rapidly during early diagenesis. However, lipids, which represent ~5–25 wt.% of these plankton, are preserved more readily during diagenesis, especially under anaerobic conditions (e.g. Tissot and Welte, 1984). Abundant evidence discussed in this book shows that lipids and lipid-derived kerogen in source rocks are the main precursors for petroleum.
- The model ignores the fact that petroleum-like oil has been commercially retorted from the organic matter in oil shales since ~1860 (Cook and Sherwood, 1991) and

that the natural process of petroleum generation can be simulated routinely in the laboratory (e.g. Lewan *et al.*, 1979).

ABIOGENIC HYDROCARBON GASES

Compound-specific isotope evidence clearly refutes an abiogenic origin for the vast majority of hydrocarbon gases (Figures 9.1 and 9.2) (e.g. Des Marais *et al.*, 1981; Jenden *et al.*, 1993a; Sherwood Lollar *et al.*, 2002). The relative amount of ^{13}C in thermogenic gases increases from methane to ethane to propane and higher homologs. Figure 6.3 gives other examples of this normal isotopic trend that characterizes thermogenic gases. In contrast, gases polymerized by spark discharge experiments on methane (Figure 9.1), some meteoritic gases (Figure 9.3), and other known or suspected abiogenic gases show the reverse trend. Trace amounts of gas from diabase and granite in the Gravberg-1 well show increasing ^{13}C-content in the order propane, ethane, methane, as expected of abiogenic gas (Castaño, 1993). However, no commercial accumulations of such gas have been found anywhere on Earth, and the concentrations of hydrocarbon gases in these intervals in the Gravberg-1 well are very low (<1000 ppm). Background gas readings in wells from producing basins commonly exceed 1000 ppm.

Gas sample 3 in Figure 6.15 has the reversed isotopic trend that might indicate an abiogenic origin. However, many of these rare gases with reversed isotopic

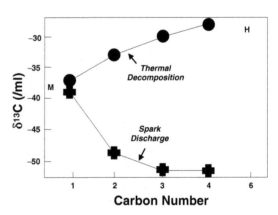

Figure 9.1. Laboratory thermal decomposition of hexane (H) and spark-discharge polymerization of methane (M) result in opposing normal and reversed stable carbon isotope trends (modified from Des Marais *et al.*, 1981).

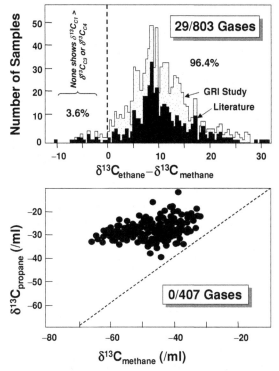

Figure 9.2. Of 803 natural gases with methane and ethane data, only 29 (3.6%) gases have methane that is isotopically enriched in ^{13}C relative to ethane, as might be expected of abiogenic gas (Jenden *et al.*, 1993a). None of these gases for which propane analyses could be made (407 samples) has methane that is isotopically enriched in ^{13}C relative to propane, suggesting that none of the C_{2+} gases could arise from abiogenic methane polymerization.

trends for methane, ethane, and propane are not abiogenic but result from heterogeneity in the source organic matter, mixing of gases from different sources, oxidation of thermogenic gas, or partial diffusive leakage of the gas reservoir (Jenden *et al.*, 1983a; Laughrey and Baldassare, 1998). Only 29 of 803 gases analyzed by Jenden *et al.* (1993) had methane that was isotopically enriched in ^{13}C compared with ethane (Figure 9.2). None of the 407 gases in this sample suite with sufficient propane for analysis had methane that was isotopically enriched in ^{13}C compared with propane, suggesting that none of these gases is abiogenic.

Geochemists do not deny the existence of limited amounts of abiogenic hydrocarbon gases (e.g. Sherwood Lollar *et al.*, 2002). Based on ^{3}He content of natural gas production, Jenden *et al.* (1993a) calculated that

<200 ppm of commercially produced gas consists of inorganic hydrocarbons. Castaño (1993) discussed the problem of concentrating and entrapping such diffuse inorganic hydrocarbons with respect to the Siljan Ring exploratory wells.

Several methane-rich gases from mining sites on the Canadian and Fennoscandian shields did not originate from bacterial or thermogenic alteration of organic matter, but they also show no evidence for a mantle-derived component (Sherwood Lollar *et al.*, 1993). These gases have methane δ^{13}C from −22.4 to −57.5‰, which is consistent with a mantle origin but is not inconsistent with methane produced by abiogenic water–rock interactions, as demonstrated in recent experimental studies by Horita and Berndt (1999). They heated dissolved bicarbonate with δ^{13}C near −4‰ in the presence of hydrothermally formed nickel–iron alloy at 200–300°C for various times to generate dry gas with methane having δ^{13}C as low as −53.5‰. Such gas compositions would normally be interpreted to indicate microbial gas. These abiogenic hydrocarbons may form by reduction of carbon dioxide in hydrothermal systems during water–rock interactions involving serpentinization of ultramafic rocks and during magma cooling as a result of Fischer–Tropsch reactions, as discussed below. Sherwood Lollar *et al.* (2002) discuss other potential mechanisms for abiogenic gas synthesis. The relative concentrations and stable carbon isotope ratios for

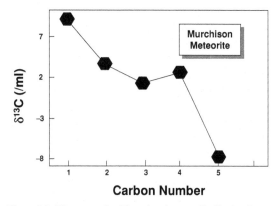

Figure 9.3. The unusual stable carbon isotope distribution for methane through pentane in a carbonaceous chondrite is consistent with an abiogenic origin of these compounds by polymerization from methane (Yuen *et al.*, 1984). Similar isotopic distributions characterize the products from spark discharge experiments on methane (see Figure 9.1) (Des Marais *et al.*, 1981).

C_1–C_4 alkanes in 11 gas samples from the Kidd Creek Mine, Ontario, Canada, provide direct evidence for inorganic synthesis of methane and higher hydrocarbons in crystalline rocks of the Canadian Shield (Westgate *et al.*, 2001). In contrast to the trend of increasing ^{13}C from C_1–C_4 in the *n*-alkanes of thermogenic gas, the Kidd Creek gases show significant depletion in ^{13}C for C_2–C_4 with respect to C_1. Furthermore, the C_1–C_4 *n*-alkanes in these gases follow a Schulz–Flory distribution (discussed below), consistent with hydrocarbons produced by polymerization of lower homologs. Depletion in ^{13}C correlates with deuterium enrichment for these gases, possibly indicating that the formation of higher hydrocarbons through stepwise addition of individual monomers involves addition of a C–C bond and elimination of C–H bonds (Sherwood Lollar *et al.*, 2002).

In general, these shield gases discharge from boreholes at rates of 1 l/min to more than 30 l/min and contain 50–90% methane, up to 10% ethane, a few percent propane, and <1% *iso*- and *n*-butane. While the shield gases are volumetrically significant compared with the trace amounts of abiogenic gases identified to date, the ephemeral nature of the gas discharges suggests little economic potential. More importantly, based on the distinct isotopic patterns identified for abiogenic gases in Sherwood Lollar *et al.* (2002, their Figure 3, p. 523), there is no evidence that such gases made any significant contribution to commercial hydrocarbon gas deposits worldwide.

Potter *et al.* (2001) propose the formation of abiogenic hydrocarbons in igneous rocks from the Khibina and Lovozero intrusions from the Kola Peninsula in Russia by late or post-magmatic processes, thus providing a link between hydrocarbon generation and mineralization. These nepheline-syenite rocks contain significant hydrocarbon concentrations up to 100–170 cm^3/kg in Khibina and 20–50 cm^3/kg in Lovozero. Fluid inclusions in the Lovozero samples contain up to 35 mole% H_2. Petrographic, microtextural, and pressure–volume–temperature modeling data suggest that the inclusions were trapped at low pressures and temperatures (~0.5–1.5 kbar, 350°C). Isotopic data are consistent with abiogenic methane ($\delta^{13}C = -12$ to $-27‰$), although it is unclear whether their isotopic values represent pure methane or mixed hydrocarbons from the inclusions. Potter *et al.* (2001) conclude that the hydrocarbons originated during post-magmatic hydrothermal alteration involving Fischer–Tropsch reactions of the type $CO_2 + 4H_2 = CH_4 + 2H_2O$, catalyzed by magnetite and iron silicates.

Fischer–Tropsch synthesis

Fischer–Tropsch synthesis (Fischer and Tropsch, 1926) is used to convert carbon monoxide or carbon dioxide generated from coal into synthetic petroleum, as follows:

$$nCO + (2n + 1)H_2 = C_nH_{2n+2} + nH_2O \qquad (9.1)$$

$$nCO_2 + (3n + 1)H_2 = C_nH_{2n+2} + 2nH_2O \qquad (9.2)$$

For example, Fischer–Tropsch synthesis generates methane from carbon dioxide based on Equation (9.2) and similar reactions can form ethane, propane, and higher homologs:

$$CO_2 + 4H_2 = CH_4 + 2H_2O \qquad (9.3)$$

Fischer–Tropsch synthesis generates complex mixtures that are always dominated by the lighter hydrocarbons. For a given Fischer–Tropsch product, the log of the hydrocarbon concentration decreases linearly with increasing carbon numbers and is called the Schulz–Flory distribution (Salvi and Williams-Jones, 1997). A Schulz–Flory distribution results in nearly constant ratios of Fischer–Tropsch hydrocarbons with successive carbon numbers, i.e. synthetic hydrocarbon mixtures have C_{n+1}/C_n below ~0.6 (e.g. Szatmari, 1989). Fischer–Tropsch synthesis forms methane under experimental conditions as low as 127°C, and both magnetite and hydrated silicates were proposed as natural catalysts for these reactions in rocks (Lancet and Anders, 1970; Anderson, 1984).

Salvi and Williams-Jones (1997) interpreted a Schulz–Flory distribution for hydrocarbons in pegmatites from Strange Lake in Canada, which have C_2/C_1, C_3/C_2, C_4/C_3, and C_5/C_4 alkane ratios of 0.08, 0.18, 0.34, and 0.34, respectively. The low ethane/methane ratio (C_2/C_1) was explained by allochthonous methane that entered the system from another source.

Metamorphism of carbonate and serpentinization

Metamorphism of graphite-bearing carbonates in crystalline rocks at temperatures below 300–400°C may generate methane rather than carbon dioxide (Holloway, 1984), as follows.

$$\text{Talc} + \text{calcite} + \text{graphite} + H_2O = \text{dolomite}$$
$$+ \text{quartz} + CH_4$$

Furthermore, heating of calcite, dolomite, and siderite at 400°C in the presence of H_2 yields methane, ethane, propane, and butane (Giardini and Salotti, 1969), suggesting that similar reactions may occur in crystalline rocks.

Serpentinization reactions are believed to be key processes accounting for hydrothermal methane at the mid-ocean ridges (Vanko and Stakes, 1991; Charlou and Donval, 1993; Berndt et al., 1996; Horita and Berndt, 1999) and have been suggested to be one mechanism for formation of the Canadian and Fennoscandian shield hydrocarbon gases (Sherwood Lollar et al., 1993). Field and experimental evidence indicates that if a carbon source is available, then methane and hydrogen gas form during hydration of olivine in ultramafic rocks (Abrajano et al., 1990), as follows:

$$\text{Olivine} + H_2O + C \text{ (or } CO_2) = \text{magnetite}$$
$$+ \text{serpentine} + \text{brucite} + CH_4 + H_2$$

More recent work suggests that the potential for abiotic formation of hydrocarbons during serpentinization may be more limited than previously believed and that mineral catalysts or vapor-phase reactions might be required to support abiogenic hydrocarbon formation in igneous rocks (McCollom and Seewald, 2001).

> Note: In the process of serpentinization, water in contact with olivine is reduced to molecular hydrogen, while iron in the olivine is oxidized to Fe^{2+}. The H_2 may be used as an electron source by chemotrophic bacteria, or it may combine with CO_2 through Fischer–Tropsch synthesis at high temperatures to produce organic compounds such as hydrocarbons and fatty acids. Holm and Charlou (2001) interpreted gas chromatography/mass spectrometry (GCMS) analyses of fluids from the Rainbow hydrothermal field on the Mid-Atlantic Ridge to indicate *de novo* synthesis of small amounts of C_{16}–C_{29} *n*-alkanes. However, they do not discuss the distinct possibility that these *n*-alkanes might represent the products of organisms growing near the vents that were swept up by circulating hydrothermal fluids. Aqueous Fischer–Tropsch

experiments have generated lipids ranging from C_2 to $>C_{35}$, including *n*-alkanols, *n*-alkanoic acids, *n*-alkylformates, *n*-alkanals, *n*-alkanes, *n*-alkenes, and *n*-alkanones (Rushdi and Simoneit, 2001).

THERMOGENIC HYPOTHESIS

Evidence supporting the thermal origin of petroleum from sedimentary organic matter (thermogenic hypothesis) is widespread in the literature (e.g. Stinnett, 1982; Tissot and Welte, 1984; Hunt, 1996) and can be divided into several categories:

- experimental
- geological
- geochemical.

Experimental evidence

Laboratory heating experiments, such as hydrous pyrolysis, can be used to generate products from potential source rocks that are physically and chemically similar to natural crude oils (e.g. Lewan et al., 1979; Lewan, 1985; Lewan, 1994; Peters et al., 1990; Ruble et al., 2001). For example, Ruble et al. (2001) completed hydrous pyrolysis experiments on two thermally immature source-rock facies of the Eocene Green River Formation, described informally as the mahogany shale and lower black shale facies. These facies represent end members in a continuum of type I kerogens from different stratigraphic levels in the formation. Low-maturity aromatic-intermediate oil from a shallow reservoir in the Altamont Field (4700 feet (1432.6 m)) (Figure 9.4) is similar to pyrolyzates from the mahogany shale and upper carbonate marker horizons (Figure 9.5). The 4700-feet (1432.6-m) oil and the lowest-temperature pyrolyzates are black-colored, heavy, and viscous and have common geochemical features, such as significant amounts of acyclic isoprenoids and β-carotene and a strong predominance of odd versus even *n*-alkanes. Paraffinic crude oils from deeper reservoirs in the Altamont Field (8569–10 217 feet (2611.8–3144.1 m)) (Figure 9.4) are similar to pyrolyzates from the basal Green River black shale facies (Figure 9.6). The paraffinic crude oils and black shale pyrolyzates are solid waxes at room temperature and are dominated by *n*-alkanes with no odd-to-even predominance. Acyclic isoprenoids are low and β-carotene is absent. The hydrous pyrolysis results support conventional oil–source rock correlations,

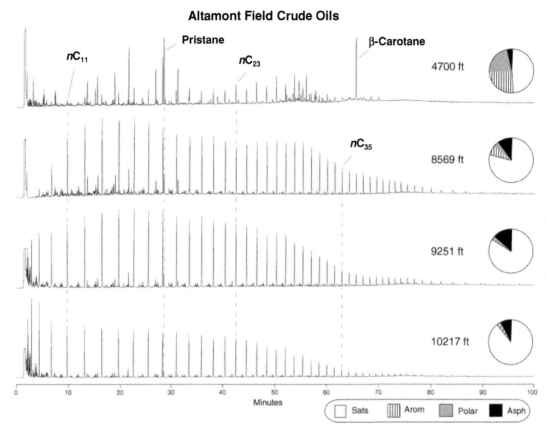

Figure 9.4. Whole-oil gas chromatograms and fractional liquid compositions for drill-stem test oils from the Altamont Field, illustrating the different types of crude oils in the Uinta Basin (Ruble *et al.*, 2001). Reprinted by permission of the AAPG, whose permission is required for further use.

which indicate genetic relationships between waxy crude oil production and source rocks in the basal Green River Formation and between shallow low-maturity oil and the upper units of the formation (Tissot *et al.*, 1978; Fouch *et al.*, 1994).

Geological evidence

More than 99% of petroleum reserves worldwide occur in sedimentary rather than basement igneous or metamorphic rocks, indicating an origin from sedimentary organic matter. Sedimentary rocks are essential for a petroleum province because they provide the source, reservoir, and seal rocks (Levorsen, 1967). No oil has ever been reported along major faults in continental shield areas where sedimentary rocks are absent. Although bitumen nodules and oil in fluid inclusions have

been reported in Archaean sandstones (see Figure 18.6) (Dutkiewicz *et al.*, 1998; Buick *et al.*, 1998), no economically significant amounts of petroleum occur in thousands of deep mines that penetrate basement rocks.

As of 1999, 5027 wildcat wells penetrate basement rocks outside of the USA, accounting for ∼5% of all non-US wildcats (109 292 wells). Many of these 5027 wells were dry (2499) or drilling results were unreported (249), some contained non-commercial hydrocarbon shows (449) or hydrocarbon gas (382), and 1448 (29%) wells contained mixed oil and gas or condensate (IHS/Petroconsultants S.A., 1999). However, the production intervals in most of these wildcat wells occur in sedimentary rocks above the basement.

Only ∼250 fields outside the USA produce at least some petroleum from basement reservoirs. Total estimated ultimate recovery (EUR) from these 250 fields,

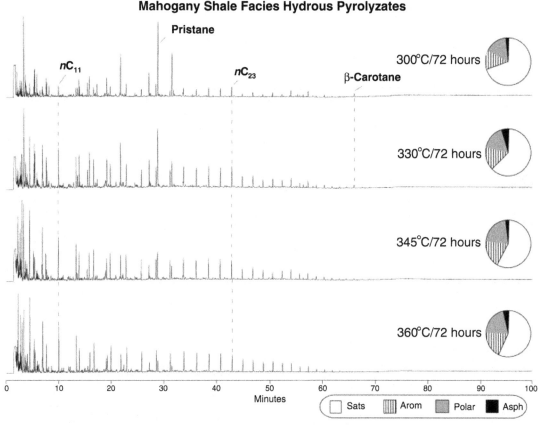

Figure 9.5. Whole-oil gas chromatograms and fractional compositions of immiscible oils from hydrous pyrolysis of mahogany shale member of the Green River Formation (Ruble *et al.*, 2001). Conditions refer to the temperature and time of hydrous pyrolysis. Reprinted by permission of the AAPG, whose permission is required for further use.

including substantial production from sedimentary rocks, is ~34 billion barrels of oil equivalent (BBOE). This amounts to less than ~0.2% of the EUR for all fields outside the USA (~17 219 BBOE) (IHS/ Petroconsultants S.A., 1999). It is difficult to quantify the relative amounts of petroleum originating from basement and sedimentary reservoirs in many of these wells because they commonly produce mixtures originating from multiple zones in the wellbore. However, nearly all of this 34 BBOE can be explained readily as originating from sedimentary source and reservoir rocks. One example is the White Tiger or Bach Ho Field from offshore Vietnam, which produces from granitic basement and accounts for nearly one billion of the 34 BBOE (Figures 18.174 and 18.175). Other examples include many of the prolific fields in the Los Angeles Basin in the USA

and the giant La Paz and Mara fields in Venezuela, as discussed below.

Petroleum in basement rocks

Petroleum occurs in basement rocks from various countries, including Algeria, Brazil, the Czech Republic, China, Egypt, Indonesia, Russia, the UK, the USA, Venezuela, and Vietnam. Petroleum in virtually all of these basement rocks can be explained by migration from adjacent organic-rich sedimentary source rocks, as in the examples below.

ALGERIA

The occurrence of petroleum in basement rocks is not evidence for an abiogenic origin of petroleum. Many

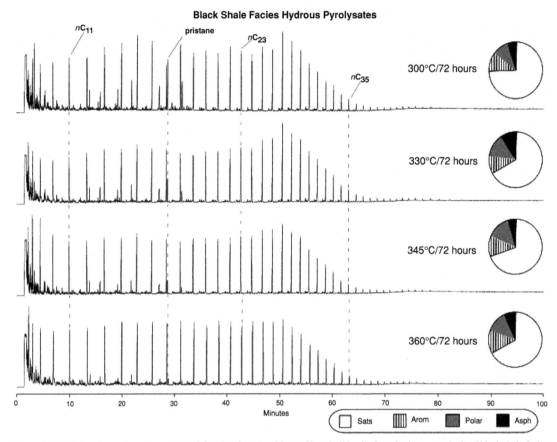

Figure 9.6. Whole-oil gas chromatograms and fractional compositions of immiscible oils from hydrous pyrolysis of black shale facies of the Green River Formation (Ruble et al., 2001). Conditions refer to the temperature and time of hydrous pyrolysis. Reprinted by permission of the AAPG whose permission is required for further use.

basement reservoirs occur below regional unconformities because paleo-exposure allowed weathering and enlargement of fractures in the basement rock, thus increasing porosity and permeability. For example, the giant Hassi R-Mel and Hassi Messaoud fields in Algeria contain crude oil that migrated from thermally mature Lower Silurian source rock into fractured Cambro-Ordovician quartzites (metamorphic basement rocks) below the regional Hercynian unconformity (Figure 9.7, top). Regional unconformities can be major pathways for petroleum migration in the subsurface (Halbouty, 1972). Where the basement and source rock are juxtaposed, fractured and porous basement rocks can be highly productive (North, 1985). The Berkine Trend in Algeria traps crude oil that migrated from thermally mature Devonian source rock to the Hercynian unconformity and then followed the unconformity to basal

Triassic sandstone reservoirs sealed by Lower Jurassic evaporites (Figure 9.7, bottom). Figure 9.7 shows how most Silurian oil from the deeper Tanezzuft source rock was lost at the unconformity, while oil generated from Devonian source rock was trapped.

THE UK

Controversy arose over the possible abiogenic origin of natural bitumens associated with Mississippi Valley-type lead–zinc (Pb–Zn) hydrothermal mineralization in Carboniferous rocks of central England (e.g. Sylvester-Bradley and King, 1963). Early geochemical analyses of one of these bitumens extracted from Caledonian age Mountsorrel Granodiorite near Leicestershire added to the controversy. Ponnamperuma and Pering (1966) inferred a likely abiogenic origin for the Mountsorrel

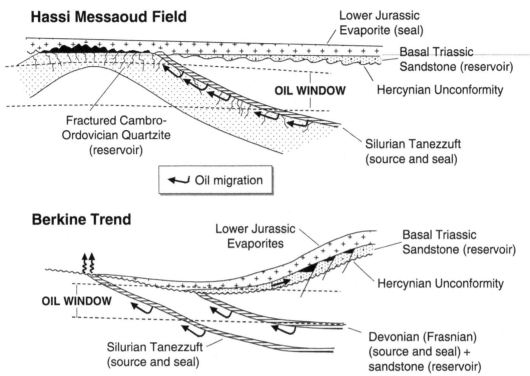

Figure 9.7. Schematic secondary migration paths for crude oils generated from Silurian and Devonian source rocks in Algeria (Peters and Creaney, 2004). Top: migration of crude oil from thermally mature Silurian Tanezzuft source rock resulted in the giant Hassi Messaoud Field in fractured Cambro-Ordovician quartzite at the Hercynian unconformity beneath the Lower Jurassic evaporite seal and additional overburden. Bottom: migration resulted in loss of Silurian oil at the Hercynian unconformity but allowed updip movement of Devonian oil (to right) along the unconformity and within the overlying Triassic sandstones, resulting in entrapment of Devonian oil in the Berkine Trend.

bitumen based mostly on (1) gas chromatography showing it to consist of an unresolved complex mixture with no *n*-alkanes, pristane, or phytane, and (2) the location of the bitumen deposits in igneous basement below sedimentary rocks. Gold and Soter (1982) cite this work as evidence for abiogenic petroleum. Ponnamperuma and Pering (1966) cautioned that they could not discount migration of oil into the Mountsorrel Granodiorite from sedimentary source rock but that, because oil floats on water, it was unlikely that the oil migrated downward into the igneous rock. Points (1) and (2) are discussed in more detail below.

Lack of *n*-alkanes (or when they show a distinct odd-to-even- or even-to-odd-number predominance), pristane, and phytane is not evidence for an abiogenic origin. The interpretations of Ponnamperuma and Pering (1966) were made before documentation that

biodegradation can cause preferential loss of these compounds from crude oil (e.g. Seifert and Moldowan, 1979; Goodwin *et al.*, 1983; Connan, 1984). Later analyses of Mountsorrel bitumen showed that it is biodegraded, accounting for the absence of *n*-alkanes, pristane, and phytane, but that it still retains abundant biomarkers, indicating a biogenic origin (Gou *et al.*, 1987). Similar bitumens from the nearby South Pennine Orefield and crude oils from the East Midlands Field have biomarker distributions that correlate with extracts of organic-rich lower Namurian (Carboniferous) mudstone source rocks that contain type II kerogen (Ewbank *et al.*, 1993). These results support the earlier conclusion that the Mountsorrel bitumens migrated into the basement rocks from overlying (now mostly eroded) Namurian mudstones during hydrothermal mineralization (Ford, 1968).

The occurrence of petroleum in igneous basement below sedimentary rocks is not evidence for an abiogenic origin. Many petroleum source rocks become overpressured due to oil and gas generation during catagenesis (Momper, 1980; Hunt, 1996). During primary migration, these generated products move toward lower-pressure regimes, which may be stratigraphically above or below the source rock, as discussed below.

Note: Expulsion of petroleum from fine-grained, low-porosity, and low-permeability source rocks (primary migration) is driven by pressure and can result in upward or downward movement of petroleum into carrier beds with higher porosity and permeability. Subsequent migration of petroleum in carrier beds (secondary migration) is driven by buoyancy (Levorsen, 1967; England and Fleet, 1991).

Pressure tends to rise gradually above hydrostatic pressure as source rocks enter the oil-generative window. Hydrostatic pressure is the pressure increase with depth for water in contact with the surface, which is about 9.8 kPa/m (kilopascals per meter) or 0.433 psi/ft. Source rocks are anisotropic and heterogeneous because much of the oil-generating kerogen occurs along approximately horizontal bedding surfaces or laminae. In argillaceous source rocks, clay and clay-sized quartz provide brittle seals between the organic-rich laminae. These seals are susceptible to fracturing caused by high fluid pressures. In carbonate-evaporite source rocks, the evaporites that seal laminae are less likely to fracture. During thermal maturation, typically 25–30 wt.% of the kerogen in the laminae is converted to bitumen, hydrocarbon gases, CO_2, and other non-hydrocarbon gases (Momper, 1980). Generation of these fluids reduces the volume of the residual kerogen. However, this is more than offset by the greater volume of generated fluids.

Abnormal overpressure (>12 kPa/m or >0.53 psi/ft.) is a departure from hydrostatic pressure that results when the pore fluids are unable to migrate out of a fine-grained rock over significant spans of geologic time. For this reason, overpressure is common in active source rocks (Hunt, 1996). If the source rock is organic-rich (Table 4.1) and sealed adequately, then early generation results in approximately horizontal migration of oil and gas along bedding laminae and some vertical movement within the source interval along microfractures that existed before or formed during generation. Expulsion occurs when high pressures cause dilation of these approximately vertical fractures and the release of petroleum and other products from the source rock into adjacent lower-pressure rock units. In this sense, source rocks behave like pressure cookers. As petroleum is expelled, the pressure drops and fractures close temporarily. Silica or calcite cement commonly precipitate along closed fractures, helping to repressurize the system until the next episodic release of petroleum. This process continues until the fluid-generation rate diminishes below the point where fluid pressures are sufficient for expulsion (Momper, 1980).

Parnell (1988) concluded that essentially all petroleum in granite plutons and other basement rocks in the UK originated from sedimentary source rocks, including those in the Mountsorrel Granodiorite. He envisioned several relationships that might explain the occurrence of petroleum in basement rocks (Figure 9.8), including (1) igneous intrusions into source rocks, (2) thermal reactivation of igneous rocks below younger source rocks, (3) hydrothermal circulation and incorporation of migrated petroleum from mature source rocks into basement fractures, (4) migration of petroleum into basement structural highs from mature source rocks in adjacent sedimentary basins, and (5) migration of petroleum into surficial porosity in basement rock after weathering and reburial. For example, intrusion of hot magma at time t_2 into source rock deposited at time t_1 might generate petroleum within the source rock that migrates into fractures in the intrusion during and after cooling (upper left in Figure 9.8).

THE LOS ANGELES BASIN

Petroleum is produced from Jurassic glaucophane-bearing schist basement along northwest–southeast anticlinal ridge trends in coastal fields of the Los Angeles Basin, including the Playa del Rey, Hyperion, El Segundo, Lawndale, Alondra, and Wilmington fields (Figure 9.9). Schist is a high-rank metamorphic rock. Exposure of the basement before Miocene time resulted in a weathered unconformity surface with considerable

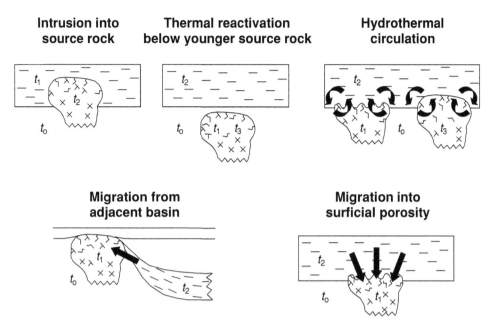

| Intrusion into source rock | Thermal reactivation below younger source rock | Hydrothermal circulation |

Figure 9.8. Some mechanisms by which petroleum becomes associated with plutonic basement rocks. Times t_0, t_1, t_2, and t_3 represent the sequence of rock ages. Modified from Parnell (1998). © Copyright 1998, with permission from Elsevier.

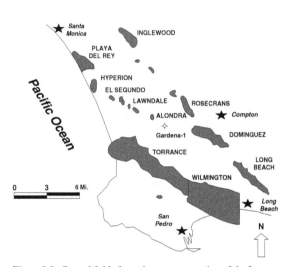

Figure 9.9. Several fields from the western portion of the Los Angeles Basin, California, produce oil from Jurassic metamorphic basement, including Playa del Rey, El Segundo, Lawndale, Alondra, and Wilmington. The Gardena-1 exploration hole contains oil from the lower Puente Formation schist-conglomerate (10 580–10 715 feet (3225–3266 m)) with geochemical characteristics typical of Miocene source rocks and related oils from California.

topographic relief. Reworked basement rocks deposited above the unconformity consist of poorly sorted schist conglomerate and sandstone with good porosity and permeability. In the Playa del Rey Field, for example, this reworked unit has average porosity near 12% and contains sand, pebbles, and cobbles, locally with boulders up to several feet in diameter (Hoots *et al.*, 1935). These redeposited basement-rock sedimentary units are thickest on the flanks of the ridge and in depressions between the highs, whereas only a thin mantle of these deposits remains on the crest of the ridge.

Subsequent burial deposition blanketed the basement and clastic units with organic-rich, laminated Miocene Nodular Shale. The Nodular Shale consists of hard black shale with tan- and gray-colored phosphatic laminations and nodules. Petrography reveals abundant algal kerogen and marine microfossils in the shale (Walker *et al.*, 1983). Composite samples yield up to 6 gallons of extractable bitumen per ton (25 l/metric ton) and 15 gallons of sulfur-rich oil per ton (62.6 l/metric ton) of rock upon retorting. Burial and thermal cracking of organic matter in the Nodular Shale resulted in expulsion and migration of petroleum into the reworked zone and the underlying weathered basement at the unconformity surface (Hoots *et al.*, 1935).

Many coastal Los Angeles Basin fields described as producing from basement reservoirs actually produce mainly from other shallower units, such as reworked basement debris immediately above the unconformity surface. Generally less production originates from fractured or weathered basement at the unconformity surface, and little or no production originates from fresh basement due to low porosity and permeability. For example, oil is produced mainly from Miocene and Pliocene sandstones and conglomerates in the giant Wilmington Field with comparatively little production from fractured, brecciated Jurassic schist (Mayuga, 1970).

Note: Land in the Wilmington Field of Los Angeles subsided by up to ~30 feet (9.1 m) in an elliptical area of ~20 square miles (51.8 km²) through 1967, apparently due to reduced pressures in the reservoirs caused by withdrawal of ~1.2 billion barrels of oil and 840 billion cubic feet (23.8×10^9 m³) of gas. A massive water-injection program increased oil recovery while reducing subsidence and the threat of inundation of low-lying harbor infrastructure (Mayuga, 1970).

The contacts between fresh basement, weathered basement, and reworked basement debris are commonly gradational (Figure 9.10). Drilling into large basement clasts in the reworked debris above the unconformity may result in the incorrect assumption that basement has been reached. For example, in the Hyperion Field the schist basement is unconformably overlain by 70 feet (21.3 m) of Upper Miocene Puente schist-conglomerate followed by 200 feet (61 m) of laminated Nodular Shale source rock. The deepest penetration into the schist basement is ~200 feet (61 m). Oil is produced from the upper 10 feet (3 m) or weathered portion of the schist, the schist-conglomerate (~70 feet (21.3 m) thick), and the lower 40 feet (12.2 m) of the Nodular Shale. However, the schist-conglomerate is the most prolific zone and yields the highest API gravity oil (~17°) (Crowder, 1960). Some wells find good production from the schist even where the schist-congolmerate is thin or missing.

The close stratigraphic relationship between the Nodular Shale and producing oil zones along the Playa del Rey–Alondra trend is strong geologic evidence for the origin of the oil from the shale. Furthermore, the occurrence of this type of oil correlates with the suspected

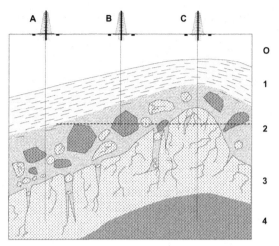

Figure 9.10. Schematic showing gradational contacts between fresh basement (4), weathered basement (3), and reworked basement debris (2). In the Playa del Rey-Alondra trend in the Los Angeles Basin, most oil is produced from reworked schist-conglomerate (2) immediately below the Nodular Shale source rock (1) and overburden rock (O). Hypothetical well A produces minor amounts of oil from thermally mature source rock zone 1 but does not produce from zones 2, 3, or 4 because the oil–water contact (dashed line) is updip. Well B produces from zones 1 and 2 but could be misinterpreted to have crossed the unconformity surface into basement. Well C produces mainly from zone 2 but also from zones 1 and 3. Zone 4 has insufficient porosity and permeability to contain significant amounts of petroleum.

regional distribution of the Nodular Shale in the Los Angeles Basin (Hoots et al., 1935).

Geochemical analyses support an origin of the oil in the Playa del Rey–Alondra trend from the Nodular Shale. Curiale et al. (1985) analyzed oil from the schist-conglomerate in the basal Puente Formation (10 580–10 715 feet (3225–3266 m)) in the Gardena-1 exploration hole (Figure 9.9). The Gardena-1 oil has geochemical characteristics, such as sterane distributions, similar to 28 extracts from stratigraphically equivalent Miocene source rocks and 22 related crude oils from California. The data are consistent with a clastic-poor marine source rock deposited under anoxic conditions (see Table 13.3). This 22°API gravity sulfur-rich oil has low diasteranes, low pristane/phytane (0.63), high 28,30-bisnorhopane, high benzothiophenes, and stable carbon isotope ratios for saturated and aromatic hydrocarbons of −23.9 and −22.7‰, respectively. All of these geochemical characteristics are remarkably similar

NW   **SE**

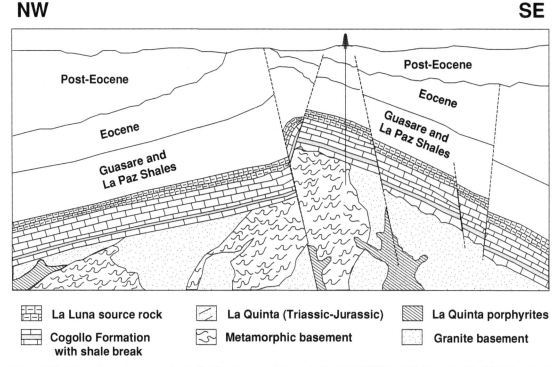

La Luna source rock   La Quinta (Triassic-Jurassic)   La Quinta porphyrites

Cogollo Formation
with shale break   Metamorphic basement   Granite basement

Figure 9.11. Schematic section across the La Paz Field, western Venezuela. Scale, 1 : 80 000 (modified from Smith, 1956). Chemically similar oil is produced from voids and fractures in Cretaceous La Luna Formation and Cogollo Formation limestones along the crest of a sharp anticline and from underlying fractured basement rocks. Reprinted by permission of the AAPG, whose permission is required for further use.

to those for other Miocene crude oils from California (Peters and Moldowan, 1991).

MARACAIBO BASIN

Oil is produced from granitic and metamorphic basement rocks in the giant La Paz and Mara fields in the Maracaibo Basin, western Venezuela. Estimated ultimate recoverable reserves of oil and gas are 900 million barrels and 2525 billion cubic feet $(17.5 \times 10^9 \text{ m}^3)$ of gas in the La Paz Field and 500 million barrels and 1000 billion cubic feet $(28.3 \times 10^9 \text{ m}^3)$ of gas in the Mara Field (IHS/Petroconsultants S.A., 1999). Oil from the basement reservoirs is compositionally similar to that from fractured Cretaceous Cogollo Formation limestones and the associated organic-rich La Luna Formation (Table 4.3). Extensive vertical faulting in both fields allowed migration of oil from the thermally mature La Luna source rock to the south of the fields into the

prolific Cogollo Formation limestones and the basement rocks (Smith, 1956). In the La Paz Field, maximum deformation of the limestone and basement occurred on the crest of a sharp anticlinal fold, coinciding with the most prolific wells (Figure 9.11).

As early as 1956, 12 wells penetrated basement at an average depth of 1650 feet (503 m) with a maximum penetration of 3087 feet (941 m) in the La Paz Field. Combined production from La Paz and Mara fields at this time amounted to ~80 000 barrels per day (Smith, 1956). Cores of basement rocks in these fields are intensely fractured, most commonly in vertical planes. Opening or dilatancy of these fractures reduced hydrostatic pressures along the crest of the anticline during deformation. The resulting pressure gradient allowed petroleum generated in the overlying organic-rich La Luna marls to migrate stratigraphically downward ~1400 feet (427 m) though brittle, heavily fractured Cogollo limestones into the basement rocks. Much of this migration may have been updip from the flanks of the anticline.

The literature describes other examples of petroleum from igneous rocks, but in virtually all of these cases an origin of the petroleum from sedimentary rocks is indicated. As another example, oil in the Central Kansas uplift is produced from fractured quartzites along the crest of buried Precambrian hills. The source rocks are flanking Cambro-Ordovician shales or overlying Pennsylvanian shales (Levorsen, 1967). Other examples of petroleum in basement rocks from the Czech Republic and Sweden are discussed in the next section.

Geochemical evidence

Much of the evidence for a biological origin of petroleum is based on geochemical data.

Biomarkers

Gold (1985; 1999) argues that biomarkers are contaminants introduced by microbes in the reservoir or by dissolution of biogenic materials during upward migration of inorganic hydrocarbons that originate by polymerization of methane in the mantle. He also argues that coal was originally biogenic (e.g. leaf fossils, woody fragments) but may have been replaced entirely by abiogenic hydrocarbons. If biomarkers were contaminants introduced at shallow depth by solubilization, then their stereochemistry should indicate the level of maturity that they achieved at their depth of origin. Contaminating biomarkers from different depths would result in mixed maturity signals, depending on which compounds were analyzed. As discussed earlier in this chapter, crude oils with mixed maturity signals are rare. For example, increased maturation favors the 18α stereoisomer of oleanane, because it is more stable than the 18β configuration (see Figure 14.9). The $18\alpha/18\beta$-oleanane ratio for most crude oils indicates thermal maturity consistent with that of the whole oil. Riva *et al.* (1988) show correlations between the oleanane epimer ratio and other maturity parameters, including Ts/Tm, vitrinite reflectance, and T_{max} with depth based on data from several wells. When mixed maturity signals occur, geologic and geochemical evidence generally supports commingling of oils from different source rocks in the reservoir. Philp and Brassell (1986) provide additional strong geochemical arguments that biomarkers are not contaminants.

Molecular mechanics calculations of stabilities for various biomarker stereoisomers are consistent with their relative abundance in crude oils generated at reasonable geologic temperatures. For example, observed $22S/(22S + 22R)$ ratios for the C_{31}–C_{35} homohopanes in crude oils (Zumberge, 1987b) are in the range of those predicted from molecular mechanics calculations at temperatures near $125°C$ (Kolaczkowska *et al.*, 1990; Peters *et al.*, 1996b).

PETROLEUM IN CRYSTALLINE ROCKS FROM THE CZECH REPUBLIC

Oil from the Zdanice-7 well in the West Carpathian thrust belt, Moravia, Czech Republic, is produced from Precambrian crystalline basement rocks (Picha and Peters, 1998), which might be interpreted to indicate an abiogenic origin. However, multiple geochemical parameters, including sterane distributions (Figure 9.12), support correlation of Zdanice-7 oil with Damborice-1 oil (produced from Carboniferous reservoirs) and Jurassic source-rock extracts. Upper Jurassic (Malm) organic-rich Mikulov marls are up to 1000 m thick in the area and contain up to 10 wt.% TOC and type II to type III kerogens. Other geochemical studies also support an origin for most oils in this area from these Upper Jurassic source rocks (e.g. Francu *et al.*, 1996).

Oils produced from Precambrian crystalline rocks in the nearby Lubna-18 and Dolni Lomna-1 wells are distinct from the Zdanice-7 oil based on many geochemical parameters, such as steranes (Figure 9.12), homohopanes (see Figure 13.85), C_{26} nordiacholestanes, and oleanane (see Figure 13.56). Although suitable source rocks for correlation were not available, the geochemical composition of the Lubna-18 and Dolni Lomna-1 oils suggests that they originated from organic-rich Menilitic Shale of the Carpathian thrust belt or autochthonous Paleogene source rocks buried below the thrust belt (Peters and Picha, 1998). For example, unlike the Zdanice-7 oil and rock extracts, the Lubna-18 and Dolni Lomna-1 oils exhibit age-related biomarker ratios consistent with a Paleogene source (see Figure 13.56). The C_{26}-nordiacholestanes and oleanane in these oils originate from diatoms and angiosperms, respectively. Furthermore, the $18\alpha/18\beta$-oleanane ratios for the Lubna-18 and Dolni Lomna-1 oils have achieved endpoint values ($\sim$0.6), indicating thermal maturity within the oil window. This excludes the possibility that

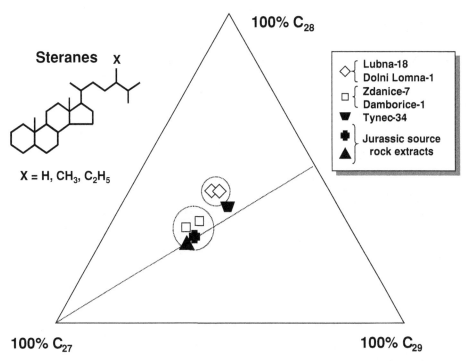

Figure 9.12. Sterane distributions support two genetic oil groups in Moravia, Czech Republic (Picha and Peters, 1998). Zdanice-7 and Damborice-1 oils correlate with extracts from two Jurassic source rocks. Repeatability approaches the diameter of each symbol.

the oleanane was a contaminant picked up during upward migration of abiotic petroleum through thermally immature carrier beds.

PETROLEUM IN THE IMPACT CRATER AT LAKE SILJAN

Near the end of the Devonian ($\sim$362 Ma) (Bottomley et al., 1978), a meteor struck Precambrian to Upper Silurian sandstones, shales, and limestones in central Sweden, resulting in an impact crater with an estimated diameter of 50–60 km. Tremendous kinetic energy transferred from the meteor to the target rocks upon impact. Temperatures were sufficient to melt or vaporize silicate minerals and induce refraction waves in the excavated cavity. The resulting transient cavity was two to three times deeper than the excavated cavity. The transient cavity floor underwent elastic rebound to form a central uplift that was surrounded by an annular depression that still exists today. Immediately after the impact, sedimentary rocks along the rim of the crater that had been tilted upward by the impact collapsed

downward into the annular depression. Some of these sedimentary rocks were tilted by up to 60–80° as they collapsed into the annular depression. The sedimentary rocks were then covered with rubble and, in certain areas, by $\sim$200 m of molten rock with initial temperatures near 2200°C that probably required $\sim$10 000 years to cool (Grieve, 1988). The present-day Lake Siljan ring structure has a 32-km diameter central core of uplifted Dala granites that represent the eroded remnants of the uplifted transient cavity floor. The 45-km diameter annular depression surrounding the granites contains lakes distributed within the down-faulted Ordovician and Silurian sedimentary rocks.

Numerous asphalts and biodegraded seep oils occur in quarries in sedimentary rocks near Lake Siljan, and some shallow boreholes contain a thin layer of free oil floating on the water table. Vlierboom et al. (1986) identified three organic-rich, oil-prone intervals in representative Paleozoic shales and limestones from the Siljan area. These rocks were never buried deeply enough to generate petroleum. Nonetheless, seep oils from the site correlate best with two facies of the Ordovician Tretaspis

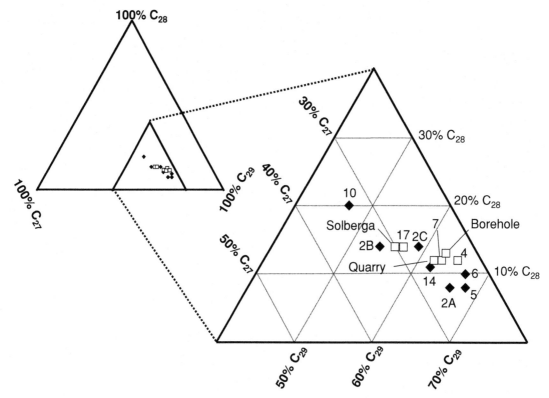

Figure 9.13. Distributions of C_{27}–C_{29} 5α 20R steranes, showing relationships among Siljan area samples. Modified from Vlierboom *et al.* (1986). © Copyright 1986, with permission from Elsevier.

Open squares = seep oils, including oil-stained limestones; solid diamonds = source-rock extracts. 2A–2C = Silurian Rastrites Shale, 5, 14 = Ordovician bituminous Tretaspis Shale; 6 = limestone interbed within Tretaspis Shale; 10 = basal Ordovician Black Tretaspis Shale.

Shale based on stable carbon isotopes, steranes, and terpanes. The data indicate that heat supplied by the meteor impact matured the source rocks near the point of impact, resulting in geologically instantaneous generation and expulsion of the seep oils. Although the amounts of crude oil at Lake Siljan are not commercial, the evidence for its origin from organic matter in sedimentary rocks is important in light of the claim by Gold (1999) that it is abiogenic.

Figure 9.13 suggests that the seep oils originated mainly by mixing of products from two organic-rich facies of the Ordovician Tretaspis Formation represented by rock samples 5, 14, and 10 (TOC = 9.3, 7.8, and 5.1 wt.%; HI = 555, 510, and 326 mg HC/g TOC, respectively). Samples 5 and 14 are from the bituminous facies, while sample 10 is from the basal black shale facies of the Tretaspis Formation. Biodegradation has not

affected steranes or triterpanes in the seep oils. The seep oils have sterane compositions similar to extracts from rock sample 14 and intermediate between extracts from samples 5 and 10. Stable carbon isotope ratios for saturated ($-30.3 \pm 0.1\%$) and aromatic ($-29.5 \pm 0.1\%$) fractions of the seep oils also support an origin mainly by mixing of products from Tretaspis source rocks. For example, the stable carbon isotope ratios for the saturated (-30.9 and -28.3%) and aromatic (-30.5 and -28.4%, respectively) hydrocarbons in samples 5 and 10 bracket the corresponding values for the seep oils. This mixing scenario is also supported by terpane data (Vlierboom *et al.*, 1986).

Optical activity
Many biological compounds are optically active, i.e. a beam of plane-polarized light that passes through

a solution containing the compound is rotated to the left (levororotary) or right (dextrorotatory). Optical activity is a consequence of asymmetric carbon centers in biological compounds (Figure 2.21). For example, nearly all amino acids in living organisms are levorotatory (Macko *et al.*, 1994). Compounds formed non-biologically in the laboratory or in interstellar space generally consist of racemic mixtures that do not rotate a polarized beam and are optically inactive. Likewise, one would not expect compounds formed by abiogenic polymerization of methane in the mantle to show optical activity.

Crude oils are optically active, strongly suggesting a biological origin. For example, the most optically active components of crude oils consist of ^{13}C-poor steranes and triterpanes derived from once-living organisms (Silverman, 1971; Whitehead, 1971). These components represent a major fraction of most crude oils. Individual biomarkers, such as gammacerane, have high optical activity, supporting the view that they account for most of the optical activity in crude oils (Hills *et al.*, 1966). This optical activity is a consequence of the biologically imposed preference of one enantiomer over another.

The optical activity of crude oils diminishes with thermal maturity because cracking results in fewer asymmetric centers (Williams, 1974). For example, pristane isolated from low-maturity rocks and crude oils consists of a non-racemic mixture that becomes racemic with increasing maturity (Patience *et al.*, 1978). A racemic mixture of farnesane isomers in two mature crude oils was attributed to loss of enantiomer preference due to high thermal maturity (Brooks *et al.*, 1997). Microbial degradation generally increases optical activity due to selective degradation of *n*-alkanes, which are not optically active (Winters and Williams, 1969).

Isomer abundance

As discussed in Chapter 2, structural isomers of various compounds that originate from terpenoid precursors in living organisms are more abundant in petroleum than might be expected if they originated by polymerization of methane. For example, 2,6-dimethyloctane and 2-methyl-3-ethylheptane are more abundant than all other 49 possible structural isomers in Ponca City crude oil (Figure 2.14). These two monoterpanes conform to the isoprene rule and can be explained as thermal breakdown products of terpenoids. Likewise, the abundance of pristane (C_{19}) and phytane (C_{20}) relative to other members of the homologous acyclic isoprenoid series (Figure 2.16) can be explained by their primary origin from chlorophyll a in plants (Figure 3.24). The C_{17} acyclic isoprenoid is low or absent in petroleum because its formation requires cleavage of two carbon–carbon bonds in any larger isoprenoid precursor. The more abundant C_{16} and C_{18} acyclic isoprenoids can form by cleavage of only one carbon–carbon bond in a larger isoprenoid precursor, such as pristane (C_{19}) or phytane (C_{20}). Similarly, C_{28} $\alpha\beta$-hopanes are rare or absent in petroleum because their formation requires cleavage of two carbon–carbon bonds attached to C-22 in the C_{35} biological hopanoid precursor (Figure 2.29). Extended tricyclic terpanes contain a regular isoprenoid side chain at C-14 as evidenced by lower abundance of the C_{22}, C_{27}, C_{32}, C_{37}, and C_{42} homologs, which require cleavage of two carbon–carbon bonds to form from higher homologs (see Figure 13.73). Polymerization of methane would not result in the major differences in the relative concentrations of the individual acyclic isoprenoids, $\alpha\beta$-hopanes, tricyclic terpanes or other terpenoid series that are evident in crude oils.

High concentrations of non-terpenoids, such as *n*-alkanes, isoalkanes, and anteisoalkanes in many crude oils, can also be explained by a biological origin. For example, 2-methyloctadecane is more abundant than many of its structural isomers because it is a common biosynthesized product of archaea. As discussed elsewhere in this book, porphyrins in petroleum clearly originate from chlorophylls and hemes in living organisms, as shown originally by Treibs (1936).

Stable carbon isotopes

As discussed in Chapter 6, carbon has two naturally occurring stable isotopes, ^{13}C and ^{12}C. Organic matter formed by photosynthetic fixation of carbon from carbon dioxide or bicarbonate is depleted in ^{13}C because plants preferentially fix ^{12}C. Kerogen in source rocks and petroleum show similar depletions in ^{13}C to those observed in living organisms (Figure 6.2), consistent with an origin of these materials by photosynthesis. No known inorganic process results in high-molecular-weight carbonaceous material that is as depleted in ^{13}C as living organic matter, kerogen, or petroleum.

Although not definitive, one suggested criterion to identify abiogenic methane is $\delta^{13}C$ more positive than $-25‰$ (Jenden et al., 1993a). However, nearly pure abiogenic methane with $\delta^{13}C$ as negative as $-53.6‰$ was recently produced under laboratory conditions resembling those in the Earth's upper crust (Horita and Berndt, 1999).

The amount of ^{12}C in natural and artificially cracked gases decreases from methane to ethane to propane and higher homologs, consistent with kinetic isotope fractionation during thermal cracking of kerogen or petroleum (Figure 6.3). The opposite trend of increasing ^{12}C from methane to ethane and higher homologs is consistent with polymerization from methane but does not occur in commercial petroleum accumulations, as discussed above. A few carbonaceous chondrites have this reversed trend in isotope composition, consistent with an origin of the small quantities of occluded hydrocarbon gases by Fischer–Tropsch type reactions in interstellar space (Figure 9.3). Limited experimental data from one study suggested that some gases produced under laboratory conditions that mimic abiogenic hydrocarbon formation have the normal isotopic trend, where methane is enriched in ^{12}C compared with a C_{2+} fraction (Lancet and Anders, 1970). They heated an equimolar mixture of CO and H_2 at 400 K under one atmosphere pressure in the presence of cobalt catalyst. After four hours in one experiment, the $\delta^{13}C$ for methane and the C_{2+} fraction was -100 and $-62‰$, respectively. However, after 30 hours, the $\delta^{13}C$ for methane and the C_{2+} fraction became -50.2 and $-61.2‰$, respectively, indicating a reversed distribution typical of abiogenic products. Similar experiments heating CO and H_2 under Fischer–Tropsch type conditions yielded hydrocarbons at temperatures above $250°C$, but no systematic isotopic variations with carbon number were observed in the major saturated hydrocarbons that formed (Yuen et al., 1990). Hu et al. (1998) conducted Fischer–Tropsch type heating experiments on CO and H_2 (2 : 3 ratio) using different catalysts. Although the generated methane had variable $\delta^{13}C$, the heavier hydrocarbons have a reversed trend, where ^{12}C content increased from ethane to propane to butane, typical of abiogenic hydrocarbons as found in the Murchison meteorite.

Compound-specific isotope analyses show that many components in petroleum can be tied directly to biological organic matter in the source rock (e.g. Hayes et al., 1987; Hayes et al., 1990; Freeman et al., 1990; Schoell et al., 1994; Guthrie et al., 1996).

Oil–source rock correlation and petroleum systems

During thermal maturation, petroleum expelled from fine-grained organic-rich source rock migrates some distance to coarse-grained or fractured reservoir rock. The migrated petroleum inherits the geochemical fingerprint of the bitumen, some of which remains in the source rock. Typical chromatographic fingerprint parameters might include the carbon preference index or pristane/phytane. Biomarker ratios might include relative distributions of C_{27}–C_{28}–C_{29} steranes or monoaromatic steroid ratios, and various terpane ratios. Other source-related parameters might include stable carbon isotope ratios for the saturated and aromatic hydrocarbons and vanadium/nickel. Secondary processes, such as migration and biodegradation, do not significantly affect these ratios. Oil–source rock correlation is based on similar geochemical fingerprints among specific crude oils and source-rock extracts. Specific source rocks for many large oil fields have been established by this method and many are discussed in this book (e.g. see Chapter 18).

Vertical and lateral variation in source-rock geochemical composition can complicate correlation, particularly for lacustrine and deltaic samples. Unlike crude oils that are generated from large volumes of mature source rock, extracts from discrete rock samples represent only the sampled interval. Vertical and lateral variations in the geochemical composition of lake sediments depend on geologically rapid changes in climate, biota, and lake chemistry. Furthermore, these variations may differ depending on the analytical method. For example, biomarker distributions might be similar over a source-rock interval, while stable carbon isotope ratios of the bulk extract or distributions of n-alkanes might show wide variations, and vice versa. This occurs because biomarker compositions are controlled largely by organic facies, while stable carbon isotope ratios are related mainly to factors controlling isotopic fractionation, such as surface water temperature and pH. Although many of these controlling factors may overlap, there are many examples where biomarker distributions change, but isotopic ratios vary, and *vice versa*. Geochemical correlations need to take these variations in source rock character into account. One approach is distributed source-rock sampling (Figure 6.14) (Curiale and Sperry, 1998). Correlations between source rocks and crude oils from lacustrine settings are generally less convincing than those between the oils.

Thermal modeling

Our ability to forecast petroleum occurrence using geochemical correlations, kinetic models, and basin thermal histories (petroleum system modeling) is sufficiently good to validate the biogenic rather than the abiogenic concept as the most reliable tool for commercial exploration (e.g. Demaison and Murris, 1984; Magoon and Dow, 1994b; Hunt, 1996; Welte et al., 1997). There are many examples in the literature where thermal models of source rock generation correctly identify the pod of active source rock. Petroleum exploration success ratios within or immediately updip of these areas exceed those of adjacent areas (e.g. see Figures 14.1, 18.1, 18.54, and 18.138).

10 · Biomarkers in the environment

This chapter explains how analyses of biomarkers and other environmental markers, such as polycyclic aromatic hydrocarbons, are used to characterize, identify, and assess the environmental impact of oil spills. The discussion covers processes affecting the composition of spilled oil, such as emulsification, oxidation, and biodegradation, as well as oil-spill mitigation and modeling. Field and laboratory procedures for sampling and analyzing spills are discussed, including program design, chemical fingerprinting, and data quality control. The chapter includes sections on smoke, natural gas, and gasoline and other light fuels as pollutants, and a detailed discussion of the controversial *Exxon Valdez* oil spill.

ENVIRONMENTAL MARKERS

After World War II, large-scale production of synthetic organic compounds, increased combustion of fossil fuels, and rapid population growth contributed to the introduction of many persistent contaminants into the environment. Examples of these contaminants include [1,1,1-trichloro-2,2-bis-(p-chlorophenyl) ethane] (DDT), polychlorinated biphenyls (PCBs), phthalates, dioxins, and polycyclic aromatic hydrocarbons (PAHs). At about the same time, rapid improvements in analytical methods, including gas chromatography, high-pressure liquid chromatography, and mass spectrometry allowed insights into the complex compositions of environmental materials. The landmark book *Silent Spring* by Rachel Carson (1962) identified some of the dangers of synthetic organic contaminants. Public awareness also increased after several dramatic events, e.g. Minimata disease (in 1953) and the Yusho incident, involving mercury and PCB poisoning, respectively, and the *Torrey Canyon* (1967), Buzzard's Bay (1969), and Santa Barbara Channel (1969) oil spills. These events led to the formation of the US Environmental Protection Agency (EPA) in 1970 and similar organizations worldwide.

Like biomarkers in petroleum, environmental markers in recent sediments, groundwater, rivers, lakes, oceans, and the atmosphere can be used to identify specific source materials. Environmental markers are compounds that are introduced into the environment as pollutants as a result of man's activities and from natural sources. These markers include biomarkers and other compounds that do not fit the conventional definition of biomarkers, such as PAHs. Our discussion includes many of these non-biomarker compounds, because they also play a critical role in evaluating environmental contamination. Environmental markers can be used to predict the behavior, fate, and effects of toxic pollutants and to monitor the progress of diagenetic processes. For simplicity, we separate environmental markers into three categories (Eganhouse, 1997): (1) contemporary biogenic markers, (2) anthropogenic markers, and (3) petroleum and fossil fuel biomarkers. These categories are not always distinct, as discussed below.

Note: The term "biomarker" as used in toxicology and the medical sciences has a different meaning to that in geochemistry. The broadest definitions include almost any measurement reflecting an interaction between a biological system and an environmental agent, which may be chemical, physical, or biological. In toxicology, biomarkers are components in biological fluids, cells, tissues, or whole organisms that indicate the presence, magnitude, and exposure of toxicants or of host response. In medicine, biomarkers refer to many compounds and genetic markers that are used to diagnose cancerous tumors, the effects of aging, and nutritional intake. In a study of the etymology of the term "biomarker" as used in the biological sciences, Benford *et al.* (2000) list no less than

17 definitions. However, they recognize that the term originated in geochemistry.

Contemporary biogenic markers

Contemporary biogenic markers are generally compounds produced by microorganisms, higher plants, or animals that occur with little to no alteration in the environment. These compounds may be diagnostic of specific organisms, classes of organism, or general biota that contribute organic matter to the atmospheric, aqueous, or sedimentary environment.

Anthropogenic markers

Anthropogenic markers are compounds that are produced by humans or human activity and are introduced intentionally or inadvertently into the atmospheric, aqueous, or sedimentary environment. These compounds may be natural or synthetic, benign or toxic. They include compounds that are classified as pollutants because they are mutagens or carcinogens or because they impose other risks to human health or ecology. Common anthropogenic markers that are toxic include PCBs and chlorinated pesticides. However, many synthetic and natural organic compounds, such as coprostanol and synthetic surfactants, are not considered pollutants but can also be introduced into the environment as a result of human activity. Coprostanol is unusual because it can fit the definitions of both contemporary biogenic and anthropogenic markers.

Petroleum markers

Petroleum markers are compounds produced during diagenesis or catagenesis of buried organic matter, which includes biomarkers in source rocks and fossil fuels in general. Many petroleum markers show clear structural relationships with contemporary biogenic markers because they have the same origins. However, petroleum markers have undergone catagenesis and thus lack certain thermally reactive structural moieties, such as hydroxyl groups or double bonds, that are common in contemporary biogenic markers.

Classification of an individual compound may depend on the situation. For example, polynuclear aromatic hydrocarbons enter the environment from various sources. When contributed directly from biota or as

pyrogenic compounds from fires, they are considered to be contemporary biogenic markers. When introduced by humans, typically by inadvertent or accidental release or as pyrogenic products, they are considered to be anthropogenic markers. Petroleum PAHs in the environment may result from natural oil seepages or the erosion of source rocks. They may also be anthropogenic markers of releases of crude oil, refined product, or pyrogenic emissions. Perylene, for example, is a common and abundant PAH in modern sediments, but its origins have long been debated (e.g. Aizenshtat, 1973; Louda and Baker, 1984; Chunqing *et al.*, 2000b; Silliman *et al.*, 2000; Silliman *et al.*, 2001).

Difficulties in classifying environmental markers

The classification of an environmental marker may change as our knowledge of its possible sources increases. For example, the fecal sterol, coprostanol, originates from bacteria in the intestinal tracts of mammals. Coprostanol occurs in waters and sediments impacted by human waste, and early work suggested its use as a tracer of human fecal pollution (Hatcher *et al.*, 1977). However, coprostanol is also generated from cholesterol under reducing conditions in sediments, and it occurs in non-human mammalian and bird feces (Venkatesan *et al.*, 1986; Leeming *et al.*, 1997; Sherblom *et al.*, 1997). Thus, high coprostanol can occur in reducing sediments not heavily impacted by mammalian feces or where non-mammalian fecal inputs are important. Nonetheless, coprostanol has been used effectively as a proxy indicator for sewage pollution (e.g. Venkatesan and Kaplan, 1990; Venkatesan and Mirsadeghi, 1992; Nichols *et al.*, 1993). Several studies have proposed using the ratios of various stanols to differentiate human from other mammalian feces (e.g. Venkatesen *et al.*, 1986; Evershed and Bethell, 1998). Using the distribution of specific stanols and bile acids, the presence of human feces can be differentiated from canine, porcine, and ruminant animals (Bull *et al.*, 2002).

Two different pathways have been proposed for the reduction of cholesterol to coprostanol by microorganisms (Figure 10.1): (1) formation of the intermediate Δ^4-cholesten-3-one, which is converted to coprostanone and coprostanol, and (2) direct reduction of the Δ^5 double bond (Venkatesen and Santiago, 1989). These authors found coprostanone in marine

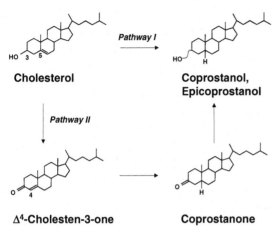

Cholesterol

Pathway I

Coprostanol, Epicoprostanol

Pathway II

Δ⁴-Cholesten-3-one

Coprostanone

Figure 10.1. Two pathways have been proposed for the reduction of cholesterol [choles-5-en-3β-ol] to coprostanol [5β-cholestan-3β-ol] and epicoprostanol [5β-cholestan-3α-ol] by intestinal microorganisms in mammals (Venkatesan and Santiago, 1989). Pathways I and II involve the intermediate formation of Δ⁴-cholesten-3-one and direct reduction of the Δ⁵ bond, respectively.

mammalian feces, suggesting that at least part of the conversion of cholesterol to coprostanol occurs via pathway 1 in the intestines of marine mammals. They determined that the type of mammalian fecal input to sediments can also be determined using the ratio of coprostanol (cop) to epicoprostanol (e-cop) and the percentage of cholesterol in the total lipid extract (%chol) in samples from (1) humans, (2) baleen whales, (3) toothed whales and penguins, and (4) pinnepeds. For example, unlike other sediment and mammalian fecal samples, the cop/e-cop ratio (~2.3) and %chol (~9%) for sludge from the Hyperion sewage treatment plant fall within the range for nearby offshore Santa Monica Basin sediments, suggesting dominantly human fecal input.

Petroleum and fossil fuel biomarkers and spills of crude oil in the marine environment are the main emphasis of this chapter. Although an extensive literature exists related to land contamination of groundwater and soil, most biomarker work has been done on marine oil spills (e.g. the *Exxon Valdez*).

OIL SPILLS

The *Torrey Canyon* spill and other major spills, such as that caused by the grounding of the *Argo Merchant* at Buzzard's Bay and blowout of Unocal Platform

A in the Santa Barbara Channel in the late 1960s, led to cooperation among the disciplines of petroleum geochemistry, oceanography, and environmental chemistry. Many field and laboratory studies investigated the fate of petroleum in aquatic environments (e.g. National Research Council, 1985; 2002). Pioneering work demonstrated the use of biomarkers to differentiate petroleum and background hydrocarbons (Blumer et al., 1972; Farrington and Meyers, 1975) and track petroleum contaminants in the environment (Blumer and Sass, 1972; Teal et al., 1992). These early applications provided the basis for more sophisticated understanding of later spills, including the *Exxon Valdez* in 1989 (Bence et al., 1996).

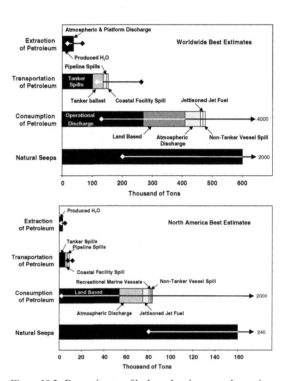

Figure 10.2. Best estimates of hydrocarbon inputs to the marine environment (National Research Council, 2002). Error bars show the minimum and maximum values estimated for each category. There is considerable uncertainty in the contribution from natural seeps and land-based run-off. Data for the worldwide contribution from recreational marine vessels are available only for North American estimates. US regulations prohibit the dumping of tanker ballast waters/oil and waste fluids from marine vessels. Consequently, these sources are negligible in North America.

Table 10.1. *Source and scale of oil pollution in the marine environment (from Patin, 1999)*

	Type of input	Source of input	Environment		Scale of impact		
			Hydrosphere	Atmosphere	Local	Regional	Global
Anthropogenic	Natural	Seeps, erosion of sediments or rocks	+	−	+	?	−
		Biosynthesis by marine organisms	+	−	+	+	+
	Marine	Marine oil transportation (e.g. accidents, operational discharges from tankers)	+	−	+	+	?
		Marine non-tanker shipping (operational, accidental, and illegal discharges)	+	−	+	?	−
		Offshore oil production (drilling discharges, accidents)	+	+	+	?	−
	Onshore	Sewage waters	+	−	+	+	?
		Oil terminals, pipelines, trucks	+	−	+	−	−
		Rivers, land run-off	+	−	+	+	?
	Combustion	Incomplete fuel combustion	−	+	+	+	?

+, presence confirmed; −, absent or negligible contribution; ?, impact uncertain.

Assessing damage to the marine environment caused by oil spills requires knowledge of the total petroleum input. Environmental damage from spills can be immediate and obvious to the casual observer, or chronic and less obvious. Numerous natural and anthropogenic sources of oil pollution are now recognized (Table 10.1). Most experts believe that chronic inputs contribute far more petroleum to the oceans than highly publicized, acute spills (National Research Council, 1985; National Research Council, 2002; Patin, 1999). Current best estimates now recognize natural marine seepage as the single largest contributor of oil to the marine environment, followed by operational discharges from large vessels (e.g. legal and illegal clearing of bilge and fuel tanks), and land-based sources (e.g. rivers with input from urban roads and municipal and industrial waste) (Figure 10.2). The contributions from these sources, particularly natural seeps and land-based effluents, are poorly constrained, and the actual values may be several times greater than the best estimates. For example, the National Research Council (2002) best estimate for worldwide land-based annual input is 140 thousand barrels, but the possible estimated range may be as low as 7000 barrels and as high as 5 million barrels. The volumes of oil released in tanker accidents, pipeline spills, and the production of oil are very well documented for US waters and are fairly reliable worldwide. These sources are minor compared with those mentioned above, particularly where marine vessel operations are regulated to prevent spillage.

Table 10.2. *The largest oil spills (DeCola, 2000)*

Rank	Incident	Millions of gallons
1	January 26, 1991; terminals, tankers; eight sources total, Sea Island installations; Kuwait; off coast in Persian Gulf and in Saudi Arabia	240.0
2	June 3, 1979; exploratory well Ixtoc-1; Mexico; Gulf of Mexico, Bahia de Campeche, 80 km northwest of Ciudad del Carmen, Campeche	140.0
3	March 2, 1992; Mingbulak-5 well; Uzbekistan; Fergana Valley	88.0
4	February 4, 1983; platform No. 3 well (Nowruz); Iran; Persian Gulf, Nowruz Field	80.0
5	August 6, 1983; tanker *Castillo de Bellver*; South Africa; Atlantic Ocean, 64 km off Table Bay	78.5
6	March 16, 1978; tanker *Amoco Cadiz*; France; Atlantic Ocean, off Portsall, Brittany	68.7
7	November 10, 1988; tanker *Odyssey*; Canada; North Atlantic Ocean, 1175 km northeast of St John's, Newfoundland	43.1
8	July 19, 1979; tanker *Atlantic Empress*; Trinidad and Tobago; Caribbean Sea, 32 km northeast of Trinidad–Tobago	42.7
9	August 1, 1980; production well D-103 concession well; 800 km southeast of Tripoli, Libya	42.0
10	April 11, 1991; tanker *Haven*; Italy; Mediterranean Sea, port of Genoa	42.0
53	March 24, 1989; tanker *Exxon Valdez*; USA; Prince William Sound, Alaska	11.0

The relative contribution of marine petroleum pollution from various sources depends on the region. Large-scale natural seepage strongly influences some areas, such as the Caspian Sea, offshore California, and the Gulf of Mexico, where seepage supports extensive chemosynthetic communities (Sassen *et al.*, 1993). In areas with extensive offshore oil production, such as the Gulf of Mexico and the North Sea, operational accidents and legal or illegal discharges may account for up to 30% of oil pollution (Corbin, 1993). Up to 50% of the oil pollution in non-petroliferous areas of the oceans may originate from tanker ballast, as in the Caribbean Sea and northern Indian Ocean (Corbin, 1993). The USA does not allow oil tankers to dump ballast, wastewater from the washing of cargo folds, fuel tanks, or engine maintenance, or contaminated bilge water. Consequently, the contributions from these inputs are negligible for North America, although they are a major contributor worldwide in areas where such regulations do not exist (National Research Council, 2002).

Although accidental oil spills may contribute only a small percentage of the total petroleum pollution on a global and time-averaged scale, they can have profound effects on regional ecology (Table 10.2, Figure 10.3). The largest release of oil occurred during the 1991 Gulf War, when ~0.5–1 million tons of oil was released into the coastal waters and over 70 million tons of oil and

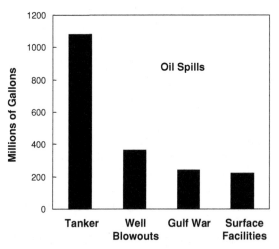

Figure 10.3. Amount of oil released during incidents from ~1950 to 2000 (>10 million gallons) (data from DeCola, 2000). Surface facilities include pipelines, refineries, and tank farms.

Figure 10.4. The Ixtoc-1 exploratory well blew out on June 3, 1979, in the Bay of Campeche off Ciudad del Carmen, Mexico. By the time the well was brought under control in February 1980, ~140 million gallons of oil had spilled into the bay. (From Office of Response and Restoration, National Ocean Service, National Oceanic and Atmospheric Administration, http://response.restoration.noaa.gov.)

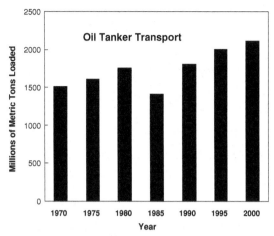

Figure 10.5. Total tonnage of oil transported around the world (data from GESAMP, 2001).

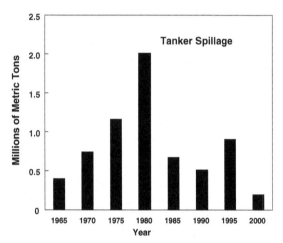

Figure 10.6. Metric tons of oil released by tanker spills (data from DeCola (2000) and ITOPF (2001)). Each bar represents the cumulative spillage over a five-year period.

combustion products were emitted into the atmosphere (Fowler *et al.*, 1993). The ecological effects were devastating but were restricted largely to within 400 km of the source (Fowler *et al.*, 1993). As a consequence of this event, leaking oil replaced transport spillage as the major source of petroleum pollution in the area (Readman *et al.*, 1996). The largest blowout was the Ixtoc-1 well, which spilled ~40 million tons (~300 million barrels) oil into the Gulf of Mexico from June 1979 to February 1980 (Figure 10.4). The largest onshore spill occurred due to a blowout of the Mingbulak-5 well near the Sar-Daryna River in Uzbekistan in March 1992. The well took months to control, releasing 35 000–150 000 barrels per day. This blowout prompted Western interest in exploration of the Fergana Valley (DOE/EIA, 1995). Although the Mingbulak well blowout was a major release, it may not have been as environmentally damaging as the chronic releases from the Fergana refinery. This refinery may have lost more than 120 million gallons of crude oil and refined products to the local environment over several decades of operation (DOE/EIA, 1995).

Incidents involving supertankers are particularly problematic because of the volumes of oil that can be released quickly and because they can affect shorelines that are ecologically sensitive or have high economic value (e.g. fishing and tourism). In terms of spill volume, tanker accidents account for more than all other catastrophic releases combined (Figure 10.3). Despite generally increasing volumes of transported petroleum, there has been a marked decrease in volumes spilled since 1980 (Figures 10.5 and 10.6).

PROCESSES AFFECTING THE FATE OF MARINE OIL SPILLS

Various processes alter crude oil in the marine environment, including spreading and aggregation, dispersion, evaporation, emulsification, dissolution, settling,

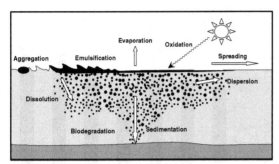

Figure 10.7. Weathering processes that affect marine oil spills (from ITOPF (2001) and Patin (1999)).

oxidation, and biodegradation (Figure 10.7). Collectively, these processes are termed "weathering." The degree to which any process affects spills depends on the chemical and physical nature of the oil or refined product, and the local environmental conditions (e.g. wind, currents, temperature, salinity, and biota) (National Research Council, 2002). Some processes occur immediately and go to completion quickly, while others may extend for years.

Spreading

When oil spills on water, it spreads over the water surface. The initial spreading is controlled by oil viscosity and wetability. Typical crude oils spread at a rate of 100–300 m/hour, while some highly refined products spread at rates of 600 m/hour. One ton of oil can form a 10-mm-thick slick with a radius of 50 m within ten minutes of being spilled (Patin, 1999). Wind, wave, and current directions and strength, temperature, and tides also affect the rate and direction of spreading. Oil slicks usually drift in the same direction as the wind, forming narrow bands. Spreading is rarely of uniform thickness.

Dispersion

When oil slicks thin to ~0.1 mm, they disaggregate into isolated patches. Eventually, the oil disperses as small droplets that remain suspended in the water column or adsorb to mineral or organic particles. Wave action and water turbulence accelerate dispersion. Dispersed droplets can travel far from the spill and can

coalesce to reform the slick. Because dispersed oil has more surface area exposed to the water, biodegradation and dissolution alter it much faster than aggregated oil. Chemical dispersants can be used to accelerate this process, e.g. the Sea Empress spill remediation (Lunel et al., 1997).

Evaporation

The amount of spilled hydrocarbons that evaporate into the atmosphere depends on the volatile content of the spilled petroleum and local environmental factors. Nearly all of a light, refined product (e.g. gasoline or kerosene) evaporates within the first day of the spill. Evaporation increases with temperature and other factors that contribute to faster spreading and dispersion (Fingas, 1995a; 1995b).

Aqueous vapor extraction

Evaporation is usually considered to be effective in the removal of volatile hydrocarbons ($<C_{15}$). Loss of hydrocarbons up to at least nC_{36} and the chrysenes can be enhanced by the aeration of oil in water. This process, termed "aqueous vapor extraction," may help to explain the observed weathering of some oil spills. Prince et al. (2002) documented aqueous vapor extraction in the analysis of oil spilled from a pipeline into the Río Desaguadero on the Bolivian Altiplano that occurred during a very turbulent flood.

Dissolution

Petroleum compounds partition between the hydrocarbon and aqueous phases (Yaws et al., 1993). The degree of dissolution depends on oil composition and solubility, water chemistry, temperature, and physical processes that allow mixing (e.g. spreading and dispersion). Because most components are more soluble in the hydrocarbon phase, only trace amounts dissolve in the water, except for the light aromatic hydrocarbons (e.g. benzene, toluene, xylenes). Most of these light compounds, however, evaporate before dissolving. Some polar compounds that form from oxidation may dissolve into the water.

Emulsification

Emulsification occurs when water droplets become suspended in the spilled oil, and vice versa. Emulsification depends on the presence of surface-active chemicals (i.e. amphipathic molecules), oil and water chemistry, temperature, and the degree of physical mixing. Viscous oils with over 0.5% asphaltenes can form stable emulsions that contain 30–80% water, resulting in a three- to four-fold increase in volume. Emulsions, commonly called mousse, slow several destructive processes and can persist in the marine environment for months. Emulsions break down when stranded on shorelines or when heated, but if buried they can persist for years (National Research Council, 2002).

Aggregation

Spilled oil can aggregate into pelagic tars or stranded tar balls. Tars can form after the dissolution and evaporation of light hydrocarbons, following emulsification, and through chemical and microbial alteration. Tar chemical compositions are highly variable, although most are rich in asphaltenes (up to 50%) and waxes. Tar balls typically have a hard outer crust that protects a softer, less weathered interior. They may persist for years in the ocean or on the seafloor, where they can provide a surface for growing microorganisms or even shelters for invertebrates. They may also serve as efficient sorbents of hydrophobic organic contaminants.

Sinking and sedimentation

Biodegraded oils and some refined products, such as asphalt, may sink in freshwater or seawater when their API gravities are <10.0° or ~6.6°, respectively. Oil residues from burning oil may also be dense enough to sink. More commonly, sedimentation occurs when oil adsorbs on to suspended clay and other minerals in nearshore environments. Oils and tars stranded on beaches can aggregate with sands and sink into nearshore sediments (Bragg et al., 1994).

Oil can be removed from the water column through biological consumption. Planktonic filter feeders and other organisms can ingest or absorb emulsified or dispersed oil. Oil accumulated within these organisms is excreted in fecal pellets or remains within the body upon death, followed by sinking through the water column to the sediments.

Oxidation

Petroleum compounds may oxidize into water-soluble compounds or tars. Most oxidation depends on the initial petroleum composition and slow photochemical reactions controlled by exposure to sunlight. For example, aromatic hydrocarbons are generally more susceptible to photo-oxidation than aliphatics. The degree of oxidation varies, resulting in water-soluble compounds that include phenols, ketones, aldehydes, and carboxylic acids. The soluble products of photo-oxidation tend to be more toxic than the parent crude oil. Oxidation of heavy oils may form a surface tar by means of photolytic reactions that initiate decomposition and polymerization of polar compounds. Rapid oxidation can occur if the spilled oil catches fire, resulting from the accident or as part of an intentional remediation (Figure 10.8). Incomplete combustion results in the formation of pyrogenic PAHs. However, these compounds are sensitive to photo-oxidation, thus limiting transport in the atmosphere and redeposition far from the original spill (Garrett et al., 1998).

Biodegradation

Many organisms, from bacteria to fungi, are capable of degrading petroleum using various enzymatic reactions

Figure 10.8. The *Mega Borg* released 5.1 million gallons of oil as the result of a lightning discharge and fire on June 8, 1990, ~60 nautical miles south-southeast of Galveston, Texas. (From Office of Response and Restoration, National Ocean Service, National Oceanic and Atmospheric Administration, http://response. restoration.noaa.gov.)

based on oxygenases, dehydrogenases, and hydrolases. Ideally, microbial degradation could totally oxidize an oil spill to carbon dioxide (i.e. remineralization). However, microbial activity occurs mainly at the oil–water interface and is controlled strongly by the rate of diffusion of petroleum components across that interface. Consequently, the rate of biodegradation depends on the type of compound and its solubility in water. Aliphatic hydrocarbons are the most readily degraded, while large, polar compounds are the least susceptible. Environmental factors, such as oxygen and nutrient availabilty, temperature, and degree of dispersion or dissolution, also strongly influence degradation rates (Prince, 1998).

Promotion of microbial degradation is one of the most effective and least costly methods to remediate oil spills. Microbes that can degrade oil are ubiquitous. In pristine areas, their proportions may be <1% of the total heterotrophic community, but in heavily contaminated areas, these microbes may account for more than 10% of the biomass (Watanabe, 2001). In most environments, oxygen and inorganic nutrients limit the rate of biodegradation. Aeration and addition of soluble nitrogen, sulfur, phosphorus, and trace metals can greatly accelerate biodegradation (Bragg *et al.*, 1994; Prince and Bragg, 1997; Alexander, 1999). However, attempts to increase marine productivity by adding nutrients have been largely unsuccessful because added nutrients dispersed rapidly.

MITIGATION OF OIL SPILLS

The techniques used to limit or mitigate damage caused by oil spills include physical removal, *in situ* burning, addition of chemical dispersants, and shoreline clean-up. The following brief discussion is included because readers need to be aware that mitigation processes may alter chemical signatures of the original spill. The effectiveness of these techniques depends on many factors, such as the composition and physicochemical properties of the spilled oil, timing of deployment, the local environment, and sensitivity of the impacted ecosystem. Depending on the expected impact of mitigation, it may be best simply to not treat a spill in some cases.

Physical methods involve containing the petroleum by using floating booms, followed by skimming the surface water. Such methods work best in calm water,

and their success depends largely on the oil properties. Chemicals may be added to minimize emulsification. A quick response is essential to limit the amount of oil spilled. For example, oil from a distressed tanker can be offloaded (lightered) before the entire cargo is spilled. About four-fifths of the 1.25 million barrels of oil aboard the grounded *Exxon Valdez* was lightered (Harrison, 1991).

In situ burning of spilled oil may result from the accident (e.g. *Haven* or *Mega Borg*) or may be part of a planned remediation effort. A controlled burn can remove more than 90% of spilled oil, but it leaves weather-resistant asphalt and can produce pyrogenic PAHs that may be more harmful than the petroleum. The efficiency and use of *in situ* burning depend on the amount of water that can mix with the oil, the volatility of the oil, the thickness of the oil layer, and wind conditions.

Dispersant can be sprayed on the water surface and shoreline to help dissipate oil spills. Dispersants are surfactants that form micelles around oil droplets. Their efficiency depends on oil chemistry and wave action.

Absorbent pads can be placed along the shoreline to intercept oil slicks. Once oil has reached the beach, several methods can be used to clean it up. Rock surfaces can be cleaned manually, and sands and soils can be removed for treatment. Surfactants can be used to lift the oil from the solid surfaces so that it can be collected using a skimmer. Bioremediation is probably the most efficient method. Addition of fertilizers containing bioavailable nitrogen or phosphorus can greatly accelerate this process.

No mitigation method results in total removal of all spilled oil, and there are substantial drawbacks for each method. *In situ* burning can produce pyrogenic PAHs that can be transported in the atmosphere far from the spill. Early use of toxic chemical additives may have had greater environmental impact than the oil, but modern chemical additives are considered to be benign. The addition of nutrients and foreign microbes to promote biodegradation alters the local ecosystem and may produce compounds that are more hazardous, but such occurrences have not been documented. Shoreline treatments using hot water may reduce the amount of oil available to contaminate other beaches. However, hot water can affect the flora and fauna, such that treated beaches recover more slowly than untreated beaches (Prince, 1993). Ultimately, the best method to limit oil spills is prevention.

MODELING MARINE OIL SPILLS

Because mitigating activities can cause more harm than good, responders to oil spills must predict rapidly and accurately their effect in different settings. The objective of spill modeling is to facilitate application of the most effective response techniques.

The processes of spreading, evaporation, dispersion, and emulsification are most important during the early stages of an oil spill, and they can be modeled reasonably well. Rapid oxidation from burning or rapid sedimentation of dense asphalt may come into play in the early stages. Photo–oxidation, dissolution, sedimentation, and biodegradation are important in the later stages. The simplest models postulate that the time it takes for an oil spill to dissipate depends mostly on oil volatility (ITOPF, 2001). Oils and refined products can be classified into one of four groups by their specific gravity, with each group having a predefined half-life based on normal marine conditions (Figure 10.9). For rough weather conditions, group III oil may follow the group II curve.

The simple models described above are only approximations and have been replaced by more complex models. For example, Automated Data Inquiry for Oil Spills (ADIOS) 2, which is available for PC and

Macintosh from National Oceanic and Atmospheric Administration (NOAA), can be downloaded at http://response.restoration.noaa.gov/software/adios/adios.html. This program estimates the expected behavior of oil spills by accounting for oil chemistry and water conditions, including temperature, wind and current speed and directions, wave heights, and salinity (Figure 10.10). The program uses a database of >1000 oils and refined products to predict changes in oil viscosity over time and the volume changes of oil and oil–water emulsions. The model accounts for most of the weathering processes, including sedimentation, and can estimate the effects of common clean-up techniques, such as the use of chemical dispersants, skimming, and burning (Figure 10.11).

OIL SPILLS ON LAND

Oil spills on land are generally easier to control than those in water. Unless the spill reaches a river or the water table, it is largely confined to the immediately affected soil. When spilled on land, oil and petroleum products enter the soil by gravity and capillary action according to their viscosity. Spread of the oil is controlled mostly by local topography, hydrodynamics, and the porosity and mineralogy of the soil.

Many processes that alter spilled petroleum at sea also occur on land, but at different rates. While evaporation and biodegradation of spilled oil are important processes on land, photo-oxidation and dissolution reactions are limited because there is no wave-based dispersion. Heavy oils that saturate soils can remain largely unaltered for years by preventing water and nutrient flow into the contaminated interior spaces.

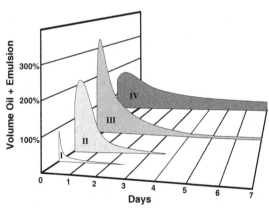

Figure 10.9. Simple model to determine the rate of removal of different oil groups from the sea surface (from ITOPF, 2001). The vertical axis is the volume of oil and oil-in-water emulsion that remains at the sea surface as a percentage of the volume of oil spilled.

I, light condensates and refined products (>45° API); II, light oils (35–45° API); III, normal oils (17–35° API); IV, heavy oils (<17° API).

UNDERGROUND LEAKAGE

Leakage of oil and processed fuels from underground storage tanks (USTs) and buried pipelines is a major potential health risk. About half of the population in the USA (most of the farm and rural communities) relies on aquifers to supply drinking water. There are over a million federally regulated USTs in the USA that contain petroleum or hazardous materials (all service stations have USTs). Many rural homes and farms have privately owned, underground fuel tanks that are not federally registered but may be under state and local control. Unlike catastrophic releases associated with

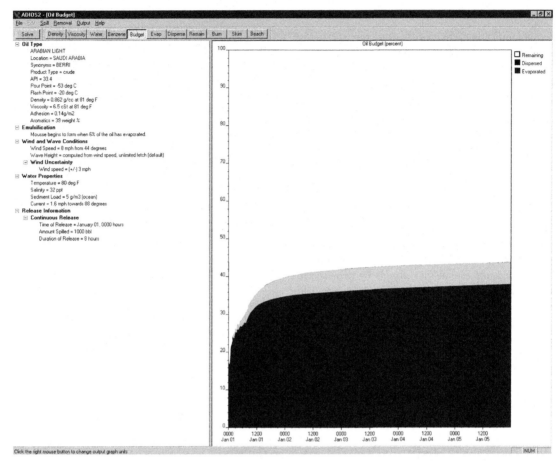

Figure 10.10. Automated Data Inquiry for Oil Spills (ADIOS) 2 results showing input parameters and modeled results for a hypothetical oil spill.

tanker or industrial accidents, most discharges from underground storage tanks and pipelines are chronic and involve slow rates of leakage. Nevertheless, the impact on groundwater quality can be substantial. A leak of 1 ml/min can release over 500 l of petroleum in one year and can contaminate $\sim 5 \times 10^8$ l of groundwater to the point where its odor and taste is no longer acceptable. One of the largest pipeline leaks was discovered under New York City, where it is estimated that more than 17 million gallons of oil was released over a 40-year period dating back to the late 1890s.

The time it takes for underground leakage to enter an aquifer depends on many factors, including soil chemistry, microbial ecology, local and regional geology, hydrology, and the composition of the leaking material. Once petroleum enters an aquifer, it tends to accumulate. Natural cleansing processes that are rapid on the surface, such as evaporation and biodegradation, may be slow in the subsurface. Remediation, such as aeration, filtration, and stimulated bioremediation, is expensive, and alternative sources of potable water, which also are costly, usually must be found before clean-up is completed.

The best practice is to prevent leakage from USTs and pipelines. Facilities are controlled by numerous federal and state regulations that mandate their construction, use, and content (EPA, 1995). The lifetime for a typical UST is $\sim$15–25 years, with the probability of leakage increasing with age.

TOXICITY OF PETROLEUM

Petroleum is a naturally occurring substance that can be assimilated by the environment when present in small

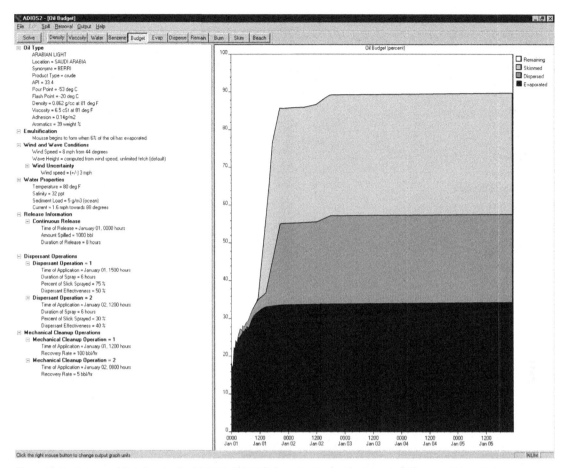

Figure 10.11. Automated Data Inquiry for Oil Spills (ADIOS) 2 results showing the amount of oil removed by *in situ* burning, chemical dispersant, and mechanical skimming. All oil and environmental conditions are the same as in Figure 10.10. These models allow spill responders to optimize remediation using available resources.

quantities. However, large volumes of spilled oil and refined products can harm organisms directly or impact their environment and cause indirect harm. Direct effects may be physical or chemical and can cause acute and/or chronic sickness or death. Physical damage can occur when oil coats feathers or fur, resulting in decreased thermal insulation and buoyancy, and by closing off respiratory passages. Chemical effects may cause acute poisoning, tissue and organ damage, and mutations or cancer. Indirect harm results from habitat destruction, interference of nutrient flow, and disruption of food chains.

Because crude oils are complex mixtures, their toxicity is difficult to assess by conventional methods. While lethal dosages have been determined for a few of the compounds in oil, the interactive effects of these compounds are poorly known. Furthermore, chemical and biological alteration of spills results in the formation of new compounds that may have different toxicities to the unaltered materials. The toxicity of oil is known to vary greatly among organisms. In general, an organism's ability to deal with petroleum is tied to evolutionary adaptations to environmental exposure. Many prokaryotes and algae tolerate or even thrive in areas affected by oil pollution. Among eukaryotic organisms, benthic invertebrates endure well, while marine planktonic species are more sensitive. Higher organisms, such as birds and fish, have much lower tolerance. These animals can only minimize their exposure by avoidance of impacted areas.

Crude oil contains hydrocarbons known to have acute toxicity. The US Environmental Protection

Table 10.3. *Acute toxicity of petroleum hydrocarbons for* Daphnia magna *(from the Environmental Protection Agency (EPA) ECOTOX (ECOTOX icology) AQUIRE (aquatic life) database, www.epa.gov/ecotox)*

Compound	48-hour LC_{50} (mg/l)*	Solubility (mg/l)	Relative toxicity**
Alkanes			
Hexane	3.9	9.5	2.4
Octane	0.37	0.66	1.8
Decane	0.028	0.052	1.9
Cycloalkanes			
Cyclohexane	3.8	55	14.5
Methylcyclohexane	1.5	14	9.3
Monoaromatics			
Benzene***	9.2	1800	195.6
Toluene***	11.5	515	44.8
Ethylbenzene***	2.1	152	72.4
p-Xylene***	8.5	185	21.8
m-Xylene***	9.6	162	16.9
o-Xylene***	3.2	175	54.7
1,2,4-Trimethylbenzene	3.6	57	15.8
1,3,5-Trimethylbenzene	6	97	16.2
Cumene	0.6	50	83.3
1,2,4,5-Tetramethylbenzene	0.47	3.5	7.4
Polyaromatics			
1-Methylnaphthalene	1.4	28	20.0
2-Methylnaphthalene	1.8	32	17.8
Biphenyl	3.1	21	6.8
Phenanthrene	1.2	6.6	5.5
Anthracene	3	5.9	2.0
9-Methylanthracene	0.44	0.88	2.0
Pyrene	1.8	2.8	1.6

*48-hour LC_{50} is the concentration of a compound necessary to cause 50% mortality in laboratory test organisms in 48 hours.
**Relative toxicity of an individual hydrocarbon is the ratio of its solubility divided by the LC_{50}.
***BTEX (benzene, toluene, ethylbenze, and xylene) compounds.

Agency has an extensive database of the concentrations needed for mortality to occur in various organisms. These values are expressed as LC_{50} or LD_{50}, representing the concentration or dose, respectively, needed to kill half of the organisms in a test population within a fixed period of time. However, to act as a toxin, a compound must be bioavailable, which, in most cases, depends on its solubility in water.

Table 10.3 shows the relative toxicity of different hydrocarbons for *Daphnia magna*, a common freshwater zooplankton. Although the light aromatics generally require higher concentrations to cause death than most other hydrocarbons, they are much more soluble and likely to be bioavailable. Light aromatics are abundant in most crude oils, and high concentrations dissolve in the water immediately following a spill. However, these compounds are highly volatile and evaporate rapidly. Compared with the light aromatics, lower concentrations of various polynuclear aromatic hydrocarbons are needed to cause death, but they are relatively insoluble

Table 10.4. *General toxicity of petroleum chemical group types in aquatic environments (modified from Krahn and Stein, 1998)*

Group type	Abundance in oil*	Bioavailability**	Persistence in tissues	Toxicity
Aliphatic hydrocarbons	High	High	Low	Low
Cyclic hydrocarbons (e.g. cycloalkanes, cycloalkenes)	High	High	Low	Low
Aromatics, alkylaromatics, N- and S-heterocycles	High	High	Species-dependent	Species-dependent
Polars (e.g. acids, phenols, thiols, thiophenols)	Low (?)	Low (?)	Low	Low
Elements (e.g. sulfur, vanadium, nickel, iron)	Low	Low	Low	Low
Insoluble components (e.g. asphaltenes, resins, tar)	Low	Low	Low	Low

*Relative abundance of various group types can be highly variable among different oils. The example assumes a typical mature, non-biodegraded crude oil.
**Bioavailability is that portion of the total concentration of a chemical that is potentially available for biological uptake by aquatic organisms.

and not immediately bioavailable. Some PAHs, such as benzo(a)pyrene, are mutagens and/or carcinogens. Because some PAHs can persist in the environment, long-term exposure is a potential environmental risk.

Table 10.4 compares the abundance of various chemical group types that occur in petroleum with their bioavailability, persistence in tissues, and toxicity. Overton *et al.* (1994) provide an extensive review of the toxicity of petroleum for different types of higher organism. Possible routes of exposure are ingestion, inhalation, dermal contact, and bioaccumulation through the food chain. Epifauna, pelagic species, birds, and marine mammals all have different responses to spilled oil.

ENVIRONMENTAL CHEMISTRY FIELD AND LABORATORY PROCEDURES

Program design and decision trees

Environmental chemical studies range from responding to an immediate crisis that threatens human safety to investigating chronic exposure of fauna and long-term ecological effects. In all cases, the design of the investigation is critical to success. Decision trees are used commonly by regulatory agencies and spill responders. Decision trees assure conformity to established

policies and procedures. Provisions and guidelines for field sampling and subsequent analyses of collected samples should be included in the plans for all oil-spill response, clean-up, and remediation programs. When litigation is involved, sample handling should be conducted under prescribed chain-of-custody regulations, so that the analytical results will be admissible as evidence by the courts.

The design of any sampling program should address the specific goals of the study. However, the initial sampling following a major spill is commonly conducted under non-ideal conditions. Sampling may be haphazard and not focused on a specific goal, limited by available resources, and statistically flawed. Rarely have prespill samples of the regionally impacted biota and sediment been collected or analyzed. Improper design of the sampling program may induce bias that leads to false conclusions. For example, sampling in the Arctic only during the summer months may yield a distorted view of oil-spill impact on the year-round ecology. Given an opportunity, the researcher can design a sampling program that is adequate to test hypotheses or answer specific questions.

After defining the goals of a study, the researcher decides what samples are needed to provide valid conclusions. A well-designed plan should have sufficient sampling density to ensure that observed differences between samples are statistically significant. Other issues include the number and type of samples (e.g. oil slicks, intertidal and subtidal sediments, benthic biota, and surface-feeding fish), and their spatial (location, depth) and temporal distribution (e.g. diurnal, tidal, and seasonal effects). Resampling because of an unrecognized issue may be costly or impossible. Sample design also must specify personnel and equipment logistics, methods to capture site information, sample-collection apparatus and containers, and post-collection sample transportation and storage. Prevention of loss of analytes or contamination of samples is essential, and field/method blanks as well as duplicate samples need to be collected. Most oil spills involve litigation. The potentially responsible party needs to have all scientific samples collected and handled under chains of custody, otherwise the data may be inadmissible in a court of law.

Tiered approach for oil-spill identification

The tiered analytical approach is designed to minimize the number of analyses while ensuring the availability of sufficient data to address the problem. The cost and the time required to complete laboratory analyses tend to increase with the level of precision needed to describe contamination at the site. Typical costs range from a few hundred to a few thousand dollars per sample, and the time needed to run a sample through an analytical scheme may be anywhere from a few hours to a week or longer. The cost to collect samples may far exceed that for their analysis. Therefore, the investigator must consider the scope of the investigation, a cost–benefit analysis, and whether the samples and site provide a reasonable opportunity for success. The most effective means to minimize analytical costs is to evaluate data from screening procedures before advancing to more specific techniques.

Wang et al. (1999a) outlined a tiered analytical scheme for oil-spill identification. The initial tier involves comparing gas chromatograms of the spilled oil and suspected sources. Weathering can alter n-alkane distributions and other diagnostic ratios, such that the spilled oil no longer resembles the suspected source. If weathering might account for a mismatch between gas chromatograms, or if the gas chromatograms match, then the samples are analyzed further by gas chromatog-

raphy/mass spectrometry (GCMS) in the second-tier analysis.

If the PAH and biomarker patterns (fingerprints) and diagnostic ratios of the second-tier GCMS analyses match, then the source of the spilled oil can be identified with a high degree of certainty. However, severe weathering and biodegradation can change the distributions of these compounds. If a mismatch cannot be attributed to such alteration, then the suspect sample can be eliminated as a possible source. If, however, the mismatch is attributed to severe weathering, then the suspect sample can be artificially weathered and subjected to a third tier of analyses. A mismatch of source-specific marker compounds and ratios between the spilled oil and the artificially weathered suspect sample eliminates that sample as a possible source. A match strongly implicates a suspect source. The tiered-analysis scheme is particularly useful in order to eliminate suspected sources using minimal analyses.

Benthic sediments may contain background hydrocarbons from multiple sources in addition to spill oil. These sources might include eroded shales and coals, combustion products, residue from natural oils seeps, and anthropogenic pollution. For these types of sediment samples, identification of a spill oil component may require the use of least-square fingerprint matching (Burns et al., 1997; Boehm et al., 2001).

Sampling of oil slicks

Most marine oil spills and natural oil seeps involve the release of small amounts of materials into a relatively large aqueous system. When oil is dispersed on surface water, it forms sheens and slicks, whose thickness can be estimated by color (Taft et al., 1995). If the oil layer is thin (<0.002 mm) and the viewing conditions are good, then oil thickness can be estimated from the presence of interference tints and rainbows. If the oil thickness is >0.002 mm, then the color is usually brown or black, and visual estimation of oil layer thickness is not possible. Table 10.5 is a rough guide to the relation between the appearance, thickness, and volume of floating oil.

Collecting sufficient volumes of oil for laboratory analysis from thin oil sheens and slicks is difficult. Either a large volume of water must be sampled or the oil must be removed selectively. One of the most efficient methods uses a highly porous tetrafluoroethylene (TFE) polymer (Teflon®) net (Figure 10.12). As

Table 10.5. *Visual properties of oil slicks are related to thickness and volume (Taft* et al., 1995)

Oil type	Appearance	Thickness (mm)	Volume (l/km^2)
Oil sheen	Silvery	>0.0001	100
Oil sheen	Iridescent	>0.0003	300
Oil slick	Dull colors	0.001	1000
Oil slick	Brown or black	0.01	10 000
Crude/fuel	Oil black/dark brown	>0.1	100 000
Water/oil emulsion	Brown/orange	>1	1 000 000

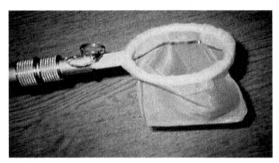

Figure 10.12. TFE-fluorocarbon net used to capture oil sheens and slicks and developed by the US Coast Guard for oil sheen/slick sampling. The netting greatly improves the original design that placed Telfon^{TM} strips on the end of a rake. The net is part of a complete oil sampling kit available from General Oceanics, Inc.

the net is swirled through sheen, the oil adheres to the Teflon^{TM} while water passes through. The oil is easily removed from the netting using solvent or mechanical shaking.

The use of passive sampling devices, such as the semipermeable membrane devices (SPMDs) developed by Huckins *et al.* (1990), is becoming more common for environmental monitoring and research. SPMDs have been used to monitor ambient concentrations of environmental contaminants in air, water, soil, and sediments. The compounds most commonly measured include PAHs, pesticides, PCBs, and dioxins. Shigenaka and Henry (1995) compared PAH accumulations by mussels and SPMD at a heavily oiled and extensively treated site on Smith Island, Prince William Sound, Alaska. While the amounts of PAHs accumulated were similar, the fingerprints differed. Luellen and Shea (2002) investigated the efficacy of a modified SPMD design to sample petroleum biomarkers in water. They exposed the SPMD to a complex hydrocarbon mixture

(Alaska North Slope Crude) for 29 days to measure the rate of uptake and to determine whether the biomarker diagnostic ratios were conserved after uptake into the device. There was little difference in the uptake rate constants or effective sampling rates for the individual di- and triterpanes, even though the number of carbon atoms ranged from 19 to 35. The average sampling rate for hopanes (1.28 l/day) was lower than for steranes (1.65 l/day). Sampling rates were normalized to the standard 400-cm^2 SPMD design of Huckins *et al.* (1990) to allow direct comparison with the standard commercially available SPMD. The commonly used diagnostic ratios were extremely well conserved for both hopanes and steranes between the oil, water, and SPMD, i.e. SPMD does not alter the saturated biomarker fingerprints found in the oil or water. Huckins *et al.* (1990) deployed these passive sampling devices in the Delta National Wildlife Refuge at the mouth of the Mississippi River as part of a study to determine possible sources of hydrocarbon contamination. More field trials are needed to test the effectiveness of these devices to sample biomarkers.

CHEMICAL FINGERPRINTING OF OIL SPILLS

Chemical analyses of oil-contaminated environmental samples are conducted to determine pollutant source and concentration to better assess the environmental impact or to monitor remediation. To achieve these goals, environmental chemists use many of the analytical methods developed by petroleum geochemists as well as those established for non-hydrocarbon pollutants (see reviews by Wang and Fingas (1999) and Wang *et al.* (1999a)). The analyses can be divided into non-specific and source-specific methods. Non-specific methods are used to screen sediments for total petroleum hydrocarbons (TPHs), to assess the extent of site contamination, and to determine the presence, type, and degree of weathering of spilled petroleum and petroleum products. Source-specific methods are used to characterize molecular and isotopic distributions for identifying the origins of oil spills, to correlate unaltered to weathered oils, to differentiate the various sources of hydrocarbons within a contaminated site, and to monitor the efficacy of remediation.

Non-specific methods

Various techniques have been used to describe general chemical characteristics of petroleum pollutants, to determine gross compositional and physical properties, to describe oil loading, and to differentiate contamination from indigenous background organic matter. These non-specific methods usually are designed to be rapid and inexpensive.

Gas chromatography with flame-ionization detection (GC/FID) of solvent-extracted samples is a routine procedure to rapidly characterize the oil in contaminated samples. Non-specific gas chromatography methods may be qualitative or semi-quantitative and can be conducted in the field using portable equipment. In rapid response situations, the GC/FID fingerprint provides a gross hydrocarbon distribution that can be referenced against a catalog of oils and refined products. Identification of the type of contamination (e.g. crude oil, diesel, bunker "c" oil) allows spill responders to quickly identify remediation procedures and potential sources of the spill. GC/FID can provide general information to map the regional, vertical, and temporal extent of contamination, to differentiate petroleum contamination from natural background, and to assess the effectiveness of biodegradation.

Several non-specific methods involve fractionation of oil into chemical group types or by molecular size distributions. Fractionation methods utilize thin-layer, liquid, supercritical fluid, and size-exclusion (gel-permeation) chromatography. Quantitation of the fractions can be obtained by gravimetry, or by ultraviolet (UV), infrared, or refractive index spectrometry. Such methods can be used to determine the relative amounts of hydrocarbons to non-hydrocarbons and the gross composition of the contaminating petroleum in sediment, soil, and water extracts.

Various methods can be used to measure the total petroleum hydrocarbons (TPHs) in environmental samples. Many of these methods are adaptations of Environmental Protection Agency (EPA) and American Society for Testing and Materials (ASTM) procedures that were originally designed for wastewaters, industrial wastes, and volatile organic compounds. All of these methods rely on extraction using a non-interfering solvent followed by non-specific hydrocarbon detection, such as ultraviolet fluorescence. Non-petrogenic organic matter in sample extracts can result in anomalously high TPH concentrations. Such bias can occur at locations with abundant organic matter, including wood-processing sites. GCMS of selected samples can help to determine whether abundant wood-related terpenes might interfere with the interpretation of data. Column chromatography can remove much of the interfering polar compounds.

Concentrations of hydrocarbons can be determined by weighing the hydrocarbon fraction in the total solvent extracts (EPA 9071). This approach was originally designed to measure oil and grease in sludge, but accurate measurements are difficult for small samples. Gravimetric measurements require removal of elemental sulfur from the extract, usually by treatment with activated copper. Furthermore, accurate weights are difficult to measure for small samples.

EPA method 418.1 involves the use of infrared spectroscopy to determine TPH. Absorption by C–H bonds between 3200 and 2700 wave numbers correlates with TPH. The method is inaccurate but was used widely. Originally designed for the analysis of TPH in wastewaters, the method was modified for soil samples by using the extraction procedure in EPA 9071. The procedure originally used Freon 113, which is no longer available. A revised method, EPA 8440, uses perchlorethane as an alternative solvent.

Several TPH methods were developed independent of the EPA. Thin-layer chromatography/flame-ionization detection (TLC/FID) provides an alternative means to determine the amount of petroleum in contaminated samples with reproducibility near 8% (Volkman and Nichols, 1991). However, TLC/FID is not accurate for light oils or condensates. Total polycyclic hydrocarbons in an extract can be estimated using ultraviolet fluorescence spectroscopy (UVF), where UV light excites PAHs, which then emit light at longer wavelengths. UVF is highly sensitive, but it is difficult to calibrate and does not provide much information on the source of the oil. Solvent extracts can be analyzed by gas chromatography using either a flame-ionization detector or a mass spectrometer. TPHs are determined by summing the total detector response for resolved and unresolved eluting compounds. Such methods are similar to EPA 8015, which was designed originally to detect non-halogenated volatile organics. Reproducibility of gas-chromatography-based TPH measurements can be improved by separating the hydrocarbon and polar fractions using column chromatography.

Many environmental projects rely solely on the TPH measurements to determine the extent of oil loading and efficacy of clean-up. However, the accuracy of any of these methods is poor, and there is a high degree of variance among laboratories. Louati *et al.* (2001) compared TPHs and total saturated hydrocarbons using various detection methods. TPH measurements using gravimetric and FTIR detection were correlated closely ($r^2 = 0.87$). Total saturated hydrocarbons using gravimetric, FTIR, and GCMS detection correlated, with r^2 ranging from 0.69 to 0.78, where values for individual samples differed by up to 500%. Given this variance, replicate TPH analyses and statistical evaluation of the significance of the data are mandatory.

Source-specific methods

Specific methods, such as GCMS and CSIA, are designed to differentiate crude oils and refined petroleum products. Methods based on GC/FID or gas chromatography equipped with element-specific detectors may or may not be source-specific. The difference between non-specific and specific methods depends on the problem. For example, if the sources of a polluting fluid can be limited to several refined products, then GC/FID may be sufficiently specific to differentiate the products based on boiling-point cuts. Other source-specific methods include trace-element composition, stable and radiogenic isotopic analyses, high-resolution mass spectroscopy, and chromatographic separations of NSO compounds (e.g. phenols, carbazoles, quinones, and porphyrins).

Many GCMS methods used for environmental studies originated from EPA or ASTM methods. Of the more than 160 priority pollutants targeted by EPA methods, only 20 are hydrocarbons in petroleum. Over half of these 20 hydrocarbons consist of four- and five-ring PAHs that occur as trace components in oil. Furthermore, the EPA methods (e.g. 610, 625, or 8270) examine parent PAHs (e.g. naphthalene, phenanthrene, fluorene, and chrysene) but ignore the alkylated analogs that are relatively abundant in oils. The volatile aromatic hydrocarbons that are priority pollutants (benzene, toluene, ethylbenzene, and xylenes) evaporate readily and are commonly absent in weathered, spilled oils. For these reasons, environmental analysts modified the EPA methods, expanding the list of target compounds to include PAHs more relevant to petroleum. Isomers of alkylated PAHs are usually summed, although they may be reported as individual compounds (e.g. C_1 naphthalenes = 1-methylnaphthalene + 2-methylnaphthalene). The target compound method developed for PAHs is often applied to the saturated biomarkers (Table 10.6).

Quality assurance and control

Because of legal and financial liabilities associated with oil spills, laboratories engaged in chemical fingerprinting need to have strict quality-assurance programs. Non-standardized procedures must be documented to the same degree as EPA and ASTM standardized methods. Quality-control measures include instrument performance specifications, instrument calibration using five-point curves covering a broad range of concentrations, surrogate spiking, procedural blanks, matrix spike recoveries, internal standards, replicate analyses, and standard laboratory procedures (e.g. ASTM standards). Information assigned to each analysis includes: a date/time stamp, designation of the specific instrumentation and method, analyst's name, data-processing procedures, and the sample's chain of custody. Assurance programs mean not necessarily that the analyses are of highest quality but that they were performed within specified limits. Failure to adhere to the quality control/quality assurance (QC/QA) program can lead to fines and, in cases of fraud, criminal prosecution.

Adherence to QC/QA specifications usually is a requirement in all contracts between a service laboratory and its clients. Unintentional errors can arise under the best circumstances, but when both the service laboratory and the client collaborate in vigilantly inspecting the data and procedures, problems can be discovered and corrected. Companies that have

Table 10.6. Target compounds for oil-spill studies

Saturated target compounds	Rings	Target ions	Aromatic target compounds	Rings	Target ions
n-Alkanes C_6–C_{40}	0	57, GC	Benzene*	1	78
Pristane, phytane	0	57, GC	Toluene*	1	91,92
C_{19}–C_{29} tricyclic terpanes	3	191	Ethylbenzene*	1	105
C_{24} tetracyclic terpane	4	191	Xylenes*	1	105
Hopanes (five-ring)	*Carbon no.*		C_3-benzenes	1	105,119
T_s;18α-22,29,30-Trisnorneohopane	27	191	Naphthalene*	2	128
17α,18α,21β(H)-25,28,30-Trisnorhopane	27	191	C_1-napthalenes	2	142
T_m:17α-22,29,30-Trisnorhopane	27	191	C_2-napthalenes	2	156
17α,18α,21β(H)-28,30-Bisnorhopane	28	191	C_3-napthalenes	2	170
17α,21β(H)-30-Norhopane	29	191	C_4-napthalenes	2	184
18α,21β(H)-30-Norneohopane	29	191	Phenanthrene	3	178
17α,21β(H)-Hopane	30	191	C_1-phenanthrenes	3	192
17β,21α(H)-Hopane	30	191	C_2-phenanthrenes	3	206
17α,21β(H)-30-Homohopane 22S	31	191	C_3-phenanthrenes	3	220
17α,21β(H)-30-Homohopane 22R	31	191	C_4-phenanthrenes	3	234
17α,21β(H)-30-Bishomohopane 22S	32	191	Dibenzothiophene*	3	184
17α,21β(H)-30-Bishomohopane 22R	32	191	C_1-dibenzothiophenes	3	198
17α,21β(H)-30-Trishomohopane 22S	33	191	C_2-dibenzothiophenes	3	212
17α,21β(H)-30-Trishomohopane 22R	33	191	C_3-dibenzothiophene	3	226
17α,21β(H)-30-Tetrakishomohopane 22S	34	191	Fluorene*	3	166
17α,21β(H)-30-Tetrakishomohopane 22R	34	191	C_1-fluorene	3	180
17α,21β(H)-30-Pentakishomohopane 22S	35	191	C_2-fluorene	3	194
17α,21β(H)-30-Pentakishomohopane 22R	35	191	C_3-fluorene	3	208
Gammacerane	30	191	Chrysene*	4	228
18α-Oleanane	30	191	C_1-chrysene	4	242
Steranes (four-ring)	*Carbon no.*		C_2-chrysene	4	256
5α,14α,17α(H)-Pregnane	20	217	C_3-chrysene	4	270
5α,14α,17α(H)-Homopregnane	21	217	Biphenyl	2	154
5α,14α,17α(H)-Bishomopregnane	22	217	Acenaphthylene*	3	152

(cont.)

Table 10.6. (cont.)

Saturated target compounds	Rings	Target ions	Aromatic target compounds	Rings	Target ions
13β,17α(H)-Diacholestane 20S	27	217, 259	Acenaphthene*	3	153
13β,17α(H)-Diacholestane 20R	27	217, 259	Anthracene*	3	178
13α,17β(H)-Diacholestane 20S + 20R	27	217, 259	Fluoranthene*	4	202
13β,17α(H)-Methyldiacholestane 20S	28	217, 259	Pyrene*	4	202
13β,17α(H)-Ethyldiacholestane 20S	29	217, 259	Benz[a]anthracene*	4	228
13β,17α(H)-Ethyldiacholestane 20R	29	217, 259	Benz[a]fluoranthene*	5	252
5α,14α,17α(H)-Cholestane 20S	27	217, 218	Benzofluoranthene*	5	252
5α,14β,17β(H)-Cholestane 20R	27	217, 218	Benzo[e]pyrene	5	252
5α,14β,17β(H)-Cholestane 20S	27	217, 218	Benzo[a]pyrene	5	252
5α,14α,17α(H)-Cholestane 20R	27	217, 218	Perylene	5	252
5α,14α,17α(H)-Methylcholestane 20S	28	217, 218	Dibenz[a,h]anthracene	5	278
5α,14β,17β(H)-Methylcholestane (20R)	28	217, 218	Indeno[1,2,3-cd]pyrene	6	276
5α,14β,17β(H)-Methylcholestane (20S)	28	217, 218	Benzo[ghi]perylene	6	276
5α,14α,17α(H)-Methylcholestane (20R)	28	217, 218	*Surrogates*		
5α,14α,17α(H)-Ethylcholestane (20S)	29	217, 218	[^{2}H$_{10}$] acenaphthene	2	164
5α,14β,17β(H)-Ethylcholestane (20R)	29	217, 218	[^{2}H$_{10}$] phenanthrene	3	188
5α,14β,17β(H)-Ethylcholestane (20S)	29	217, 218	[^{2}H$_{12}$] benz[a]anthracene	4	240
5α,14α,17α(H)-Ethylcholestane (20R)	29	217, 218	[^{2}H$_{12}$] perylene	5	264
Surrogates and internal standards			o-Terphenyl	3	GC
5α-Androstane	19	GC, 217	*Internal standards*		
5β-Cholane	24	217	[^{2}H$_{14}$] terphenyl	3	244
17β,21β(H)-Hopane	29	191	[^{2}H$_{10}$] anthracene	3	188
Triaromatic steroids (TAS)	*Target ions*				*Target ions*
C$_{20}$ TAS (pregnane-derived)	231		C$_{28}$ TAS 20S (24S + 24R) (ethylcholestane-derived)		231
C$_{21}$ TAS (homopregnane-derived)	231		C$_{27}$ TAS 20R (24S + 24R)(methylcholestane-derived)		231
C$_{22}$ TAS (bishomopregnane-derived)	231		C$_{28}$ TAS 20R (24S + 24R) (ethylcholestane-derived)		231
C$_{26}$ TAS 20S (Cholestane-derived)	231		C$_{29}$ TAS 20R (24S + 24R) (24-n-propylcholestane-derived)		231
C$_{26}$ TAS 20R + C$_{27}$ TAS 20S	231				

* EPA priority pollutant.
Modified from Wang et al. (1999a).

full internal analytical capabilities still use licensed service laboratories in environmental studies to obtain unbiased, legally presentable data. These companies have an advantage of conducting their own independent checks, whereas companies without their own facilities must either rely on the service laboratory or have multiple contracts to crosscheck results.

ANALYSIS OF BIOMARKERS AND POLYCYCLIC AROMATIC HYDROCARBONS IN OIL-SPILL STUDIES

Biomarkers are used routinely to investigate oil and refined products in the environment and to determine the nature and origin of spills, the impact on the ecosystem of released petroleum, and the effectiveness of spill clean-up and remediation programs. As with petroleum geochemical studies, biomarker interpretations are most reliable when supported by other analyses in the context of models that include site geology and hydrology.

Biomarker parameters in oil-spill characterization

It is not surprising that the compound distributions and ratios in environmental studies are similar to those used by petroleum geochemists. Distributions of n-alkanes and isoprenoids from gas chromatography of samples can be compared with reference oils or refined products. Common ratios derived from gas chromatography include pristane/phytane, pristane/nC_{17}, phytane/nC_{18}, carbon preference index (CPI) (C_{15}–C_{25}) for recent marine input, and CPI (C_{26}–C_{35}) for recent terrigenous input.

The broad and pronounced rise of the baseline on some gas chromatograms represents an unresolved complex mixture (UCM) (Figure 4.22). The UCM consists mainly of linear hydrocarbon chains connected at one or more branch points (Gough and Rowland, 1990) and is strong evidence for petroleum contamination in sediment or water samples. There is no evidence for UCM in recent organic matter, although some UCM may originate by weathering of ancient rocks (Rowland and Maxwell, 1984). Accurate measurement of UCM is difficult, especially when the baseline does not return to zero at the maximum gas chromatography oven temperature. To obtain semi-quantitative UCM concentrations for a saturate fraction, the chromatogram is processed twice: first to obtain the total detector response,

and second to integrate all resolved peaks. UCM is quantified by assuming a response factor of 1.0 based on n-alkanes. The total area of resolved peaks and the total area obtained previously from a stored column compensation (blank) run are subtracted from the total detector response. The remaining area is assigned to UCM and reported either as relative concentration or as UCM/Σ(resolved n-alkanes) or UCM/Σ(resolved peaks).

The aromatic hydrocarbon fraction also has a so-called hump of unresolved compounds. Monoaromatic components in the aromatic UCM can be toxic and may impair the health of marine mussels (Rowland *et al.*, 2001). Aromatic UCM can be quantified in the same manner as the saturated hydrocarbon UCM, but generally yields greater measurement errors, partly due to the common lack of suitable standards.

Environmental geochemists treat saturated and aromatic biomarkers like other target compounds, compiling histogram distributions for visual or statistical comparisons of samples, with particular emphasis on distinguishing recent from thermogenic organic matter. The cyclic biomarkers resist biodegradation, making them ideal compounds for identifying the origin and fate of spill oils. Commonly used parameters derived from saturated and aromatic biomarkers are nearly identical to those used in petroleum geochemistry; they include Ts/Tm, C_{26} tricyclic terpanes/C_{24} tetracyclic terpane, C_{23}/C_{24} tricyclic terpanes, C_{23} and C_{24} tricyclic terpane to hopane, C_{28}/C_{30} hopane, C_{29}/C_{30} hopane, oleanane/C_{30} hopane, gammacerane/C_{30} hopane, C_{35}/C_{30} hopane, normalized C_{27}–C_{29} steranes, C_{27}–C_{35} hopane distributions, and C_{31}–C_{35} S/(S + R) hopanes.

Steranes, triterpanes, and aromatized steroids are highly resistant to biodegradation within the timeframe of most environmental studies. As such, ratios based on these biomarkers are used routinely to identify and correlate oiled beach and soil samples to their source. The approach is valid for fresh spills, but biomarkers will degrade in areas with long-term exposure. Biomarkers in soils contaminated by leaking pipelines are particularly prone to biodegradation.

Conservation of hopane during oil-spill weathering

The extent of biodegradation can be determined for crude oil, chemical groups, and individual compounds using $17\alpha,21\beta$(H)-hopane (hopane) as a conserved internal standard. Early studies used pristane/nC_{17} and

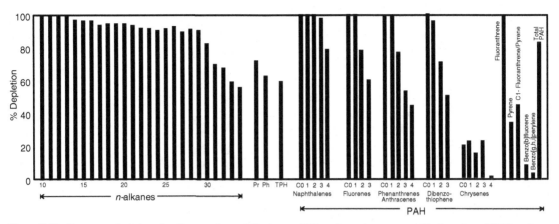

Figure 10.13. Percentage depletion of *n*-alkanes, pristane (Pr), phytane (Ph), total petroleum saturated hydrocarbons (TPH), and polycyclic aromatic hydrocarbons (PAHs) in a shoreline sediment sample collected from Prince William Sound 16 months after the *Exxon Valdez* oil spill. Depletion was calculated assuming that 17α,21β(H)-hopane is conserved and serves as an internal standard. Reprinted with permission from Douglas *et al.* (1996). © Copyright 1996, American Chemical Society.

phytane/nC_{18} as measures of the extent of alteration. Although these ratios are easily obtained from gas chromatography, they are of limited use because pristane and phytane can be removed during advanced stages of biodegradation. Research relating to the 1989 *Exxon Valdez* oil spill showed that 17α,21β(H)-hopane is essentially conserved and can be used as an internal standard (Prince *et al.*, 1994; Douglas *et al.*, 1994). Although there are field indications that hopane can be degraded (Prince *et al.*, 1995), its use as a conserved internal standard over the timeframe and conditions of most environmental studies is justified (Le Dréau *et al.*, 1997; Venosa *et al.*, 1997).

Calculations of the amount of depletion of an individual compound, summed components, or the total oil are simple (Douglas *et al.*, 1996). Assuming that hopane is conserved, then the amount of oil degraded is the ratio of the concentration (measured on an oil-weight basis) of initial hopane (H_o) relative to the amount of hopane in the weathered oil (H_w):

$$\% \text{ oil depleted} = [1 - (H_o/H_w)] \times 100$$

For individual compounds, the ratio of the weathered (C_w) and initial (C_o) concentrations can be normalized to the amount of hopane:

$$\% \text{ compound depleted}$$
$$= [1 - (C_w/C_o)/(H_o/H_w)] \times 100$$

Because hopane is known to degrade under some reservoir conditions (Peters and Moldowan, 1993), the calculations of percentage oil and percentage compound depletions are considered to be minimum estimates (Figure 10.13).

Sasaki *et al.* (1998) found that vanadium is conserved to the same degree as hopane during aerobic biodegradation. By simulating nutrient-enhanced biodegradation of an artificially weathered crude oil, they showed that the concentrations of vanadium and hopane correlated linearly ($R^2 = 0.99$). Recovery in the biodegraded samples, however, was not 100% of the initial measured values. This difference was attributed to decreased extraction efficiency, although biodegradation of the hopane and vanadium ligands could not be ruled out.

Polycyclic aromatic hydrocarbon parameters for oil-spill characterization

PAHs occur in crude oil, coal, and refined petroleum products. PAHs also form during incomplete combustion of fossil fuels, wood, garbage, and other organic substances, such as tobacco and grilled meats, and occur in numerous hazardous chemicals, e.g. creosote. Although they comprise only a few percent of the typical crude oil, PAHs are the most acutely toxic components and are associated with many chronic and carcinogenic effects in animals. PAHs are generally more resistant to biodegradation than many saturated biomarkers and tend to persist in contaminated water and sediments (Alexander, 1999). Because of these characteristics, many environmental studies use PAHs to identify spilled oil, distinguish between sources of contaminants, provide information on the extent of oil weathering and degradation, and evaluate the degree of ecological impact.

Frequently, PAH measurements include the so-called total polycyclic aromatic hydrocarbons (TPAH), which must be defined carefully to be useful because some studies include only the 16 EPA priority pollutants in TPAH, while others include many more compounds. PAHs are hydrocarbons that are composed of two or more fused aromatic rings. However, PAH analyses usually include other compounds that are not strictly PAHs (e.g. biphenyl, C_0–C_3 dibenzothiophenes).

Distributions or "fingerprints" of PAH concentrations are often used to compare petroleum sources. The most abundant PAHs in crude oils are C_0–C_4 naphthalenes, C_0–C_4 phenanthrenes, C_0–C_4 fluorenes, and the C_0–C_4 chrysenes. Alkylated C_1–C_2 species are more abundant than the non-alkylated (C_0) parent compounds and are usually more abundant than the C_{3+} homologs. Ring-number distributions depend on initial oil composition and weathering. Lighter oils are enriched in the naphthalenes, whereas heavy, degraded, or weathered oils are enriched in larger-ring compounds. The concentration of C_0–C_3 dibenzothiophenes in oils depends on the source-rock facies. High-sulfur kerogens from anoxic carbonate-evaporite source rocks yield crude oils rich in C_0–C_3 dibenzothiophenes, whereas low-sulfur kerogens from clastic source rocks yield crude oils with low C_0–C_3 dibenzothiophenes (Hughes, 1984). Dibenzothiophene concentrations may decrease during refining. Other PAHs occur in crude oils, but in lower concentrations. Anthracene, for example, is a trace component in most crude oils, whereas C_0–C_4 phenanthrene is relatively abundant. Other naturally occurring PAHs, such as retene, perylene, and 3,4-benzopyrene, are geochemically significant but are not used widely in environmental studies because of low concentrations or uncertain origins.

Douglas et al. (1996) proposed various PAH ratios as indicators of source and weathering. Oiled beach sands from the Exxon Valdez spill in Prince William Sound that were collected 16 months after the accident exhibited a wide range of weathering. Several PAH ratios varied greatly and were used to indicate the degree of weathering. Other ratios varied only slightly and were considered to be source-specific (Figure 10.14).

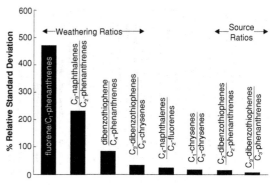

Figure 10.14. Comparison of the relative standard deviations for selected polycyclic aromatic hydrocarbon (PAH) ratios measured for oiled beach samples from Prince William Sound as part of the bioremediation and monitoring program following the *Exxon Valdez* oil spill. Ratios that show large variations can be used to indicate weathering, whereas those that are relatively constant can be used to indicate source. Reprinted with permission from Douglas *et al.* (1996). © Copyright 1996, American Chemical Society.

Source-specific ratios are based on compounds that resist weathering or that weather at the same rate, such that their ratio remains constant until the compounds can no longer be detected. Three-ring PAHs are well suited as source parameters because they are abundant in most crude oils, are relatively resistant to biodegradation, are not prone to evaporation, and have similar solubility in water. Comparisons of the concentrations of C_2-phenanthrenes (C_2-P) to C_2-dibenzothiophenes (C_2-D) and C_3-phenanthrenes (C_3-P) to C_3-dibenzothiophenes (C_3-D), or ratio plots of C_2-D/C_2-P versus C_3-D/C_3-P, are used frequently to differentiate sources with similar chemical composition (Figure 10.15). Laboratory (Douglas *et al.*, 1996) and field studies (Wang *et al.*, 1994) show that these ratios remain constant up to ~70% depletion of the total petroleum hydrocarbons. Severely degraded samples from old oil spills do not have consistent C_2-D/C_2-P and C_3-D/C_3-P ratios. Combustion also alters these ratios, because alkylated dibenzothiophenes decreased faster than alkylated phenanthrenes during a controlled burn (Wang *et al.*, 1999b).

Hostettler *et al.* (1999) proposed the ratio of C_{26}(20R) + C_{27}(20S) triaromatic steroids (co-eluting compounds that are a major peak on m/z 231 traces) to methylchrysene as a means to discriminate background sources of hydrocarbons in the Prince William Sound region. This ratio, termed the refractory PAH index (PAH-RI), is based on the observation that both

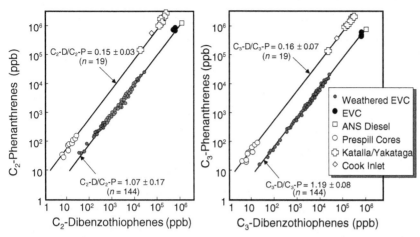

Figure 10.15. Plots of polycyclic aromatic hydrocarbon (PAH) concentrations show source relationships for samples from the area near the *Exxon Valdez* oil spill. *Exxon Valdez* crude (EVC), weathered EVC, and Alaskan North Slope (ANS) diesel plot on the same lines, indicating a genetic relationship. Samples from prespill Prince William Sound cores, the Katalla and Yakataga oil seeps, and Cook Inlet core correlate with each other and are distinct from the EVC samples. Reprinted from Page *et al.* (1995) and Bence *et al.* (1996). © Copyright 1995, 1996, with permission from Elsevier.

alkylchrysenes and triaromatic steroids are highly resistant to weathering. The ratio varies widely among samples that might contribute to background hydrocarbons in sediments from Prince William Sound. However, Bence *et al.* (2000) questioned whether the thermal dependency of PAH-RI (decrease with increased maturity) limits its application as a source-specific parameter in areas where potential sources span a wide range of thermal maturities. Resolution of this disagreement would require a survey of the PAH-RI ratio for all potential point sources for comparison with Prince William Sound sediments.

Weathering PAH parameters are based on compounds that have different rates of evaporation, water solubility, absorption on to clay and mineral surfaces, photo-oxidation, and/or biodegradation. Weathering alters PAH distributions as a function of ring number and degree of alkylation. Naphthalenes decrease faster than other PAHs due to greater evaporation and water solubility. Alkylation imparts a degree of stability, such that the distribution in a given class of PAHs shifts toward the C_{3+}-isomers during weathering. Chrysenes are particularly refractory and increase relative to smaller PAHs.

Several parameters are useful as indicators of the degree of weathering. The C_3-naphthalenes to C_3-phenanthrenes ratio is used for lighter crude oils in the early stages of weathering. For late-stage weathering of heavy crude oils, the C_3-dibenzothiophenes to

C_3-chrysenes ratio (C_3-D to C_3-C) is preferred. Weathering PAH parameters can be combined with source-specific PAH parameters to provide an overview of the spilled crude, oiled sediments, and regional background (Figure 10.16). Other PAH ratios are indicators of

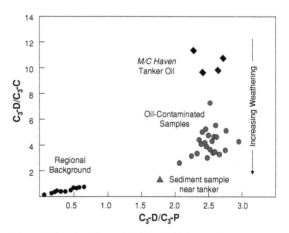

Figure 10.16. C_3-dibenzothiophenes/C_3-phenanthrenes (C_3-D/C_3-P) (polycyclic aromatic hydrocarbon (PAH) source-specific ratio) versus C_3-dibenzothiophenes/ C_3-chrysenes (C_3-D/C_3-C) (PAH weathering ratio) for samples associated with oil spilled from the *M/C Haven*. The incident occurred in 1991 near Genoa, Italy, and involved the release and combustion of heavy Iranian crude oil. Reprinted with permission from Douglas *et al.* (1996). © Copyright 1996, American Chemical Society.

photochemical decay that can occur during aqueous transport. These ratios, such as benzo[e]pyrene to benzo[a]pyrene or benz[a]anthracene to chrysene compare labile PAHs with more resistant species (Maldonado et al., 1999). Using data from controlled weathering experiments, Short and Heintz (1997) constructed a kinetic model for weathering of PAHs to explain data collected following the *Exxon Valdez* oil spill.

Several characteristic compounds and distributions can be used to differentiate pyrogenic from petrogenic PAHs. Pyrogenic PAHs form during the incomplete combustion of fossil fuels, wood, tires, and other organic matter. Pyrogenic PAHs are characterized by a predominance of C_0-parent species over alkylated forms, and by increased concentrations of four- to six-ring PAHs over the two- to three-ring hydrocarbons.

The distinction between biogenic, pyrogenic, and petrogenic origin is difficult because PAHs can form from each of these sources. For example, benzo[a]pyrene can be biosynthesized by some bacteria and plants, formed by incomplete combustion of organic matter (e.g. forest fires), and it occurs in crude oils and refined products (e.g. asphalts, and coal tar pitches). Creosote is a high-temperature (200–300°C) distillate of coal or wood that is used widely as a wood preservative. Creosote contains ~85% PAHs, which may occur as a contaminant in soils and marine sediments (Fowler et al., 1994).

Many ratios have been proposed as indicators of pyrogenic hydrocarbons. These ratios typically relate PAHs that are produced preferentially during combustion (e.g. anthracene, fluoranthene, benz[a]anthracene, benzofluoranthenes, and benzopyrenes) to PAHs that occur preferentially in petroleum (e.g. alkylated PAH). Some pyrogenic ratios used in environmental studies include anthracene/phenanthrene, phenanthrene/methylphenanthrene, fluoranthene/pyrene, benzo[a]anthracene/chrysene, benzo[e]pyrene/benzo[a]pyrene, and indeno[1,2,3-cd]pyrene/benzo[ghi] perylene. A more general approach is to compare the sum of several pyrogenic PAHs with a group of petrogenic PAHs. Maldonado et al. (1999) proposed one such pyrogenic PAH/fossil PAH ratio:

$$\frac{\text{anthracene} + \text{fluoranthene} + \text{benz[a]anthracene} + \text{benzofluoranthenes} + \text{benzopyrenes}}{(C_1 + C_2 \text{ phenanthrenes}) + (C_1 + C_2 \text{ dibenzothiophenes})}$$

Wang et al. (1999b) proposed a more inclusive ratio called the pyrogenic index:

$$\frac{\begin{array}{l}\text{acenaphthylene} + \text{acenaphthene} + \text{anthracene} \\ + \text{fluoranthene} + \text{pyrene} + \text{benz[a]anthracene} \\ + \text{benzofluoranthenes} + \text{benzopyrenes} + \text{perylene} \\ + \text{indeno[1,2,3-}cd\text{]pyrene} + \text{dibenz[a,h]anthracene} \\ + \text{benzo[}ghi\text{]perylene}\end{array}}{\begin{array}{l}(C_0 \text{ to } C_4 \text{ naphthalenes}) + (C_0 \text{ to } C_4 \text{ phenanthrenes}) + \\ (C_0 \text{ to } C_3 \text{ dibenzothiophenes}) \\ + (C_0 \text{ to } C_3 \text{ fluorenes}) + (C_0 \text{ to } C_3 \text{ chrysenes})\end{array}}$$

These ratios are particularly diagnostic when pyrogenic hydrocarbons derived from the combustion of creosote or coal tars are abundant. The quantity and distribution of pyrogenic PAHs depend on starting material, combustion conditions, mode of transport (e.g. aerosols, soot), and mineral adsorption. Consequently, it is often difficult to reconstruct the contribution of pyrogenic PAHs to contaminated sediment. The input of pyrogenic hydrocarbons to soils or sediments is difficult to interpret when these components are mixed with natural or petrogenic PAH.

Note: The discovery of PAHs in the Allen Hills meteorite illustrates some of the ambiguity of assigning a specific source to PAHs. McKay et al. (1996) argued for extraterrestrial PAHs rather than terrestrial contamination (e.g. glacially eroded Antarctic coal) in this meteorite from Mars. More than 90% of the PAHs consist of parent compounds, yet dibenzothiophene is absent, typical of terrestrial pyrogenic PAHs. They argued that the PAHs provided direct evidence for carbon derived from Martian microorganisms. For a more complete discussion of extraterrestrial biomarkers, see Figure 19.22 and the related discussion in Chapter 19.

APPLICATIONS OF BIOMARKERS AND POLYCYCLIC AROMATIC HYDROCARBONS TO OIL-SPILL STUDIES

Petroleum biomarkers to identify the source of spilled oil

In most cases of an accidental oil release, the offender notifies the appropriate government agencies, voluntarily admits culpability, aids in the clean-up, and accepts

Table 10.7. *Diagnostic ratios used to identify the source of oil spilled into Lachine Canal in 1998 (from Wang et al., 2000)*

Diagnostic ratio	March 17 canal spill sample	March 23 canal spill sample	Pumping station diesel	No. 2 Ref. Diesel
Gas chromatography total hydrocarbons (mg/g oil)	861	828	865	841
Total saturates (mg/g oil)	723	654	729	705
Saturates (% of total hydrocarbons)	84	81	84	84
Resolved peaks/total hydrocarbons	0.27	0.27	0.27	0.32
Resolved saturates/total saturates	0.30	0.30	0.31	0.35
Total n-alkanes (mg/g oil)	135	122	133	156
nC_{17}/pristane	2.73	2.70	2.76	5.37
nC_{18}/pristane	1.72	1.72	1.75	2.74
Pristane/phytane	0.87	0.88	0.89	0.76
C_2-dibenzothiophene/C_2-phenanthrene	0.22	0.21	0.22	1.84
C_3-dibenzothiophene/C_3-phenanthrene	0.36	0.34	0.35	1.51
$(2+3)/4$ Methyldibenzothiophenes	0.78	0.78	0.78	0.70
$1/4$ Methyldibenzothiophenes	0.24	0.24	0.24	0.13
$(3+2)/(4+9+1)$ Methylphenanthrenes	1.52	1.51	1.54	1.07
C_{23}-terpane/C_{24}-terpane	2.40	2.45	2.43	1.52

financial liability. Unfortunately, some offenders may not be aware of the spillage or may hope to avoid penalties by remaining silent. When spilled oil of unknown origin is discovered, it is important to identify the source as rapidly as possible. Commonly, investigators identify the offending source by physical evidence or eyewitness accounts. Many instances of marine pollution can be traced to a specific vessel by correlating the timing of the spill with the passage of the vessel. Chemical analyses may be required to identify the source of so-called mystery oils when physical evidence is inconclusive and multiple sources are possible. A positive chemical correlation may be sufficient to convince responsible parties or their insurance carriers to assume liability for the release of oil. Geochemical data are used in most instances of illegal dumping, where prosecution requires that the source of the spilled oil be identified beyond reasonable doubt. The Coast Guard Marine Safety Laboratory (Groton, Connecticut) and the Environmental Technology Center (Ottawa, Ontario) routinely analyze oils spilled in US and Canadian waters, respectively.

Wang *et al.* (2000) identified the source of mystery oil in the Lachine Canal located in Quebec, Canada. On March 17 and March 23, 1998, the oil flowed from a sewer pipe into the canal, possibly from a nearby pumping station. Diesel fuel from the station's reservoir was sampled and compared with the mystery oil. Diagnostic ratios from gas chromatography of the whole oil and GCMS of saturated biomarkers and PAHs proved that the pumping station was the source of the mystery oil (Table 10.7). Furthermore, the mystery oil was unaltered by weathering, indicating that the spill had occurred recently. In this case, diagnostic saturated biomarkers were low in abundance and limited to low-molecular-weight tricyclic terpanes. PAH analysis included isomeric resolution of the methylphenanthrenes and dibenzothiophenes.

Wang *et al.* (2001a) identified the source of oil spilled into the St Lawrence River, Quebec. On February 28, 1999, ~10 tons of mystery oil was reported floating in the river ~50 km northwest of Montreal. The oil was unusual because of its strong, unpleasant odor. Samples were collected from storage tanks at a nearby processing factory that was the likely source. Diagnostic ratios and compound distributions based on gas chromatography of the whole crude and GCMS of saturated biomarkers and PAHs proved that the factory oil was identical to that in the St Lawrence River. This oil,

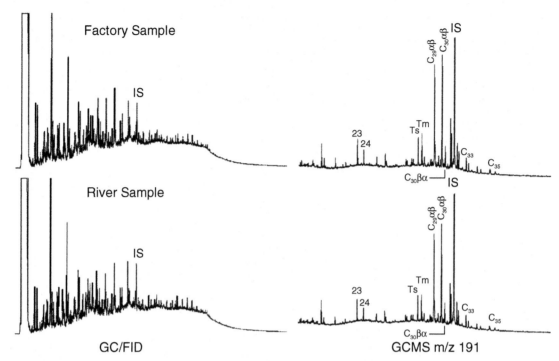

Figure 10.17. Correlation of oil spilled into the St Lawrence River with oil from a nearby factory. This unusual oil contains pyrogenic polycyclic aromatic hydrocarbons (PAHs) far in excess of natural crude oils, possibly because it was produced by recycling waste oils and old tires. Saturated biomarkers in the spill and factory oil samples were nearly identical. Low biomarker abundance and the relatively high proportion of 17β,21α(H)-hopane (C$_{30}$ βα) are unusual for thermally mature crude oil. Reprinted from Wang *et al.* 2001a. © Copyright 2001, with permission from Elsevier. IS, internal standard.

however, was unlike anything encountered previously (Figure 10.17). Gas chromatograms of the whole oil exhibited an unusual UCM or hump and a bimodal distribution of *n*-alkanes. Saturated biomarkers were much less abundant than expected for heavy oil, and abundant 17β,21α(H)-hopane was inconsistent with mature crude oil. The high pyrogenic index for these samples (0.11–0.13) exceeds that for normal crude oils, refined fuels, and highly degraded tars (Figure 10.18). Based on the identification of the source, the owners of the factory agreed to pay for the clean-up. The unusual character of the spilled oil resulted from processing of recycled waste oil and old tires at the factory.

Compound-specific isotope analysis to identify oil spills

Compound-specific isotopic analysis (CSIA) is emerging as an alternative means to identify the source of spilled oil. Mansuy *et al.* (1997) demonstrated that CSIA could be used to correlate oils over a wide range of weathering. The δ^{13}C of individual *n*-alkanes and isoprenoids maintained their initial values. Furthermore, when biodegraded to the extent that the *n*-alkanes were no longer present, the oils could still be correlated using *n*-alkanes liberated by asphaltene pyrolysis. O'Malley *et al.* (1994) used CSIA of PAHs in sediments from St John's Harbor and Conception Bay, Newfoundland, to identify contamination by crankcase oil. Despite some interference by co-elution and the presence of UCM, the PAHs were found to maintain their initial isotopic values after extensive weathering.

Mazeas and Budzinski (2001) described a simple and rapid sample preparation procedure suitable for CSIA of PAHs in petroleum and sediments. Sediment collected off the North Atlantic coast of France that was suspected of being contaminated had δ^{13}C of PAHs identical to the oil spilled by the *Erika*

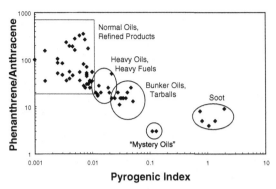

Figure 10.18. "Mystery oil" spilled into the St Lawrence River was produced from recycled waste oils and tires, resulting in an unusually high pyrogenic index (data and figure modified from Wang *et al.* (1999b; 2001a)). The pyrogenic index for light, non-biodegraded oils is <0.01 and can increase to 0.03 in some heavily degraded crude oils and tar balls. Controlled burning of diesel fuel with an initial pyrogenic index of <0.004, yielded pyrogenic indices of 0.009–0.019 for the residual fuel and 0.81–1.94 for the soot. Data and figure modified from Wang *et al.* (1999b; 2001a). © Copyright 1999, American Chemical Society.

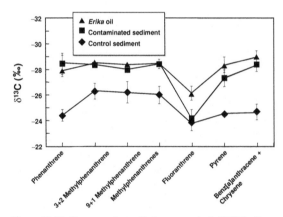

Figure 10.19. Compound-specific isotope analysis (CSIA) of polycyclic aromatic hydrocarbons (PAHs) in sediment from Traict du Croizic, France, North Atlantic coast, indicating contamination by petroleum spilled from the tanker *Erika* (from Mazeas and Budzinski, 2001).

tanker (Figure 10.19). The exception was fluoranthene, which had the same isotopic value in the contaminated and uncontaminated sediments, apparently because fluoranthene is much more abundant in the uncontaminated sediment than in the oil.

The coupling of CSIA with molecular analyses of major hydrocarbons, PAHs, and even trace

components offers environmental researchers another data dimension for oil-spill identification. It is likely that this technique, as well as other advanced molecular measurements, such as of metalloporphyrins, may become routine in oil-spill studies.

Biomarkers and polycyclic aromatic hydrocarbons to identify the impact of spilled oil

In order to assess the ecological impact of a marine oil spill, one must know the full extent and the level of contamination of the impacted area. Shoreline and sediments directly in the path of spilled oil are easily identified as contaminated. However, remote areas, where spilled oil may be transported by water, suspended particles, or biota, may also be affected adversely by the spill. The challenge is to distinguish the spilled oil signature from the background hydrocarbons in the soil or sediments before the spill. Biomarkers, PAHs, and other molecular and isotopic characterizations can be used to identify and quantify the sources of background hydrocarbons.

Before a petroleum spill, uncontaminated sediments always have some level of background hydrocarbons. Because natural background concentrations tend to follow log-normal distributions, even high concentrations of a compound might originate solely from non-anthropogenic sources. Consequently, total petroleum hydrocarbons (TPHs) are not a reliable indicator of oil contamination. To use TPHs, a statistically significant number of background samples must be measured to determine the natural distribution before contamination. These measurements should be done for both surface and core samples to provide both spatial and temporal variations. TPH concentrations that are statistically greater than background indicate contamination but do not identify the source.

There are many possible origins for background hydrocarbons. Biogenic hydrocarbons may be generated by local biota or transported into the area by wind or water. Both natural and anthropogenic sources can contribute petrogenic hydrocarbons. Petroleum seeps or eroded coals and other organic-rich rocks can contribute organic matter to the sediments. Past oil spills and chronic or accidental industrial discharges may contribute to the petrogenic hydrocarbon background. Multiple releases of contaminants are common near ship channels, harbors, petrochemical refineries, and other heavy industry. Pyrogenic hydrocarbons also have natural and anthropogenic sources. Such compounds are produced during the incomplete combustion of organic matter (e.g. forests, grassland, landfills, coal, tires, and fossil fuels). Most background pyrogenic hydrocarbons in marine sediments are produced onshore and are transported through the atmosphere. High concentrations of pyrogenic hydrocarbons may be contributed to marine sediments and soils when spilled oil catches fire. There are numerous other anthropogenic sources of background hydrocarbons. Road run-off may contribute asphalt, motor oil, gasoline, and exhaust. Creosote in treated timbers, which are commonly used to construct marine docks and piers, can leach into the environment. Dumping of used oil, household chemicals, and other wastes into sewers can result in their discharge into the environment. Discerning trace amounts of petrogenic hydrocarbons in sediments containing predominantly biogenic or anthropogenic hydrocarbons can be particularly difficult. Complexity increases when the background hydrocarbons are petrogenic or when the background petrogenic hydrocarbons have the same origin as the spilled oil.

Faure *et al.* (2000) used biomarkers in samples collected from different rivers and watersheds to determine the regional sources of hydrocarbons for river sediments in Alsace-Lorraine. Several minor rivers in the study had few potential contaminant inputs (Rosselle, Fensch, Sarre, and Thur). For example, the Thur River is located in a deep glacially carved valley in the Vosges Mountains and is lined by conifers and beech trees with no adjacent industrial activities. The major rivers and their tributaries (Rhine, Moselle, Meurhe, Ill, and Meuse) can receive contaminants from various industrial activities (e.g. coal extraction and processing, petroleum refining, and polymer manufacturing) and from more diffuse sources.

Biomarker analysis of the river sediment extracts revealed that the background hydrocarbons were nearly identical in almost every sample. The saturated hydrocarbons are dominated by C_{23}–C_{33} *n*-alkanes, with a pronounced odd carbon preference, which are attributed to input from leaf waxes. The saturated biomarkers, however, were clearly petrogenic, having nearly identical distributions of triterpanes and steranes with thermally mature isomeric ratios. The exceptions were samples from the Thur River, which contained only a single peak on the *m/z* 191 ion trace that was identified

as hop(22)29-ene. The widespread diffusion of nearly identical petrogenic hydrocarbons suggests that none of the industrial point sources contributed significantly. The pristine condition of the Thur River suggested that the petrogenic contamination was not atmospheric or carried by wind-blown dust. The only remaining contributors of the hydrocarbon pollution common to all Alsace-Lorraine river systems, except the Thur, were nearby roadways.

Vehicular traffic and roadways can contribute several different types of hydrocarbon pollutant: gasoline, engine exhaust, used motor oil, and road asphalt. Gasoline and engine exhaust can be eliminated as contributing biomarkers to the Thur River because they lack the observed biomarkers. Used motor oil can contain steranes and hopanes, but they are associated with a large hump or UCM, which was not observed in the river sediment extracts. This leaves road asphalt as the most likely possible contributor of petrogenic hydrocarbons. The study showed a strong biomarker correlation between the river sediments and road asphalt (Figure 10.20).

Chemical markers to monitor biodegradation and bioremediation

Bioremediation is the most cost-effective and environmentally acceptable treatment of oiled shorelines in temperate climates (Prince, 1993; Prince,1998; Swannell *et al.*, 1996). Most hydrocarbons in crude oils are readily biodegraded, and oil-degrading bacteria are ubiquitous. However, biodegradation is limited in most marine environments by suboptimal levels of biologically available nitrogen, phosphorus, and other trace nutrients. Non-optimal growth temperatures and limited oxygen availability may further hinder the effectiveness of the clean-up in Arctic waters and soils. The addition of fertilizers to oiled shorelines increases the number of hydrocarbon-degrading microorganisms, their hydrocarbon mineralization potentials, and the rate of hydrocarbon degradation (Lindstrom *et al.*, 1991). However, an important limitation of bioremediation applied to oil-contaminated shorelines is the difficulty in formulating treatment strategies that will produce a specific outcome regarding degradation rate and residual contaminant concentration (Head and Swannell, 1999). Moldowan *et al.* (1995) suggest that it is possible to adapt the current scale for ranking the biodegradation of subsurface crude oils (e.g. see Figure 16.11) to monitor

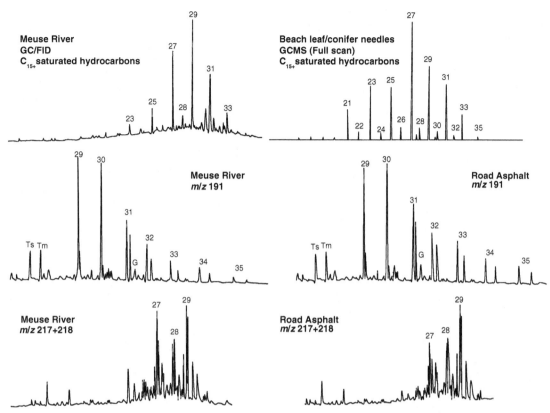

Figure 10.20. Saturated hydrocarbons from Meuse River (Alsace-Lorraine) sediments proved to be mixtures that originated from leaf waxes and petrogenic hydrocarbons from road asphalt. Normal alkanes, hopanes, and $\alpha\beta\beta$-steranes indicated by their carbon number. Reprinted with permission from Faure *et al.* (2000). © Copyright 2000, American Chemical Society.

G, gammacerane; GC/FID, gas chromatography/flame-ionization detection; GCMS, gas chromatography/mass spectrometry.

the bioremediation of soils contaminated with refined petroleum products and wastes.

In fresh oil spills, $17\alpha,21\beta$(H)-hopane is considered to be non-biodegradable and conserved. Consequently, it can be used as an internal standard to monitor the amount of total oil removed by bioremediation (Prince *et al.*, 1994). The technique was used widely to monitor the effectiveness of bioremediation efforts for the *Exxon Valdez* oil spill (Bragg *et al.*, 1994), and it continues to be applied in monitoring the effectiveness of clean-up efforts at other contaminated sites (e.g. Venosa *et al.*, 1996). If hopane is conserved, then bioremediation of other compounds can be assessed. PAHs are particularly important because of their bioavailability and toxicity. Although it is generally accepted that the lower-molecular-weight *n*-alkanes are the most biodegradable hydrocarbons, there is some

evidence that PAHs can be degraded preferentially under exceptional conditions. Jones *et al.* (1983) reported that alkylaromatic hydrocarbons were preferentially degraded over *n*-alkanes in oil-contaminated sediments that were incubated aerobically. Preferential biodegradation of aromatic hydrocarbons was also reported to occur naturally in asphalts in the southern Aquitaine Basin (Connan, 1981), but the *n*-alkanes in the asphalts could have entered the reservoir after biodegradation during a recharge event.

A recent study of the Nipisi pipeline oil spill illustrates the use of biomarkers to monitor biodegradation and remediation (Wang *et al.*, 1998). From 1970 to 1972, ~60 000 barrels of oil spilled from a pipeline in the Nipisi area in northern Alberta, Canada, contaminating over 25 acres (0.1 km^2). Several remediation methods were used, including burning, tilling, and addition of

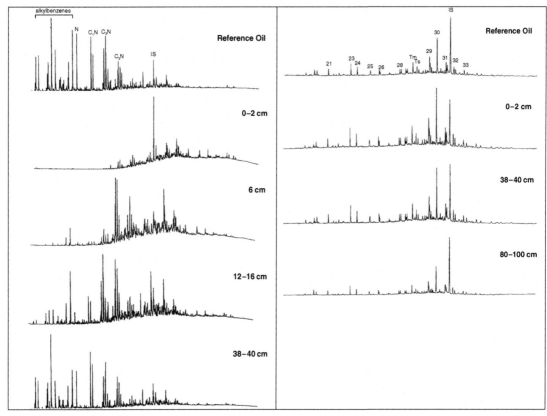

Figure 10.21. Gas chromatography/mass spectrometry (GCMS) analysis of polycyclic aromatic hydrocarbons (PAHs) (summed ions, left) and triterpanes (m/z 191, right) in soils from the Nipisi pipeline oil spill. The chromatograms show near complete removal of alkylbenzenes (not PAHs), naphthalenes (N), and alkylnaphthalenes (C_1N, C_2N, C_3N) in the near surface but no sign of degradation below ~40 cm depth. The saturated biomarkers remain unchanged except for their relative abundance, which increases as other hydrocarbons are degraded. Reprinted with permission from Wang *et al.* (1998). © Copyright 1998, American Chemical Society. IS, internal standard.

fertilizer. Samples for a geochemical assessment were collected in 1995. Although 25 years had passed and there was clear evidence that oil had been removed from the immediate surface, soil deeper than 10 cm still retained large quantities of oil. The degree of *n*-alkane and PAH degradation decreased with sample depth. Furthermore, this distribution was observed regardless of treatment, suggesting that the initial efforts had little long-term effect.

The distribution of the saturated biomarkers in the Nipisi samples remained unchanged regardless of the degree of total hydrocarbon degradation (Figure 10.21). The only difference between the samples collected at the surface and at depth is their relative concentration. This effect was attributed to their conservation and en-

richment as other hydrocarbons were removed. Close inspection of the ion chromatograms reveals that the soil sample from 80 to 100 cm depth is slightly different from the reference oil, which had been stored in a glass jar since 1972. Either the reference oil is slightly weathered or it is not completely characteristic of the oil spilled from the pipeline.

Biomarkers as toxicological indicators

As noted above, toxicologists and medical scientists use the term "biomarker" to denote anything that can be used as a surrogate measure of exposure. Several studies examine the effect of petroleum exposure on organisms, such as fish, by measuring changes in the levels of various enzymes (e.g. Kurelec *et al.*, 1977; Martin and Black, 1996; Cajaraville *et al.*, 1997). Detection of petroleum hydrocarbons in body fluids of exposed animals also suggests petroleum exposure. For example, fish living near natural seepage in the Santa Barbara Channel contained high levels of naphthalene in their bile and showed physiological and enzymatic evidence of exposure (Spies *et al.*, 1996).

PAHs are the most commonly studied petrogenic hydrocarbons in exposed biota. For example, Bence and Burns (1993) report PAH distributions used to assess the exposure of various organisms to the *Exxon Valdez* oil spill. These samples were collected in 1989, 1990, and 1991 by the Oil Spill Health Task Force to assess the spill impact on subsistence foods and by the State and Federal Trustees to evaluate the impact on biota. Except for shellfish, which bioaccumulate oil, only a few biological samples (mainly from 1989) contained evidence of the *Exxon Valdez* pipeline crude oil. Most of these samples were associated with external surfaces (e.g. eggshells, skin, hair) or with the gastrointestinal tract (e.g. stomach contents, intestines). *Exxon Valdez* oil was found rarely in internal tissues (e.g. liver) and never in body fluids (e.g. blood, milk). However, mussels and clams from some of the heavily oiled shorelines initially contained high concentrations of *Exxon Valdez* oil. Those concentrations dropped by approximately an order of magnitude annually and approached background concentrations by 1991, two years after the spill.

Saturated biomarkers are used less frequently as toxicological markers of exposure. Kaplan *et al.* (1996) used petroleum biomarkers in kangaroo rat livers as indicators of exposure to crude oil contamination in soil and seeds. They showed that steranes and hopanes were useful because they are more resistant to metabolic degradation than alkanes and aromatic hydrocarbons. In this case, the steranes and hopanes serve as both petroleum and toxicological biomarkers. Kangaroo rats from uncontaminated areas contained no petroleum biomarkers.

Porte *et al.* (2000) provide an example of the current use of aliphatic and aromatic petroleum biomarkers as toxicological biomarkers. In 1992, the *Aegean Sea* tanker ran aground off the Galicia coast of northwestern Spain and began to leak oil. A series of explosions set the oil on fire, releasing both petroliferous and pyrolytic hydrocarbons into the environment. Porte *et al.*

(2000) showed that petroleum uptake by bivalves along 200 km of affected coastline could be monitored by the presence of the aliphatic unresolved complex mixture, triterpanes, and both aliphatic and aromatic steroidal hydrocarbons. The markers proved useful in order to assess the spatial and temporal distribution of the spilled oil. Presence of the spilled oil (Brent-type, North Sea) was confirmed easily using the triterpane signature, where 28,30-bisnorhopane was particularly diagnostic. Curiously, sterane profiles in bivalves collected three years after the spill exhibited strong depletion of the 20R stereoisomers and enrichment of pregnane derivatives. Such enrichments were not observed in the surrounding sediments. Because bivalves have only a low ability to metabolize hydrocarbons, the enrichment in pregnanes may result from selective transport across cellular membranes. PAHs in the bivalve tissue reflected the presence of both spilled oil and combustion products. Monitoring of biomarkers in the bivalves revealed a significant decline in hydrocarbon content within three to six months after the spill. There were subsequent, incidental increases at some monitoring stations over a three-year period, probably from resuspension of polluted sediments by storms. Porte *et al.* (2000) also showed that aliphatic and aromatic hydrocarbons bioaccumulate at different rates.

BIOMARKERS AND THE *EXXON VALDEZ* OIL SPILL

On March 24, 1989, the *Exxon Valdez* tanker ran aground on Bligh Reef in Prince William Sound, Alaska. About 11 million gallons (258 000 barrels) (Harrison, 1991) of North Slope pipeline crude oil spilled, impacting nearly 800 km of shoreline. The slick drifted in the currents for days, reaching as far as the Alaska Peninsula (Figure 10.22). The *Exxon Valdez* oil spill was the largest accident from a vessel to occur in US waters. (Larger spills occurred along the east coast of the US as a result of torpedo attacks by German submarines in World War II.) In an international comparison, the *Exxon Valdez* oil spill ranks 53 out of 66 major spills from 1960 to 1999 (DeCola, 2000). Assessment of the environmental impact began almost immediately after the spill and continues today. This incident prompted the most comprehensive studies of the fate and effects of spill oil, including its acute and chronic impact on the local biota and ecology and the sources of hydrocarbons

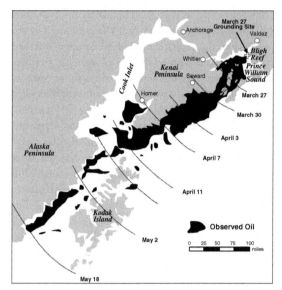

Figure 10.22. Map showing where the *Exxon Valdez* ran aground and the areas where spilled oil was observed (Alaska Department of Environmental Conservation, 1993). The map does not reflect the degree of shore contamination or the areas that were environmentally impacted.

to the sediments of Prince William Sound and the Gulf of Alaska (e.g. Wells *et al.*, 1995; Rice *et al.*, 1996).

Many oil spills occur in areas that are already heavily polluted from past events. The *Exxon Valdez* release appeared to provide an opportunity to study the effect of an oil spill on biota and the environment in a pristine setting removed from industrial activity. Short and Babcock (1996) argued that baseline studies showed that prespill mussels and intertidal sediments from Prince William Sound were free of hydrocarbon pollution. Kvenvolden *et al.* (1993a; 1993b; 1995) provided early evidence that Prince William Sound shorelines were not pristine before the *Exxon Valdez* oil spill. While the stable carbon isotope ratios of some oil residues from the impacted shoreline could be correlated directly to North Slope oil, other tars and tar balls clearly were from a different source. The stable carbon isotopes of the tar balls were all closely grouped ($\delta^{13}C = -23.7 \pm 0.2‰$) and distinct from the *Exxon Valdez* oil ($\delta^{13}C = -29.4 \pm 0.1‰$). The ^{13}C-rich isotopic signature of the tar balls suggested that their parent oil was Miocene or younger. Correlation with saturated biomarkers proved that these tars were from crude oil generated from the Monterey Formation of California (Figure 10.23). For example,

the tars contained 28,30-bisnorhopane, which is absent in North Slope oils but abundant in many Monterey oils (Jeffrey *et al.*, 1991). Wooley (2001) used archaeological samples collected from throughout the Prince William Sound area to conclude that hydrocarbons from numerous historic industrial sites were already present before the spill.

Kvenvolden *et al.* (1995) suggested that most of the Monterey tars were probably released in a single event that occurred almost exactly 25 years before the tanker spill. On March 27, 1964, the Great Alaska Earthquake caused asphalt and fuel oil to spill from coastal storage tanks near the old town of Valdez (Figure 10.24). Monterey crude oil was used routinely for asphalt before 1970 in southern Alaska. The isotopic ratios and biomarker distributions of these historic asphalts correlated to Monterey crude oil but were not as similar to each other as those in Prince William Sound. The high degree of similarity of the Monterey-derived tars

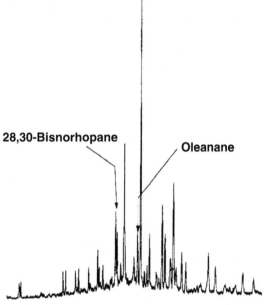

Figure 10.23. Ion *m/z* 191 trace shows the terpane distribution for a tar ball from a beach in Prince William Sound. The presence of 28,30-bisnorhopane and oleanane proves that the tar is not North Slope oil spilled by the *Exxon Valdez*. The biomarker distribution and the stable carbon isotopic ratios indicate that the oil originated from Monterey Formation source rock in California. Reprinted from Bence *et al.* (1996). © Copyright 1996, with permission from Elsevier.

Figure 10.24. Asphalt from California remains at the abandoned site of an asphalt storage plant destroyed in 1964 near Valdez, Alaska. Photo from USGS.

in the sound suggests that they originated from one event.

Shoreline tars unrelated to the *Exxon Valdez* oil spill are common in Prince William Sound. Other possible sources of contamination from industrial sites include Port Ashton, Thumb Bay, and the Latouche mine (Figure 10.25). These sites, mostly old herring reduction plants that operated in the 1920s, have abandoned and leaking oil tanks and barrels containing low-sulfur Monterey oil that was used to fire the boil-

ers (Page *et al.*, 1999a). Shoreline tars were found as far away as the Kodiak Archipelago and the Alaska Peninsula. Relatively fresh, waxy tars occur on Kodiak and nearby islands (Figure 10.22). These tars contain saturated biomarkers, including 18α(H)-oleanane, bicadinane, and taraxastane, that are characteristic of Tertiary Southeast Asian crude oils and are probably related to small spills of imported oil from ship traffic (Bence *et al.*, 1996).

Although the *Exxon Valdez* spill had significant short-term impacts on shorelines, marine wildlife, and the commercial fishing industry in 1989, most of the oil vanished from the shoreline within a year. This removal is attributed partly to clean-up efforts (11 000 personnel at work during peak clean-up) but mostly to wave action during winter storms and other processes, such as biodegradation. By 1993, there was little evidence on the shorelines that the spill had occurred. Shoreline oil residues occurred largely in the subsurface and in isolated locations protected from the wave action of winter storms. The rapid removal of the spill oil from the shoreline raised concerns that the oil was not degraded but had been transported into deeper water, where it could be a long-term risk to the health of the ecosystem. Exxon and oil-spill trustee scientists conducted independent studies of bottom sediments in the

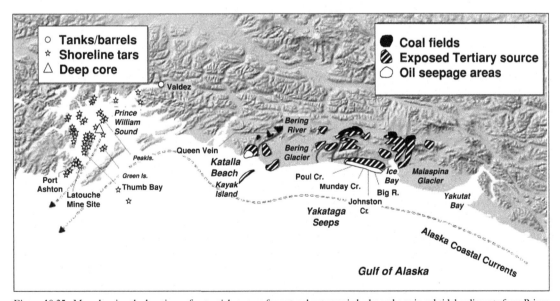

Figure 10.25. Map showing the locations of potential sources for natural petrogenic hydrocarbons in subtidal sediments from Prince William Sound and the Gulf of Alaska (from Bence *et al.* (1996) and Boehm *et al.* (2001)). The deep core is described in the text and in Figure 10.26. Reprinted with permission from Boehm *et al.* (2001). © Copyright 2001, American Chemical Society.

region to evaluate this possibility. The trustee scientists included representatives of the US Departments of Interior and Agriculture, the US Forest Service, the Alaska Region National Marine Fisheries Service, the National Oceanic and Atmospheric Administration (NOAA), and the State of Alaska Departments of Law, Environmental Conservation, and Fish and Game.

Examination of subtidal sediments indicated that there is a substantial regional background of hydrocarbons in Prince William Sound and in adjacent areas (Page *et al.*, 1995; 1996a). Age-dated sediment cores indicate that this regional background existed for at least 160 years and likely much longer. The background is mostly petrogenic, with minor pyrogenic and biogenic-diagenetic hydrocarbons. Pyrogenic PAHs arise from past human activities involving combustion of wood and fossil fuels, creosote leaching and forest fires (Page *et al.*, 1999a). Pyrogenic PAHs in typical subtidal sediments, including fluoranthene and pyrene, account for only a few percent of the total PAHs. In a few locations associated closely with human activities, pyrogenic PAHs exceed 90% of total PAHs. Biogenic hydrocarbons are mostly plant waxes characterized by C_{25}–C_{33} *n*-alkanes with a strong odd-carbon preference and perylene. Using a compositional mixing model, Page *et al.* (1995; 1996a) argued that petrogenic PAHs in the older subtidal and deep offshore sediments came from pre-existing natural sources. Nearshore subtidal surface sediments along coastlines heavily oiled in 1989 exhibited mixed patterns consisting of a minor component of *Exxon Valdez* spill oil superimposed on the natural background. The major distinguishing trait of the contaminated surficial sediments is elevated dibenzothiophenes, which are abundant in the *Exxon Valdez* spill oil.

Possible sources for the natural petrogenic hydrocarbon background in Prince William Sound include oils from seepages, eroded coals, and eroded shales. Modern allochthonous sources, such as the Monterey tars and *Exxon Valdez* spill oil, can be excluded as contributors to the petrogenic background in sediment samples that predate industrial activity in the area. Page *et al.* (1995; 1996b) identified oil seeps and petroleum source rocks as the major natural petrogenic sources. There are no known natural oil seepages in Prince William Sound, but numerous seeps occur to the east (Martin, 1908), where oil from these seeps could be picked up and transported by Alaska coastal currents (Figure 10.25). Page *et al.* (1995; 1996b) proposed that

the Katalla oil seep was a likely source based mainly on saturated and aromatic biomarker analyses. Oleanane, a saturated petroleum biomarker found in Prince William Sound prespill sediments, occurs in the Katalla seep oil but not in *Exxon Valdez* oil (Figure 10.26). Source-specific PAH ratios also implied a correlation between the prespill hydrocarbons and the Katalla seep (Figure 10.15). Seeps around Cape Yakataga, east of Katalla, were later identified as additional sources (Bence *et al.*, 1996).

Coal versus seep hypotheses for background petrogenic hydrocarbons

Short and Heintz (1997) question the interpretations of Page *et al.* (1996a) and Bence *et al.* (1996) regarding the source of background petrogenic hydrocarbons in Prince William Sound. They used first-order kinetics to model PAH weathering in sediments and mussels directly in the spill path of the *Exxon Valdez* oil. Surprisingly, prespill mussels and mussels taken from depth along the spill path lacked petrogenic PAHs. Short and Heintz (1997) concluded that petrogenic PAHs in the sediments are not available to biota and suggested that coals transported by the Bering River were the source of the PAHs. The coal-source hypothesis is mentioned only briefly in this publication without supporting data. This publication, however, initiated a series of papers that supported or discounted a coal source. Short and Heintz were later joined by others who argued that coals were the source of petrogenic PAHs in Prince William Sound (Short and Heintz, 1998; Short *et al.*, 1999; Short *et al.*, 2000; Hostettler *et al.*, 1999; Kvenvolden *et al.*, 2000). A series of papers followed that reject coals as the source while defending the position that seeps and/or source rock were the source of the petrogenic background hydrocarbons (Page *et al.*, 1997; Page *et al.*, 1998; Page *et al.*, 1999b; Boehm *et al.*, 1997; Boehm *et al.*, 1998; Boehm *et al.*, 2000; Boehm *et al.*, 2001; Bence *et al.*, 2000).

The coal versus seep hypotheses may affect determination of the relative impact of the *Exxon Valdez* spill. If the background petrogenic hydrocarbons in Prince William Sound are from oil seeps, then exposure to petroleum occurred over geological time and the biota have long since adapted to the presence of PAHs. PAHs from the *Exxon Valdez* oil spill must then be considered as only a partial contributor to the overall long-term

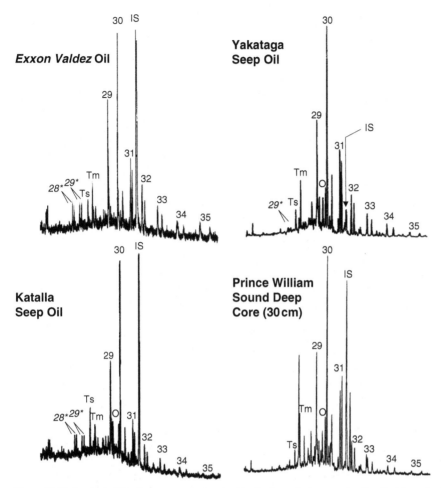

Figure 10.26. Ion (*m/z* 191) chromatograms showing distributions of triterpanes in oil spilled from the *Exxon Valdez*, oil seeps from Katalla and Yakataga, and a sediment core from Prince William Sound (from Bence *et al.*, 1996) (see also Figure 10.25). Sediments in the core predate industrial activity, eliminating spilled North Slope oil or Monterey asphalt as sources of the extracted petrogenic hydrocarbons. The presence of oleanane (O) and the correlation with source-specific polycyclic aromatic hydrocarbon (PAH) parameters indicate that the two seeps were likely sources for background hydrocarbons in the sediment core. Reprinted from Bence *et al.* (1996). © Copyright 1996, with permission from Elsevier.
IS, internal standard; *italics** = tricyclic terpanes; normal font = Ts, Tm, oleanane, and 17α-hopanes.

health risk from exposure to PAHs. On the other hand, PAHs from coals are considered to be unavailable to biota (Neff, 1979). If the PAH background petrogenic hydrocarbons are from eroded coals, then the PAHs may not be available to biota. In this scenario, *Exxon Valdez* oil and Monterey tars would have contributed bioavailable PAHs to which local biota were not previously exposed.

Arguments for and against various sources evolved as additional samples and data were analyzed. The following claims were made in support of the coal-source

hypothesis (Short and Heintz, 1998; Short *et al.*, 1999; Short *et al.*, 2000; Hostettler *et al.*, 1999; Kvenvolden *et al.*, 2000):

1. Prespill Prince William Sound mussels exhibit no evidence for bioaccumulation of PAHs from past exposure.
2. Diagnostic PAH (C_2-dibenzothiophene/C_2-phenanthrene = DPI) and biomarker (oleanane/hopane) ratios are consistent for either the Katalla

oil seep or coal samples. Mismatches in PAH and biomarker ratios between the measured coal samples and benthic sediments are minor and due to sampling inadequacies.

3. Sterane and triaromatic steroid distributions of the oil seeps at Katalla and Yakataga are unlike those from Gulf of Alaska and Prince William Sound benthic sediments.

4. The (PAH-RI) (C_{26} + C_{27} triaromatic steroid/ methylchrysene) for Prince William Sound sediment (<0.1–0.5) is similar to that for Bering River coals (<0.1) but unlike the Katalla oil seep (11–13).

5. The PAH pattern for the coal samples resembles the pattern in Prince William Sound sediment more than that in weathered Katalla seep oil.

6. The geological setting is consistent with erosion and transport of coals to Prince William Sound. Active glaciers transport coal eroded from the Kulthieth Formation to the northern Gulf of Alaska, as evidenced by the black particulate strand line on the beach at Katalla.

7. The observed distribution of PAHs is consistent with coal influx into the Gulf of Alaksa.

8. In contrast to coals, it is not clear how seep oil might be transported to Prince William Sound without being weathered. There is no evidence of oil seep association with fine-grained particles either at the site of the seeps or elsewhere.

9. Oils slicks are rarely seen at Katalla and Yakataga, indicating minor leakage.

10. Transport of sediment from the Yakataga area is disrupted by intervening landforms before they could enter Prince William Sound.

11. Coal particles occur in Gulf of Alaska and Prince William Sound sediments.

12. Mass-balance calculations equating TOC to PAH or biomarker influx are influenced by particle-size effects of the transported coals on PAH extraction efficiencies.

The following claims were made in support of the seep hypothesis (Page *et al.*, 1997; Page *et al.*, 1998; Page *et al.*, 1999a; Boehm *et al.*, 1997; Boehm *et al.*, 1998; Boehm *et al.*, 2000; Boehm *et al.*, 2001; Bence *et al.*, 2000).

1. Diagnostic PAH (DPI) and biomarker (oleanane/ hopane) ratios for contaminated sediments are consistent with an origin from the Katalla or Yakataga oil seeps and eroded Tertiary rocks, but not from coal samples.

2. The sterane and triaromatic steroid distributions of the Gulf of Alaska and Prince William Sound sediments are consistent with mixtures of Yakataga oil seeps and eroded Tertiary source rocks, with minor contributions from coal (generally <1%).

3. PAH-RI for the Yakataga oil seeps are consistent with their contributing petrogenic PAHs to the sediments.

4. PAH patterns and abundances are consistent with oil seep and eroded source-rock sources.

5. Comparison of seep oil and sediments shows that PAHs are associated and transported with fine-grained sediments, accounting for the lack of obvious oil slicks.

6. Mass balance places an upper limit on the amount of PAHs that coals can contribute (<1%). If coal particles (average TOC ~80%) accounted for all carbon in the sediments of Prince William Sound (average TOC ~0.6%), then they would contribute <15% of the observed PAHs (~1100 ng/g) and <1% of selected biomarkers. Sources with much higher PAH + biomarker/TOC ratios, such as seeps, must contribute most of these compounds to the sediments

7. High-maturity coals along the Bering River and the Queen Vein coal (>1.5% R_o) cannot have supplied the biomarkers or thermally sensitive PAHs in the sediments. However, Tertiary Katalla and Yakataga Formation shales are in the oil window (0.4–1.3% R_o).

The above arguments highlight the complexity of claims that biomarker and PAH distributions may or may not verify coals versus seeps and organic shales as sources of background hydrocarbons in Prince William Sound. PAH distributions are particularly problematic because they can be altered severely by weathering. Correlations based on PAHs rest on the degree of weathering and the interpretive model for how weathering alters composition. Additional confusion can be traced to inadequate sampling. Sediments from Prince William Sound were sampled extensively, particularly from areas impacted by the *Exxon Valdez* oil spill. However, few initial samples were used to formulate either the oil seep or the coal source hypotheses. Data for Tertiary source rocks, although mentioned by Page *et al.* (1995), were not reported until their

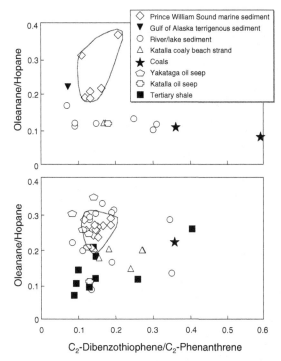

Figure 10.27. Comparison of dibenzothiophene phenanthrene index (DPI) versus oleanane ratios from Short *et al.* (1999) (top) and Boehm *et al.* (2000) (bottom). The dotted lines enclose most background values for Prince William Sound and the Gulf of Alaska. The bottom portion of the figure contains one additional Tertiary shale sample from Page *et al.* (1999b) that plots within the background area. The top figure shows that the coal samples do not match the background, while the bottom figure indicates that the seep oils and some eroded Tertiary shale samples are a better match to the background than the coal. Reprinted with permission from Boehm *et al.* (2000). © Copyright 2000, American Chemical Society.

paper in 2001. With limited data, mismatches between specific biomarker ratios and distributions can always be attributed to sampling inadequacy. The available samples are not really representative, but they are close enough to argue a given point of view. For example, Figure 10.27 plots the DPI and oleanane ratios for comparable samples from two different sources. Although there are some differences due to analytical precision, the general trends for the same samples remain relatively constant. Interpretations of the significance of the sample variance and the resulting conclusions, however, differ radically. The controversy extends to the character of collected samples. For example, Bence *et al.* (2000) assert

that while the black strand line at Katalla Beach is a mixture of kerogen, bitumen, rock fragments, coals, and natural coke, mica accounts for most of the black coloration and TOC is <12 wt.%. Hostettler *et al.* (2000) refute this claim, asserting that the Katalla Beach stranding consists of organic-rich fragments (74–82 wt.% TOC) with the optical properties of coal. These different observations may be due to differences in sampling technique and/or natural heterogeneity.

Boehm *et al.* (2001) provide data on sources of hydrocarbons that might enter Prince William Sound with the Alaska coastal current. They sampled sediments along the Gulf of Alaska between Prince William Sound and potential sources of background hydrocarbons, including stream sediments from the Copper to Malaspina Glacier, seep oils and Tertiary shales from the Katalla and Yakataga areas, and Bering River coals. They used biomarker and PAH analyses and a least-squares mixing model (Burns *et al.*, 1997) to apportion source inputs and the changing nature of the background hydrocarbons in these marine sediments (Figure 10.28). According to their interpretation, eroded shales were the major contributor of hydrocarbons to Prince William Sound. Seep oils and sediments from Copper River were minor contributors, while coal input is less than 1% of the contributed hydrocarbons. Sediments near oil seeps show high input from the seeps (~20–80%). Bottom sediments near the Bering and Duktoth Rivers have elevated coal input (>5%).

Mudge (2002) conducted an independent multivariate statistical analysis using data provided by both Page (PAHs and biomarkers) and Short (PAHs only). He found that the variability in the background sediments from Prince William Sound and the Gulf of Alaska could be attributed to mixed inputs of coal, seep oil, eroding shales, and rivers, whose relative contributions vary significantly across the sampling area. The seeps were best defined by naphthalene and methyl- and dimethylnaphthalene, whereas coals and shales are best defined by the larger PAHs, such as benzo[*ghi*]perylene.

One of the major unsolved issues is the significance of background hydrocarbons relative to those of the Exxon Valdez oil spill (Figure 10.28). Transport of oil adsorbed on fine-grained minerals or the organic matter associated with eroded source rocks (bitumen and kerogen) may limit bioavailabilty to a similar extent to PAHs in coals. If adsorption limits bioavailability, then adaptation of the biota in Prince William

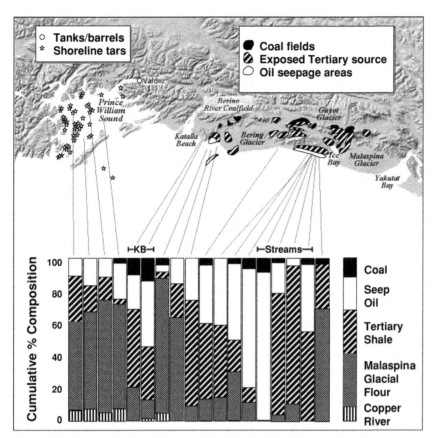

Figure 10.28. Sources of background petrogenic hydrocarbons in offshore, beach, and stream sediments in Prince William Sound. The study indicates that most background hydrocarbons in sediments came from eroded Tertiary shale and Malaspina glacial flour. Minor hydrocarbons came from oil seeps and Copper River sediments, with coals contributing <1%. The bioavailability of polycyclic aromatic hydrocarbons (PAHs) in eroded shales and oil absorbed on fine-grained particles is unknown. Reprinted with permission from Boehm *et al.* (2001). Copyright 2001, American Chemical Society.

Sound to hydrocarbons before the spill may not have occurred. Boehm *et al.* (2001) argue that PAHs associated with clay-sized minerals can be consumed and solubilized by organisms. For example, deposit feeders solubilize sediment-bound PAHs in their digestive fluids, thereby facilitating entry of the PAHs into the food chain (Voparil and Mayer, 2000). Furthermore, PAH-metabolizing enzymes in fish may be activated, even when background hydrocarbons are tightly bound on minerals (Arthur D. Little, Inc., 1999). It is not known whether organisms can extract PAHs from consumed coal particles. Whether the background hydrocarbons are bioavailable may be a moot question. Boehm *et al.* (1998) showed that the contributions of PAHs from the spill are now lower than the background PAHs in most

sediment samples, including those from areas that were impacted heavily. All sediments contained <4000 ng TPAH per gram of sediment, the lower limit for sediment toxicity effects. These studies suggest that the environment in Prince William Sound has effectively recovered from the *Exxon Valdez* spill.

GASOLINE AND OTHER LIGHT FUELS AS POLLUTANTS

The accidental release of gasoline, kerosene, and other light, refined fuels may involve large, discrete events, but it is more likely to occur as chronic, slow leakage from storage tanks and pipelines. The EPA (http://www.epa.

gov/ada/csmos/models.html) lists over 400 000 leakages from underground storage tanks, most of which have been cleaned up or are undergoing remediation. Most hydrocarbons evaporate from spilled gasoline and other light fuels. However, when the soil becomes saturated, plumes of released fuel can migrate vertically and laterally. If refined fuels reach the water table, then soluble light aromatics have the potential to contaminate local water supplies.

Forensic environmental chemistry identifies contaminants, such as spilled gasoline and light fuels, using chemical analyses (Kaplan, 1989; Murphy and Morrison, 2002). Characterization of light, refined petroleum products involves many of the techniques described in this chapter (e.g. gas chromatography and GCMS of product, soil, and water samples), but biomarkers and most PAHs, except naphthalenes, are low or absent. Distributions of hydrocarbons are not particularly diagnostic because there is little variation between commercial brands of refined products, and water washing, evaporation, and biodegradation readily alter them. As discussed below, non-petroleum additives, including dyes, organometallics, halogenated hydrocarbons, detergents, and oxygenates, are commonly used to identify origins of contaminants and as markers for plume delineation and remediation monitoring. Kaplan et al. (1997) reviewed many of these techniques.

Dyes added by refiners to brand their products can be used to distinguish ownership and product grade. For example, different dyes are commonly added to premium and regular gasolines. Dyes are easily separated and characterized by thin-layer chromatography and UV–visible absorption spectrometry. However, dyes are relatively unstable and do not persist in the environment, so their use is limited to fresh spills.

Metals may exist in gasoline as organometallic additives that improve performance or in lubricating oils as a result of engine wear. Various formulations of tetraethyl- and tetramethyl-lead additives used to be added to gasoline since 1923, until government regulations phased out leaded gasoline from ~1985 to 1995. Manganese compounds (e.g. 2-methyl cyclopentadienyl manganese tricarbonyl, MMT) also were added to prevent engine knock. Other metal compounds containing phosphorus, boron, nickel, zinc, and barium have been added to control various properties of refined fuels.

Halogenated organics were added to leaded gasoline to prevent deposits of lead oxides in the combustion chamber. Ethylene dichloride and ethylene dibromide were used in automotive fuels, and ethylene dibromide was used in aviation fuels. Organo-lead compounds are absorbed strongly by soil and hydrolyzed by water. Halogenated organic additives may be the only organic residual indication of contamination of soils by leaded gasoline.

Oxygenates, such as ethanol and methyl-tert-butyl ether (MTBE), may be added to gasoline to improve octane ratings and reduce emissions. Oxygenates were used in the early 1980s and became major additives in reformulated and premium-grade gasoline with the phasing out of lead additives in the late 1980s. However, oxygenates that were originally touted as the environmentally friendly alternative to lead have become a new water-quality issue. Oxygenates are more soluble in water than most other gasoline components and tend to move freely in flowing groundwater. Partitioning of MTBE and other volatile oxygenates into groundwater can be used to evaluate the timing, extent, and dispersion rates of lead-free gasolines.

Detergents, gum inhibitors, and cetane enhancers are added to gasolines and other fuels in trace amounts. Many of these compounds are readily biodegraded, limiting their use as markers for contamination.

Age-dating fuel spills

Knowledge of the timing of a fuel spill is essential to assess liability and to develop a remediation program. This is usually not an issue in acute point releases, but it can be problematic for chronic subsurface leakages. Gasolines are difficult to age-date because they contain abundant volatile and water-soluble compounds that are lost readily. Age-dating techniques for gasoline rely on measuring the degree of weathering of the more volatile components, changes in the availability of alkylated lead and other additives, lead isotopes, and changes in refinery formulations.

In many cases, the approximate age of manufacture of the fuel in a spill can be estimated to within about one decade by using historical information on changes in fuel formulations and refining practices (Figure 10.29). These changes in composition were necessary to maintain octane ratings during the mandated removal of lead from the gasoline to meet increasingly stringent air-quality regulations. For example, regular and mid-grade gasolines were formulated with progressively greater

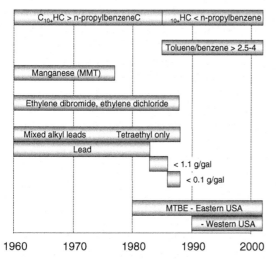

Figure 10.29. Chronology of US gasoline formulations can be used to approximate the age of certain spilled gasolines with resolution of ~5–10 years. Reprinted with permission from Hurst *et al.* (1996). © Copyright 1996, American Chemical Society.
MMT, 2-methyl cyclopentadienyl manganese tricarbonyl.
MTBE, methyl-tert-butyl ether.

amounts of the aromatics and less normal alkanes since the 1970s (Schmidt *et al.*, 2002). Oxygenates, such as MTBE, were added after 1980 and 1990 to eastern and western US gasolines, respectively. Anti-knock additives, such as alkyl leads and manganese compounds (2-methylcylcopendienyl manganese tricarbonyl, MMT) also have age significance. MMT can be age-diagnostic, but its absence does not prove a post-1978 gasoline because some refiners did not use it. Tetraethyl lead was the only alkyl-lead additive in leaded fuels after 1980, but it was phased out in the late 1980s. Ethylene dibromide and ethylene dichloride are lead scavengers that were added to leaded fuels to prevent the accumulation of metallic lead in engines. Permitted lead contents in gasolines dropped significantly in the 1980s, but the ranges shown in Figure 10.29 are not used readily to date gasolines because of secondary processes that affect lead content, such as selective adsorption by soils.

In special cases, even better age discrimination is possible. For example, gasohol (gasoline with 10% ethanol or 3% methanol) was prevalent during the 1972 oil embargo. Further refinement of the spill date may involve monitoring product weathering and changes

in specific hydrocarbons due to water washing and biodegradation. Changes in the concentration of light aromatic hydrocarbons in contaminated soil and associated groundwater and in nC_{17}/pristane have been calibrated in specific sites to fuel residence times. In silty soils saturated with gasoline, over half of the benzene, toluene, ethylbenzene, and xylenes (BTEX) may partition into the groundwater within five years (Hurst *et al.*, 1996).

Lead isotopes provide a direct method to date gasoline spills (Hurst *et al.*, 1996; Hurst *et al.*, 2001; Hurst, 2002). The technique relies on constructing a lead isotope model for the source of ore used in the manufacture of alkylated lead additives over time. The model is based on the fact that the major producers used similar starting mixtures of lead ore for their alkyl-lead additives. Hurst *et al.* (1996) proposed that their comparison of the lead isotopes in soils, waters, and aerosols to the isotope model could be used to date a gasoline spill with an accuracy of $\sim \pm 2$ years (Figure 10.30). However, most of the pre-1985 data used in the model are from southern California. Hurst *et al.* (1996) extrapolated these data nationwide based on questionable calibrations, including marine sediments that could contain lead from multiple unknown sources. It is likely that the calibration in the figure can be applied only to southern

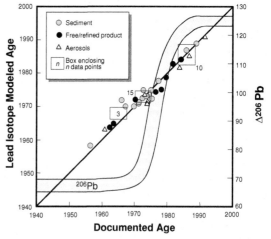

Figure 10.30. Comparison of documented age of anthropogenic lead sources to calculated age determined by a lead isotope model (shaded S-curve from Hurst *et al.*, 1996). $\Delta^{206}Pb = k[(^{206}Pb/^{207}Pb)_{sample}]/[(^{206}Pb/^{207}Pb)_{standard}]$. Reprinted with permission from Hurst *et al.* (1996). © Copyright 1996, American Chemical Society.

California samples dating from ~1970 to 1983 (I. R. Kaplan, personal communication, 2002).

Compound-specific isotopic analysis of gasoline hydrocarbons

Compound-specific isotope analysis of individual hydrocarbons in gasoline is useful in order to determine the source of groundwater contamination. Hydrocarbons and additives inherit their $\delta^{13}C$ from the feedstock and refinery processes. Variations in the pattern of isotopic values for common components in gasolines are usually sufficient to determine their regional origin (Smallwood *et al.*, 2002). In general, the $\delta^{13}C$ of selected gasoline hydrocarbons is not sufficiently fractionated by water washing or evaporation to obscure their source.

It is often possible to determine specific sources for hydrocarbons after they reach the water table using isotopic methods. This is particularly true of BTEX because they are water-soluble and move with groundwater flow. CSIA of dissolved BTEX provides a means to determine the source of these light aromatic hydrocarbons in water (Dempster *et al.*, 1997). The technique requires that the possible sources have sufficiently different $\delta^{13}C$ signatures for BTEX to be distinguished in contaminated water. Differentiation of BTEX sources can be improved using a combination of hydrogen and carbon isotope measurements. For example, Hunkeler *et al.* (2001) successfully monitored the aerobic biodegradation of benzene and distinguished benzenes from different manufacturers. Two of the four analyzed benzenes had similar $\delta^{13}C$, but all four benzenes showed distinct δD-$\delta^{13}C$ values, where δD ranged over 66.5‰.

Modeling gasoline leakage

Computer models of fuel release into the near subsurface can provide estimates of the age of the release and the dispersion rates of various hydrocarbons. The results can be used to evaluate weathering effects over time, to test hypotheses concerning the release and subsurface movements of contaminants, and to help to determine the most effective remediation. Models, however, are based on various assumptions and require knowledge of site geology, hydrology, and topography that may not be available. The EPA Center for Subsurface Modeling Support (CSMoS) provides public-domain software for groundwater and vadose zone modeling (www.epa.gov/ada/csmos/models.html).

NATURAL GAS AS A POLLUTANT

Although an asphyxiation and explosion hazard, methane is considered non-toxic. Accidental release of methane into the atmosphere produces no obvious environmental damage, but it is a greenhouse gas and may contribute to global warming. Natural gas, of which methane is a significant component, is released from both natural (e.g. gas seepage and swamps) and anthropogenic sources. Anthropogenic natural gas sources include thermogenic gases from oil, gas, and coal production, leakage during transportation, and incomplete burning, and release of microbial methane from farming, sewage, and landfills. Sources for methane emissions in the USA in 1995 included landfills (36%), enteric (cattle and other ruminants) (20%), natural gas (18%), coal (11%), manure (9%), petroleum (1%), and other (5%) (EPA, 1999). Global atmospheric methane concentration has more than doubled since pre-industrial times and is responsible for ~20% of the change in direct radiative forcing due to anthropogenic greenhouse-gas emissions (Dlugokencky *et al.*, 1998).

Several non-hydrocarbon constituents of natural gas pose substantially greater environmental hazards than methane. Hydrogen sulfide can be produced by bacterial sulfate reduction in low-temperature reservoirs (<80°C) or thermochemical sulfate reduction in high-temperature (>100°C) non-clastic reservoirs. Both bacterial and thermochemical sulfate reduction produce CO_2 and H_2S as primary inorganic products and solid bitumen as a common organic product. Gas chromatography and/or stable carbon or sulfur isotope ratios can be used to distinguish these processes (Machel *et al.*, 1995b; Machel, 2001). Alkali earth metals in the surrounding rocks combine with CO_2 to form carbonates, particularly calcite and dolomite. Iron sulfides, galena, and sphalerite may form by reaction of H_2S with transition or base metals. H_2S is acutely toxic in the atmosphere and hydrosphere and may cause death, even at low concentrations. In non-lethal concentrations below 0.1 ppt in air, H_2S rapidly dulls the sense of smell, so that increasing concentrations go unnoticed. A 0.1%

concentration of H_2S in air is fatal in <30 minutes. Because it is dangerous, H_2S is monitored during drilling and gas treatment. Many areas in the former Soviet Union, such as the lower Volga River near the Astrakhan gas/condensate field, have experienced prolonged H_2S exposure, resulting in health problems for nearby inhabitants and disruption of the ecology (Patin, 1999).

Microbial oxidation and various organic and mineral reactions generate subsurface carbon dioxide. Produced CO_2 is commonly pumped into subsurface reservoirs to maintain pressure. The main health risk of CO_2 is asphyxiation. Because CO_2 is the greenhouse gas most responsible for global warming, CO_2 removal and sequestration from natural gas is now an area of intense research. Mercury is a trace component in natural gas that poses health risks. Mercury ore (mostly cinnabar with some elemental mercury) is commonly associated with bitumen and hydrocarbon gas, probably due to focusing of CO_2-rich buoyant fluids into traps during migration (Peabody, 1993). Mercury vapor is oxidized to Hg^{2+} in the atmosphere and returned to Earth in rainwater. Bacteria can convert inorganic mercury to methylmercury, which is more biologically active. It bioaccumulates up the food chain, particularly in aquatic settings, where small fish eat plankton and are eaten by larger fish. Methylmercury can damage the brain and central nervous system.

Microbial and thermogenic natural gases

Near-surface hydrocarbon gases can arise from various microbial and thermogenic sources (Schoell, 1988). Microbial methane is produced at low temperatures (<80° C) by microbial degradation of organic matter (i.e. fermentation) in anoxic, near-surface environments, including natural aquatic systems, such as oceans, rivers, lakes, marshes, and swamps, and manmade settings, such as sewers, waste-treatment facilities, and landfills (Rice and Claypool, 1981). Rice fields and composting manure release high concentrations of microbial methane. Thermogenic methane can migrate to the surface from natural reservoirs, thermally mature source rocks, and coals. Thermogenic methane can also leak from oil and gas wells with faulty casing or damaged subsurface pipelines and storage tanks. Several remote-sensing techniques can identify trace amounts of thermogenic gases in soils, hydrates, and sediment cores.

Consortia of methanogenic archaea and bacteria produce microbial methane by decomposition of organic matter. There are at least three interacting groups. Various fermentative bacteria consume complex organic compounds and excrete volatile fatty acids, H_2, and CO_2. Aceticlastic bacteria can oxidize the higher acids to acetate or formate and H_2. Methanogens use several enzymatic pathways to form microbial methane. Methylotrophs are eubacteria with a suite of unique enzymes and metabolic pathways that allow them to utilize various one-carbon compounds (e.g. methanol, methylamine, methylbromide, and, in some cases, methane) as their sole carbon and energy source. They occur in members of the α-, β-, and γ-proteobacteria and in Gram-positive groups and are widespread in natural habitats (Haber *et al.*, 1983). Aceticlastic methanogens use acetate, formate, or small alcohols as both the electron acceptor and electron donor, liberating CH_4 and CO_2 in equal quantities. Collectively, these reactions of microbial consortia are frequently called acetic acid or acetate fermentation, because acetate is the most important methyl species. This fermentation process dominates in sanitary landfills and some organic-rich freshwater environments, such as marshes and swamps, where abundant organic acids are generated. The second major enzymatic pathway for microbial methane is carbon dioxide reduction, where CO_2 or $CO_3{}^{2-}$ is converted to CH_4 using electrons from H_2 or formate. Methane produced by CO_2 reduction dominates most marine and estuarine environments where the production of labile acids is limited.

Methane consumption also occurs in anoxic marine sediments, where consortia of archaea and sulfate-reducing bacteria operate CO_2 reduction in reverse (Hinrichs *et al.*, 1999; Boetius *et al.*, 2000). Orphan *et al.* (2001a) studied symbiotic cell aggregates that anaerobically oxidize methane in marine sediments from the Eel River Basin, offshore California. The aggregates consist of methanotrophic archaea belonging to the *Methanosarcinales* surrounded by sulfate-reducing bacteria related to *Desulfosarcina*. *Methanosarcinales* in the core of each aggregate are highly depleted in ^{13}C (e.g. $\delta^{13}C = -96\%o$) compared with the surrounding outer shell of *Desulfosarcina*. The ^{13}C-poor biomass of *Methanosarcinales* is consistent with assimilation and fractionation of carbon from methane in nearby seeps ($\delta -63$ to $-35\%o$).

Methanogenesis has characteristic carbon and hydrogen isotopic fractionations that differentiate microbial from thermogenic methane (Figure 10.31).

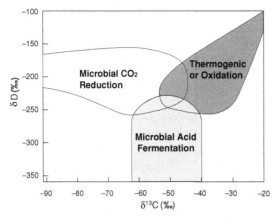

Figure 10.31. Microbial and thermogenic methane can be differentiated using $\delta^{13}C$ and δD measurements (from Whiticar *et al.*, 1986).

Near-surface methane with $\delta^{13}C < -60‰$ is almost certainly microbial, although Rowe and Muehlenbachs (1999) reported $\delta^{13}C$ for methane of $-63‰$, which they believe represents early thermogenic gas produced from low-maturity shale. Several secondary processes alter the primary $\delta^{13}C$ and δD of methane, including bacterial oxidation and molecular diffusion into air and across the water–air interface. Each of these processes favors methane containing the light isotope (^{12}C or ^{1}H). Bacterial oxidation imparts the greatest isotopic fractionation of these reactions. Consequently, methane with $\delta^{13}C > -40‰$, usually considered to be thermogenic, may be oxidized microbial methane. Bacterially oxidized methane can be distinguished from thermogenic methane by determining whether the associated CO_2 is isotopically depleted in ^{13}C or by determining its ^{14}C age.

Microbial and thermogenic gases can be distinguished further by the molecular and isotopic composition of associated higher hydrocarbons. Microbial hydrocarbon gases are typically very dry (>99% methane by volume). Minor amounts of olefins and ethane (<0.1% by volume) and trace propane and butane (<10 ppm) may be present. Thermogenic natural gases associated with oil are usually much wetter, containing >5% C_{2+} alkanes. Most high-maturity thermogenic gases still contain more than 1% ethane but contain no ethene. Most gases can be characterized as either microbial or thermogenic based on interpretation of both the gas composition and the $\delta^{13}C$ of the methane.

Methane in soils associated with petroleum spills

Although methane is not present in refined petroleum products or spilled oils, elevated concentrations of CH_4 and CO_2 occur in contaminated soils (Lundegard *et al.*, 2000). Because of the potential legal liability, it is important to determine whether soil gas methane originated from an oil spill or another source. Near-surface methane concentrations can readily exceed the explosive limit (>5% by volume).

Soil gas methane resulting from oil spills can originate by microbial degradation of petroleum or by alteration of the associated subsurface organic matter under anoxic conditions. To prove direct degradation, one must differentiate between microbial methane produced by degradation of the oil and methane generated by degradation of indigenous organic matter. Simple correlations between the spatial distributions of soil gas concentration in the area affected by the oil spill may be coincidental and insufficient to identify the source.

Radiogenic carbon isotopic analysis can be used to determine whether methane originated from spilled oil or recent organic matter. Methane resulting from degradation of recent organic matter contains radiogenic carbon, whereas that from degradation of spilled oil does not. Living organisms incorporate ^{14}C that is produced in the upper atmosphere by cosmic rays and by nuclear bomb tests. After methanogenesis ceases, no additional ^{14}C is incorporated, and the ^{14}C in the organic matter begins to decay with a half-life of 5730

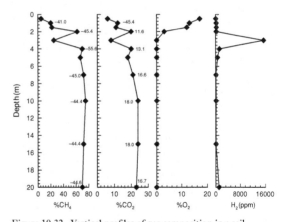

Figure 10.32. Vertical profiles of gas composition in a soil column heavily contaminated by crude oil. $\delta^{13}C$ for CH_4 and CO_2 are posted next to their concentration data points (from Lundegard *et al.*, 2000). Used with the permission of AEHS, 150 Fearing St, Amherst, MA 01002, USA.

Table 10.8. *Key source-specific tracers for organic components in ambient aerosol particles (from Simoneit, 2002)*

Compound or compound class	Major source	Emission process*
n-Alkanes		
C_{15}–C_{20} (odd/even)	Microbial	Direct/resuspension
C_{20}–C_{37} (odd/even)	Plant waxes	Direct/biomass burning
C_{15}–C_{37} (CPI = 1)	Urban	Vehicle exhaust
n-Alkenes, C_{15}–C_{37}	Biomass/coal	Burning
n-Alkanones, C_{15}–C_{35}	Biomass/coal	Biodegradation/burning
n-Alkanals, C_{15}–C_{35}	Biomass/coal	Biodegradation/burning
n-Alkanoic acids		
C_{15}–C_{37}	Microbial/biomass	Direct/resuspension/burning
C_{20}–C_{36}	Higher plants	Direct/burning
n-Alkanoic acid salts C_{15}–C_{20}	Marine biomass	Sea slick resuspension
n-Alkanols, C_{14}–C_{36}	Biomass	Direct
Alkanedioic acids, C_6–C_{28}	Various	Photo-oxidation/combustion
Wax esters	Plant waxes	Biomass burning/direct
Triterpenyl alkanoates	Tropical vegetation	Biomass burning
Triacylglycerides	Flora/fauna	Biomass burning/cooking
Methoxyphenols	Biomass with lignins	Burning
Levoglucosan (mannosan, galactosan)	Biomass with cellulose	Burning
Cholesterol	Urban/algae	Cooking/direct
Phytosterols	Higher plants	Burning (direct)
Triterpenoids	Higher plants	Burning (direct)
Diterpenoids (resin acids)	Higher plants, i.e. gymnosperms	Burning (direct)
Hopanes/steranes	Petroleum	Urban (e.g. vehicle exhaust)
Unresolved complex mixture	Petroleum	Urban (e.g. vehicle exhaust)
Alkylpicenes/alkylchrysenes	Coals	Urban (burning/heating)
Polycyclic aromatic hydrocarbons	Ubiquitous	All pyrogenic processes

* Listed in order of importance.

years. After ~50 000 years, insufficient ^{14}C remains for reliable measurements.

Lundegard *et al.* (1998) described a case study of a large gasoline spill (~80 000 gallons) from an urban service station. Remediation of the contaminated soil by vapor extraction revealed high concentrations of methane, apparently resulting from biodegradation of the gasoline. Methane concentrations remained high during remediation (>25% by volume) and there was little spatial correlation between methane concentration and the area affected by the spill. Exploratory cores penetrated a zone that contained abundant sawdust and wood, consistent with historic records of land use. The methane had an average δ^{13}C of −49.5‰ and an average δD of −310‰. Associated CO_2 had an average δ^{13}C of −17.1‰. The stable isotopic analyses proved that the methane resulted from microbial acetic acid fermentation but did not identify the source organic matter because the gasoline and wood had about the same δ^{13}C (~−25.9‰). The methane contained radiogenic ^{14}C with most ages ranging from 200 to 1100 years, proving that the wood rather than the gasoline was the carbon source. As a result, the owner of the service station was not liable for the high concentrations of near-surface methane.

Lundegard *et al.* (2000) showed how isotopic and compositional depth profiles support an origin of microbial methane from carbon in spilled oil at an unidentified industrial facility. Sediments from the facility contained ^{13}C-depleted methane (~−45‰)

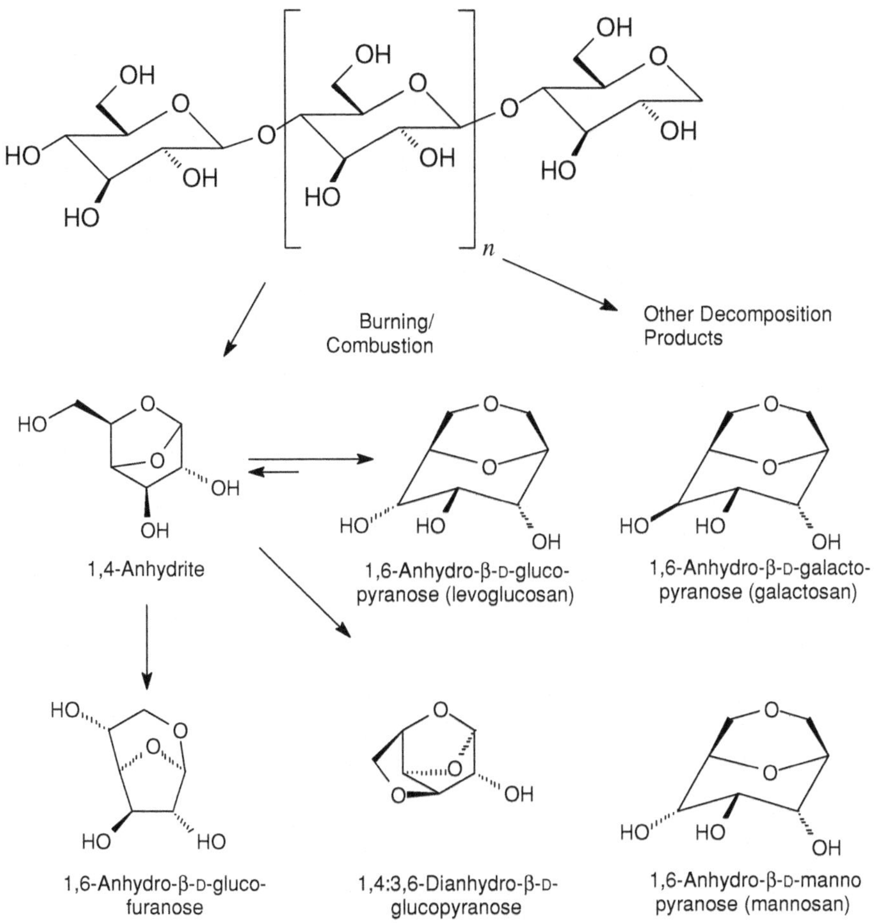

Figure 10.33. The major decomposition products from the burning of cellulose (from Simoneit, 2002).

with no higher hydrocarbons (dry gas) that was interpreted to be microbial. However, the gas contained no radiogenic carbon, indicating that the carbon source was more than 50 000 years old and, therefore, most likely to be thermogenic oil. The identity of this source was unclear because the facility was near subsurface petroleum reservoirs and several documented oil spillages. Detailed vertical profiles of soil gas concentrations and isotopic compositions revealed evidence for the source of the methane (Figure 10.32). Anoxic conditions occur below depths of 3 m. Hydrogen concentration decreases sharply below 3 m, whereas CO_2 is ^{13}C-rich ($\delta^{13}C > +10\%_0$). These conditions suggest that methanogenesis by CO_2 reduction occurs within the oil-contaminated soil immediately below the oxic/anoxic

boundary. Microbial oxidation of methane within the oxic zone accounts for the ^{13}C-depleted CO_2 ($-45.4\%_0$). The fact that methanogenesis occurs in the contaminated soil suggests strongly that the spilled oil, rather than hydrocarbons migrating from deeper reservoirs, is the carbon source.

BIOMARKERS IN SMOKE

The atmosphere contains a complex mixture of organic compounds, as either volatiles or aerosol particles (Table 10.8). These organic compounds originate either from natural materials (e.g. wind-blown plant waxes, vegetation, and soils) or from manmade emissions (vehicle exhaust, cooking fumes, industrial soot).

Table 10.9. *Major molecular tracers identified in smoke particles from biomass burning (from Simoneit, 2002)*

Compound group	Molecular tracers	Biomass source
Monosaccharide derivatives	Levoglucosan (mannosan, galactosan)	All with cellulose
	Vanillin, vanillic acid	Conifers
Methoxyphenols	Syringaldehyde, syringic acid	Angiosperm
	p-Hydroxybenzaldehyde, *p*-hydroxybenzoic acid	Gramineae (grasses)
Diterpenoids	Abietic, pimaric, isopimaric, sandaracopimaric acids	Conifers
	Dehydroabietic acid	Conifers
	Pimanthrene, retene	Conifers
Triterpenoids	α-Amyrin, β-amyrin, lupeol	Angiosperm
Phytosterols	β-Sitosterol, stigmasterol	All
	Campesterol	*Gramineae*
Sterols	Cholesterol	Meat cooking/algae
Chitin derivatives	Anhydroacetamidodeoxyglucose	Seafood cooking

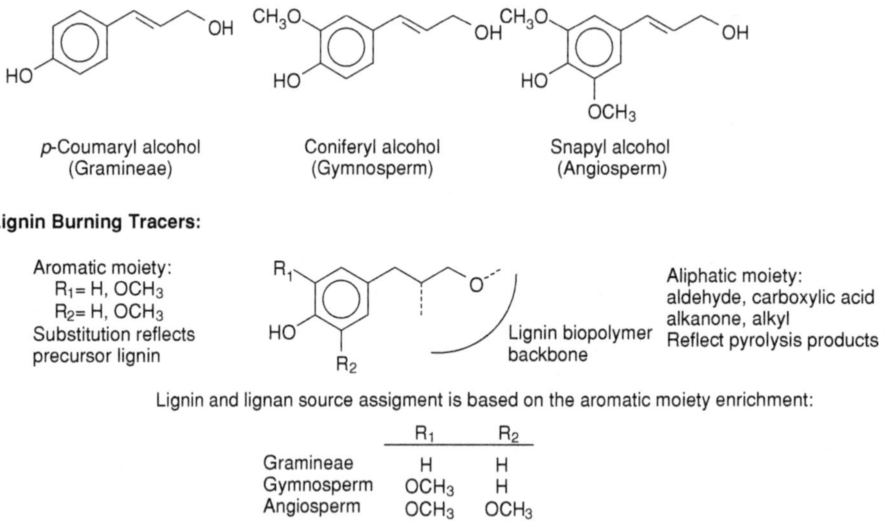

Figure 10.34. Biochemical precursors for lignins and lignans (top), and of lignin burning products (bottom), as tracers for biomass sources (from Simoneit, 2002).

Combustion of fossil fuels, such as petroleum and coal, introduces biomarkers, whereas incomplete combustion of biomass, from either natural wildfires or anthropogenic sources, introduces a host of pyrogenic and biogenic compounds that can be used to trace the material to its source.

Simoneit (2002) provided a comprehensive review of biomarkers in aerosol particles and their sources (Table 10.8). Burning biomass produces suites of compounds on smoke particles that are similar to natural background emissions (Table 10.9). Monosaccharide derivatives from the incomplete combustion of cellulose are most abundant but are non-diagnostic of origin (Figure 10.33). The methoxyphenols from lignin are, however, specific markers (Figure 10.34) that can be used to distinguish the major vegetation classes:

angiosperms (flowering plants), gramineae (grasses), and gymnosperms (conifers). Terpanoids liberated from biomass burning also may be quite specific. The source of some compounds depends on location. In open marine settings, cholesterol arises from direct algal input. In urban environments, cholesterol is produced mostly from the cooking of meat. Emissions from food preparations (e.g. frying and grilling) may account for ~20% of the airborne organic carbon over cities. 1,6-Anhydro-2-acetamido-2-deoxyglucose arises from the burning of chitin and can be traced to the grilling of shellfish.

11 · Biomarkers in archeology

This chapter provides examples of the growing use of biomarker and isotopic analyses to evaluate organic materials in archeology. Some of the topics include bitumens in Egyptian mummies, such as Cleopatra, archeological gums and resins, and biomarkers in art and ancient shipwrecks. The discussion covers the use of biomarkers and isotopes in studies of paleodiet and agricultural practices, including studies of ancient wine and beeswax. Other topics include archeological DNA, proteins, and evidence for ancient narcotics.

Human remains and artifacts provide critical evidence of mankind's origin, diversification, migration, interactions, and cultures. They also contain chemical fossils that provide important clues to the past. Consequently, the molecular methods developed in other sciences are becoming new and important tools in archeology. The same molecular and isotopic techniques used in petroleum geochemistry are applied to analyze bitumen and other natural substances used by ancient man. Biomarkers identified in modern agricultural and environmental studies can be used to investigate ancient dietary and agricultural practices. DNA amplification and sequencing techniques are being used increasingly on ancient human remains, and genome studies have led to new theories of man's origins. Results from these chemical studies not only complement and refine but frequently challenge long-accepted theories.

THE AGES OF MAN

Arguably, one of the distinguishing characteristics of humanoids is their use of tools. The ages of man are named after the most commonly used technological material for tools of the period: stone, bronze, and iron (Table 11.1). The Stone Age is divided further into Paleolithic, Mesolithic, and Neolithic Periods, reflecting increasing sophistication of stone tool fabrication

and diversity. The Bronze Age is marked by the development of metallurgy, initially copper and then bronze (copper–tin). Eventually, the Bronze Age gave way to the Iron Age. The time when one age ends and the other begins varies for each culture. For example, the Bronze Age in the Middle East and in Britain began in ~6500 BC and 1900 BC, respectively.

ORIGINS AND TRANSPORT OF PETROLIFEROUS MATERIALS IN ANTIQUITY

Slime is a translation of the Hebrew word hemar, which means asphalt. *And they had brick for stone, and slime had they for mortar. (Genesis 11:3.)*

And when she could not longer hide him, she took for him an ark of bulrushes, and daubed it with slime and with pitch . . . (Exodus 2:3.)

Petroleum-based materials were used widely in the ancient world, particularly in the Middle East, where oil, solid bitumens, and source rocks commonly occur at the surface. Bitumen was used to attach flint tools to wooden handles, as an adhesive to repair broken statues and pottery, and to attach decorative façades of mother-of-pearl, lapis lazuli, and other minerals to ornamental items. When mixed with sand, clay, and straw, bitumen served as mortar in both common buildings and ceremonial palaces and temples, as in the construction of the Tower of Babel. The hydrophobic nature of bitumen was exploited as a waterproofing agent in the ancient Near East. Bitumen was used as caulking in ships; as a sealant for various containers, palm mats, and baskets; as roofing material, as in the Hanging Gardens of Babylon; and in plumbing systems.

The methods developed by petroleum geochemists for oil–oil correlations are directly applicable in order to determine the presence, origins, and transport of

Table 11.1. *The ages of man are based on the materials used to construct tools*

			~Years ago	Technological characteristics
Stone	Paleolithic	Lower	2 500 000–200 000	Simple pebble tools, chopping tools
		Middle	700 000–40 000	Hand axes, carefully flaked flint tools
		Upper	40 000–10 000	More complex and specialized stone tools, distinctive regional and artistic traditions
	Mesolithic		10 000–5000	Refers mainly to northwestern Europe; finely made, chipped stone tools, microliths (small stone tools)
	Neolithic		10 000–5000	Stone tools shaped by polishing or grinding; domestication of plants and animals, permanent villages, pottery and weaving
	Bronze		8500–2000	Copper and, later, bronze tools, the wheel, the ox-drawn plow
	Iron		3000–present	Iron replaces bronze

archeological bitumens. Marschner and Wright (1978) were first to correlate various ancient bitumens from the Middle East to their likely sources using the techniques then available to the petroleum industry. Venkatesan *et al.* (1982) and Connan (1988) started the modern era of these interdisciplinary investigations with biomarker and isotopic analyses of bituminous artifacts. For example, Venkatesan *et al.* (1982) showed that archeological charcoal samples from an ancient dumpsite at Terqa, Syria, gave much older apparent ^{14}C ages than expected because of contamination by native asphalt. The triterpanes in the contaminated charcoal samples and associated asphalt have similar distributions, including greater C_{29} than C_{30} 17α-hopane. In the third millennium in Syria, native asphalt was commonly heated to seal cracks in ceramic vessels and to fasten tool heads on to wooden handles. Broken storage pots sealed with asphalt and dumped in the refuse heap could easily have contaminated the charcoal samples.

Bitumens in Egyptian mummies

Archeologists have long debated the materials and procedures used by ancient Egyptians in mummification (David, 2000). Ritual mummification began in the Old Kingdom (~2600 BC) and continued until the Arab conquest in the seventh century AD (Koller *et al.*, 1998). The secrets of mummification were closely guarded. We have only the written accounts of Herodotus in the fifth century BC and Diodorus in the first century AD. The organic materials used in the mummification process involved complex recipes that changed over time. Various resins, gums, waxes, bitumen, and honey are mentioned in these ancient texts. Because many of the materials were imported, their use also varied depending on the prestige and wealth of the deceased. Archeologists long described these balms in general terms such as "resinous" or "bituminous," but their compositions and origins have only been revealed by modern chemical analyses. These analyses revealed much about

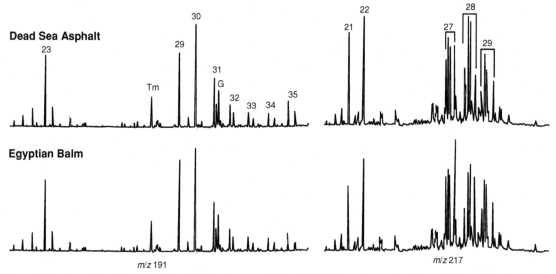

Figure 11.1. Comparison of terpane (left, m/z 191) and sterane (right, m/z 217) biomarkers in Dead Sea asphalt and balm from an Egyptian mummy (from Connan, 1999).

mummification, its theological significance, the health of these people, and ancient trade routes.

Rullkötter and Nissenbaum (1988) and Connan and Dessort (1989a; 1989b) were first to apply biomarker analyses to ancient bitumens as components of balms used to prepare Egyptian mummies. Biomarkers in the balms correlated with Dead Sea asphalt, proving sustained trade between cultures (Figure 11.1). Subsequent studies, mostly on mummies dating from 1000 BC–400 AD, found that bitumens from other sources, possibly Iraq, also were used. The balms were complex mixtures with inconsistent compositions. They contain varying amounts of bitumen, ranging from ∼0 to 30% (Connan and Dessort, 1991). Biomarkers from conifer resins, such as longifolene (structure in Figure 11.8), beeswax, and fats prove that these materials were also major ingredients (Proefke et al., 1992; Connan, 1999; Maurer et al., 2002). The organic residues from other mummies contain neither bitumen nor resins but a complex suite of altered lipids and proteinaceous compounds, suggesting that mixtures of oils and gelatin also were used as embalming agents (Buckley et al., 1999).

Rullkötter and Nissenbaum (1988) analyzed biomarkers in asphalts from four Egyptian mummies, including that of Cleopatra, dating from the ninth century BC to the early second century AD. The three younger mummy asphalts show sterane and triterpane biomarker distributions almost identical to those from a modern floating block of asphalt from the Dead Sea. These samples have low diasteranes and high gammacerane, characteristic of petroleum from clay-poor carbonate or evaporitic source rocks. Floating asphalt blocks from the Dead Sea are believed to originate from Upper Cretaceous (Senonian) limestone deposited under hypersaline conditions (Rullkötter et al., 1985) based on the following characteristics: even-to-odd n-alkane predominance, low pristane/phytane (∼0.5), lack of diasteranes, high sulfur content, $\delta^{13}C$ in the range −29.7 to −27.6‰, predominance of C_{27}–C_{29} steranes, and high gammacerane. The oldest mummy (Pasenhor) has biomarker distributions indicating a different origin, possibly because trade of Dead Sea asphalt into Egypt began only after 900 BC.

Harrell and Lewan (2002) compared five mummies with seeps from the Dead Sea and Abu Durba and Gebel Zeit from the Gulf of Suez (Figure 11.2). Four of the mummies were the same as those analyzed by Rullkötter and Nissenbaum (1988), including Cleopatra and Soter from western Thebes (early second century AD), Djedoler from Akhmin (∼200 BC), and Pasenhor from western Thebes (∼900 BC). The fifth mummy is that of an unknown priest from western Thebes (∼800 BC). The Egyptian seeps from Abu Durba and Gebel Zeit were not included in the earlier study.

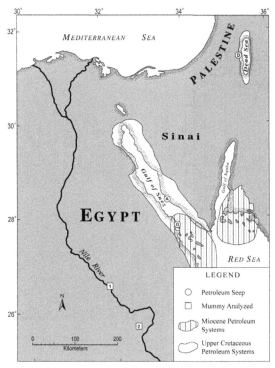

Figure 11.2. Map shows locations of Miocene and Upper Cretaceous petroleum systems, petroleum seeps (D = Dead Sea, A = Abu Durba, G = Gebel Zeit, also Jebel Zeit in Figure 11.5), and collection localities for ancient mummies (1 = Akhmin, 2 = western Thebes) (from Harrell and Lewan, 2002).

The Abu Durba seep and Dead Sea asphalt blocks originated from the same regional Upper Cretaceous marine carbonate source rock (Brown Limestone in the Gulf of Suez or Ghareb Formation near the Dead Sea) and show only minor differences in biomarker distributions due to local variations in organic facies. These samples have high gammacerane and C_{35} hopanes, low diasteranes, low 18α-30-norneohopane, and little or no oleanane (Figure 11.3). The Cleopatra, Soter, Djedoler, and priest mummies have biomarkers similar to these seeps (Figure 11.4). Although the Gulf of Suez Abu Durba seep was closer to mummification sites in Egypt, it appears that the ancient Egyptians preferred the Dead Sea bitumen. This may have been due to a better established coastal trade route from the Dead Sea compared with routes from the Gulf of Suez or because the semi-solid Dead Sea bitumen was more readily transported than a liquid.

The Gebel Zeit seep originated from Miocene siliciclastic source rocks in the Rudeis, Kareem, and Belayim formations (Alsharhan and Salah, 1997). This seep has many characteristics of the oldest mummy (Pashenor, ~900 BC), including low gammacerane and C_{35} hopanes and high diasteranes, 18α-30-neonorhopane, and oleanane. The Pasenhor mummy had not been correlated previously with any petroleum seep (Rullkötter and Nissenbaum, 1988).

Maurer et al. (2002) analyzed the embalming materials in four mummies from the Kellis 1 tomb at the Dakhleh Oasis in the Western Desert ~300 km west of the Nile Valley. Artifacts with the mummies were dated at about the fourth century AD, but some radiocarbon measurements yielded ages up to 1000 years older. The admixture of petroleum bitumen to the mummy material could cause such anomalous ages. The balms from the Kellis 1 tomb were found to be mixtures of petroleum bitumen, various contemporaneous plant materials, and possibly beeswax. C_{25}–C_{31} odd-carbon-number n-alkanes dominate the hydrocarbon distributions but are mixed with an envelope of C_{17}–C_{35} n-alkanes with no carbon preference. Steranes with petroliferous distributions are mixed with ster-2-enes. The triterpane distribution was dominated by petroleum components. One interesting finding was that bitumen in the thorax samples correlated well with Dead Sea asphalt, whereas bitumen in the cranium samples was from a different, unidentified source.

Bitumens in Middle Eastern artifacts

Bitumen was traded throughout the ancient Middle East (Figure 11.5). Correlating excavated bitumens to their sources can reveal former trade routes. Bitumens from Tell el'Oueili, one of the earliest settlements in southern Mesopotamia, reveal changing trade routes through three millennia (Connan, 1999). Numerous bituminous artifacts were recovered from five levels representing a discontinuous history of settlement from 5800 to 3200 BC. Some of the earliest bitumens have biomarker distributions that correlate to the Middle Jurassic Sargelu Formation source rocks from the Lurestan region (see Figure 18.87). Other early bitumens contain oleanane and correlate with the Tertiary Pabdeh Formation source rocks from the Khuzestan area in modern Iran, consistent with other evidence for close contact between the Susian and Mesopotamian cultures. Bitumens

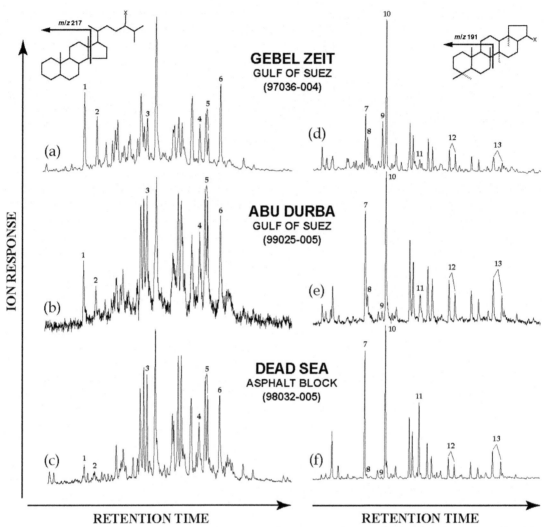

Figure 11.3. Sterane (a–c) and terpane (d–f) mass chromatograms for the saturate fractions of Gebel Zeit, Abu Durba, and Dead Sea petroleum seeps (from Harrell and Lewan, 2002). Labeled peaks include (1) 13β,17α(H)-diacholestane 20S, (2) 13β,17α(H)-diacholestane 20R, (3) 5α,14β,17β(H)-cholestane 20S, (4) 5α,14α,17α(H)-stigmastane 20S, (5) 5α,14β,17β(H)-stigmastane 20R and 20S, (6) 5α,14α, 17α(H)-stigmastane 20R, (7) 17α,21β(H)-30-norhopane, (8) 18α-30-norneohopane, (9) oleanane, (10) 17α,21β(H)-hopane, (11) gammacerane, (12) 17α,21(H)β-29-trishomohopane 22S and 22R, and (13) 17α,21(H)β-29-pentakishomohopane 22S and 22R.

from higher levels, dating from 4550–3700 BC, correlate with sources from the Kirkuk region of northern Iraq. During this time, southern Mesopotamian settlements expanded northward. Bitumens from the youngest levels, dating from 3500–3200 BC, correlate to yet another source, the Hit-Ramadi-Abu Jir asphalts, a major source of bitumen found in excavations along the Euphrates River.

Hummal and Umm El Tiel, two Middle Paleolithic sites in Syria (~40 000 BC), are among the oldest examples of the use of bitumen (Boëda et al., 1996). Among the thousands of artifacts recovered from these excavations, a few stone tools have traces of adhering black material. In some specimens, this black material proved to be inorganic manganese oxides, but on other stone artifacts the black matter is bitumen used to attach

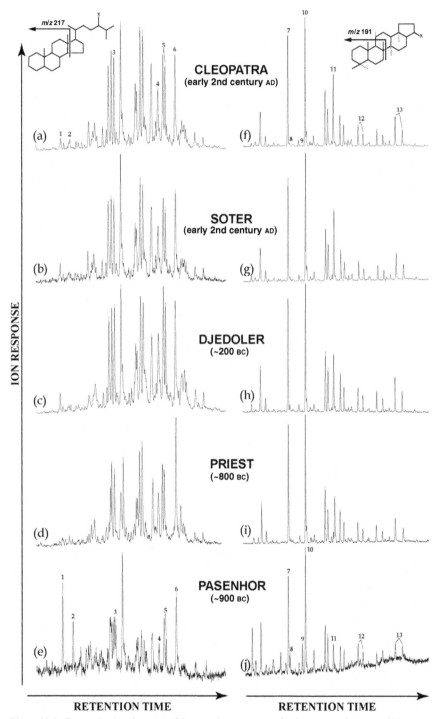

Figure 11.4. Sterane (a–e) and terpane (f–j) mass chromatograms for the saturate fractions of bitumen extracted with chloroform from Cleopatra, Soter, Djedoler, priest, and Pasenhor mummies (from Harrell and Lewan, 2002). Labeled reference peaks are the same as those in Figure 11.3.

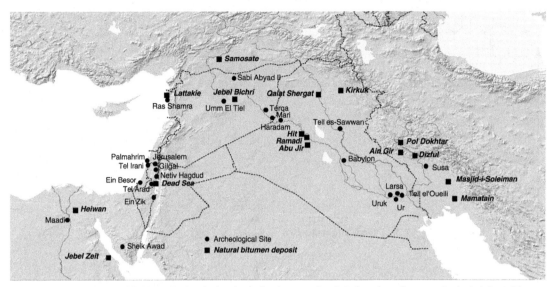

Figure 11.5. Map of the Middle East, showing the locations of major natural asphalt deposits and some archeological sites with bitumen artifacts (adapted from Connan, 1999).

flint tools to wooden handles. This bitumen contains linear and T-branched alkylbenzenes, fluoranthenes, pyrenes, and other compounds characteristic of pyrolytic reactions, suggesting thermal treatment. However, it was discovered later that the samples were heavily contaminated during collection and archiving procedures and that these compounds may be modern artifacts.

Special care was taken to minimize contamination in the 1996 excavations of Umm El Tiel. Unfortunately, no new stone artifacts that had bitumen adhesive were found, but lumps of bituminous sands were discovered in layers associated with other artifacts. Another oil-stained sandstone was found in a wadi 15 km north of the site. Biomarkers and isotopic ratios relate these samples to oil sands that outcrop in Jebel Bichri, ~50 km east of Umm El Tiel. Bitumen from these oil sands is enriched in tricyclic terpanes, unlike that from Hit-Abu Jir, the most extensively used bitumen source along the Euphrates River. The displaced sample in the wadi may have been lost during transport along a Paleolithic trade route from Jebel Bichri to Umm El Tiel (Boëda et al., 1996; Connan, 1999).

Biomarkers and isotopes correlate seep and floating-block asphalts in the Dead Sea with archeological bitumens from ancient Canaan, the Sinai, and Egypt dating from 3900 to 2200 BC (Connan et al., 1992).

This correlation provided the first evidence for extensive trade and export of Dead Sea asphalt into Egypt before its documented use in mummies younger than 1000 BC. Older mummies rarely are available for geochemical study because of their scarcity and historic significance. The Dead Sea asphalts and related archeological bitumens originated from Senonian bituminous chalk source rock deposited in a hypersaline environment, as discussed above (Figure 11.3).

Bitumen and oil shales were worked into jewelry and sculptures. Some of the most famous are sculptures and carved cylinder and stamp seals from Susa (Shushan in the Bible). Susa, located in present-day western Iran, is the ruined capital city of ancient Elam and, later, Suslana, where the stele containing the Babylonian Code of Hammurabi was discovered. The Susian carvings are composed of a bituminous-mineral mixture called bitumen mastic. In their book Le bitumen à Susa, which documented many of the Susian masterpieces in the Louvre museum collection, Connan and Deschesne (1996) proposed that the bitumen mastic was an artificial material formed by moderate heating of bitumen and minerals. Connan and Deschesne (2001) discovered a carbonate source rock near Susa that is likely to be the source of the bitumen used in the carvings. This hypothesis explains why the Susian sculptures are unique to the area and why these objects were carved rather than molded.

Although bitumen use is most widespread in the Near East, where natural occurrences are plentiful, petroliferous bitumens have been identified in archeological sites in prehistoric Europe. Koller and Baumer (1993) identified a bituminous substance on an early Bronze Age knife from Xanten-Wardt, Germany. Regert *et al.* (1998b) identified bitumen used as hafting (adhesive) for an artifact from prehistoric lake dwellings at Chalain, France. There are no local sources for these bitumens, and their origins remain unknown.

Jet rock, cannel coal, lignite, and torbanite are some of the materials that were used in the past to make black shiny ornaments. Archeologists had difficulty in identifying these different materials. Pyrolysis/gas chromatography/mass spectrometry (py-GCMS) proves that ancient samples of workable black lithic materials can be readily distinguished (Watts *et al.*, 1999).

Note: Slabs of bitumen covered with barnacles and with impressions of reed bundles and rope may be the remains of the oldest boat (Lawler, 2002). The bitumen slabs consist of a composite of petroleum asphalt, crushed coral, and fish oils and were found in the ruins of a 7000-year-old stone building at As-Sabiyah, Kuwait. Some archeologists conclude that the bitumen slabs were part of a seagoing craft made of reeds and tar that was disassembled for storage or repair. If this is correct, then marine transport and trade could explain the dispersal of Mesopotamian artifacts from throughout the Persian Gulf from the Ubaid period.

ARCHEOLOGICAL GUMS AND RESINS

The resinous secretions of certain plants were highly valued in ancient times, and trade routes were established over land and sea for their transport. Frankincense, myrrh, and terebinth are secretions of *Boswellia*, *Commiphora*, and *Pistacia* terebinth trees, respectively. These natural products contain complex mixtures of fragrant terpenoids, mainly mono- and sesquiterpenes (Figure 11.6), which prompted their use in incense, medicines, and perfumes and as preservatives and flavorings for wine. Later, dammar, mastic, and sandarach resins along with Congo, Kauri, and Manila copals were used widely as varnishes in oil paintings.

Natural resins are still used widely today. Gum arabic is a water-soluble, gummy exudate from the stems and branches of *Acacia senegal* and any tree or shrub belonging to the *Acacia* genus. An extremely versatile substance, gum arabic is used as a food thickener and emulsifier and in inks and medications. Rosin is a distilled pine resin used on various string instruments to create suitable friction contact between the bow and string. Shellac, a common varnish, is a secretion of the insect *Coccus lacca*, which exudes a resinous material that builds up on the bark of host trees. Modern synthetic polymers, however, have replaced many of the traditional natural resins.

The natural resins can be characterized by their distributions of terpenoid hydrocarbons and acids (Figures 11.6 and 11.7). Most resins contain a few abundant terpenoids with a host of accessory terpenoids in minor or trace concentrations. For example, the acid terpenoid content of sandarach resin is nearly all sandaracopimaric acid, a compound that is absent or occurs only in trace amounts in other common resins. Myrrh is characterized by myrrhoric acid. These terpenoids perform various functions, including hormonal regulation, chemical and physical defense against insects, and antibacterial and antifungal activity. There are thousands of these types of terpenoids, all constructed from isoprenoid units. Even the complex multi-ring structures are condensed forms of isoprenoids (Figure 11.8).

Comparisons of ancient and modern resins must account for the fact that volatile terpenes can be lost and non-volatile terpenoids can be altered by heating and oxidation. For example, modern *Pinus* resins contain abundant α-pinene, minor amounts of β-pinene and limonene, and trace amounts of other terpenoids. Most archeological samples lack these volatile monoterpenoids, and less volatile terpanoid hydrocarbons and acids or their partially oxidized species characterize ancient coniferous resins. Hence, the detection of biomarkers, such as longifolene (Connan, 1999), dehydroabietic acids, and oxo-dehydroabietic acids (Koller *et al.*, 1998; Buckley and Evershed, 2001; Maurer *et al.*, 2002) in Egyptian mummies provides evidence for the use of conifer resins in embalming.

For example, Weser *et al.* (1998) documented the use of smoked pine resins in an Egyptian skeleton dating from ~2200 BC. Evidence was found for ritual mummification practices during the Old Kingdom that differed substantially from later times. Pyrrole derivatives, which are produced by smoldering bones, were found in extracts, suggesting that the skeleton may have

Monoterpenes

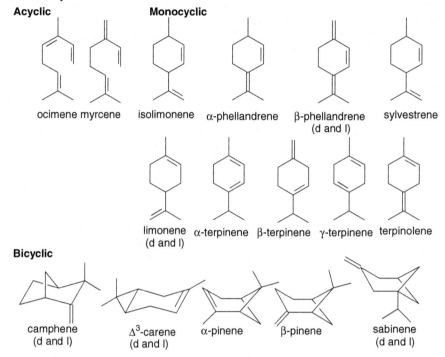

Acyclic   **Monocyclic**

ocimene myrcene   isolimonene   α-phellandrene   β-phellandrene   sylvestrene
(d and l)

limonene α-terpinene β-terpinene γ-terpinene terpinolene
(d and l)

Bicyclic

camphene   Δ³-carene   α-pinene   β-pinene   sabinene
(d and l)   (d and l)   (d and l)

Sesquiterpenes

Acyclic   **Monocyclic**

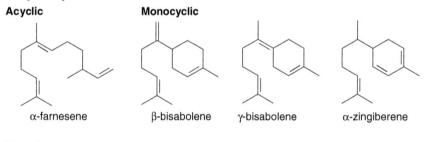

α-farnesene   β-bisabolene   γ-bisabolene   α-zingiberene

Bicyclic

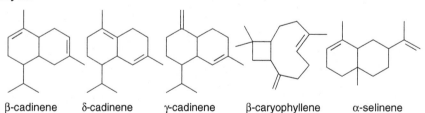

β-cadinene   δ-cadinene   γ-cadinene   β-caryophyllene   α-selinene

Figure 11.6. Common monoterpenes and sesquiterpenes in natural resins.

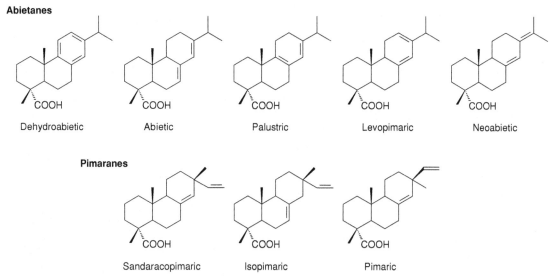

Abietanes

Dehydroabietic   Abietic   Palustric   Levopimaric   Neoabietic

Pimaranes

Sandaracopimaric   Isopimaric   Pimaric

Figure 11.7. Common acids found in plant resins.

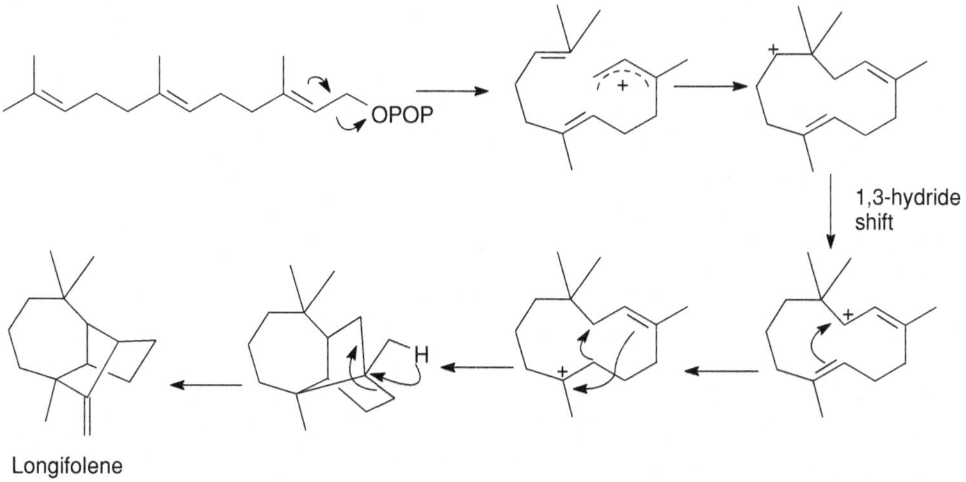

1,3-hydride shift

Longifolene

Figure 11.8. Biosynthesis of longifolene, a minor terpenoid in conifer resins.

been flamed. The bone extracts also contained high concentrations of diterpenoid resin acids (mainly dehydroabietic acids and its oxidized forms and methyl esters of these acids (Figure 11.9). Liquid chromatography was used to detect the natural methyl esters because conventional esterification of the acids before gas chromatographic separation would obscure their presence. Dehydroabietic acid results from the heating and oxidation of abietane- and pimarane-type acids found in pine resins. Methyl esters are not very abundant in untreated pine

resins and are not formed by heating alone. However, methyl esters form when the resins are burned. Weser *et al.* (1998) also noted that aromatic hydrocarbons, such as retene, which are indicative of pine tars produced by oxygen-deficient heating (see below), are absent. They concluded that a smoking process had altered the pine resins used in the preparation of these Old Kingdom remains.

Cadinanes, polycadinanes, and diagnostic resin acids (e.g. darmarenolic acid) on prehistoric potsherds

methyl dehydroabietate   methyl dehydrodehydroabietate   methyl 7-oxo-dehydroabietate   methyl 7-oxo-dehydrodehydroabietate   7-oxo-dehydroabietic acid

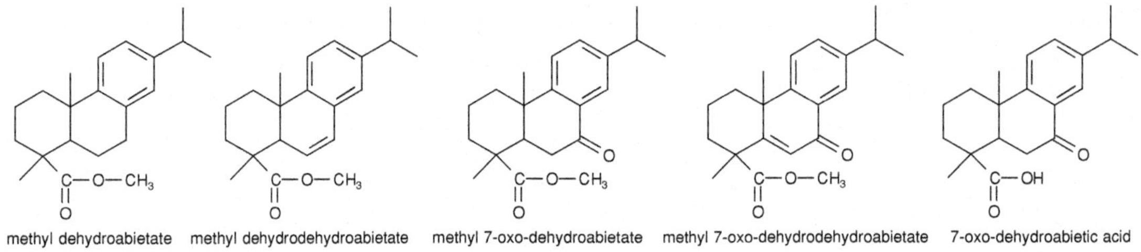

Figure 11.9. Oxidized and esterified pine resin acids identified within a mummified Old Kingdom skeleton (Weser *et al.*, 1998). High relative concentrations of dehydroabietic acid and dehydrodehydroabietic acid were present.

document the use of *Dipterocarpaceae* resins as coatings or adhesives in southeast Asia (Lampert *et al.*, 2001; 2002). The resin coatings were applied soon after firing as a waterproofing material or as an adhesive. Because the resins are contemporaneous within the functional lifetime of the pot, ^{14}C abundance can be used to age date the ceramics (Lampert *et al.*, 2003a; 2003b).

Biomarkers for frankincense

Frankincense was one of the most valued aromatic resins used by ancient Near Eastern cultures. Mixed in medicines and ritual incenses throughout history, frankincense originates from the gum of *Boswellia* trees, which are limited to areas of northern Somalia and southern Arabia. Its spiritual significance is recorded in biblical passages and historic records. Frankincense and other aromatic resins are amorphous materials that are rarely recovered from archeological sites, although incense burners are common artifacts.

Biomarker analyses identified the origins of amorphous resins recovered in the excavation of Qasr Ibrîm, an Egyptian settlement in Nubia dating from ~400–500 AD (Evershed *et al.*, 1997c; van Bergen *et al.*, 1997). Some of these materials contained tricyclic diterpenoid acids (e.g. isopimaric acid, abietic acid, and dehydroabietic acid) that originated from conifer resins. Other resinous remains contained α- and β-boswellic acids, identified as biomarkers specific to frankincense (Figure 11.10). The boswellic acids can be detected directly by GCMS using m/z 292, or as O-acetyl derivatives using m/z 352 as a diagnostic fragment. Pyrolysis of frankincense forms 24-noroleana-3,12-diene and 24-noruras-3,12-diene.

Biomarkers for terebinth resin

Terebinth is a yellowish, semi-solid resin secreted by several species of *Pistacia*, which are known collectively as terebinth or turpentine trees. These trees grow throughout the Mediterranean area, but only in the eastern areas do winter temperatures drop sufficiently to cause the tree to produce resin. Several classical authors considered terebinth resin to be the most valuable of all resins, and they describe its use in perfumed oils and incense (Peachey, 1995) and in the preparation of resinate wines (McGovern *et al.*, 1995). A late Bronze Age shipwreck at Ulu Burun, Turkey, confirms the importance of terebinth (Bass, 1986; Bass and Pulak, 1987; Pulak, 1988). The ship's cargo consisted of the most luxurious and expensive items of the time, including tons of copper, tin and glass ingots, gold and silver jewelry, elephant and hippopotamus ivory, and Egyptian ebony logs (*Dalbergia melanoxylon*). With these treasures were ~130 Canaanite amphoras containing terebinth resin (Mills and White, 1989; Hairfield and Hairfield, 1990). Late Bronze Age Egyptian and other Near Eastern rulers used these items as royal tribute.

The oldest known use of terebinth is based on chemical analysis of residues in a pottery jar from Hajji Firuz, a Neolithic (5400–5000 BC) village in the northern Zagros Mountains, Iran (McGovern *et al.*, 1996). The analyses did not monitor the characteristic terpenoids of terebinth resin but relied on HPLC,

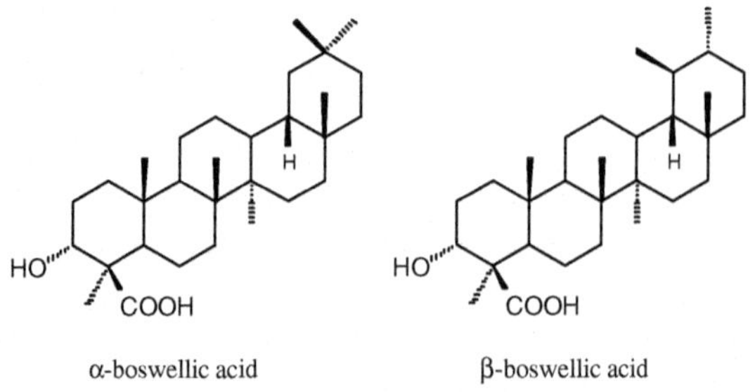

α-boswellic acid   β-boswellic acid

Figure 11.10. α- and β-boswellic acid.

UV-spectrometry, and diffuse reflectance IR-spectrometry to compare extracts of the pottery jar with modern reference materials. The jar contained mainly the calcium salt of tartaric acid, which occurs naturally in large amounts only in grapes and terebinth resin. Presumably, the resin was added to inhibit the growth of bacteria (*Acetobacter*) that convert wine to vinegar. The addition of terebinth proves that wine was made deliberately and did not result from unintentional fermentation of grape juice.

Archeologists have long debated the material and procedures used by ancient Egyptians in mummification rituals. The recipes for the balms used to anoint the body were a lost secret, but recent applications of biomarker analysis have rediscovered some of the ingredients. Terebinth has a suite of diagnostic triterpanoids and hydroxylaromatic acids, including isomasticadienoic, masticadienonic, moronic, and oleanonic acids. These compounds, their oxidized forms, and dammaranes that formed in highly degraded terebinth resins were identified in several Egyptian mummies (Figure 11.11) (Colombini *et al.*, 2000; Buckley and Evershed, 2001). Black substances, which were indiscriminately termed bitumen, proved to be mixtures of

beeswax and resins. The organic residues from other mummies contain neither bitumen nor resins but a complex suite of altered lipids and proteinaceous compounds, suggesting that mixtures of oils and gelatin also were used as embalming agents (Buckley *et al.*, 1999).

BIOMARKERS IN ART

Petroleum asphalts have been used as pigment materials in European paintings since the seventeenth century, reaching widespread application in the eighteenth and nineteenth centuries. Recipes varied from suspension in cold oil to hot solvent blends. When applied in thin layers, the asphaltic paint imparted a warm, mellow brown tone with glaze-like transparency. In addition to petroleum asphalt, soot, lampblack, roasted pitch, and tar coals were some of the ingredients collectively termed "asphalts." Mills and White (1994) used hopane distributions as indicators of asphaltic pigments. The saturated biomarkers may not be ideal for this purpose because they are only trace components in asphalt and may be removed by solvent cleaning. Languri *et al.* (2002) compared a nineteenth-century paint pigment from the Hafkenscheid Collection (one of

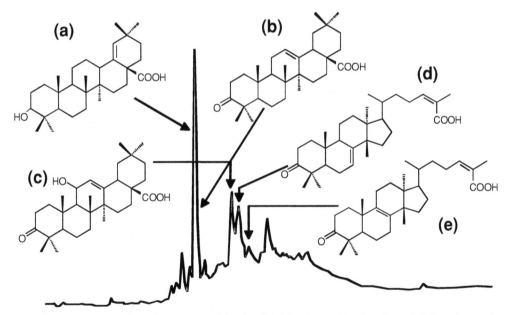

Figure 11.11. Reconstructed ion chromatogram of the trimethylsilylated neutral fraction of an embalming substance from an Egyptian mummy from the Ptolemaic Period (332–330 BC) (from Buckley and Evershed, 2001). The substance consisted mostly of beeswax with minor amounts of plant oils and resins. The extract contained a mixture of morolic acid (a), oleanonic acid (b), hydroxyoleanonic acid (c), isomasticadienonic acid (d), and masticadienonic acid (e).

the few surviving collections of pigments and paint materials from the early nineteenth century) with Dead Sea asphalt using pyrolysis/MSMS and pyrolysis/GCMS. Gammacerane, $\alpha\beta$-hopanes, and C-ring monoaromatic steroidal hydrocarbons were some of the biomarkers detected with similar distributions. Languri et al. (2002) also found that the Dead Sea and the paint pigment yielded similar pyrolyzate distributions of homologous series of non-biomarker compounds (e.g. alkylaromatics and alkylbenzothiophenes) and had similar isotopic values ($\delta^{13}C \sim -29.3\%$). They concluded that provenance of the asphaltic pigment from the Hafkenscheid Collection could be established.

Artists applied varnishes prepared from resins to oil paintings as a protective coating and to impart a uniform surface that improved the optical qualities of the underlying pigments by enhancing color saturation and gloss. An ideal varnish is transparent, durable, fast-drying, easily removed, and free of changes in coloration over time. The Old Masters used various resins as varnish. Since the ninth century, varnishes were prepared mostly using mastic resins from *Anacardiaceae* trees belonging to the genus *Pistacia* (mainly *P. lentiscus*). The mastic resins were mixed with heated linseed, colophony (pine), and sandarac (juniper) oils. Because mastic varnishes yellow rapidly, they were largely replaced by ~1850 by mixtures using dammar resin (*Dipterocarpaceae* trees), which are less prone to degradation. Dammar varnishes were also about nine times cheaper than mastic. Congo, Manila, and Kauri copals also were used widely in varnishes. These varnishes are very hard, but their coloration, stiffness, and difficulty in removal discouraged their use. Resins from multiple sources were frequently mixed with a variety of solvents, resulting in different properties and chemical changes with age. Complicating the situation further, Old Master paintings were commonly restored many times using different materials and varnishes that may interact with each other (White and Kirby, 2001).

Modern conservation and restoration of museum art employs some of the most sophisticated chemical analyses available. MOLART is a multi-year cooperative project between art historians, restorers, and analytical chemists funded by the Netherlands Organization for Scientific Research to investigate the chemical aging of paints and varnishes. As varnishes age, they become yellow and brittle. Molecular alteration can change their solubility, leading to damage of the underlying pigments. The MOLART (molecular aspects of aging in painted works of art) program was established to develop a scientific framework for the conservation of painted art at the molecular level.

The chemistry of varnish yellowing and degradation is not known fully, but it involves oxidation, polymerization, and possibly isomerization reactions (Mills and White, 1994; Van der Doelen, 1999). These authors observed that the triterpenoid fraction in varnishes decreases during aging and that the varnishes contain similar oxidation products but with different distributions. These variations are due to several factors, including the age of the varnish, the environmental conditions in the museum, and the restoration history of the painting. For dammarane-type molecules, oxidation leads to a side chain of the ocotillone-type (e.g. 20,24-epoxyl-25-hydroxyl-dammaran-3-ol) and to a γ-lactone in the side chain (3-oxo-25,26,27-trisnor-darmmarano-24,20-lactone) (Figure 11.12). In dammarane-type compounds, oxidation of the A-ring produces dammarenolic acid. In the oleanane and ursane skeleton types, oxidation of C-28 in the side chain occurs readily, leading to an alcohol, an aldehyde, and finally a carboxylic acid group (Figure 11.13). Artificial aging using high-intensity UV light does not accurately reproduce natural oxidation of paintings. Low-level UV light mimics the natural aging process.

Van der Berg et al. (1996) investigated the oxidation of pine resins in varnishes and museum pieces. The aging of *Pinaceae* diterpenoid resins results predominantly from oxidation of abietane-diterpenoid acids to form dehydroabietic acid (DHA). Further oxidation yields mainly de-7-oxo-DHA and 15-hydroxyl-7-oxo-DHA (Figure 11.14). The ratio of these oxidation products can be used to express aging effects.

Molecular studies of resin-based finishes are not restricted to European paintings. Niiumura et al. (1999) used two-stage pyrolysis/GCMS to analyze Oriental lacquer films. Japanese or Chinese (urushiol-based), Vietnamese (laccol-based), and Burmese lacquers (thitsiol-based), as well as added components, such as linseed oil, can be differentiated.

ARCHEOLOGICAL WOOD TARS (PITCH)

Make thee an ark of gopher wood, rooms shalt thou make in the ark, and shalt pitch it within and without with pitch. (Genesis 6:15.)

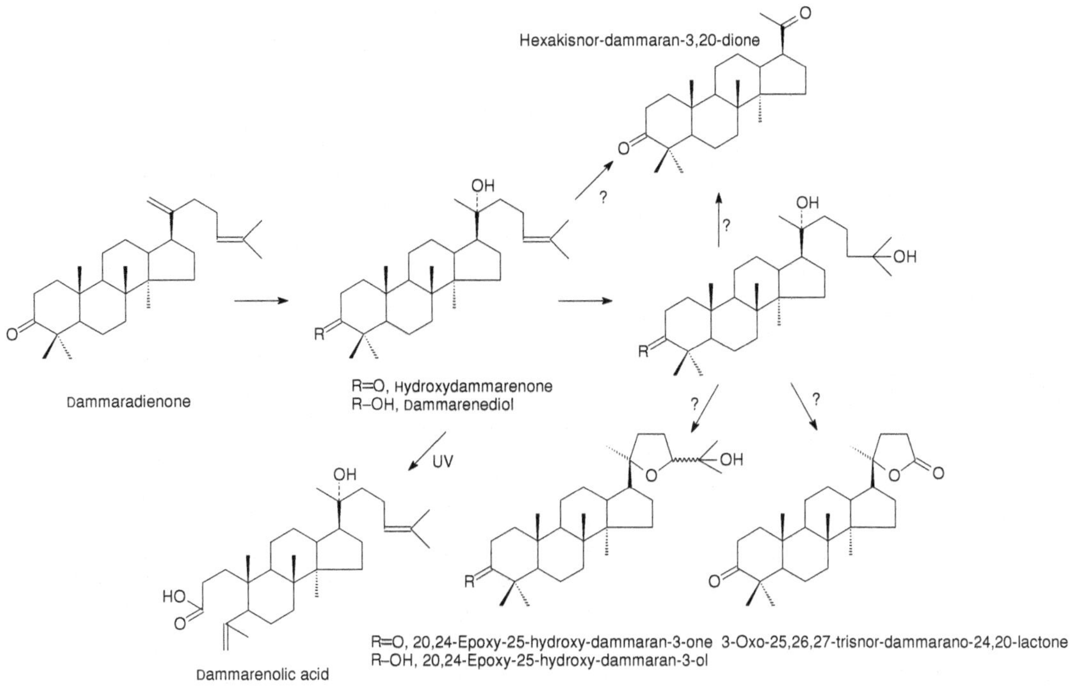

Figure 11.12. Different stages of side–chain oxidation in dammarane-type molecules. Formation of dammarenolic acid occurs only under artificial aging with intense ultraviolet light (van der Doelen, 1999).

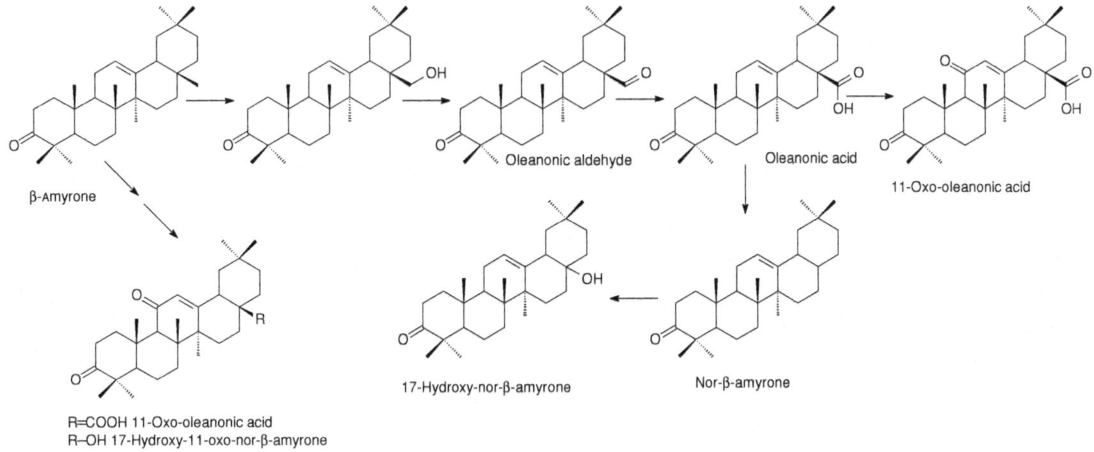

Figure 11.13. Oxidation of an oleanane-type molecule. C-11 can be oxidized to a conjugated keto group at any point in the reaction sequence (van der Doelen, 1999).

Wood tars or pitches were used widely in Europe, much like petroliferous bitumens in the Middle East. Unlike bitumens, a high level of technology is required to produce wood tars. After the bark or wood from resinous trees, such as pine, fir, and birch, is heated in air, a few small droplets of blackened resin remain. When these materials are heated with limited oxygen, however, large quantities of wood tar are produced by destructive

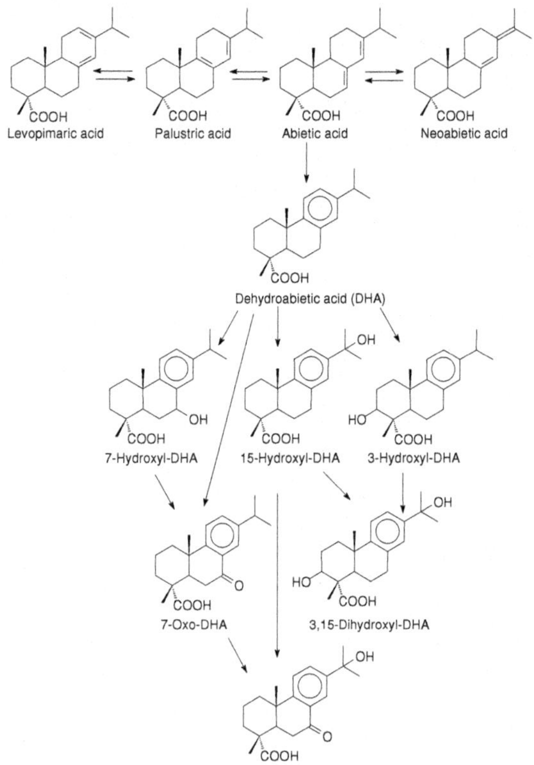

Figure 11.14. Oxidation reactions of abietic acid (van der Berg *et al.*, 1996).

pyrolysis. Control of temperature and oxygen are critical in order to obtain maximum yields of tar. The earliest fractions, turpentine and creosote, are enriched in mono- and sesquiterpenoids. The techniques of ancient tar manufacture are known only from literary sources and from scant archeological remains. Beck *et al.* (1998; 1999) investigated several ancient pitch ovens using pyrolysis and biomarker techniques to prove that relatively high temperatures were used. At one site, mono- and sesquiterpenoids in tar-soaked sand near a pitch oven were not considered to indicate a low operating temperature but may be the discarded first fraction of the pyrolysis process.

Wood tars and shipwrecks

One of the first biomarker studies of archeological pitch was conducted on barrels containing tar and on tarred caulking, ropes, and luting recovered from the *Mary Rose*, Henry VIII's sunken flagship (Evershed *et al.*,

1985). The tar from this shipwreck contained a series of tricyclic diterpenoid hydrocarbons nearly identical to modern Stockholm tar produced by destructive distillation of pine tree wood (Figure 11.15). In addition to the diterpenoid hydrocarbons, more than 80% of the ester fraction was methyldehydroabietate and more than 90% of the acid fraction was dehydroabietic acid. These compounds form from abietic acid, which is abundant in pine resin, during retorting. Beck and Borromeo (1990) found similar distributions identifying pine tars in an ancient Hellenistic shipwreck. Connan and Nissenbaum (2003) also identified pine tars in keel and hull planking from a shipwreck dating to ~500 BC, offshore Israel.

Paleolithic-Mesolithic birch bark tar

Wood tars, particularly from birch bark, have been recovered from many Mesolithic and Neolithic sites throughout northern Europe, where they were used as adhesives and as hafting, sealant, and waterproofing materials (Aveling, 1998; Pollard and Heron, 1996; Urem-Kotsou *et al.*, 2002). Birch bark tar was identified on the copper ax and arrows found with the frozen remains of the Ice Man, a Neolithic body recovered from a glacier in the Otztal Alps on the Austrian-Italian border (Sauter *et al.*, 1992). Many prehistoric birch bark tars have well-defined tooth impressions, suggesting their use as medicine or chewing gum (Aveling, 1997). Birch bark tar was used in two remarkable ~80 000-year-old Neanderthal stone tools recovered from an open-pit lignite mine in the Harz Mountains, Germany (Koller *et al.*, 2001). The pitch on one flint tool had a fingerprint and the imprint of wood cells, indicating that the pitch was used as an adhesive to secure a wooden shaft to the flint blade.

How these tars were produced in the Paleolithic-Mesolithic age remains a mystery. Experiments show that tar begins to form at ~340°C but is produced efficiently only at much higher temperatures. Unless the bark is heated in a sealed vessel with limited air, it chars and no tar is produced. Ceramic vessels were available since the Neolithic age, but no archeological evidence indicates how the process was achieved in earlier times. Modern attempts to produce tar by combining birch bark with heated stones in a pit were unsuccessful (Aveling and Heron, 1999). Because birch bark tars do not form naturally, their use as far back as

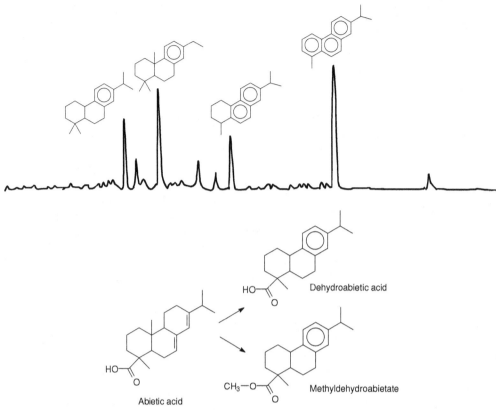

Figure 11.15. Partial gas chromatogram of the hydrocarbon fraction of tar from the *Mary Rose* (top) (Evershed *et al.*, 1985). Modern Stockholm tar produced by the destructive distillation of pine wood has nearly identical composition, including abundant dehydroabietic acid and methyldehydroabietate (bottom).

80 000 years ago suggests that ancient man had a higher degree of manufacturing sophistication than was once believed possible.

The biomarker signature of birch bark tar is diagnostic (Ruthenberg *et al.*, 2001). Birch bark is particularly enriched in biomarkers with a lupane skeleton (Figure 11.6) and contains none of the diterpenoids characteristic of resinous substances in the *Pinaceae*

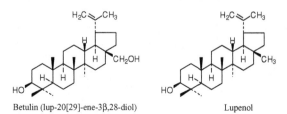

Figure 11.16. Betulin and lupenol are common pyrolysis products of birch bark.

family, such as pine, spruce, and fir. Up to 20–30% of the bark can be converted to betulin, lupenol, and lupenone by pyrolysis. Detection of this suite of triterpenoids is routine using GCMS (Binder *et al.*, 1990; Hayek *et al.*, 1990; Hayek *et al.*, 1991). Although pine-wood tars were identified in prehistoric pottery fragments (Heron *et al.*, 1991) and as hafting glue (Sheldrick *et al.*, 1997), positive identification is more difficult than for birch-bark tars, where the presence of betulin alone is considered diagnostic (Charters *et al.*, 1993; Heron *et al.*, 1991; Regert *et al.*, 1998b; Reunanen *et al.*, 1993). Koller *et al.* (2001) found that the biomarker distribution in ~80 000-year-old birch-bark tar was remarkably well preserved. Gas chromatography and GCMS analyses indicate that the pitch from the open-pit mine in the Harz Mountains consists predominantly of pentacyclic triterpenoids of the lupane series with betulin forming the major component.

Mankind used various organic substances other than bitumen, resins, and wood tars as adhesives. Baumer and Koller (2002) analyzed the material used to bond gold leaf to the surface of a cult tree discovered at a third-century Celtic oppidum (a fortified site on an elevated location) in Manching, Germany. Based on GCMS analysis, which revealed the presence of wax esters, wax alcohols, sterols, and fatty acids, they concluded that the organic adhesive was lanolin. Lanolin, also known as wool fat and wool oil, can be extracted from sheep wool with boiling water.

PALEODIETS AND AGRICULTURAL PRACTICES

The availability of food influenced much of mankind's history and culture. A stable food source was necessary for a rapidly increasing population and the advent of cities. Food was and continues to be a driving force for trade and wars between cultures. Surviving documents and relics provide much of our knowledge of paleodiets from historic time. Prehistoric paleodiets, however, are revealed mostly from biomarkers and isotopes in archeological remains.

Stable isotopes and paleodiets

The adage "you are what you eat" is the principle behind the use of stable carbon and nitrogen isotopes to indicate paleodiet. The ranges of $\delta^{13}C$ and $\delta^{15}N$ for various foods are known, as are isotopic fractionations that take place after consumption, the changing nature of the isotopic pool through time, post-depositional alterations, and the influence of soil contaminants. Therefore, measurements of $\delta^{13}C$ and $\delta^{15}N$ of organic matter from human remains can be used to indicate their diet. Isotopic composition is a time-averaged value of total food consumption, thus providing clearer evidence for long-term paleodiets than associated artifacts (e.g. faunal, floral, and skeletal remains).

Most archeological stable isotopic studies use organic matter extracted from bones and teeth, specifically collagen. Bones and teeth are the most common human remains, and their hard apatite matrix provides a measure of protection from degradation and contamination of the entrained collagen. In adults, bone collagen is continually built from consumed protein and is used to replace older collagen. It takes about ten years for adult humans to replace collagen in a large bone, such as a femur. Therefore, adult human bone collagen provides an average record of all of the protein eaten by a person over about a ten-year period.

The factors that determine stable carbon isotopic values in fossil remains are understood well (Ambrose, 1993; DeNiro and Epstein, 1978; Koch et al., 1992a; Schoeninger and DeNiro, 1984; van der Merwe and Vogel, 1978). The $\delta^{13}C$ of collagen can be used to distinguish the amount of marine (fish and shellfish) versus terrigenous (e.g. grains, meat, and milk) protein and the amount of C3 versus C4 plants consumed (Figure 11.17). The $\delta^{13}C$ of organic matter increases by $\sim 1‰$ upon consumption. Hence, the collagen of herbivores is $\sim 1‰$ heavier (^{13}C-enriched) than that of the plants they eat, and the collagen of carnivores is $\sim 1‰$ heavier than that of the herbivores or $\sim 2‰$ heavier than that of the plants. Omnivores that eat both plants and animals are difficult to distinguish from either herbivores or carnivores because their $\delta^{13}C$ values are intermediate, at $\sim 1.5‰$ heavier than those of the plants. In addition, because of the canopy effect, species that live in forest environments can have $\delta^{13}C$ values that are more negative than those of species that live in open environments.

Primary marine plankton and land plants can have a wide range of $\delta^{15}N$ depending on mean annual precipitation and their ability to fix nitrogen. Consumption results in $\sim 2–4‰$ enrichment in $\delta^{15}N$ (Figure 11.18). This large fractionation permits the use of $\delta^{15}N$ to distinguish the relative amounts of plant and animal foods in paleodiets and to identify the trophic levels organisms within ecosystems. Herbivores are $\sim 3‰$ heavier, omnivores $\sim 4.5‰$ heavier, and carnivores $\sim 6–7‰$ heavier than the primary plants. Nursing infants are at the highest trophic level, and their proteins are $\sim 3‰$ heavier than their mothers' (Fogel et al., 1997).

Reconstruction of paleodiets can be refined using stable isotopic analysis of amino acids, typically obtained by degrading collagen (Hare et al., 1991). Humans retain the ability to synthesize some amino acids de novo, while other amino acids must be obtained from the diet. These essential amino acids reflect paleodiet directly, and trophic effects can be defined using $\delta^{13}C$. Ancient human hair may be better than bone collagen for such studies (Macko et al., 1999).

The first archeological studies of paleodiet using stable carbon isotopes were conducted on the remains

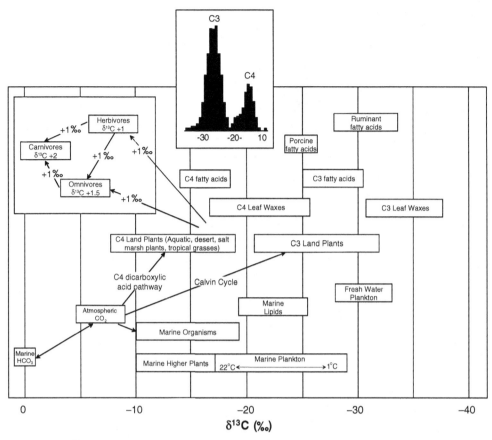

Figure 11.17. $\delta^{13}C$ ranges for primary modern producers and consumer lipids (modified from Stahl (1979) and Evershed *et al.* (1999)). Distribution of $\delta^{13}C$ for land plants from Deines (1980b). Diagram of $\delta^{13}C$ fractionation in the food web from Fogel *et al.* (1997).

of northeastern Native American populations to determine when maize was introduced. A shift toward heavier $\delta^{13}C$ occurred $\sim$1000 AD, consistent with initial consumption of maize, a C4 plant with $\delta^{13}C \sim -12.5\%o$ (Vogel and van der Merwe, 1977; van der Merwe and Vogel, 1978). Tauber (1981) conducted one of the first studies of Mesolithic and Neolithic humans. Bone collagen extracted from Mesolithic remains from coastal Denmark had $\delta^{13}C$, indicating a marine diet ($\delta^{13}C \simeq -12\%o$). Bone collagen extracted from Neolithic remains, however, was more depleted in ^{13}C ($\delta^{13}C \simeq -20\%o$), indicting a shift from marine foods. This Mesolithic/Neolithic transition from a predominantly marine to a terrigenous diet has been observed in other coastal areas, such as Portugal (Lubell *et al.*, 1994) and Britain (Richards and Hedges, 1999). There is now a considerable database of $\delta^{13}C$ for collagen from

post-Mesolithic human bones that also have radiocarbon dates. Since the Neolithic, the $\delta^{13}C$ for bone collagen in Europeans has remained fairly constant at $\sim -20\%o$.

Recent studies of Neanderthal remains illustrate the utility of coupling $\delta^{13}C$ and $\delta^{15}N$ to determine trophic level (Figure 11.19). Bone collagen from Neanderthals at Marillac, France ($\sim$40 000–45 000 years before Present (BP), Scladina Cave, Belgium ($\sim$80 000–130 000 years BP) (Bocherens *et al.*, 1999), and Vindija Cave, Croatia ($\sim$28 000–29 000 years BP) (Richards *et al.*, 2000b), as well as Upper Paleolithic humans at Gough's Cave, UK (Richards *et al.*, 2000a), indicate that these populations were at the top trophic level and obtained nearly all of their dietary protein from animal sources. The $\delta^{15}N$ for these Neanderthal and early human remains are nearly identical to those for bone

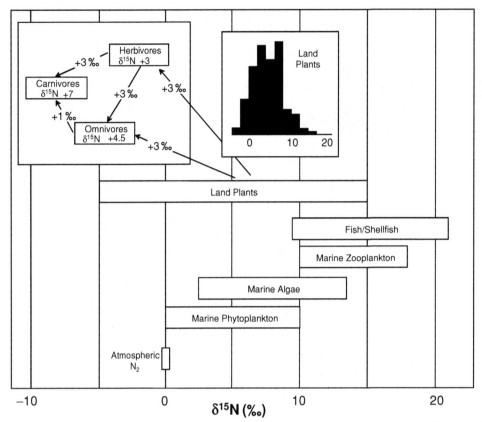

Figure 11.18. $\delta^{15}N$ ranges for primary modern producers and consumer lipids (modified from Kaplan, 1983). Distribution of $\delta^{15}N$ for land plants and diagram of $\delta^{15}N$ fractionation in the food web from Fogel *et al.* (1997).

collagen from other top carnivores, such as wolves and panthers, from the same sites or nearby sites of similar age.

Isotopic values for some faunal remains shown in Figure 11.19 illustrate remaining issues in understanding paleodiet. Bison and cervid samples from Vindija Cave have $\delta^{15}N$ consistent with those for other European Holocene specimens. However, cave bear samples are puzzling because they have very low $\delta^{15}N$ that are not consistent with omnivores. These low values suggest a high degree of herbivory, but they may be the result of unusual metabolism associated with hibernation. Wooly mammoth $\delta^{15}N$ values, however, are higher than those of other herbivores and may result from consumption of specific plant species. Other herbivores may have been less selective for the plants that they consumed. The high $\delta^{15}N$ for Neanderthals is consistent with mammoths as their main dietary protein source.

Isotopic analysis of individual amino acids in bone collagen is providing another level of refinement in paleodietary reconstructions (Jones *et al.*, 2001). Laboratory studies showed that in animal collagen, the essential amino acids are used directly from the diet with little isotopic fractionation, while the non-essential amino acids are synthesized and reflect the $\delta^{13}C$ values of total dietary carbon (e.g. carbohydrates and lipids). The potential of this technique was demonstrated in studies that identified regional dietary preferences of late Stone Age populations along the South African coast (Corr *et al.* 2002) and of prehistoric North Americans (Fogel and Tuross, 2003)

By combining bulk and compound-specific isotopic values with dietary lipid analyses (see below), it may be possible to reconstruct detailed descriptions of changes in diet and farming practices. The dry desert air of the Qasr Ibrîm site, a fortified citadel in Nubian Egypt that

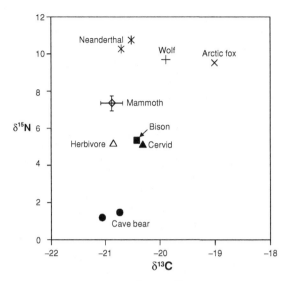

Figure 11.19. Bone collagen $\delta^{13}C$ and $\delta^{15}N$ for Neanderthals and associated fauna from Vindija Cave, Croatia, dated at 28 500 years BP and other faunal values from 22 000–26 000 years before Present (BP) for sites in the Czech Republic. Reprinted from Richards *et al.* (2000b). © Copyright 2000, with permission from Elsevier.

was occupied from ∼1000 BC to 1812 AD, provided conditions favorable for exceptional preservation of plant biopolymers and plant and animal lipids (Evershed *et al.*, 1997d; Regert *et al.*, 1998a). By examining the $\delta^{13}C$ of $C_{16:0}$ and $C_{18:0}$ lipids found in livestock bones and potsherds spanning the time of site occupation, Copley *et al.* (2002) found that the diet of the cattle changed over time, with higher quantities of C4 plants being prevalent during later times. This conclusion was supported by independent measurements of the $\delta^{13}C$ and $\delta^{15}N$ of bulk collagen and the $\delta^{13}C$ for apatite, cholesterol, and individual amino acids from bones. In contrast, the diet of sheep and goats did not change over time.

While numerous studies of human organic remains focus on bone collagen, amino acid racemization, and proteins (Bada *et al.*, 1999; Collins *et al.*, 1999; Flannery *et al.*, 1999), few publications examine human remains for their biomarker lipid signatures. Gülaçar *et al.* (1990) examined ancient Egyptian mummies and Evershed (1992) and Evershed and Connolly (1994) investigated the Lindow Man bog body. In the latter study, comparison of sterols in the body tissues and the associated peat shows that mobility of sterols in the aqueous peat bog was restricted because of their hydrophobic

properties. Evidence supported postmortem molecular transformations mediated by endogenous and possibly exogenous bacteria.

Stott and Evershed (1996) proved that $\delta^{13}C$ measurements could be made reliably on cholesterol extracted from the skeletal remains of individuals excavated from coastal cemeteries in northern Lincolnshire, UK. Most of the bones, dating from Saxon times to the eighteenth century, contained cholesterol with $\delta^{13}C$ ∼−22‰, indicating a strong preference for marine foods by the community over the past ∼1500 years. A few individuals exhibited $\delta^{13}C$ as negative as −26‰, indicating preferences for terrigenous rather than marine foodstuffs. Stott *et al.* (1997) proved that $\delta^{13}C$ of cholesterol isolated from fossil whales, dated at 9735 ± 160 years and 75 000 ± 15 000 years BP, were within the range expected for the bulk fat of marine mammals. The isotopic depletion between the cholesterol and collagen was similar for both ancient and modern whale bones, indicating reliable preservation of the $\delta^{13}C$ signal between ancient cholesterol and the protein constituents. Stott *et al.* (1999) extended $\delta^{13}C$ analysis of cholesterol extracted from bones to a wide range of human and animal remains dating to the Mesolithic Period.

Cholesterol $\delta^{13}C$ provides information complementary to that from collagen and apatite, which are used more commonly in paleodietary studies. Stott *et al.* (1999) showed the potential advantages of using $\delta^{13}C$ of biomarker lipids, such as cholesterol. Cholesterol $\delta^{13}C$ retains isotopic integrity through diagenesis, provided that the carbon skeleton is not altered. Cholesterol $\delta^{13}C$ reflects the original isotopic composition of the carbohydrates and fats in the diet and is turned over more rapidly than collagen, so that the $\delta^{13}C$ of cholesterol represents a shorter timescale than collagen.

Note: The Nahal Heimar Cave, on a cliff near the Dead Sea just northwest of Mount Sedom in Israel, contains evidence for the oldest glue, carbon dated at 6310–6110 BC (Walker, 1998). The blackish glue was used as a waterproofing agent and an adhesive for many items from daily life and on ritual objects, such as stone masks and decorated skulls. The substance was thought to be asphalt, but chemical analyses proved it to be collagen glue extracted from animal skins (Connan, 1996). While ancient Egyptians used collagen-based glues ∼4000 years ago, the sophistication of the Neolithic people of

Nahal Heimar Cave was unexpected because they had not yet begun to produce pottery.

Dietary lipids

The molecular and isotopic compositions of food residues in ancient potsherds, cooking vessels, and soils provide archeologists with insights into the ancient dietary, agricultural, and culinary practices. These residues can be classified as biological n-alkanes and functionalized lipids, pyrolyzates from cooking, and proteinaceous material. Natural lipids are the best preserved and most diagnostic of the food residues. They are absorbed readily into the surfaces of fired and unfired clay pots, which were used by all ancient civilizations to store and prepare food. Charred food remains on the surfaces of cooking pots can be identified by visual inspection but provide less chemical information on their origins (Oudemans and Boon, 1991; Regert et al., 1998a). Proteinaceous material can survive in archeological potsherds and associated soils (Evershed and Tuross, 1996). However, proteins degrade rapidly, and

fats (Condamin et al., 1976; Patrick et al., 1985), but these distributions were insufficient to differentiate the sources of archeological lipids. Advances in high-temperature gas chromatography allowed Evershed et al. (1990) to characterize the free fatty acids and n-alkanes as well as the non-degraded triacylglyercols, wax esters, and long-chained n-alkanols and ketones. When coupled with compound-specific isotopic analyses, the origins of the archeological lipids can be correlated to specific animal and plant sources.

Degraded and intact animal fats are common lipid residues in potsherds that are recognized easily by the distribution of saturated C_{16} and C_{18} fatty acids. Adipose fats from ruminant and non-ruminant animals can be distinguished by the distribution of monounsaturated isomers of fatty acids and by compound-specific isotopic analysis (Evershed et al., 1999; Evershed et al., 1997b; Mottram et al., 1999). Potsherds record evidence that animal fats were heated to high temperatures during cooking. The fatty acids undergo a ketonic decarboxylation, resulting in self- and cross-head-to-head condensation and the formation of ketones, as follows (Evershed et al., 1995):

$$CH_3(CH_2)_nCO_2H + CH_3(CH_2)_mCO_2H \xrightarrow[\substack{-CO_2 \\ -H_2O}]{>300°C} CH_3(CH_2)_nCO(CH_2)_mCH_3$$

gelatin/collagen amino acid patterns occur only in the best preserved samples. Hydrolysis of potsherds liberates substantial quantities of amino acids, but their distribution is not sufficiently diagnostic to contribute to paleodietary analysis. In plant remains excavated from Qasr Ibrîm, Evershed et al. (1997e) found evidence for specific diagenetic reactions that occurred since burial. Abundant alkyl pyrzaines, which are characteristic byproducts of the Maillard condensation reactions of proteins and carbohydrates, were trapped within networks of structural and storage macromolecules. In contrast, fatty acid distributions ($C_{12:0}$ and $C_{14:0}$) that are diagnostic for the processing of palm fruit were detected in closed-form vessels from this location (Copley et al., 2001).

Use of fatty acids as dietary markers in potsherds requires knowledge of their biosynthetic origins, their diagenesis through natural burial and aging, and alteration under cooking conditions. Early gas chromatographic studies compared the distribution of saturated fatty acids extracted from potsherds with those in reference

The reaction is catalyzed by metal oxides and occurs at temperatures >300°C. For animal fats, the subscripts n and m range from 13 to 16 (Evershed et al., 1995).

Polycyclic biomarkers also may be diagnostic of dietary plant lipids. Decavallas et al. (2002) analyzed ceramic vessels from the Neolithic site of Bercy, Paris, France, where the amount and different species of preserved seeds indicated a wide use of plant materials. Extracts from 2 of 22 potsherds contained about four times the amount of palmitic ($C_{16:0}$) acid compared with stearic ($C_{18:0}$) acid and significant amounts of unsaturated C_{18} fatty acids. Campesterol, 5α-campestanol, sitosterol, and 5α-stigmastanol, which are higher-plant sterols formed by direct biosynthesis or diagenesis, also were detected in these potsherds. These were accompanied by triterpanoid compounds in the oleanane and lupane series, including a very high-mass (m/z 664) species that is probably indicative of a triterpanoid palmitate.

Investigations of biomarkers in cooking pots excavated from a late Saxon medieval settlement in England illustrate the progress made in characterizing archeological lipids (Evershed *et al.*, 1999). Three lipids dominated extracts from a cooking pot: *n*-nonacosane, nonacosan-15-one, and nonacosan-15-ol. They occurred in proportions nearly identical to those in the epicuticular leaf wax of *Brassica* (cabbage) (Evershed *et al.*, 1991). Isotopic analysis of the lipids proved that they were from a C3 plant (Evershed *et al.*, 1994) with $\delta^{13}C$ (-33.1 to $-34.8‰$) that was ^{13}C-rich compared with lipids from modern wild cabbage ($-35.8‰$). These results are consistent with the isotopic shift in atmospheric CO_2 toward more negative $\delta^{13}C$ caused by the modern burning of fossil fuels. Concentrations of the cabbage lipids varied within the cooking pot, decreasing from the top toward the base. Laboratory experiments show that this distribution can be duplicated in replica ceramic jars by boiling cabbage leaves (Charters *et al.*, 1997). Boiling causes non-selective mobilization of the epicuticular wax and its incorporation into the porous ceramic.

Dairy practice biomarkers

Milk production from domesticated animals is documented well in ancient cultures. Historic records from North Africa and the Near East prove organized dairying began by ~4000–2900 BC. Milk and milk products were major economic commodities for the Romans, who had large, well-organized dairy farms throughout their empire. Milk lipids can be distinguished from lipids in adipose fats by the presence of C_4–C_{14} fatty acids and their corresponding triacylgylcerols.

Selective degradation of the short-chained C_4–C_{14} fatty acids in milk results in their absence in ancient potsherds (Figure 11.20) (Dudd and Evershed, 1998). Simulation experiments show that milk lipid triacylglycerols hydrolyze faster than their long-chain analogs. Once released, the short-chain fatty acids are readily biodegraded and are far more soluble in water than the longer-chain fatty acids. Degraded milk lipids yield fatty acid distributions nearly identical to those produced by animal fats (Figure 11.21). Thus, although milk lipids may be common in ancient potsherds, they may not be recognized.

By examining the isotopic compositions of fatty acids in milk and apidose fats, Dudd and Evershed (1998) and Dudd *et al.* (1998) found that milk lipids could be distinguished in ancient potsherds. The $\delta^{13}C$

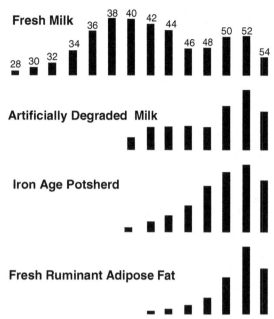

Figure 11.20. Triacylglycerol distributions in fresh milk, artificially degraded milk, an Iron Age potsherd from a Romano-British vessel excavated at Stanwick, and fresh ruminant adipose fat. Reprinted with permission from Dudd and Evershed (1998). © Copyright 1998, American Association for the Advancement of Science.

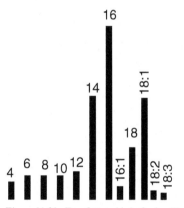

Figure 11.21. Free fatty acids in fresh milk.

of the $C_{16:0}$ and $C_{18:0}$ fatty acids vary, reflecting different biosynthetic pathways (Figure 11.22). $C_{16:0}$ fatty acid in milk is synthesized in the mammary gland, whereas the $C_{18:0}$ fatty acid originates, in part, directly from diet. The $C_{16:0}$ and $C_{18:0}$ fatty acids in ruminant adipose fats also originate from both diet and *de novo* synthesis. During lactation, however, the dietary fatty acids are

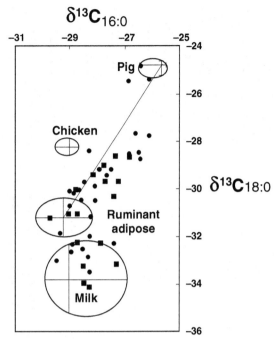

$\delta^{13}C_{16:0}$

Figure 11.22. Carbon isotopic ratios of $C_{16:0}$ and $C_{18:0}$ n-alkanoic acids in lipids extracted from potsherds from a late Saxon/early medieval site in Northamptonshire, UK (●) and an Iron Age Romano-British site in Stanwick, UK (■). Many potsherds contain fatty acids with $\delta^{13}C$ that fall along a mixing curve between porcine (pigs) and ovine/bovine (sheep/cows) sources. However, some potsherds have ^{13}C-depleted $C_{18:0}$ fatty acids and exhibit triacylglycerol distributions characteristic of degraded milk. There was no obvious correlation between the type of archeological vessel and the lipid origins. Reprinted with permission from Dudd and Evershed (1998). © Copyright 1998, American Association for the Advancement of Science.

routed directly to milk, giving rise to more ^{13}C-depleted $C_{18:0}$ fatty acid compared with $C_{16:0}$ fatty acid within the adipose fats in the same ruminant animal.

> Note: Studies of archeological lipids require reference lipids from living animals for comparison. Modern agricultural feeds are mixtures of C3 and C4 plants, fish oils and meals, cane sugars, animal parts, and supplements. Consequently, the stable isotopic ratios of most modern animals are anomalous. To obtain isotopic reference samples, domesticated animals are raised on organic farms with controlled C3 diets without supplements or commercial feeds. A correction for the effects of fossil-fuel burning on the isotopic ratio of

atmospheric carbon dioxide also must be applied (Evershed et al., 1999).

Copley et al. (2003) extended this work to potsherds from prehistoric Britain, dating from the Neolithic Age, Bronze Age, and Iron Age. Evidence for milk lipids was found at all 14 sites, showing that dairying was an established practice throughout British prehistory. These results are consistent with the hypothesis that the exploitation of animals for milk was already an established farming practice by the time organized agriculture arrived in Britain ~7000 years ago.

Biomarkers of wine

Several compound classes are biomarkers for wine and/or grapes. Tartaric acid (McGovern et al., 1996; 1999) and gallic acid (3,4,5-trihydroxybenzoic acid) (Condamin et al., 1976) provide circumstantial evidence for grape fermentation. Both acids occur in high concentrations in grapes but also are common as both free acids and as part of tannins in many plants. Gallic acid, for example, can be found in gallnuts, sumach, tea leaves, and oak bark. More specific biomarkers can be released from grape tannins using a soft depolymerization technique developed by Hernes and Hedges (2000) for fresh produce. Polyphenols (m/z 354 and 367) form as degraded derivatives of catechin, a three-ring flavanol. Using this procedure, Garnier and Regert (2002) demonstrated that these compounds could be recovered in archeological samples, such as a closed amphora of Roman wine from a shipwreck (70 BC) and from grape seeds from several locations (400–100 BC).

ARCHEOLOGICAL BEESWAX

Beeswax occurs with a wide variety of archeological artifacts, including Neolithic pottery (Heron et al., 1994), Egyptian mummification balms (Figure 11.23) (Connan, 1999; Colombini et al., 2000; Buckley and Evershed, 2001; Maurer et al., 2002), Minoan lamps (Evershed et al., 1997d), and varnishes from oil paintings (van der Doelen, 1999). Beeswax is a mixture of long-chain (C_{24}–C_{32}) alcohols, C_{25}–C_{33} odd–carbon-numbered n-alkanes, palmitic wax esters (C_{40}–C_{52}), and hydroxyl wax esters (C_{42}–C_{54}). This chemical fingerprint is distinct, and beeswax can be identified even when mixed with animal and plant lipids (Charters et al.,

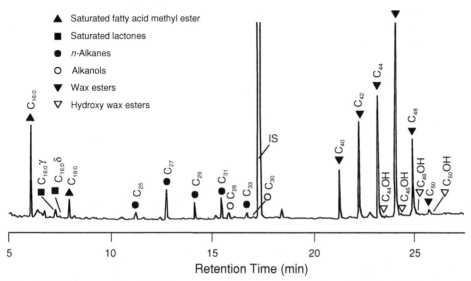

Figure 11.23. Reconstructed total-ion chromatogram of the trimethylsilylated neutral fraction of balm extracted from a Late Period (664–404 BC) mummy. Beeswax in this sample comprised 38% of the embalming resin (from Buckley and Evershed, 2001).

1995). A series of diesters also are diagnostic for beeswax. However, these compounds are heat- sensitive and require liquid-phase chromatographic analysis. Using a combination of SIM/GCMS and MRM/GCMS, Garnier *et al.* (2002) identified over 50 biomarkers for beeswax in an Etruscan cup at parts-per-million levels. They concluded that the beeswax was used either for waterproofing or as a fuel.

The use of beeswax as fuel for lamps was proved through chemical analysis. Prehistoric lamps are rare in the Neolithic Age but appear with increasing frequency beginning in the early Bronze Age, particularly in Aegean cultures. It was long assumed that these lamps burned olive oil because they originated at about the same time as the cultivation of olives. Kimpe *et al.* (2001) found direct evidence for the use of olive oil in late Roman to early Byzantine ceramic oil lamps discovered at the archeological site of Sagalassos, southwest Turkey. They used LCMS techniques to detect triacylglycerols, characteristic of olive oil, as the predominant lipids in extracts from the lamps. The presence of multiple unsaturated triacylglycerols and traces of saturated triacylglycerols indicated that other oils and animal fat were added.

Other ancient oil lamps may have used numerous other natural materials as fuel, including other plant oils and animal fats, resins, bitumens, and natural waxes.

Chemical analysis of terracotta lamps from late Minoan (~1600–1450 BC) settlements and eastern Crete revealed complex mixtures of wax esters, long-chain alcohols, and *n*-alkanes, suggesting beeswax (Evershed *et al.*, 1997d). The distribution of these compounds, however, was different from that in fresh beeswax, presumably due to degradation. The long-chain alcohols have an ambiguous origin. Compound-specific isotopic analysis proved that the lamps contained beeswax residues. $\delta^{13}C$ of the wax esters, long-chain alcohols, and degraded palmitic acid moieties derived from the wax esters correlated closely and eliminated the possibility that these lipids could be from C3 plants.

Excavations from Isthmia, Greece, uncovered combed pottery resembling ceramics that are still in use on the Cycladic islands and Crete as beehives. Chemical analysis of the residues from these excavated coarse-ware vessels found evidence for *n*-alkanes, wax esters, fatty acids, and long-chain alcohols with $\delta^{13}C$ values diagnostic of beeswax (Evershed *et al.*, 2003).

Beeswax was identified in artifacts excavated from the Tumulus Mount at Gordion, central Turkey, which is believed to be the tomb of Phrygian King Midas of the eighth century BC (McGovern *et al.*, 1999). The site contained hundreds of drinking vessels, mixing bowls, and serving dishes encrusted with food residues. Molecular techniques identified many animal fats, drink residues, and spices, which collectively reflect a royal feast. Among the many drinking vessels were some that contained long-chain wax esters diagnostic of beeswax. These residues are believed to be evidence for mead, the waxes being residues that were not filtered out during preparation. Other markers for fermented beverages included tartaric acid and its salts, which are abundant only in grapes and wine, and calcium oxalate, also known as beerstone, the main precipitate of barley beer. Triacylglycerols and fatty acids characteristic of lamb were found in food vessels along with polycyclic aromatic hydrocarbons (e.g. phenanthrene) and alkylated phenol derivatives (e.g. cresol), typical of barbecued meat. Other compounds indicative of a banquet included anisic acid (from anise or fennel), chondrillasterol (lentil), elaidic acid (the *trans* isomer of oleic acid, characteristic of olive oil), and terpineol and terpenoid signatures of spices.

Organic artifacts in the tomb were severely degraded by a soft-rot fungus. Filley *et al.* (2001) used nitrogen isotopes to show that the decay patterns provide clues as to the king's diet. The worst decomposition within the chamber is around the cedar tomb and on a small table nearby. This observation is atypical because soft-rot fungus needs large amounts of fixed nitrogen, which wood typically lacks. The nitrogen must have come from different sources, most likely the decaying body of King Midas and the remains of his funerary feast left on the table. The rot from the table and tomb was enriched in ^{15}N, indicating that food offerings were mostly meat and that King Midas consumed a meat-rich diet.

BIOMARKERS AND MANURING PRACTICES

Soil contains biomarkers that reflect historic agricultural use. For example, by examining organic matter in a Yorkshire, UK lacustrine sedimentary sequence that spanned the past 3000 years, Fisher *et al.* (2003) identified periods of deforestation at ~600 BC and 1200 AD. One higher-plant triterpenoid, identified as 28-carboxyursen-12-enol (Figure 11.24), was particularly diagnostic, appearing only in sediments where tree and scrub pollen dominated.

Several soil studies focused on the use of manure as fertilizer and in tracing sewage systems. As discussed in Chapter 11, coprostanol (5β-cholestan-3β-ol) is used

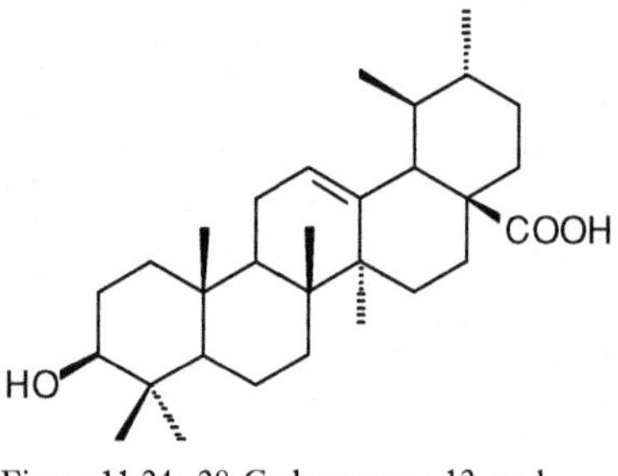

Figure 11.24. 28-Carboxyursen-12-enol.

routinely as a marker for sewage pollution (Grimalt *et al.*, 1990). It is the major sterol in omnivorous mammalian feces and forms by microbial catabolism of cholesterol (5α-cholestan-3β-ol) in the gut. Bethell *et al.* (1994) tested the hypothesis that coprostanol could be used as a marker of manure in archeological soils. Samples from a modern latrine, a seventeenth-century garderobe, a medieval garderobe, and two soils suspected to be from Roman cesspits contained coprostanols. Although low levels of coprostanol also occur in control soil samples, the relative proportion of the 5β-stanols provided a distinct chemical signature for fecal matter. Soil lipids, micromorphology, and stable carbon isotopic analyses show that a dark-colored loam horizon near Tofts Ness, Scotland, was cultivated with a mixture of grassy turf and animal feces (Simpson *et al.*, 1998). The horizon is associated with Bronze Age cultural landscape activity that may have started during the late Neolithic Period. Simpson *et al.* (1999) further examined agricultural soils from West Mainland Orkney dating from the twelfth to the nineteenth century. Soil lipid concentrations reflected the areas used for farming. Campesterol, sitosterol, and 5β-stigmastanol confirmed the use of ruminant animal manure, while coprostanol and hyodeoxycholic acid confirmed the use of pig manure. Bull *et al.* (1999a) reported on manure usage in Crete during the Minoan Age, combining biomarker and compound-specific isotopic analyses. The archeological community is rapidly becoming aware of the use of biomarkers to study ancient agricultural practices (Bull *et al.*, 1999b).

Modern experimental agricultural sites provide natural, controlled conditions in order to examine the fate of biomarker lipids from manure and compost. Evershed *et al.* (1997a) showed that 5β-stanol markers, characteristic of farmyard manure, persist where manure was applied at concentrations above those in

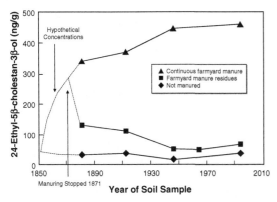

Figure 11.25. Concentration of 24-ethyl-5β-cholestan-3β-ol in the three test plots from the Hoosfield Spring Barley experiment. 5β-Stanols still persist above background concentrations in an intensely cultivated plot where manure use ended in 1871. Modified from Bull et al., (1998). © Copyright 1998, with permission from Elsevier.

non-manured and control areas. Analysis of soils from the Rothamsted Experimental Station, Hertfordshire, UK, proved particularly diagnostic (Bull et al., 1998; 2000). Soils were collected over time from experimental plots where (1) barley was cultivated continually with farmyard manure, (2) manure use was suspended in 1871, and (3) no manure was used. The manuring signal from 5β-stanols still persisted above background levels in soil that was cultivated intensely and received no manure since 1871 (Figure 11.25). 5β-Stanyl esters are even more persistent than free stanols and may better indicate manure use in archeological soils (Bull et al., 2000).

Isaksson (1998) traced animal products in soils through the analysis of sterols and their corresponding 5α-stanols. Analyses of samples from different cultural layers at Vendel, Uppland, Sweden, allowed statistical identification of culinary and non-culinary areas, essentially mapping the kitchen areas used by German aristocracy.

Bile acids provide further evidence for the origins of feces in both modern and archeological sites (Elhmmali et al., 2000; Bull et al., 2002). Bile acids are C_{24}, C_{27}, and C_{28} steroidal acids produced in the digestive tracts of animals. The carboxylic acid group is attached to C-23 on the steroidal side chain, which may be modified further by the addition of NSO moieties (e.g. keto or amino groups). Bile acids retain the steroidal 3α- or 3β-hydroxyl group on the A-ring and may have one or more

additional hydroxyl groups on the B- and C-rings. By combining sterol and bile acid analyses, it is possible to distinguish whether feces originated from humans, ruminants, pigs, or dogs.

High concentrations of bile acids occurred in a Roman latrine ditch in Bearsden, Scotland (Knights et al., 1983) and in a Roman sewage culvert in Agora, Athens (Elhmmali, 1998; Bull et al., 2003). Bile acids have also been detected in 2000-year-old human coprolites (Lin et al., 1978) and in the tissues of a 4000-year-old Nubian mummy (Gülaçar et al., 1990).

ARCHEOLOGICAL DNA

Modern methods of amplification and sequencing allow characterization of ancient deoxyribonucleic acid (DNA), even when it is partially degraded. The technique has been used to study ancient diseases (Rollo and Marota, 1999) and the development of agriculture and animal husbandry practices (Brown, 1999; MacHugh et al., 1999). Ancient DNA combined with genetic information from living populations is helping to uncover patterns in human migration (Hagelberg et al., 1999; Merriwether, 1999), genetic diversity (Stone and Stoneking, 1999), and the origin of the species. The nature of these genetic compounds places them beyond the scope of this book. However, a brief discussion of recent work on Neanderthal DNA illustrates the utility and controversy of such studies.

There are two major, competing hypotheses for the origins of modern humans: recent replacement and multi-regional evolution (Figure 11.26). The recent replacement (recent African, or "out of Africa") hypothesis proposes that there is a sharp break between archaic and modern humans. In this hypothesis, early modern humans speciated in Africa and then migrated into Europe and Asia, replacing archaic Homo populations. The hypothesis predicts that the earliest modern anatomical forms originated in Africa ~100 000 years ago, that interbreeding with archaic Homo population was limited, and that hybrid populations did not arise. The replacement hypothesis also suggests that the early modern humans methodically exterminated the archaic humans. The multi-regional hypothesis proposes that Homo erectus populations in Africa, Europe, and Asia evolved independently into Homo sapiens. Sufficient gene flow occurred across continents to maintain species continuity, while geographic distance preserved

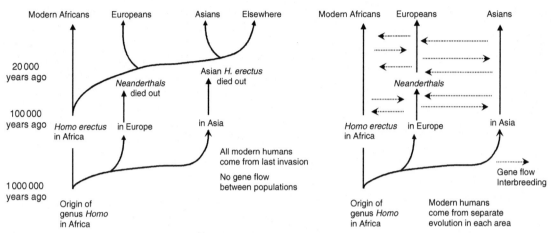

Figure 11.26. Two hypotheses attempting to explain the origin of modern humans.

regional differences. The multi-regional hypothesis predicts that no single definition of modern humans applies to all regions due to regional diversity, that the earliest modern humans did not necessarily originate in Africa, and that early modern humans should have some anatomical features derived from archaic *Homo* species in their region.

The relationship between the Neanderthals and modern humans is complex because the populations co-existed for thousands of years. Stone tools and personal ornaments believed to be made by modern humans were found with 34 000-year-old Neanderthal remains at Arcy-sur-Cure, France, suggesting some cultural exchange. Whether the modern and archaic humans interbred, however, cannot be determined from the fossil evidence. Anatomic data are equivocal, and selective interpretations can be made supporting either the recent replacement or the multi-regional hypothesis.

Molecular genetics entered into the controversy when studies of the living human population indicated little diversity in mitochondrial DNA (mtDNA). Assumptions on the rates of mutations led to the conclusion that modern man evolved recently and that all humans descended from a small African gene pool, perhaps even a single female, called the mitochrondrial Eve (Cann *et al.*, 1987; Stoneking and Cann, 1989). These conclusions obviously favored the replacement hypothesis and were attacked by those favoring the multi-regional hypothesis. Critics argued that there were flaws

in the mathematical treatment of the mtDNA data and that an underestimation of the variability in mutation rates could push the human split back to earlier times, far enough to include *Homo erectus*. Using a different set of assumptions, the mtDNA data can support the multi-regional hypothesis just as well as the replacement hypothesis (Relethford, 1998). DNA from the living human population is inconclusive, but DNA from the archaic and earliest modern humans might resolve the issue.

DNA begins to degrade immediately after cell death and is absent after 50 000–100 000 years. Under the most optimistic conditions, DNA remaining in a Neanderthal might consist of trace amounts of highly damaged, short fragments. The damaged state of this ancient DNA would make amplification by the polymerase chain reaction (PCR) far more prone to errors, including contamination from modern DNA. The first mtDNA for a Neanderthal (Krings *et al.*, 1997; 1999) was discovered in a sample from the original Neanderthal 1856 Feldhofer type specimen. Neanderthal mtDNA proved to be distinctly different from that of all living human populations, showing no affinity to European DNA. The researchers calculated that Neanderthals split from the human lineage ~500 000 years ago, indicating that they were an early offshoot of *Homo erectus* that did not lead to modern humans. A second mtDNA analysis on a Neanderthal fossil from the northern Caucasus gave similar results (Ovchinnikov *et al.*, 2000). This specimen is ~29 000 years old, making it one of the last living Neanderthals. Phylogenetic analysis placed the two Neanderthals from the Caucasus and western Germany together in a single group distinct from modern humans, suggesting that their mtDNA did not contribute to the modern human mtDNA pool. Caramelli *et al.* (2003) extracted and sequenced mtDNA from Cro-Magnon bones from ~25 000–23 000 years ago. The extracted sequences are nearly identical to modern human mtDNA and are very different from the Neanderthal specimens. Caramelli *et al.* (2003) concluded that the genomic discontinuity indicates that the Neanderthals contributed little if anything to the modern European gene pool. Many anthropologists now consider the Neanderthal and Cro-Magnon mtDNA studies as proof of the replacement hypothesis.

However, supporters of the multi-regional hypothesis have not yet capitulated (Wong, 1999). In addition to all of the arguments on assumptions of mutation rates, Paleolithic populations, and genetic drift, new physical evidence in support of multi-regionalism has been discovered. A 24 500-year-old skeleton of a child in Portugal appears to have hybrid features of Neanderthals and modern humans (Duarte *et al.*, 1999) and a 60 000-year-old skeleton from Australia has modern human features with no indication of an African origin (Thorne *et al.*, 1999). Advocates of the rapid replacement hypothesis counter that these anatomical features are adaptations to local environments. Furthermore, the mtDNA evidence is not accepted universally because it may be impossible to rule out modern contamination (Abbott, 2003).

Analysis of ancient DNA from the fossils of archaic and early modern man now seems technically feasible, and protocols have been established to minimize the likelihood of modern contamination (Cooper *et al.*, 1997; Cooper and Poinar, 2002). Global understanding of genetic diversity in the Paleolithic age may be improved by additional data from other specimens. Perhaps a more complex theory that accommodates both the rapid replacement and the multi-regional hypotheses will emerge. The Paleo DNA Laboratory at Lakehead University maintains a comprehensive reference list on this topic at www.ancientdna.com.

Although intact DNA degrades within about 10 000 years, molecular hybridization methods show that portions of DNA may remain within much older fossils (i.e. 465 000-year-old animal bones) (Geigl, 2002). However, these DNA strands appear to be cross-linked to each other and to other organic compounds, thus rendering them insoluble and unavailable for PCR amplification.

ANCIENT PROTEINS

Biochemical methods for protein analysis, including amplification, immunology, and sequencing, are applied widely to archeological artifacts and animal remains. Proteins have limited use in petroleum geochemical studies because, like DNA, they are rarely preserved in samples older than one million years (Endo *et al.*, 1995; Muyzer *et al.*, 1984; Muyzer *et al.*, 1986). As such, these compounds are beyond the scope of this book, but a discussion of recent work on evidence of cannibalism by the Anasazi illustrates the present level of application.

An anthropologic controversy appears to have been resolved by detection of ancient protein. Many bones excavated from Anasazi (Puebloan, ~900–1300 AD) archeological sites in the Four Corners region, southwestern USA, show evidence of being cut up and roasted or boiled, which suggests cannibalism. Critics argue that the bone evidence does not prove that human flesh was eaten and that the cut marks and heat-alteration could result from ritualistic burial or even the execution of witches. However, Marlar *et al.* (2000) proved that human flesh was consumed. The disarticulated, cooked bones of at least five people were found at a Puebloan site in Cowboy Wash, southwestern Colorado, dated at 1150 AD. The bones were from both males and females and of various ages. The site was hastily abandoned, with many household goods, valuable items, and salvageable construction materials being left behind. Two cutting tools tested positive for human blood. Based on an immunological detection assay (enzyme-linked immunosorbent assay, ELISA), cook-pot shards contained a protein confined to human heart and skeletal muscle. This evidence proved that humans were defleshed and their meat cooked. Proof of cannibalism emerged when an antibody assay of a coprolite recovered from the site also proved positive for myogloblin. This human feces was in the ashy fill of the hearth, with the cooking pot shards, but its unburned condition indicates that it was deposited after the last use of the hearth. The human coprolite contained no plant remains, which is unusual for ancient Puebloan coprolites that typically contain starch granules indicative of maize. Myoglobin is not found in blood, intestinal tissues, or the smooth-muscle cells of the digestive system. Consequently, its presence in a coprolite proves consumption of human skeletal and cardiac muscle tissue. The ELISA technique is taxon-specific, and human myoglobin can be distinguished from that of other animals. Although previous arch-eological studies have showed strong circumstantial ev-idence for cannibalism among the ancient Anasazi, the results of the human myoglobin ELISA analyses of the human coprolite and shards from a ceramic vessel pro-vide the first indisputable proof.

Note: Under ideal diagenetic conditions, proteins remain reasonably intact for ∼500 000 years. However, under the conditions of fossilization, most proteins become too degraded to be sequenced and identified in a few thousand years. The exception may be osteocalcin, a small (∼50 amino acid) protein that is abundant in vertebrate bones. It is the only ancient protein to have been sequenced, from 6000-year-old Moa bones (Huq *et al.*, 1990) and from >55 000-year-old bison bones (Nielsen-Marsh *et al.*, 2002). Experimental evidence suggests that the survival of osteocalcin is tied to mineral-phase diagenetic stabilization, which occurs in the hydroxyapatite of bones during early diagenesis (Child *et al.*, 1997; Collins *et al.*, 1998).

Muyzer *et al.* (1992) presented evidence for osteocalcin preserved in Cretaceous dinosaur bones using an indirect immunological assay method. Monoclonal antibodies produced by a rabbit injected with ostecalcin from alligator bones were reacted with various fossil materials. Strong positive reactions were seen for several Pleistocene vertebrate fossils to powdered dinosaur bones from species identified as Lambeosaurus F38 (∼75.5 Ma), Pachyrhinosaurus F39 (∼73.25 Ma), and a third dinosaur sample identified only as F33. Two other dinosaur fossils and three control materials tested negative. Additional evidence for osteocalcin in the dinosaur bones was provided with an amino–acid analysis of the degraded protein. The proportion of an unusual amino acid, γ-carboxyl glutaminic acid, was consistent with osteocalcin. The presence of proteins in such old fossils is controversial. Contamination is always possible and difficult to eliminate. Bada *et al.* (1999) question the

conclusions of Muyzer and co-workers, but concede that although dinosaur bones contain mainly exogenous amino acids, traces of endogenous amino acids may occur in some cases.

ARCHEOLOGICAL NARCOTICS

Mankind has long made extensive medicinal, religious, and recreational use of botanical drugs. Detection of drugs and drug metabolites in ancient tissues and arti-facts yields insight into social behavior, cultivation prac-tices, and trade.

Opium is one of the oldest and most widely used narcotics. The drug occurs in the latex sap of the plant *Papaver somniferum*. The Greeks called this sap *opion*. In antiquity, the narcotic was extracted by steeping the unripe poppy capsule in a honey or wine solution or by slicing the capsule and scraping off the exuded sap. The latex was then drunk, eaten, or burned and inhaled to produce a psychoactive effect. The ancient preparations were far less potent than modern refined heroin, but they were still addictive (Booth, 1999).

The earliest botanical remains of *Papaver som-niferum* occur in frozen peat from Switzerland near archeological sites dating to the third millennium BC (Merlin, 1984). Archeological evidence suggests that the cultivation of the opium poppy did not reach the Near East until much later. The oldest known representation is a limestone carving from the Central Court of the Palace of Knossos in Crete, dating to the late Minoan I (1600–1450 BC) (Evans, 1928). A new, distinctive type of pottery call Base-ring Ware emerged on Crete around this time. This pottery consisted of small juglets 10–15 cm tall, with a ring-like foot at the base. The ring at the bottom formed a flat surface that supported the jugs in an upright position, unlike earlier amphora, which were tapered at the base. Chemical analyses prove that Base-ring Ware was used to store opium and was prob-ably created in connection with its cultivation and com-merce. The principal narcotics in opium are morphine and codeine. Traces of the drugs are present in Cypriot Base-ring juglets from the fifteenth century BC that were unearthed in Tell el-Ajjul near Gaza and in Palestine (Bisset *et al.*, 1996; Koschel, 1996).

The case of the cocaine mummies

Use of narcotics from poppy seeds and lotus plants by ancient Egyptians is documented in hieroglyphs and papyrus writings. More controversial is the reported detection of cocaine, nicotine, and hashish (tetrahydrocannabinol) in hair, tissue, and bone samples from Egyptian mummies (Balabanova *et al.*, 1992; Parsche and Nerlich, 1995). The mummies were from German museum collections and included seven heads, one complete adult female, and one incomplete adult male mummy dating from the Third Intermediate Period (~1070 BC) to the Ptolemaic/Roman Period (~395 AD). These results are most unexpected, as cocaine originates from *Erthroxylon* and nicotine from *Nicotiana*, two plants that grew only in the Americas. The findings were criticized immediately with arguments suggesting that the drugs were misidentified, that nicotine could be a contaminant from cigarette smoke, and that the mummies may have been fakes (Bisset and Zenk, 1993; Schäfer, 1993; Bjorn, 1993). The latter possibility is particularly curious because mummies were once in high demand as a medicine. Between the fifteenth and eighteenth centuries, mummies were ground up for their bitumen, which was purported to cure various illnesses. The word "mummy" is derived from the Persian *mummia*, which means bitumen. The demand for mummies in Europe was so great that it grew into the business of creating new "mummies" using freshly dead people or animals dipped into pitch or asphalt and dried in the sun. The use of mummies in medicines became unfashionable as the supply of real mummies diminished and the public became aware that suppliers had been using the bodies of recently dead subjects instead of ancient ones.

The controversy of cocaine and nicotine in Ancient Egypt may have languished within the archeological community, but for a 1994 British television documentary, *The Mystery of the Cocaine Mummies* (Equinox, Channel 4). The documentary featured Balabanova, her critics, and Rosalie David, a respected Egyptologist at The Manchester Museum in England. David was unable to examine the mummies used by Balabanova, but she was able to tentatively verify the pedigree of at least one of the specimens. More interesting, David found evidence for nicotine but not cocaine in mummies.

The presence of cocaine and nicotine in Egyptian mummies can be interpreted as evidence for pre-Colombian trans-oceanic trade. Such early contact between the Old and New Worlds is highly controversial. An alternative possibility is that the drugs were present in plants native to the Old World that are undiscovered or are now extinct. Tropane alkaloids that are structurally related to cocaine are present in henbane, mandrake, and nightshade and may have been altered during the mummification process into a cocaine-like compound (Schäfer, 1993). Nicotine may also have much wider occurrence than just in tobacco and may have been present in Old World plants (Bisset and Zenk, 1993). Such plants may no longer exist because many medicinal plants have become extinct through overuse. For example, the demand for *silphium*, a plant prized for its medicinal and contraceptive properties, was so great in ancient Greece that it was extinct by the third or fourth century AD.

However, the problems associated with organic contamination of archived archeological materials are only now being addressed. Wischmann *et al.* (2002) conducted a test by exposing a nicotine-free human femur from the Bronze Age to environmental tobacco smoke for six weeks. They analyzed portions of the bone before and after washing it. Surprisingly, the unwashed sample contained 11.6 ng nicotine per gram of bone, while the washed sample contained 35.9 ng/g. They attributed this increase to tobacco smoke deposits being rinsed from the surface into the bone's interior during the washing step, thus concentrating the nicotine. As part of this study, the femurs from eighteenth-century citizens of Goslar, Germany, were examined for nicotine. Approximately two-thirds of the 34 samples contained trace amounts of nicotine, suggesting widespread use of tobacco in this community. However, no traces of cotinine, the major metabolite of nicotine, were detected. Wischmann *et al.* (2002) concluded that the nicotine in the Goslar samples was probably contamination, but they acknowledged that little information exists on the long-term persistence of cotinine.

Note: Organic residues in seventeenth-century clay pipes from England, including samples from William Shakespeare's residence, contain not only nicotine but also cocaine and possibly residues of cannabis (Thackeray *et al.*, 2001). These findings are surprising because the introduction of cocaine into Europe from South America was thought to have occurred only ~200 years ago. Although Shakespeare may not have used the pipes, there is some evocative evidence for drug use in his Sonnet 76 that contains the phrases "noted weed" and "compounds strange" (Wong, 2001). Whether these analyses found endogenous residues of

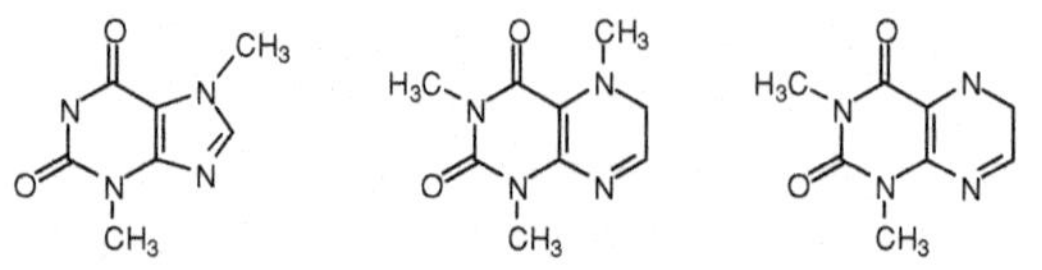

Theobromine
3,7-dihydro-3,7-dimethyl-
1H-purine-2,6-dione

Caffeine
3,7-dihydro-1,3,7-trimethyl-
1H-purine-2,6-dione

Theophylline
3,7-dihydro-1,3-dimethyl-
1H-purine-2,6-dione

Figure 11.27. Theobromine, a member of the methylxanthines, is the primary alkaloid in products from the cocoa tree *Theobroma cacao*. Other members of the methylxanthines include caffeine and theophylline, the primary alkaloids in coffee and tea, respectively.

narcotics or are experimental artifacts or measures of contamination during sample treatment or storage has yet to be determined.

Cacao in Mesoamerica

Cacao (cocoa) was a key beverage and spice for the Aztec and Mayan civilizations (Coe and Coe, 2000). Its use during the Classic Period (250–900 AD) was well documented, but recent chemical analysis has now established its use as far back as 600 BC (Hurst *et al.*, 2002). The only Mesoamerican plant to contain theobromine (Figure 11.27) as its primary alkaloid is *Theobroma cacao*. This compound was detected in the residues in several spouted jars from the Preclassic Period using HPLC coupled with atmospheric chemical ionization mass spectrometry. The jars may have been used to pour the cacao from one vessel to another to generate a froth, which was considered by the Mayans and Aztecs to be the most desirable part of the drink.

The ritual smoking of tobacco was an important part of Native American culture. Pictorial and ethnobotanical sources indicate that the practice was widespread and existed for thousands of years. Direct evidence for early tobacco use in eastern North America, however, was lacking until the detection of tobacco residues in pipes from the Early Woodland Period (~1000 –200 BC) (Rafferty, 2002).

BIOMARKERS AND INTERDISCIPLINARY STUDIES

Chemical analyses of archeological artifacts have greatly improved our understanding of ancient man. Detection of biomarkers in bitumens, resins, and other natural products shows that these substances were used in much more diverse and sophisticated ways than was once believed. The present studies, however, represent only a small fraction of what could be done. Few bituminous artifacts have been examined chemically outside of the Middle East, yet petroleum seeps occur worldwide.

Interdisciplinary biomarker studies can extend into other sciences as well. For example, the techniques developed to characterize source-rock extracts and kerogen pyrolyzates were applied in a biomedical study of cerumen (earwax) (Burkhart *et al.*, 2001). Cerumen proved to be a complex mixture of hydrocarbons, diterpenoids, sterols, and a host of unidentified compounds.

Biomarker geochemistry is a highly interdisciplinary science, drawing upon a broad knowledge base in geology, chemistry, biology, and ecology. Nevertheless, geochemists rarely venture far from the geosphere. We urge our colleagues to be proactive in establishing cross-disciplinary studies. Routine geochemical procedures and instrumentation could be used to support other disciplines with little additional laboratory effort or expense. More important is a commitment to provide the needed expertise for interpreting the data. Many archeologists and museum curators would welcome cooperative studies that utilized geochemical expertise.

Appendix: geologic time charts

Table 1 (Phanerozoic — Cenozoic/Mesozoic)

EONOTHEM EON	ERATHEM ERA	SYSTEM PERIOD	SERIES EPOCH	STAGE AGE	AGE Ma	+/-
PHANEROZOIC	CENOZOIC	NEOGENE (Quaternary)	HOLOCENE			
			PLEISTOCENE		0.01	
				Calabrian	1.81	
			PLIOCENE	Gelasian	2.58	
				Piacenzian	3.60	
				Zanclean	5.32	
			MIOCENE	Messinian	7.12	
				Tortonian	11.2	
				Serravallian	14.8	
				Langhian	16.4	
				Burdigalian	20.5	
				Aquitanian	23.8	
		PALEOGENE	OLIGOCENE	Chattian	28.5	
				Rupelian	33.7	
			EOCENE	Priabonian	37.0	
				Bartonian	41.3	
				Lutetian	49.0	
				Ypresian	55.0	
			PALEOCENE	Thanetian	57.9	
				Selandian	61.0	
				Danian	65.5	0.1
	MESOZOIC	CRETACEOUS	UPPER/LATE	Maastrichtian	71.5	0.5
				Campanian	83.5	0.5
				Santonian	85.8	0.5
				Coniacian	89.0	0.5
				Turonian	93.5	0.2
				Cenomanian	98.9	0.6
			LOWER/EARLY	Albian	112.2	1.1
				Aptian	121.0	1.4
				Barremian	127.0	1.6
				Hauterivian	132.0	1.9
				Valanginian	136.5	2.2
				Berriasian	142.0	2.6
		JURASSIC	UPPER/LATE	Tithonian	150.7	3.0
				Kimmeridgian	154.1	3.3
				Oxfordian	159.4	3.6
			MIDDLE	Callovian	164.4	3.8
				Bathonian	169.2	4.0
				Bajocian	176.5	4.0
				Aalenian	180.1	4.0
			LOWER/EARLY	Toarcian	189.6	4.0
				Pliensbachian	195.3	3.9
				Sinemurian	201.9	3.9
				Hettangian	205.1	4.0
		TRIASSIC	UPPER/LATE	Rhaetian	209.6	4.0
				Norian	220.7	4.4
				Carnian	227.4	4.5
			MIDDLE	Ladinian	234.3	4.6
				Anisian	241.7	4.7
			LOWER/EARLY	Olenekian	244.8	4.8
				Induan	250.0	4.8

Table 2 (Phanerozoic — Paleozoic)

EONOTHEM EON	ERATHEM ERA	SYSTEM PERIOD	SERIES EPOCH	STAGE AGE	AGE Ma	+/-
PHANEROZOIC	PALEOZOIC	PERMIAN	LOPINGIAN	Changhsingian	251.4	3.6
				Wuchiapingian	253.4	
			QUADALUPIAN	Capitanian	265	
				Wordian		
				Roadian		
			CISURALIAN	Kungurian		
				Artinskian	283	
				Sakmarian		
				Asselian	292	
		CARBONIFEROUS	PENNSYLVANIAN	Gzhelian		
				Kazimovian		
				Moscovian		
				Bashkirian	320	
			MISSISSIPPIAN	Serpukhovian	327	
				Visean	342	3.6
				Tournaisian	354	4
		DEVONIAN	UPPER/LATE	Famennian	364	
				Frasnian	370	
			MIDDLE	Givetian	380	
				Eifelian	391	
			LOWER/EARLY	Emsian	400	
				Pragian	412	
				Lochkovian	417	
		SILURIAN	PRIDOLI		419	
			LUDLOW	Ludfordian	423	
				Gorstian		
			WENLOCK	Homerian	428	
				Sheinwoodian		
			LLANDOVERY	Telychian		
				Aeronian	440	
				Rhuddanian		
		ORDOVICIAN	UPPER/LATE			
			MIDDLE	Darriwilian	467.5	3
			LOWER/EARLY	Tremadocian	495	
		CAMBRIAN	UPPER/LATE		500	
			MIDDLE		520	
			LOWER/EARLY		545	

Table 3 (Precambrian)

EONOTHEM EON	ERATHEM ERA	SYSTEM PERIOD	AGE Ma
PRECAMBRIAN	PROTEROZOIC NEOPROTEROZOIC	NEOPROTEROZOIC III	540
		CYROGENIAN	650
		TONIAN	850
	MESOPROTEROZOIC	STENIAN	1000
		ECTASIAN	1200
		CALYMMIAN	1400
	PALEOPROTEROZOIC	STATHERIAN	1600
		OROSIRIAN	1800
		RHYACIAN	2050
		SIDERIAN	2300
	ARCHEAN NEOARCHEAN	No subdivisions into periods	2500
	MESOARCHEAN		2800
	PALEOARCHEAN		3200
	EOARCHEAN		3600

PHANEROZOIC

CENOZOIC

AGE (Ma)	PERIOD	EPOCH	AGE	Boundary Ages (Ma)
		HOLOCENE / PLEISTOCENE	Calabrian	0.01 / 1.8
5		L PLIOCENE E	Piacenzian	1.8
			Zanclean	3.6
	NEOGENE		Messinian	5.3
10		L	Tortonian	7.1
		MIOCENE M	Serravallian	11.2
15			Langhian	14.8
		E	Burdigalian	16.4
20				20.5
			Aquitanian	
25				23.8
		L OLIGOCENE	Chattian	
30	TERTIARY			28.5
		E	Rupelian	
35				33.7
	PALEOGENE	L	Priabonian	37.0
40		EOCENE	Bartonian	41.3
45		M	Lutetian	
50		E	Ypresian	49.0
55				54.8
		L PALEOCENE	Thanetian	57.9
60			Selandian	61.0
		E	Danian	
65				65.0

MESOZOIC

AGE (Ma)	PERIOD	EPOCH	AGE	Boundary Ages (Ma)
70			Maastrichtian	65
				71.3
80		LATE	Campanian	
			Santonian	83.5
90			Coniacian	85.8
	CRETACEOUS		Turonian	89.0
			Cenomanian	93.5
100				99.0
			Albian	
110				112
		EARLY	Aptian	
120				121
			Barremian	127
130		NEOCOMIAN	Hauterivian	132
			Valanginian	137
140			Berriasian	144
			Tithonian	
150		LATE	Kimmeridgian	151
			Oxfordian	154
160			Callovian	159
	JURASSIC		Bathonian	164
170		MIDDLE	Bajocian	169
			Aalenian	176
180				180
			Toarcian	
190		EARLY		190
			Pliensbachian	195
200			Sinemurian	
			Hettangian	202
			Rhaetian	206
210				210
		LATE	Norian	
220				221
			Carnian	
230	TRIASSIC			227
			Ladinian	234
240		MIDDLE	Anisian	
				242
		EARLY	Olenekian	245
			Induan	248

PALEOZOIC

AGE (Ma)	PERIOD	EPOCH	AGE	Boundary Ages (Ma)
			Tatarian	248
		LATE	Ufimian-Kazanian	252
260			Kungurian	256 / 260
	PERMIAN		Artinskian	269
280		EARLY	Sakmarian	
				282
			Asselian	290
300			Gzelian	296
		PENNSYLVANIAN LATE	Kasimovian	303
320	CARBONIFEROUS		Moscovian	311
			Bashkirian	323
		MISSISSIPPIAN EARLY	Serpukhovian	327
340			Visean	342
			Tournaisian	
360		LATE	Famennian	354
			Frasnian	364
		MIDDLE	Givetian	370
380	DEVONAN		Eifelian	380
			Emsian	391
400		EARLY	Praghian	400
			Lockhovian	412 / 417
420		LATE	Pridolian	423
	SILURIAN		Ludlovian	428
		EARLY	Wenlockian	
440			Llandoverian	
		LATE	Ashgillian	443
460			Caradocian	449
	ORDOVICIAN	MIDDLE	Llandeilian	458
			Llanvirnian	464
480		EARLY	Arenigian	470
			Tremadocian	485
		D	Sunwaptan	490
500			Steptoean	495
		C	Marjuman	500
	CAMBRIAN		Delamaran	506
520		B	Dyeran	512
			Montezuman	516 / 520
540		A		
				543

PRECAMBRIAN

AGE (Ma)	EON	ERA	ERA	Boundary Ages (Ma)
				543
750	NEOPROTEROZOIC	LATE	SINIAN	
				800
1000				900
1250	PROTEROZOIC	MIDDLE	RIPHEAN	
1500				1600
1750				1700
2000		EARLY	ANIMIKEAN	
2250			HURO-NIAN	2200
2500				2500
2750		LATE	RAN-DIAN	
3000	ARCHEAN		SWAZIAN	2800
3250		MIDDLE		3000
				3400
3500		EARLY	ISUAN	3500
3750				3800

Glossary

abiogenic   see Abiotic.

abiotic   non-biological, not related to living organisms.

abyssal   oceanic depths in the range 3500–6000 m.

accretionary wedge   a piece of continental crust that has been accreted or attached to a larger continental mass.

acetogenic bacteria   prokaryotic organisms that use carbonate as a terminal electron acceptor and produce acetic acid as a waste product.

acidophile   an organism that grows best under acid conditions (as low as pH = 1).

acritarch   a unicellular, alga-like eukaryotic microfossil of uncertain biological origins. Acritarchs may be related to dinoflagellates.

activation energy   the energy necessary for chemical transformations to take place, usually expressed in kcal/mol. Activation energy distributions are used in maturation modeling to calculate the degree of transformation of kerogen to oil and gas.

active margin   a continental margin characterized by uplifted mountains, earthquakes, and igneous activity caused by convergent or transforming plate motion.

active source rock   a source rock that is currently generating petroleum due to thermal maturation or microbial activity (e.g. methanogenesis). Source rock that was active in the past is either inactive or spent today.

acyclic   straight or branched carbon – carbon linkage in a compound without cyclic structures. See Aliphatic.

adaptation   a modification of an organism that improves its ability to exist in its environment or enables it to live in a different environment.

adaptive radiation   diversity that develops among species as they adapt to different sets of environmental conditions.

adenosine diphosphate (ADP)   a natural product in living cells formed by the hydrolysis of adenosine triphosphate, which is accompanied by release of energy and organic phosphate.

adenosine triphosphate (ATP)   adenosine $5'$-triphosphoric acid. A natural product in living cells that serves as a source of energy for biochemical reactions.

adsorbent   packing used in adsorption chromatography. Silica gel and alumina are the most commonly used adsorbents in liquid chromatography.

adsorption   the interaction between the solute and the surface of an adsorbent. The forces in adsorption can be strong (e.g. hydrogen bonds) or weak (van der Waals' forces). For silica gel, the silanol group is the driving force for adsorption, and any solute functional group that can interact with this group can be retained by liquid–solid chromatography on silica.

adsorption chromatography   a type of liquid chromatography that relies on adsorption to separate compounds. Silica gel and alumina are the most commonly used supports. Molecules are retained by the interaction of their polar functional groups with the surface functional groups (e.g. silanols of silica).

aeolian   wind-deposited, e.g. desert sands.

aerobe   an organism that requires molecular oxygen (terminal electron acceptor) to carry out respiratory processes.

aerobic   refers to metabolism under oxic conditions using molecular oxygen, as in aerobic respiration (see Table 1.5). Most organic matter is destroyed or severely altered by aerobic microbes when exposed to oxic conditions for extended periods.

alcohol   a compound that contains one or more hydroxyl groups (–OH), e.g. methanol, ethanol, bacteriohopanetetrol.

aldehyde   a compound that contains the –CHO group.

alga (*pl*. algae)   a non-vascular uni- or multicellular eukaryotic plant (thallophyta) that contains chlorophyll and is capable of photosynthesis. Blue-green algae are not true algae but belong to a group of bacteria called cyanobacteria.

alginite   an oil-prone maceral of the liptinite group composed of morphologically recognizable remains of algae.

aliphatic   hydrocarbons in petroleum that contain saturated and/or single unsaturated bonds that elute during liquid chromatography using non-polar solvents. Aliphatic hydrocarbons include normal and branched alkanes and alkenes, but not aromatics.

alkane (paraffin)   a saturated hydrocarbon that may be straight (normal), branched (isoalkane or isoparaffin), or cyclic (naphthene).

alkaline   having pH > 7.3.

alkene (olefin)   a straight- or branched-chain unsaturated hydrocarbon (C_nH_{2n}) containing only hydrogen and carbon and with one or more double bonds, but not aromatic. Alkenes are not abundant in crude oils, but they can occur as a result of rapid heating, as during turbo-drilling, steam treatment, refinery cracking, and laboratory pyrolysis. Ethene (C_2H_4) is the simplest alkene. Refined olefins, such as ethene and propene, are made from oil or natural gas liquids and are commonly used to manufacture plastics and gasoline.

alkylation   a refining process that converts light olefins into high-octane gasoline components, i.e. the reverse of cracking.

alkyl group   a hydrocarbon substituent with the general formula C_nH_{2n+1} obtained by dropping one hydrogen from the fully saturated compound, e.g. methyl (–CH_3), ethyl (–CH_2CH_3), propyl (–$CH_2CH_2CH_3$), or isopropyl [($CH_3)_2CH$–]. The letter "R" typically symbolizes alkyl groups attached to biomarkers, such as the alkyl side chain in steranes. In the text, we prefer to use the letter "X" so as not to confuse the reader with the R designation for stereochemistry (e.g. see Figure 2.29).

allochthonous   not indigenous, derived from elsewhere. For example, sediments in a marine basin might receive autochthonous macerals from aquatic organisms in the overlying water column and allochthonous macerals from terrigenous plants that were transported into the basin from distal swamps and soils.

alumina   a common adsorbent in liquid chromatography. Aluminum oxide (Al_2O_3) is a porous adsorbent that is available with a slightly basic surface.

aluminosilicate   a silicate mineral where aluminum substitutes for silicon in the SiO_4 tetrahedra.

amino acid   organic compounds that contain amino (–NH_2) and carboxylic acid (–COOH) groups. Certain amino acids, such as alanine, $CH_3CH(NH_2)COOH$, are the building blocks of proteins in living organisms.

amino group   an –NH_2 group attached to a carbon skeleton, as in the amines and amino acids.

ammonification   liberation of ammonium (ammonia) from organic nitrogenous compounds by the action of microorganisms.

ammonite   an extinct group of cephalopods with coiled, chambered shells having septa with crenulated margins.

amorphous   non-crystalline, lacking a crystal structure; a solid, such as glass, opal, wood, or coal, that lacks an ordered internal arrangement of atoms or ions.

amorphous kerogen   insoluble particulate organic matter that lacks distinct form or shape as observed in microscopy. Amorphous kerogen that fluoresces in ultraviolet light has petroleum-generative potential. However, highly mature oil-prone kerogen and degraded gas-prone organic matter may appear amorphous but do not fluoresce and do not act as a source for petroleum.

amphibians   cold-blooded vertebrates with gills for respiration in early life but with air-breathing lungs as adults.

amphipathic   refers to organic compounds with polar (hydrophilic) and non-polar (hydrophobic) ends, e.g. cholesterol, bacteriohopanetetrol, phospholipids.

anaerobe   an organism that does not require oxygen for respiration but uses other processes, such as fermentation, to obtain energy. Anaerobic bacteria grow in the absence of molecular oxygen.

anaerobic   refers to metabolism under anoxic conditions without molecular oxygen, as in anaerobic respiration (see Table 1.5).

anaerobic respiration   a metabolic process in which electrons are transferred from organic or, in some cases, inorganic compounds to an inorganic acceptor molecule other than oxygen. The most common acceptors are nitrate, sulfate, and carbonate.

analyte   a compound to be analyzed.

angiosperms   flowering or higher land plants having floral reproductive structures and seeds in an ovary that became prominent on land during Early Cretaceous time. Oleanane is a biomarker believed to indicate angiosperm input to petroleum from source rocks of Late Cretaceous or younger age.

angular unconformity   an unconformity between two groups of rocks where the bedding planes of rocks above and below are not parallel. Usually, the older, underlying rocks dip at a steeper angle than the younger, overlying strata.

anhydrite   see Evaporite.

anion   a negatively charged atom that has gained one or more electrons.

annulus   the space between the drill string and the exposed wall of the well bore, or between the production tubing and the casing.

anoxic   water column or sediments that lack oxygen (see Table 1.5). Demaison and Moore (1980) define anoxic sediments as containing <0.5 ml oxygen/l water, which is the threshold below which the activity of multicellular deposit feeders is significantly depressed. Between ~0.1 and 0.5 ml oxygen/l, epifauna (non-burrowing) can still survive above the sediment–water interface, but bioturbation by deposit feeders virtually ceases. Organic matter in sediments below anoxic water stands a better chance of preservation and is commonly more hydrogen-rich, more lipid-rich, and more abundant than that under oxic water.

anoxygenic photosynthesis   a type of photosynthesis in green and purple bacteria in which oxygen is not produced.

anteisoalkanes   straight-chain alkanes that have a methyl group attached to the third carbon atom (i.e. 3-methyl alkanes).

anthracene   an aromatic hydrocarbon consisting of three fused benzene rings; an isomer of phenanthrene.

anthracite   coal of the highest thermal maturity. See also Coal.

anthropogenic   derived from human activities.

antibody   protein that is produced by animals in response to the presence of an antigen and that can combine specifically with that antigen.

anticline   a concave-downward fold where strata are bent into an arch.

API gravity   a scale of the American Petroleum Institute that is related inversely to the density of liquid petroleum: API gravity = [141.5°/(specific gravity at 16°C) −131.5°]. The unusual form of the equation results in a convenient scale; higher API indicates lighter oil. Fresh water has a gravity of 10°API. Heavy oils are <25°API; medium oils are 25–35°API; light oils are 35–45°API; condensates are >45°API.

aquifer   permeable rock that carries moving subsurface water.

aquitard   a relatively impermeable rock unit that retards the flow of groundwater.

archaea   prokaryotic organisms that contain ether-linked lipids built from phytanyl chains and are common in extreme environments. They include extreme halophiles (hypersaline environments), thermoacidophiles (hot, acidic environments), thermophiles (hot environments), and methanogens (methane generated as a product of metabolism). Although generally considered prokaryotes due to their similarities to bacteria, archaea show a number of eukaryotic features indicating that they may represent a third distinct group (Woese *et al.*, 1978).

archaebacteria   an old term for the archaea.

archaeocyathids   extinct marine organisms with double, perforated, calcareous, conical to cylindrical walls.

Archean   part of Precambrian time beginning 3.8 and ending 2.5 billion years ago.

archosaurs   advanced reptiles of a group called diapsids, which includes thecodonts, dinosaurs, pterosaurs, and crocodiles.

argillaceous   composed largely of clay-sized particles or clay minerals, e.g. argillaceous marls or shales contain appreciable clay.

argillite   a low-grade metamorphic rock composed of shaly sedimentary rock and characterized by irregular fractures and lack of foliation.

aromaticity ratio (B)   toluene/*n*-heptane, as in Figure 7.21 (Thompson, 1987).

aromatics   organic compounds with one or more benzene rings in their structure, including pure aromatics, such as benzene and polycyclic aromatic hydrocarbons, plus cycloalkanoaromatics, such as monoaromatic steroids, some cyclic sulfur compounds, such as benzothiophenes, and porphyrins. Because of their high solubility, low-molecular-weight aromatics such as benzene and toluene are readily washed from petroleum by circulating groundwater. Light aromatics are components in unleaded gasoline and are used as feedstocks for petrochemicals.

aromatic steroids   aromatic hydrocarbons that originate from sterols, including monoaromatic and triaromatic steroids.

aromatization parameter (monoaromatic steroid aromatization)   a biomarker maturity parameter based on the hypothesis of irreversible, thermal conversion of C_{29} monoaromatic (MA) to C_{28} triaromatic (TA) steroids [TA/(MA + TA)].

arthropod   Invertebrate with jointed body and limbs (includes insects, arachnids, and crustaceans).

asphalt   dark brown to black, solid to semi-solid heavy ends of petroleum that gradually liquefy when heated and are generally soluble in carbon disulfide but insoluble in *n*-heptane. Natural and refined asphalts contain resins, asphaltenes, and heavy waxes, are dominated by carbon and hydrogen, but are also rich in nitrogen, sulfur, oxygen, and complexed metals, such as vanadium and nickel. Refinery asphalts are residues of crude oil distillations and air oxidation of crude oil feedstock.

asphaltenes   a complex mixture of heavy organic compounds precipitated from oils and extracts by natural processes or in the laboratory by addition of excess *n*-pentane, *n*-hexane, or *n*-heptane. After precipitation of asphaltenes, the remaining oil or bitumen consists of saturates, aromatics, and NSO compounds. See Asphalt.

associated gas   natural gas that occurs with oil, either dissolved in the oil or as a gas cap above the oil.

asteroid   small planetary bodies (<800 km in diameter), mostly orbiting the sun between Mars and Jupiter. Asteroids in Earth-crossing orbits can collide with Earth, resulting in major-impact structures and possible extinction events.

asthenosphere   the layer of Earth below the lithosphere where isostatic adjustments take place, magmas can be generated, and seismic waves are strongly attenuated, suggesting that the asthenosphere is a zone of convective flow.

ASTM   American Society for Testing Materials.

asymmetric carbon (center)   a carbon atom surrounded by four different substituents (see Figure 2.21), thus having no plane of symmetry.

atom   the smallest unit of an element that retains the physicochemical properties of that element.

atomic mass   the mass of the number of neutrons and protons in an atomic nucleus. A Dalton is one atomic mass unit (amu).

atomic number   the number of protons in the nucleus of an atom.

atomic weight   the average mass number of an element; the mass (in grams) of Avogadro's number of atoms of an element.

autochthonous   indigenous, derived from nearby. For example, sediments in a marine basin might receive autochthonous macerals from aquatic organisms in the overlying water column and allochthonous macerals from terrigenous plants that were transported into the basin from distal swamps and soils.

autotroph   an organism that uses carbon dioxide as the source for cellular carbon.

azeotrope   a mixture of two solvents that boils at a constant boiling point, as if it were a pure compound, e.g. the azeotrope of ethanol : water is a 95 : 5 mixture.

bacillus   a bacterium with an elongated, rod shape.

backflushing   a column-switching technique whereby a four-way valve placed between the injector and the column allows for mobile phase flow in either direction. Backflushing is used to elute strongly held compounds from the head of the column.

bacteria   a general term that includes unicellular, prokaryotic (archaea or eubacteria) organisms and microorganisms.

bacterial sulfate reduction (BSR)   bacterially mediated reduction of sulfate to sulfide, accompanied by oxidation of organic matter to carbon dioxide.

bacteriochlorophyll   light-absorbing pigment in green sulfur and purple sulfur bacteria.

bacteriohopanetetrol   a C_{35}-hopanoid containing four hydroxyl groups found in the lipid membranes of prokaryotes (see Figure 2.29). This compound is presumed to be a major precursor for the hopanes in petroleum.

bacteriophage   a type of virus that infects bacteria, often with destruction or lysis of the host cell.

barrel (bbl)   the standard unit of oil in the petroleum industry; 42 US standard gallons.

barrel of oil equivalent (BOE)   the volume of gas divided by about six and added to volumes of crude oil and natural gas.

basalt   an extrusive igneous rock that is rich in ferromagnesian minerals and low in silica.

basement   the oldest rocks in an area; commonly igneous or metamorphic rocks of Precambrian or Paleozoic age that underlie other sedimentary formations. Basement generally does not contain significant oil or gas, unless it is fractured and in a position to receive these materials from sedimentary strata.

base peak   the largest peak in the mass spectrum of a compound. Typically all other peaks in the spectrum are normalized to the base peak, which is assigned an intensity of 100%. For example, the mass spectrum of cholestane (a C_{27} sterane) has a base peak at m/z 217 (see Figure 8.27), typical of many steranes.

basin   a low-lying or depressed area that collects sediments or where sediments collected in the past; also refers to the basin fill.

bathyal   oceanic depths in the range 200–3500 m.

BBOE   billions of barrels of oil equivalent.

bedding   sedimentary rock layers of varying thickness and character that represent the original sediments that were deposited.

bedrock   solid rock that underlies unconsolidated surface materials.

belemnites   mollusks of the class Cephalopoda that have straight internal shells.

benthic   refers to a bottom-dwelling marine or lacustrine organism.

benzene   the simplest aromatic hydrocarbon, consisting of a flat, six-member ring with the formula C_6H_6. Some important products manufactured from benzene include phenol, styrene, nylon, and synthetic detergent.

B–F diagram   a plot of aromaticity ratio (toluene/n-heptane, B) versus paraffinity ratio (n-heptane/methylcyclohexanes, F) that has trends associated with several reservoir alteration processes (Thompson, 1987).

bioaccumulation   build-up of pollutants, such as heavy metals, in organisms at higher trophic levels in a food chain.

biodegradable   capable of being decomposed by biological processes.

biodegradation   microbial alteration of organic matter. Petroleum can be biodegraded during migration, while in the reservoir, and at surface seep locations. Biodegradation of petroleum occurs at low temperatures (not more than ~80°C) and, hence, at shallow depths. Conditions may be anoxic or involve circulating groundwater with dissolved oxygen where water washing accompanies biodegradation. Microorganisms typically degrade petroleum by attacking the less complex, hydrogen-rich compounds first. For example, n-alkanes and acyclic isoprenoids are attacked before steranes and triterpanes.

biofacies   distinct spatial or temporal distributions of assemblages of organisms that are controlled by environmental factors in the surrounding environment. For example, biofacies can vary within the same lithologic unit due to differences in water depth, oxicity, or salinity during deposition. Organic facies are analogous to biofacies, except that they refer to distinctive organic matter composition (e.g. including geochemical characteristics) rather than only assemblages of recognizable organisms.

biofilm   an adhesive, usually a polysaccharide, that encases some microbial material and assists with attachment to a surface.

biogenic   related to, or originating from, organisms. The terms "biogenic," "abiogenic," and "microbial" can be confusing. Although some methane is abiogenic (originating abiotically), most originates as a result of microbial or thermal degradation of (biogenic) organic matter. For this reason, we prefer to use the terms "biogenic" and "abiogenic" when describing the biotic or abiotic origins of methane, and the terms "microbial" and "thermogenic" when describing the origin of most petroleum.

biogeochemistry   the study of microbially mediated chemical transformations of geochemical interest, such as nitrogen or sulfur cycling.

biological marker (biomarker, molecular fossil)   complex organic compounds composed of carbon, hydrogen, and other elements that are found in petroleum, rocks, and sediments and show little or no change in structure from their parent organic molecules in living organisms. These compounds are typically analyzed using gas chromatography/mass spectrometry. Most, but not all, biomarkers are isopentenoids, composed of isoprene subunits. Biomarkers include pristane, phytane, steranes, triterpanes, and porphyrins.

biological oxygen demand (BOD)   the tendency to consume dissolved oxygen during the biological breaking down of organic matter.

biomarker   see Biological marker.

biomass   the amount of living organic matter in a particular environment or habitat.

bioremediation   the process by which organisms remove or detoxify hazardous organic contaminants in soils, water, and the subsurface.

biosphere   zones of the atmosphere, water, sediments, and rocks that are capable of supporting life. Most life in these zones depends on energy from the sun, although some organisms living in the deep subsurface or at deep-sea hydrothermal vents obtain their energy from other sources, such as carbon dioxide and hydrogen.

biosynthesis   production of cellular constituents from simpler molecules.

biotic   caused or induced by living organisms.

bioturbation   disturbance of sediment by burrowing, boring, and sediment-ingesting multicellular organisms.

28,30-bisnorhopane a desmethylhopane (two methyl groups removed compared with hopane) that contains 28 carbon atoms and is common in sulfur-rich source rocks, such as the Miocene Monterey Formation. Also called 28,30-BNH and 28,30-dinorhopane (28,30-DNH).

bitumen organic matter extracted from fine-grained rocks using common organic solvents, such as methylene chloride. Unlike oil, bitumen is indigenous to the rock in which it is found (i.e. it has not migrated). If mistaken for bitumen, migrated oil impregnating a proposed source rock could result in an erroneous oil–source rock correlation. Bitumen differs from pyrobitumen, which is less soluble. In archeology, bitumen refers to heavy oil or asphalt.

bitumen ratio (transformation ratio) the ratio of extractable bitumen to total organic carbon (Bit/TOC) in fine-grained, non-reservoir rocks. The ratio varies from near-zero in shallow sediments to ~250 mg/g TOC at peak generation and decreases because of conversion of bitumen to gas at greater depths. Anomalously high values for the bitumen ratio in immature sediments can be used to help show contamination by migrated oil or manmade products.

bituminous coal coal with maturity between lignite and anthracite; see also Coal.

bivalves a class of the phylum Mollusca (also known as the class Pelecypoda).

bleaching whitening of red-colored sandstones along faults by migrating saline brines that convert the immobile Fe^{3+} in hematite to mobile Fe^{2+} (e.g. Chan et al., 2000). These brines were reducing due to interaction with petroleum, organic acids, or hydrogen sulfide.

blue-green algae see Cyanobacteria.

blueschist a high-pressure, low-temperature metamorphic rock characteristic of subduction zones.

BOE barrels of oil equivalent.

boghead coal a liptinite-rich, oil-prone (type I) coal that is dominated by algal remains.

bolide an extraterrestrial object, such as an asteroid or comet, that explodes upon entering the atmosphere or striking the Earth.

bonded-phase chromatography the most popular mode of liquid chromatography, in which a stationary phase chemically bonded to a support is used for the separation. The most popular support and bonded phase are silica gel and an organosilane, such as octadecyl (for reversed-phase chromatography), respectively.

BOPD barrels of oil per day.

botryals C_{52}–C_{64} even-carbon-numbered α-branched, α-unsaturated aldehyldes that represent up to 45% of the lipids in some strains of living *Botryococcus braunii* race A (Metzger et al., 1989).

botryococcane a saturated, irregular isoprenoid biomarker formed from precursors in *Botryococcus braunii*, a fresh/brackish water, colonial Chlorophycean alga found in ancient rocks and still living today. Concentrated deposits of the alga may result in boghead coals or oil shales.

bottoms the heavy fraction of crude oil that does not vaporize during distillation.

bottom-simulating reflector (BSR) a seismic response that marks the interface between higher-sonic-velocity, gas hydrate-bearing sediment (above) and lower-sonic-velocity, free gas-bearing sediment (below). BSRs at depth in the sediment commonly parallel the overlying seafloor topography.

boulder a rock >256 mm in diameter.

BPD barrels per day.

brachiopod marine bivalve (double-shelled) invertebrates that were widespread during the Paleozoic Era but are less common today.

brackish mixed fresh and marine waters with intermediate salinities (<35 parts per thousand).

branched hydrocarbon branched alkanes containing carbon atoms that are linked to more than two other carbon atoms.

breccia clastic sedimentary rock composed mainly of angular fragments of various sizes.

Bryozoa a phylum of attached and encrusting colonial marine invertebrates.

BSR see Bacterial sulfate reduction and Bottom-simulating reflector (BSR is a common abbreviation for both terms; they can be distinguished by the context of the discussion).

BTEX water-soluble aromatic hydrocarbons, including benzene, toluene, ethylbenzene, and xylenes.

Burgess Shale fauna preserved fossil fauna of soft-bodied Cambrian animals discovered in 1910 by Charles Walcott in Kicking Horse Pass, Alberta, Canada.

burial history the depth of a sedimentary layer versus time, usually corrected for compaction.

burial history chart a diagram that shows the time interval for hydrocarbon generation, the essential elements, and the critical moment for a petroleum system.

butadiene   a widely used raw material in the manufacture of synthetic rubber.

butane   flammable, gaseous hydrocarbons consisting of a mixture of isobutane and normal butane.

^{14}C carbon-14. A radioactive isotope of carbon with atomic mass 14 that can be used to determine the age of materials <50000 years old.

C3 plant   most green plants, eukaryotic algae, and autrotrophic bacteria use the C3 or Calvin pathway to fix carbon from carbon dioxide during photosynthesis. All resin-producing plants, including gymnosperms and angiosperms, use the C3 pathway (Murray *et al.*, 1998). C3 and C4 plants are isotopically distinct (see Figure 6.2).

C4 plant   Most tropical grasses use C4 photosynthesis, which originated during the Paleocene Epoch, to fix carbon from carbon dioxide. C3 and C4 plants are isotopically distinct (see Figure 6.2).

$C_{15}+$ fraction   the fraction of oil or bitumen composed of compounds eluting after the C_{15} *n*-alkane from a gas chromatographic column (boiling point $nC_{15} \sim 271°C$). This fraction is not altered significantly by evaporation or sample preparation and is the most reliable for correlations involving oils and bitumens. For example, after extraction of bitumen from rock, removal of the solvent by rotoevaporation results in loss of the $<C_{15}$ fraction. It is for this reason that oils are commonly topped (distilled) to remove the $<C_{15}$ fraction so that the remaining material ($C_{15}+$) can be compared with bitumens more reliably.

C_{30} sterane index   ratio of parts per thousand C_{30} steranes (proposed markers of marine algal input) to total C_{27}–C_{30} steranes.

C_{35} homohopane index   percentage of C_{35} 17α-homohopanes among the total C_{31}–C_{35} homohopanes. High values indicate selective preservation of the C_{35} homolog due to low redox potential conditions or complexation of bacteriohopanoids with sulfur.

CAI   see Color alteration index.

calcite   see Evaporite.

Caledonian Orogeny   an early Paleozoic mountain-building episode in Europe that created the Caledonides orogenic belt, extending from Ireland and Scotland northeastward through Scandinavia.

Calvin cycle   the biochemical route of carbon dioxide fixation in many autotrophic organisms.

CAM plants   Succulents that can use C3 or C4 pathways to fix carbon from carbon dioxide during photosynthesis. CAM plants show $\delta^{13}C$ that covers the range for both C3 and C4 plants.

cannel coal   oil-prone (type II) coal that consists mainly of liptinite macerals (i.e. spores and pollen) with little or no alginite.

canonical variable (CV)   a statistical parameter based on stable carbon isotopic compositions of the saturated and aromatic hydrocarbons for a collection of terrigenous ($\delta^{13}C = 1.12\delta^{13}C_{sat} + 5.45$) and marine oils ($\delta^{13}C = 1.10\delta^{13}C_{sat} + 3.75$). Based on stepwise discriminant analysis of 339 non-biodegraded oils, $CV = -2.53\delta^{13}C_{sat} + 2.22\delta^{13}C_{aro} - 11.65$. $CV >0.47$ indicates mainly waxy terrigenous oils, while $CV<0.47$ indicates mainly non-waxy marine oils (Sofer, 1984).

capillary tubing   metal, plastic, or silica tubing used to connect various parts of a chromatograph. Tubing with inner diameters ranging from 0.2 to 0.5 mm is in common use.

carbohydrate   carbohydrates are a class of organic compounds consisting of polyhydroxy aldehydes and ketones or substances that yield these materials when hydrolyzed. These range from monosaccharides (i.e. simple sugars such as glucose, $C_6H_{12}O_6$) to polysaccharides (e.g. cellulose), which are polymers of many monosaccharide units. Complex carbohydrates decompose to water-soluble sugars in shallow sediments.

carbon   the main element in all hydrocarbons. Capable of combining with hydrogen to form huge numbers of compounds.

carbonaceous shale   organic-rich shales that contain less total organic carbon (TOC) than coals (50 wt.% TOC).

carbonado   microcrystalline black diamond thought to originate in the crust rather than the mantle of the Earth.

carbonate carbon   carbon present in a rock as carbonate minerals. Commonly measured by converting the carbonate to carbon dioxide.

carbonate rock   a sedimentary rock dominated by calcium carbonate ($CaCO_3$). In the text, we define carbonate rocks as containing at least 50 wt.% carbonate minerals, the consolidated equivalent of limy mud, shell fragments, or calcareous sand.

carbonates   chemical precipitates formed when carbon dioxide dissolved in water combines with oxides of calcium, magnesium, potassium, sodium, and iron. Carbonate rocks, such as limestone and dolostone, consist mainly of the carbonate minerals calcite ($CaCO_3$) and dolomite $CaMg(CO_3)_2$, respectively.

carbon cycle   the geochemical cycle by which carbon dioxide is fixed to form organic matter by photosynthesis or chemosynthesis, cycled through various tropic levels in the biosphere, partially lost to sediments, and finally returned to its original state through respiration or combustion.

carbon fixation   conversion of carbon dioxide or other single-carbon compounds to organic matter, such as carbohydrate.

carbonization   an antiquated term that refers to the concentration of carbon during fossilization.

carbon isotope ratio   see Stable carbon isotope ratio.

carbon preference index (CPI)   ratio of peak heights or peak areas for odd- to even-numbered n-alkanes in the range nC_{24}–nC_{34} (Bray and Evans, 1961). If other ranges of n-alkanes are used, then the results are usually referred to as odd-to-even predominance (OEP) (Scalan and Smith, 1970). Petroleum that originates from terrigenous organic matter has high CPI that decreases toward 1.0 with increasing maturation.

carboxyl group   a functional group (–COOH) attached to a carbon skeleton, as in fatty acids.

carboxylic acid   an acid that contains a carboxyl (−COOH) group, e.g. fatty acids and various naphthenic and aromatic acids.

carcinogen   a substance that causes cancer.

carotane   (β-carotane, perhydro-β-carotene) a saturated tetraterpenoid biomarker ($C_{40}H_{78}$) typical of petroleum from saline, lacustrine environments.

carotenoid   yellow to black pigments that include a class of hydrocarbons (carotenes) and their oxygenated derivatives (xanthophylls). They are characterized by eight isoprenoid units (tetraterpanoid, C_{40}) that are linked so that the two central methyl groups are in 1,6-positional relationship (tail-to-tail) while the remaining non-terminal methyl groups are in a 1,5-positional relationship (head-to-tail).

carrier bed   a permeable rock that acts as a conduit for migrating petroleum.

casing   steel pipe that is cemented into a well bore to prevent caving and loss of circulating fluids into the formation. In a completed well, the casing is perforated

at the proper depths to allow oil or natural gas to flow into the production tube.

catabolism   the biochemical breakdown of organic compounds, resulting in the production of energy.

catagenesis   thermal alteration of organic matter by burial and heating in the range of $\sim$50–150°C, typically requiring millions of years. Thermal maturity equivalent to the range 0.6–2.0% vitrinite reflectance includes the oil window or principal zone of oil formation and wet gas zone (e.g. see Figure 1.2).

catalyst   a substance that promotes a chemical reaction by lowering the activation energy but does not enter into the reaction, e.g. enzymes in living organisms or platinum catalysts in refineries.

catalytic cracking   a petroleum-refining process whereby heavy hydrocarbons and other compounds break down (crack) into lighter molecules in the presence of heated catalysts.

cation   a positively charged atom that has lost one or more electrons.

cation exchange capacity   the total exchangeable cations that a material can adsorb at a given pH.

cation-exchange chromatography   a form of ion-exchange chromatography that uses resins or packings with functional groups that can separate cations.

cell   the basic unit of living matter.

cell membrane   a selectively permeable membrane that surrounds the cytoplasm.

cellulose   a complex carbohydrate or glucose polysaccharide with beta-1,4-linkage in the cell walls of woody plants.

cell wall   layer outside the cell membrane that supports and protects the membrane and gives the cell shape.

chalk   a soft, white, fine-grained variety of limestone that is composed mainly of the calcium carbonate skeletal remains of marine plankton.

channeling   occurs when voids in the packing material of a chromatographic column cause mobile phase and solutes to move more rapidly than the average flow velocity, resulting in band broadening. These voids can be caused by poor packing or by erosion of the packed bed.

charge   the volume of expelled petroleum available for entrapment.

cheilanthanes   see Tricyclic terpanes.

chelate   a chemical structure that allows a metal to be held between two or more atoms.

chemoautotroph   an organism that uses inorganic compounds as the source for cellular carbon and energy.

chemocline   the depth at which hydrogen sulfide first appears and oxygen disappears (Sinninghe Damsté *et al.*, 1993b). For example, the present-day chemocline in the Black Sea occurs at 80–100 m.

chemoheterotroph   an organism that uses organic compounds as the source for cellular carbon and energy.

chemosynthesis   the process by which certain microbes create biomass (e.g. carbohydrate) by chemical oxidation of simple inorganic compounds. For example, chemosynthetic bacteria create biomass from carbon dioxide and water by mediating inorganic chemical reactions at hydrothermal vents in the deep sea.

chemotroph   an organism that uses inorganic or organic substances for energy and as a source of carbon. Includes all non-photosynthetic microorganisms, animals, and fungi.

chert   a dense, hard, sedimentary rock composed of chemically or biologically precipitated silica (SiO_2) as microcrystalline quartz. Unless colored by impurities, chert is white while flint is black.

chiral   many molecules containing asymmetric carbon atoms are chiral, i.e. they have the potential to exist as two non-superimposable structures that are mirror images. These mirror-image structures differ in the spatial relationship between atoms in the same manner as right- and left-handed gloves. In solution, chiral molecules rotate plane-polarized light either clockwise or counterclockwise. Certain chiral molecules are not asymmetric.

chiral stationary phase   stationary chromatographic phase designed to separate enantiomeric compounds.

Chlorobiaceae   photosynthetic green sulfur bacteria in which carbon fixation takes place by the reversed tricarboxylic acid cycle, resulting in ^{13}C-rich organic matter (e.g. Grice *et al.*, 1997). *Chlorobiaceae* are obligate anaerobes that require light and hydrogen sulfide (Clifford *et al.*, 1997). Presence of isorenieratane in crude oil or rock extracts indicates that the source rock was deposited under conditions of photic zone anoxia.

chlorophylls   green tetrapyrrole pigments in higher plants, algae, and certain bacteria that are required for photosynthesis. Most porphyrins in petroleum originate from the pyrrole ring in chlorophyll, while much of the pristane and phytane originate from the isoprenoid side chain.

chloroplast   an organelle in eukaryotes that contains chlorophyll, which is necessary for photosynthesis.

cholestane C_{27} saturated sterane derived from cholesterol. A major biomarker in petroleum.

cholesterol (choles-5-en-3β-ol)   a sterol (steroid alcohol) containing 27 carbon atoms that is found in the lipid membranes of eukaryotes (see Figure 2.30). This compound is a precursor for the cholestane in petroleum.

chondrites   stony meteorites that contain rounded silicate grains or chondrules, believed to have formed by crystallization of liquid silicate droplets from the solar nebula.

chordates   animals that develop a notochord during some stage of their development, including vertebrates.

chromatogram   a plot of the detector signal for separated components as peaks versus time or elution volume during chromatography (e.g. see Figure 8.9). Each peak on the chromatogram may represent more than one compound. Comparison of petroleum samples is typically accomplished using peak-height or peak-area ratios. However, internal standards allow direct comparisons of concentrations for specific peaks between samples.

chromatography   separation of mixtures of compounds or ions based on their physicochemical properties. Liquid chromatography and gas chromatography are used routinely to separate petroleum components based on their partitioning between a mobile and a stationary phase in a chromatographic column.

chromosome   the genetic template for a living organism that consists of a threadlike, microscopic body that contains genes. Occurs in the nucleus of eukaryotes but within the cytoplasm in prokaryotes. The number of chromosomes is normally constant for each species.

chronostratigraphic unit   rocks formed during a specific geologic time. Also known as time-stratigraphic and time-rock unit.

cilia (*sing*. cilium)   short, whiplike appendages on some protozoa that are used for propulsion.

ciliate   a protozoan that moves by means of cilia found on the surface of the cell.

circulation   (1) water movement. (2) Movement of drilling mud pumped down through the drill string to the drill bit and back up the annulus during rotary drilling.

cladistic phylogeny   biologic classification in which organisms are grouped according to similar characteristics that were inherited from a common ancestor.

clastic   sedimentary rock that consists of fragments or clasts of pre-existing rock, such as sandstone or shale.

clastic wedge   an accumulation of mainly clastic sediments deposited adjacent to uplifted areas. Sediments in the wedge become finer and the section thinner in a direction away from the upland source area.

clay   a particle <0.004 mm (<4 μm) in diameter, or a group of layer-silicate minerals characterized by poor crystallinity and fine particle size.

claystone   a non-fissile sedimentary rock composed mainly of clay-sized particles.

cleavage   (1) bond-breaking. (2) Tendency of a mineral or rock to break in preferred directions.

coal   a rock containing >50 wt.% organic matter. Most, but not all, coals are of higher-plant origin, plot along the type III (gas-prone) pathway on a van Krevelen diagram, and are dominated by vitrinite-group macerals. Increasing burial maturation of peat results in lignite, bituminous, and anthracite coals.

coalbed methane   a methane-rich, sulfur-free natural gas from coal beds.

cobble   a rock particle 64–256 mm in diameter.

coccolithophorids   planktonic, marine, golden-brown algae that typically secrete discoidal calcareous coverings called coccoliths.

coccus   spherical bacterial cell.

coenzyme   a chemical that participates in an enzymatic reaction by accepting and donating electrons or functional groups.

co-injection   chromatographic technique used to help identify unknown compounds. A synthesized or isolated standard (some are available commercially) is mixed with the sample (a process called spiking) containing the compound to be identified. If the standard and unknown compounds co-elute, then the relative peak intensity of the unknown compound on chromatograms of the mixture is higher than that for the neat (unspiked) sample (e.g. see Figure 13.96). Co-elution supports, but does not prove, the idea that the compounds are identical. Proof of structure might include co-elution of the unknown and standard on various chromatographic columns, identical mass spectra, and nuclear magnetic resonance (NMR) or X-ray structural verification.

coke   solid carbonaceous residues that can form during refinery processes.

collinite   a structureless vitrinite group maceral that originates from woody organic matter.

collision-activated decomposition (CAD)   dissociation of a projectile ion into fragment ions due to collision with a target neutral species. For example, in a typical application of triple quadrupole mass spectrometry, parent ions separated in the first quadrupole collide with an inert collision gas (e.g. argon) and decompose into daughter ions in the second quadrupole (collision cell).

colloid   a gas, liquid, or solid that is dispersed so finely as to settle only very slowly.

color alteration index (CAI)   measure of thermal stress based on color changes in fossil conodonts (tiny fossil cone-shaped teeth of the extinct eel-like vertebrate). Also called condodont alteration index.

column chromatography   use of a column or tube to hold the stationary phase, e.g. open-column chromatography, high-performance liquid chromatography (HPLC), and open-tubular capillary chromatography.

column performance   refers to the efficiency of a column as measured by the number of theoretical plates for a test compound.

column switching   use of multiple columns connected by switching valves to effect better chromatographic separations or for sample cleanup.

comet   an extraterrestrial body consisting of a nucleus of frozen gases (mostly water, but also CO, CO_2, and others) and organic- and silicate-based dust.

commingled production   a mixture of petroleum from two or more reservoirs at different depth.

competition   rivalry between two or more species or groups for a limiting environmental factor in the environment.

compost   a mixture of moist organic materials, with or without added fertilizer and lime, that is allowed to undergo thermophilic decomposition until the original organic materials are altered substantially.

compound-specific isotope analysis (CSIA)   a technique that measures isotopic compositions of individual components separated by a gas chromatograph. Compounds that elute from the gas chromatograph are combusted to CO_2 or pyrolyzed to form H_2 or N_2 for measurement of carbon, hydrogen, or nitrogen isotopes, respectively.

compression   lateral force or stress (e.g. tectonic) that tends to decrease the volume of, or shorten, a substance.

condensate   a light oil with an API gravity greater than 45° API. The term was used originally to indicate petroleum that is gaseous under reservoir conditions but is liquid at surface temperatures and pressures.

condensed rings   naphthenic or aromatic molecules containing multiple rings of carbon atoms where adjacent rings are joined by two shared carbon atoms (e.g. cholestane, benzopyrene).

condensed section   a thin marine stratigraphic unit, typically deep-water shale, that may be rich in organic matter and/or phosphate that was deposited at very slow sedimentation rates (<1–10 mm/1000 years). High total organic carbon (TOC)-condensed sections are time-synchronous markers that can be correlated across sedimentary basins (Creaney and Passey, 1993).

configuration   refers to the arrangement of atomic groups around an asymmetric carbon atom.

conformation   the three-dimensional arrangements that an organic molecule can assume by rotating carbon atoms or their substituents around single covalent carbon bonds. Although not fixed, one conformation may be more likely to occur than another. For example, cyclohexane exists in the preferred chair formation in addition to the boat and twisted conformations (see Figure 2.11).

conglomerate   a sedimentary rock dominated by rounded pebbles, cobbles, or boulders.

conjugation   the process of mating and exchange of DNA between bacterial cells.

conodonts   small, toothlike fossils, probably from primitive chordates, that are composed of calcium phosphate and that occur in Cambrian to Triassic rocks. Conodonts are useful for stratigraphic correlation and maturity assessment.

contact metamorphism   changes in the mineralogy and texture of a rock caused by heat and pressure from a nearby igneous intrusion.

continental crust   the part of the Earth's crust that underlies the continents and continental shelves and ranges in thickness from ~35 km to as much as 60 km under mountain ranges.

continental drift   the horizontal movement or rotation of continents relative to one another. The theory proposes that continents move away from each other by seafloor spreading along a median ridge or rift, producing new oceanic areas between the continents. Materials moving away from the ridges consist of thick plates of continental and oceanic crust, which move independently of each other.

continental margin   part of the ocean floor that extends from the shoreline to the landward edge of the abyssal plain, including the continental shelf, slope, and rise.

continental rise   a broad, gentle slope that rises from the abyssal plain to the continental slope.

continental shelf   the gently seaward-sloping submerged edge of a continent that extends to the edge of the continental slope, typically near 200 m depth.

continental slope   steeply sloping seafloor between the continental shelf and the continental rise.

convection   circulation of a fluid in response to uneven heat distribution.

convergent evolution   a process in which genetically different organisms develop similar characteristics during their adaption to similar habitats.

coorongite   a low-maturity equivalent of torbanite or boghead coal from Australia that consists mainly of *Botryococcus braunii.*

cordaites   a primitive order of treelike plants with long, blade-like leaves and clusters of naked seeds that may be evolutionary intermediates between seed ferns and conifers.

cordillera   an extensive mountain range or system of mountain ranges.

Coriolis effect   deflection of winds and water currents to the right in the Northern Hemisphere and to the left in the Southern Hemisphere due to the Earth's rotation.

cosmic rays   high-energy particles, mostly protons, that move through the space and frequently penetrate the Earth's atmosphere.

covalent (sigma) bond   a non-ionic chemical bond in which valence electrons are shared equally between the bonding atoms (for example, C–C or C–H bonds). Because of equally shared electrons, covalent bonds are stronger than ionic bonds (as in NaCl), which are characterized by electrical asymmetry.

CPI   see Carbon preference index (CPI).

cracking   a thermal process in which large molecules break down into smaller molecules, which occurs during burial maturation and in refineries. Kerogen in source rocks and oil in reservoirs are cracked to generate lighter petroleum products at high temperatures. Catalytic cracking increases the rate and efficiency of generation of light hydrocarbons in refineries.

crater   a bowl-shaped depression, commonly surrounding the central vent of a volcano.

craton   the old, geologically stable interior of a continent. Commonly composed of Precambrian rocks at the surface or covered only thinly by younger sedimentary rocks.

crinoids   stalked marine echinoderms with a calyx composed of regularly arranged plates and radiating arms for gathering food.

critical moment   the time that best depicts the generation–migration–accumulation of hydrocarbons in a petroleum system. The geographic and stratigraphic extents of the system are best evaluated using a map and cross-section drawn at the critical moment.

cross-linking   during copolymerization of resins, a difunctional monomer is added to form cross-linkages between adjacent polymer chains. The degree of cross-linking is determined by the amount of this monomer added to the reaction. For example, divinylbenzene is a typical cross-linking agent for polystyrene ion-exchange resins.

crude oil   natural oil that consists of a mixture of hydrocarbons and other compounds that have not been refined. It commonly contains solution gas and closely resembles the original oil from the reservoir rock. Hydrocarbons are the most abundant compounds in crude oils, but crude oils also contain NSO compounds and widely varying concentrations of trace elements, such as vanadium, nickel, and iron (e.g. National Research Council, 1985).

crust   outer part of the Earth, from the surface to the Mohorovicic discontinuity (Moho). Continental crust is thick (30–60 km), light, silica-rich, and older, while oceanic crust is thin (<20 km), dense, silica-poor, and younger.

Crustacea   a subphylum of the phylum Arthropoda that includes lobsters and crayfish.

cryogenic trap   a cold trap used to remove contaminants, e.g. in compound-specific isotope analysis, a cryogenic trap is commonly used to remove water and other combustion products from the carbon dioxide analyte in the effluent from the gas chromatograph/combustion furnace (see Figure 6.12).

CSIA   see Compound-specific isotope analysis.

culture   (1) a population of microbes cultivated in an artificial growth medium. Pure cultures grow from one cell, while mixed cultures consist of two or more strains or species growing together. (2) A human population with a distinct set of customs and traditions.

cuticle   the waxy layer on the outer walls of epidermal plant cells that represents the precursor to cutinite.

cutinite   a maceral derived from the waxy remains of land plants.

CV   see Canonical variable (CV).

cyanobacteria   prokaryotic, photosynthetic microorganisms with chlorophyll and phycobilins that produce oxygen. Formerly called blue-green algae.

cycadales   seed plants that were common during the Mesozoic Era and were characterized by palm-like leaves and coarsely textured trunks marked by leaf scars.

cycloalkane (naphthene, cycloparaffin)   a saturated, cyclic compound containing only carbon and hydrogen. One of the simplest cycloalkanes is cyclohexane (C_6H_{12}). Steranes and triterpanes are branched cycloalkanes consisting of multiple condensed five- or six-carbon rings.

cyclohexane   the cyclic form of hexane; used as a raw material to manufacture nylon.

cyclothem   a vertical package of sedimentary units that reflects environmental events that occurred in a constant order. Multiple cyclothems are particularly characteristic of the Pennsylvanian System.

cyst   the resting stage of some bacteria, nematodes, and protozoa, during which the cell becomes surrounded by a protective layer.

cytochrome   respiratory pigment that consists of an iron-containing porphyrin ring (e.g. heme) complexed with proteins and serves to transfer electrons from a substrate to a terminal electron acceptor, such as oxygen.

cytoplasm   the contents of a cell inside the cell membrane, but excluding the nucleus.

28,30-dinorhopane   see 28,30-Bisnorhopane.

Dalton   one atomic mass unit (amu).

data processing   storage and manipulation of gas chromatography/mass spectrometry (GCMS) data using an electronic computer.

daughter ion   an electrically charged product of the reaction of a particular parent ion.

Dean Stark extraction   a method used to measure fluid saturations in core samples by distillation extraction. Water in the sample is vaporized by boiling the solvent, condensed, and measured in a calibrated trap. The solvent is condensed repeatedly and flows back through the sample to extract oil. Extraction continues until the condensed solvent is clean or the sample lacks residual fluorescence. The sample is weighed before and after extraction, and the volume of oil can be calculated from the loss in weight of the sample. Water (S_w) and oil (S_o) saturations are calculated from these data.

deasphalting   the process by which asphaltenes precipitate from crude oil. Laboratory or refinery deasphalting is used to remove complex components from oil by adding light hydrocarbons, such as pentane or hexane. This process can occur in nature when methane and other gases that escape from deep reservoirs enter a shallower oil reservoir.

decarboxylation   loss of one or more carboxyl (–COOH) groups from a compound. For example, decarboxylation of acetic acid (CH_3COOH) results in methane (CH_4) and carbon dioxide (CO_2). Decarboxylation of phytanic acid ($C_{18}H_{36}O_2$) results in pristane ($C_{17}H_{36}$).

Deccan Traps   a thick (3200 m) sequence of Upper Cretaceous basaltic lava flows that cover $\sim$500 000 km^2 in India.

deformation   folding, faulting, shearing, compression, or extension of rocks due to the Earth's forces.

dehydration   loss of one or more molecules of water from a compound. For example, dehydration of ethyl alcohol (CH_3CH_2OH) results in ethylene ($CH_2=CH_2$) and water.

delta   a low, nearly flat area near the mouth of a river, commonly forming a fan-shaped plain that can extend beyond the coast into deep water. Deltas form in lakes and oceans when sediment supplied by a stream or river overwhelms that removed by tides, waves, and currents.

delta log R   a method used to estimate total organic carbon (TOC) content in rocks based on the difference between the scaled transit-time and resistivity curves from conventional well logs (Creaney and Passey, 1993).

demethylated hopanes   demethylated hopanes (25-norhopanes) resulting from microbial removal of the methyl group attached to C-10 during heavy biodegradation of hopanes (Peters *et al.*, 1996b).

denitrification   reduction of nitrate or nitrite to molecular nitrogen or nitrogen oxides by microbial activity or by chemical reactions involving nitrite.

deoxyribonucleic acid (DNA)   a biomacromolecule that consists of nucleotides connected by a phosphate–deoxyribose sugar backbone. DNA contains the genetic information of the cell and controls protein synthesis.

deposition   sedimentation of any material, as in the mechanical settling of sediment from suspension in water, precipitation of mineral matter by evaporation from solution, and accumulation of organic material.

detrital   refers to sediment or organic matter deposited by moving water from a source within or outside the depositional basin.

detritus   loose rock, mineral, or organic material that is worn off or removed by mechanical means, as by disintegration or abrasion and deposited in a basin, e.g. sand, silt, and clay derived from older rocks and moved from its place of origin.

development   a phase in which newly discovered oil or gas fields are put into production by drilling and completing production wells.

development well   a well drilled with the intent to produce petroleum from a proven field.

diagenesis   chemical, physical, and biological changes that affect sediment during and after deposition and lithification but before significant changes caused by heat. Alteration processes occurring at maturity levels up to an equivalent of 0.6% vitrinite reflectance (e.g. see Figure 1.2). Diagenesis occurs before oil generation but includes the formation of microbial gas.

diahopane   hopane in which the methyl group attached to C-14 has been rearranged to C-15, e.g. 17α-15α-methyl-27-norhopane (Dasgupta *et al.*, 1995).

diamondoid   fused-ring cycloalkane with a diamond structure that shows high thermal stability and commonly precipitates directly from the gas to the solid phase during production.

diasterane (rearranged sterane)   rearrangement product from sterol precursors (e.g. see Figure 2.15) through diasterenes (see Figure 13.49). The rearrangement involves migration of C-10 and C-13 methyl groups to C-5 and C-14 and is favored by acidic conditions, clay catalysis, and/or high temperatures. Diasteranes increase relative to steranes with thermal maturation and they are low in clay-poor carbonate source rocks and related oils.

diasterane index   the ratio of rearranged (diasteranes) to regular steranes. Diasterane concentrations in petroleum depend on anoxicity and pH of the depositional environment and clay content and thermal maturity in the source rock.

diastereomer   a steroisomer that is not an enantiomer (mirror image). Diastereomers show different physical and chemical properties. An epimer is one type of diastereomer.

diatom   microscopic unicellular or colonial alga (chrysophyte) with a siliceous cell wall called a frustule that persists after death.

diatomaceous   refers to geologic deposits of fine siliceous material composed mainly of the remains of diatoms.

diazotroph   organism that can use diatomic nitrogen as a sole nitrogen source, i.e. capable of N_2 fixation.

diesel   a straight-run refinery distillation cut that contains hydrocarbons with ~14–18 carbon atoms. Diesel is a common additive in oil-based drilling muds and can interfere with geochemical analyses. See also Straight-run refinery products.

dike   a small, discordant (injected into massive igneous, metamorphic, or across layers of sedimentary rock) body of intrusive igneous rock.

dinoflagellates   unicellular marine algae typified by two flagella and a cellulose wall.

diploid   describes eukaryotic organism or cell with two chromosome complements, one derived from each haploid gamete.

disproportionation   cracking of organic matter resulting in lighter and heavier products, e.g. cracking of petroleum in the reservoir generates gases and a pyrobitumen residue.

distillation   heating of crude oil or other materials to separate components according to their relative volatility (see Table 5.1).

diterpanes   a class of biomarkers containing four isoprene subunits (e.g. see Figure 2.15) that are mainly bi- and tricyclic compounds. Many diterpanes originate from higher plants.

DNA   see Deoxyribonucleic acid (DNA).

DNA fingerprinting   molecular genetic techniques for assessing the differences between DNA in samples.

dolomite   a rhombohedral carbonate mineral with the formula $CaMg(CO_3)_2$.

dolostone   a carbonate sedimentary rock that contains over 50% of the mineral dolomite [$CaMg (CO_3)_2$].

domain   a major taxonomic division ranking higher than a kingdom. The three domains of organisms consist of the archaea, eubacteria, and eukarya.

dome   a fold in rocks shaped like an inverted bowl. Strata in a dome dip outward and downward in all directions from a central area.

downstream   all petroleum activities from refining of crude oil into petroleum products to the distribution, marketing, and shipping of these products.

DPEP   deoxophylloerythroetioporphyrins. Common tetrapyrrole pigments in sediments and petroleum. The DPEP/etio porphyrin ratio is used as a thermal maturity parameter.

drill bit   a drilling tool that cuts through rock by a combination of crushing and shearing.

drilling mud   a mixture of clay, water, and chemicals that is pumped in and out of the well bore during drilling. Circulation of drilling mud flushes rock cuttings produced by the drill bit at the bottom of the well bore to the surface and maintains pressure in the well bore.

dry gas   gas that is dominated by methane (>95% by volume) with little or no natural gas liquids and is generally microbial or highly mature. Microbial gas is depleted in ^{13}C compared with highly mature gas.

dry hole   a well lacking oil or gas in commercial quantities.

dwell time   the time spent by the detector at a given mass during scanning by the mass spectrometer. Longer dwell times result in more accurate measurements of peaks by increasing the signal-to-noise ratio. Dwell times for selected ion monitoring/gas chromatography/mass spectrometry (SIM/GCMS) and full-scan GCMS are ~0.04 and <0.0075 s/ion, respectively.

dysaerobic   refers to metabolism under dysoxic conditions with limited molecular oxygen (see Table 1.5).

dysoxic   refers to water column or sediments with molecular oxygen contents of 0.2–2 ml/l of interstitial water (see Table 1.5).

early-mature   early-mature source rocks have begun to generate some petroleum but have not yet reached the main stage of oil generation. Early mature is commonly assumed to represent ~0.5–0.6% vitrinite reflectance.

earthquake   shaking of the ground caused by a sudden release of energy stored in the rocks beneath the Earth's surface, commonly caused by movement along faults.

EASY %R_o   a simple model that uses Arrhenius first-order kinetics for a distribution of activation energies to calculate vitrinite reflectance for given time/temperature conditions (Sweeney and Burnham, 1990). The model applies in the range 0.3–4.5% mean random reflectance.

echinoderms   marine invertebrates of the phylum Echinodermata, showing fivefold symmetry and with exoskeletons and spines commonly consisting of calcite. Cystoids, blastoids, crinoids, and echinoids are examples of echinoderms.

ecology   the science of the interrelationships among organisms and between organisms and their environment.

ecosystem   the community of organisms and the environment in which they live.

Ediacaran fauna   Late Proterozoic fauna first discovered in Australia but subsequently found in rocks ~600 million years old in many continents.

effective source rock   a source rock that is generating and expelling or has generated and expelled oil and gas.

effluent   mobile phase used to carry out a chromatographic separation. See also Eluate.

Eh   see Redox potential.

electron   a subatomic particle with an electric charge of -1 and mass of 9.11×10^{-28} g.

electron acceptor   a small organic or inorganic compound that accepts electrons during a redox reaction and is reduced to complete an electron-transport chain.

electron donor   a small organic or inorganic compound that donates electrons during a redox reaction and is oxidized to provide an electron for the electron transport chain.

electron impact   a gas chromatography/mass spectrometry (GCMS) ionization technique. Compounds eluting from the gas chromatograph enter the ion source of the mass spectrometer through a transfer line, where they are bombarded with energetic electrons, typically at 70 eV (electron-volts). This causes the compounds to ionize into a series of fragment ions characteristic of the structure of the particular compound, e.g. $M + e- \rightarrow M + \cdot 2e-$, where M is an eluting compound.

electron-transport chain   the sequence of reactions in biological oxidation that moves electrons through a series of oxidizing agents arranged in order of increasing strength and terminating in oxygen.

eluate   combined mobile phase and solute exiting a chromatographic column. Also called effluent.

eluent   a liquid used as the mobile phase to carry components to be separated through a chromatographic column. For example, in liquid chromatographic fractionation of petroleum, hexane can be used as the eluent for the saturate fraction while methylene chloride can be used for the polar fraction.

elution   the process of passing mobile phase through a chromatographic column to transport solutes.

enantiomers   a pair of stereoisomers of a molecule that have the same molecular formula but differ in the arrangement of substituents around every asymmetric carbon atom, thus representing two mirror-image structures (e.g. see Figure 2.21). When only one asymmetric carbon atom is present, the mirror-image structure can be obtained only by transposing (inverting) any two of the four substituents on the asymmetric carbon atom (isomerization). In a molecule containing more than one asymmetric center, inversion of all the centers leads to the enantiomer. Enantiomers are isomers that are non-superimposable mirror images.

endosymbiosis   a permanent or long-lasting intracellular incorporation of one or more genetically autonomous biological systems. For example, mitochondria and chloroplasts in eukaryotic cells may have originated as aerobic non-photosynthetic bacteria and photosynthetic cyanobacteria, respectively.

enhanced oil recovery   displacement and recovery of oil or gas by injecting gas or water into the reservoir.

enzyme   a protein in a living organism that catalyzes specific reactions.

eon   the largest division of geologic time consisting of several eras. The Phanerozoic Eon consists of all geologic periods, from the Cambrian to the Quaternary.

EPA   United States Environmental Protection Agency.

epifaunal   organisms living on the sediment surface.

epimer   in a molecule containing more than one asymmetric center, inversion of all the centers leads to the enantiomer (mirror image). Inversion of less than all the asymmetric centers yields an epimer or diastereomer.

epiric sea   a shallow water body formed during major transgressions of continental shelf or landmass. Source rocks with high petroleum-generative potential can be deposited in shallow basins within epiric seas. The name originates from the Greek *Epirus*, a region of northwestern Greece bordering the Ionian Sea.

epoch   a subdivision of a geologic period. Rocks deposited during an epoch represent the series for that epoch.

era   a division of geologic time that is less than an eon and that consists of several periods.

ergostane   the C_{28} sterane; also known as methylcholestane.

erosion   the movement of unconsolidated materials at the Earth's surface by wind, water, or glacial ice.

essential elements   the source, reservoir, seal, and overburden rocks in a petroleum system. Together with the processes of generation–migration–accumulation and trap formation, essential elements control the distribution of petroleum in the lithosphere.

ethane (C_2H_6)   a gaseous hydrocarbon that occurs in petroleum.

ethylcholestane   see Stigmastane.

etio   a class of porphyrins. The deoxyphylloerythroetio-porphyrin (DPEP)/etio porphyrin ratio is used as a maturity parameter.

ETR   see Extended tricyclic terpane ratio.

eubacteria   prokaryotic organisms that contain ester-linked lipids and are not archaea.

eukarya   see Eukaryotes.

eukaryotes   a phylogenetic domain that contains all higher organisms (eukarya or eukaryotes) with a membrane-bound nucleus and organelles and well-defined chromosomes. Includes virtually all organisms except the prokaryotes (bacteria and cyanobacteria) and the archaea.

euphotic zone   see Photic zone.

eustatic   refers to worldwide simultaneous changes in sea level, such as might result from change in the volume of continental glaciers.

eutrophic   describes a lake containing abundant nutrients that support high primary productivity. Eutrophic lakes commonly have low dissolved oxygen due to high biological oxygen demand. They are generally last in the age sequence oligotrophic, mesotrophic, and eutrophic, before complete infilling of the lake basin and subaerial conditions.

euxinic   anoxic, restricted depositional conditions with free hydrogen sulfide (H_2S). Present-day examples include anoxic bottom water and sediments in Lake Tanganyika, the Black Sea, the coast of Peru, and the northeastern Pacific. Anoxic sediments with organic matter and sulfate become euxinic through the activity of sulfate-reducing bacteria. The term originates from the classical Greek name for the Black Sea, *Euxynos*. While the deep Black Sea is euxinic, the name is contradictory because it means "hospitable" or "good to foreigners."

evaporation   the process in which a substance passes from the liquid or solid state to the vapor state. Also called vaporization.

evaporite   various minerals that precipitate from a water solution during evaporation, e.g. anhydrite ($CaSO_4$), gypsum [($CaSO_4 \cdot 2(H_2O)$)], and halite (NaCl). A typical sequence of minerals deposited with increasing extent of evaporation is (1) calcite ($CaCO_3$), (2) gypsum, (3) halite, and (4) sylvite (KCl).

events chart   a petroleum system chart (also called timing-risk chart) that shows the timing of essential elements and processes and the preservation time and critical moment of the system.

evolution   genetic adaptation of organisms or species to the environment.

exinite   the liptinite group, sometimes called exinite, consists of hydrogen-rich macerals that generate significant quantities of oil during maturation. These macerals include alginite, resinite, sporinite, and cutinite (from algae, terrigenous plant resins, spores, and cuticle, respectively). In our simplified classification, exinite consists of all liptinite macerals, except alginite.

exobiology   the branch of biology that deals with the effects of extraterrestrial environments on living organisms.

expulsion   the process of primary migration, whereby oil or gas escapes from the source rock due to increased pressure and temperature. Generally involves short distances (meters to tens of meters).

extended tricyclic terpane ratio $ETR = (C_{28} + C_{29})/(C_{28} + C_{29} + Ts)$. Early results by Holba *et al.* (2001) suggest that ETR can be used to differentiate crude oils generated from Triassic, Lower Jurassic, and Middle–Upper Jurassic source rocks (see Table 13.8).

extract   bitumen or oil removed from rocks using organic solvents.

extractable organic matter (EOM)   see Extract.

extrusive   igneous rock that erupted on to the surface of the Earth, such as lava flows, tuff (consolidated pyroclastic material), and volcanic ash.

facies   the characteristics of a rock unit that reflect the conditions of its depositional environment.

facultative   describes an organism that can carry out both options of mutually exclusive processes (e.g. aerobic and anaerobic metabolism).

facultative anaerobes   microbes that can grow under both oxic and anoxic conditions but do not shift from one mode of metabolism to another as conditions change. They obtain energy by fermentation.

fatty acids   saturated or unsaturated organic compounds with a terminal carboxyl group (–COOH) that are common membrane components in organisms and are converted to *n*-alkanes and other hydrocarbons during diagenesis and thermal maturation. Fatty acids commonly contain an even number of carbon atoms in the range $\sim C_{14}$–C_{24}, which are synthesized by condensation of malonyl coenzyme A (see Figure 3.5).

fault a fracture or zone of fractures in the Earth's crust along which rocks on one side were displaced relative to those on the other side.

FC43 a standard compound (perfluorotributylamine) used to calibrate the mass scale of a mass spectrometer.

feedstock crude oil, natural gas liquids, natural gas, and refined materials that are used to make gasoline, other refined products, and chemicals.

fermentation an anaerobic oxidation–reduction by which some prokaryotes and eukaryotes (e.g. yeasts) break down organic compounds to yield energy for growth, alcohol, and lactic acid. Some atoms of the energy source become more reduced, while others become more oxidized.

FID see Flame ionization detector.

field an area under which a producing or prospective oil and/or natural gas reservoir lies.

filter feeders animals that filter water to obtain small particles of food.

fingerprint a chromatographic signature used in oil–oil or oil–source rock correlations. Mass chromatograms of steranes or terpanes are examples of fingerprints that can be used for qualitative or quantitative comparison of oils and source-rock extracts.

Fischer–Tropsch an industrial process used to convert carbon monoxide and carbon dioxide from coal into synthetic petroleum. Fischer–Tropsch reactions may occur naturally in the presence of magnetite or other metal oxide catalysts.

fissile the tendency of certain rocks, such as shale, to split into thin plates or lamina.

fissility a property of rocks, mainly shales, that causes them to split into thin pieces parallel to bedding.

fission tracks tiny damage trails in minerals produced by high-energy particles emitted from radioactive elements (e.g. uranium) during spontaneous fission.

fjord a long, U-shaped, glacier-carved inlet from the ocean.

flagellum (*pl.* **flagella**) a whip-like structure attached to certain microbial cells that allows motility.

flame ionization detector (FID) a detector for gas chromatographs or Rock-Eval pyrolysis that measures ions generated by combustion of eluting compounds. FID responds to all compounds that combust (e.g. hydrocarbons, but not carbon dioxide).

flash point the lowest temperature to which liquid fuel must be heated to produce a vapor/air mixture above the liquid that ignites when exposed to an open flame.

flood basalt a regionally extensive plateau consisting of layers of basalt that originated from repeated fissure eruptions, e.g. the Columbia Plateau in the western USA and the Deccan Traps in India.

floodplain a flat area of unconsolidated sediment near a stream channel that is submerged only during high-flow periods.

fluorescence light emission by organic matter when excited by energy, such as ultraviolet (UV) light. UV fluorescence is diagnostic of low to moderately mature, oil-prone kerogen. UV-excitation fluorescence can be used to identify crude oil in conventional cores.

fluvial sediments or other geologic features formed by streams.

flux the rate of emission, sorption, or deposition of a material from one pool to another.

fold a curve or bend of a formerly planar structure, such as rock strata or bedding planes, that generally results from deformation.

footwall the underlying side of a fault, below the hanging wall.

foraminifera an order of mostly marine, unicellular protozoans that secrete tests (shells) composed of calcium carbonate and that account for most carbonate sediments in the modern oceans.

formation a distinct stratigraphic unit of rock that shares common lithologic features and is large enough to be mapped.

formation water water that occurs naturally in sedimentary rocks.

fossil remains or indications of an organism that lived in the geologic past.

fragment ion an electrically charged product of the fragmentation of a molecule or ion. Fragment ions can dissociate further to generate other, lower-molecular-weight fragments.

fragmentogram see Mass chromatogram.

free energy energy that is, or that can be made, available to do useful work. Defined as the difference between the internal energy of a system and the product of its absolute temperature and entropy. $\Delta G = \Delta H - T\Delta S$. Also called Gibbs free energy, after Josiah Willard Gibbs, who developed the concept.

frustule the siliceous wall and protoplast of a diatom.

fuel oil heavy distillates from oil refining that are used mainly for heating and fueling industrial processes, ships, locomotives, and power stations.

full scan   an operational mode in mass spectrometry in which a range of ion mass/charge ratios is monitored repeatedly at a specified rate.

fulvic acid   yellow organic material that remains in solution after removal of humic acid by acidification.

functional group   a site of chemical reactivity in a molecule that arises from differences in electronegativity or from a pi bond. For example, alcohol (–OH), thiol (–SH), and carboxyl (–COOH) groups, and double bonds (C=C), are common functional groups in organic compounds.

fungus (*pl.* fungi)   non-phototrophic, eukaryotic microorganisms that contain rigid cell walls.

fusinite   an inert maceral of the inertinite group that is rich in carbon and has high reflectance.

fusulinids   mainly spindle-shaped foraminifers with coiled, calcareous tests separated into many complex chambers. Fusulinids were abundant during the Pennsylvanian and Permian periods.

gammacerane   a C_{30} pentacyclic triterpane in which each ring contains six carbon atoms. Source rocks deposited in stratified anoxic water columns (commonly hypersaline) and related crude oils commonly have high gammacerane indices (gammacerane/hopane).

gamma ray   high-frequency electromagnetic wave.

gas cap   part of a petroleum reservoir that contains free gas.

gas chromatography   an analytical technique designed to separate compounds (e.g. see Figure 8.6), whereby a mobile phase (inert carrier gas) passes through a column containing immobile stationary phase (high-molecular-weight liquid). The stationary phase can be coated on the walls of the column or on a solid support packing material. Compounds separate based on their relative tendencies to partition into the stationary or mobile phases as they move through the column. Various detectors can be used to measure the separated components as they elute from the column. Most gas chromatographs are equipped with flame ionization detectors (FIDs).

gas chromatography/mass spectrometry   see GCMS.

gas hydrate   a crystalline phase of water that contains gas (mainly methane) in arctic and deep-water settings.

gas injection   an enhanced recovery method, whereby natural gas is injected under pressure into a producing reservoir through an injection well to drive oil to the well bore.

gasoline   a light fuel used to drive automobiles and other vehicles. The gasoline distillation cut from a refinery includes the approximate range C_5–C_{10}, while commercial gasoline may also contain synthetic additives. See also Straight-run refinery products.

gasoline range   the fraction of crude oil boiling between $\sim$15°C and 200°C, including low-molecular-weight compounds, usually containing fewer than 12 carbon atoms; e.g. the C_5–C_{10} hydrocarbons represent the gasoline-range compounds that are low or absent in uncontaminated, thermally immature sediments.

gas-prone   describes organic matter that generates predominantly gases with only minor oil during thermal maturation. Type III organic matter is gas-prone.

gas-to-oil ratio (GOR)   the amount of hydrocarbon gas relative to oil in a reservoir, commonly measured in cubic feet per barrel.

gas wetness   can be expressed in different ways, but generally represents the amount of methane (C_1) relative to the total hydrocarbon gases ($C_1 + C_{2+}$) in a sample. For example, $C_1/(C_1$–$C_5)$ ratios above and below 98% are dry and wet gases, respectively.

Gaussian curve   a symmetrical, bell-shaped error curve; e.g. ideal chromatographic peaks are Gaussian curves.

GC   see Gas chromatography. Most gas chromatographs are equipped with a flame ionization detector (FID);. see Flame ionization detector. As used in the text, the term "gas chromatography" implies use of FID unless another detector is specified.

GCMS (gas chromatography/mass spectrometry)   the gas chromatograph separates organic compounds while the mass spectrometer is used as a detector to provide structural information (see Figure 8.5). Biomarker interpretations depend mainly on GCMS analysis.

GC/MSD (gas chromatograph/mass selective detector)   a benchtop instrument that provides separation and detection of organic compounds, including biomarkers.

gel   the solid packing used in gel permeation chromatography for separation of macromolecules (e.g. nucleic acids or proteins), which consists of a dispersed medium (solid portion) and dispersing medium (the solvent).

gel permeation chromatography   a type of chromatography used to separate and characterize polymers.

gene   the unit of heredity in a chromosome consisting of a segment of DNA that codes for a single t-RNA, r-RNA, or protein.

gene pool   all of the genes in the living population of a species.

generation–accumulation efficiency (GAE)   the percentage of the total volume of trapped (in-place) petroleum for a petroleum system relative to the total volume of petroleum generated from the corresponding pod of active source rock (Magoon and Valin, 1994).

generation–migration–accumulation   a petroleum system process that includes the generation and movement of petroleum from the pod of active source rock to the petroleum seep, show, or accumulation.

genetic code   information for the synthesis of proteins contained in the nucleotide sequence of a DNA molecule (or, in certain viruses, of an RNA molecule).

genome   a complete set of genes for an organism. All of the chromosomal genes in a haploid cell (e.g. prokaryotes and archaea) or the haploid part of chromosomes in eukaryotes.

genotype   the complete genetic constitution of an organism.

genus (*pl.* genera)   a subdivision of a taxonomic family of plants or animals that usually consists of more than one species; the first name (e.g. *Homo*) of the scientific name (e.g. *Homo sapiens*).

geochemical fossil   see Biological marker.

geochromatography   chromatographic separation of organic compounds, in which clay minerals and organic matter in the source rock and carrier beds retard heavier, more polar compounds.

geochronology   the study of geologic time.

geographic extent   the area of occurrence of a petroleum system as defined by a line that encloses the pod of active source rock and all discovered petroleum shows, seeps, and accumulations that originated from that pod. The geographic extent is mapped at the critical moment.

geologic range   the span of geologic time between the origin and extinction of an organism.

geology   the science of the history of the Earth, including the materials that comprise the planet, the physicochemical changes that occur on and within the Earth, and the evolution of life as recorded in the rocks.

geopolymer   a term commonly used to describe fulvic acids, humic acids, and kerogen in rocks and sediments.

However, these organic materials are not strictly polymers because they are not composed of repeating subunits like those in organic (e.g. cellulose) or inorganic (e.g. nylon) polymers.

geosphere   the solid Earth, including continental and oceanic crust, mantle, and deep interior. The geosphere does not include the oceans, atmosphere, or life.

geothermal gradient   the rate of increase in temperature with depth in the Earth, usually measured in $°C/km$ or $°F/100$ ft. Gradients are sensitive to lithology, circulating groundwater, and the cooling effect of drilling fluids. Worldwide average geothermal gradients range from 24 to $41°C/km$ ($1.3–2.2°F/100$ ft), with extremes outside this range.

Gibb's free energy   see Free energy.

gilsonite   solid bitumen composed mainly of nitrogen, sulfur, and oxygen (NSO) compounds and asphaltenes that occurs as dike complexes in the Uinta Basin, Utah. It is a common drilling mud additive.

glacier   a massive body of ice, formed from recrystallized snow, that flows slowly under the influence of gravity.

glauconite   a green clay mineral common in marine sandstones and believed to have formed at the site of deposition.

Gloeocapsomorpha prisca   a microorganism peculiar to Middle Ordovician rocks that appears responsible for the strong odd n-alkane ($<C_{20}$) predominance in oils and source-rock extracts of this age. Ordovician rocks that contain abundant *G. prisca*, such as the Estonian kukersites and the Guttenberg Member of the Decorah Formation (North America), contain a polymeric material consisting mainly of C_{21} and C_{23} n-alkenyl resorcinol building blocks that represent selectively preserved cell-wall or sheath components or materials that polymerized during diagenesis (Blokker *et al.*, 2001).

***Glossopteris* flora**   an assemblage of late Paleozoic to early Mesozoic fossil plants from South Africa, India, Australia, and South America named after the associated seed fern, *Glossopteris*.

glucose   a common monosaccharide sugar ($C_6H_{12}O_6$) that occurs in honey and fruit juices. Glucose is also called dextrose because it is optically active and rotates the plane of polarized light in a clockwise direction.

glycerides   esters formed between one or more acids and glycerol. For example, fatty-acid esters with glycerol occur in plant oils and animal fats.

Gondwana a large continent in the southern hemisphere during Permo-Carboniferous time composed of parts of modern Australia, South Africa, India, Africa–Arabia, and Antarctica.

GOR see Gas-to-oil ratio.

GPA Gas Processors Association.

graben a down-dropped block of crust bounded by parallel normal faults.

Gram stain a stain that divides bacteria into two groups, Gram-positive and Gram-negative, depending on their ability to retain crystal violet when decolorized with an organic solvent, such as ethanol. The cell wall of Gram-positive bacteria consists chiefly of peptidoglycan and lacks the outer membrane of Gram-negative cells.

granite an intrusive igneous rock with high silica (SiO_2) content typical of continental regions.

graphite a mineral composed mainly of staggered flat layers of carbon atoms with minor amounts of hydrogen. Individual layers are bound weakly to each other and are composed of strongly bonded carbon atoms at the vertices of a network of regular hexagons.

graptolites extinct colonial marine invertebrates, possibly protochordates that range from Late Cambrian to Mississippian in age.

gravity segregation a process whereby heavier and lighter petroleum components accumulate near the bottom and top of the reservoir, respectively, possibly due to movement of gas toward the top of the reservoir.

greenhouse effect a process whereby short-wavelength solar radiation that impinges on the Earth is re-radiated from Earth and cannot escape back into space because the Earth's atmosphere is not transparent to the re-radiated energy, which is in the form of infrared radiation (heat).

greenschist schist containing the minerals chlorite and epidote (which are green) and formed by low-pressure, low-temperature metamorphism of mafic volcanic rocks.

groundwater water that lies below the surface in fractures and pore space in rocks.

growth fault a slump fault sedimentary rock that grows continuously during deposition of sediments, such that strata on the downthrown side of the fault are thicker than equivalent strata on the upthrown side.

GRZ three organic-rich Cretaceous intervals on the North Slope of Alaska, including the pebble shale (informal name), gamma-ray zone (GRZ) (also called the highly radioactive zone, HRZ), and the lower part of the Torok Formation, are considered to be one source rock

because of similar kerogen composition (Magoon and Bird, 1988).

guard column a small column placed between the injector and the analytical chromatographic column that protects the latter from contamination by sample particulates and strongly retained species.

gymnosperms flowerless seed plants in which the seeds are not enclosed (naked seeds), e.g. Coniferales, Ginkgoales, Bennettitales, and Cycadales, which were the dominant land plants after their origin in the middle Paleozoic Era until the advent of angiosperms during the Cretaceous Period.

habitat the environment where an organism lives.

half-life the time required for one-half of an original amount of radioactive material to decay to daughter products. The half-life of $_{14}C$ is ~5730 years.

halite see Evaporite.

halocline a layer of water in which the salinity increase with depth is greater than that of the underlying or overlying water.

halophile organism that grows readily in a highly saline environment, e.g. purple halophilic bacteria in modern lakes.

hanging wall the overlying side of a fault, above the footwall.

haploid describes cells with a single set of chromosomes, as in gametes.

HBIs highly branched isoprenoids in petroleum probably originate from diatom precursors (Nichols *et al.*, 1988) and could be markers of Jurassic or younger source rock. The C_{25} HBI is generally more abundant than C_{20} and C_{30} HBI and is also called 2,6,10,14-tetramethyl-7-(3-methylpentyl)-pentadecane.

H/C the atomic hydrogen to carbon ratio, typically used to describe kerogen type on van Krevelen diagrams.

HRZ see GRZ.

headspace gas analysis gas chromatographic analysis of light hydrocarbon gases that collect in the contained space above canned cuttings.

heavy crude thick, viscous crude oil showing <20°API.

heavy oil crude oil that shows low API gravity (<25°API).

hemipelagic describes ocean waters near land masses. Hemipelagic sediments consist of large amounts of lithic materials transported from land by turbidity currents, volcanic activity, and other processes.

heptane ratio a thermal maturity parameter based on the abundance of *n*-heptane versus other gasoline-range hydrocarbons and commonly used with the isoheptane ratio (Thompson, 1983).

herbaceous organic matter   cuticular, spore, and pollen components in kerogen (type II).

heterocompounds   see NSO compounds.

heterocyst   a specialized cyanobacterial cell that carries out fixation of diatomic nitrogen.

heterotroph   an organism that uses organic compounds as a source for cellular carbon.

hexane   a petroleum liquid found in natural gasoline.

HGT   see Horizontal gene transfer.

HI   see Hydrogen index.

higher-plant index (HPI)   (retene + cadalene + iHMN)/1,3,6,7-TeMN, where iHMN is 1-isohexyl-2-methyl-6-isopropylnaphthalene and TeMN is tetramethylnaphthalene (van Aarssen *et al.*, 1996).

high-performance liquid chromatography (HPLC)   a chromatographic method used to separate petroleum into saturate, aromatic, and polar fractions (see Figure 8.3). The method is typically automated. Larger amounts of materials can be separated by column chromatography.

high-sulfur oil   see Sour crude.

holocene   the time since the last major episode of glaciation. Equivalent to Recent.

homohopane index   see C_{35} Homohopane index.

homohopane isomerization   a biomarker maturity ratio [22S/(22S + 22R)] describing the conversion of the biological 22R to the geological 22S configuration of homohopane molecules. Typically, this ratio is calculated for the C_{32} 17α-homohopanes, but other carbon numbers in the range C_{31}–C_{35} are used sometimes.

homohopanes C_{31}–C_{35} $17\alpha,21\beta$(H)-hopanes (pentacyclic triterpanes) that elute as isomeric doublets after the C_{30} hopane on m/z 191 mass chromatograms.

homolog   one member of a series of organic compounds of the same chemical type that are constructed from discrete chemical subunits. For example, nC_{17} is the seventeenth homolog in the series of n-alkanes (n-paraffins) that begins with methane (nC_1). Each homolog in the n-alkane series differs from the previous and subsequent homolog by one methylene group ($-CH_2-$). Likewise, the homologous series of acyclic isoprenoids includes various compounds, such as norpristane (iC_{18}), pristane (iC_{19}), and phytane (iC_{20}).

homologous series   compounds showing similar structures but differing by the number of methylene ($-CH_2-$) groups (homologs). For example, the homologous series of n-alkanes can be described by the formula C_nH_{2n+2}, where $n = 1, 2, 3, 4$, etc. (methane, ethane, propane, butane, etc.). Many biomarkers consist of homologous series.

hopane   the C_{30} homolog in the hopane series of pentacyclic hydrocarbons (see Hopanes).

hopanes C_{27}–C_{35} pentacyclic triterpanes that originate from bacteriohopanoids in bacterial membranes and generally dominate the triterpanes in petroleum. Various isomeric series of hopanes include diahopanes, neohopanes, moretanes, demethylated hopanes, and homohopanes. The C_{30} hopane has $17\alpha,21\beta$-stereochemistry and commonly is more abundant than later-eluting homohopanes on m/z 191 mass chromatograms.

horizontal gene transfer (HGT) (also called lateral gene transfer)   the acquisition of new genes by organisms. For example, bacteria can incorporate genes during conjugation with other genetically distinct bacteria, or they can absorb genes introduced into the surrounding environment by dead bacteria.

host   an organism capable of supporting the growth of a virus or other parasite.

HPI   see Higher-plant index.

HPLC   see High-performance liquid chromatography.

HTB   "high TOC at the base, decreasing upward" units (Creaney and Passey, 1993). Most marine organic-rich shales are composed of these discrete sedimentary units, with TOC values that decrease upward from maxima near their bases.

humic acids   complex, high-molecular-weight organic acids that originate from decomposing organic matter and can be extracted from sediments and low-maturity rocks using weak base (e.g. dilute NaOH) and precipitated by acid (pH 1–2). Some kerogen originates from humic acids during diagenesis.

humic coal   gas-prone (type III) coal that consists mainly of higher-plant detritus, including vitrinite and inertinite group macerals, with little or no liptinite macerals.

humic substances   high-molecular-weight, brown to black substances formed in sediments by secondary synthesis reactions during diagenesis. The generic term describes the colored material or its fractions obtained on the basis of solubility characteristics, such as humic acid or fulvic acid.

hydration   a chemical reaction in which water or hydrogen is added to a compound or mineral.

hydrocarbons   various organic compounds composed of hydrogen and carbon atoms that can exist as solids, liquids, or gases. Sometimes this term is used loosely to refer to petroleum.

hydrogenation   any reaction of hydrogen with an organic compound. Typically, hydrogen reacts with double bonds of unsaturated compounds, resulting in a saturated product. For example, hydrogen reacts with ethylene ($CH_2=CH_2$) to produce ethane.

hydrogen bond   a chemical bond between a hydrogen atom of one molecule and two unshared electrons of another molecule.

hydrogen index (HI)   a Rock-Eval pyrolysis parameter defined as (S2/TOC) × 100 (TOC is total organic carbon) and measured in mg hydrocarbon/g TOC (see Figure 4.5). The HI is used in van Krevelen-type plots of HI versus oxygen index (OI) to determine organic matter type (see Figure 4.4).

hydrologic cycle   the movement of water from the ocean by evaporation, precipitation on land, and transport back to the ocean as surface and ground water.

hydrology   study of the distribution and movement of water in the atmosphere and upon and within the Earth.

hydrophilic   refers to water-soluble molecules and chromatographic stationary phases that are compatible with water.

hydrophobic   refers to molecules with little affinity to water and chromatographic stationary phases that are not compatible with water.

hydrostatic   equal pressure in all directions, as under a column of water extending from the surface.

hydrothermal   formed by precipitation from hot aqueous solutions.

hydrotreating   a refinery process that removes sulfur and nitrogen from crude oil and other feedstocks.

hydrous pyrolysis   a laboratory technique in which potential source rocks are heated without air, under pressure, and with water to artificially increase the level of thermal maturity. Oils generated during hydrous pyrolysis may be used for oil–source rock correlation when natural source rock extracts of suitable maturity are not available (see Figure 19.16).

hypercapnia   carbon dioxide poisoning, generally caused by rapid overturn of CO_2-rich bottom waters and release of dissolved gases to the atmosphere.

hypersaline   refers to brines that show salinities greater than ~35 parts per thousand.

hypha (*pl.* hyphae)   a long, commonly branched tubular filament that serves as the vegetative body for many fungi and bacteria of the order Actinomycetes.

icecap   a mountain glacier that flows outward in several directions.

ice sheet   a large mass of ice covering a significant portion of a continent, as in Greenland and Antarctica.

ichnofossils   trace fossils, e.g. tracks, trails, burrows, borings, castings, and other markings made in sediments.

igneous rock   a rock formed by cooling or solidification of magma.

illite   a general name for three-layer, mica-like clay minerals in argillaceous rocks, especially marine shales and related soils.

immature   refers to conditions too cool (or of too short duration) for thermal generation of petroleum, i.e. vitrinite reflectance <0.6%.

I_{max}   a thermal maturity parameter based on the maximum wavelength of the fluorescence intensity, e.g. acritarch fluorescence

inactive source rock   a source rock with remaining generative potential that has stopped generating petroleum (e.g. due to uplift and reduced thermal stress).

inertinite   a maceral group composed of inert, hydrogen-poor organic matter with little or no petroleum-generative potential. Type IV kerogen is dominated by inertinite.

infauna   organisms that live within bottom sediments.

infrared (IR)   the electromagnetic spectrum with wavelengths from ~0.1 to 10 μm.

inlet   the initial part of a chromatographic column, where the solvent and sample enter.

inspissation   drying-up. For example, inspissation of seep oils can result in loss of gases and the lighter oil fractions due to reduced pressure (compared with that in the reservoir), evaporation and drying by exposure to sunlight, oxidation, and other related processes.

internal standard   a compound that shows similar behavior to the unknown compounds in a mixture to be analyzed but that is absent in the original mixture.

Addition of the internal standard to the mixture allows more reliable measurement of the unknown compounds. For example, 5β-cholane is added to the saturate fractions of oils as an internal standard. This compound is not found in significant amounts in natural oils, but its mass spectral characteristics are similar to those of steranes and other compounds in saturate fractions.

interstitial   refers to the pores of a host rock.

intracellular   inside the cell.

intrusion   the process of emplacement of magma into pre-existing rock.

ion   a charged atom or compound that is electrically charged as a result of the loss of electrons (to produce cations) or the gain of electrons (to produce anions).

ionic bond   a type of chemical bond in which atoms exchange electrons.

IRM/GCMS   see Compound-specific isotope analysis.

island arc   a linear or arc-shaped chain of volcanic islands formed at a convergent plate boundary, characterized by volcanic and earthquake activity, and usually near deep oceanic trenches.

isoalkanes   straight-chain alkanes that have a methyl group attached to the second carbon atom (i.e. 2-methyl alkanes).

isoheptane ratio   a thermal maturity parameter based on the abundance of several C_7 hydrocarbons and commonly used with the heptane ratio (Thompson, 1983).

isomerization (configurational isomerization, stereoisomerization)   any rearrangement of the atoms in a molecule to form a different structure. During stereoisomerization, a hydride or hydrogen radical is removed from an asymmetric carbon atom, resulting in formation of a planar carbocation or radical intermediate, followed by reattachment of the hydride or hydrogen radical. Reattachment can occur on the same side of the planar intermediate as removal (thus resulting in no change in configuration) or on the opposite side (resulting in an inverted or mirror-image configuration). Stereoisomerization occurs only when cleavage and reassembly of bonds results in an inverted configuration compared with the starting asymmetric center.

isomers   compounds with the same molecular formula but different arrangements of their structural groups (e.g. *n*-butane and isobutane).

isopach   a contour on a map that shows equal thickness of a sedimentary unit.

isopentenoids   compounds composed of isoprene subunits (see Figure 2.15). Most biomarkers are isopentenoids. Some biomarkers that are not isopentenoids include certain normal, iso-, and anteisoalkanes.

isoprene (isopentadiene)   the basic structural unit composed of five carbon atoms found in most biomarkers, including mono-, sesqui-, di-, sester-, and triterpanes, steranes, and polyterpanes (see Figure 2.15). The saturated analog of isoprene is 2-methylbutane.

isoprenoids   hydrocarbons composed of, or derived from, polymerized isoprene units. Typical acyclic isoprenoids include pristane (iC_{19}) and phytane (iC_{20}).

isostacy   gravitational balance or equilibrium, comparable to floating, of the units of the lithosphere above the asthenosphere.

isotopes   atoms whose nuclei contain the same number of protons but different numbers of neutrons. For example, all carbon atoms have six protons, but there are isotopes containing 6, 7, and 8 neutrons resulting in atomic masses 12, 13, and 14. ^{12}C is the principal naturally occurring isotope (98.89%). ^{13}C is the stable carbon isotope (1.11%) and ^{14}C is the unstable (radioactive) carbon isotope ($\sim 1 \times 10^{-11}$%). Carbon isotopes are commonly used in oil–oil and oil–source rock correlation.

IUPAC   the International Union of Pure and Applied Chemistry.

jet rock   an organic-rich black gem material of uncertain origin that takes on a bright luster when polished. It is believed to consist of plant remains (possibly related to the *Araucaria* tree) that were deposited under anoxic conditions.

joint   a fracture in rock on which no movement has taken place.

kerogen   insoluble (in organic solvents) particulate organic matter preserved in sedimentary rocks that consists of various macerals originating from components of plants, animals, and bacteria. Kerogen can be isolated from ground rock by extracting bitumen with solvents and removing most of the rock matrix with hydrochloric and hydrofluoric acids.

kerogen type I, II, IIS, III, and IV   see Type I kerogen, Type II kerogen, Type IIS kerogen, Type III kerogen, and Type IV kerogen.

kerosene   a distillate from crude oil refining that is used for heating and fuel for aircraft engines. See also Straight-run refinery products.

ketone   an organic compound with a carbonyl group (–CO).

kinetics   study of the rates of biological, physical, and chemical changes.

Krebs cycle   see Tricarboxylic acid cycle.

lacustrine   refers to lakes; e.g. lacustrine oils were generated from organic matter originally deposited in lake sediments.

lamella (*pl*. lamellae)   a thin plate-like arrangement or membrane.

lamination   fine stratification in shale or siltstone, where layers commonly 0.05–1.0 mm thick differ from adjacent layers.

Laurasia   a supercontinent that was composed of parts of modern Europe, Asia, Greenland, and North America.

lava   magma that reaches the Earth's surface.

leaching   the removal of materials in solution; e.g. extraction of soluble materials by water or crude oil percolating through rock.

level of certainty   a measure of confidence that petroleum originated from a specific pod of active source rock. Three levels include known (!), hypothetical (.), and speculative (?), depending on the level of geochemical, geophysical and geological evidence.

lichen   a symbiotic association of fungi and algae or cyanobacteria.

light hydrocarbons   gases that are volatile liquids at standard temperature and pressure (STP) and range from methane to octane, including normal, iso-, and cyclic alkanes, and aromatic compounds.

light oil   crude oil that has 35–45°API gravity.

lignans   in woody plants, dimers of p-coumaryl, coniferyl, and sinapyl alcohols that serve mainly as supportive fillers or toxins.

lignin   a phenolic organic polymer found with cellulose in the cell walls of many plants. Composed primarily of three aromatic precursors, p-coumaryl, coniferyl, and sinapyl alcohols.

lignite   a low-rank coal ($\sim$0.2–0.4% R_o) that is commonly used as a drilling mud additive. See also Coal.

limestone   see Carbonate rock.

linked-scan mode   a gas chromatography/mass spectrometry (GCMS) mode in which two or more quadrupole, electrostatic, and/or magnetic fields are scanned simultaneously, thus, allowing detection of specific parent, daughter, or neutral loss relationships between ions.

lipids   oil-soluble, water-insoluble organic compounds, including fatty acids, waxes, pigments, sterols, and hopanoids, and substances related biosynthetically to these compounds. Lipids are major precursors for petroleum (Silverman, 1971).

liptinite   a maceral group composed of oil-prone, fluorescent, hydrogen-rich kerogen. Both structured (e.g. resinite, sporinite, and cutinite) and unstructured or amorphous liptinites, sometimes called amorphinite, can occur.

liquid chromatography   a preparative column method used to separate oils and bitumens into saturate/aromatic and porphyrin/polar fractions before high-performance liquid chromatography (HPLC).

lithification   the solidification of loose sediments to form sedimentary rocks.

lithofacies map   a map that shows the lateral variation in lithologic attributes of a stratigraphic unit.

lithosphere   the outermost shell of the solid Earth, consisting of $\sim$100 km of crust and upper mantle, that lies above the asthenosphere.

lithotroph   an organism that oxidizes an inorganic substrate, such as ammonia or hydrogen, to use it as an electron donor in energy metabolism. Also known as chemoautotroph.

low-sulfur oil   see Sweet oil.

Lycopsida   leafy plants with simple, closely spaced leaves that carry sporangia on their upper surfaces. Lycopsida include living club mosses and many extinct late Paleozoic Lepidodendrales (scale trees). Also called lycopsids and lycophytes.

lysis   rupture of a cell, which results in the loss of cell contents.

maceral   microscopically recognizable, particulate organic component of kerogen showing distinctive physicochemical properties that change with thermal maturity. The three main maceral groups are liptinite, vitrinite, and inertinite.

macromolecule   a large molecule formed from the connection of a number of small molecules.

mafic   igneous rock composed mainly of dark-colored ferromagnesian minerals.

magma   molten rock below the Earth's surface.

magnetic sector   a magnetic (versus electric or quadrupole) mass analyzer in mass spectrometers. Ions follow a curved path down the flight tube in the magnetic field (e.g. see Figure 8.15). The degree of curvature is related to the mass and velocity of the ion. Only those ions with a given m/z reach the detector for a given magnetic field strength or accelerating voltage. All other ions collide with the walls of the flight tube.

maleimides   biomarkers for bacteriochlorophylls c, d, and e in green sulfur bacteria and chlorophyll a in phytoplankton (Grice, 2001).

mammoth   extinct elephant of the Pleistocene Epoch.

mantle   the region of the Earth composed mainly of solid silicate rock that extends from the base of the crust (Moho) to the core–mantle boundary at a depth of ~2900 km.

marl   a sedimentary rock containing calcareous clay.

marsupials   mammals of the order Marsupialia. Females have mammary glands and carry their young in a stomach pouch.

mass chromatogram (fragmentogram)   intensity of a specific ion versus gas–chromatographic retention time (e.g. see Figure 8.18). Allows identification of carbon number and isomer distributions for selected compound types.

mass number   the number of protons plus neutrons in the nucleus of an isotope of an element.

mass spectrometer   an instrument that separates ions of different mass but equal charge and measures their relative quantities.

mass spectrometry   a method used to supply information on the molecular structure of compounds, particularly biomarkers. Molecules in the gaseous state (inserted directly into the mass spectrometer or eluting from a gas chromatograph after separation) are ionized, usually by high-energy electrons. The resulting molecular and fragment ions are detected and displayed on the basis of increasing mass/charge ratio (m/z) in a mass spectrum.

mass spectroscopy   see Mass spectrometry.

mass spectrum   depiction of a beam of ions on a plot of the mass/charge (m/z) ratio versus intensity. Mass spectra can be used for provisional compound identification because they represent a fingerprint that is often diagnostic of specific structures.

MA steroid   see Monoaromatic steroid.

mature   describes organic matter that is in the oil-generative window, i.e. at a maturity equivalent to vitrinite reflectance in the approximate range 0.6–1.4%

(e.g. see Figure 14.3). Organic matter can also be mature with respect to the gas window (~0.9–2.0% vitrinite reflectance).

maturity   see Thermal maturity.

meiosis   the process of nuclear division in eukaryotes, in which the change from diploid to haploid occurs, resulting in daughter cells with half the number of chromosomes of the original cell.

mercaptan   sulfur-containing compounds that occur in sour crude and gas and are added to odorless natural gas and natural gas liquids to give them a characteristic smell and thus allow them to be detected. Light mercaptans have a strong, repulsive odor like that of rotten eggs.

meso compound   molecule with two or more asymmetric centers that also has planes of symmetry and thus cannot exist as enantiomers. For example, the 6(R),10(S)- and 6(S),10(R)-configurations of pristane are identical and are called mesopristane.

mesophile   an organism with optimal growth in the approximate range 15–40°C.

mesotrophic   describes a lake at a stage of evolution intermediate between oligotrophic and eutrophic.

metabolism   all biochemical reactions in a cell.

metagenesis   the thermal destruction of organic molecules by cracking to gas, which occurs after catagenesis but before greenschist metamorphism (>200°C) in the range of ~150–200°C (e.g. see Figure 1.2). The level of maturity equivalent to the range 2.0–4.0% vitrinite reflectance.

metamorphic rock   a rock formed from pre-existing rock due to high temperature and pressure in the Earth's crust, but without complete melting.

metamorphism   the solid-state transformation of pre-existing rock into texturally or mineralogically distinct new rock due to high temperatures, pressures, and chemically reactive fluids at depth, as occurs in the roots of mountain chains and adjacent to large intrusive igneous bodies.

metastable ion   an ion that has been accelerated from the ion source and decomposes in one of the field-free regions of the mass spectrometer, producing a broad, diffuse peak in the mass spectrum.

metastable reaction (metastable transition)   decomposition of a metastable ion. In metastable reaction monitoring/gas chromatography/ mass spectrometry (MRM/GCMS, also called metastable reaction monitoring/gas chromatography/

mass spectrometry/mass spectrometry, MRM/GCMS/MS) parent mode, all or selected metastable ions (usually molecular ions) in the mass spectrometer decomposing to a single daughter ion are monitored using a linked-scan method (see Linked-scan mode). A daughter mode is also possible, i.e. monitoring daughter ions from a given parent ion.

metazoa multicellular animals whose cells differentiate to form tissues, i.e. all animals except protozoa.

meteorites stony or metallic extraterrestrial bodies that have impacted the Earth's surface.

meteors solid extraterrestrial materials, mostly small particles, that are heated frictionally to incandescence upon entering the Earth's atmosphere. Most disintegrate, but some reach the surface of the Earth to become meteorites.

methane the main compound in natural gas.

methanogen a methane-producing prokaryote and member of the archaea.

methanogenesis the generation of methane by bacterial fermentation. Occurs only under anoxic conditions, where little sulfate is available (e.g. water column or sediments).

methanotroph a methane-oxidizing microbe.

methylcholestane see Ergostane.

methylotrophic bacteria heterotrophs that use reduced carbon substrates with no carbon–carbon bonds (e.g. methane, methanol, methylated amines, and methylated sulfur species) as their sole source of carbon and energy.

methylphenanthrene index (MPI) a thermal maturity parameter based on the relative abundances of phenanthrene and methylphenanthrenes (three-ring aromatic hydrocarbons). MPI depends on organic facies but is generally most reliable for petroleum generated from type III kerogen.

methylsteranes steranes with a methyl group attached to C-2, C-3, or C-4 on the A-ring. The 4-methylsteranes in petroleum generally originated from dinoflagellates.

microbe a living organism too small to be seen with the naked eye (<0.1 mm), e.g. algae, bacteria, fungi, and protozoans.

microbial biomass total mass of living microorganisms in a volume or mass of sediment or water.

microbial gas microbial gases are dominated by methane (typically $>99\%$) produced by bacteria in shallow sediments. Microbial methane is generally depleted in ^{13}C compared with thermogenic gas.

microbiology the study of microorganisms.

microenvironment the close physical and chemical surroundings of a microorganism.

micrometer 10^{-6} m (one millionth of a meter). A common unit for measuring microorganisms.

microorganism an organism of microscopic size.

microscopic organic analysis (MOA) petrographic analysis of kerogen or rock in transmitted and/or reflected light. General use of the term includes both maceral (relative percentages of phytoclast types) and thermal maturity (vitrinite reflectance, thermal alteration index (TAI), transmittance color index (TCI), color alteration index (CAI)) analysis. MOA provides critical support for interpretations based on elemental analysis (Jones and Edison, 1978) and biomarkers.

MID (multiple ion detection) see Selected ion monitoring.

migration the process whereby petroleum moves from source rocks toward reservoirs or seep sites. Primary migration consists of movement of petroleum to exit the source rock. Secondary migration occurs when oil and gas move along a carrier bed from the source to the reservoir or seep. Tertiary migration is where oil and gas move from one trap to another or to a seep.

Milankovitch effect the proposed long-term effect on world climate caused by three components of the Earth's motion. The combination of these components provides a possible explanation for repeated glacial to interglacial climatic swings.

mineral an element (e.g. gold) or compound (e.g. calcite, $CaCO_3$) that has a definite chemical composition or range of compositions with distinctive properties and form that reflect its atomic structure.

mineralization conversion of an element from an organic to an inorganic state due to microbial decomposition.

mitochondrion (*pl.* mitochondria) the eukaryotic organelle responsible for respiration and oxidative phosphorylation.

mitosis the reproductive process of cell division in eukaryotes in which each of the two daughter nuclei receives exactly the same complement of chromosomes as had existed in the parent nucleus.

MMCFD million cubic feet per day.

mobile phase see Gas chromatography.

Moho the Mohorivicic discontinuity (seismic reflector) at the base of the crust.

Mohorovicic discontinuity a zone that separates the crust of the Earth from the underlying mantle. The so-called Moho occurs at ~70 km below the surface of continents and 6–14 km below the floor of the oceans.

molecular fossil   see Biological marker.

molecular ion (M^+ or M^-)   an ion formed by the removal (or addition) of one or more electrons from a molecule without fragmentation.

molecular weight   the sum of the atomic weights of all the atoms in a molecule. A compound's mass may be expressed by its nominal, accurate, or average molecular weight. Nominal molecular weight is calculated from integer masses (e.g. H = 1, C = 12, O = 16). Accurate molecular weight uses the exact monoisotopic mass values (e.g. ^{1}H = 1.00783, ^{12}C = 12, ^{16}O = 15.99491). Average molecular weight uses the average of the accurate weight for isotopes as they occur in their natural abundance (e.g. H = 1.01, C = 12.01 O = 16.00). For example, $C_{35}H_{62}$ has a nominal mass of 482, an accurate mass of 482. 485152, and an average mass of 482.88 Daltons (atomic mass unit, amu).

molecule   two or more atoms combined by chemical bonding.

mollusk   a member of the invertebrate phylum Mollusca, which includes cephalopods, pelecypods, gastropods, scaphopods, and chitons.

monoaromatic steroid   a class of biomarkers that contain one aromatic ring (usually the C-ring) (e.g. see Figure 13.107), probably derived from sterols.

monoaromatic steroid triangle   plot of C_{27}, C_{28}, and C_{29} monoaromatic steroids on a ternary diagram (e.g. see Figure 13.105) used for correlation similar to the sterane triangle (e.g. see Figure 13.38). However, monoaromatic steroids are probably derived from different sterol precursors compared with the steranes.

monomer   a simple molecular unit, such as ethylene or styrene, from which a polymer can be made.

monoterpane   a class of saturated biomarkers constructed from two isoprene subunits ($^\sim C_{10}$).

monotremes   egg-laying mammals.

montmorillonite   a smectite clay that is cation-exchangeable and occurs as a major component in bentonite clay deposits, soils, sedimentary and metamorphic rocks, and some mineral deposits. Catalytic activity of montmorillonite and other clays is believed to account for the origin of diasterenes and disasteranes from sterols (see Figure 13.49).

moraine   a deposit of rock and soil (till) left by a retreating glacier.

moretanes C_{27}–C_{35} pentacyclic triterpanes that are stereoisomers of the hopanes and have $17\beta,21\alpha$-stereochemistry (see Figure 2.29). Because of lower stability, moretanes decrease relative to hopanes with increasing thermal maturation (see Figure 14.5).

mosasaurs   large marine lizards from the Late Cretaceous Period.

MPI   see Methylphenanthrene index.

MRM   see Metastable reaction monitoring.

mucilage   a gelatinous secretion produced by many microorganisms and plant roots.

mud gas   gases in the drilling mud that originate from formations in a well.

mudrock   see Mudstone.

mudstone   a general term for sedimentary rock made up of clay-sized particles, typically massive and not fissile.

multidimensional chromatography   use of two or more chromatographic columns (e.g. two-dimensional gas chromatography) to improve compound separation, either off-line by collecting fractions and re-injecting on to a second column, or on-line by the use of a switching valve.

multiple ion detection (MID)   see Selected ion monitoring.

murein (peptidoglycan)   an aminosugar polymer that consists of two hexoses, N-acetylglucosamine and N-acetyl-muramic acid, linked by a β-1,4-bond and short oligopeptides, which cross-link the polysaccharide chains in bacterial cell walls.

mutagen   a substance that causes genes to mutate.

mutant   an organism, gene, or chromosome that differs from the corresponding wild type by one or more base pairs.

mutation   an inheritable change in a gene of the DNA of an organism.

mycoplasmas   very small bacteria that lack a cell wall and have a single triple-layered membrane that is stabilized by small amounts of sterols. These sterols are apparently not synthesized by mycoplasmas but are obtained from the surrounding environment.

mycorrhiza   the symbiotic association between specific fungi with the fine roots of higher plants.

m/z   the mass/charge ratio of an ion in mass spectrometry measured in units of Daltons per charge, with positive or negative values denoting cations or anions, respectively, e.g. m/z 217 is a characteristic fragment of steranes (e.g. see Figure 8.15). Older publications may use the term m/e instead of m/z. Incorrect use of mass and Daltons as synonyms for m/z is especially confusing when applied to multiply charged ions.

NADH, NADPH   NADH is the reduced form of nicotinamide adenine dinucleotide (NAD). NADH and NADPH (additional phosphate) are key coenzymes for biochemical reactions in living cells (e.g. Figures 3.5 and 3.21).

nafion R   a family of perfluorosulfonate ion-exchange membranes (Dupont, Inc.) used in many component-specific isotope analysis (CSIA) instruments to remove H_2O from CO_2 during continuous combustion of the gas chromatography effluent.

naphtha   a distillation cut of petroleum showing volatility between gasoline and kerosene ($\sim C_7$–C_{10}). Used as a manufacturing solvent, a dry-cleaning fluid, and a gasoline-blending stock.

naphthalene   an aromatic hydrocarbon ($C_{10}H_8$) consisting of two fused benzene rings.

naphthenes   see Cycloalkane.

natural gas   gaseous petroleum that can consist of a mixture of C_1–C_5 hydrocarbons, CO_2, N_2, H_2, H_2S, and He. When natural gas occurs with oil, it is called associated gas.

NBS-22   a Pennsylvania No. 30 lubricating oil supplied by the National Bureau of Standards as one international standard for stable carbon isotope ratios. The most common standard for carbon is Peedee belemnite (PDB).

NDR C_{26} $24/(24 + 27)$ nordiacholestanes greater than 0.25 and 0.55 typify oils from Cretaceous or younger and Oligocene or younger (generally Neogene) source rocks, respectively (see Figure 13.55).

Neogene Period   a subdivision of the Cenozoic Era that consists of the Miocene and Pliocene epochs.

neohopane   a hopane in which the methyl group at C-18 is rearranged to C-17 (i.e. hopane II). Ts is the C_{27}-neohopane.

neritic   describes oceanic depths in the range between low tide and 200 m or the approximate edge of the continental shelf.

neutron   an electrically neutral subatomic particle of matter with a rest mass of 1 Dalton that occurs with protons in the atomic nucleus of all elements except the mass 1 isotope of hydrogen.

niche   the physical and biologic conditions under which an organism can live and reproduce.

nitrogen fixation   the conversion of molecular nitrogen (N_2) to ammonia and finally to organic nitrogen.

NMR   see Nuclear magnetic resonance.

non-associated gas   natural gas in gas accumulations.

non-hydrocarbon gases   mainly carbon dioxide (CO_2), nitrogen (N_2), and hydrogen sulfide (H_2S), but also including helium (He), argon (Ar), and hydrogen (H_2).

non-hydrocarbons   see NSO compounds.

non-polar   not easily dissolved in water due to hydrophobic (water repelling) characteristics.

25-norhopanes   see Demethylated hopanes.

normal fault   a fault in which the hanging wall appears to have moved downward relative to the footwall, normally occurring in areas of crustal tension.

normal-phase chromatography   chromatography using a polar stationary phase and a non-polar mobile phase, e.g. adsorption on silica gel using hexane as a mobile phase.

notochord   a rod-shaped cord of cartilage cells forming the primary axial structure of the chordate body. In vertebrates, the notochord is present in the embryo and is later supplanted by the vertebral column.

NSO compounds (resins)   a pentane-soluble fraction (or individual compounds) of petroleum that contains various elements in addition to hydrogen and carbon, including nitrogen, sulfur, and/or oxygen. Compounds in this fraction are sometimes called heterocompounds or non-hydrocarbons. Other fractions include saturates, aromatics, and asphaltenes.

nuclear magnetic resonance (NMR)   proton NMR (^1H NMR) is a spectroscopic method for determining molecular structure. The high resolution of the method commonly distinguishes stereoisomers that cannot be distinguished by mass spectroscopy. ^{13}C NMR requires more sample than proton NMR and can identify the types of carbon atoms (e.g. aromatic versus saturated). New ^{13}C NMR methods allow determination of numbers of hydrogen atoms bound to individual carbon atoms (i.e. methyl, methylene, methine, or quaternary). New two-dimensional NMR techniques allow correlation of ^{13}C and proton NMR data and provide improved resolution and in some cases can rival X-ray diffraction crystallography in structural elucidation.

nucleic acids   organic acid polymers of nucleotides that control hereditary processes within cells and make possible the manufacture of proteins from the amino acids ingested by the cells as food.

nucleotide   a monomeric unit of nucleic acid that consists of a pentose sugar, a phosphate, and a nitrogenous base.

nucleus   a membrane-enclosed structure that contains the genetic material (DNA) organized in chromosomes.

obligate   refers to an environmental factor (e.g. oxygen) that is always required for growth.

O/C   the atomic oxygen/carbon ratio from elemental analysis of kerogen, typically used to describe kerogen type on van Krevelen diagrams. Because oxygen is not measured readily on kerogens, atomic O/C is not always used directly on van Krevelen plots (Jones and Edison, 1978).

oceanic crust   the Earth's crust underlying the ocean basins, which is $\sim$5–10 km thick.

octane number   a measure of the resistance of a fuel to engine knock (pre-ignition) when burned in an internal combustion engine. In tests using pure hydrocarbons in gasoline engines, *n*-heptane and isooctane caused the most and least engine knock and were assigned octane ratings of 0 and 100, respectively. Higher octane numbers for fuels on this scale correspond to less engine knock.

odd-to-even predominance (OEP)   the ratio of odd- to even-numbered *n*-alkanes in a given range (Scalan and Smith, 1970). Immature rock extracts can show high or low OEP, but most mature oils and source rocks show OEP near 1.0.

OEP   see Odd-to-even predominance.

OI   see Oxygen index.

oil   a mixture of liquid hydrocarbons and other compounds of different molecular weights.

oil deadline   the depth at which oil no longer exists as a liquid phase in petroleum reservoirs, generally corresponding to gas-to-oil ratio (GOR) >5000 standard cubic feet/barrel or temperatures >150°C (typically in the range 165–185°C).

oil field   an area with an underlying oil reservoir.

oil–oil correlation   a comparison of chemical compositions to describe the genetic relationships among crude oils based on source-related geochemical data that might include biomarkers, isotopes, and metal distributions, although the source rock may not be defined.

oil-prone   organic matter that generates significant quantities of oil at optimal maturity. For example, type I and II kerogens are highly oil-prone and oil-prone, respectively. Oil-prone organic matter is typically also more gas-prone than gas-prone kerogen.

oil shale   organic-rich shale that contains significant amounts of oil-prone kerogen and liberates crude oil upon heating, as might occur during laboratory pyrolysis or commercial retorting.

oil–source rock correlation   a comparison of chemical compositions to describe the genetic relationships among crude oils and source-rock extracts based on source-related geochemical data that might include biomarkers, isotopes, and metal distributions.

oil–water contact (OWC)   the boundary between oil and underlying water in a reservoir.

oil window   the maturity range in which oil is generated from oil-prone organic matter ($\sim$0.6–1.4% vitrinite reflectance), i.e. within the catagenesis zone ($\sim$0.5–2.0% vitrinite reflectance) (e.g. see Figure 14.3).

oleanane index   the ratio of oleanane (a C_{30} triterpane marker of angiosperms) to 17α-hopane (a bacterial marker). Oleanane occurs mainly in Late Cretaceous or younger rocks, but its absence cannot be used to prove age.

olefin   see Alkene.

oligotrophic   describes a lake with few nutrients and abundant dissolved oxygen that supports little primary productivity. Oligotrophic lakes are earliest in the age sequence oligotropic, mesotrophic, and eutrophic, before complete infilling of the lake basin and subaerial conditions.

OPEC   the Organization of Petroleum Exporting Countries, which includes Algeria, Indonesia, Iran, Iraq, Kuwait, Libya, Nigeria, Qatar, Saudi Arabia, United Arab Emirates, and Venezuela.

open-tubular column   a chromatographic column with small internal diameter, e.g. fused-silica tubing for capillary gas chromatography.

optical activity   the tendency of some organic compounds to rotate monochromatic plane-polarized light to the right or left. Biomarkers and other organic materials contain asymmetric carbon atoms resulting in left- or right-handed molecules (enantiomers). Enzymes in living organisms tend to produce one or the other of these optically active compounds. Solutions of these compounds rotate plane-polarized light to varying degrees. Solutions of the same compounds produced without enzymes or living organisms are not optically active because they represent a racemic (50:50) mixture of enantiomers.

organelles   various membrane-enclosed bodies specialized for carrying out certain functions in eukaryotic cells.

organic acids   organic molecules that contain at least one functional group capable of releasing a proton. Carboxyl groups (–COOH) account for the acidity of most organic acids, e.g. fatty, naphthenic, and aromatic acids. Common monocarboxylic acids include formic ($HCOOH$), acetic (CH_3COOH), and proprionic (CH_3CH_2COOH) acids (methanoic, ethanoic, and propanoic acids, respectively, in International Union of Pure and Applied Chemistry (IUPAC) nomenclature). Oxalic acid ($HOOCCOOH$) is the simplest dicarboxylic acid (ethanedioic acid in IUPAC nomenclature).

organic carbon   carbon in compounds derived from living organisms rather than inorganic sources.

organic facies   a mappable rock unit containing a distinctive assemblage of organic matter without regard to the mineralogy (Jones, 1987). Biofacies are analogous to organic facies, except that biofacies refer specifically to distinctive assemblages of recognizable organisms.

organic matter   biogenic, carbonaceous materials. Organic matter preserved in rocks includes kerogen, bitumen, oil, and gas. Different types of organic matter can have different oil-generative potential. See also Biomass.

organic yield   a crude estimate (total organic carbon (TOC) is more accurate) of the amount of organic matter in a rock based on the volume of organic material that survives demineralization of the rock with acids during preparation of kerogen.

organotroph   an organism that obtains reducing equivalents (stored electrons) from organic substrates. Also known as chemoheterotroph.

orogenic belt   extensive tracts of deformed rocks, primarily developed near continental margins by compressional forces accompanying mountain building.

orogeny   the process of mountain building; the process whereby structures within fold-belt mountainous areas formed.

ostracodes (also ostracods)   small ($\sim$0.4–30 mm) aquatic crustaceans with a calcified bivalve carapace. Range from Lower Cambrian to present.

ostracoderms   extinct jawless fish from the early Paleozoic Era.

outcrop   an exposure of bedrock at the surface.

outer continental shelf   the portion of a continental land mass that constitutes the slope down to the ocean floor. The outer continental shelves may contain much of the world's remaining undiscovered oil and gas.

overburden rock   sedimentary or other rock that compresses and consolidates the underlying rock. Overburden rock is an essential element of petroleum systems because it contributes to the thermal maturation of the underlying source rock.

overload   in chromatography, the mass of sample injected on to the column at which efficiency and resolution begin to be affected adversely if the sample size is increased further.

overmature   see Postmature.

oxic   refers to water column or sediments with molecular oxygen contents of 2–8 ml/l of interstitial water. Used in reference to a microbial habitat (see Table 1.5). Maximum oxygen saturation in seawater is in the range 6–8.5 ml/l, depending on salinity and water temperature.

oxidation   the process of (1) loss of one or more electrons by a compound, or (2) addition of oxygen to a compound, or (3) loss of hydrogen (dehydrogenation). Oxidation and reduction reactions are always coupled (redox reactions). For example, phytol (an alcohol) can be oxidized to phytanic acid, or iron can be oxidized from 2^+ (ferrous) to 3^+ (ferric).

oxidation–reduction (redox) reaction   a coupled pair of reactions, in which one compound is oxidized while the other is reduced and takes up the electrons released in the oxidation reaction. See also Redox potential.

oxidation state   the number of electrons to be added or removed from an atom in a combined state to convert it to the elemental form.

oxidative phosphorylation   the synthesis of adenosine triphosphate, involving a membrane-associated electron-transport chain.

oxygenic photosynthesis   the use of light energy to synthesize ATP and NADPH by acyclic photophosphorylation with the production of oxygen from water.

oxygen index   a Rock-Eval parameter defined as (S3/TOC) $\times$ 100 (where TOC is total organic carbon) and measured in mg carbon dioxide/g TOC (see Figure 4.5). The OI is used on van Krevelen-type plots of hydrogen index (HI) versus OI to describe organic matter type (see Figure 4.4), and there is a general relationship between OI and atomic oxygen/carbon ratio (O/C).

PAH   see Polynuclear or polycyclic aromatic hydrocarbons.

PAL   present atmospheric level of molecular oxygen (21%).

paleoecology   the study of the relationship of ancient organisms to their environment.

Paleogene Period   a subdivision of the Cenozoic Era that consists of the Paleocene, Eocene, and Oligocene epochs.

paleogeography   the geography as it existed in the geologic past.

paleolatitude   the latitude of a location at a particular time in the geologic past.

paleomagnetism   remnant magnetization of iron minerals in ancient rocks that allows reconstruction of the orientation of Earth's ancient magnetic field and the former positions of the continents relative to the magnetic poles.

paleontology   the study of all ancient forms of life, their fossils, interactions, and evolution.

palynomorph   organic-walled, acid-resistant microfossils useful in providing information on age, paleoenvironment, and thermal maturity (e.g. thermal alteration index (TAI)).

Pangea   a late Paleozoic supercontinent that included all major present-day continents, which broke up in the Mesozoic Era.

Panthalassa   a great ocean that surrounded the supercontinent Pangea before its breakup.

paraffin   see Alkane.

paraffinicity ratio (F)   *n*-heptane/methylcyclohexane, as in Figure 7.21 (Thompson, 1987).

parasequence   a conformable succession of genetically related progradational beds bounded by marine-flooding surfaces that mark an abrupt increase in water depth.

parasitic   a relationship in which an organism lives in intimate association with another, from which it obtains food and other benefits.

paraxylene   an aromatic compound used to make polyester fibers and plastic bottles.

partition coefficient   the amount of solute in the chromatographic stationary phase relative to that in the mobile phase.

passive margin   a continental margin characterized by thick, flat-lying shallow-water sediments and limited tectonic activity.

pathogen   an organism that can damage or kill a host that it infects.

pay zone   the layer of rock in which significant oil and/or gas are found.

PDB   see Peedee belemnite.

peak mature   the level of maturity associated with the maximum rate of petroleum generation from the kerogen.

peat   unconsolidated, slightly decomposed organic matter accumulated under moist conditions. Related to the early stages of coal formation.

pebble   a rock particle 2–64 mm in diameter.

Peedee belemnite (PDB)   an international primary stable isotope standard for carbon obtained from the carbonate fossil of a cephalopod, *Belemnitella americana*, in the Cretaceous Peedee Formation. Measurements are given in parts per thousand (‰ or per mil) relative to the standard using the standard delta notation. On the PDB scale, a secondary standard, NBS-22 oil, measures $\sim-29.81‰$. Accurate conversion of PDB to NBS-22 values is not straightforward, especially for gases, and requires a correction.

pelagic   describes deep ocean areas far removed from land masses. Pelagic sediments consist of sea-surface materials that settle to the deep ocean floor, including biogenic oozes and aeolian (wind-blown) clays.

pelycosaurs   early mammal-like reptiles, e.g. the sail-back animals of the Permian Period.

period   the most commonly used unit of geologic time, representing one subdivision of an era.

permeability   the capacity of a rock layer to allow water or other fluids, such as oil, to pass through it.

petrochemical   a chemical derived from petroleum, hydrocarbon liquids, or natural gas, e.g. ethylene, propylene, benzene, toluene, and xylene.

petroleum   a mixture of organic compounds composed predominantly of hydrogen and carbon and found in the gaseous, liquid, or solid state in the Earth, including hydrocarbon gases, bitumen, migrated oil, pyrobitumen, and their refined products, but not kerogen. In European usage, the term is sometimes restricted to refined products only.

petroleum province   a geographic area where petroleum occurs. Also referred to as a petroleum basin or basin. For example, the Western Canada Basin and the Zagros Thrust Belt are well-known petroleum provinces.

petroleum system   the essential elements and processes and all genetically related petroleum that occurs in shows, seeps, and accumulations and originated from one pod of active source rock. Also called a hydrocarbon system.

petroleum system name   a name that includes the pod of active source rock, the reservoir rock containing the largest volume of corresponding petroleum, and the level of certainty in the petroleum system, e.g. the Mandal-Ekofisk(!) petroleum system.

petroleum system processes   trap formation and generation–migration– accumulation. Combined with the essential elements, the processes control the distribution of petroleum in the lithosphere.

pH   a measure of acidity; the tendency of an environment to supply protons (hydrogen ions) to a base or to take up protons from an acid (negative logarithm of the hydrogen ion concentration). A pH of 7.0 under standard conditions is neutral, while lower and higher values are acidic and basic, respectively. In nature, most environments fall in the pH range 4–9.

Phanerozoic Eon   the eon of geologic time when Earth became populated by abundant and diverse life. The Phanerozoic Eon is divided into the Paleozoic, Mesozoic, and Cenozoic eras.

phenanthrene   an aromatic hydrocarbon consisting of three fused aromatic rings; an isomer of anthracene.

phenol   hydroxybenzenes; oxygen-containing aromatic compounds that are acidic due to the effect of the aromatic ring on the hydroxyl group.

phosphate rock   a rock that contains abundant apatite or other phosphatic minerals. Most phosphate minerals are mined from phosphorites.

phospholipid   lipids that contain a substituted phosphate group and two fatty acid chains on a glycerol backbone.

phosphorite   a sedimentary rock composed mainly of microcrystalline carbonate fluorapatite.

photic zone   the uppermost layer of water or soil that receives enough sunlight to permit photosynthesis, e.g. for water, typically <200 m depending on water turbidity.

photoautotroph   an organism that uses light as the source of energy and carbon dioxide as the source for cellular carbon.

photoheterotroph   an organism that uses light as the source of energy and organic compounds as the source for cellular carbon.

photosynthesis   a process in plants that uses light energy captured by chlorophyll to synthesize carbohydrates from carbon dioxide and water. Photosynthesis requires chlorophyll and other light-trapping pigments, such as carotenoids and phycobilins. Photosynthetic eukaryotes include higher green plants, multicellular green, brown, and red algae, and unicellular organisms, such as dinoflagellates and diatoms. Photosynthetic prokaryotes include cyanobacteria, green bacteria, and purple bacteria.

phototroph   an organism that uses light as the energy source to drive the electron flow from the electron donors, such as water, hydrogen and sulfide. Chlorophyll and other light-trapping pigments, such as carotenoids and phycobilins, are used by phototrophs in photosynthesis. Phototrophs include green cells of higher plants, cyanobacteria, photosynthetic bacteria, and non-sulfur purple bacteria.

phthalate   an ester or salt of phthalic acid common in plasticizers.

phycobilins   water-soluble compounds that occur in cyanobacteria and function as the light-harvesting pigments for photosystem II.

phylogeny   classification of species into higher taxa and evolutionary trees based on genetic relationships.

phytane (Ph)   a branched acyclic (no rings) isoprenoid hydrocarbon containing 20 carbon atoms (see Figure 2.15); a prominent peak eluting immediately after the C_{18} n-alkane in petroleum on most gas chromatographic columns (e.g. see Figure 8.9).

phytoclast   an identifiable particle or maceral in kerogen, e.g. phytoclasts of vitrinite are used for measurement of vitrinite reflectance.

phytol   a branched acyclic (no rings) isoprenoid alcohol containing 20 carbon atoms (e.g. see Figure 2.26).

phytoplankton   unicellular photosynthetic planktonic plants, such as algae, diatoms, and dinoflagellates that inhabit the photic zone of water bodies.

PI   see Production index.

placoderms   extinct jawed fish from the Paleozoic Era.

plasmids   circular DNA molecules that are separate from chromosomal DNA and are exchanged between bacteria during conjugation. They may provide a selective advantage to bacteria, e.g. resistance to antiobiotics. Plasmids are common in bacteria, but they also occur in eukaryotes.

plastid   specialized cell organelle that contains pigments.

plates   segments of the Earth's lithosphere that ride as distinct units over the asthenosphere.

platform   part of a craton that is covered by layered sedimentary rocks and characterized by relatively stable tectonic conditions; or an offshore facility where development wells are drilled.

plankton   small, free-floating aquatic organisms.

plateau   an elevated area with little internal relief.

plate tectonics   a theory that explains the tectonic behavior of the crust of the Earth by means of moving but rigid crustal plates that form by volcanic activity at oceanic ridges and are destroyed along ocean trenches.

playa   a desert lake bed that is dry for most of the year.

PMP   see Porphyrin maturity parameter.

PNA   see Polynuclear or polycyclic aromatic hydrocarbons.

pod of active source rock   a contiguous volume of organic-rich rock that generates and expels petroleum at the critical moment and accounts for genetically related petroleum shows, seeps, and accumulations in a petroleum system. A pod of mature source rock may be active, inactive, or spent.

polar compound   an organic compound with distinct regions of partial positive and negative charge. Polar compounds include alcohols, such as sterols, and aromatics, such as monoaromatic steroids. Because of their polarity, these compounds are more soluble in polar solvents, including water, compared with non-polar compounds of similar molecular weight.

polymer   a macromolecule composed of a large number of repeating subunits (monomers). Biopolymers include proteins (composed of amino acids) and polysaccharides (composed of sugars). Rubber is a polyunsaturated polymer composed of isoprene $[(C_5H_8)_n]$ subunits. Kerogen is not a polymer.

polymerase chain reaction   a method used to amplify DNA *in vitro* that involves the use of DNA polymerase to copy target genetic sequences.

polynuclear or polycyclic aromatic hydrocarbons (PAHs or sometimes PNAs)   organic compounds containing more than two aromatic rings.

polysaccharide   long chains of monosaccharides (simple sugars) that are linked by glycosidic bonds.

polysiloxane   see Siloxane.

polyterpanes   a class of saturated biomarkers constructed of more than eight isoprene subunits ($\sim C_{40}+$).

pool   see Reservoir.

population   a group of individuals of the same species that occupy an area at the same time.

porosity   the volume percentage of a rock that consists of open or pore space.

porphyrin maturity parameter (PMP)   a biomarker maturity parameter for petroleum based on generation. $PMP = C_{28}$ etio/$(C_{28}$ etio $+ C_{32}$ DPEP).

porphyrins   complex biomarkers characterized by a tetrapyrrole ring, usually containing vanadium or nickel (e.g. see Figure 3.24), which originate from various sources including chlorophyll and heme.

postmature   a high level of maturity at which no further oil generation can occur, i.e. vitrinite reflectance $>1.3\%$.

potential source rock   a gas- or oil-prone, organic-rich rock that has not yet generated petroleum. A potential source rock becomes an effective source rock when it generates microbial gas at low temperatures or when it reaches the level thermal maturity necessary to generate petroleum.

pour point   the temperature below which a crude oil does not flow freely as a liquid. High API gravity oils can show low pour points due to abundant wax content.

ppb   parts per billion.

ppm   parts per million – the scale for measuring many components in oils, gases, and petrochemicals.

Precambrian Era   refers to all of geologic time before the Paleozoic Era.

precipitate   to drop out of a saturated solution.

precision   the degree of agreement of repeated measurements of a quantity, which may or may not be accurate; i.e. a measurement can be precise but inaccurate.

precolumn   a small chromatographic column placed between the pump and the injector to remove particulate matter that may be in the mobile phase and chemically adsorb substances that might interfere with the separation.

prenols   alcohols having the general formula $H\text{-}[CH_2C(CH_3)\text{=}CHCH_2]_nOH$, where the carbon skeleton is composed of one or more isoprene units.

preparative chromatography   use of liquid chromatography to isolate sufficient amounts of material for other purposes.

preservation time   the time after generation–migration–accumulation of petroleum, including time during which petroleum might be exposed to secondary processes, such as remigration and inspissation, biodegradation, and water washing.

primary migration   see Expulsion and Migration.

primary producer   an organism that adds biomass to the ecosystem by synthesizing organic molecules from carbon dioxide and simple inorganic nutrients.

pristane (Pr)   a branched acyclic (no rings) isoprenoid hydrocarbon containing 19 carbon atoms (see Figure 2.15); a prominent peak eluting immediately after the C_{17} *n*-alkane in petroleum on most gas chromatography columns (e.g. see Figure 8.9).

production index (PI)   a Rock-Eval parameter useful for describing thermal maturity of source rocks or indicating contamination. PI = S1/(S1 + S2) (Peters, 1986).

production platform   a platform from which development wells are drilled and that carries all of the processing plants and other equipment needed to maintain production of a field.

production string   the pipes in a production well through which oil or gas flows from the reservoir to the wellhead.

production well   a well used to remove oil or gas from a reservoir.

progradation   movement of the shoreline into a sedimentary basin when clastic input exceeds the accommodation space, as might occur due to reduced basinal subsidence or increased erosion and sediment supply.

prokaryotes   a phylogenetic domain that consists of organisms (bacteria and cyanobacteria) that lack membrane-bound nuclei and organelles and that maintain their genome dispersed throughout the cytoplasm. Prokarya is one of the three domains of life.

proteins   large molecules in all living cells that are composed of chains of amino acids.

protista   an old taxonomic term that refers to algae, fungi, and protozoa (eukaryotic protists), and the prokaryotes.

proton   a subatomic particle in the nucleus of all atoms with an electric charge of +1 and a mass similar to that of a neutron.

protoplasm   all cellular contents, cytoplasmic membrane, cytoplasm, and nucleus. Usually considered the living portion of the cell, thus excluding those layers peripheral to the cell membrane.

protozoan (*pl.* protozoa)   unicellular eukaryotic microorganisms.

pseudohomologous series   compounds showing similar structures but differing in the length of a branched alkyl group. For example, the tricyclic terpanes (cheilanthanes) represent a pseudohomologous series.

psychrophile   an organism able to grow at low temperatures. Optimal growth occurs below 15°C.

pterosaur   a flying reptile of the Jurassic and Cretaceous periods.

pure culture   a population composed of a single strain of organism.

pycnocline   a layer of water in which the density increase with depth is greater than that of the underlying or overlying water. For example, a pycnocline or halocline in an estuary might separate shallow, fresh water from deeper, saline water with higher density.

pyrobitumen   thermally altered, solidified bitumen that is insoluble in common organic solvents and has an atomic hydrogen/carbon ratio (H/C) of 0.5 or less.

pyrogram   a graph showing detector response to products generated by pyrolysis (e.g. see Figure 4.5). For the Rock-Eval pyroanalyzer, a pyrogram has S1, S2, S3, T_{max}, the programmed temperature trace, and other information (Peters, 1986).

pyrolysis   the breakdown of organic matter during heating in the absence of oxygen; as in Rock-Eval pyrolysis (for source-rock evaluation) or hydrous pyrolysis (for simulating oil generation in source rocks). Rock-Eval pyrolysis employs programmed-temperature pyrolysis because the temperature is programmed to increase at a selected rate during analysis. Hydrous pyrolysis typically employs a constant temperature for each experiment.

quadrupole rods   an electrical (versus magnetic) mass analyzer in a mass spectrometer (e.g. see Figure 8.17). Using a combination of direct current (DC) and radio frequency (RF) fields on the quadrupole rods serves as a mass filter for ions. Only ions of a given m/z reach the detector for a given magnitude of the DC and RF fields. All other ions collide with the rods before reaching the detector.

quasi-dysaerobic   refers to metabolism under suboxic conditions with limited molecular oxygen (see Table 1.5).

R or S configuration   the stereochemical arrangement of atomic groups around a asymmetric carbon atom can be assigned by orienting the molecule according to specific conventions and finding whether a circle passes through the remaining groups surrounding the center clockwise (R, rectus) or counterclockwise (S, sinister).

racemic mixture   a 50:50 or racemic mixture or racemate of enantiomers in solution results in no optical activity. Racemic mixtures of enantiomers are typical of the non-biologic (abiotic) synthesis of molecules containing asymmetric centers. Many biologically formed compounds are optically active because they consist of only one enantiomer.

radioactive   the spontaneous emission of a particle from the atomic nucleus, thus transforming the atom from one element to another.

radiolaria   protozoa that secrete a skeleton of opaline silica.

radiometric dating   techniques used to determine the age of geologic materials utilizing the known rates of decay of radioactive isotopes.

rearranged hopane   see Diahopane and Neohopane.

rearranged sterane   see Diasterane.

recharge   the replacement of groundwater by infiltrating rain or stream water or the replenishment of water in a lake or ocean.

reconstructed ion chromatogram (RIC)   the magnitude of the total ion current during a gas chromatography/mass spectrometry (GCMS) analysis plotted versus scan or retention time. RIC and gas chromatographic traces of petroleum samples are nearly identical. Also known as total ion chromatogram (TIC).

recoverable reserves   the amount of oil and/or gas in a reservoir that can be removed using currently available techniques.

recovery   the amount of solute (sample) that elutes from a chromatographic column relative to that injected.

red beds   red-colored, usually clastic sedimentary deposits.

redox potential (oxidoreduction potential)   a measure of the ability of an environment to supply electrons to an oxidizing agent or to take up electrons from a reducing agent. In redox reactions, there is transfer of electrons from an electron donor (the reducing agent or reductant) to an electron acceptor (the oxidizing agent or oxidant). Expressed as the tendency of a reducing agent to lose electrons relative to the standard reduction potential. Standard reduction potential is the electromotive force in volts given by a half-cell in which the reductant and oxidant are present at 1.0-M concentration, 25°C, and pH 7.0, in equilibrium with an electrode, which can reversibly accept electrons from the reductant species. The standard of reference is the reduction potential of the reaction: $H_2 = 2H^+ + 2e^-$, which is set at 0.0 V under conditions in which the pressure of H_2 gas is 1.0 atmosphere, $[H^+]$ is 1.0 M, pH is 7.0, and temperature is 25°C. Systems having a more negative standard reduction potential than the H_2–$2H^+$ couple have a greater tendency to lose electrons than hydrogen, and vice versa. Most redox potentials of seawater, for example, lie in the range between +0.3 V for aerated water to −0.6 V for oxygen-depleted (anoxic) bottom water.

reduction   the process of (1) acceptance of one or more electrons by a compound, or (2) removal of oxygen from a compound, or (3) addition of hydrogen (hydrogenation). Reduction and oxidation always occur as coupled reactions (redox reactions). For example, alkenes can be reduced to alkanes by hydrogenation or iron can be reduced from 3^+ (ferric) to 2^+ (ferrous) state by gaining electrons.

reef   a wave-resistant organic structure built by calcareous organisms that stands in relief above the surrounding seafloor.

refinery   a facility used to separate various components from crude oil and convert them into fuel products or feedstock for other processes.

reflectance   see Vitrinite reflectance.

refractory polycyclic aromatic hydrocarbon index (PAH-RI)   ratio of $C_{26}(20R) + C_{27}(20S)$ triaromatic steroids (co-eluting compounds that are a major peak on m/z 231 traces) to methylchrysene (Hostettler *et al.*, 1999).

regression   withdrawal of the sea from land areas that shifts the boundary between marine and non-marine deposition (or between deposition and erosion) toward the center of a marine basin. Regressions can be caused by tectonic uplift of the land, eustatic lowering of sea level, and progradation of sediments into the basin.

relief   the maximum regional difference in elevation.

replication   conversion of one double-stranded DNA molecule into two identical double-stranded DNA molecules.

reservoir   a porous, permeable sedimentary rock formation that contains oil and/or natural gas enclosed or surrounded by layers of less permeable rock. An oil pool consists of a reservoir or group of reservoirs. However, the term is misleading because petroleum exists not in pools but in pores between rock grains.

reservoir characterization   integrating and interpreting geological, geophysical, petrophysical, fluid, and performance data to describe a reservoir.

reservoir rock   any porous and permeable rock that contains petroleum, such as porous sandstone, vuggy carbonate, and fractured shale.

resin   a solid or semi-solid mixture of complex organic substances with significant nitrogen, sulfur, and oxygen. Resins can be divided into five classes based on chemical composition (Anderson and Muntean, 2000), but classes I and II are the most common. Class I resins originate from gymnosperms (i.e. conifers), especially the family Araucariacea, while class II resins originate mainly from angiosperms, especially tropical hardwoods of the family *Dipterocarpaceae*.

resinite   macerals of the liptinite group derived from plant resins, e.g. amber.

resolution   the efficiency of a column to separate adjacent chromatographic peaks. Baseline separation indicates that two adjacent peaks are resolved (separated) completely and that the valley between the peaks reaches background (low) levels. In mass spectrometry, resolution refers to the ratio of mass to the difference between two adjacent masses ($M/\Delta M$) that a mass spectrometer can just separate completely. Low and high resolutions are $\sim$1000 and >2000 (no units).

respiration   oxidative catabolic reactions in living organisms that yield energy (as ATP), where either organic or inorganic substrates are primary electron donors and oxygen or other compounds are the ultimate electron acceptors.

retention time   the time required for an injected compound to pass through a chromatographic column. An unknown compound can be identified provisionally if it has the same retention time as a standard compound when injected into the same chromatographic column under the same conditions.

retinoids   oxygenated derivatives of 3,7-dimethyl-1-(2,6, 6-trimethylcyclohex-1-enyl)nona-1,3,5,7-tetraene.

retrograde condensate   a liquid petroleum formed by condensation due to reduced pressure and temperature.

reverse fault   a fault in which the hanging wall appears to have moved upward relative to the footwall. Common in compressional regimes.

reverse-phase chromatography   the most common high-performance liquid chromatography (HPLC) mode, which uses hydrophobic packing, such as octadecyl- or octylsilane phases bonded to silica, or neutral polymeric beads. The mobile phase is usually water and a water-miscible organic solvent, such as methanol or acetonitrile.

rhizoid   a root-like structure that anchors an organism to a substrate.

ribonucleic acid (RNA)   a polymer of nucleotides involved in protein synthesis.

ribonucleic acid (RNA)   a biochemical consisting of long usually single-strand chains of alternating phosphate and ribose units with the nitrogen bases adenine, guanine, cytosine, and uracil bound to the ribose. Found in all organisms, RNA is used in protein synthesis, transmission of genetic material, and regulation of cellular processes.

ribosomes   small organelles consisting of rRNA and enzymes where proteins are manufactured from amino acids in the cytoplasm of living organisms.

ribosomal RNA (rRNA)   a type of ribonucleic acid (RNA) found in the ribosome that participates in protein synthesis. 16S rRNA is a large polynucleotide ($\sim$1500 bases) that is part of the ribosome of prokaryotes. Comparisons of 16S rRNA base sequences are used to show evolutionary relationships among organisms. The eukaryotic counterpart is 18s rRNA.

RIC   see Reconstructed ion chromatogram.

rift   a narrow depression in a rock caused by cracking or splitting.

rift valley   a valley formed by faulting, usually involving a central fault block that moves downward in relation to adjacent blocks.

R_m   see Vitrinite reflectance.

RNA   see Ribonucleic acid.

R_o   see Vitrinite reflectance.

Rock-Eval   a commercially available pyrolysis instrument used as a rapid screening tool in evaluating the quantity, quality, and thermal maturity of rock samples.

root nodule   a specialized structure in plant roots, such as legumes, where bacteria fix diatomic nitrogen, making it available for the plant.

rRNA   see Ribosomal RNA.

rudists   specialized Mesozoic bivalves, commonly having one valve in the shape of a horn coral and covered by the other valve in the form of a lid.

ruminant   a herbivorous ungulate, e.g. cow.

ruthenium tetroxide degradation   oxidation of organic matter by RuO_4 disrupts aromatic systems by producing a carboxyl group at the point of attachment of the side chains or groups. Gas chromatography/mass spectrometry (GCMS) or other analysis of the products of RuO_4 oxidation provides evidence on principal structural components in these complex materials (e.g. Blokker *et al.*, 2001).

S configuration   see R or S configuration.

S1, S2, S3   Rock-Eval pyrolysis parameters, where S1 is volatile organic compounds (mg HC/g TOC), S2 is organic compounds generated by cracking of the kerogen (mg HC/g TOC), and S3 is organic carbon dioxide generated from the kerogen up to 390°C (mg CO_2/g TOC) (TOC is total organic carbon) (see Figure 4.5).

sabkha   an evaporitic environment of sedimentation formed under arid to semi-arid conditions, usually on restricted coastal plains. Characterized by evaporite-salt, tidal-flood, and aeolian deposits along many modern coastlines, e.g. the Persian Gulf and the Gulf of California.

salt dome   a structural dome in sedimentary strata resulting from the upward flow of a large body of salt.

sand   particles measuring 0.05–2.0 mm in diameter.

sandstone   sedimentary rock composed of sand-sized particles, typically quartz.

sapropel   putrefying organic matter, mostly algae and microbes, deposited in an anoxic setting. The word "sapropel" originates from the Greek *sapros*, meaning rotten.

saprophyte   an organism that feeds on dead organic material.

Sarcopterygii   lobe-finned bony fish, including air-breathing crossopterygian fish.

saturates (saturate fraction)   non-aromatic organic compounds in petroleum. Includes normal and branched alkanes (paraffins) and cycloalkanes (naphthenes).

Saurischia   an order of dinosaurs with triradiate pelvic structures, including both the gigantic herbivorous sauropods and the carnivorous theropods.

scan mode   a gas chromatography/mass spectrometry (GCMS) operating mode in which the detector records the entire mass range per unit of time (e.g. 50–600 amu/3 s), resulting in a spectrum of masses for each analyzed peak.

schist   a foliated metamorphic rock rich in mica.

Schulz–Flory distribution   a mixture of hydrocarbons dominated by lighter components, which shows a linear decrease on a semi-log plot of hydrocarbon concentration (e.g. mole%) versus carbon number.

screening   rejection of inappropriate samples using rapid, inexpensive analyses to allow upgrading of other samples for more detailed analysis. Large numbers of potential source rocks can be screened using Rock-Eval pyrolysis and total organic carbon (TOC) analysis before further study. Similarly, benchtop gas chromatography/mass spectrometry (GCMS) in selected ion monitoring (SIM) mode can be used to screen petroleum samples for their general biomarker composition before more detailed analysis, such as gas chromatography/mass spectrometry/mass spectrometry (GCMS/MS).

seafloor spreading   the process in which oceanic crust rises from deep in the Earth by convective upwelling of magma along mid-oceanic ridges or rift systems and spreads outward at ∼1–10 cm/year, carrying tectonic plates.

seal rock (cap rock)   an impervious layer of rock that overlies a reservoir rock, thus preventing leakage of petroleum to the surface.

secondary migration   migration of petroleum through permeable rocks (carrier beds) after expulsion from the source rock. Unlike primary migration (expulsion), secondary migration generally involves long distances from tens of meters to hundreds of kilometers.

secondary recovery   enhanced recovery of oil or gas from a reservoir beyond the oil or gas that could be recovered by normal pumping operations. Secondary recovery techniques involve maintaining or enhancing reservoir pressure by injecting water, gas, or other substances into the formation.

sediment   various materials deposited by water, wind, or glacial ice, or by precipitation from water by chemical or biological action, e.g. clay, sand, carbonate.

sedimentary rock   rock formed by lithification of sediment transported or precipitated at the Earth's surface and accumulated in layers. These rocks can contain fragments of older rock transported and deposited by water, air, or ice, chemical rocks formed by precipitation from solution, and remains of plants and animals.

sedimentation   the process of deposition and accumulation of sedimentary layers.

seismic exploration   an exploration technique that provides information on subsurface structures by monitoring seismic waves from a surface energy source (e.g. dynamite or vibrating heavy equipment), which reflect from the subsurface structures. Used to determine the best places to drill for hydrocarbons.

seismic reflection profiling   three-dimensional analysis of structures beneath the Earth's surface using seismic waves. See also Seismic exploration.

selected ion monitoring (SIM)   mass spectrometric monitoring of a specific mass/charge (m/z) ratio or limited number of ratios. For example, SIM for m/z 217 results in a mass chromatogram dominated by steranes. Using the SIM method of monitoring one or a few masses results in better sensitivity than can be obtained using the full-scan mode. Also known as multiple ion detection (MID).

series   the time-rock term representing rocks deposited or emplaced during a geologic epoch. A series is a subdivision of a system.

serpentinization   low-temperature hydration and alteration of mafic minerals, such as olivine, to form serpentine-group minerals.

sesquiterpanes   a class of saturated biomarkers constructed from three isoprene subunits ($\sim$C$_{15}$).

sesterterpanes   a class of saturated biomarkers constructed from five isoprene subunits ($\sim$C$_{25}$).

shale   a fine-grained sedimentary rock formed by lithification of mud that is fissile or fractures easily along bedding planes and is dominated by clay-sized particles. In this text, shales, calcareous shales, and limestones contain <25%, 25–50%, and >50 wt.% carbonate, respectively.

shear   a frictional force that tends to cause contiguous parts of a body to slide relative to each other in a direction parallel to their plane of contact.

sheath   tubular structure formed around a chain of cells or around a bundle of filaments.

shelf   the physiographic area between the shoreline and the slope.

silica gel   an amorphous, porous packing material common in liquid chromatography composed of siloxane and silanol groups.

silicalite   a term that has been used incorrectly to designate high Si/Al ZSM-5 zeolite (Flanigen *et al.*, 1978). ZSM-5 and so-called silicalite are equivalent (Fyfe *et al.*, 1982). Budiansky (1982) describes the patent dispute that arose due to the use of this term. It is now generally recognized that the Si/Al ratio for synthetic ZSM-5 is limited only by Al impurities in the starting materials.

silicate   a compound whose crystal structure contains SiO$_4$ tetrahedra, either isolated or joined through one or more of the oxygen atoms to form groups, chains, sheets, or three-dimensional structures with metallic elements.

silicic   rich in silica.

siliciclastic   clastic, non-carbonate rocks dominated by quartz or silicate minerals.

silicon tetrahedron   an atomic structure in silicate minerals that consists of a central silicon atom linked to four oxygen atoms placed symmetrically around the silicon at the corners of a tetrahedron (SiO$_4$).

sill   (1) a small, concordant (injected between layers of sedimentary rock) body of intrusive igneous rock; (2) a subaqueous elevated area that partially isolates the water in two basins.

siloxane   the Si–O–Si bond as in polysiloxane; a principal bond in silica gel or for attachment of a silylated compound or bonded phase.

silt   a rock fragment or particle 0.004–0.063 mm in diameter.

SIM   see Selected ion monitoring.

slime mold   non-phototrophic eukaryotic microbes that lack cell walls.

soil   weathered, unconsolidated minerals and organic matter on the surface of the Earth that serves as a growth medium for organisms.

solid bitumen   includes pyrobitumen and bitumen formed by non-thermal processes, such as deasphalting and biodegradation (Curiale, 1986).

solute   the dissolved component of a mixture that is to be separated in a chromatographic column.

sorting   a measure of the size uniformity of particles in a sediment or sedimentary rock.

source rock   an essential part of the petroleum system that consists of fine-grained, organic-rich rock that could generate (potential source rock) or has already generated (effective source rock) significant amounts of thermogenic or microbial petroleum. An effective source rock must satisfy requirements as to quantity, quality, and thermal maturity of the organic matter.

sour crude   crude oil with high sulfur content (>0.5 wt.%).

sour gas   natural or associated gas with abundant hydrogen sulfide.

Soxhlet extraction   a common method for removing soluble organic compounds from crushed rock using a hot organic solvent (e.g. dichloromethane) or solvent mixture that is refluxed through the sample.

species   a population of individuals sufficiently different from others to be recognized as a distinct group and that in nature breed only with one another.

specific gravity   a measure of the density of a material compared with that of water (g/cm^3).

spent source rock   source rock that generated and expelled petroleum but that is now postmature, with no further generative potential.

sphenopsids (also called sphenophytes)   sponge-bearing plants that were common during the late Paleozoic Era and were characterized by articulated stems with leaves borne in whorls at the nodes.

sphingolipids   membrane lipids that function in cell-to-cell communication, signal transduction, immunorecognition, Ca^{2+} mediation, and the physical state of membranes and lipoproteins. Sphingolipids are derivatives of sphingosine, a long-chain unsaturated amino alcohol in nervous tissue and cell membranes. It has a similar structure to a glycerol-based phospholipid, having a polar head group and two hydrophobic hydrocarbon chains (one is the sphingosine and the other is a fatty acid chain).

spiking see Co-injection.

spore coloration index see Thermal alteration index.

spores specialized asexual reproductive cells that germinate without uniting with other cells, as occurs in bacteria, ferns, and mosses.

stable carbon isotope ratio relative amount of ^{13}C versus ^{12}C (non-radioactive isotopes) in organic matter. Generally used to show relationships between oils or between oils and source rocks.

stage the time-rock unit equivalent to an age. A subdivision of a series.

standard temperature and pressure (STP) $0°$C and one atmosphere pressure.

stationary phase see Gas chromatography.

sterane isomerization stereochemical conversions between the biological and geological configuration at several asymmetric centers, which are used as indicators of thermal maturity. Includes $20S/(20S + 20R)$ and $\beta\beta/(\beta\beta + \alpha\alpha)$ parameters.

steranes a class of tetracyclic, saturated biomarkers constructed from six isoprene subunits ($\sim$C$_{30}$) (see Figure 2.15). Steranes originate from sterols, which are important membrane and hormone components in eukaryotic organisms. Most commonly used steranes are in the range C$_{26}$–C$_{30}$ and are detected using m/z 217 mass chromatograms.

sterane triangle plot of %C$_{27}$, C$_{28}$, and C$_{29}$ steranes on a ternary diagram used for correlation of oils and bitumens (e.g. see Figure 13.38). Relative location on plot can be used to infer organic matter input. For example, abundant %C$_{29}$ usually, but not always, indicates major input from higher-plant sterols.

stereochemistry the three-dimensional relationship between atoms in molecules.

stereoisomers compounds that have the same molecular formula and the same linkage between atoms but different spatial arrangements of the atoms, typically around an asymmetric carbon atom. Stereoisomers include enantiomers (mirror-image structures) and diastereomers (epimers), which differ at certain asymmetric centers but are identical at others.

steroids similar to steranes, except that other elements in addition to carbon and hydrogen occur, especially oxygen in alcohol groups. Certain structural elements may be rearranged or missing compared with sterols. For example, unlike sterols (see Figure 2.31) triaromatic steroid hydrocarbons lack the α-methyl group at C-10. Further, the methyl group at C-13 in sterols is rearranged to C-17 in triaromatic steroids (see Figure 13.109).

stigmastane the C$_{29}$ ethylcholestane.

stoichiometry a branch of science that deals with the laws of definite proportions and conservation of matter during chemical reactions, e.g. the predictable quantities of substances that enter into and are produced by chemical reactions. For example, carbon in saturated organic compounds is always bound to four substituents.

STP see Standard temperature and pressure.

straight-run refinery products distillation cuts from crude-oil feedstock that include gasoline (C$_5$–C$_{10}$), kerosene (C$_{11}$–C$_{13}$), diesel (C$_{14}$–C$_{18}$), heavy gas oil (C$_{19}$–C$_{25}$), lubricating oil (C$_{26}$–C$_{40}$), and residuum (>C$_{40}$) (Hunt, 1996). The carbon number ranges are approximate and differ depending on specific distillation conditions (e.g. see Table 5.1).

strain (1) change in the shape of a body as a result of stress; a change in relative configuration of the particles of a substance. (2) A population of cells that descended from a single pure isolate.

stratification layering in sedimentary rocks that results from changes in texture, color, or rock type.

stratigraphy the study of sedimentary rock strata, including their age, form, distribution, lithology, fossil content, and other characteristics useful in interpreting their environment of origin and geologic history.

stratum a tabular or sheet-like bed of sedimentary rock that is distinct from other layers above and below.

stromatolites Precambrian to Recent laminated accumulations of calcium carbonate with rounded, branching, or frondose shape that form as a result of the metabolic activity of marine cyanobacterial mats, commonly in the high intertidal to low supratidal zones.

stromatoporids extinct reef-building organisms believed to have affinities with the Porifera and noted for the large, often laminated masses constructed by the colonies.

structural isomers molecules that have the same molecular formula but different linkages between atoms, e.g. n-butane and isobutane (see Figure 2.7), and

2-methyl,3-ethyl-heptane and 2,6-dimethyloctane (see Figure 2.14). Stereoisomers are a special form of structural isomer.

subaerial   exposed near to or at a sediment surface above sea level.

subduction zone   an inclined planar zone defined by a high frequency of earthquakes that is thought to locate the descending leading edge of a moving oceanic plate.

sublimation   passing from a gaseous to a solid state without going through the liquid state.

suboxic   refers to a water column or sediment with molecular oxygen contents of 0–0.2 ml/l of interstitial water (see Table 1.5).

subsalt   refers to rock formations lying beneath long, horizontal layers of salt. These rock formations may contain hydrocarbons.

subsidence   sinking or gradual downward settling of the Earth's surface with little or no horizontal motion.

substrate   (1) a nutrient or medium required by an organism to live and grow. (2) The surface or underlying rocks to which organisms are attached.

supercritical fluid chromatography   a method that uses a supercritical fluid as the mobile phase. This allows separation of compounds that cannot be handled by liquid chromatography because of detection problems or by gas chromatography because of the lack of volatility.

sweet crude   crude oil with low sulfur content (<0.5 wt.%).

sweet gas   natural gas that contains hydrocarbons with little or no hydrogen sulfide.

sweet oil   a low-sulfur (<0.5 wt.%) crude oil.

symbiosis   a permanent, mutually beneficial association of different organisms. Symbiosis differs from parasitism, in which one organism benefits to the detriment of its host. See also Endosymbiosis.

syncline   a concave-upward fold in rock that contains stratigraphically younger strata toward the center.

synthetic fuels   combustible fluids made from coal or other hydrocarbon-containing substances.

synthetic natural gas   gases made from coal or other hydrocarbon-containing substances.

synthetic oil   liquid fuel made from coal and other hydrocarbon-containing substances.

TAI   see Thermal alteration index.

tailing   the chromatographic phenomenon where, unlike a normal Gaussian peak, the peak has a trailing edge.

tar   a black liquid or solid residue produced by petroleum refining. The terms "tar mat" and "tar sand" refer to natural bitumens.

taxon (*pl.* taxa)   a unit in the taxonomic classification, e.g. phylum, class, order, family.

taxonomy   the science of classifying and naming organisms.

TBR   see Trimethylnaphthalene ratio.

tectonics   the structural behavior of the Earth's crust.

TCI   see Transmittance color index.

teleosts   the most advanced of the bony fish, characterized by thin, rounded scales, completely bony internal skeleton, and symmetric tail. Teleosts range from Cretaceous to Recent.

telinite   a maceral of the vitrinite group that has a distinct cellular structure of woody tissue.

temperature programming   changing the column temperature with time during the gas chromatographic separation.

terminal electron-acceptor   an external oxidant, usually oxygen, that accepts electrons as they exit from the electron-transport chain.

terpanes   a broad class of complex branched, cyclic alkane biomarkers, including hopanes and tricyclic compounds, commonly monitored using m/z 191 mass chromatograms.

terpenoids (isopentenoids)   a broad class of complex branched, cyclic biomarkers composed of isoprene (C_5) subunits, including mono- (two isoprene units), sesqui- (three), di- (four), sester- (five), tri- (six), tetra- (eight), and higher terpenoids. Terpenoids include the saturated terpanes (hydrocarbons) and compounds that may contain double bonds or other elements (in addition to carbon and hydrogen), such as oxygen. Certain structural units may be rearranged or missing compared with terpenes and terpanes.

terrane   an area of crust with a distinct assemblage of rocks (as opposed to terrain, which implies topography, such as rolling hills or rugged mountains).

terrestrial   pertaining to the Earth. The term "terrestrial" is sometimes used to refer to dry land, rather than marine, fluvial, or lacustrine settings. However, this is confusing because the antonym of terrestrial is extraterrestrial. Organic matter that originated on land is best described using the term "terrigenous."

terrigenous   refers to land rather than marine, fluvial, or lacustrine settings. Terrigenous organic matter can be deposited in any of these settings. The term "terrestrial" is commonly used as a synonym for terrigenous. However, this is confusing because the antonym of terrestrial is extraterrestrial.

Tethys Seaway an east–west trending seaway between Laurasia and Gondwana during Paleozoic and Mesozoic time from which arose the Alpine–Himalayan mountain ranges.

tetrapyrrole pigments compounds required for photosynthesis that contain a macrocyclic nucleus composed of four linked pyrrole (nitrogen-containing) rings, i.e. chlorophylls. Porphyrins contain a tetrapyrrole nucleus and are degradation products of these pigments.

tetraterpanes a class of saturated biomarkers constructed from eight isoprene subunits ($\sim$C$_{40}$).

texture a property of rocks relating to size, size variability, rounding or angularity, and orientation of mineral grains.

thecodonts an order of primarily Triassic reptiles considered to be the ancestral archosaurians.

theoretical plate a concept that relates chromatographic separation to the theory of distillation and is a measure of column efficiency. The length of column relating to this concept is called height equivalent to a theoretical plate (HETP).

therapods the carnivorous saurischian dinosaurs.

therapsids an order of mammal-like reptiles.

thermal alteration index (TAI) various maturity scales based on changes in the color of spores and pollen from yellow to brown to black with thermal maturity. Although this TAI scale does not correspond to that of Staplin (1969), it was designed to correlate linearly with vitrinite reflectance, e.g. 0.4% and 0.7% vitrinite reflectances are 2.4 and 2.7, respectively, on the TAI scale.

thermal maturity the extent of heat-driven reactions that convert sedimentary organic matter into petroleum and finally to gas and graphite. Three levels of maturity are early or low, mid or peak, and late or high. Different geochemical scales including vitrinite reflectance, pyrolysis T$_{max}$, and biomarker maturity ratios indicate the level of thermal maturity of organic matter. Because common thermal maturity parameters are measured using irreversible reactions, uplifted source rocks may show higher maturity indices than expected based on their current burial depth.

thermocline a layer of water in a lake or ocean where the temperature decrease with depth is greater than that of

the underlying and overlying water. Thermoclines may be seasonal or permanent.

thermodynamics the physics and chemistry of reactions and chemical stability.

thermogenic gas hydrocarbon gases generated by the thermal breakdown of organic matter.

thermophile an organism whose optimum temperature for growth is between 45 and 85°C.

thin-layer chromatography (TLC) separation of organic compounds by movement of a solvent through a thin layer of stationary phase coated on a glass plate. The glass plate is positioned vertically in a tray of solvent (mobile phase) so that the solvent rises upward by capillary action. The sample is placed near the bottom edge of the plate, above the original level of solvent. Upward movement of solvent results in partitioning of components between the mobile and stationary phases. Because it is less time-intensive, most laboratories use column chromatography and high-performance liquid chromatography (HPLC) rather than TLC.

thrust fault a low-angle reverse fault, with inclination of fault plane generally <45°.

TIC see Reconstructed ion chromatogram.

till a glacial deposit of unconsolidated, unsorted, unstratified angular fragments of rock.

tillite lithified rock composed mainly of poorly sorted fragments ranging in size and shape (e.g. clay, silt, sand, gravel, and boulders) that were deposited by a glacier.

TLC see Thin-layer chromatography.

Tm see Ts/Tm

T$_{max}$ a Rock-Eval pyrolysis thermal maturity parameter based on the temperature at which the maximum amount of pyrolyzate (S2) is generated from the kerogen in a rock sample. The beginning and end of the oil-generative window approximately correspond to T$_{max}$ of 435°C and 470°C, respectively (see Figure 4.5) (Peters, 1986).

TNR see Trimethylnaphthalene ratio.

TOC see Total organic carbon.

toluene methylbenzene. A key petrochemical and an organic solvent. Toluene and xylene are important components in unleaded gasoline.

topography the configuration of the land surface, including its relief and the position of natural and man-made features.

torbanite an oil-prone coal that consists mainly of algal remains, e.g. Permian torbanites in Australia and Carboniferous torbanites in the Midland Valley,

Scotland. See also Boghead coal. Although many torbanites contain recognizable remains of *Botryococcus braunii*, no biomarkers from *B. braunii* lipids have been reported, possibly due to intense microbial reworking or transformation of reactive precursors (Audino *et al.*, 2001).

total ion chromatogram   see Reconstructed ion chromatogram.

total organic carbon (TOC)   the quantity of organic carbon (excluding carbonate carbon) expressed as percentage weight of the rock. For rocks at a thermal maturity equivalent to vitrinite reflectance of 0.6% (beginning of oil window), TOC can be described as follows: poor, TOC <0.5 wt.%; fair, TOC = 0.5–1 wt.%; good, TOC = 1–2 wt.%; very good, TOC >2 wt.% (see Table 4.1). TOC decreases with maturity. TOC × 1.22 is commonly equated to total organic matter (TOM). The resulting TOM is approximate because the 1.22 factor assumes certain amounts of oxygen, nitrogen, and sulfur are present.

total organic matter (TOM)   see Total organic carbon.

trace fossils   see Ichnofossils.

transformation ratio   the difference between the original hydrocarbon potential of a sample before maturation and the measured hydrocarbon potential divided by the original hydrocarbon potential. Ranges from 0 to 1.0.

transgression   advance of the sea on to land areas that shifts the boundary between marine and non-marine deposition (or between deposition and erosion) toward the edges of a marine basin. Transgressions can be caused by tectonic subsidence of the land or a eustatic rise in sea level.

transmittance color index (TCI)   an optical maturity measurement of amorphous kerogen that complements other optical maturity parameters, such as vitrinite reflectance, thermal alteration index (TAI), and CAI. TCI covers the range of petroleum generation and preservation (Robison *et al.*, 2000). TCI is obtained by analysis of white light from a 100-W 6-V tungsten lamp attached to a photometric microscope.

transpiration   evaporation of water from plant leaves.

trap   a geometric arrangement of reservoir rock and overlying or updip seal rock that allows the accumulation of oil or gas, or both, in the subsurface. Traps may be structural (e.g. domes, anticlines), stratigraphic (pinchouts, permeability changes), or combinations of both (e.g. Levorsen, 1967).

triaromatic steroids   a class of biomarkers containing three fused aromatic rings and one five-member naphthene ring (naphthenophenanthrenes). Probably derived from monoaromatic steroids during maturation (e.g. see Figure 13.109).

tricarboxylic acid cycle (Krebs cycle)   a series of metabolic reactions in which pyruvate is oxidized to carbon dioxide, also forming NADH, which allows ATP production.

tricyclic terpanes   the most prominent tricyclic terpanes (cheilanthanes, 14-alkyl-13-methylpodocarpanes) range from C_{19} to more than C_{54} and consist of three six-member rings with an isoprenoid side chain (Peters, 2000). Abundant tricyclic terpanes commonly correlate with high paleolatitude *Tasmanites*-rich rocks, suggesting an origin from these algae, although other sources are possible. Tricyclic terpane/hopane increases with thermal maturity of petroleum.

trilobites   Paleozoic marine arthropods of the class Crustacea that show longitudinal and transverse division of the carapace into three parts, or lobes.

trimethylnaphthalene ratio   a maturity parameter defined by different authors as follows: TBR = 1,3,6-TMN/1,2,4-DMN (Fisher *et al.*, 1998); TNR-1 = 2,3,6-TMN/(1,4,6-TMN + 1,3,5-TMN) (Alexander *et al.*, 1985).

triterpanes   a class of saturated biomarkers constructed from six isoprene subunits ($\sim C_{30}$) (see Figure 2.15).

trophic level   refers to nutrients in various organisms along a food chain, ranging from the primary autotrophs to the predatory carnivorous animals.

Ts/Tm   ratio of two C_{27} hopanes: 18α-22,29,30-trisnorneohopane (Ts) (previously called trisnorhopane-II) and 17α-22,29,30-trisnorhopane (Tm). Ts/Tm and Ts/(Ts + Tm) depend on both source and maturity (Moldowan *et al.*, 1986).

tsunami   a large sea wave generated by an earthquake or underwater landslide.

tuff   volcanic ash that was consolidated into rock.

turbidites   sediment deposited from a turbidity current and characterized by graded bedding and generally poor sorting.

turbidity   scattering of light due to fine, suspended particulate matter, such as clay, in water.

turbidity current   a submarine flow of water and suspended sediment that is denser than surrounding water and thus moves downward.

two-dimensional gas chromatography   see Multidimensional chromatography.

type I kerogen   highly oil-prone organic matter showing Rock-Eval pyrolysis hydrogen indices over 600 mg hydrocarbon/g total organic carbon (TOC) when thermally immature. Contains algal and bacterial input dominated by amorphous liptinite macerals. Common in, but not restricted to, lacustrine settings.

type II kerogen   oil-prone organic matter showing Rock-Eval pyrolysis hydrogen indices in the range 400–600 mg hydrocarbon/g total organic carbon (TOC) when thermally immature. Contains algal and bacterial organic matter dominated by liptinite macerals, such as exinite and sporinite. Common in, but not restricted to, marine settings.

type IIS kerogen   composition similar to type II kerogen, but sulfur-rich. Type IIS kerogens contain unusually high organic sulfur (8–14 wt.%, atomic S/C ≥ 0.04) and appear to begin to generate oil at lower thermal exposure than typical type II kerogens with <6 wt.% sulfur (Orr, 1986).

type III kerogen   gas-prone organic matter showing Rock-Eval pyrolysis hydrogen indices in the range 50–150 mg hydrocarbon/g total organic carbon (TOC) when thermally immature. Contains higher-plant organic matter dominated by vitrinite macerals. Common in, but not restricted to, paralic marine settings.

type IV kerogen   inert organic matter showing Rock-Eval pyrolysis hydrogen indices below 50 mg hydrocarbon/g total organic carbon (TOC) in immature rocks. Contains predominantly higher-plant organic matter that was recycled or oxidized extensively during deposition.

UCM   see Unresolved complex mixture.

ultralaminae   bundles of very thin lamellae ($\sim$10–30 nm) observed under transmission electron microscopy that are widespread in so-called amorphous kerogens (Largeau *et al.*, 1990).

ultramafic   igneous rock dominated by mafic minerals, such as augite and olivine.

unassociated gas   natural gas in reservoirs that do not contain crude oil.

unconformity   a surface that separates an overlying younger rock formation from an underlying formation and represents a break in deposition (non-deposition) or erosion of formerly deposited material, or both (e.g. see Figure 9.10). Unconformities represent gaps in the geologic record.

uniformitarianism   a principle that suggests that Earth history can be interpreted using natural laws, i.e. the Earth is old and its geologic features can be explained by the action of present-day geologic processes operated over long periods of time.

unit cell   the smallest repeating unit of a crystal that has the properties and symmetry of that crystal.

unresolved complex mixture (UCM)   the hump on gas chromatograms of biodegraded (e.g. see Figure 4.22) or low-maturity crude oils (see Figure 4.25). Consists of complex organic compounds not resolved by the column (Gough and Rowland, 1990). Becomes more pronounced in biodegraded oils, probably because it contains abundant T-branched alkanes, such as 2,6,10,14-tetramethyl-7-(3-methylpentyl)-pentadecane (C_{25} highly branched isoprenoid).

unsaturated   describes compounds that contain one or more double or triple bonds, such as olefins (alkenes). Saturation of the double bond in ethylene (C_2H_4) results in ethane (C_2H_6). Most unsaturated organic compounds are unstable in natural petroleum, except for aromatics such as benzene.

upstream   describes oil and natural gas exploration and production activities.

upwelling   upward movement of shallow waters due to wind. Upwelling can carry nutrients required for algal blooms, especially on the western coast of continents (Demaison and Moore, 1980).

urea adduction   a procedure used to separate *n*-alkanes from other organic compounds in the saturate fraction of petroleum. Used to concentrate biomarkers for more reliable gas chromatography/mass spectrometry (GCMS) analysis (e.g. see Figure 13.26). The method is typically applied to samples with low biomarkers, such as extremely waxy oils and condensates.

valence   the combining capacity of an element for other elements based on the number of electrons available for chemical bonding. For example, carbon has a valence of four, indicating that it will form four covalent bonds with other elements.

van Krevelen diagram   a plot of atomic H/C versus O/C originally used to characterize the compositions of coals but now also used to describe different types of kerogen (see Type I kerogen, Type II kerogen, Type III kerogen, and Type IV kerogen) in rocks (see Figure 4.4). Van

Krevelen-type plots can be made using hydrogen index (HI) versus oxygen index (OI) from Rock-Eval pyrolysis (Peters, 1986).

varve   a lamina or pair of laminae (thin sedimentary layers) that represents the depositional record of a single year.

vascular plants   higher land plants that have a system of vessels and ducts for distributing moisture and nutrients.

virus   a submicroscopic infective agent capable of growth only within a living cell. Viruses typically consist of a protein coat surrounding a nucleic acid core.

viscosity   the resistance of a fluid to flow. Viscosity decreases with increasing temperature, as measured in centipoise at a given temperature.

vitrinite   a group of gas-prone macerals derived from land-plant tissues. Particles of vitrinite are used for vitrinite reflectance (R_o) determinations of thermal maturity.

vitrinite reflectance (R_o)   a parameter for determining maturation of organic matter in fine-grained rocks (see Figure 14.3). Average R_o is based on measurements of at least 50–100 randomly oriented vitrinite phytoclasts in a polished slide of kerogen, coal, or whole rock (see Figure 4.12). Each measurement represents the percentage of incident light (546 nm) reflected from a phytoclast under an oil-immersion microscope objective, as measured by a photometer. Some microscopes have rotating stages that allow measurements of anisotropy. Thus, R_m, R_{max}, R_{min}, and R_r indicate mean, maximun, minimum, and random vitrinite reflectance, respectively.

void volume   the total volume of mobile phase in the column, where the remainder of the column is taken up by packing material.

V/(V + Ni) porphyrin   the ratio of vanadyl to nickel porphyrins, used for correlation of oils and bitumens. The ratio is higher for oils and bitumens derived from marine source rocks deposited under anoxic compared with oxic or suboxic conditions (see Figure 4.11). Low ratios may indicate oxic to suboxic or lacustrine deposition. The ratio is sometimes expressed as Ni/(Ni + V).

waterflooding   a method of secondary recovery, whereby water is injected into an oil reservoir to force additional oil into the well bores of producing wells.

water injection   a method of enhanced recovery in which water is injected into an oil reservoir to increase pressure and maintain or improve oil production.

water table   the upper surface of the zone of water saturation in sediments. Below this surface, the pore space is filled by liquid water.

water washing   contact by formation waters in reservoirs or during migration that causes removal of light hydrocarbons, aromatics, and other soluble compounds from petroleum. Biodegradation commonly accompanies water washing of petroleum because bacteria can be introduced from the water (Palmer, 1984; 1993).

wax   the solid *n*-alkanes in petroleum. The term "waxy oil" generally indicates oil that contains abundant *n*-alkanes greater than nC_{25} (see Figure 4.21).

weathering   the breakdown of rocks and other materials at the Earth's surface caused by mechanical action and reactions with air, water, and organisms. Weathering of seep oils or improperly sealed oil samples by exposure to air results in evaporative loss of light hydrocarbons. Biodegradation and water washing commonly accompany weathering.

well   a hole drilled for the purpose of obtaining water, oil, gas, or other natural resources.

wet gas   natural gas that contains ethane, propane, and heavier hydrocarbons and <98% methane/total hydrocarbons. The wet-gas zone occurs during catagenesis below the bottom of the oil window and above the top of the gas window (1.4–2.0% vitrinite reflectance).

wildcat   an exploration well drilled in unproven territory to discover petroleum, without direct evidence of the contents of the underlying rock structure.

xanthophylls   a class of carotenoids consisting of the oxygenated carotenes.

xylene   an aromatic compound consisting of a benzene ring with methyl groups at C-1 and C-2 (orthoxylene), C-1 and C-3 (metaxylene), or C-1 and C-4 (paraxylene).

yeast   a fungus consisting of single cells that multiply by budding or fission.

zygote   the single diploid cell in eukaryotes that results from the fusion of two haploid gametes.

References

Abbott, A. (2003) Anthropologists cast doubt on human DNA evidence. *Nature*, 423, 468.

Abe, I., Rohmer, M. and Prestwich, G. D. (1993) Enzymatic cyclization of squalene and oxidosqualene to sterols and triterpenes. *Chemical Reviews*, 93, 2189–206.

Abelson, P. H. (1963) Organic geochemistry and the formation of petroleum. *6th World Petroleum Congress Proceedings*, Section 1, John Wiley & Sons, Chichester, pp. 397–407.

Abrajano, T. A., Sturchio, N. C., Kennedy, B. M., *et al.* (1990) Geochemistry of reduced gas related to serpentinization of the Zambales ophiolite, Philippines. *Applied Geochemistry*, 5, 625–30.

Adams, J. A. S. and Weaver, C. E. (1958) Thorium-to-uranium ratios as indicators of sedimentary processes: example of concept of geochemical facies. *American Association of Petroleum Geologists Bulletin*, 42, 387–430.

Aiello, A., Fattorusso, E. and Menna, M. (1999) Steroids from sponges: recent reports. *Steroids*, 64, 687–714.

Aizenshtat, Z. (1973) Perylene and its geochemical significance. *Geochimica et Cosmochimica Acta*, 37, 559–67.

Alaska Department of Environmental Conservation (1993) *The Exxon Valdez Oil Spill. Final Report, State of Alaska Response*. Alaska Department of Environmental Conservation, Juneau, Alaska.

Albaigés, J. (1980) Identification and geochemical significance of long chain acyclic isoprenoid hydrocarbons in crude oils. In: *Advances in Organic Geochemistry 1979* (A. G. Douglas and J. R. Maxwell, eds.), Pergamon, New York, pp. 19–28.

Albaigés, J., Borbon, J. and Salagre, P. (1978) Identification of a series of C_{25}–C_{40} acyclic isoprenoid hydrocarbons in crude oils. *Tetrahedron Letters*, 6, 595–8.

Albaigés, J., Borbon, J. and Walker, W., II (1985) Petroleum isoprenoid hydrocarbons derived from catagenetic degradation of archaebacterial lipids. *Organic Geochemistry*, 8, 293–7.

Alberdi, M., López, C. E. and Galarraga, F. (1996) Genetic classification of crude oil families in the Eastern Venezuela Basin. *Boletin de la Sociedad Venezolana de Geológos*, 21, 7–21.

Alexander, M. (1999) *Biodegradation and Bioremediation*, 2nd edn. Academic Press, San Diego, CA.

Alexander, R., Kagi, R. and Woodhouse, G. W. (1981) Geochemical correlation of Windalia oil and extracts of Winning Group (Cretaceous) potential source rocks, Barrow Subbasin, Western Australia. *American Association of Petroleum Geologists Bulletin*, 65, 235–50.

Alexander, R., Kagi, R. I., Roland, S. J., Sheppard, P. N. and Chirila, T. V. (1985) The effects of thermal maturity on distributions of dimethylnaphthalenes and trimethylnaphthalenes in some ancient sediments and petroleum. *Geochimica et Cosmochimica Acta*, 49, 385–95.

Alexander, R., Bastow, T. P., Kagi, R. I. and Singh, R. K. (1992) Identification of 1,2,2,5-tetramethyltetralins and 1,2,2,5,6-pentamethyltetralins as racemates in petroleum. *Journal of the Chemical Society*, Chemical Communications, 23, 1712–14.

Alsharhan, A. S. and Salah, M. G. (1997) A common source rock for Egyptian and Saudi hydrocarbons in the Red Sea. *American Association of Petroleum Geologists Bulletin*, 81, 1640–59.

Alvarez, H. M. and Steinbüchel, A. (2002) Triacylglycerols in prokaryotic microorganisms. *Applied Microbiology and Biotechnology*, 60, 367–76.

Ambrose, S. H. (1993) Isotopic analysis of paleodiets: methodological and interpretive considerations. In: *Investigations of Ancient Human Tissue* (M. K. Stanford, ed.), Gordon and Breach Science Publishers, Langhorne, PA, pp. 59–130.

American Society for Testing and Materials (1992) Detailed analysis of petroleum naphthas through *n*-nonane by capillary gas chromatography. Procedure ASTM D 5134–92.

Andersen, N., Paul, H. A., Bernasconi, S. M., *et al.* (2001) Large and rapid climate variability during the Messinian salinity crisis: evidence from deuterium concentrations of individual biomarkers. *Geology*, 29, 799–802.

Anderson, R. B. (1984) *The Fischer-Tropsch Synthesis*. Academic Press, New York.

Anderson, K. B. and Muntean, J. V. (2000) The nature and fate of natural resins in the geosphere. Part X. Structural characteristics of the macromolecular constituents of modern dammar resin and Class II ambers. *Geochemical Transactions*, 1, 1–9.

Andrusevich, V. E., Engel, M. H., Zumberge, J. E. and Brothers, L. A. (1998) Secular, episodic changes in stable carbon isotope composition of crude oils. *Chemical Geology*, 152, 59–72.

Andrusevich, V. E., Engel, M. H. and Zumberge, J. E. (2000) Effects of paleolatitude on the stable carbon isotope composition of crude oils. *Geology*, 28, 847–50.

Aplin, A. C., Larter, S. R., Bigge, M. A., *et al.* (2000) PVTX history of the North Sea's Judy oilfield. *Journal of Geochemical Exploration*, 69–70, 641–4.

Apps, J. A. and van de Kamp, P. C. (1993) Energy gases of abiogenic origin in the Earth's crust. U.S. Geological Survey Professional Paper 1570, pp. 81–132.

Armanios, C. (1995) *Molecular sieving, analysis and geochemistry of some pentacyclic triterpanes in sedimentary organic matter*. Ph.D. thesis, Curtin University of Technology, School of Applied Chemistry, Perth, Australia.

Armanios, C., Alexander, R. and Kagi, R. I. (1992) Shape-selective sorption of petroleum hopanoids by ultrastable Y zeolite. *Organic Geochemistry*, 18, 399–406.

Armanios, C., Alexander, R. and Kagi, R. I. (1994) Fractionation of higher-plant derived triterpanes using molecular sieves. *Organic Geochemistry*, 21, 531–43.

Armanios, C., Alexander, R., Sosrowidjojo, I. M. and Kagi, R. I. (1995) Identification of bicadinanes in Jurassic organic matter from the Eromanga Basin, Australia. *Organic Geochemistry*, 23, 837–43.

Arnold, R. and Anderson, R. (1907) Geology and oil resources of the Santa Maria oil district, Santa Barbara County, California. *U. S. Geological Survey Bulletin*, 322.

Arthur D. Little, Inc. (1999) *Sediment Quality in Depositional Areas of Shelikof Strait and Outermost Cook Inlet*, draft final report, U.S. Department of the Interior, Minerals Management Service, contract no. 1435-01-97-CT-30830.

Arthur, M. A., Dean, W. E. and Pratt, L. M. (1988) Geochemical and climatic effects of increased marine organic carbon burial at the Cenomanian/Turonian boundary. *Nature*, 335, 714–17.

Audino, M., Grice, K., Alexander, R., Boreham, C. J. and Kagi, R. I. (2001) Unusual distribution of monomethylalkanes in *Botryococcus braunii*-rich samples. Origin and significance. *Geochimica et Cosmochimica*, 65, 1995–2000.

Aveling, E. M. (1997) Chew, chew, that ancient chewing gum. A slovenly modern habit? Or one of the world's oldest pastimes. *British Archaeology*, 21, 6.

Aveling, E. M. (1998) Characterisation of natural products from the Mesolithic of Northern Europe. Ph. D. thesis, University of Bradford, Bradford, UK.

Aveling, E. M. and Heron, C. (1999) Chewing tar in the early Holocene: an archaeological and ethnographic evaluation. *Antiquity*, 73, 579–84.

Ayres, M. G., Bilal, M., Jones, R. W., *et al.* (1982) Hydrocarbon habitat in main producing areas, Saudi Arabia. *American Association of Petroleum Geologists Bulletin*, 66, 1–9.

Azevedo, D. A., Aquino Neto, F. R., Simoneit, B. R. T. and Pinto, A. C. (1992) Novel series of tricyclic aromatic terpanes characterized in Tasmanian tasmanite. *Organic Geochemistry*, 18, 9–16.

Bada, J. L., Wang, X. S. and Hamilton, H. (1999) Preservation of key biomolecules in the fossil record: current knowledge and future challenges. *Philosophical Transactions of the Royal Society of London, Biological Sciences*, 354, 77–88.

Baelocher, C., Meier, W. M. and Olson, D. H. (2001) *Atlas of Zeolite Framework Types*, 5th edn, Elsevier, Amsterdam.

Bailey, N. J. L., Burwood, R. and Harriman, G. E. (1990) Application of pyrolyzate carbon isotope and biomarker technology to organofacies definition and oil correlation problems in North Sea basins. *Organic Geochemistry*, 16, 1157–72.

Baker, E. W. and Louda, J. W. (1983) Thermal aspects of chlorophyll geochemistry. In: *Advances in Organic Geochemistry 1981* (M. Bjorøy, C. Albrecht, C. Cornford, *et al.*, eds.), John Wiley & Sons, New York, pp. 401–21.

Baker, E. W. and Louda, J. W. (1986) Porphyrins in the geological record. In: *Biological Markers in the Sedimentary Record* (R. B. Johns, ed.), Elsevier, New York, pp. 125–224.

Balabanova, S., Parsche, F. and Pirsig, W. (1992) First identification of drugs in Egyptian mummies. *Naturwissenschaften*, 79, 358.

Ballentine, C. J., Schoell, M., Coleman, D. and Cain, B. A. (2001) 300-Myr-old magmatic CO_2 in natural gas reservoirs of the west Texas Permian Basin. *Nature*, 409, 327–31.

Balogh, B., Wilson, D. M., Christiansen, P. and Burlingame, A. L. (1973) 17α(H)-hopane identified in oil shale of the Green River Formation (Eocene) by carbon-13 NMR. *Nature*, 242, 603–5.

Barker, C. and Smith, M. P. (1986) Mass spectrometric determination of gases in individual fluid inclusions in natural minerals. *Analytical Chemistry*, 58, 1330–33.

Barwise, A. J. G. (1990) Role of nickel and vanadium in petroleum classification. *Energy & Fuels*, 4, 647–52.

Barwise, A. J. G. and Whitehead, E. V. (1980) Separation and structure of petroporphyrins. In: *Advances in Organic Geochemistry 1979* (A. G. Douglas and J. R. Maxwell, eds.), Pergamon, New York, pp. 181–92.

Baskin, D. K. (1979) A method of preparing phytoclasts for vitrinite reflectance analysis. *Journal of Sedimentary Petrology*, 49, 633–5.

(1997) Atomic H/C ratio of kerogen as an estimate of thermal maturity and organic matter conversion. *American Association of Petroleum Geologists Bulletin*, 81, 1437–50.

Baskin, D. K. (2001) Comparison between atomic H/C and Rock-Eval hydrogen index as an indicator of organic matter quality. In: *The Monterey Formation: From Rocks to Molecules* (C. M. Isaacs and J. Rullkötter, eds.), Columbia University Press, New York, pp. 230–40.

Baskin, D. K. and Jones, R. W. (1993) Prediction of oil gravity prior to drill-stem testing in Monterey Formation reservoirs, offshore California. *American Association of Petroleum Geologists Bulletin*, 77, 1479–87.

Baskin, D. K. and Peters, K. E. (1992) Early generation characteristics of a sulfur-rich Monterey kerogen. *American Association of Petroleum Geologists Bulletin*, 76, 1–13.

Bass, G. (1986) A Bronze Age shipwreck at Ulu Burun (Kas) 1984 Campaign. *American Journal of Archaeology*, 90, 269–96.

Bass, G. and Pulak, C. (1987) A Late Bronze Age shipwreck at Ulu Burun: 1986. *American Journal of Archaeology*, 93, 1–29.

Bastow, T. P. (1998) Sedimentary processes involving aromatic hydrocarbons. Ph. D. thesis, Curtin University of Technology, Perth, Australia.

Bauer, P. E., Dunlap, N. K., Arseniyadis, S., *et al.* (1983) Synthesis of biological markers in fossil fuels. 1. 17α and 17β isomers of 30-norhopane and 30-normoretane. *Journal of Organic Chemistry*, 48, 4493–7.

Baumer, U. and Koller, J. (2002) The gold tree from the Celtic oppidum at Manching: investigation of the organic adhesive used for the gilding procedure. Presented at the 33rd International Symposium on Archaeometry, April 22–25, 2002, Amsterdam.

Beato, B. D., Yost, R. A., van Berkel, G. J., Filby, R. H. and Quirke, M. E. (1991) The Henryville bed of the New Albany Shale. III: tandem mass spectrometric analyses of geoporphyrins from the bitumen and kerogen. *Organic Geochemistry*, 17, 93–105.

Beck, C. W. and Borromeo, C. (1990) Ancient pine pitch: technological perspectives from a Hellenistic shipwreck. In: *Organic Content of Ancient Vessels: Materials Analysis and Archaeological Investigation*. Vol. 7 (A. R. Biers and P. E. McGovern, eds.), University of Pennsylvania Press, Philadelphia, pp. 51–8.

Beck, C. W., Stout, E. C. and Janne, P. A. (1998) The pyrotechnology of pine tar and pitch inferred from quantitative analysis by gas chromatography/mass spectrometry and carbon-13 nuclear magnetic resonance spectroscopy. In: *Proceedings of the First International Symposium on Wood Tar and Pitch* (W. Brzeinski, and W. Piotrowski, eds.), Biskupin, Poland, pp. 181–90.

Beck, C. W., Stout, E. C., Bingham, J., Lucas, J. and Purohit, V. (1999) Central European pine tar technologies. *Ancient Biomolecules*, 2, 281–93.

Beesley, T. E. and Scott, P. W. (1998) *Chiral Chromatography*. John Wiley & Sons, New York.

Bellamine, A., Mangla, A. T., Nes, W. D. and Waterman, M. R. (1999) Characterization and catalytic properties of the sterol 14α-demethylase from *Mycobacterium tuberculosis*. *Proceedings of the National Academy of Science, USA*, 96, 8937–8942.

BeMent, W. O., Levey, R. A. and Mango, F. D. (1995) The temperature of oil generation as defined with C_7 chemistry maturity parameter (2,4-DMP/2,3-DMP ratio). In: *Organic Geochemistry: Development and Applications to Energy, Climate, Environment and Human History* (J. O. Grimalt and C. Dorronsoro, eds.), AIGOA, Donostia-San Sebastián, Spain, pp. 505–7.

BeMent, W. O., McNeil, R. I. and Lippincott, R. G. (1996) Predicting oil quality from sidewall cores using PFID, TEC, and NIR analytical techniques in sandstone reservoirs, Rio Del Rey Basin, Cameroon. *Organic Geochemistry*, 24, 1173–8.

Bence, A. E. and Burns, W. A. (1993) Fingerprinting hydrocarbons in the biological resources of the *Exxon Valdez* spill area. In: *Exxon Valdez Oil Spill: Fate and Effects in Alaskan Waters (3rd ASTM Environmental Toxicology and Risk Assessment Symposium)* (P. G. Wells, J. N. Butler, and J. S. Hughes, eds.) American Society for Testing and Materials, STP 1219, Philadelphia, p. 84–140.

Bence, A. E., Kvenvolden, K. A. and Kennicutt, M. C., II (1996) Organic geochemistry applied to environmental assessments of Prince William Sound, Alaska, after the *Exxon Valdez* oil spill – a review. *Organic Geochemistry*, 24, 7–42.

Bence, A. E., Burns, W. A., Mankiewicz, P. J., Page, D. S. and Boehm, P. D. (2000) Comment on "PAH refractory index as a source discriminant of hydrocarbon input from crude oil and coal in Prince William Sound, Alaska" by F. D. Hostettler, R. J. Rosenbauer, K. A. Kvenvolden. *Organic Geochemistry*, 31, 931–8.

Benford, D. J., Hanley, A. B., Bottrill, K., *et al.* (2000) Biomarkers as predictive tools in toxicity testing. *Alternatives to Laboratory Animals*, 28, 119–31.

Berkert, U. and Allinger, N. L. (1982) *Molecular Mechanics*, monograph 177, American Chemical Society, Washington, DC.

Berndt, M. E., Allen, D. E. and Seyfried, W. E. J. (1996) Reduction of CO_2 during serpentinization of olivine at 300°C and 500 bar. *Geology*, 24, 351–4.

Berner, R. A. (1984) Sedimentary pyrite formation: an update. *Geochimica et Cosmochimica Acta*, 48, 605–15.

Berner, R. A. and Raiswell, R. (1983) Burial of organic carbon and pyrite sulfur in sediments over Phanerozoic time: a new theory. *Geochimica et Cosmochimica Acta*, 47, 605–15.

Berry, A. M., Harriott, O. T., Moreau, R. A., *et al.* (1993) Hopanoid lipids compose the *Frankia* vesicle envelope, presumptive barrier of oxygen diffusion to nitrogenase. *Proceedings of the National Academy of Science, USA*, 90, 6091–4.

Berthelot, M.-P. (1860) *Chimie organique fondée sur la synthèse*. Mallet-Bachelier, Paris.

Bertsch, W. (1999) Two-dimensional gas chromatography. Concepts, instrumentation, and applications. Part 1. Fundamentals, conventional two-dimensional gas chromatography, selected applications. *Journal of High Resolution Chromatography*, 22, 647–65.

(2000) Two-dimensional gas chromatography. Concepts, instrumentation, and applications. Part 2.

Comprehensive two-dimensional GC. *Journal of High Resolution Chromatography*, 23, 167–81.

Bethell, P. H., Goad, L. J., Evershed, R. P. and Ottaway, J. (1994) The study of molecular markers of human activity: the use of coprostanol in the soil as an indicator of human faecal material. *Journal of Archaeological Science*, 21, 619–32.

Bhullar, A. G., Karlsen, D. A., Backer-Owe, K., Seland, R. T. and Le Tran, K. (1999) Dating reservoir filling – a case history from the North Sea. *Marine Petroleum Geology*, 16, 581–603.

Bidigare, R. R., Kennicutt, M. C., II, Ondrusek, M. E., Keller, M. D. and Guillard, R. R. L. (1990) Novel chlorophyll-related compounds in marine phytoplankton: distributions and geochemical implications. *Energy & Fuels*, 4, 653–7.

Bidigare, R. R., Kennicutt, M. C., II, Keeney-Kennicutt, W. L. and Macko, S. A. (1991) Isolation and purification of chlorophylls a and b for the determination of stable carbon and nitrogen isotope compositions. *Analytical Chemistry*, 63, 130–33.

Binder, D., Bourgois, G., Benoist, F. and Votry, C. (1990) Identification de brai de bouleau (*Betula*) dans le Neolithique de Giribaldi (Nice, France) par la spectrometrie de masse. *Revue d'Archeometrie*, 14, 37–42.

Bird, C. W., Lynch, J. M., Pirt, F. J., *et al.* (1971) Steroids and squalene in *Methylococcus capsulatus* grown on methane. *Nature*, 230, 473–4.

Bisset, N. G. and Zenk, M. H. (1993) Responding to 'First identification of drugs in Egyptian mummies'. *Naturwissenschaften*, 80, 244–5.

Bisset, N., Bruhn, J. G. and Zenk, M. H. (1996) Was opium known in the 18th dynasty in Egypt? An examination of materials from the tomb of the chief royal architect Kha, the presence of opium in a 3,500 year old Cypriote base-ring juglet. *Ägypten und Levante*, 6, 199–204.

Biswas, S. K., Rangaraju, M. K., Thomas, J. and Bhattacharya, S. K. (1994) Cambay-Hazad(!) petroleum system in South Cambay Basin, India. In: *The Petroleum System – From Source to Trap* (L. B. Magoon and W. G. Dow, eds.), American Association of Petroleum Geologists, Tulsa, OK, pp. 615–24.

Bjorn, L. O. (1993) Responding to 'First identification of drugs in Egyptian mummies'. *Naturwissenschaften*, 80, 244.

Bjorøy, M., Hall, K., Gillyon, P. and Jumeau, J. (1991) Carbon isotope variations in *n*-alkanes and isoprenoids of whole oils. *Chemical Geology*, 93, 13–20.

Bjorøy, M., Hall, P. B. and Moe, R. P. (1994) Variation in the isotopic composition of single components in the C_4–C_{20} fraction of oils and condensates. *Organic Geochemistry*, 21, 761–76.

Blankenship, R. (1992) Origin and early evolution of photosynthesis. *Photosynthetic Research*, 33, 91–111.

Blankenship, R. E. and Hartman, H. (1998) The origin and evolution of oxygenic photosynthesis. *Trends in Biochemical Sciences*, 23, 94–7.

Bloch, K. (1983) Sterol structure and membrane function. *CRC Critical Reviews of Biochemistry*, 4, 47–92.

Blokker, P., Van Bergen, P., Pancost, R., *et al.* (2001) The chemical structure of *Gloeocapsomorpha prisca* microfossils: implications for their origin. *Geochimica et Cosmochimica Acta*, 65, 885–900.

Blomberg, J., Schoenmakers, P. J. and Brinkman, U. A. (2002) Gas chromatographic methods for oil analysis *Journal of Chromatography A*, 972, 137–73.

Blumer, M. and Sass, J. (1972) Oil pollution: persistence and degradation of spilled fuel oil. *Science*, 176, 1120–2.

Blumer, M., Guillard, R. R. L. and Chase, T. (1971) Hydrocarbons of marine plankton. *Marine Biology*, 8, 183–9.

Blumer, M., Blokker, P. C., Cowell, E. B. and Duckworth, D. F. (1972) Petroleum. In: *A Guide to Marine Pollution* (E. D. Goldberg, ed.), Gordon and Breach, New York, pp. 19–40.

Bocherens, H., Billiou, D., Mariotti, A., *et al.* (1999) Palaeoenvironmental and palaeodietary implications of isotopic biogeochemistry of last interglacial Neanderthal and mammal bones in Scladina Cave (Belgium). *Journal of Archaeological Science*, 26, 599–607.

Boëda, E., Connan, J., Dessort, D., *et al.* (1996) Bitumen as a hafting material on Middle Paleolithic artefacts. *Nature*, 380, 336–8.

Boehm, P. D., Douglas, G. S., Burns, W. A., *et al.* (1997) Application of petroleum hydrocarbon chemical fingerprinting and allocation techniques after the *Exxon Valdez* oil spill. *Marine Pollution Bulletin*, 34, 599–613.

Boehm, P. D., Page, D. S., Gilfillan, E. S., *et al.* (1998) Study of the fates and effects of the *Exxon Valdez* oil spill on benthic sediments in two bays in Prince William Sound, Alaska. 1. Study design, chemistry, and source fingerprinting. *Environmental Science & Technology*, 32, 567–76.

Boehm, P. D., Douglas, G. S., Borwn, J. S., *et al.* (2000) Comment on "Natural hydrocarbon background in benthic sediments of Prince William Sound, Alaska: oil vs. coal". *Environmental Science & Technology*, 34, 2064–5.

Boehm, P. D., Page, D. S., Burns, W. A., *et al.* (2001) Resolving the origin of the petrogenic hydrocarbon background in Prince William Sound, Alaska. *Environmental Science & Technology*, 35, 471–9.

Boetius, A., Raveschlag, K., Schubert, C. J., *et al.* (2000) A marine microbial consortium apparently mediating anaerobic oxidation of methane. *Nature*, 407, 577–9.

Booth, M. (1999) *Opium: A History*. St Martin's Griffin, New York.

Bordenave, M. L. (1993) *Applied Petroleum Geochemistry*. Editions Technip, Paris.

Bordovskiy, O. K. and Takh, N. I. (1978) Organic matter in the Recent carbonate sediments of the Caspian Sea. *Oceanology*, 18, 673–8.

Boreham, C. J. and Powell, T. G. (1993) Petroleum source rock potential of coal and associated sediments: qualitative and quantitative aspects. In: *Hydrocarbons from Coal* (D. D. Rice, ed.), American Association of Petroleum Geologists, Tulsa, OK, pp. 133–57.

Boreham, C. J., Fookes, C. J. R., Popp, B. N. and Hayes, J. M. (1989) Origins of etioporphyrins in sediments: evidence from stable carbon isotopes. *Geochimica et Cosmochimica Acta*, 53, 2451–5.

Borgund, A. E. and Barth, T. (1994) Generation of short-chain organic acids from crude oil by hydrous pyrolysis. *Organic Geochemistry*, 21, 943–52.

Bostick, N. H. (1979) Microscopic measurement of the level of catagenesis of solid organic matter in sedimentary rocks to aid exploration for petroleum and to determine former burial temperatures – a review. In: *Aspects of Diagenesis* (P. A. Schdle and P. R. Schulger, eds.), Society for Sedimentary Geology, Houston, TX, pp. 17–43.

Bostick, N. H. and Alpern, B. (1977) Principles of sampling, preparation and constituent selection for microphotometry in measurement of maturation of sedimentary organic matter. *Journal of Microscopy*, 109, 41–7.

Botneva, T. A., Eremenko, N. A. and Pankina, R. G. (1984) Isotopic composition of carbon, hydrogen, nitrogen, and sulphur in crude oils, gases, and organic matter of rocks. In: *Handbook on Oil and Gas Geology* [in Russian], Nedra, Moscow, pp. 78–97.

Bottomley, R. J., York, D. and Grieve, R. A. F. (1978) ^{40}Ar–^{39}Ar ages of Scandinavian impact structures:

I. Mien and Siljan. *Contributions to Mineralogy and Petrology*, 68, 79–84.

Bouvier, P., Rohmer, M., Benveniste, P. and Ourisson, G. (1976) Δ8,14-Steroids in the bacterium *Methylococcus capsulatus*. *Biochemistry Journal*, 159, 267–71.

Bragg, J. R., Prince, R. C., Harner, E. J. and Atlas, R. M. (1994) Effectiveness of bioremediation for the *Exxon Valdez* oil spill. *Nature*, 368, 413–8.

Brassell, S. C. (1987) Natural gas from the mantle. Book review. *Power from the Earth* by Thomas Gold. *New Scientist*, 116, 54–5.

Brassell, S. C., Wardroper, A. M. K., Thompson, I. D., Maxwell, J. R. and Eglinton, G. (1981) Specific acyclic isoprenoids as biological markers of methanogenic bacteria in marine sediments. *Nature*, 290, 693–6.

Brassell, S. C., Eglinton, G. and Fu, J. M. (1985) Biological marker compounds as indicators of the depositional history of the Maoming oil shale. *Organic Geochemistry*, 10, 927–41.

Bray, E. E. and Evans, E. D. (1961) Distribution of *n*-paraffins as a clue to recognition of source beds. *Geochimica et Cosmochimica Acta*, 22, 2–15.

Breck, D. W. (1974) *Zeolite Molecular Sieves*. John Wiley & Sons, New York.

Britton, G. (1998) Overview of carotenoid biosynthesis. In: *Carotenoids*, Vol. 3 (G. Britton, S. Leeaen-Jensen and H. Pfander, eds.), Birkhauser Verlag, Basel, pp. 13–147.

Brock, T. D. and Madigan, M. T. (1991) *Biology of Microorganisms*. Prentice-Hall, Englewood Cliffs, NJ.

Brocks, J. J., Logan, G. A., Buick, R. and Summons, R. E. (1999) Archean molecular fossils and the early rise of eukaryotes. *Science*, 285, 1033–6.

Bromley, B. W. and Larter, S. R. (1986) Biogenic origin of petroleums. *Chemical and Engineering News*, August 25, 3, 43.

Brooks, P. W., Maxwell, J. R., Cornforth, J. W. Butlin, A. G. and Milne, C. B. (1977) Stereochemical studies of acyclic isoprenoid compounds. VI. The stereochemistry of farnesane from crude oil. In: *Advances in Organic Geochemistry 1975* (R. Campos and J. Goni, eds.), Pergamon, Oxford, pp. 91–97.

Brooks, J. M., Kennicutt, M. C., II and Carey, B. D., Jr. (1986) Offshore surface geochemical exploration. *Oil and Gas Journal*, 84, 66–72.

Brown, T. A. (1999) How ancient DNA may help in understanding the origin and spread of agriculture. *Philosophical Transactions of the Royal Society of London, Biological Sciences*, 354, 89–98.

Buck, S. P. and McCulloh, T. H. (1994) Bampo-Peutu(!) petroleum system, North Sumatra, Indonesia. In: *The Petroleum System – From Source to Trap* (L. B. Magoon and W. G. Dow, eds.), American Association of Petroleum Geologists, Tulsa, OK, pp. 624–38.

Buckley, S. A. and Evershed, R. P. (2001) Organic chemistry of embalming agents in Pharaonic and Graeco-Roman mummies. *Nature*, 413, 837–41.

Buckley, S. A., Stott, A. W. and Evershed, R. P. (1999) Studies of organic residues from ancient Egyptian mummies using high temperature gas chromatography mass spectrometry and sequential thermal desorption gas chromatography mass spectrometry and pyrolysis gas chromatography mass spectrometry. *Analyst*, 124, 443–52.

Budiansky, S. (1982) Research article triggers dispute on zeolite. *Nature*, 300, 309.

Buick, R., Rasmussen, B. and Krapez, B. (1998) Archean oil: evidence for extensive hydrocarbon generation and migration 2.5–3.5 Ga. *American Association of Petroleum Geologists Bulletin*, 82, 50–69.

Bull, I. D., van Bergen, P. F., Poulton, P. R. and Evershed, R. P. (1998) Organic geochemical studies of soils from the Rothamsted classical experiments – II. Soils from the Hoosfield spring barley experiment treated with different quantities of manure. *Organic Geochemistry*, 28, 11–26.

Bull, I. D., Betancourt, P. P. and Evershed, R. P. (1999a) Chemical evidence supporting the existence of structured agricultural manuring regime on Pseira Island, Crete during the Minoan Age. *Malcolm Wiener Fertschrift Volume, Aegeum*, 20, 69–74.

Bull, I. D., Simpson, I. A., van Bergen, P. F., Poulton, P. R. and Evershed, R. P. (1999b) Muck 'n' molecules: organic geochemical methods for detecting ancient manuring. *Antiquity*, 73, 86–96.

Bull, I. D., van Bergen, P. F., Nott, C. J., Poulton, P. R. and Evershed, R. P. (2000) Organic geochemical studies of soils from the Rothamsted classical experiments – V. The fate of lipids in different long-term experiments. *Organic Geochemistry*, 31, 389–408.

Bull, I. D., Lockheart, M. J., Elhummali, M. M., Roberts, D. J. and Evershed, R. P. (2002) The origin of faeces by means of biomarker detection. *Environment International*, 27, 647–54.

Bull, I. D., Elhmmali, M. M., Roberts, D. J. and Evershed, R. P. (2003) The application of steroidal biomarkers to track the abandonment of a Roman wastewater course at the Agora (Athens, Greece). *Archaeometry*, 45, 149–62.

Bullock, C. (2000) The archaea – a biochemical perspective. *Biochemistry and Molecular Biology Education*, 28, 186–91.

Burke, E. A. J. (2001) Raman microspectrometry of fluid inclusions. *Lithos*, 55, 139–58.

Burkhart, C. N., Kruge, M. A., Burkhart, C. G. and Black, C. (2001) Cerumen composition by flash pyrolysis-gas chromatography/mass spectrometry. *Otology and Neurotology*, 22, 715–22.

Burlingame, A. L., Haug, P., Belsky, T. and Calvin, M. (1965) Occurrence of biogenic steranes and pentacyclic triterpanes in an Eocene shale (52 million years) and in an early Precambrian shale (2.7 billion years): a preliminary report. *Proceedings of the National Academy of Sciences, USA*, 54, 1406–12.

Burlingame, A. L., Baillie, T. A., Derrick, P. G. and Chizhov, O. S. (1980) Mass spectrometry. *Analytical Chemistry*, 52, 214–58R.

Burns, W. A., Mankiewicz, P. J., Bence, A. E., Page, E. S. and Parker, K. R. (1997) A principal-component and least-squares method for allocating polycyclic aromatic hydrocarbons in sediment to multiple sources. *Environmental Toxicological Chemistry*, 16, 1119–31.

Burruss, R. C., Cercone, K. R. and Harris, P. M. (1983) Fluid inclusion petrography and tectonic-burial history of the Al Ali no. 2 well: evidence for the timing of diagenesis and oil migration, northern Oman foredeep. *Geology*, 7, 567–70.

Burwood, R., Cornet, P. J., Jacobs, L. and Paulet, J. (1990) Organofacies variation control on hydrocarbon generation: a Lower Congo Coastal Basin (Angola) case history. *Organic Geochemistry*, 16, 325–38.

Cahn, R. S., Ingold, C. and Prelog, V. (1966) Specification of molecular chirality. *Angewandte Chemie International Edition*, 5, 385–415.

Cajaraville, M. P., Orbea, A., Mrigomez, I. and Cncio, I. (1997) Peroxisome proliferation in the digestive epithelium of mussels exposed to the water accommodated fraction of three oils. *Comparative Biochemistry and Physiology C: Pharmacology, Toxicology and Endocrinology*, 117C, 233–42.

Callot, H. J., Ocampo, R. and Albrecht, P. (1990) Sedimentary porphyrins: correlations with biological precursors. *Energy & Fuels*, 4, 635–9.

Calvert, S. E. (1987) Oceanographic controls on the accumulation of organic matter in marine sediments. In: *Marine Petroleum Source Rocks* (J. Brooks and A. J. Fleet, eds.), Blackwell, London, pp. 137–51.

Calvert, S. E. and Pederson, T. (1992) Organic carbon accumulation and preservation in marine sediments: how important is anoxia? In: *Productivity, Accumulation and Preservation of Organic Matter in Recent and Ancient Sediments* (J. Whelan and J. W. Farrington, eds.), Columbia University Press, New York, pp. 231–63.

Calvert, S. E., Karlin, R. E., Toolin, L. J., *et al.* (1991) Low organic carbon accumulation rates in Black Sea sediments. *Nature*, 350, 692–5.

Cameron, N. R., Brooks, J. M. and Zumberge, J. E. (1999) Deepwater petroleum systems in Nigeria: their identification and characterization ahead of the drill bit using SGE technology. www.tdi-bi.com (accessed 2 October, 1999).

Cane, R. F. (1969) Coorongite and the genesis of oil shale. *Geochimica et Cosmochimica Acta*, 33, 257–65.

Cann, R. L., Stoneking, M. and Wilson, A. C. (1987) Mitochondrial DNA and human evolution. *Nature*, 325, 31–6.

Caplan, M. L. and Bustin, R. M. (1996) Factors governing organic matter accumulation and preservation in a marine petroleum source rock from the Upper Devonian to Lower Carboniferous Exshaw Formation, Alberta. *Bulletin of Canadian Petroleum Geology*, 44, 474–94.

Caramelli, D., Lalueza-Fox, C., Vernesi, C., *et al.* (2003) Evidence for a genetic discontinuity between Neandertals and 24,000-year-old anatomically modern Europeans. *Proceedings of the National Academy of Sciences, USA*, 100, 6593–7.

Carlson, R. M. K., Croasmun, W. R. and Chamberlain, D. E. (1995) Transformations of cholestane useful for probing processing chemistry. Presented at the 210th National Meeting of the American Chemical Society, August 20–25, 1995, Chicago, IL.

Carpentier, B., Ungerer, P., Kowalewski, I., *et al.* (1996) Molecular and isotopic fractionation of light hydrocarbons between oil and gas phases. *Organic Geochemistry*, 24, 1115–39.

Carrigan, W. J., Tobey, M. H., Halpern, H. I., *et al.* (1998) Identification of reservoir compartments by geochemical methods: Jauf reservoir, Ghawar. *Saudi Aramco Journal of Technology*, Summer, 28–32.

Carroll, A. R. and Bohacs, K. M. (2001) Lake-type controls on petroleum source rock potential in nonmarine basins. *American Association of Petroleum Geologists Bulletin*, 85, 1033–53.

Carson, R. (1962) Silent Spring. Houghton Mifflin, Boston, MA.

Casagrande, D. J. (1987) Sulfur in peat and coal. In: *Coal and Coal-bearing Strata: Recent Advances* (A. C. Scott, ed.), Geological Society, London, pp. 87–105.

Castagna, J. P. and Backus, M. M. (eds.) (1997) *Offset-dependent Reflectivity – Theory and Practice of AVO Analysis.* Society of Exploration Geophysicists, Tulsa, OK.

Castaño, J. R. (1993) *Prospects for Commercial Abiogenic Gas Production: Implications from the Siljan Ring Area, Sweden.* U.S. Geological Survey Professional Paper 1570.

Cazier, E. C., Hayward, A. B., Espinosa, G., *et al.* (1995) Petroleum geology of the Cusiana Field, Llanos Basin Foothills, Colombia. *American Association of Petroleum Geologists Bulletin*, 79, 1444–62.

Chan, M. A., Parry, W. T. and Bowman, J. R. (2000) Diagenetic hematite and manganese oxides and fault-related fluid flow in Jurassic Sandstones, Southeastern Utah. *American Association of Petroleum Geologists Bulletin*, 84, 1281–310.

Chapelle, F. H., O'Neill, K., Bradley, P. M., *et al.* (2002) A hydrogen-based subsurface microbial community dominated by methanogens. *Nature*, 415, 312–5.

Chapman, D. J. and Gest, H. (1983) Terms used to describe biological energy conversions, electron transport processes, interactions of cellular systems with molecular oxygen, and carbon nutrition. In: *Earth's Earliest Biosphere* (J. W. Schopf, ed.), Princeton University Press, Princeton, NJ, pp. 459–63.

Chapman, D. J. and Schopf, J. W. (1983) Biological and biochemical effects of the development of an aerobic environment. In: *Earth's Earliest Biosphere* (J. W. Schopf, ed.), Princeton University Press, Princeton, NJ, pp. 302–20.

Chappe, B., Michaelis, W. and Albrecht, P. (1980) Molecular fossils of archaebacteria as selective degradation products of kerogen. In: *Advances in Organic Geochemistry 1979* (A. G. Douglas and J. R. Maxwell, eds.), Pergamon Press, Oxford, pp. 265–74.

Charlou, J.-L. and Donval, J.-P. (1993) Hydrothermal methane venting between 12°N and 26°N along the Mid-Atlantic Range. *Journal of Geophysical Research*, 98, 9625–42.

Charters, S., Evershed, R. P., Goad, L. J., Heron, C. and Blinkhorn, P. W. (1993) Identification of an adhesive used to repair a Roman jar. *Archaeometry*, 35, 211–23.

Charters, S., Evershed, R. P., Blinkhorn, P. W. and Denham, V. (1995) Evidence for the mixing of fats and waxes in archaeological ceramics. *Archaeometry*, 37, 113–27.

Charters, S., Evershed, R. P., Quye, A., Blinkhorn, P. W. and Reeves, V. (1997) Simulation experiments for determining the use of ancient pottery vessels: the behaviour of epicuticular leaf wax during boiling of a leafy vegetable. *Journal of Archaeological Science*, 24, 1–7.

Chen, J., Fu, J., Sheng, G., Liu, D., and Zhang, J. (1996) Diamondoid hydrocarbon ratios: novel maturity indices for highly mature crude oils. *Organic Geochemistry*, 25, 179–90.

Chicarelli, M. I., Kaur, S. and Maxwell, J. R. (1987) Sedimentary porphyrins: unexpected structures, occurrence, and possible origins. In: *Metal Complexes in Fossil Fuels* (R. H. Filby and J. F. Branthaven, eds.), American Chemical Society, Washington, DC, pp. 41–67.

Child, A. M., Collins, M. J., Vermeer, C., *et al.* (1997) Osteocalcin – a "long-term" protein. In: *Archaeological Sciences Conference Proceedings, 2–4 September 1997* (A. Millard, ed.), British Archaeological Reports International Series No. 939, University of Durham, Durham, UK.

Chung, H. M., Brand, S. W. and Grizzle, P. L. (1981) Carbon isotope geochemistry of Paleozoic oils from Big Horn Basin. *Geochimica et Cosmochimica Acta*, 45, 1803–15.

Chung, H. M., Gormly, J. R. and Squires, R. M. (1988) Origin of gaseous hydrocarbons in subsurface environments: theoretical considerations of carbon isotope distribution. *Chemical Geology*, 71, 97–103.

Chung, H. M., Rooney, M. A., Toon, M. B. and Claypool, G. E. (1992) Carbon isotope composition of marine crude oils. *American Association of Petroleum Geologists Bulletin*, Vol. 76, p. 1000–1007.

Chung, H. M., Walters, C. C., Buck, S. and Bingham, G. (1998) Mixed signals of the source and thermal maturity for petroleum accumulations from light hydrocarbons: an example of the Beryl Field. *Organic Geochemistry*, 29, 381–96.

Chunqing, J., Li, M. and van Duin, A. C. T. (2000a) Inadequate separation of saturate and monoaromatic hydrocarbons in crude oils and rock extracts by alumina column chromatography. *Organic Geochemistry*, 31, 751–6.

Chunqing, J., Alexander, R., Kagi, R. I. and Murray, A. P. (2000b) Origin of perylene in ancient sediments and its geological significance. *Organic Geochemistry*, 31, 1545–59.

Clarke, F. W. (1916) Data of geochemistry, third edition. *US Geological Survey Bulletin*, 616.

Claus, H., Akca, E., Debaerdemaeker, T., *et al.* (2002) Primary structure of selected archaeal mesophilic and extremely thermophilic outer surface layer proteins. *Systematic and Applied Microbiology*, 25, 3–12.

Claypool, G. E. and Kaplan, I. R. (1974) The origin and distribution of methane in marine sediments. In: *Natural Gases in Marine Sediments* (I. R. Kaplan, ed.), Plenum Press, New York, pp. 99–140.

Claypool, G. E. and Magoon, L. B. (1985) Comparison of oil-source rock correlation data for Alaskan North Slope: techniques, results, and conclusions. In: *Alaska North Slope Oil/Source Rock Correlation Study* (L. B. Magoon and G. E. Claypool, eds.), American Association of Petroleum Geologists, Tulsa, OK, pp. 49–81.

Claypool, G. E. and Mancini, E. A. (1989) Geochemical relationships of petroleum in Mesozoic reservoirs to carbonate source rocks of Jurassic Smackover Formation, Southwestern Alabama. *American Association of Petroleum Geologists Bulletin*, 73, 904–24.

Claypool, G. E., Love, A. H. and Maughan, E. K. (1978) Organic geochemistry, incipient metamorphism, and oil generation in black shale members of Phosphoria Formation, western interior United States. *American Association of Petroleum Bulletin*, 62, 98–120.

Clayton, C. (1991) Carbon isotope fractionation during natural gas generation from kerogen. *Marine and Petroleum Geology*, 8, 232–40.

Clayton, C. J. and Bjorøy, M. (1994) Effect of maturity on $^{13}C/^{12}C$ ratios of individual compounds in North Sea oils. *Organic Geochemistry*, 21, 737–50.

Clifford, D. J., Clayton, J. L. and Sinninghe Damsté, J. S. (1997) 3,4,5–2,3,6 Substituted diaryl carotenoid derivatives (*Chlorobiaceae*) and their utility as indicators of photic zone anoxia in sedimentary environments. In: *Abstracts from the 18th International Meeting on Organic Geochemistry, September 22–26, 1997, Maastricht, The Netherlands* (B. Horsfield, ed.), Forschungszentrum Jülich, Jülich, Germany, pp. 685–6.

Coe, S. D. and Coe, M. D. (2000) *The True History of Chocolate*. Thames and Hudson, London.

Coleman, I. W. M. and Lawrence, B. M. (2000) Examination of the enantiomeric distribution of certain monoterpene hydrocarbons in selected essential oils by automated solid-phase microextraction–chiral gas chromatography-mass selective detection. *Journal of Chromatographic Science*, 38, 95–9.

Collins, M. J., Child, A. M., van Duin, A. T. C. and Vermeer, C. (1998) Ancient osteocalcin; the most stable bone protein? *Ancient Biomolecules*, 2, 223–38.

Collins, M. J., Waite, E. R. and van Duin, A. C. T. (1999) Predicting protein decomposition: the case of aspartic-acid racemization kinetics. *Philosophical Transactions of the Royal Society London, Biological Sciences*, 354, 51–64.

Colombini, M. P., Modugno, C., Silvano, F. and Onor, M. (2000) Characterization of the balm of an Egyptian mummy from the seventh century B.C. *Studies in Conservation*, 45, 19–29.

Condamin, J., Formenti, F., Metais, M. O., Michel, M. and Blond, P. (1976) The application of gas chromatography to the tracing of oil in ancient amphorae. *Archaeometry*, 18, 195–201.

Connan, J. (1981) Un exemple de biodegradation preferentielle des hydrocarbures aromatique dans des asphaltes du bassin Sud-Aquitain (France). *Bulletin des Centres de Recherches Exploration Production Elf Aquitaine*, 5, 151–71.

(1984) Biodegradation of crude oils in reservoirs. In: *Advances in Petroleum Geochemistry*, Vol. 1 (J. Brooks and D. H. Welte, eds.), Academic Press, London, pp. 299–335.

(1988) Quelques secrets des bitumens archéologiques de Mésopotamie révélés par les analyses de Géochimi Organique Pétrolière. *Bulletin des Centres de Recherches Exploration Production Elf Aquitaine*, 12, 759–87.

(1996) La colle au collagène, innovation du Néolithique. *La Recherche*, 284, 33–4.

(1999) Use and trade of bitumen in antiquity and prehistory: molecular archaeology reveals secrets of past civilizations. *Philosophical Transactions of the Royal Society, Biological Sciences*, 354, 33–50.

Connan, J. and Deschesne, O. (1996) *Le Bitumen à Suse (Bitumen at Susa), Réunion des musées nationaux (Collection du musée du Louvre)*. Elf Aquitaine Production, Pau, France.

(2001) Matériau artificiel ou roche naturelle? [Artificial material or natural rock?] *La Recherche*, 347, 46–7.

Connan, J. and Dessort, D. (1989a) Du bitume dans les baumes de momies égyptienne (1295 av. J-C.-300 ap. J. C.): détermination de son origine et évaluation de sa quantité. *Comptes Rendus de l'Academie des Sciences, Paris*, 312, 1445–52.

(1989b) Du bitume de la Mer Morte dans les baumes d'une momie égyptienne: identification par critèrese moléculaires. *Comptes Rendus de l'Academie des Sciences, Paris*, 309, 1665–72.

(1991) Le bitume dans l'Antiquité. *La Recherche*, 229, 152–9.

Connan, J. and Lacrampe-Couloume, G. (1993) The origin of the Lacq Superieur heavy oil accumulation and of the giant Lacq Inferieur gas field (Aquitaine Basin, SW France). In: *Applied Petroleum Geochemistry* (M. L. Bordenave, ed.), Editions Technip, Paris, pp. 465–87.

Connan, J. and Nissenbaum, A. (2003) Conifer tar on the keel and hull planking of the Ma'agan Mikhael Ship (Israel, 5th century BC): identification and comparison with natural products and artefacts employed in boat construction. *Journal of Archaeological Science*, 30, 709–19.

Connan, J., Nissenbaum, A. and Dessort, D. (1992) Molecular archaeology: export of Dead Sea asphalt to Canaan and Egypt in the Chalcolithic-Early Bronze Age (4th–3rd millennium BC). *Geochimica et Cosmochimica Acta*, 56, 2743–59.

Cook, A. C. and Sherwood, N. R. (1991) Classification of oil shales, coals and other organic-rich rocks. *Organic Geochemistry*, 17, 211–22.

Cooles, G. P., Mackenzie, A. S. and Quigley, T. M. (1986) Calculation of petroleum masses generated and expelled from source rocks. *Organic Geochemistry*, 10, 235–45.

Cooper, A. and Poinar, H. N. (2002) Ancient DNA: do it right or not at all. *Science*, 289, 1139.

Cooper, A., Poinar, H. N., Paabo, S., *et al.* (1997) Neandertal genetics. *Science*, 277, 1021–5.

Coplen, T. B. (1996) New guidelines for reporting stable hydrogen, carbon, and oxygen isotope-ratio data. *Geochimica et Cosmochimica Acta*, 60, 3359–60.

Copley, M. S., Rose, P. J., Clapham, A., *et al.* (2001) Processing palm fruits in the Nile Valley – biomolecular evidence from Qasr Ibrim. *Antiquity*, 75, 538–42.

Copley, M., Jones, V., Rose, P., *et al.* (2002) Biomolecular analysis of pottery and palaeoenvironmental material from Qasr Ibrim as indicators of changing economy. Presented at the 33rd International Symposium on Archaeometry, April 22–25, 2002, Amsterdam.

Copley, M. S., Berstan, R., Dudd, S. N., *et al.* (2003) Direct chemical evidence for widespread dairying in prehistoric Britain. *Proceedings of the National Academy of Sciences, USA*, 100, 1524–9.

Corbin, C. J. (1993) Petroleum contribution of the coastal environment of St Lucia. *Marine Pollution Bulletin*, 26, 579–80.

Cornford, C. (1994) Mandal-Ekofisk(!) petroleum system in the Central Graben of the North Sea. In: *The Petroleum System – From Source to Trap* (L. B. Magoon and W. G. Dow, eds.), American Association of Petroleum Geologists, Tulsa, OK, pp. 537–71.

Corr, L. T., Sealy, J., Jones, V. and Evershed, R. P. (2002) Carbon isotopic analysis of individual collagenous amino acids and coastal diets in the Late Stone Age of South Africa. Presented at the 33rd International Symposium on Archaeometry, April 22–25, 2002, Amsterdam.

Cox, R. M. and Gallois, R. W. (1981) *The Stratigraphy of the Kimmeridge Clay of the Dorset Type Area and its Correlation With Some Other Kimmeridgian Sequences*. Institute of Geological Sciences Report 80/4.

Cragg, B. A., Parkes, R. J., Fry, F. C., *et al.* (1996) Bacterial populations and processes in sediments containing gas hydrates (ODP Leg 146: Cascadia Margin). *Earth and Planetary Science Letters*, 139, 497–507.

Craig, H. (1953) The geochemistry of the stable carbon isotopes. *Geochimica et Cosmochimica Acta*, 3, 53–92.

Cranwell, P. A., Eglinton, G. and Robinson, N. (1987) Lipids of aquatic organisms as potential contributors to lacustrine sediments. II. *Organic Geochemistry*, 11, 513–27.

Creaney, S. and Passey, Q. R. (1993) Recurring patterns of total organic carbon and source rock quality within a sequence stratigraphic framework. *American Association of Petroleum Geologists Bulletin*, 77, 386–401.

Croasmun, W. R. and Carlson, R. M. K. (1987) Two-dimensional NMR spectroscopy – applications for chemists and biochemists. In: *Methods in Stereochemical Analysis*, Vol. 9 (W. R. Croasmun and R. M. K. Carlson, eds.), VCH Publishers, New York, pp. 1–534.

Cronin, J., Pizzarello, S. and Cruikshank, D. P. (1988) Organic matter in carbonaceous chondrites, planetary satellites, asteroids, and comets. In: *Meteorites and the Early Solar System* (J. F. Kerridge and M. S. Mathews, eds.), University of Arizona Press, Tempe, AZ, pp. 819–57.

Crowder, R. E. (1960) Hyperion oil field. In: *Summary of Operations, California Oil Fields*, Vol. 46, Department of Natural Resources, Division of Oil and Gas, San Francisco, pp. 86–91.

Curiale, J. A. (1986) Origin of solid bitumens, with emphasis on biological marker results. *Organic Geochemistry*, 10, 559–80.

(1995) Saturated and olefinic terrigenous triterpenoid hydrocarbons in a biodegraded Tertiary oil of northeast Alaska. *Organic Geochemistry*, 23, 177–82.

Curiale, J. A. and Sperry, S. W. (1998) An isotope-based oil-source rock correlation in the Camamu-Almada

Basin, offshore Brazil. *Revista Latino Americana de Geoquimica Organica*, 4, 51–64.

Curiale, J. A., Cameron, D. and Davis, D. V. (1985) Biological marker distribution and significance in oils and rocks of the Monterey Formation, California. *Geochimica et Cosmochimica Acta*, 49, 271–88.

Curiale, J., Morelos, J., Lambiase, J. and Mueller, W. (2000) Brunei Darussalam – characteristics of selected petroleums and source rocks. *Organic Geochemistry*, 31, 1475–93.

Curran, R., Eglinton, G., Maclean, I., Douglas, A. G. and Dungworth, G. (1968) Simplification of complex mixtures of alkanes using 7A molecular sieve. *Tetrahedron Letters*, 14, 1669–73.

Curtis, C. D. (1987) Inorganic geochemistry and petroleum exploration. In: *Advances in Petroleum Geochemistry*, Vol. 2 (J. Brooks and D. Welte, eds.), Academic Press, London, pp. 91–140.

Dahl, J. E., Moldowan, J. M., Peters, K. E., *et al.* (1999) Diamondoid hydrocarbons as indicators of natural oil cracking. *Nature*, 399, 54–7.

Dahl, J. E., Liu, S. G. and Carlson, R. M. K. (2002) Isolation and structure of higher diamondoids, nanometer-sized diamond molecules. *Science*, 299, 96–9.

Dahl, J. E., Moldowan, J. M., Peakman, T. M., *et al.* (2003) Isolation and structural proof of the large diamond molecule, cyclohexamantane ($C_{26}H_{30}$). *Angewandte Chemie International Edition*, 42, 2040–4.

Dai, J. (1992) Identification and distribution of various alkane gases. *Science in China. Series D, Earth Sciences*, 35, 1246–57.

Dasgupta, S., Tang, Y., Moldowan, J. M., Carlson, R. M. K. and Goddard, W. A., III (1995) Stabilizing the boat conformation of cyclohexane rings. *Journal of the American Chemical Society*, 117, 6532–4.

David, R. (2000) Mummification. In: *Ancient Egyptian Materials and Technology* (P. Nicholson and I. Shaw, eds.), Cambridge University Press, Cambridge, pp. 372–89.

David, R. and Sandra, P. (1999) Use of hydrogen as carrier gas in capillary GC. *American Laboratory*, 9, 18–9.

Dean, R. A. and Whitehead, E. V. (1961) The occurrence of phytane in petroleum. *Tetrahedron Letters*, 21, 768–70.

Decavallas, O., Garnier, N. and Regert, M. (2002) Chemical characterisation of plant commodities in archaeological ceramic vessels. Presented at the 33rd International Symposium on Archaeometry, April 22–25, 2002, Amsterdam.

Decola, E. (2000) *International Oil Spill Statistics: 1999*. Cutter Information Corporation, Arlington, MA.

Deines, E. T. (1980a) Biogeochemistry of stable carbon isotopes. In: *Organic Geochemistry* (G. Eglinton and M. T. J. Murphy, eds.), Springer-Verlag, New York, pp. 306–29.

Deines, P. (1980b) The isotopic composition of reduced organic carbon. In: *Handbook of Environmental Isotope Geochemistry*, Vol. 1 (P. Fritz and J. C. Fontes, eds.), Elsevier, Amsterdam, pp. 329–406.

De Leeuw, J. W., Cox, H. C., van Graas, G., *et al.* (1989) Limited double bond isomerization and selective hydrogenation of sterenes during early diagenesis. *Geochimica et Cosmochimica Acta*, 53, 903–9.

Del Río, J. C. and Philp, R. P. (1992) High molecular weight hydrocarbon ($>C_{40}$) in source rock extracts. *American Association of Petroleum Geologists Bulletin, Annual Meeting Abstracts*, 76, 1097.

(1999) Field ionization mass spectrometric study of high molecular weight hydrocarbons in a crude oil and a solid bitumen. *Organic Geochemistry*, 30, 279–86.

Demaison, G. J. and Huizinga, B. J. (1994) Genetic classification of petroleum systems using three factors: charge, migration, and entrapment. In: *The Petroleum System – From Source to Trap* (L. B. Magoon and W. G. Dow, eds.), American Association of Petroleum Geologists, Tulsa, OK, pp. 73–89.

Demaison, G. J. and Moore, G. T. (1980) Anoxic environments and oil source bed genesis. *American Association of Petroleum Geologists Bulletin*, 64, 1179–209.

Demaison, G. and Murris, R. J. (1984) *Petroleum Geochemistry and Basin Evaluation*. American Association of Petroleum Geologists, Tulsa, OK.

Demaison, G., Holck, A. J. J., Jones, R. W. and Moore, G. T. (1983) Predictive source bed stratigraphy; a guide to regional petroleum occurrence. In: *Proceedings of the 11th World Petroleum Congress*, Vol. 2, John Wiley & Sons, London, pp. 1–13.

Demirel, I. H., Yurtsever, T. S. and Guneri, S. (2001) Petroleum systems of the Adiyaman region, Southeastern Anatolia, Turkey. *Marine and Petroleum Geology*, 18, 391–410.

Dempster, H. S., Sherwood Lollar, B. and Feenstra, S. (1997) Tracing organic contaminants in groundwater: a new methodology using compound-specific isotopic analysis. *Environmental Science & Technology*, 31, 3193–7.

DeNiro, M. J. and Epstein, J. (1978) Influence of diet on the distribution of carbon isotopes in animals. *Geochimica et Cosmochimica Acta*, 42, 495–506.

Derenne, S., Largeau, C. and Taulelle, F. (1993) Occurrence of non-hydrolysable amides in the macromolecular constituent of *Scenedesmus quadricauda* cell wall as revealed by ^{15}N NMR: origin of *n*-alkylnitriles in pyrolysates of ultralaminae-containing kerogens. *Geochimica et Cosmochimica Acta*, 57, 851–7.

Derenne, S., Largeau, C. and Behar, F. (1994) Low polarity pyrolysis products of Permian to Recent *Botryococcus*-rich sediments; first evidence for the contribution of an isoprenoid to kerogen formation. *Geochimica et Cosmochimica Acta*, 58, 3703–11.

De Rosa, M., Gambacorta, A., Nicolaus, B., Sodano, S. and Bu'lock, J. D. (1980) Structural regulations in tetraetherlipids of *Caldariella* and their biosynthetic and phyletic implications. *Phytochemistry*, 19, 833–6.

De Rosa, M., Trincone, A., Nicolaus, B. and Gambacorta, A. (1991) Achaebacteria: lipids, membrane structures and adaptation to environmental stresses. In: *Life Under Extreme Conditions* (G. di Prisco, ed.), Spinger-Verlag, Berlin, pp. 61–87.

Des Marais, D. J., Donchin, J. H., Nehring, N. L. and Truesdell, A. H. (1981) Molecular carbon isotopic evidence for the origin of geothermal hydrocarbons. *Nature*, 292, 826–8.

Des Marais, D. J., Stallard, M. L., Nehring, N. L. and Truesdell, A. H. (1988) Carbon isotope geochemistry of hydrocarbons in the Cerro Prieto geothermal field, Baja California Norte, Mexico. *Chemical Geology*, 71, 159–67.

Devon, T. K. and Scott, A. I. (1972) *Handbook of Naturally Occurring Compounds*, Vol. II. Academic Press, New York.

De Vivo, B. and Frezzotti, M. L. (1994) *Fluid Inclusions in Minerals: Methods and Applications*. Short Course of the IMA Working Group "Inclusions in Minerals", Virginia Tech., Blacksberg, VA, p. 376.

Dias, R. F., Freeman, K. H. and Franks, S. G. (2002) Gas chromatography-pyrolysis-isotope ratio mass spectrometry: a new method for investigating intramolecular isotopic variation in low molecular weight organic acids. *Organic Geochemistry*, 33, 161–8.

Dimmler, A. and Strausz, O. P. (1983) Enrichment of polycyclic terpenoid, saturated hydrocarbons from petroleum by adsorption on zeolite NaX. *Journal of Chromatography*, 270, 219–25.

Dlugokencky, E. J., Masarie, K. A., Lang, L. M. and Tans, P. M. (1998) Continuing decline in the growth rate of the atmospheric methane burden. *Nature*, 393, 447–50.

DOE/EIA (1995) *Oil and Gas Resources of the Fergana Basin (Uzbekistan, Tadzhikistan, Kyrgyzstan)*. US Department of Energy/Energy Information Administration, Report No. DOE/EIA-TR/0575, Washington, D.C.

Dott, R. H. (1969) Hypotheses for an organic origin. In: *Sourcebook for Petroleum Geology, Part 1. Genesis of Petroleum* (R. H. Dott and M. J. Reynolds, eds), American Association of Petroleum Geologists, Tulsa, OK, pp. 1–244.

Douglas, A. G., Sinninghe Damsté, J. S., Fowler, M. G., Eglinton, T. I. and de Leeuw, J. W. (1991) Unique distributions of hydrocarbons and sulphur compounds released by flash pyrolysis from the fossilized alga *Gloecapsomorpha prisca*, a major constituent in one of four Ordovician kerogens. *Geochimica et Cosmochimica Acta*, 55, 275–91.

Douglas, G. S., Prince, R. C., Butler, E. L. and Steinhauer, W. G. (1994) The use of internal chemical indicators in petroleum and refined products to evaluate the extent of biodegradation. In: *Hydrocarbon Bioremediation* (R. E. Hinchee, B. C. Alleman, R. E. Hoeppel, and R. N. Miller, eds.), Lewis Publishers, Ann Arbor, MI, pp. 59–72.

Douglas, A. G., Bence, A. E., McMillen, S. J., Prince, R. C. and Butler, E. L. (1996) Environmental stability of selected petroleum hydrocarbon source and weathering ratios. *Environmental Science & Technology*, 30, 2332–9.

Douka, E., Koukkou, A., Drainas, C., Grosdemange-Billiard, C. and Rohmer, M. (2001) Structural diversity of the triterpenic hydrocarbons from the bacterium *Zymomonas mobilis*: the signature of defective squalene cyclization by the squalene/hopene cyclase. *FEMS Microbiology Letters*, 199, 247–51.

Dow, W. G. (1977) Kerogen studies and geological interpretations. *Journal of Geochemical Exploration*, 7, 79–99.

Duarte, C., Maurício, J., Pettitt, P. B., *et al.* (1999) The early Upper Paleolithic human skeleton from the Abrigo do Lagar Velho (Portugal) and modern human emergence in Iberia. *Proceedings of the National Academy of Science, USA*, 96, 7604–9.

Dudd, S. N. and Evershed, R. P. (1998) Direct demonstration of milk as an element of archaeological economies. *Science*, 282, 1478–81.

Dudd, S. N., Regert, M. and Evershed, R. P. (1998) Assessing microbial lipid contributions during laboratory degradations of fats and oils and pure triacylglycerols absorbed in ceramic potsherds. *Organic Geochemistry*, 29, 1345–54.

Durand, B. (1980) *Kerogen. Insoluble Organic Matter From Sedimentary Rocks*. Editions Technip, Paris.

(1983) Present trends in organic geochemistry in research on migration of hydrocarbons. In: *Advances in Organic Geochemistry 1981* (M. Bjorøy, C. Albrecht, C. Cornford, *et al.*, eds.), John Wiley & Sons, New York, pp. 117–28.

Durand, B. and Monin, J. C. (1980) Elemental analysis of kerogens (C,H,O,N,S,Fe). In: *Kerogen. Insoluble Organic Matter from Sedimentary Rocks* (B. Durand, ed.), Editions Technip, Paris, pp. 113–42.

Dutkiewicz, A., Rasmussen, B. and Buick, R. (1998) Oil preserved in fluid inclusions in Archean sandstones. *Nature*, 395, 885–8.

Dzou, L. I. P. and Hughes, W. B. (1993) Geochemistry of oils and condensates, K Field, offshore Taiwan: a case study in migration fractionation. *Organic Geochemistry*, 20, 437–62.

Eganhouse, R. P. (1997) *Molecular Markers in Environmental Geochemistry*. American Chemical Society, Washington, DC.

Eglinton, G. and Calvin, M. (1967) Chemical fossils. *Scientific American*, 216, 32–43.

Eglinton, G. and Hamilton, R. J. (1967) Leaf epicuticular waxes. *Science*, 156, 1322–35.

Eglinton, T. I. and Douglas, A. G. (1988) Quantitative study of biomarker hydrocarbons released from kerogens during hydrous pyrolysis. *Energy & Fuels*, 2, 81–8.

Eglinton, G., Scott, P. M., Besky, T., Burlingame, A. L. and Calvin, M. (1964) Hydrocarbons of biological origin from a one-billion-year-old sediment. *Science*, 145, 263–4.

Eglinton, T. I., Curtis, C. D. and Rowland, S. J. (1987) Generation of water-soluble organic acids from kerogen during hydrous pyrolysis: implications for porosity development. *Mineralogical Magazine*, 51, 495–503.

Ekweozor, E. M., and Daukoru, E. M. (1994) Northern delta depobelt portion of the Akata-Agbada(!) petroleum system, Niger Delta, Nigeria. In: *The Petroleum System – From Source to Trap* (L. B. Magoon and W. G. Dow, eds.), American Association of Petroleum Geologists, Tulsa, OK, pp. 599–614.

Elhmmali, M. M. (1998) Complementary use of bile acids and sterols as sewage pollution indicators. Ph. D. thesis, University of Bristol, Bristol, UK.

Elhmmali, M. M., Roberts, D. J. and Evershed, R. P. (2000) Combined analysis of bile acids and sterols/stanols from riverine particulates to assess sewage discharges and other fecal sources. *Environmental Science & Technology*, 34, 39–46.

Ellis, L. (1995) Aromatic hydrocarbons in crude oil and sediments: Molecular sieve separations and biomarkers. Ph. D. thesis, Curtin University of Technology, Perth, Australia.

Ellis, L. and Fincannon, A. L. (1998) Analytical improvements in IRM-GC/MS analyses: advanced techniques in tube furnace design and sample preparation *Organic Geochemistry*, 29, 1101–17.

Ellis, L., Kagi, R. I. and Alexander, R. (1992) Separation of petroleum hydrocarbons using dealuminated mordenite molecular sieve. I. Monoaromatic hydrocarbons. *Organic Geochemistry*, 18, 587–93.

Ellis, L., Alexander, R. and Kagi, R. I. (1994) Separation of petroleum hydrocarbons using dealuminated mordenite molecular sieve. II. Alkylnaphthalenes and alkylphenanthrenes. *Organic Geochemistry*, 21, 849–55.

Elvert, M., Suess, E., Greinert, J. and Whiticar, M. J. (2000) Archaea mediating anaerobic methane oxidation in deep-sea sediments at cold seeps of the eastern Aleutian subduction zone. *Organic Geochemistry*, 31, 1175–87.

Emerson, S. (1985) Organic carbon preservation in marine sediments. In: *The Carbon Cycle and Atmospheric CO_2: Natural Variations from Archean to Present* (E. T. Sundquist and W. S. Broecker, eds.), American Geophysical Union, Washington, DC, pp. 78–86.

Endo, K., Walton, D., Urry, G. B. C. and Reyment, R. A. (1995) Fossil intra-crystalline biomolecules of brachiopod shells: diagenesis and preserved geo-biological information. *Organic Geochemistry*, 23, 661–73.

Engel, M. H. and Maynard, R. J. (1989) Preparation of organic matter for stable carbon isotope analysis by sealed tube combustion: a cautionary note. *Analytical Chemistry*, 61, 1996–8.

England, W. A. (1990) The organic geochemistry of petroleum reservoirs. *Organic Geochemistry*, 16, 415–25.

England, W. A. and Fleet, A. J. (1991) *Petroleum Migration*, Geological Society, London.

Engelhardt, G. and Michel, D. (1987) *High Resolution Solid State NMR of Silicates and Zeolites*. John Wiley & Sons, New York.

Ensminger, A. (1977) Evolution de composes polycycliques sedimentaires [in French]. Doctorate thesis, University Louis Pasteur, Strasbourg, France.

EPA (1995) *Musts for USTs: A Summary of the Federal Regulations for Underground Storage Tank Systems*. EPA 510-K-95–002, July 1995, Environmental Protection Agency, Washington, DC.

   (1999) *Estimates of Methane Emissions from the US Oil Industry*. Final draft, October 1999, Prepared by ICF Consulting for the US Environmental Protection Agency, Washington, DC.

Epstein, A. G., Epstein, J. B. and Harris, L. D. (1977) *Conodont Color Alteration: An Index to Organic Metamorphism*. Geological Survey Professional Paper 995, US Geological Survey, Washington, DC.

Erdman, J. G. and Morris, D. A. (1974) Geochemical correlation of petroleum. *American Association of Petroleum Geologists Bulletin*, 58, 2326–37.

Espitalié, J., Madec, M., Tissot, B. and Leplat, P. (1977) Source rock characterization method for petroleum exploration. In: *Proceedings of the Offshore Technology Conference, May 2–5, 1977*, OTC, Houston, TX, pp. 439–44.

Espitalié, J., Marquis, F. and Sage, L. (1987) Organic geochemistry of the Paris Basin. In: *Petroleum Geology of Northwest Europe* (J. Brooks and K. Glennie, eds.), Graham and Trotman, London, pp. 71–86.

Etminan, H. and Hoffmann, C. F. (1989) Biomarkers in fluid inclusions: a new tool in constraining source regimes and its implications for the genesis of Mississippi Valley-type deposits. *Geology (Boulder)*, 17, 19–22.

Evans, S. A. (1928) *The Palace of Minos: A Comparative Account of the Successive Stages*. Macmillan, London.

Evershed, R. P. (1992) Chemical composition of bog body adipocere. *Archaeometry*, 34, 253–65.

Evershed, R. P. and Bethell, P. H. (1998) Application of multimolecular biomarker techniques to the identification of fecal material in archaeological soils and sediments. In: *Archaeological Chemistry, 23, ACS Symposium Series 625* (M. V. Orna, ed.), American Chemical Society, Washington, DC, pp. 157–72.

Evershed, R. P. and Connolly, R. C. (1994) Post-mortem transformations of sterols in bog body tissues. *Journal of Archaeological Science*, 21, 577–83.

Evershed, R. P. and Tuross, N. (1996) Proteinaceous material from potsherds and associated soils. *Journal of Archaeological Science*, 23, 429–36.

Evershed, R. P., Jerman, K. and Eglinton, G. (1985) Pine wood origin for pitch from the *Mary Rose*. *Nature*, 314, 528–30.

Evershed, R. P., Heron, C. and Goad, L. J. (1990) Analysis of organic residues of archaeological origin by high temperature gas chromatography/mass spectrometry. *Analyst*, 115, 1339–42.

   (1991) Epicuticular wax components preserved in potsherds as chemical indicators of leafy vegetables in ancient diets. *Antiquity*, 65, 540–4.

Evershed, R. P., Arnot, K. I., Collister, J., Eglinton, G. and Charters, S. (1994) Application of isotope ratio monitoring gas chromatography-mass spectrometry to the analysis of organic residues of archaeological origin. *Analyst*, 119, 909–14.

Evershed, R. P., Stott, A. W., Raven, A., *et al.* (1995) Formation of long-chain ketones in ancient pottery vessels by pyrolysis of acyl lipids. *Tetrahedron Letters*, 36, 8875–8.

Evershed, R. P., Bethell, P. H., Reynolds, R. J. and Walsh, N. J. (1997a) 5β-Stigmastanol and related 5β-stanols as biomarkers of manuring: analysis of modern experimental material and assessment of the archaeological potential. *Journal of Archaeological Science*, 24, 485–95.

Evershed, R. P., Mottram, H. R., Dudd, S. N., *et al.* (1997b) New criteria for the identification of animal fats preserved in archaeological pottery. *Naturwissenschaften*, 84, 402–6.

Evershed, R. P., van Bergen, P. F., Peakman, T. M., *et al.* (1997c) Archaeological frankincense. *Nature*, 390, 667–8.

Evershed, R. P., Vaughan, S. J., Dudd, S. N. and Soles, J. S. (1997d) Fuel for thought? Beeswax in lamps and conical cups from Late Minoan Crete. *Antiquity*, 71, 979–85.

Evershed, R. P., Bland, H. A., van Bergen, P. F., *et al.* (1997e) Volatile compounds in archaeological plant remains and the Maillard reaction during decay of organic matter. *Science*, 278, 432–3.

Evershed, R. P., Dudd, S. N., Charters, S., *et al.* (1999) Lipids as carriers of anthropogenic signals from prehistory. *Philosophical Transactions of the Royal Society, Biological Sciences*, 354, 19–32.

Evershed, R. P., Dudd, S. N., Anderson-Stojanovic, V. R. and Gebhard, E. R. (2003) New chemical evidence for the use of combed ware pottery vessels as beehives in ancient Greece. *Journal of Archaeological Science*, 30, 1–12.

Ewbank, G., Manning, D. A. C. and Abbott, G. D. (1993) An organic geochemical study of bitumens and their potential source rocks from the South Pennine Orefield, Central England. *Organic Geochemistry*, 20, 579–98.

Farrimond, P., Eglinton, G., Brassell, S. C. and Jenkyns, H. C. (1989) Toarcian anoxic event in Europe: an organic geochemical study. *Marine and Petroleum Geology*, 6, 136–47.

Farrimond, P., Head, I. M. and Innes, H. E. (2000) Environmental influence on the biohopanoid composition of Recent sediments. *Geochimica et Cosmochimica Acta*, 64, 2985–92.

Farrington, J. W. and Meyers, P. A. (1975) Hydrocarbons in the marine environment. In: *Environmental Chemistry* (G. Eglinton, ed.), The Chemical Society, London, pp. 109–36.

Faulon, J. L., Carlson, G. A. and Hatcher, P. G. (1993) Statistical model for bituminous coal: a three-dimensional evaluation of structural and physical properties based on computer-generated structures. *Energy & Fuels*, 7, 1062–72.

Faure, P., Landais, P., Schlepp, L. and Michels, R. (2000) Evidence for diffuse contamination of river sediments by road asphalt particles. *Environmental Science & Technology*, 34, 1174–81.

Feazel, C. T. and Aram, R. B. (1990) Interpretation of discontinuous reflectance profiles. Discussion. *American Association of Petroleum Geologists Bulletin*, 74, 91–3.

Filby, R. H. and Berkel, G. J. V. (1987) Geochemistry of metal complexes in petroleum, source rocks, and coals: an overview. In: *Metal Complexes in Fossil Fuels* (R. H. Filby and J. F. Branthaven, eds.), American Chemical Society, Washington, DC, pp. 2–39.

Filby, R. H. and Branthaven, J. F. (1987) *Metal Complexes in Fossil Fuels*. American Chemistry Society, Washington, DC.

Filley, T. R., Blanchette, R. A., Simpson, E. and Fogel, M. L. (2001) Nitrogen cycling by wood decomposing soft-rot fungi in the "King Midas tomb," Gordion, Turkey *Proceedings of the National Academy of Science, USA*, 98, 13346–50.

Fingas, M. F. (1995a) A literature review of the physics and predictive modeling of oil spill evaporation. *Journal of Hazardous Materials*, 42, 157–75.

(1995b) The evaporation of oil spills: variations with temperature and correlation with distillation data. *Journal of Hazardous Materials*, 42, 29–72.

Fischer, F. and Tropsch, H. (1926) Über die direkte synthese von erdöl-kohlenwasserstoffen bei gewöhnlichem druck. *Berichte der Deutschen Chemischen Gesellschaft*, 59, 830–1.

Fischer, P., Aichholz, R., Boelz, S., Juza, M. and Krimmer, S. (1990) Chiral recognition in capillary gas chromatography. 3. Polysiloxane-bound permethyl-β-cyclodextrin – a chiral stationary phase with broad application in gas-chromatographic enantiomer separation. *Angewandte Chemie International Edition*, 29, 427.

Fisher, K., Largeau, C. and Derenne, S. (1996a) Can oil shales be used to produce fullerenes? *Organic Geochemistry*, 24, 715–23.

Fisher, S. J., Alexander, R., Ellis, L. and Kagi, R. I. (1996b) The analysis of dimethylphenanthrenes by direct deposition gas chromatography-Fourier transform infrared spectroscopy (GC-FTIR). *Polycyclic Aromatic Compounds*, 9, 257–64.

Fisher, S. J., Alexander, R., Kagi, R. I. and Oliver, G. A. (1998) Aromatic hydrocarbons as indicators of biodegradation in north Western Australian reservoirs. In: *Sedimentary Basins of Western Australia: West Australian Basins Symposium* (P. G. Purcell and R. R. Purcell, eds.), Petroleum Exploration Society of Australia, WA Branch, Perth, Australia, pp. 185–94.

Fisher, E., Oldfield, F., Wake, R., *et al.* (2003) Molecular marker records of land use change. *Organic Geochemistry*, 34, 105–19.

Fishman, N. S., Ridgley, J. L., Hall, D. L. and Lillis, P. G. (2002) Timing of biogenic methane generation in Cretaceous rocks of the Northern Great Plains, Southeastern Alberta and Southwestern Saskatchewan: petrologic and fluid inclusion evidence. Presented at the Annual Meeting of the American Association of Petroleum Geologists, March 10–13, 2002, Houston, TX.

Flanigen, E. M., Bennett, J. M., Grosee, R. W., *et al.* (1978) Silicalite, a new hydrophobic crystalline silica molecular sieve. *Nature*, 271, 512–6.

Flannery, M. B., Stankiewicz, B. A., Hutchins, J. C., White, C. W. and Evershed, R. P. (1999) Chemical and morphological changes in human skin during preservation in waterlogged and desiccated environments. *Ancient Biomolecules*, 3, 37–50.

Fogel, M. L. and Tuross, N. (2003) Extending the limits of paleodietary studies of humans with compound specific carbon isotope analysis of amino acids. *Journal of Archaeological Science*, 30, 535–45.

Fogel, M. L., Tuross, N., Johnson, B. J. and Miller, G. H. (1997) Biogeochemical record of ancient humans. *Organic Geochemistry*, 27, 275–87.

Ford, T. B. D. (1968) Field meeting to Charnwood Forest, Leicestershire. *Proceedings of the Yorkshire Geological Society*, 45, 67–9.

Forsman, J. P. and Hunt, J. M. (1958) Insoluble organic matter (kerogen) in sedimentary rocks of marine origin. In: *Habitat of Oil: A Symposium* (L. G. Weeks, ed.) American Association of Petroleum Geologists, Tulsa, OK, pp. 747–78.

Fouch, T. D., Nuccio, V. F., Anders, D. E., *et al.* (1994) Green River (!) petroleum system, Uinta Basin, Utah, USA. In: *The Petroleum System-From Source to Trap* (L. B. Magoon and W. G. Dow, eds.), American Association of Petroleum Geologists, Tulsa, OK, pp 399–421.

Fowler, M. G. and Brooks, P. W. (1990) Organic geochemistry as an aid in the interpretation of the history of oil migration into different reservoirs at the Hibernia K-18 and Ben Nevis I-45 wells, Jeanne d'Arc Basin, offshore eastern Canada. *Organic Geochemistry*, 16, 461–75.

Fowler, M. G. and McAlpine, K. D. (1995) The Egret Member, a prolific Kimmeridgian source rock from offshore eastern Canada. In: *Petroleum Source Rocks* (B. Katz, ed.), Springer-Verlag, Berlin, pp. 111–30.

Fowler, S. W., Readman, J. W., Oregioni, B., Villeneuve, J. P. and McKay, K. (1993) Petroleum hydrocarbons and trace metals in nearshore Gulf sediments and biota before and after the 1991 war: an assessment of temporal and spatial trends. *Marine Pollution Bulletin*, 27, 171–82.

Fowler, M. G., Brooks, P. W., Northcott, M., *et al.* (1994) Preliminary results from a field experiment investigating the fate of some creosote components in a natural aquifer. *Organic Geochemistry*, 22, 641–9.

Fox, P. A., Carter, J. F. and Farrimond, P. (1998) Analysis of bacteriohopanepolyols in sediment and bacterial extracts by high performance liquid chromatography/atmospheric pressure chemical ionization mass spectrometry. *Rapid Communications in Mass Spectrometry*, 12, 1–4.

Francois, R. (1987) A study of sulphur enrichment in the humic fraction of marine sediments during early diagenesis. *Geochimica et Cosmochimica Acta*, 51, 17–27.

Francu, J., Radke, M., Schaefer, R. G., *et al.* (1996) Oil-oil and oil-source rock correlations in the northern Vienna Basin and adjacent Carpathian Flysch Zone (Czech and Slovak area). In: *Oil and Gas in Alpidic Thrustbelts and Basins of Central and Eastern Europe* (G. Wessely and W. Liebl, eds.), Geological Society of London, London, pp. 343–53.

Frank, H. A., Young, A. J., Britton, G. and Cogdell, R. J. (2000) *The Photochemistry of Carotenoids*, Kluwer Academic Publishers, Dordrecht.

Franks, S. G., Dias, R. F., Freeman, K. H., *et al.* (2001) Carbon isotopic composition of organic acids in oil field waters, San Joaquin Basin, California, USA. *Geochimica et Cosmochimica Acta*, 65, 1301–10.

Freedman, P. A., Gillyon, E. C. P. and Jumeau, E. J. (1998) Design and application of a new instrument for GC-isotope ratio MS. *American Laboratory*, 20, 114–9.

Freeman, K. H. and Colarusso, L. A. (2001) Molecular and isotopic records of C_4 grassland expansion in the late Miocene. *Geochimica et Cosmochimica Acta*, 65, 1439–54.

Freeman, K. H., Hayes, J. M., Trendel, J. M. and Albrecht, P. (1990) Evidence from carbon isotope measurements for diverse origins of sedimentary hydrocarbons. *Nature*, 343, 254–6.

Frenkel, D. and Smit, B. (2001) *Understanding Molecular Simulation*, 2nd ed. Academic Press, San Diego, CA.

Froelich, P. N., Klinkhammer, G. P., Bender, M. L., *et al.* (1979) Early oxidation of organic matter in pelagic sediments of the eastern equatorial Atlantic: suboxic diagenesis. *Geochimica et Cosmochimica Acta*, 43, 1075–90.

Frolov, E. B., Smirnov, M. B., Melikhov, V. A. and Vanyukova, N. A. (1998) Olefins of radiogenic origin in crude oils. *Organic Geochemistry*, 29, 409–20.

Frysinger, G. S. and Gaines, R. B. (1999) Analysis of petroleum fuels by comprehensive two-dimensional gas chromatography with mass spectrometry detection (GC×GC/MS). Presented at the Pittsburgh Conference on Analytical Chemistry and Applied Spectroscopy, March 8, 1999, Orlando, FL.

(2001) Separation and identification of petroleum biomarkers by comprehensive two-dimensional gas chromatography. *Journal of Separation Science*, 24, 87–96.

Fu, J., Sheng, G., Peng, P., *et al.* (1986) Peculiarities of salt lake sediments as potential source rocks in China. *Organic Geochemistry*, 10, 119–26.

Fuex, A. N. (1977) The use of stable carbon isotopes in hydrocarbon exploration. *Journal of Geochemical Exploration*, 7, 155–88.

Futrell, J. H. (2000) Development of tandem mass spectrometry: one perspective. *International Journal of Mass Spectrometry*, 200, 495–508.

Fyfe, C. A., Gobbi, G. C., Klinowski, J., Thomas, J. M. and Ramdas, S. (1982) Resolving crystallographically distinct tetrahedral sites in silicalite and ZSM-5 by solid-state NMR. *Nature*, 296, 530–3.

Gaffney, J. S., Premuzic, E. T. and Manowitz, B. (1980) On the usefulness of sulfur isotope ratios in crude oil correlations. *Geochimica et Cosmochimica Acta*, 44, 135–9.

Gaines, R. B., Frysinger, G. S., Hendrick-Smith, M. S. and Stuart, J. D. (1999) Oil spill source identification by comprehensive two-dimensional gas chromatography. *Environmental Science & Technology*, 33, 2106–12.

Galimov, E. M. (1973) *Carbon Isotopes in Oil – Gas Geology* (translation from Russian). National Aeronautics and Space Administration, Washington, DC.

Galimov, E. M., Lopatin, N. V. and Espitalié, J. (1988) Oil-source properties of the Bazhenovskaya suite at Salym area, Western Siberia. *Geokhimiya*, 4, 467–78.

Gallegos, E. J. (1976) Analysis of organic mixtures using metastable transition spectra. *Analytical Chemistry*, 48, 1348–51.

Gallegos, E. J. and Moldowan, J. M. (1992) The effect of hold time on GC resolution and the effect of collision gas on mass spectra in geochemical "biomarker" research. In: *Biological Markers in Sediments and Petroleum* (J. M. Moldowan, P. Albrecht and R. P. Philp, eds.), Prentice-Hall, Englewood Cliffs, NJ, pp. 156–81.

Garcia-Asua, G., Lang, H. P., Cogdell, R. J. and Hunter, C. N. (1998) Carotenoid diversity: a modular role for the phytoene desaturase step. *Trends in Plant Science*, 3, 445–9.

Garnier, N. and Regert, M. (2002) Development of a new methodology to detect polyphenols, biomarkers of archaeological wine and grape seeds. Presented at the 33rd International Symposium on Archaeometry, April 22–25, 2002, Amsterdam.

Garnier, N., Cren-Olivé, C., Rolando, C. and Regert, M. (2002) Characterization of archaeological beeswax by electron ionization and electrospray ionization mass spectrometry. *Analytical Chemistry*, 74, 4868–77.

Garrett, R. M., Pickering, I. J., Haith, C. E. and Prince, R. C. (1998) Photooxidation of crude oils. *Environmental Science & Technology*, 32, 3719–23.

Gas Processors Association (1995) *Tentative Method for the Extended Analysis of Hydrocarbon Liquid Mixtures Containing Nitrogen and Carbon Dioxide by Temperature Programmed Gas Chromatography*. GPA Standard 2186–95.

Geigl, E. M. (2002) DNA preservation in 500,000 year-old fossils: hibernation in molecular niches? Presented at the 33rd International Symposium on Archaeometry, April 22–25, 2002, Amsterdam.

Gelin, F, De Leeuw, J. W., Sinninghe Damsté, J. S., *et al.* (1994) The similarity of chemical structures of soluble aliphatic polyaldehyde and insoluble algaenan in the green microalga *Botryococcus braunii* race A as revealed by analytical pyrolysis. *Organic Geochemistry*, 21, 423–35.

Gelpi, V., Schneider, H., Mann, J. and Oró, J. (1970) Hydrocarbons of geochemical significance in microscopic algae. *Phytochemistry*, 9, 603–12.

George, S. C., Krieger, F. W., Eadington, P. J., *et al.* (1997) Geochemical comparison of oil-bearing fluid inclusions and produced oil from the Toro sandstone, Papua New Guinea. *Organic Geochemistry*, 26, 155–73.

George, S. C., Eadington, P. J., Lisk, M. and Quezada, R. A. (1998a) Geochemical comparison of oil trapped in fluid inclusions and reservoired oil in Blackback Oilfield, Gippsland Basin, Australia. *PESA (Petroleum Exploration Society of Australia) Journal*, 26, 64–81.

George, S. C., Lisk, M., Summons, R. E. and Quezada, R. A. (1998b) Constraining the oil charge history of the South Pepper oilfield from the analysis of oil-bearing fluid inclusions. *Organic Geochemistry*, 29, 631–48.

George, S. C., Ruble, T. E., Dutkiewicz, A. and Eadington, P. J. (2001) Assessing the maturity of oil trapped in fluid inclusions using molecular geochemistry data and visually-determined fluorescence colours. *Applied Geochemistry*, 16, 451–73.

GESAMP (2001) *A Sea of Troubles*. Joint Group of Experts on the Scientific Aspects of Marine Environmental Protection and Advisory Committee on Protection of the Sea, United Nations Environment Program, Report 70.

Gest, H. (1993) Photosynthetic and quasi-photosynthetic bacteria. *FEMS Microbiology Letters*, 112, 1–6.

Giardini, A. A. and Salotti, C. A. (1969) Kinetics and relations in the calcite-hydrogen reaction and relations in the dolomite-hydrogen and siderite-hydrogen systems. *American Mineralogist*, 54, 1151–72.

Gibbison, R., Peakman, T. M. and Maxwell, J. R. (1995) Novel porphyrins as molecular fossils for anoxygenic photosynthesis. *Tetrahedron Letters*, 36, 9057–60.

Gogou, A., Stratigakis, N., Kanakidou, M. and Stephanou, E. G. (1996) Organic aerosols in Eastern Mediterranean: components source reconciliation by using molecular markers and atmospheric back trajectories. *Organic Geochemistry*, 25, 79–96.

Gold, T. (1985) The origin of natural gas and petroleum and the prognosis for future supplies. *Annual Review of Energy*, 10, 53–77.

(1999) *The Deep Hot Biosphere*. Copernicus, New York.

Gold, T. and Soter, S. (1980) The deep-earth gas hypothesis. *Scientific American*, 242, 154–62.

(1982) Abiogenic methane and the origin of petroleum. *Energy Exploration and Exploitation*, 1, 89–104.

Goldhaber, M. B. and Orr, W. L. (1995) Kinetic controls on thermochemical sulfate reduction as a source of sedimentary H_2S. In: *Geochemical Transformations of Sedimentary Sulfur* (M. A. Vairavamurthy and M. A. A. Schoonen, eds.), American Chemical Society, Washington, DC, pp. 412–25.

Goldstein, T. P. and Aizenshtat, Z. (1994) Thermochemical sulfate reduction. A review. *Journal of Thermal Analysis*, 42, 241–90.

Goldstein, R. H. and Reynolds, T. J. (1994) *Systematics of Fluid Inclusions in Diagenetic Minerals*. Society for Sedimentary Geology, Tulsa, OK.

Goodwin, N. S., Park, P. J. D., and Rawlinson, T. (1983) Crude oil biodegradation. In: *Advances in Organic Geochemistry 1981* (M. Bjorøy, C. Albrecht, C. Cornford, *et al.*, eds.), John Wiley & Sons, New York, pp. 650–8.

Goossens, H., de Leeuw, J. W., Schenck, P. A. and Brassell, S. C. (1984) Tocopherols as likely precursors of pristane in ancient sediments and crude oils. *Nature*, 312, 440–2.

Goth, K., de Leeuw, J. W., Püttmann, W. and Tegelaar, E. W. (1988) Origin of Messel oil shale kerogen. *Nature*, 336, 759–61.

Gou, X., Fowler, M. G., Comet, P. A., *et al.* (1987) Investigation of three natural bitumens from central England by hydrous pyrolysis and gas chromatography-mass spectrometry. *Chemical Geology*, 64, 181–95.

Gough, M. A. and Rowland, S. J. (1990) Characterization of unresolved complex mixtures of hydrocarbons in petroleum. *Nature*, 344, 648–50.

Gransch, J. A. and Posthuma, J. (1974) On the origin of sulfur in crudes. In: *Advances of Organic Geochemistry 1973* (B. Tissot and F. Bienner, eds.), Editions Technip, Paris, pp. 727–39.

Grantham, P. J., Posthuma, J. and DeGroot, K. (1980) Variation and significance of the C_{27} and C_{28} triterpane content of a North Sea core and various North Sea crude oils. In: *Advances in Organic Geochemistry 1979* (A. G. Douglas and J. R. Maxwell, eds.), Pergamon Press, Oxford, UK, pp. 29–38.

Grice, K. (2001) $\delta^{13}C$ as an indicator of paleoenvironments: a molecular approach. In: *Application of Stable Isotope Techniques to Study Biological Processes and Functioning Ecosystems* (M. Unkovich, J. Pate, A. McNeill and J. Gibbs, eds.), Kluwer Scientific, Dordrecht, The Netherlands, pp. 247–81.

Grice, K., Schaeffer, P., Schwark, L. and Maxwell, J. R. (1997) Changes in palaeoenvironmental conditions during deposition of the Permian Kupferschiefer (Lower Rhine Basin, northwest Germany) inferred from molecular and isotopic compositions of biomarker components. *Organic Geochemistry*, 26, 677–90.

Grice, K., Schouten, S., Peters, K. E. and Sinninghe Damsté, J. S. (1998a) Molecular isotopic characterisation of hydrocarbon biomarkers in Palaeocene-Eocene evaporitic, lacustrine source rocks from the Jianghan Basin, China. *Organic Geochemistry*, 29, 1745–64.

Grice, K., Alexander, R. and Kagi, R. I. (2000) Diamondoid hydrocarbon ratios as indicators of biodegradation levels in Australian crude oils. *Organic Geochemistry*, 31, 67–73.

Grieve R. A. F. (1988) The formation of large impact structures and constraints on the nature of Siljan. In: *Deep Drilling in Crystalline Bedrock* (A. Boden and K. G. Eriksson, eds. Vol. 1, Springer-Verlag, New York, pp. 328–48.

Grimalt, J. O., Torras, E. and Albaigés, J. (1988) Bacterial reworking of sedimentary lipids during sample storage. *Organic Geochemistry*, 13, 741–6.

Grimalt, J. O., Fernandez, P., Bayona, J. M. and Albaigés, J. (1990) Assessment of faecal sterols and ketones as indicators of urban sewage inputs to coastal waters. *Environmental Science & Technology*, 24, 357–63.

Grob, K. (2001) *Split and Splitless Injection for Quantitative Gas Chromatography*. Wiley-VCH, New York.

Guadalupe, M. F. M., Castello branco, V. A. and Schmid, J. C. (1991) Isolation of sulfides in oils. *Organic Geochemistry*, 17, 355–61.

Guilhaumou, N., Ellouz, N., Jaswal, T. M. and Mougin, P. (2001) Genesis and evolution of hydrocarbons entrapped in the fluorite deposit of Koh-i-Maran, (North Kirthar Range, Pakistan). *Marine and Petroleum Geology*, 17, 1151–64.

Gülaçar, F. O., Susini, A. and Koln, M. (1990) Preservation of post-mortem transformation of lipids in samples from a 4000-year-old Nubian mummy. *Journal of Archaeological Science*, 17, 651–9.

Guthrie, J. M., Trindade, L. A. F., Eckardt, C. B. and Takaki, T. (1996) Molecular and carbon isotopic analysis of specific biological markers: evidence for distinguishing between marine and lacustrine depositional environments in sedimentary basins of Brazil. Presented at the Annual Meeting of the American Association of Petroleum Geologists, 1996, San Diego, CA.

Guthrie, J. M., Walters, C. C. and Peters, K. E. (1998) Comparison of micro-techniques used for analyzing oils in sidewall cores to model viscosity, API gravity and sulfur content. *American Association of Petroleum Geologists Bulletin*, 82, 1883–4.

Haber, C. L., Allen, L. N., Zhao, S. and Hanson, R. S. (1983) Methylotrophic bacteria: biochemical diversity and genetics. *Science*, 221, 1147–53.

Haddon, W. F. (1979) Computerized mass spectrometry linked scan system for recording metastable ions. *Analytical Chemistry*, 51, 983–8.

Hagelberg, E., Kayser, M., Nagy, M., *et al.* (1999) Molecular genetic evidence for the human settlement of the Pacific: analysis of mitochondrial DNA, Y chromosome and HLA markers. *Philosophical Transactions of the Royal Society London, Biological Sciences*, 354, 141–52.

Hairfield, H. H. and Hairfield, E. M. (1990) Identification of a late Bronze Age resin. *Analytical Chemistry*, 62, 41–5.

Halbouty, M. T. (1972) Rationale for deliberate pursuit of stratigraphic, unconformity, and paleogeomorphic traps. *American Association of Petroleum Geologists Bulletin*, 56, 537–41.

Hall, D. L., Bigge, M. A. and Jarvie, D. M. (2002a) Fluid inclusion evidence for alteration of crude oils. *2002 American Association of Petroleum Geologists Annual Convention, March 10–13, 2002, Houston, Texas, Abstract, p. A70.*

Hall, D. L., Sterner, S. M., Shentwu, W. and Bigge, M. A. (2002b) Applying fluid inclusions to petroleum exploration and production. American Association of Petroleum Geologists, Search and Discovery, article #40042, www.searchanddiscovery.net/documents/donhall/index.htm.

Halpern, H. I. (1995) Development and applications of light-hydrocarbon-based star diagrams. *American Association of Petroleum Geologists Bulletin*, 79, 801–15.

Hanin, S., Adam, P., Kowalewski, I., *et al.* (2002) Bridgehead alkylated 2-thiaadamantanes: novel markers for sulfurisation processes occurring under high thermal stress in deep petroleum reservoirs. *Chemical Communications – Royal Society of Chemistry*, 16, 1750–1.

Hare, P. E., Fogel, M. L., Stafford, T. W., Mitchell, A. and Hoering, T. C. (1991) The isotopic composition of carbon and nitrogen in individual amino acids isolated from modern and fossil proteins. *Journal of Archaeological Science*, 18, 277–92.

Harrell, J. A. and Lewan, M. D. (2002) Sources of mummy bitumen in ancient Egypt and Palestine. *Archaeometry*, 44, 285–93.

Harris, N. B., Freeman, K. H., Pancost, R. D., *et al.* (1998) The origin of lacustrine petroleum source rocks, Congo Basin, West Africa: preliminary results of a multidisciplinary study. *American Association of Petroleum Geologists Bulletin*, 82, 1922–3.

Harrison, O. R. (1991) An overview of the *Exxon Valdez* oil spill. In: Proceedings of the 1991 International Oil Spill Conference (Prevention, Behavior, Control, Cleanup), March 4–7, 1991, San Diego, California, American Petroleum Institute, Washington, DC, pp. 313–9.

Harvey, H. R. and McManus, G. B. (1991) Marine ciliates as a widespread source of tetrahymanol and hopan-3β-ol in sediments. *Geochimica et Cosmochimica Acta*, 55, 3387–90.

Haskell, N., Nissen, S., Hughes, M., *et al.* (1999) Delineation of geologic drilling hazards using 3-D seismic attributes. *The Leading Edge*, 18, 373–4, 376, 378, 381–2.

Hatch, J. R., Jacobson, J. R., Witzke, B. J., *et al.* (1987) Possible Middle Ordovician organic carbon isotope excursion: evidence from Ordovician oils and hydrocarbon source rocks, Mid-Continent, and East-Central United States. *American Association of Petroleum Geologists Bulletin*, 71, 1342–54.

Hatcher, P. G., Keister, L. E. and McGillivary, P. A. (1977) Steroids as sewage specific indicators in New York bight sediments. *Bulletin of Environmental Contamination and Toxicology*, 17, 491–8.

Hawes, I. and Schwarz, A.-M. (1995) Photosynthesis in benthic cyanobacterial mats from Lake Hoare, Antarctica. *Antarctic Journal of the United States* 30, 296–7.

Hayek, E. W. H., Krenmayer, P., Lonhinger, H., *et al.* (1990) Identification of archaeological and recent wood tar pitches using gas chromatography/mass spectrometry and pattern recognition. *Analytical Chemistry*, 62, 2038–43.

(1991) Gas chromatography/mass spectrometry and chemometrics in archaeometry. Investigation of glue on Copper Age arrowheads. *Fresenius' Journal of Analytical Chemistry*, 340, 153–6.

Hayes J. M., Kaplan, I. R. and Wedeking, K. M. (1983) Precambrian organic geochemistry, preservation of the record. In: *Earth's Earliest Biosphere, Its Origin and Evolution* (J. W. Schopf, ed.), Princeton University Press, Princeton, NJ, pp. 93–134.

Hayes, J. M., Takigiku, R., Ocampo, R., Callot, H. J. and Albrecht, P. (1987) Isotopic compositions and probable origins of organic molecules in the Eocene Messel shale. *Nature*, 329, 48–51.

Hayes, J. M., Freeman, K. H., Popp, B. N. and Hoham, C. H. (1990) Compound-specific isotopic analyses: a novel tool for reconstruction of ancient biogeochemical processes. *Organic Geochemistry*, 16, 1115–28.

Head, I. M. and Swannell, R. P. J. (1999) Bioremediation of petroleum hydrocarbon contaminants in marine habitats. *Current Opinion in Biotechnology*, 10, 234–9.

Hedberg, H. D. (1988) *The 1740 Description by Daniel Tilas of Stratigraphy and Petroleum Occurrence at Osmundsberg in the Siljan Region of Central Sweden*. American Association of Petroleum Geologists, Tulsa, OK.

Hedges, J. I. and Keil, R. G. (1995) Sedimentary organic matter preservation: an assessment and speculative synthesis. *Marine Chemistry*, 49, 81–115.

Helgesen, H. C., Knox, A. M., Owens, E. E. and Shock, E. L. (1993) Petroleum, oil field waters, and authigenic mineral assemblages: are they in metastable equilibrium in hydrocarbon reservoirs? *Geochimica et Cosmochimica Acta*, 57, 3295–339.

Henley, D. and Hoffmann, C. (1987) Complex hydrocarbons in fluid inclusion in gold and tin deposits; a new frontier for mineral exploration. *BMR Research Newsletter*, 6, 1–2.

Hermans, M. A. F., Neuss, B. and Sahm, H. (1991) Content and composition of hopanoids in *Zymomonas mobilis* under various growth conditions. *Journal of Bacteriology*, 173, 5592–5.

Hernes, P. J. and Hedges, J. I. (2000) Determination of condensed tannin monomers in environmental samples by capillary gas chromatography of acid depolymerization extracts. *Analytical Chemistry*, 72, 5115–24.

Heron, C., Evershed, R. P., Chapman, B. and Pollard, A.-M. (1991) Glue, disinfectant and 'chewing gum' in prehistory. In: *Archaeological Sciences 1989: Proceedings of a Conference on the Application of Scientific Techniques to Archaeology* (P. Budd, B. Chapman, C. Jackson, R. Janaway and B. Ottaway, eds.), Oxbow, Oxford, UK, pp. 325–31.

Heron, C., Nemcek, N., Bonfield, K. M., Dixon, D. and Ottaway, B. S. (1994) The chemistry of Neolithic beeswax. *Naturwissenschaften*, 81, 266–9.

Herrmann, D., Bisseret, P., Connan, J. and Rohmer, M. (1996) A non-extractable triterpanoid of the hopane series in *Acetobacter xylinum*. *FEMS Microbiology Letters*, 135, 323–6.

Hills, I. R. and Whitehead, E. V. (1966) Triterpanes in optically active petroleum distillates. *Nature*, 209, 977–9.

Hills, I. R., Whitehead, E. V., Anders, D. E., Cummins, J. J. and Robinson, W. E. (1966) An optically active triterpane, gammacerane in Green River, Colorado, oil shale bitumen. *Journal of the Chemical Society, Chemical Communications*, 20, 752–4.

Hinrichs, K.-U., Haver, J. M., Sylva, S. P., Brewer, P. G. and Delong, E. F. (1999) Methane-consuming archaebacteria in marine sediments. *Nature*, 398, 802–5.

Ho, T. Y., Rogers, M. A., Drushel, H. V. and Kroons, C. B. (1974) Evolution of sulfur compounds in crude oils. *American Association of Petroleum Geologists Bulletin*, 58, 2338–48.

Hoefs, J. (1997) *Stable Isotope Geochemistry*. Springer-Verlag, New York.

Hoering, T. C. and Freeman, D. H. (1984) Shape-selective sorption of monomethylalkanes by silicalite, a zeolite form of silica. *Journal of Chromatography*, 316, 333–41.

Hoffmann, C. F., Mackenzie, A. S., Lewis, C. A., *et al.* (1984) A biological marker study of coals, shales, and oils from the Mahakam Delta, Kalimantan, Indonesia. *Chemical Geology*, 42, 1–23.

Holba, A. G., Dzou, L. I. P., Masterson, W. D., (1998) Application of 24-norcholestanes for constraining source age of petroleum. *Organic Geochemistry*, 29, 1269–83.

Holba A. G., Ellis, L., Dzou, I. L., *et al.* (2001) Extended tricyclic terpanes as age discriminators between Triassic, Early Jurassic and Middle-Late Jurassic oils. Presented at the 20th International Meeting on Organic Geochemistry, 10–14 September, 2001, Nancy, France.

Hollander, D. J. and Mckenzie, J. A. (1991) CO_2 control on carbon-isotope fractionation during aqueous photosynthesis: a paleo-pCO_2 barometer. *Geology*, 19, 929–32.

Holloway, J. R. (1984) Graphite-CH_4-H_2O-CO_2 equilibria at low-grade metamorphic conditions. *Geology*, 12, 455–8.

Holm, N. G. and Charlou, J. L. (2001) Initial indications of abiotic formation of hydrocarbons in the Rainbow ultramafic hydrothermal system, Mid-Atlantic Ridge. *Earth and Planetary Science Letters*, 191, 1–8.

Honghan, C., Sitian, L., Yongchuan, S., and Qiming, Z. (1998) Two petroleum systems charge the YA13-1 gas field in Yinggehai and Qiongdongnan basins, South China Sea. *American Association of Petroleum Geologists Bulletin*, 82, 757–72.

Hoots, H. W., Blount, A. L. and Jones, P. H. (1935) Marine oil shale, source of oil in Playa del Rey Field, California. *American Association of Petroleum Geologists Bulletin*, 19, 172–205.

Horita, J. and Berndt, M. E. (1999) Abiogenic methane formation and isotopic fractionation under hydrothermal conditions. *Science*, 285, 1055–7.

Horsfield, B., Schenk, H. J., Mills, N. and Welte, D. H. (1992) An investigation of the in-reservoir conversion of oil to gas: compositional and kinetic findings from closed-system programmed-temperature pyrolysis *Organic Geochemistry*, 19, 191–204.

Horstad, I., Larter, S. R., Dypvik, H., *et al.* (1990) Degradation and maturity controls on oil Field petroleum column heterogeneity in the Gullfaks field, Norwegian North Sea. *Organic Geochemistry*, 16, 497–510.

Hostettler, F. D., Rosenbauer, R. J. and Kvenvolden, K. A. (1999) PAH refractory index as a source discriminant of hydrocarbon input from crude oil and coal in Prince William Sound, Alaska. *Organic Geochemistry*, 30, 873–9.

(2000) Response to comment by Bence *et al.* on "PAH refractory index as a source discriminant of hydrocarbon input from crude oil and coal in Prince William Sound, Alaska." *Organic Geochemistry*, 31, 939–943.

Hoyle, F. (1955) *Frontiers of Astronomy*. Heinemann, London.

Hu, G., Ouyang, Z., Wang, Z. and Wen, Q. (1998) Carbon isotopic fractionation in the process of Fischer–Tropsch reaction in primitive solar nebula. *Scientia Sinica*, 41, 202–7.

Huang, W.-Y. and Meinschein, W. G. (1978) Sterols in sediments from Baffin Bay, Texas. *Geochimica et Cosmochimica Acta*, 42, 1391–6.

(1979) Sterols as ecological indicators. *Geochimica et Cosmochimica Acta*, 43, 739–45.

Huc, A. Y. (1988a) Aspects of depositional processes of organic matter in sedimentary basins. *Organic Geochemistry*, 13, 263–72.

(1988b) Sedimentology of organic matter. In: *Humic Substances and Their Role in the Environment* (F. H. Frimmel and R. F. Christman, eds.), John Wiley & Sons, New York, pp. 215–43.

Huckins, J. N., Tubergen, M. W. and Manuweera, G. K. (1990) Semipermeable membrane devices containing model lipid: a new approach to monitoring the bioavailability of lipophilic contaminants and estimating their bioconcentration potential. *Chemosphere*, 20, 533–52.

Hughes, W. B. (1984) Use of thiophenic organosulfur compounds in characterizing crude oils derived from carbonate versus siliciclastic sources. In: *Petroleum Geochemistry and Source Rock Potential of Carbonate Rocks* (J. G. Palacas, ed.), American Association of Petroleum Geologists, Tulsa, OK, pp. 181–196.

Hughes, W. B., Holba, A. G., Mueller, D. E. and Richardson, J. S. (1985) Geochemistry of greater Ekofisk crude oils. In: *Geochemistry in Exploration of the Norwegian Shelf* (B. M. Thomas, ed.), Graham and Trotman, London, pp. 75–92.

Hughey, C. A., Rodgers, R. P., Marshall, A. G., Qian, K. and Robbins, W. K. (2002a) Identification of acidic NSO compounds in crude oils of different geochemical origins by negative ion electrospray Fourier transform ion cyclotron resonance mass spectrometry. *Organic Geochemistry*, 33, 743–59.

Hughey, C. A., Rodgers, R. P. and Marshall, A.G. (2002b) Resolution of 11 000 compositionally distinct components in a single electrospray ionization Fourier transform ion cyclotron resonance mass spectrum of crude oil. *Analytical Chemistry*, 36, 4145–9.

Hulen, J. B. and Collister, J. W. (1999) The oil-bearing, Carlin-type gold deposits of Yankee Basin, Alligator Ridge district, Nevada. *Economic Geology and the Bulletin of the Society of Economic Geologists*, 94, 1029–49.

Hunkeler, D., Andersen, N., Aravena, R., Bernasconi, S. M. and Butler, B. J. (2001) Hydrogen and carbon isotope fractionation during aerobic biodegradation of benzene. *Environmental Science & Technology*, 35, 3462–7.

Hunt T. S. (1863) *Report on the Geology of Canada.* Canadian Geological Survey report: progress to 1863.

Hunt, J. M. (1984) Generation and migration of light hydrocarbons. *Science*, 1226, 1265–70.

(1996) *Petroleum Geochemistry and Geology.* W. H. Freeman, New York.

Hunt, J. M., Miller, R. J. and Whelan, J. K. (1980a) Formation of C_4–C_7 hydrocarbons from bacterial degradation of naturally occurring terpenoids. *Nature*, 288, 577–8.

Hunt, J. M., Whelan, J. K. and Huc, A.-Y. (1980b) Genesis of petroleum hydrocarbons in marine sediments. *Science*, 209, 403–4.

Huq, N. L., Tseng, A. and Chapman, G. E., (1990) Partial amino acid sequence of osteocalcin from an extinct species of ratite bird. *Biochemistry International*, 21, 491–6.

Hurst, R. W. (2002) Lead isotopes as age-sensitive genetic markers in hydrocarbons. 3. Leaded gasoline, 1923–1990 (ALAS Model). *Environmental Geosciences*, 9, 43–50.

Hurst, R. W., Davis, T. E. and Chinn, B. D. (1996) The lead fingerprints of gasoline contamination. *Environmental Science & Technology*, 30, 304–7A.

Hurst, R. W., Barron, D., Washington, M. and Bowring, S. A. (2001) Lead isotopes as age-sensitive, genetic markers in hydrocarbons. 1. Copartitioning of lead with MTBE into water and implications for MTBE-source correlations. *Environmental Geosciences*, 8, 242–50.

Hurst, W. J., Tarka, S. M., Jr, Powis, T. G., Valdez, F., Jr and Hester, T. R. (2002) Cacao usage by the earliest Maya civilization. *Nature*, 418, 289–90.

Hutton, A. C. (1987) Petrographic classification of oil shales. *International Journal of Coal Geology*, 8, 203–31.

Hutton, A. C. and Cook, A. C. (1980) Influence of alginite on the reflectance of vitrinite from Joadja, NSW, and some other coals and oils shales containing alginite. *Fuel*, 59, 711–4.

Hutton, A. C., Kantsler, A. J., Cook, A. C. and Mckirdy, D. M. (1980) Organic matter in oil shales. *Journal of the Australian Petroleum Exploration Association*, 20, 44–67.

Hwang, R. J. (1990) Biomarker analysis using GC-MSD. *Journal of Chromatographic Science*, 28, 109–13.

Hwang R J., Sundararaman, P., Teerman, S. C. and Schoell, M. (1989) Effect of preservation on geochemical properties of organic matter in immature lacustrine sediments. Presented at the 14th International Meeting on Organic Geochemistry, September 18–22, 1989, Paris, France.

Ibach, L. E. J. (1982) Relationship between sedimentation rate and total organic carbon content in ancient marine sediments. *American Association of Petroleum Geologists Bulletin*, 66, 170–88.

IHS (Information Handling Services)/Petroconsultants S. A. (1996–99) Petroleum exploration and production database. Available from Petroconsultants, Inc., PO Box 740619, Houston, TX 77274–0619, USA.

Isaacs, C. M. (2001) Statistical evaluation of interlaboratory data from the cooperative Monterey organic geochemistry study. In: *The Monterey Formation: From Rocks to Molecules* (C. M. Isaacs and J. Rullkötter, eds.), Columbia University Press, New York, pp. 461–524.

Isaksen, G. H. and Bohacs, K. M. (1995) Geological controls on source rock geochemistry through relative sea level; Triassic, Barents Sea. In: *Petroleum Source Rocks* (B. J. Katz, ed.), Springer-Verlag, New York, pp. 25–50.

Isaksen, G. H., Pottorf, R. J. and Jenssen, A. I. (1998) Correlation of fluid inclusions and reservoired oils to infer trap fill history in the South Viking Graben, North Sea. *Petroleum Geoscience*, 4, 41–55.

Isaksen, G. H., Aliyev, A. A., Mamedova, S. A., *et al.* (1999) Geochemistry of organic-rich rocks from mud-volcano ejecta in Azerbaijan – a novel approach for regional assessment of source rock quality. Presented at the Geodynamics of the Black Sea–Caspian Segment of the Alpine Folded Belt International Conference, Baku, Azerbaijan, June 9–10, 1999.

Isaksson, S. (1998) A kitchen entrance to the aristocracy – analysis of lipid biomarkers in cultural layers. *Journal of Nordic Archaeological Science*, 10–11, 289–93.

Itoh, Y. H., Sugai, A., Uda, I. and Itoh, T. (2001) The evolution of lipids. *Advances in Space Research: the Official Journal of the Committee on Space Research (COSPAR)*, 28, 719–24.

ITOPF (2001) *ITOPF Handbook 2001/2002.* International Tanker Owners Pollution Federation Ltd, London.

Jacob, S. M., Quann, R. J., Sanchez, E. and Wells, M. E. (1998) Composition modeling reduces crude-analysis time, predicts yield. *Oil and Gas Journal*, 96, 51–6.

Jacobson, S. R., Hatch, J. R., Teerman, S. C. and Askin, R. A. (1988) Middle Ordovician organic matter assemblages and their effect on Ordovician-derived oils. *American Association of Petroleum Geologists Bulletin*, 72, 1090–100.

Jaffé, R., Albrecht, P. and Oudin, J. L. (1988a) Carboxylic acids as indicators of oil migration. I. Occurrence and geochemical significance of C-22 diastereoisomers of the $17\beta(H)$, $21\beta(H)$ C_{30} hopanoic acid in geological samples. *Organic Geochemistry*, 13, 483–8.

(1988b) Carboxylic acids as indicators of oil migration. II. Case of the Mahakam Delta, Indonesia. *Geochimica et Cosmochimica Acta*, 52, 2599–607.

James, A. T. (1983) Correlation of natural gas by use of carbon isotopic distribution between hydrocarbon components. *American Association of Petroleum Geologists Bulletin*, 67, 1176–91.

Jarvie, D. M. (1991) Total organic carbon (TOC) analysis. In: *Source and Migration Processes and Evaluation Techniques* (R. K. Merril, ed.), American Association of Petroleum Geologists, Tulsa, OK, pp. 113–8.

(2001) Williston Basin petroleum systems: inferences from oil geochemistry and geology. *Mountain Geologist*, 38, 19–42.

Jarvie, D. M. and Walker, P. R. (1997) Correlation of oils and source rocks in the Williston Basin using classical correlation tools and thermal extraction very high resolution C_7 gas chromatography. In: *Abstracts from the 18th International Meeting on Organic Geochemistry, September 22–26, 1997, Maastricht, the Netherlands* (B. Horsfield, ed.). Forschungszentrum Jülich, Germany, pp. 51–2.

Jarvie, D. M. and Lundell, L. L. (2001) Kerogen type and thermal transformation of organic matter in the Miocene Monterey Formation. In: *The Monterey Formation: From Rocks to Molecules* (C. M. Isaacs and J. Rullkötter, eds.), Columbia University Press, New York, pp. 268–95.

Jarvie, D. M., Morelos, A. and Zhiwen, H. (2001a) Detection of pay zones and pay quality, Gulf of Mexico: application of geochemical techniques. *Gulf Coast Association of Geological Societies Transactions*, 51, 151–60.

Jarvie, D. M., Claxton, B. L., Henk, F. and Breyer, J. T. (2001b) Oil and shale gas from the Barnett Shale, Fort Worth Basin, Texas. *American Association of Petroleum Geologists Bulletin*, 85, A100.

(1999) The increasing use of stable isotopes in the pharmaceutical industry. *Pharmaceutical Technology*, 23, 106–14.

(2001) Quantitative estimates of precision for molecular isotopic measurements. *Rapid Communications in Mass Spectrometry*, 15, 1554–7.

Jasra, R. V. and Bhat, S. G. (1987) Sorption kinetics of higher *n*-paraffins on zeolite molecular sieve 5A. *Indian Engineering and Chemical Research*, 26, 2544–6.

Jeffrey, A. W. A. and Kaplan, I. R. (1989) Drilling fluid additives and artifact hydrocarbon shows: examples from the Gravberg-1 well, Siljan Ring, Sweden. *Scientific Drilling*, 1, 63–70.

Jeffrey, A. W. A., Alimi, H. M. and Jenden, P. D. (1991) Geochemistry of Los Angeles Basin oil and gas systems. In: *Active Margin Basins* (K. T. Biddle, ed.), American Association of Petroleum Geologists, Tulsa, OK, pp. 197–219.

Jenden, P. D., Hilton, D. R., Kaplan, I. R. and Craig, H. (1993a) *Abiogenic Hydrocarbons and Mantle Helium in Oil and Gas Fields*. US Geological Survey Professional Paper 1570.

Jenden, P. D., Drozan, D. J. and Kaplan, I. R. (1993b) Mixing of thermogenic natural gases in northern Appalachian Basin. *American Association of Petroleum Geologists Bulletin*, 77, 980–98.

Jetten, M. S. M., Wagner, M., Fuerst, J. A., *et al.* (2001) Microbiology and application of the anaerobic ammonium oxidation ("anammox") process. *Current Opinion in Biotechnology*, 12, 283–8.

Johathan, D., l'Hote, G. and du Rochet, J. (1975) Analyse géochimiques des hydrocarbures léger per thermovaporisation. *Review Institut Français du Petrolé*, 30, 65–88.

Jomaa, H., Wiesner, J., Sanderbrand, S., *et al.* (1999) Inhibitors of the nonmevalonate pathway of isoprenoid biosynthesis as antimalarial drugs. *Science*, 285, 1573–6.

Jones, R. W. (1987) Organic facies. In: *Advances in Petroleum Geochemistry* (J. Brooks and D. Welte, eds.), Academic Press, New York, pp. 1–90.

Jones, R. W. and Edison, T. A. (1978) Microscopic observations of kerogen related to geochemical parameters with emphasis on thermal maturation. In: *Low Temperature Metamorphism of Kerogen and Clay Minerals* (D. F. Oltz, ed.), Society of Economic Paleontologists and Mineralogists, Los Angeles, pp. 1–12.

Jones, D. M. and Macleod, G. (2000) Molecular analysis of petroleum in fluid inclusions: a practical methodology. *Organic Geochemistry*, 31, 1163–73.

Jones, D. M., Douglas, A. G., Parkes, R. J., *et al.* (1983) The recognition of biodegraded petroleum-derived aromatic hydrocarbons in recent marine sediments. *Marine Pollution Bulletin*, 14, 103–8.

Jones, D. M., Macleod, G., Larter, S. R., *et al.* (1996) Characterization of the molecular composition of included petroleum. In: *PACROFI VI: Sixth Biennial Pan-American Conference on Research on Fluid Inclusions: Program and Abstracts.* (P. E. Brown and St. G. Hagemann, eds.), University of Wisconsin Press, Madison, WI pp. 64–5.

Jones, V., Ambrose, S. and Evershed, R. P. (2001) Tracing the routing and synthesis of amino acids using gas chromatography-combustion-isotope ratio mass spectrometry in palaeodietary reconstruction. Presented at the 221st National Meeting of the American Chemical Society, San Diego, CA, April 1–5, 2001.

Juvancz, Z., Alexander, B. and Bzejtll, S. (1987) Permethylated β-cyclodextrin as stationary phase in capillary gas chromatography. *Journal of High Resolution Chromatography*, 10, 105–7.

Kamioka, H., Shibata, K., Kajizuka, I. and Ohta, T. (1996) Rare-earth element patterns and carbon isotopic composition of carbonados: implications for their crustal origin. *Geochemistry Journal*, 30, 189–94.

Kannenberg, E. L. and Poralla, K. (1999) Hopanoid biosynthesis and function in bacteria. *Naturwissenschaften*, 86, 168–76.

Kaplan, I. R. (1975) Stable isotopes as a guide to biogeochemical processes. *Proceedings of the Royal Society of London*, 189, 183–211.

(1983) Stable isotopes of sulfur, nitrogen, and deuterium in recent marine environments. In: *Stable Isotopes in Sedimentary Geology, Society of Economic Paleontologists and Mineralogists (SEPM) Short Course 10* (M. A. Arthur, ed.), Society of Economic Paleontologists and Mineralogists, Tulsa, OK pp. 1–108.

(1989) Forensic geochemistry methods to trace sources of oil and gasoline pollution. *American Association of Petroleum Geologists Bulletin*, 73, 543.

Kaplan, I., Lu, S.-T., Lee, R.-P. and Warrick, G. (1996) Polycyclic hydrocarbon biomarkers confirm selective incorporation of petroleum in soil and kangaroo rat liver samples near an oil well blowout site in the western San Joaquin Valley, California. *Environmental Toxicology and Chemistry*, 15, 696–707.

Kaplan, I. R., Galperin, Y., Lu, S.-T. and Lee, R.-P. (1997) Forensic environmental geochemistry: differentiation of

fuel-types, their sources and release time. *Organic Geochemistry*, 27, 289–317.

Karlsen, D. A. and Larter, S. (1990) A rapid correlation method for petroleum population mapping within individual petroleum reservoirs: applications to petroleum reservoir description. In: *Correlation in Hydrocarbon Exploration* (J. D. Collinson, ed.), Graham and Trotman, London, pp. 77–85.

Karlson, D. A., Nedvitne, T., Larter, S. R. and Bjørlkke, K. (1993) Hydrocarbon composition of authigenic inclusions: application to elucidation of petroleum reservoir filling history. *Geochimica et Cosmochimica Acta*, 57, 3641–59.

Karner, M. B., Delong, E. F. and Karl, D. M. (2001) Archaeal dominance in the mesopelagic zone of the Pacific Ocean. *Nature*, 409, 507–10.

Katz, B. J. and Dawson, W. C. (1997) Pematang-Sihapas petroleum system of Central Sumatra. In: *Petroleum Systems of SE Asia and Australasia* (J. V. C. Howes and R. A. Noble, eds.), Indonesian Petroleum Association, Jakarta, pp. 685–98.

Katz, B. J., Pheifer, R. N. and Schunk, D. J. (1988) Interpretation of discontinuous vitrinite reflectance profiles. *American Association of Petroleum Geologists Bulletin*, 72, 926–31.

Katz, B. J., Robison, V. D., Dawson, W. C. and Elrod, L. W. (1994) Simpson-Ellenburger(.) petroleum system of the Central Basin Platform, West Texas, USA. In: *The Petroleum System – From Source to Trap* (L. B. Magoon and W. G. Dow, eds.), American Association of Petroleum Geologists, Tulsa, OK, pp. 453–62.

Katz, B. J., Dittmar, E. E. and Ehret, G. E. (2000a) A geochemical review of carbonate source rocks in Italy. *Journal of Petroleum Geology*, 23, 399–424.

Kaufman, R. L., Ahmed, A. S. and Hempkins, W. B. (1987) A new technique for the analysis of commingled oils and its application to production allocation calculations. In: *Proceedings of the Sixteenth Annual Convention of the Indonesian Petroleum Association*, Indonesian Petroleum Association Jakarta, Indonesia, pp. 247–68.

Kaufman, R. L., Ahmed, A. S. and Elsinger, R. J. (1990) Gas chromatography as a development and production tool for fingerprinting oils from individual reservoirs: applications in the Gulf of Mexico. In: *Proceedings of the 9th Annual Research Conference of the Society of Economic Paleontologists and Mineralogists* (D. Schumacher and B. F. Perkins, eds.), Society of Paleontologists and Mineralogists, Tulsa, OK, pp. 263–82.

Keely, B. J., Prowse, W. G. and Maxwell, J. R. (1990) The Treibs hypothesis: an evaluation based on structural studies. *Energy & Fuels*, 4, 628–34.

Kenig, F., Popp, B. N. and Summons, R. E. (2000) Preparative HPLC with ultrastable-Y zeolite for compound-specific carbon isotopic analyses. *Organic Geochemistry*, 31, 1087–94.

Kennedy, M. J., Pevear, D. R. and Hill, R. J. (2002) Mineral surface control of organic carbon in black shale. *Science*, 295, 657–60.

Kenney, J. F. (1996) Considerations about recent predictions of impending shortages of petroleum evaluated from the perspective of modern petroleum science. *Energy World*, 240, 16–18.

Kenney, J. F., Kutcherov, V. A., Bendeliani, N. A. and Alekseev, V. A. (2002) The evolution of multicomponent systems at high pressures. VI. The thermodynamic stability of the hydrogen–carbon system: the genesis of hydrocarbons and the origin of petroleum. *Proceedings of the National Academy of Science, USA*, 99, 10 976–81.

Kerr, G. T. (1989) Synthetic zeolites. *Scientific American*, 261, 82–7.

Kerr, R. A. (1990) When a radical experiment goes bust. *Science*, 247, 1177.

Kessler, A. and Baldwin, I. I. (2001) Defensive function of herbivore-injured plant volatile emissions in nature. *Science*, 291, 2141–4.

Kihle, J. (1995) Adaptation of fluorescence excitation-emission micro-spectroscopy for characterization of single hydrocarbon fluid inclusions. *Organic Geochemistry*, 23, 1029–42.

Killops, S. D. and Al-Juboori, M. A. H. A. (1990) Characterization of the unresolved complex mixture (UCM) in the gas chromatograms of biodegraded petroleums. *Organic Geochemistry*, 15, 147–60.

Kimpe, K., Jacobs, P. A. and Waelkens, M. (2001) Analysis of oil used in late Roman cooking lamps with different mass spectrometric techniques revealed in presence of predominantly olive oil together with traces of animal fat. *Journal of Chromatography A*, 937, 87–95.

King, W. J. (1988) Operating problems in the Hanlan Swan Hills gas field. Presented at the Society of Petroleum Engineers Gas Technology Symposium, June 13–15, 1988, Dallas, TX.

Kitson, F. G., Larsen, B. S. and McEwen, C. N. (1996) *Gas Chromatography and Mass Spectrometry*. Academic Press, New York.

Klemme, H. D. (1994) Petroleum systems of the world involving Upper Jurassic source rocks. In: *The Petroleum System – From Source to Trap* (L. B. Magoon and W. G. Dow, eds.), American Association of Petroleum Geologists, Tulsa, OK, pp. 51–72.

Klemme, H. D. and Ulmishek, G. F. (1991) Effective petroleum source rocks of the world: stratigraphic distribution and controlling depositional factors. *American Association of Petroleum Geologists Bulletin*, 75, 1809–51.

Knauss, K. G., Copenhaver, S. A., Braun, R. L. and Burnham, A. K. (1997) Hydrous pyrolysis of New Albany and Phosphoria shales: production kinetics of carboxylic acids and light hydrocarbons and interactions between the inorganic and organic chemical systems. *Organic Geochemistry*, 27, 477–96.

Knights, B. A., Dickson, C. A., Dickson, J. H. and Breeze, D. J. (1983) Evidence concerning the Roman military diet at Bearsden, Scotland, in the 2nd century A.D. *Journal of Archaeological Science*, 10, 139–52.

Knöss, W. and Reuter, B. (1998) Biosynthesis of isoprenic units via different pathways: occurrence and future prospects. *Pharmaceutica Acta Helvetiae*, 73, 45–52.

Koch, P. L., Fogel, M. L. and Tuross, N. (1992a) Tracing the diets of fossil animals using stable isotopes. In: *Methods in Ecology* (K. Lajtha and B. Michener, eds.), Blackwell Scientific Publishing, Oxford, UK, pp. 63–92.

Kohl, W., Gloe, A. and Reichenbach, H. (1983) Steroids from the myxobacterium *Nannocystis exedens*. *Journal of General Microbiology*, 129, 1629–35.

Kohnen, M. E. L., Sinninghe Damsté, J. S., Kock-Van Dalen, A. C. and De Leeuw, J. W. (1991) Di- or polysulfide-bound biomarkers in sulfur-rich geomacromolecules as revealed by selective chemolysis. *Geochimica et Cosmochimica Acta*, 55, 1375–94.

Kolaczkowska, E., Slougui, N.-E., Watt, D. S., Marcura, R. E. and Moldowan, J. M. (1990) Thermodynamic stability of various alkylated, dealkylated, and rearranged 17α- and 17β-hopane isomers using molecular mechanics calculations. *Organic Geochemistry*, 16, 1033–8.

Koller, J. and Baumer, U. (1993) Analyse einer Kittprobe aus dem Griff des Messers von Xanten-Wardt. *Praehistorica et Archaeologica Acta*, 25, 129–31.

Koller, J., Baumer, U., Kaup, Y., Etspuler, H. and Weser, U. (1998) Embalming was used in Old Kingdom. *Nature*, 391, 343–4.

Koller, J., Baumer, U. and Mania, D. (2001) High-tech in the Middle Palaeolithic: Neandertal-manufactured pitch identified. *European Journal Archaeology*, 4, 385–97.

König, W. A. (1992) *Gas Chromatographic Enantiomeric Separation with Modified Cyclodextrins*. Hütig, Buch Verlag, Heidelberg.

Kontorovich, A. E. (1984) Geochemical methods for the quantitative evaluation of the petroleum potential of sedimentary basins. In: *Petroleum Geochemistry and Basin Evaluation* (G. Demaison and R. J. Murris, eds.), American Association of Petroleum Geologists, Tulsa, OK, pp. 79–109.

Kontorovich, A. E., Danilova, V. P., Kostyreva, E. A., *et al.* (1998a) Main marine oil source formations of the West Siberian petroleum megabasin and their genetic relations to oils. Presented at the Annual Meeting of the American Association of Petroleum Geologists, Salt Lake City, UT, May 17–20, 1998.

Koonin, E. V., Makarova, K. S. and Aravind, L. (2001) Horizontal gene transfer in prokaryotes: quantification and classification. *Annual Review of Microbiology*, 55, 709–42.

Koopmans, M. P., Schouten, S., Kohnen, M. E. L., and Sinninghe Damsté, J. S. (1996) Restricted utility of aryl isoprenoids as indicators for photic zone anoxia. *Geochimica et Cosmochimica Acta*, 60, 4467–96.

Kornacki, A. S. (1993) C_7 chemistry and origin of Monterey oils and source rocks from the Santa Maria Basin, California. Presented at the Annual Meeting of the American Association of Petroleum Geologists, April 25– 28, 1993.

Kornacki, A. S. and Mango, F. D. (1996) C_7 chemistry of biodegraded Monterey oils from the southwestern margin of the Los Angeles Basin, California. Presented at the Annual Meeting of the American Association of Petroleum Geologists, 1996.

Koschel, K. (1996) Opium alkaloids in a Cypriote base ring I vessel (Bilbil) of the Middle Bronze Age from Egypt. *Ägypten und Levante*, 6, 159–66.

Krahn, M. M. and Stein, J. E. (1998) Assessing exposure of marine biota and habitats to petroleum compounds. *Analytical Chemistry News and Features*, 70, 186–92A.

Krings, M., Stone, A., Schmitz, R. W., *et al.* (1997) Neanderthal DNA sequences and the origin of modern humans. *Cell*, 90, 19–30.

Krings, M., Geisert, H., Schmitz, R. W., Krainitzki, H. and Pääbo, S. (1999) DNA sequence of the mitochondrial hypervariable region II from the Neanderthal type specimen. *Proceedings of the National Academy of Science, USA*, 96, 5581–5.

Krouse, H. R., Viau, C. A., Eliuk, L. S., Ueda, A. and Halas, S. (1989) Chemical and isotopic evidence of thermo-chemical sulphate reduction by light hydrocarbon gases in deep carbonate reservoirs. *Nature*, 333, 415–9.

Kudryavtsev, N. A. (1951) Against the organic hypothesis of the origin of petroleum. *Petroleum Economy [Neftianoye Khozyaistvo]*, 9, 17–29.

Kurelec, B., Britvic, S., Rijavec, M., Muller, W. E. G. and Zahn, R. K. (1977) Benzo(a)pyrene monooxygenase induction in marine fish – molecular response to oil pollution. *Marine Biology*, 44, 211–6.

Kvenvolden, K. A. (1993) Gas hydrates – geological persepctive and global change. *Reviews of Geophysics*, 31, 173–87.

(2002) History of the recognition of organic geochemistry in geoscience. *Organic Geochemistry*, 33, 517–21.

Kvenvolden, K. A. and Lorenson, T. D. (2001) The global occurrence of natural gas hydrate. In: *Natural Gas Hydrates: Occurrence, Distribution, and Detection* (C. K. Paull and W. P. Dillon, eds.), American Geophysical Union, Washington, DC, pp. 3–18.

Kvenvolden, K. A., Carlson, P. R., Threlkeld, C. N. and Warden, A. (1993a) Possible connection between two Alaskan catastrophies occurring 25 yr apart (1964 and 1989). *Geology*, 21, 813–6.

Kvenvolden, K. A., Hostettler, F. D., Rapp, J. B. and Carlson, P. R. (1993b) Hydrocarbons in oil residue on beaches of islands of Prince William Sound, Alaska. *Marine Pollution Bulletin*, 26, 24–9.

Kvenvolden, K. A., Hostettler, F. D., Carlson, P. R., *et al.* (1995) Ubiquitous tar balls with a California-source signature on the shorelines of Prince William Sound, Alaska. *Environmental Science & Technology*, 29, 2684–94.

Kvenvolden, K. A., Carlson, P. R., Hostettler, F. D. and Rosenbauer, R. J. (2000) Response to Comment on "Natural hydrocarbon background in benthic sediments of Prince William Sound, Alaska: oil vs. coal". *Environmental Science & Technology*, 34, 2066–7.

Lafargue, E., Marquis, F. and Pillot, D. (1998) Rock-Eval 6 applications in hydrocarbon exploration, production, and soil contamination studies. *Revue de l'Insitut Francais du Petrole*, 53, 421–37.

Lampert, C. D., Heron, C., Thompson, J., *et al.* (2001) Sticky links to the past: characterization and radiocarbon dating of archaeological resins from Southeast Asia. Presented at the 222nd ACS National Meeting, August 26–30, 2001, Chicago, IL.

Lampert, C. D., Glover, I. C., Heron, C. P., *et al.* (2002) The characterization and radiocarbon dating of archaeological resins from Southeast Asia. In: *Archaeological Chemistry: Material, Methods and Meaning* (K. A. Jakes, ed.), American Chemical Society, Washington, DC, pp. 84–109.

Lampert, C. D., Glover, I. C., Hedges, R. E. M., *et al.* (2003a) Dating resin coating on pottery: the Spirit Cave early dates revised. *Antiquity*, 77, 126–33.

Lampert, C. D., Glover, I. C., Heron, C. P., *et al.* (2003b) Resinous residues on prehistoric pottery from Southeast Asia: characterisation and radiocarbon dating. In: *Proceedings of the 9th International Conference of the European Association of Southeast Asian Archaeologists* (A. Kallen & A. Karlstrom, eds.), Museum of Far Eastern Antiquities, Stockholm, in press.

Lancet, H. S. and Anders, E. (1970) Carbon isotope fractionation in the Fischer-Tropsch synthesis of methane. *Science*, 170, 980–2.

Languri, G. M., Van der Horst, J. and Boon, J. J. (2002) Characterisation of a unique "asphalt" sample from the early 19th century Hafkenscheid painting materials collection by analytical pyrolysis MS and GC/MS. *Journal of Analytical and Applied Pyrolysis*, 63, 171–96.

Langworthy, T. A. and Mayberry, W. R. (1976) A 1,2,3,4-tetrahydroxy pentane-substituted pentacyclic triterpene from *Bacillus acidocaldarius*. *Biochimica et Biophysica Acta*, 431, 570–7.

Largeau, C., Derenne, S., Casadevall, E., *et al.* (1990) Occurrence and origin of ultralaminar structures in amorphous kerogens of various source rocks and oils shales. *Organic Geochemistry*, 16, 889–95.

Larter, S. R., Bowler, F., Li, M., *et al.* (1996) Benzocarbazoles as molecular indicators of secondary oil migration distance. *Nature*, 383, 593–7.

Laughrey, C. D. and Baldassare, F. J. (1998) Geochemistry and origin of some natural gases in the Plateau Province, Central Appalachian Basin, Pennsylvania and Ohio. *American Association of Petroleum Geologists Bulletin*, 82, 317–35.

Law, B. E. and Rice, D. D. (1993) *Hydrocarbons from Coal*. American Association of Petroleum Geologists, Tulsa, OK.

Lawler, A. (2002) Report of oldest boat hints at early trade routes. *Science*, 296, 1791–2.

Laws, E. A., Popp, B. N., Bidigare, R. R., Kennicutt, M. C. and Macko, S. A. (1995) Dependence of phytoplankton carbon isotopic composition on growth rate and

[CO_2]aq: theoretical considerations and experimental results. *Geochimica et Cosmochimica Acta*, 59, 1131–8.

Le Dréau, Y., Gilbert, F., Doumenq, P., *et al.* (1997) The use of hopanes to track *in situ* variations in petroleum composition in surface sediments. *Chemosphere*, 34, 1663–72.

Leeming, R., Latham, V., Rayner, M. and Nichols, P. (1997) Detecting and distinguishing sources of sewage pollution in Australian inland and coastal waters and sediments. In: *Molecular Markers in Environmental Geochemistry*, Vol. 67 (R. P. Eganhouse, ed.), American Chemical Society, Washington, DC, pp. 306–19.

Lesquereux, L. (1866) Report on the fossil plants of Illinois: Illinois *Geological Survey*, 2, 425–70.

Levorsen, A. I. (1967) *Geology of Petroleum*. W. H. Freeman and Company, San Francisco.

Lewan, M. D. (1984) Factors controlling the proportionality of vanadium to nickel in crude oils. *Geochimica et Cosmochimica Acta*, 48, 2231–8.

  (1985) Evaluation of petroleum generation by hydrous pyrolysis experimentation. *Philosophical Transactions of the Royal Society of London, A*, 315, 123–34.

  (1987) Petrographic study of primary petroleum migration in the Woodford Shale and related rock units. In: *Migration of Hydrocarbons in Sedimentary Basins* (B. Doligez, ed.), Editions Technip, Paris, pp. 113–30.

  (1994) Assessing natural oil expulsion from source rocks by laboratory pyrolysis. In: *The Petroleum System – From Source to Trap* (L. B. Magoon and W. G. Dow, eds.), American Association of Petroleum Geologists, pp. 201–10.

Lewan, M. D. and Fisher, J. B. (1994) Organic acids from petroleum source rocks. In: *Organic Acids in Geological Processes* (E. D. Pittman and M. D. Lewan, eds.), Springer-Verlag, New York, pp. 70–114.

Lewan, M. D., Winters, J. C. and McDonald, J. H. (1979) Generation of oil-like pyrolyzates from organic-rich shales. *Science*, 203, 897–9.

Leythaeuser, D., Schaefer, R. G. and Weiner, B. (1978) Generation of low molecular weight hydrocarbons from organic matter in source beds as a function of temperature. *Chemical Geology*, 25, 95–108.

Leythaeuser, D., Schaefer, R. G., Cornford, C. and Weiner, B. (1979) Generation and migration of light hydrocarbons (C_2–C_7) in sedimentary basins. *Organic Geochemistry*, 1, 191–204.

Li, M., Larter, S. R., Stoddart, S. and Bjorøy, M. (1992) Practical liquid chromatographic separation schemes for pyrrolic and pyridinic nitrogen heterocyclic fractions from crude oils suitable for rapid characterization of geological samples. *Analytical Chemistry*, 64, 1337–44.

Li, J. G., Philp, R. P. and Cui, M. Z. (2000) Methyl diamantane index (MDI) as a maturity parameter for Lower Palaeozoic carbonate rocks at high maturity and overmaturity. *Organic Geochemistry*, 31, 267–72.

Liberti, A., Cartoni, G. P. and Bruner, F. (1965) Isotope effect in gas chromatography. In: *Gas Chromatography 1964* (A. Goldup, et.), Elsevier, Amsterdam, pp. 301–12.

Lichtenthaler, H. K. (2000) Non-mevalonate isoprenoid biosynthesis: enzymes, genes and inhibitors. *Biochemical Society Transactions*, 28, 785–9.

Lijmbach, G. W. M. (1975) On the origin of petroleum. *Proceedings of the 9th World Petroleum Congress*, 2, 357–69.

Lijmbach, G. W. M., van der Veen, F. M. and Englehardt, E. D. (1983) Geochemical characterisation of crude oils and source rocks using field ionisation mass spectrometry In: *Advances in Organic Geochemistry 1981* (M. Bjorøy, C. Albrecht, C. Cornford, *et al.*, eds.), John Wiley & Sons, New York, pp. 788–98.

Lin, R. (1995) An interlaboratory comparison of vitrinite reflectance measurement. *Organic Geochemistry*, 22, 1–9.

Lin, R. and Wilk, Z. A. (1995) Natural occurrence of tetramantane ($C_{22}H_{28}$), pentamantane ($C_{26}H_{32}$) and hexamantane ($C_{30}H_{36}$) in a deep petroleum reservoir. *Fuel*, 74, 1512–21.

Lin, D. S., Connor, W. E., Napton, L. K. and Heizer, R. F. (1978) The steroids of 2000-year-old human coprolites. *Journal of Lipid Research*, 19, 215–21.

Lindstrom, J. E., Prince, R. C., Clark, J. C., *et al.* (1991) Microbial populations and hydrocarbon biodegradation potential in fertilized shoreline sediments affected by the T/V *Exxon Valdez* oil spill. *Applied and Environmental Microbiology*, 57, 2514–52.

Lorant, F. and Behar, F. (2002) Late generation of methane from mature kerogens. *Energy & Fuels*, 16, 412–27.

Losh, L., Eglinton, L., Schoell, M. and Wood, J. (1999) Vertical and lateral fluid flow related to a large growth fault, South Eugene Island Block 330 Field, Offshore Louisiana. *American Association of Petroleum Geologists Bulletin*, 83, 244–76.

Louati, A., Elleuch, B., Kallel, M., *et al.* (2001) Hydrocarbon contamination of coastal sediments from the Sfax Area (Tunisia), Mediterranean Sea. *Marine Pollution Bulletin*, 42, 445–52.

Louda, J. W. and Baker, E. W. (1984) Perylene occurrence, alkylation and possible sources in deep-ocean sediments. *Geochimica et Cosmochimica Acta*, 48, 1043–58.

(1986) The biogeochemistry of chlorophyll. In: *Organic Marine Chemistry* (M. L. Sohn, ed.), Vol. 305, American Chemical Society, Washington, DC, pp. 107–41.

Loutit, T. S., Hardenbol, J., Vail, P. R. and Baum, G. R. (1988) Condensed sections: the key to age determination and correlation of continental margin sequences. In: *Sea-level Changes – An Integrated Approach* (C. K. Wilgus, B. S. Hastings, C. G. St C. Kendall, *et al.*, eds.), Society of Economic Paleontologists and Mineralogists, Tulsa, OK, pp. 109–24.

Lubell, D., Jackes, M., Schwarcz, H., Knyf, M. and Meiklejohn, C. (1994) The Mesolithic-Neolithic transition in Portugal: isotopic and dental evidence of diet. *Journal of Archaeological Science*, 21, 201–16.

Luellen, D. R. and Shea, D. (2002) Calibration and field verification of semipermeable membrane devices for measuring polycyclic aromatic hydrocarbons in water. *Environmental Science & Technology*, 36, 1791–7.

Lundegard, P. D., Haddad, R. and Brearley, M. (1998) Methane associated with a large gasoline spill: forensic determination of origin and source. *Environmental Geoscience*, 5, 69–78.

Lundegard, P. D., Sweeney, R. E. and Ririe, G. T. (2000) Soil gas methane at petroleum contaminated sites: forensic determination of origin and source. *Environmental Forensics*, 1, 3–10.

Lunel, T., Rusin, J., Halliwell, C. and Davies, L. (1997) The net environmental benefit of a successful dispersant operation at the *Sea Empress* incident. In *Proceedings of the 1997 International Oil Spill Conference, April 7–10, 1997, Fort Lauderdale, FL*, American Petroleum Institute, Washington, DC, pp. 185–94.

Luzzati, V., Gulik, A., de Rosa, M. and Gambacorta, A. (1987) Lipids from *Sulfolobus solfatarius*, life at high temperature and the structure of membranes. *Chemica Scripta*, 27B, 211–9.

MacDonald, I. R. (1998) Natural oil spills. *Scientific American*, 11, 31–5.

Machel, H. G. (2001) Bacterial and thermochemical sulfate reduction in diagenetic settings-old and new insights. *Sedimentary Geology*, 140, 143–75.

Machel, H. G., Krouse, H. R., Riciputi, L. R. and Cole, D. R. (1995a) Devonian Nisku sour gas play, Canada: a unique natural laboratory for study of thermochemical sulfate reduction. In: *Geochemical Transformations of Sedimentary Sulfur* (M. A. Vairavamurthy and M. A. A. Schoonen, Tedse), American Chemical Society, Washington, DC pp. 439–54.

Machel, H. G., Krouse, H. R. and Sassen, R. (1995b) Products and distinguishing criteria of bacterial and thermochemical sulfate reduction. *Applied Geochemistry*, 10, 373–89.

MacHugh, D. E., Troy, C. S., McCormick, F., *et al.* (1999) Early medieval cattle remains from a Scandinavian settlement in Dublin: genetic analysis and comparison with extant breeds. *Philosophical Transactions of the Royal Society London, Biological Sciences*, 354, 99–110.

Mackenzie, A. S., Brassell, S. C., Eglinton, G. and Maxwell, J. R. (1982) Chemical fossils: the geological fate of steroids. *Science*, 217, 491–504.

Mackenzie, A. S., Disko, U. and Rullkötter, J. (1983a) Determination of hydrocarbon distributions in oils and sediment extracts by gas chromatography–high resolution mass spectrometry. *Organic Geochemistry*, 5, 57–63.

Mackenzie, A. S., Li, R.-W., Maxwell, J. R., Moldowan, J. M. and Seifert, W. K. (1983b) Molecular measurements of thermal maturation of Cretaceous shales from the Overthrust Belt, Wyoming, USA. In: *Advances in Organic Geochemistry 1981* (M. Bjorøy, C. Albrecht, C. Cornford, *et al.* eds.), John Wiley & Sons, New York, pp. 496–503.

Mackenzie, A. S., Rullkötter, J., Welte, D. H. and Mankiewicz, P. (1985) Reconstruction of oil formation and accumulation in North Slope, Alaska, using quantitative gas chromatography-mass spectrometry. In: *Alaska North Slope Oil/Source Rock Correlation Study* (L. B. Magoon and G. E. Claypool, eds.), American Association of Petroleum Geologists, Tulsa, OK, pp. 319–77.

Mackenzie, A. S., Leythaeuser, D., Altebäumer, F.-J., Disko, U. and Rullkötter, J. (1988) Molecular measurements of maturity for Lias δ shales in N. W. Germany. *Geochimica et Cosmochimica Acta*, 52, 1145–54.

Macko, S. A. and Quick, R. S. (1986) A geochemical study of oil migration at source rock reservoir contacts: stable isotopes. *Organic Geochemistry*, 10, 199–205.

Macko, S. A., Engel, M. H. and Qian, Y. (1994) Early diagenesis and organic matter preservation – a molecular stable carbon isotope perspective. *Chemical Geology*, 114, 365–79.

Macko, S. A., Engel, M. H., Andrusevich, V., *et al.* (1999) Documenting the diet in ancient human populations through stable isotope analysis of hair. *Philosophical Transactions of the Royal Society London, Biological Sciences*, 354, 65–77.

Magoon, L. B. (1994) Tuxedni-Hemlock(!) petroleum system in Cook Inlet, Alaska, U.S.A. In: *The Petroleum System – From Source to Trap.* (L. B. Magoon and W. G. Dow, eds.), American Association of Petroleum Geologists, Tulsa, OK, pp. 359–70.

Magoon, L. B. and Bird, K. J. (1988) Evaluation of petroleum source rocks in the National Petroleum Reserve in Alaska using organic-carbon content, hydrocarbon content, visual kerogen, and vitrinite reflectance. In: *Geology and Exploration of the National Petroleum Reserve in Alaska, 1974 to 1982* (C. Gryc, ed.), U.S. Geological Survey, Washington, DC, 1399, pp. 381–450.

Magoon, L. B. and Claypool, G. E. (1981) Two oil types on the North Slope of Alaska – Implications for future exploration. *American Association of Petroleum Geologists Bulletin*, 65, 644–52.

(1983) Petroleum geochemistry of the North Slope of Alaska: time and degree of thermal maturity. In: *Advances in Organic Geochemistry 1981* (M. Bjorøy, C. Albrecht, C. Cornford, *et al.*, eds.) John Wiley & Sons, New York, pp. 28–38.

(1984) The Kingak shale of north Alaska-Regional variations in organic geochemical properties and petroleum source rock quality. *Organic Geochemistry*, 6, 533–42.

Magoon, L. B. and Dow, W. G. (1994) *The Petroleum System – From Source to Trap.* American Association of Petroleum Geologists, Tulsa, OK.

Magoon, L. B. and Valin, Z. C. (1994) Overview of petroleum system case studies. In: *The Petroleum System – From Source to Trap* (L. B. Magoon and W. G. Dow, eds.), American Association of Petroleum Geologists, Tulsa, OK, pp. 329–38.

Mahato, S. B. and Sen, S. (1997) Advances in terpenoid research 1990–1994. *Phytochemistry*, 44, 1185–236.

Mair, B. J., Ronen, Z., Eisenbraun, E. J. and Horodysky, A. G. (1966) Terpenoid precursors of hydrocarbons from the gasoline range of petroleum. *Science*, 154, 1339–41.

Maldonado, C., Bayona, J. M. and Bodineau, L. (1999) Sources, distribution, and water column processes of aliphatic and polycyclic aromatic hydrocarbons in the

Northwestern Black Sea water. *Environmental Science & Technology*, 33, 2693–702.

Mancini, E. A., Mink, R. M. and Bearden, B. L. (1989) Integrated geological, geophysical, and geochemical interpretation of Upper Jurassic petroleum in the eastern Gulf of Mexico. *Transactions – Gulf Coast Association of Geological Societies*, 36, 309–20.

Mango, F. D. (1987) Invariance in the isoheptanes of petroleum. *Science*, 247, 514–7.

(1990) The origin of light cycloalkanes in petroleum. *Geochimica et Cosmochimica Acta*, 54, 24–7.

(1991) The stability of hydrocarbons under the time-temperature conditions of petroleum genesis. *Nature*, 352, 146–8.

(1992) Transition metal catalysis in the generation of petroleum and natural gas. *Geochimica et Cosmochimica Acta*, 56, 553–5.

(1994) The origin of light hydrocarbons in petroleum: ring preference in the closure of carbocyclic rings. *Geochimica et Cosmochimica Acta*, 58, 895–901.

(1996) Transition metal catalysis in the generation of natural gas. *Organic Geochemistry*, 24, 977–84.

(1997) The light hydrocarbons in petroleum: a critical review. *Organic Geochemistry*, 26, 417–40.

(1998) Some evidence supporting catalysis in the decomposition of oil to natural gas. Presented at the 215th National Meeting of the American Chemical Society, Dallas, TX, March 29–April 2, 1998.

(2000) The origin of light hydrocarbons. *Geochimica et Cosmochimica Acta*, 64, 1265–77.

Mango, F. D. and Elrod, L. W. (1998) The carbon isotopic composition of catalytic gas: a comparative analysis with natural gas. *Geochimica et Cosmochimica Acta*, 63, 1097–106.

Mango, F. D. and Hightower, J. (1997) The catalytic decomposition of petroleum into natural gas. *Geochimica et Cosmochimica Acta*, 61, 5347–50.

Mango, F. D., Hightower, J. W. and James, A. T. (1994) Role of transition-metal catalysis in the formation of natural gas. *Nature*, 368, 536–8.

Mansfield, C. T., Barman, B. N., Thomas, J. V., Mehrotra, A. K. and McCann, J. M. (1999) Petroleum and coal. *Analytical Chemistry*, 71, 81–107R.

Mansuy, L., Philp, R. P. and Allen, J. (1997) Source identification of oil spills based on the isotopic composition of individual components in weathered oil samples. *Environmental Science & Technology*, 31, 3417–25.

Manzano, B. K., Fowler, M. G. and Machel, H. G. (1997) The influence of thermochemical sulfate reduction on hydrocarbon composition in Nisku reservoirs, Brazeau River area, Alberta, Canada. *Organic Geochemistry*, 27, 507–21.

Marlar, R. A., Leonard, B. L., Billman, B. R., Lambert, P. M. and Marlar, J. E. (2000) Biochemical evidence of cannibalism at a prehistoric Puebloan site in southwestern Colorado. *Nature*, 407, 74–8.

Marschner, R. F. and Wright, H. T. (1978) Asphalts from Middle Eastern archaeological sites. *Archaeological Chemistry*, 21, 51–171.

Martin, G. C. (1908) Geology and Mineral Resources of the Controller Bay Region, Alaska. *U.S. Geological Survey Bulletin*, 335, 3–141.

Martin, L. K., Jr and Black, M. C. (1996) Biomarker assessment of the effects of petroleum refinery contamination on channel catfish. *Ecotoxicology and Environmental Safety*, 33, 81–7.

Martin, R. L., Winters, J. C. and Williams, J. A. (1963) Composition of crude oils by gas chromatography: geological significance of hydrocarbon distribution. Presented at the Sixth World Petroleum Congress Proceedings, Frankfurt am Main, June 1963.

Mason, G. M., Rudell, L. G. and Branthaver, J. F. (1990) Review of the stratigraphic distribution and diagenetic history of abelsonite. *Organic Geochemistry*, 14, 585–94.

Masterson, W. D., Dzou, L. I. P., Holba, A. G., Fincannon, A. L. and Ellis, L. (2001) Evidence for biodegradation and evaporative fractionation in West Sak, Kuparuk and Prudhoe Bay field areas, North Slope, Alaska. *Organic Geochemistry*, 32, 411–41.

Matthews, D. E. and Hayes, J. M. (1978) Isotope-ratio monitoring gas chromatgraphy-mass spectrometry. *Analytical Chemistry*, 50, 1465–73.

Mauch, D. H., Nägler, K., Schumacher, S., *et al.* (2001) CNS synaptogenesis promoted by glia-derived cholesterol. *Science*, 294, 1354–7.

Maughan, E. K. (1993) Phosphoria Formation (Permian) and its resource significance in the Western Interior, USA. Presented at the CSPG Pangeo: Global Environment and Resources Conference, Calgary, August 15–19, 1993.

Maurer, J., Möhring, T., Rullkötter, J. and Nissenbaum, A. (2002) Plant lipids and fossil hydrocarbons in embalming material of Roman Period mummies from the Dakhleh Oasis, Western Desert, Egypt. *Journal of Archaeological Science*, 29, 751–62.

Maxwell, J. R., Cox, R. E., Eglinton, G., *et al.* (1973) Stereochemical studies of acyclic isoprenoid compounds – 2. The role of chlorophyll in the derivation of isoprenoid-type acids in a lacustrine sediment. *Geochimica et Cosmochimica Acta*, 37, 297–313.

Mayuga, M. N. (1970) Geology and development of California's giant – Wilmington oil field. In: *Geology of Giant Petroleum Fields* (M. T. Halbouty, ed.), American Association of Petroleum Geologists, Tulsa, OK, pp. 158–84.

Mazeas, L. and Budzinski, H. (2001) Polycyclic aromatic hydrocarbon $^{13}C/^{12}C$ ratio measurement in petroleum and marine sediments: application to standard reference materials and a sediment suspected of contamination from the *Erika* oil spill. *Journal of Chromatography A*, 923, 165–76.

McCaffrey, M. A., Farrington, J. W. and Repeta, D. J. (1989) Geochemical implications of the lipid composition of *Thioploca* spp. from the Peru upwelling region – 15 S. *Organic Geochemistry*, 14, 61–8.

McCaffrey, M. A., Moldowan, J. M., Lipton, P. A., *et al.* (1994) Paleoenvironmental implications of novel C_{30} steranes in Precambrian to Cenozoic age petroleum and bitumen. *Geochimica et Cosmochimica Acta*, 58, 529–32.

McCaffrey, M. A., Legarre, H. A. and Johnson, S. J. (1996) Using biomarkers to improve heavy oil reservoir management; an example from the Cymric Field, Kern County, California. *American Association of Petroleum Geologists Bulletin*, 80, 898–913.

McCollom, T. M. (2003) Formation of meteorite hydrocarbons from thermal decomposition of siderite ($FeCO_3$). *Geochimica et Cosmochimica Acta*, 67, 311–7.

McCollom, T. M. and Seewald, J. S. (2001) A reassessment of the potential for reduction of dissolved CO_2 to hydrocarbons during serpentinization of olivine. *Geochimica et Cosmochimica Acta*, 65, 3769–78.

McCusker, L. B. (1994) Advances in powder diffraction methods for zeolite structure analysis. In: *Zeolites and Related Microporous Materials: State of the Art 1994*, Vol. 84 (J. W. Weitkamp, H. G. Karge, H. Pfeifer and W. Hölderich, eds.), Elsevier, Amsterdam, pp. 341–356.

McFadden W. H. (1973) *Techniques of Combined Gas Chromatography Mass Spectrometry*. Wiley-Interscience, New York.

McGovern, P. E., Fleming, S. J. and Katz, S. H. (1995) *The Origins and Ancient History of Wine. Food and Nutrition in History and Anthropology* Vol. 11. Gordon and Breach, New York.

McGovern, P. E., Glusker, D. L., Exner, L. J. and Voigt, M. W. (1996) Neolithic resinate wine. *Nature*, 381, 480–1.

McGovern, P. E., Glusker, D. L., Moreau, R. A., *et al.* (1999) A funerary feast fit for King Midas. *Nature*, 402, 863–4.

McKay, D. S., Gibson, E. K., Jr, Thomas-Keprta, K. L., *et al.* (1996) Search for past life on Mars: possible relic biogenic activity in Martian meteorite ALH84001. *Science*, 273, 924–30.

McKirdy, D. M., Aldrige, A. K. and Ypma, P. J. M. (1983) A geochemical comparison of some crude oils from Pre-Ordovician carbonate rocks. In: *Advances in Organic Geochemistry 1981* (M. Bjorøy, C. Albrecht, C. Cornford, *et al.*, eds.), John Wiley & Sons, New York, pp. 99–107.

McLafferty, F. W. (1980) *Interpretation of Mass Spectra*, 3rd edn., University Science Books, Mill Valley, CA.

McLimans, R. K. (1987) The application of fluid inclusions to migration of oil and diagenesis in petroleum reservoirs. *Applied Geochemistry*, 2, 585–603.

McNeil, R. H. and Bement, W. O. (1996) Thermal stability of hydrocarbons: laboratory criteria and field examples. *Energy & Fuels*, 10, 60–7.

Mearns, E. W. and McBride, J. J. (1999) Hydrocarbon filling history and reservoir continuity of oil fields evaluated using $^{87}Sr/^{86}Sr$ isotope ratio variations in formation water, with examples from the North Sea. *Petroleum Geoscience*, 5, 17–27.

Meganathan, R. (2001) Ubiquinone biosynthesis in microorganisms. *FEMS Microbiology Letters*, 203, 131–9.

Meissner, F. F., Woodward, J. and Clayton, J. L. (1984) Stratigraphic relationships and distribution of source rocks in the Greater Rocky Mountain Region. In: *Hydrocarbon Source Rocks of the Greater Rocky Mountain Region* (J. Woodward, F. F. Meissner and J. L. Clayton, eds.), Rocky Mountain Association of Geologists, Denver, CO, pp. 1–34.

Mello, M. R., Koutsoukos, E. A. M., Mohriak, W. U. and Bacoccoli, G. (1994) Selected petroleum systems in Brazil. In: *The Petroleum System – From Source to Trap* (L. B. Magoon and W. G. Dow, eds.), American Association of Petroleum Geologists, Tulsa, OK, pp. 499–512.

Mendeleev, D. (1877) L'origine du petrole. *Revue Scientifique, 2e Ser.*, VIII, 409–16.

(1902) *The Principles of Chemistry*, Vol. 1. Second English edition translated from the sixth Russian edition. Collier, New York.

Mercadante, A. Z. (1999) New carotenoids: recent progress. *Pure and Applied Chemistry*, 71, 2263–73.

Merlin, M. D. (1984) *On the Trail of the Ancient Opium Poppy*. Fairleigh Dickinson University Press (Associated University Presses), Cranbury, NJ.

Merriwether, D. A. (1999) Freezer anthropology: new uses for old blood. *Philosophical Transactions of the Royal Society London, Biological Sciences*, 354, 121–30.

Metzger, P., Villarreal-Rosales, E., Casadevall, E. and Coute, A. (1989) Hydrocarbons, aldehydes and tricylglycerols in some strains of the A race of the green alga *Botryococcus braunii*. *Phytochemistry*, 28, 2349–53.

Meyers, P. A. (1994) Preservation of elemental and isotopic source identification of sedimentary organic matter. *Chemical Geology*, 144, 289–302.

(1997) Organic geochemical proxies of paleooceanographic, paleolimnlogic, and paleoclimatic processes. *Organic Geochemistry*, 27, 213–50.

Michaelis, W., Seifert, R., Nauhaus, K., *et al.* (2002) Microbial reefs in the Black Sea fueled by anaerobic oxidation of methane. *Science*, 297, 1013–5.

Michalczyk, G. (1985) Determination of *n*- and iso-paraffins in hydrocarbon waxes – comparative results of analyses by gas chromatography, urea adduction, and molecular sieve adsorption [in German]. *Fette-Seifen-Anstrichmittel*, 87, 481–6.

Miller, R. G. (1995) A future for exploration geochemistry. In: *Organic Geochemistry: Developments and Applications to Energy, Climate, Environment and Human History* (J. O. Grimalt and C. Dorronsoro, eds.), AIGOA, Donostia-San Sebastián, Spain, pp. 412–4.

Mills, J. S. and White, R. (1989) The identity of the resins from the Late Bronze Age shipwreck at Ulu Burun (Kas). *Archaeometry*, 31, 37–44.

(1994) *The Organic Chemistry of Museum Objects, Arts and Archaeology*. Butterworths, London.

Milner, C. W. D., Rogers, M. A. and Evans, C. R. (1977) Petroleum transformations in reservoirs. *Journal of Geochemical Exploration*, 7, 101–53.

Mislow, K. (1965) *Introduction to Stereochemistry*. W. A. Benjamin, New York.

Moldowan, J. M. and Seifert, W. K. (1979) Head-to-head linked isoprenoid hydrocarbons in petroleum. *Science*, 204, 169–71.

Moldowan, J. M., Seifert, W. K., Arnold, E. and Clardy, J. (1984) Structure proof and significance of stereoisomeric 28,30-bisnorhopanes in petroleum and petroleum source rocks. *Geochimica et Cosmochimica Acta*, 48, 1651–61.

Moldowan, J. M., Seifert, W. K. and Gallegos, E. J. (1985) Relationship between petroleum composition and depositional environment of petroleum source rocks. *American Association of Petroleum Geologists Bulletin*, 69, 1255–68.

Moldowan, J. M., Sundararaman, P. and Schoell, M. (1986) Sensitivity of biomarker properties to depositional environment and/or source input in the Lower Toarcian of S. W. Germany. *Organic Geochemistry*, 10, 915–26.

Moldowan, J. M., Fago, F. J., Lee, C. Y., *et al.* (1990) Sedimentary 24-*n*-propylcholestanes, molecular fossils diagnostic of marine algae. *Science*, 247, 309–12.

Moldowan, J. M., Dahl, J., Huizinga, B. J., *et al.* (1994) The molecular fossil record of oleanane and its relation to angiosperms. *Science*, 265, 768–71.

Moldowan, J. M., Dahl, J., McCaffrey, M. A., Smith, W. J. and Fetzer, J. C. (1995) Application of biological marker technology to bioremediation of refinery by-products. *Energy & Fuels*, 9, 155–62.

Momper, J. A. (1980) Oil expulsion – a consequence of oil generation. Abstract. *American Association of Petroleum Geologists Bulletin*, 64, 1279.

Mommessin, P. R., Castaño, J. R., Rankin, J. G. and Weiss, M. L. (1981) *Process for Determining API Gravity of oil by FID*. United States Patent 4 248 599.

Mook, W. G. (2001) Abundance and fractionation of stable isotopes. In: *Environmental Isotopes in the Hydrological Cycle. Principles and Applications* Vol. 1 (W. G. Mook, ed.), UNESCO/IAEA, Paris, pp. 31–48.

Morrison, R. T. and Boyd, R. N. (*1987*) *Organic Chemistry*. Allyn and Bacon, Boston, MA.

Mottram, H. R., Dudd, S. N., Lawrence, G. J., Stott, A. W. and Evershed, R. P. (1999) New chromatographic, mass spectrometric and stable isotope approaches to the classification of degraded animal fats preserved in archaeological pottery. *Journal of Chromatography A*, 833, 223–9.

Mudge, S. M. (2002) Reassessment of the hydrocarbons in Prince William Sound and the Gulf of Alaska: identifying the source using partial least-squares. *Environmental Science & Technology*, 36, 2354–60.

Müller, P. J. and Suess, E. (1979) Productivity, sedimentation rate, and sedimentary organic matter in the oceans. I. Organic carbon preservation. *Deep Sea Research*, 26A, 1347–62.

Munz, I. A. (2001) Petroleum inclusions in sedimentary basins: systematics, analytical methods and applications. *Lithos*, 55, 195–212.

Murray, A. P., Edwards, D., Hope, J. M., *et al.* (1998) Carbon isotope biogeochemistry of plant resins and derived hydrocarbons. *Organic Geochemistry*, 29, 1199–214.

Murphy, M. T. K. (1969) Analytical methods. In: *Organic Geochemistry* (G. Eglinton and M. T. J. Murphy, eds.), Springer-Verlag, Berlin, pp. 74–88.

Murphy, B. L. and Morrison, R. D. (2002) *Introduction to Environmental Forensics*. Academic Press, San Diego, CA.

Murphy, M. T. J., McCormick, A. and Eglinton, G. (1967) Perhydro-β-carotene in Green River Shale. *Science*, 157, 1040–2.

Muyzer, G., Westbroek, P., de Vrind, H. P. M., *et al.* (1984) Immunology and organic geochemistry. *Organic Geochemistry*, 6, 847–55.

Muyzer, G., Dekoster, S., van Zijl, Y., Boon, J. J. and Westbroek, P. (1986) Immunological studies on microbial mats from Solar Lake (Sinai): a contribution to the organic geochemistry of sediments. *Organic Geochemistry*, 10, 697–704.

Muyzer, G., Sandberg, P. A., Knapen, M. H. A., *et al.* (1992) Preservation of the bone protein osteocalcin in dinosaurs. *Geology*, 20, 871–4.

Mycke, B., Narjes, F. and Michaelis, W. (1987) Bacteriohopanetetrol from chemical degradation of an oil shale kerogen. *Nature*, 326, 179–81.

Myers, K. J. and Wignall, P. B. (1987) Understanding Jurassic organic-rich mudrock – new concepts using gamma-ray spectrometry and paleoecology: examples from the Kimmeridge Clay of Dorset and the Jet Rock of Yorkshire. In: *Marine Clastic Environments: Concepts and Case Studies* (J. K. Legget, ed.), Graham and Trotman, London, pp. 175–92.

National Research Council (1985) *Oil in the Sea: Input, Fates, and Effect*. National Academy Press, Washington, DC.

(2002) *Oil in the Sea III: Inputs, Fates, and Effects*. National Acadamy Press, Washington, DC.

Nederlof, P. J. R., Gijsen, M. A. and Doyle, M. A. (1994) Application of reservoir geochemistry to field appraisal. In: *Geo '94: The Middle East Petroleum Geosciences. Selected Middle East Papers from the Middle East Geoscience Conference* (M. I. Al Husseini, ed.), Gulf PetroLink, Manama, Bahrain, pp. 709–722.

Neff, J. M. (1979) *Polycyclic Aromatic Hydrocarbons in the Aquatic Environment: Sources, Fates and Biological Effects*. Applied Science Publishers, London.

Nes, W. R. and McKean, M. L. (1977) *Biochemistry of Steroids and Other Isopentenoids*. University Park Press, Baltimore.

Nes, W. D. and Venkatramesh, M. (1994) Molecular assymetry and sterol evolution. In: *Isopentenoids and Other Natural Products: Evolution and Function, ACS Symposium Series 562* (W. D. Nes, ed.), American Chemical Society, Washington, DC, pp. 55–89.

Newberry, J. S. (1873) *The General Geological Relations and Structure of Ohio*. Ohio Geological Survey Report 1, Part 1. Division of Geological Survey, Columbus, OH.

Nichols, P. D., Volkman, J. K., Palmisano, A. C., Smith, G. A. and White, D. C. (1988) Occurrence of an isoprenoid C_{25} diunsaturated alkene and high neutral lipid content in Antarctic sea-ice diatom communities. *Journal of Phycology*, 24, 90–6.

Nichols, P. D., Leeming, R., Rayner, M. S. and Latham, V. (1993) Comparison of the abundance of the fecal sterol coprostanol and fecal bacterial groups in inner-shelf waters and sediments near Sydney, Australia. *Journal of Chromatography*, 643, 189–95.

Nielsen-Marsh, C. M., Ostrom, P. H., Gandhi, H., *et al.* (2002) Sequence preservation of osteocalcin protein and mitochondrial DNA in bison bones older than 55 ka. *Geology*, 30, 1099–102.

Niklas, K. J. (1996) *The Evolutionary Biology of Plants*. University of Chicago Press, Chicago, IL.

Nisbet, E. G., Cann, J. R. and van Dover, C. L. (1995) Origins of photosynthesis. *Nature*, 373, 479–480.

Nissenbaum, A., Baedecker, M. J. and Kaplan, I. R. (1972) Studies on dissolved organic matter from interstitial water of a reducing marine fjord. In: *Advances in Organic Geochemistry 1971* (H. R. von Gaertner and H. Wehner, eds.), Pergamon Press, New York, pp. 427–40.

Nolte, D. G. (1991) *Separation of a Mixture of Normal Paraffins, Branched Chain Paraffins, and Cyclic Paraffins*. United States Patent 4 982 052, January 1, 1991.

North, F. K. (1985) *Petroleum Geology*. Allen and Unwin, London.

Oakwood, T. S., Shriver, D. S., Fall, H. H., McAleer, W. J. and Wunz, P. R. (1952) Optical activity of petroleum. *Industrial and Engineering Chemistry*, 44, 2568–70.

Obermajer, M., Stasiuk, L. D., Fowler, M. G. and Osadetz, K. G. (1997) Acritarch fluorescence as a new thermal maturity indicator. *American Association of Petroleum Geologists Bulletin*, 81, 1561.

Obermajer, M., Osadetz, K. G., Fowler, M. G. and Snowdon, L. R. (2000b) Light hydrocarbon (gasoline range) parameter refinement of biomarker-based oil-oil correlation studies: an example from Williston Basin. *Organic Geochemistry*, 31, 959–76.

Ocampo, R., Callot, H. J. and Albrecht, P. (1989) Different isotope compositions of C_{32} DPEP and C_{32} etioporphyrin III in oil shale. *Naturwissenschaften*, 76, 419–21.

Odden, W., Patience, R. L. and van Graas, G. W. (1998) Application of light hydrocarbons (C_4–C_{13}) to oil/source rock correlations. *Organic Geochemistry*, 28, 823–47.

Olson, D. H., Haag, W. O. and Lago, R. M. (1980) Chemical and physical properties of the ZSM-5 substitutional series. *Journal of Catalysis*, 61, 390–6.

O'Malley, V. P., Abrajano, T. A., Jr and Hellou, J. (1994) Determination of the $^{13}C/^{12}C$ ratios of individual PAH from environmental samples: can PAH sources be apportioned? *Organic Geochemistry*, 21, 809–22.

O'Neil, J. R. (1986) Theoretical and experimental aspects of isotopic fractionation. *Mineralogical Society of America Reviews in Mineralogy*, 16, 1–40.

Ong, R. C. Y. and Marriott, P. J. (2002) A review of basic concepts in comprehensive two-dimensional gas chromatography. *Journal of Chromatographic Science*, 40, 276–91.

Orphan, V. J., Hinrichs, K.-U., Ussler, W., III, *et al.* (2001a) Comparative analysis of methane-oxidizing archaea and sulfate-reducing bacteria in anoxic marine sediments. *Applied and Environmental Microbiology*, 67, 1922–34.

Orphan, V. J., House, C. H., Hinrichs, K.-U., McKeegan, K. D. and Delong, E. F. (2001b) Methane-consuming archaea revealed by directly coupled isotopic and phylogenetic analysis. *Science*, 293, 479–81.

Orr, W. L. (1974) Changes in sulfur content and isotopic ratios of sulfur during petroleum maturation. Study of Big Horn Basin Paleozoic oils. *American Association of Petroleum Geologists Bulletin*, 58, 2295–318.

(1977) Geologic and geochemical controls on the distribution of hydrogen sulfide in natural gas. In:

Advances in Organic Geochemistry (R. Campos and J. Goni, eds.), Enadisma, Madrid, Spain, pp. 571–97.

(1986) Kerogen/asphaltene/sulfur relationships in sulfur-rich Monterey oils. *Organic Geochemistry*, 10, 499–516.

Orr, W. L. and Gaines, A. G. (1974) Observations on rate of sulfate reduction and organic matter oxidation in the bottom waters of an estuarine basin: the upper basin of the Pettaquamscutt River (Rhode Island). In: *Advances in Organic Geochemistry 1973* (B. Tissot and F. Bienner, eds.), Editions Technip, Paris, pp. 791–812.

Othman, R. and Ward, C. R. (2002) Thermal maturation pattern in the southern Bowen, northern Gunnedah and Surat basins, northern New South Wales, Australia. *International Journal of Coal Geology*, 51, 145–67.

Oudemans, T. F. M. and Boon, J. J. (1991) Molecular archaeology: analysis of charred (food) remains from prehistoric pottery by pyrolysis-gas chromatography/mass spectrometry. *Journal of Analytical and Applied Pyrolysis*, 20, 197–227.

Ourisson, G. (1987) Bigger and better hopanoids. *Nature*, 326, 126–7.

Ourisson, G. and Nakatani, Y. (1994) The terpenoid theory of the origin of cellular life: the evolution of terpanoids to cholesterols. *Chemistry and Biology*, 1, 11–23.

Ourisson, G., Albrecht, P. and Rohmer, M. (1984) The microbial origin of fossil fuels. *Scientific American*, 251, 44–51.

Ourisson, G., Rohmer, M. and Poralla, K. (1987) Prokaryotic hopanoids and other polyterpenoid sterol surrogates. *Annual Review of Microbiology*, 41, 301–33.

Ovchinnikov, I. V., Götherström, A., Romanova, G. P., *et al.* (2000) Molecular analysis of Neanderthal DNA from the northern Caucasus. *Nature*, 404, 490–3.

Overton, E. B., Sharp, W. D. and Roberts, P. (1994) Toxicity of petroleum. In: *Basic Environmental Toxicology* (L. G. Cockerham and B. S. Shane, eds.), CRC Press, Boca Raton, FL, pp. 133–56.

Page, D. S., Boehm, P. D., Douglas, G. S. and Bence, A. E. (1995) Identification of hydrocarbon sources in the benthic sediments of Prince William Sound and the Gulf of Alaska following the *Exxon Valdez* oil spill. In: Exxon Valdez *Oil Spill: Fate and Effects in Alaskan Waters* (P. G. Wells, J. N. Butler and J. S. Hughes, eds.), American Society for Testing and Materials, Philadelphia, PA, pp. 41–83.

Page, D. S., Boehm, P. D., Douglas, G. S., *et al.* (1996a) The natural petroleum hydrocarbon background in subtidal sediments of Prince William Sound, Alaska, USA.

Environmental Toxicology and Chemistry, 15, 1266–81.

Page, D. S., Boehm, P. D., Gilifillan, E. S., *et al.* (1996b) Effects of the *Exxon Valdez* oil spill on the subtidal organic geochemistry of two bays in Prince William Sound, Alaska. In: *19th Arctic and Marine Oilspill Program (AMOP) Technical Seminar, Calgary, Alberta, June 12–14, 1996. Proceedings*, Vol. 2, Environment Canada, Emergencies Science Division, Ottawa, pp. 1195–209.

Page, D. S., Boehm, P. D., Douglas, G. S., *et al.* (1997) An estimate of the annual input of natural petroleum hydrocarbons to seafloor sediments in Prince William Sound, Alaska. *Marine Pollution Bulletin*, 34, 744–9.

(1998) Petroleum sources in the western Gulf of Alaska/Shelikoff Strait Area. *Marine Pollution Bulletin*, 36, 1004–12.

(1999a) Pyrogenic polycyclic aromatic hydrocarbons in sediments record past human activity: a case study in Prince William Sound, Alaska. *Marine Pollution Bulletin*, 38, 247–60.

(1999b) Sources of background hydrocarbons in subtidal sediments of Prince William Sound and the Eastern Gulf of Alaska: part 2. Discriminating among multiple sources. Presented at the SETAC 20th Annual Meeting, November 14–18, 1999, Philadelphia, pp. 261.

Page, D. S., Burns, W. A., Bence, A. E., *et al.* (2001) Resolving the origin of the petrogenic hydrocarbon background in Prince William Sound, Alaska. *Environmental Science & Technology*, 35, 471–9.

Palenik, B. (2002) The genomics of symbiosis: hosts keep the baby and the bath water. *Proceedings of the National Academy of Sciences*, 99, 11996–7.

Palmer, S. E. (1984) Effect of water washing on $C_{15}+$ hydrocarbon fraction of crude oils from northwest Palawan, Phillipines. *American Association of Petroleum Geologists Bulletin*, 68, 137–49.

(1993) Effect of biodegradation and water washing on crude oil composition. In: *Organic Geochemistry* (M. H. Engel and S. A. Macko, eds.), Plenum Press, New York, pp. 511–33.

Pan, C., Fu, J. and Sheng, G. (2000) Sequential extraction and compositional analysis of oil-bearing fluid inclusions in reservoir rocks from Kuche Depression, Tarim Basin. *Chinese Science Bulletin*, 45, 60–6.

Pan, C., Yang, J., Fu, J. and Sheng, G. (2003) Molecular correlation of free oil and inclusion oil of reservoir rocks in the Junggar Basin, China. *Organic Geochemistry*, 34, 357–74.

Parnell, J. (1988) Migration of biogenic hydrocarbons into granites: a review of hydrocarbons in British plutons. *Marine and Petroleum Geology*, 5, 385–96.

Parnell, J., Middleton, D., Honghan, C. and Hall, D. (2001) The use of integrated fluid inclusion studies in constraining oil charge history and reservoir compartmentation: examples from the Jeanne d'Arc Basin, offshore Newfoundland. *Marine and Petroleum Geology*, 18, 535–49.

Parsche, F. and Nerlich, A. (1995) Presence of drugs in different tissues of an Egyptian mummy. *Fresenius' Journal of Analytical Chemistry*, 352, 380–4.

Paseshnichenko, V. A. (1998) A new alternative non-mevalonate pathway for isoprenoid biosynthesis in eubacteria and plants. *Biochemistry (Biokhimiia)*, 63, 139–48.

Passey, Q. R., Creaney, S., Kulla, J. B., Moretti, F. J. and Stroud, J.D (1990) A practical model for organic richness from porosity and resistivity logs. *American Association of Petroleum Geologists Bulletin*, 74, 1777–94.

Patience, R. L., Rowland, S. J. and Maxwell, J. R. (1978) The effect of maturation on the configuration of pristane in sediments and petroleum. *Geochimica et Cosmochimica Acta*, 42, 1871–6.

Patience, R. L., Yon, D. A., Ryback, G. and Maxwell, J. R. (1980) Acyclic isoprenoid alkanes and geochemical maturation. In: *Advances in Organic Geochemistry* 1979 (A. G. Douglas and J. R. Maxwell, eds.), Pergamon Press, New York, pp. 287–94.

Patience, R. L., Pedersen, V. B., Hanesand, T., *et al.* (1993) *The Norwegian Industry Guide to Organic Geochemical Analyses*, edn 4.0. The Norwegian Petroleum Directorate, Stavanger, Norway.

Patin, S. (1999) *Environmental Impact of the Offshore Oil and Gas Industry*. EcoMonitor Publisher, East Northport, New York.

Patrick, M., Koning, A. J. and Smith, A. B. (1985) Gas-liquid chromatographic analysis of fatty acids in food residues in ceramics found in Southwestern Cape. *Archaeometry*, 27, 231–6.

Patt, T. E. and Hanson, R. S. (1978) Intracytoplasmic membrane, phospholipid, and sterol content of *Methylobacterium organophilum* cells grown under different conditions. *Journal of Bacteriology*, 134, 636–44.

Patterson, G. W. (1994) Phylogenetic distribution of sterols. In: *Isopentenoids and Other Natural Products: Evolution and Function, ACS Symposium Series* 562 (W. D. Nes, ed.), American Chemical Society, Washington, DC, pp. 90–108.

Payzant, J. D., Montgomery, D. S. and Strausz, O. P. (1986) Sulfides in petroleum. *Organic Geochemistry*, 9, 357–69.

Peabody, C. E. (1993) The association of cinnabar and bitumen in mercury deposits of the California Coast Ranges. In: *Bitumens in Ore Deposits* (J. Parnell, H. Kucha and P. Landais, eds.), Springer-Verlag, New York, pp. 178–209.

Peachey, C. P. (1995) Terebinth resin in antiquity: Possible uses in the Late Bronze Age Aegean region. M. A. thesis, Texas A&M University, College Park, TX.

Pedersen, T. F. and Calvert, S. E. (1990) Anoxia vs. productivity: what controls the formation of organic-carbon-rich sediments and sedimentary rocks. *American Association of Petroleum Geologists Bulletin*, 74, 454–66.

Pelet, R. (1987) A model of organic sedimentation on present-day continental margins. In: *Marine Petroleum Source Rocks* (J. Brooks and A. J. Fleet, eds.), Geological Society, London, pp. 167–80.

Pepper, A. and Dodd, T. A. (1995) Single kinetic models of petroleum formation. Part II: oil-gas cracking. *Marine and Petroleum Geochemistry*, 12, 321–40.

Peters, K. E. (1986) Guidelines for evaluating petroleum source rock using programmed pyrolysis. *American Association of Petroleum Geologists Bulletin*, 70, 318–29.

(1999a) Rock-Eval pyrolysis. In: *Encyclopedia of Geochemistry* (C. P. Marshall and R. W. Fairbridge, eds.), Kluwer Academic Publishers, Boston, MA, pp. 551–5.

(1999b) The Deep Hot Biosphere; Thomas Gold. Book review. *Organic Geochemistry*, 30, 473–5.

(2000) Petroleum tricyclic terpanes: predicted physicochemical behavior from molecular mechanics calculations. *Organic Geochemistry*, 31, 497–507.

Peters, K. E. and Cassa, M. R. (1994) Applied source rock geochemistry. In: *The Petroleum System – From Source to Trap* (L. B. Magoon and W. G. Dow, eds.), American Association of Petroleum Geologists, Tulsa, OK, pp. 93–117.

Peters, K. E. and Creaney, S. (2004) Geochemical differentiation of Silurian and Devonian oils from Algeria. *Geochemical Investigations: A Tribute to Isaac R. Kaplan* (R. J. Hill, J. Leventhal, Z. Aizenshtat, *et al.*, eds.), Geological Society of America, Boulder, CO, pp. 287–301

Peters, K. E. and Fowler, M. G. (2002) Applications of petroleum geochemistry to exploration and reservoir management. *Organic Geochemistry*, 33, 5–36.

Peters, K. E. and Moldowan, J. M. (1991) Effects of source, thermal maturity, and biodegradation on the distribution and isomerization of homohopanes in petroleum. *Organic Geochemistry*, 17, 47–61.

(1993) *The Biomarker Guide. Interpreting Molecular Fossils in Petroleum and Ancient Sediments*. Prentice-Hall, Englewood Cliffs, NJ.

Peters, K. E. and Nelson, D. A. (1992) REESA – an expert system for geochemical logging of wells. *American Association of Petroleum Geologists Annual Meeting Abstracts*, 103.

Peters, K. E. and Simoneit, B. R. T. (1982) Rock-Eval pyrolysis of Quaternary sediments from Leg 64, Sites 479 and 480, Gulf of California. *Initial Reports Deep Sea Drilling Project*, 64, 925–31.

Peters, K. E., Rohrback, B. G. and Kaplan, I. R. (1981) Carbon and hydrogen stable isotope variations in kerogen during laboratory simulated thermal maturation. *American Association of Petroleum Geologists Bulletin*, 65, 501–8.

Peters, K. E., Whelan, J. K., Hunt, J. M. and Tarafa, M. E. (1983) Programmed pyrolysis of organic matter from thermally altered Cretaceous black shales. *American Association of Petroleum Geologists Bulletin*, 67, 2137–46.

Peters, K. E., Moldowan, J. M., Schoell, M. and Hempkins, W. B. (1986) Petroleum isotopic and biomarker composition related to source rock organic matter and depositional environment. *Organic Geochemistry*, 10, 17–27.

Peters, K. E., Moldowan, J. M., Driscole, A. R. and Demaison, G. J. (1989) Origin of Beatrice oil by cosourcing from Devonian and Middle Jurassic source rocks, Inner Moray Firth, UK. *American Association of Petroleum Geologists Bulletin*, 73, 454–71.

Peters, K. E., Moldowan, J. M. and Sundararaman, P. (1990) Effects of hydrous pyrolysis on biomarker thermal maturity parameters: Monterey Phosphatic and Siliceous Members. *Organic Geochemistry*, 15, 249–65.

Peters, K. E., Scheuerman, G. L., Lee, C. Y., *et al.* (1992) Effects of refinery processes on biological markers. *Energy & Fuels*, 6, 560–77.

Peters, K. E., Kontorovich, A. E., Moldowan, J. M., *et al.* (1993) Geochemistry of selected oils and rocks from the central portion of the West Siberian Basin, Russia. *American Association of Petroleum Geologists Bulletin*, 77, 863–87.

Peters, K. E., Elam, T. D., Pytte, M. H. and Sundararaman, P. (1994) Identification of petroleum systems adjacent to the San Andreas Fault, California, USA. In: *The Petroleum System – From Source to Trap* (L. B. Magoon and W. G. Dow, eds.), American Association of Petroleum Geologists, Tulsa, OK, pp. 423–36.

Peters, K. E., Clark, M. E., das Gupta, U., McCaffrey, M. A. and Lee, C. Y. (1995) Recognition of an Infracambrian source rock based on biomarkers in the Bagehwala-1 oil, India. *American Association of Petroleum Geologists Bulletin*, 79, 1481–94.

Peters, K. E., Cunningham, A. E., Walters, C. C., Jiang, J. and Fan, Z. (1996a) Petroleum systems in the Jiangling-Dangyang area, Jianghan Basin, China. *Organic Geochemistry*, 24, 1035–60.

Peters, K. E., Moldowan, J. M., McCaffrey, M. A. and Fago, F. J. (1996b) Selective biodegradation of extended hopanes to 25-norhopanes in petroleum reservoirs. Insights from molecular mechanics. *Organic Geochemistry*, 24, 765–83.

Peters, K. E., Wagner, J. B., Carpenter, D. G. and Conrad, K. T. (1997) World class Devonian potential seen in eastern Madre de Dios Basin. *Oil and Gas Journal*, 95, 61–65, 84–87.

Peters, K. E., Fraser, T. H., Amris, W., Rustanto, B. and Hermanto, E. (1999) Geochemistry of crude oils from eastern Indonesia. *American Association of Petroleum Geologists Bulletin*, 83, 1927–42.

Petrov, A. A., Pustil'Nikova, S. D., Abriutina, N. N. and Kagramonova, G. R. (1976) Petroleum steranes and triterpanes. *Neftekhimiia*, 16, 411–27.

Petrov, A. A., Vorobyova, N. S. and Zemskova, Z. K. (1990) Isoprenoid alkanes with irregular "head-to-head" linkages. *Organic Geochemistry*, 16, 1001–5.

Petsch, S. T., Eglinton, T. I. and Edwards, K. J. (2001) ^{14}C-dead living biomass: evidence for microbial assimilation of ancient organic carbon during shale weathering. *Science*, 292, 1127–31.

Philippi, G. T. (1975) The deep subsurface temperature controlled origin of gaseous and gasoline-range hydrocarbons of petroleum. *Geochimica et Cosmochimica Acta*, 39, 1355–73.

Phillips, J. B. and Beens, J. (1999) Comprehensive two-dimensional gas chromatography: a hyphenated method with strong coupling between the two dimensions. *Journal of Chromatography A*, 856, 331–47.

Philp, R. P. (1985) *Fossil Fuel Biomarkers*. Elsevier, New York.

Philp, R. P. and Gilbert, T. D. (1982) Unusual distribution of biological markers in an Australian crude oil. *Nature*, 299, 245–7.

Philp, R. P. and Brassell, S. (1986) Arguments against abiogenic origin for hydrocarbons. *Chemical and Engineering News*, 64, 2–3, 48, 59.

Philp, R. P., Oung, J. and Lewis, C. A. (1988) Biomarker determinations in crude oils using a triple-stage quadrupole mass spectrometer. *Journal of Chromatography*, 446, 3–16.

Picha, F. J. and Peters, K. E. (1998) Biomarker oil-to-source rock correlation in the Western Carpathians and their foreland, Czech Republic. *Petroleum Geoscience*, 4, 289–302.

Pironon, J., Thiéry, R., Teinturier, S. and Walgenwitz, F. (2000) Water in petroleum inclusions: evidence from Raman and FT-IR measurements, PVT consequences. *Journal of Geochemical Exploration*, 69–70, 663–8.

Pollard, A. M. and Heron, C. (1996) *Archeological Chemistry*. Royal Society of Chemistry, Cambridge, UK.

Pompeckj, J. F. (1901). Die Juraablagerungen zwischen Regensburg und Regenstauf. *Geologisches Jahrbuch*, 14, 139–220.

Ponnamperuma, C. and Pering, K. (1966) Possible abiogenic origin of some naturally occurring hydrocarbons. *Nature*, 209, 979–82.

Poole, C. F. and Schuette, S. A. (1984) *Contemporary Practice of Chromatography*. Elsevier, New York.

Popp, B. N., Laws, E. A., Bidigare, R. R., *et al.* (1998) Effect of phytoplankton cell geometry on carbon isotope fractionation. *Geochimica et Cosmochimica Acta*, 62, 69–77.

Poralla, K., Muth, G. and Härtner, T. (2000) Hopanoids are formed during transition from substrate to aerial hyphae in *Streptomyces coelicolor* A3(2). *FEMS Microbiology Letters*, 189, 93–5.

Porte, C., Biosca, X., Pastor, D., Sole, M. and Albaigés, J. (2000) Aegean Sea oil spill. 2. Temporal study of the hydrocarbons accumulation in bivalves. *Environmental Science & Technology*, 34, 5067–75.

Posamentier, H. W., Jervey, M. T. and Vail, P. R. (1988) Eustatic controls on clastic deposition. I – conceptual framework. In: *Sea-level Changes – An Integrated Approach* (C. K. Wilgus, B. S. Hastings, C. G. St. C. Kendall, *et al.*, eds.), Society of Economic Paleontologists and Mineralogists, Tulsa, OK, pp. 109–124.

Potter, J., Rankin, A. H. and Treloar, P. J. (2001) The nature and origin of abiogenic hydrocarbons in the alkaline igneous intrusions, Khibina and Lovozero in the Kola Peninsula, N. W. Presented at the Geological Society London Meeting on Hydrocarbons in Crystalline Rocks, February 13–14, 2001, London.

Powell, T. G. and McKirdy, D. M. (1973) Relationship between ratio of pristane to phytane, crude oil composition and geological environment in Australia. *Nature*, 243, 37–9.

Premuzic, E. T., Gaffney, J. S. and Manowitz, B. (1986) The importance of sulfur isotope ratios in the differentiation of Prudhoe Bay crude oils. *Journal of Geochemical Exploration*, 26, 151–9.

Price, L. C. (1992) Thermal stability of hydrocarbons in nature: limits, evidence, characteristics, and possible controls. *Geochimica et Cosmochimica Acta*, 57, 3261–80.

Price, L. C. and Barker, C. E. (1985) Suppression of vitrinite reflectance in amorphous rich kerogen – a major unrecognized problem. *Journal of Petroleum Geology*, 8, 59–84.

Price, L. C. and Schoell, M. (1995) Constraints on the origins of hydrocarbon gas from compositions of gases at their site of origin. *Nature*, 378, 368–71.

Prince, R. C. (1993) Petroleum spill bioremediation in marine environments. *Critical Reviews Microbiology*, 19, 217–42.

  (1998) Crude oil biodegradation. In: *The Encyclopedia of Environmental Analysis and Remediation 2* (R. A. Meyers, ed.), John Wiley & Sons, New York, pp. 1327–42.

Prince, R. C. and Bragg, J. R (1997) Shoreline bioremediation following the *Exxon Valdez* oil spill in Alaska. *Journal of Bioremediation*, 1, 97–104.

Prince, R. C., Elmendorf, D. L., Lute, J. R., *et al.* (1994) 17α(H),21β(H)-hopane as a conserved internal standard for estimating the biodegradation of crude oil. *Environmental Science & Technology*, 28, 142–5.

Prince, R. C., Drake, E. N., Madden, P. C. and Douglas, G. S. (1995) Biodegradation of polycyclic aromatic hydrocarbons in a historically contaminated soil. In: *In Situ and On-site Bioremediation (4–2)* (B. C. Alleman and A. Leeson, eds.), Battelle Press, Columbus, OH, pp. 205–10.

Prince, R. C., Stibrany, R. T., Hardenstine, J., Douglas, G. S. and Owens, E. H. (2002) Aqueous vapor extraction: a previously unrecognized weathering process affecting oil spills in vigorously aerated water. *Environmental Science & Technology*, 36, 2822–5.

Proefke, M. L. and Rinehart, K. L. (1992) Analysis of an Egyptian mummy resin by mass spectrometry. *Journal of the American Society for Mass Spectrometry*, 3, 582–9.

Proefke, M. L., Rinehart, K. L., Mastura, R., Ambrose, S. H. and Wisseman, S. U. (1992) Probing the mysteries of ancient Egypt. Chemical analysis of a Roman period Egyptian mummy. *Analytical Chemistry*, 64, 106A–111A.

Prowse, W. G., Keely, B. J. and Maxwell, J. R. (1990) A novel sedimentary metallochlorin. *Organic Geochemistry*, 16, 1059–65.

Pulak, C. (1988) The Bronze Age shipwreck at Ulu Burun, Turkey: 1985 campaign. *American Journal of Archaeology*, 92, 1–38.

Pursch, M., Sun, K., Winniford, B., *et al.* (2002) Modulation techniques and applications in comprehensive two-dimensional gas chromatography (GC × GC). *Analytical and Bioanalytical Chemistry*, 373, 356–67.

Pustil'Nikova, S. D., Abryutina, N. N., Kayukova, G. P. and Petrov, A. A. (1980) Equilibrium composition and properties of epimeric cholestanes. *Neftekhimia*, 20, 26–33.

Quann, R. J. (1998) Modeling the chemistry of complex petroleum mixtures. *Environmental Health Perspectives*, 106, 1441–8.

Quann, R. J. and Jaffe, S. B. (1992) Structured Oriented Lumping: describing the chemistry of complex hydrocarbon mixtures. *I&EC Research*, 31, 2483–97.

Quigley, T. M. and McKenzie, A. S. (1988) The temperature of oil and gas formation in the subsurface. *Nature*, 333, 549–52.

Quirke, J. M. E., Cuesta, L. L., Yost, R. A., Johnson, J. and Britton, E. D. (1989) Studies on high carbon number geoporphyrins by tandem mass spectrometry. *Organic Geochemistry*, 14, 43–50.

Rafferty, S. M. (2002) Identification of nicotine by gas chromatography/mass spectroscopy analysis of smoking pipe residue. *Journal of Archaeological Science*, 29, 897–907.

Raiswell, R. and Berner, R. A. (1985) Pyrite formation in euxinic and semi-euxinic sediments. *American Journal of Science*, 285, 710–24.

Ran X., Fazio, G. C. and Matsuda, S. P. T. (2004) On the origins of triterpenoid skeletal diversity. *Phytochemistry*, 65, 261–91.

Raymond, J., Zhaxybayeva, O., Gogarten, J. P., Gerdes, S. Y. and Blankenship, R. E. (2003) Whole-genome analysis of photosynthetic prokaryotes. *Science*, 298, 1616–20.

Readman, J. W., Bartocci, J., Tolosa, I., *et al.* (1996) Recovery of the coastal marine environment in the Gulf following the 1991 war-related oil spills. *Marine Pollution Bulletin*, 32, 493–8.

Redfield, A. C. (1942) The processes determining the concentrations of oxygen, phosphate and other organic derivatives within the depths of the Atlantic Ocean. *Papers on Physical Oceanography and Meteorology*, 9, 1–22.

Reed, J. D., Illich, H. A. and Horsfield, B. (1986) Biochemical evolutionary significance of Ordovician oils and their sources. *Organic Geochemistry*, 10, 347–58.

Regert, M., Bland, H. A., Dudd, S. N., van Bergen, P. F. and Evershed, R. P. (1998a) Free and bound fatty acid oxidation products in archaeological ceramic vessels. *Proceedings of the Royal Society of London, Series B*, 265, 2027–32.

Regert, M., Delacotte, J.-M., Menu, M., Petrequin, P. and Rolando, C. (1998b) Identification of haftling adhesives from two lake dwellings at Chalain (Jura, France). *Ancient Biomolecules*, 2, 156–63.

Relethford, J. H. (1998) Genetics of modern human origins and diversity. *Annual Review of Anthropology*, 27, 1–7.

Requejo, A. G. (1992) Quantitative analysis of triterpane and sterane biomarkers: methodology and applications in molecular maturity studies. In: *Biological Markers in Sediments and Petroleum* (J. M. Moldowan, P. Albrecht and R. P. Philp, eds.), Prentice-Hall, Englewood Cliffs, NJ, pp. 222–40.

Reunanen, M., Holmbom, B. and Edgren, T. (1993) Analysis of archaeological birch bark pitches. *Holzforschung*, 47, 175–7.

Rhodes, D. C. and Morse, J. W. (1971) Evolutionary and ecologic significance of oxygen-deficient marine basins. *Lethaia*, 4, 413–28.

Rice, D. D. and Claypool, G. E. (1981) Generation, accumulation, and resource potential of biogenic gas. *American Association of Petroleum Geologists Bulletin*, 65, 5–25.

Rice, D. D., Law, B. E. and Clayton, J. L. (1993) Coalbed gas – an undeveloped resource. In: *The Future of Energy Gases* (D. G. Howell, K. Wiese, M. Fanelli, L. Zimk, and F. Cole, eds.), U.S. Geological Survey Professional Paper 1570, U.S. Geological Survey, Washington, DC, pp. 389–404.

Rice, S. D., Spies, R. B., Douglas, D. A. and Wright, B. A. (1996) *Proceedings of the Exxon Valdez Oil Spill Symposium, Anchorage Alaska, 2–5 February, 1993.* American Fisheries Society, Alpharetta, GA.

Richards, M. P. and Hedges, R. E. M. (1999) A Neolithic revolution? New evidence of diet in the British Neolithic. *Antiquity*, 73, 891–7.

Richards, M. P., Jacobi, R., Currant, A., Stringer, C. and Hedges, R. E. M. (2000a) Gough's Cave and Sun Hole Cave human stable isotope values indicate a high animal protein diet in the British Upper Palaeolithic. *Journal of Archaeological Science*, 27, 1–3.

Richards, M. P., Pettitt, P. B., Trinkaus, E., *et al.* (2000b) Neanderthal diet at Vindija and Neanderthal predation: the evidence from stable isotopes. *Proceedings of the National Academy of Science*, 97, 7663–6.

Riebesell, U., Revill, A. T., Holdsworth, D. G. and Volkman, J. K. (2000) The effects of varying CO_2 concentration on lipid composition and carbon isotope fractionation in *Emiliania huxleyi. Geochimica et Cosmochimica Acta*, 64, 4179–92.

Rieley, G., Collier, R. J., Jones, D. M., *et al.* (1991) Sources of sedimentary lipids deduced from stable carbon-isotope analyses of individual compounds. *Nature*, 352, 425–7.

Riva, A., Caccialanza, P. G. and Quagliaroli, F. (1988) Recognition of $18\beta(H)$-oleanane in several crudes and Tertiary-Upper Cretaceous sediments. Definition of a new maturity parameter. *Organic Geochemistry*, 13, 671–5.

Robison, C. R., van Gijzel, P. and Darnell, L. M. (2000) The transmittance color index of amorphous organic matter: a thermal maturity indicator for petroleum source rocks. *International Journal of Coal Geology*, 43, 83–103.

Rodrigues, D. C., Koike, L., De, A. M., *et al.* (2000) Carboxylic acids of marine evaporitic oils from Sergipe-Alagoas Basin, Brazil. *Organic Geochemistry*, 31, 1209–22.

Roedder, E. (1984) Fluid inclusions. *Reviews in Mineralogy*, 12, 1–644.

Rohdich, F., Kis, K., Bacher, A. and Eisenreich, W. (2001) The non-mevalonate pathway of isoprenoids: genes, enzymes and intermediates. *Current Opinion in Chemical Biology*, 5, 535–40.

Rohmer, M. (1987) The hopanoids, prokaryotic triterpenoids and sterol surrogates. In: *Surface Structures of Microorganisms and their Interactions with the Mammalian Host* (E. Schriner *et al.*, eds.), VCH Publishing, Weinlein, Germany, pp. 227–42.

(1993) The biosynthesis of triterpenoids of the hopane series in eubacteria: a mine of new enzyme reactions. *Pure and Applied Chemistry*, 65, 1293–8.

(1999) A mevalonate-independent route to isopentenyl diphosphate. In: *Comprehensive Natural Product Chemistry 2* (D. Barton and K. Nakanishi, eds.), Pergamon Press, Oxford, UK, pp. 45–68.

Rohmer, M. and Bisseret, P. (1994) Hopanoid and other polyterpenoid biosynthesis in eubacteria: phyologenetic significance. In: *Isopentenoids and Other Natural Products: Evolution and Function*, ACS Symposium Series 562 (W. D. Nes, ed.), American Chemical Society, Washington, DC, pp. 31–43.

Rohmer, M. and Ourisson, G. (1976a) Structure des bactériohopanetétrols d' *Acetobacter xylinum*. *Tetrahedron Letters*, 17, 3633–6.

(1976b) Dérivés du bactériohopane: variations structurales et répartition. *Tetrahedron Letters*, 17, 3637–40.

Rohmer, M., Bouvier, P. and Ourisson, G. (1979) Molecular evolution of biomembranes: structural equivalents and phylogenetic precursors of sterols. *Proceedings of the National Academy of Sciences USA*, 76, 847–51.

Rohmer, M., Knani, M., Simonin, P., Sutter, B. and Sahm, H. (1993) Isoprenoid biosynthesis in bacteria: a novel pathway for the early steps leading to isopentenyl diphosphate. *Biochemical Journal*, 295, 121–9.

Rollo, F. and Marota, I. (1999) How microbial ancient DNA, found in association with human remains, can be interpreted. *Philosophical Transactions of the Royal Society London, Biological Sciences*, 354, 111–20.

Rontani, J.-F. and Volkman, J. K. (2003) Phytol degradation products as biogeochemical tracers in aquatic environments. *Organic Geochemistry*, 34, 1–35.

Rooney, M. A. (1995) Carbon isotope ratios of light hydrocarbons as indicators of thermochemical sulfate reduction. In: *Organic Geochemistry: Developments and Applications to Energy, Climate, Environment and Human History* (J. O. Grimalt and C. Dorronsoro, eds.), AIGOA, Donostia-San Sebastian, Spain, pp. 523–5.

Rosell-Melé, A., Carter, J. F. and Maxwell, J. R. (1999) Liquid chromatography/tandem mass spectrometry of free base alkyl porphyrins for the characterization of the macrocyclic substitutents in components of complex mixtures. *Rapid Communications in Mass Spectrometry*, 13, 568–73.

Rowan, E. L. and Goldhaber, M. B. (1995) Duration of mineralization and fluid-flow history of the Upper Mississippi Valley zinc-lead district. *Geology (Boulder)*, 23, 609–12.

(1996) *Fluid Inclusions and Biomarkers in the Upper Mississippi Valley Zinc-Lead District – Implications for the Fluid-Flow and Thermal History of the Illinois Basin.* U.S. Geological Survey Bulletin 2094-F, U.S. Geological Survey, Washington, DC.

Rowan, E. L., Goldhaber, M. B. and Hatch, J. R. (1994a) Biomarker and fluid inclusion measurements as constraints on the time-temperature and fluid-flow history of the northern Illinois Basin and Upper Mississippi Valley zinc district. In: *Proceedings of the Illinois Basin Energy and Mineral Resources Workshop* (J. L. Ridgley, J. A. Drahovzal, B. D. Keith and D. R. Kolata, eds.), U.S. Geological Survey, Washington, DC, pp. 40–1.

(1994b) Regional fluid flow and thermal history of the Illinois Basin: evidence from fluid inclusions and biomarkers in the Upper Mississippi Valley zinc district. *Eos, Transactions, American Geophysical Union*, 75, 675.

(1995) Duration of mineralization in the Upper Mississippi Valley zinc-lead district: implications for the thermal-hydrologic history of the Illinois Basin. Presented at the Annual Meeting of the Geological Society of America, November 4–6, 1995, New Orleans, LA.

Rowe, D. and Muehlenbachs, K. (1999) Isotopic fingerprints of shallow gases in the Western Canadian Sedimentary Basin: tools for remediation of leaking heavy oil wells. *Organic Geochemistry*, 30, 861–71.

Rowland, S. J. and Maxwell, J. R. (1984) Reworked triterpenoid and steroid hydrocarbons in a Recent sediment. *Geochimica et Cosmochimica Acta*, 48, 617–24.

Rowland, S., Donkin, P., Smith, E. and Wraige, E. (2001) Aromatic hydrocarbon "humps" in the marine environment: unrecognized toxins? *Environmental Science & Technology*, 35, 2640–4.

Rubinstein, I., Strausz, O. P., Spyckerelle, C., Crawford, R. J. and Westlake, D. W. S. (1977) The origin of oil sand bitumens of Alberta. *Geochimica et Cosmochimica Acta*, 41, 1341–53.

Ruble, T. E., Lisk, M., Ahmed, M., *et al.* (2000) Geochemical appraisal of palaeo-oil columns: implications for petroleum systems analysis in the Bonaparte Basin, Australia. Presented at the Annual Meeting of the American Association of Petroleum Geologists, April 16–19, 2000, New Orleans, LA.

Ruble, T. E., Lewan M. D. and Philp, R. P. (2001) New insights on the Green River petroleum system in the Unita Basin from hydrous pyrolysis experiments. *American Association of Petroleum Geologists Bulletin*, 85, 1333–71.

Rullkötter, J. and Nissenbaum, A. (1988) Dead Sea asphalt in Egyptian mummies: molecular evidence. *Naturwissenschaften*, 75, 618–21.

Rullkötter, J., Aizenshtat, Z. and Spiro, B. (1984) Biological markers in bitumens and pyrolyzates of Upper Cretaceous bituminous chalks from the Ghareb Formation (Israel). *Geochimica et Cosmochimica Acta*, 48, 151–7.

Rullkötter, J., Spiro, B. and Nissenbaum, A. (1985) Biological marker characteristics of oils and asphalts from carbonate source rocks in a rapidly subsiding graben, Dead Sea, Israel. *Geochimica et Cosmochimica Acta*, 49, 1357–70.

Rullkötter, J., Meyers, P. A., Schaefer, R. G. and Dunham, K. W. (1986) Oil generation in the Michigan Basin: a biological marker and carbon isotope approach. *Organic Geochemistry*, 10, 359–75.

Rushdi, A. I. and Simoneit, B. R. T. (2001) Lipid formation by aqueous Fischer-Tropsch-type synthesis over a temperature range of 100 to 400°C. *Origins of Life and Evolution of the Biosphere*, 31, 103–18.

Ruthenberg, K. A., Beck, C. W. and Stout, E. C. (2001) Betulin – fate of a birch tar biomarker. In: *Archaeological Sciences Conference Proceedings, Durham 97* (A. Millard, ed.), British Archaeological Reports, Oxford, UK, pp. 91–5.

Ruthven, D. M. (1988) Zeolites as selective adsorbents. *Chemical and Engineering Progress*, 84, 42–50.

Ryan, C. G. and Griffin, W. L. (1993) The nuclear microprobe as a tool in geology and mineral exploration. *Nuclear Instruments and Methods in Physics Research B*, 77, 381–98.

Sahm, H., Rohmer, M., Bringer-Meyer, S., Sprenger, G. A. and Welle, R. (1993) Biochemistry and physiology of hopanoids in bacteria. *Advances in Microbial Physiology*, 35, 243–73.

Salvi, S. and Williams-Jones, A. E. (1997) Fischer-Tropsch synthesis of hydrocarbons during sub-solidus alteration of the Strange Lake peralkaline granite, Quebec/Labrador, Canada. *Geochemica et Cosmochimica Acta*, 61, 83–99.

Santos Neto, E. V. and Hayes, J. M. (1999) Use of hydrogen and carbon stable isotopes characterizing oils from the Potiguar Basin (onshore), northeastern Brazil. *American Association of Petroleum Geologists Bulletin*, 83: 496–518.

Santos Neto, E. V., Hayes, J. M. and Takaki, T. (1998) Isotopic biogeochemistry of the Neocomian lacustrine and Upper Aptian marine-evaporitic sediments of the Potiguar Basin, northeastern Brazil. *Organic Geochemistry*, 28, 361–81.

Sasaki, T., Maki, H., Ishihara, M. and Harayama, S. (1998) Vanadium as an internal marker to evaluate microbial degradation of crude oil. *Environmental Science & Technology*, 33, 3618–21.

Sassen, R., Roberts, H. H., Aharon, P., *et al.* (1993) Chemosynthetic bacterial mats at cold hydrocarbon seeps, Gulf of Mexico continental slope. *Organic Geochemistry*, 20, 77–89.

Sauter, F., Jordis, U. and Hayek, E. (1992) Chemische Untersuchungen der Kittschaftungs-Materialien. In: *Der Mann im Eis, Band 1, Veroffentlichungen der Universitat Innsbruck 187* (F. Hopfel, W. W. Platzer and K. Spindler, eds.), Universitat Innsbruck, Innsbruck, Austria, pp. 435–41.

Savrda, C. E. (1995) Ichnologic applications in paleoceanographic, paleoclimatic, and sea-level studies. *Palaios*, 10, 565–77.

Savrda, E. E. and Bottjer, D. J. (1986) Trace-fossil model for reconstruction of paleo-oxygenation in bottom waters. *Geology*, 14, 3–6.

Scalan, R. S. and Smith, J. E. (1970) An improved measure of the odd-to-even predominance in the normal alkanes of sediment extracts and petroleum. *Geochimica et Cosmochimica Acta*, 34, 611–20.

Schaefer, R. G. (1992) Zur Geochemie niedrigmolekularer Kohlenwasserstoffe im Posidonienschiefer der Hilsmulde. *Erdos and Kohle – Erdgas Petrochemie*, 45, 73–8.

Schaefer, R. G. and Littke, R. (1988) Maturity-related compositional changes in the low-molecular-weight hydrocarbon fraction of Toarcian shales. *Organic Geochemistry*, 13, 887–92.

Schäfer, T. (1993) Responding to "First identification of drugs in Egyptian mummies". *Naturwissenschaften*, 80, 243–4.

Schenk, J. E. A., Herrmann, R. G., Jeon, K. W., Muller, N. E. and Schwemmler, W. (1997) *Eukaryotism and Symbiosis*. Springer, New York.

Schildowski, M. and Aharon, P. (1992) Carbon cycle and carbon isotope record: geochemical impact of life over 3.89 Ga of Earth history. In: *Early Organic Evolution:*

Implications for Mineral and Energy Resources
(M. Schildowski, S. Golubic, M. M. Kimberley and
P. A. Trudinger, eds.), Springer-Verlag, Berlin,
pp. 147–75.

Schildowski, M., Matzigkeit, U. and Krumbein, W. E. (1984)
Superheavy organic carbon from hypersaline microbial
mats. *Naturwissenschaften*, 71, 303–8.

Schimmelmann, A., Lewan, M. D. and Wintsch, R. P. (1999)
D/H isotope ratios of kerogen, bitumen, oil and water
in hydrous pyrolysis of source rocks containing kerogen
types I, II, IIS, and III. *Geochimica et Cosmochimica
Acta*, 63, 3751–66.

Schleyer, P. (1957) A simple preparation of adamantane.
Journal of the American Chemical Society, 79,
3292.

Schleyer, P. von R. (1990) My thirty years in hydrocarbon
cages: from adamantane to dodecahedrane. In: *Cage
Hydrocarbons* (G. A. Olah, ed.), John Wiley & Sons,
New York, pp. 1–38.

Schmid, J. C., Connan, J. and Albrecht, P. (1987) Occurrence
and geochemical significance of long-chain
dialkylthiocyclopentanes. *Nature*, 329, 54–6.

Schmidt, G. W., Beckmann, D. D. and Torkelson, B. E.
(2002) A technique for estimating the age of
regular/mid-grade gasolines released to the subsurface
since the early 1970s. *Environmental Forensics*, 3,
145–62.

Schmoker, J. W. (1981) Determination of organic matter
content of Appalachian Devonian shales from
gamma-ray logs. *American Association of Petroleum
Geologists Bulletin*, 65, 1285–98.

Schoell, M. (1983) Genetic characteristics of natural gases.
American Association of Petroleum Geologists Bulletin, 67,
2225–38.

(1984) Stable isotopes in petroleum research. In: *Advances
in Petroleum Geochemistry*, Vol. 1 (J. Brooks and D. H.
Welte, eds.), Academic Press, London, pp. 215–45.

(1988) Multiple origins of methane in the Earth. *Chemical
Geology*, 71, 1–10.

Schoell, M. and Hayes, J. M. (1994) Compound-specific
isotope analysis in biogeochemistry and petroleum
research. *Organic Geochemistry*, 21, 1–827.

Schoell, M. and Wellmer, F.-W. (1981) Anomalous ^{13}C
depletion in early Precambrian graphites from Superior
Province, Canada. *Nature*, 290, 696–9.

Schoell, M., McCaffrey, M. A., Fago, F. J. and Moldowan,
J. M. (1992) Carbon isotopic compositions of
23,30-bisnorhopanes and other biological markers in a

Monterey crude oil. *Geochimica et Cosmochimica Acta*,
56, 1391–9.

Schoell, M., Hwang, R. J., Carlson, R. M. K. and Welton,
J. E. (1994) Carbon isotopic composition of individual
biomarkers in gilstonites (Utah). *Organic Geochemistry*,
21, 673–83.

Schoell, M., Dias, R. F., Carlson, R. M. K., *et al.* (1997)
Carbon isotope systematics in diamondoid
hydrocarbons. Presented at the 18th Meeting on
Organic Geochemistry, September 22–26, 1997,
Maastricht, the Netherlands.

Schouten, S., Bowman, J. P., Rijpstra, W. I. C. and Sinninghe
Damsté, J. S. (2000a) Sterols in a psychrophilic
methanotroph, *Methylosphaera hansonii. FEMS
Microbiology Letters*, 186, 193–5.

Schouten, S., van Kaam-Peters, H. M. E., Rijpstra, W. I. C.,
Schoell, M. and Sinninghe Damsté, J. S. (2000b) Effects
of an oceanic anoxic event on the stable carbon isotope
composition of Early Toarcian carbon. *American Journal
of Science*, 300, 1–22.

Schubert, K., Rose, G., Wachtel, H., Horhold, C. and
Ikekawa, N. (1968) Zum vorkommen von sterinen in
bacterien. *European Journal of Biochemistry*, 5, 246.

Schuchert, C. (1915) The conditions of black shale
deposition as illustrated by Kuperschiefer and Lias of
Germany. *Proceedings of the American Philosophical
Society*, 54, 259–69.

Schulz, H. D., Dahmke, A., Schinzel, U., Wallman, K. and
Zabel, M. (1994) Early diagenetic processes, fluxes, and
reaction rates in sediments of the South Atlantic.
Geochimica et Cosmochimica Acta, 58, 2041–60.

Schulz, L. K., Wilhelms, A., Rein, E. and Steen, A. S. (2001)
Application of diamondoids to distinguish source rock
facies. *Organic Geochemistry*, 32, 365–75.

Schurig, V. (1994) Review: enantiomer separation by gas
chromatography on chiral stationary phases. *Journal of
Chromatography*, A666, 111–29.

Schurig, V. and Nowotny, P. (1988) Separation of
enantiomers on diluted permethylated β-cyclodextrin
by high-resolution gas chromatography. *Journal of
Chromatography*, 441, 155–63.

Scotchman, I. C., Griffith, C. E., Holmes, A. J. and Jones,
D. M. (1998) The Jurassic petroleum system north and
west of Britain: a geochemical oil-source correlation
study. *Organic Geochemistry*, 29, 671–700.

Scott, A. R., Kaiser, W. R. and Ayers, W. B., Jr (1994)
Thermogenic and secondary biogenic gases, San Juan
Basin, Colorado and New Mexico – implications for

coalbed gas producibility. *American Association of Petroleum Geologists Bulletin*, 78, 1186–209.

Scrimgeour, C. M., Begley, I. S. and Thomason, M. L. (1999) Measurements of deuterium incorporation into fatty acids by gas chromatography/isotope ratio mass spectrometry. *Rapid Communications in Mass Spectrometry*, 13, 271–74.

Seewald, J. S. (2001) Model for the origin of carboxylic acids in basinal brines. *Geochimica et Cosmochimica Acta*, 65, 3779–89.

Seifert, W. K. (1975) Carboxylic acids in petroleum in sediments. *Fortschritte der Chemie Organischer Naturstoffe*, 32, 1–49.

(1977) Source rock/oil correlations by C_{27}–C_{30} biological marker hydrocarbons. In: *Advances in Organic Geochemistry 1974* (R. Campos and J. Goni, eds.), ENADIMSA, Madrid, pp. 21–44.

(1978) Steranes and terpanes in kerogen pyrolysis for correlation of oils and source rocks. *Geochimica et Cosmochimica Acta*, 42, 473–84.

Seifert, W. K. and Moldowan, J. M. (1979) The effect of biodegradation on steranes and terpanes in crude oils. *Geochimica et Cosmochimica Acta*, 43, 111–26.

(1980) The effect of thermal stress on source-rock quality as measured by hopane stereochemistry. *Physics and Chemistry of the Earth*, 12, 229–37.

(1986) Use of biological markers in petroleum exploration. In: *Methods in Geochemistry and Geophysics* Vol. 24 (R. B. Johns, ed.), Elsevier, Amsterdam, pp. 261–90.

Seifert, W. K., Moldowan, J. M. and Jones, R. W. (1980) Application of biological marker chemistry to petroleum exploration. In: *Proceedings of the Tenth World Petroleum Congress*, Heyden & Son, Inc., Philadelphia, PA pp. 425–40.

Seifert, W. K., Carlson, R. M. K. and Moldowan, J. M. (1983) Geomimetic synthesis, structure assignment, and geochemical correlation application of monoaromatized petroleum steranes. In: *Advances in Organic Geochemistry 1981* (M. Bjorøy, C. Albrecht, C. Cornford, *et al.*, eds.), John Wiley & Sons, New York, pp. 710–24.

Sessions, A. L., Burgoyne, T. W., Schimmelmann, A. and Hayes, J. M. (1999) Fractionation of hydrogen isotopes in lipid biosynthesis. *Organic Geochemistry*, 30, 1193–200.

Seufferheld, M., Vieira, M. C. F., Ruiz, F. A., *et al.* (2003) Identification of organelles in bacteria similar to

acidocalcisomes of unicellular eukaryotes. *Journal of Biological Chemistry*, 278, 299, 971–8.

Shah, R., Gale, J. D., Payne, M. C. and Lee, M.-H. (1996) Understanding the catalytic behaviour of zeolites: first principles study of adsorption of methanol. *Science*, 271, 1395–7.

Sheldrick, C., Lowe, J. J. and Reynier, M. J. (1997) Palaeolithic barbed point from Gransmoor, East Yorkshire, England. *Proceedings of the Prehistoric Society*, 63, 359–70.

Shelkov, D., Verkhovsky, A. B., Milledge, H. J. and Pillinger, C. T. (1997) Carbonado: a comparison between Brazilian and Ubangui sources with other forms of microcrystalline diamond based on carbon and nitrogen isotopes [in Russian]. *Geologiya i Geofizika*, 38, 315–22.

Shellie, R. A., Marriott, P. J. and Morrison, P. (2001) Concepts and preliminary observations on the triple-dimensional analysis of complex volatile samples by using GC×GC-TOFMS. *Analytical Chemistry*, 73, 4861–7.

Sherblom, P. M., Henry, M. S. and Kelly, D. (1997) Questions remain in the use of coprostanol and epicoprostanol as domestic waste markers: examples from coastal Florida. In: *Molecular Markers in Environmental Geochemistry* (R. P. Eganhouse, ed.), American Chemical Society, Washington, DC, pp. 320–31.

Sherwood Lollar, B. S., Frape, S. K., Weise, S. M., *et al.* (1993) Abiogenic methanogenesis in crystalline rocks. *Geochimica et Cosmochimica Acta*, 57, 5087–97.

Sherwood Lollar, B., Westgate, T. D., Ward, J. A., *et al.* (2002) Abiogenic formation of alkanes in the Earth's crust as a minor source for global hydrocarbon reservoirs. *Nature*, 416, 522–4.

Shigenaka, G. and Henry, C. B., Jr (1995) Use of mussels and semipermeable membrane devices to assess bioavailability of residual polynuclear aromatic hydrocarbons three years after the *Exxon Valdez* oil spill. In: Exxon Valdez *Oil Spill: Fate and Effects in Alaskan Waters* (P. G. Wells, J. N. Butler and J. S. Hughes, eds.), American Society for Testing and Materials, PA, p. 239–60.

Shock, E. L. (1988) Organic acid metastability in sedimentary basins. *Geology*, 16, 886–90.

(1994) Application of thermodynamic calculations to geochemical processes involving organic acids. In: *Organic Acids in Geological Processes* (E. D. Pittman and M. D. Lewan, eds.), Springer-Verlag, New York, pp. 270–318.

Shoeninger, M. J. and DeNiro, M. J. (1984) Nitrogen and carbon isotopic composition of bone-collagen from marine and terrestrial animals. *Geochimica et Cosmochimica Acta*, 48, 625–39.

Short, J. W. and Babcock, M. M. (1996) Prespill and postspill concentrations of hydrocarbons in mussels and sediments in Prince William Sound. In: *Proceedings of the* Exxon Valdez *Oil Spill Symposium, Anchorage, 1993. American Fisheries Society Symposium 18* (S. S. Rice, R. B. Spies, D. A. Wolfe and B. A. Wright, eds.), American Fisheries Society, Bethesda, MD, pp. 149–68.

Short, J. W. and Heintz, R. A. (1997) Identification of *Exxon Valdez* oil in sediments and tissues from Prince William Sound and the Northwestern Gulf of Alaska based on a PAH weathering model. *Environmental Science & Technology*, 31, 2375–84.

 (1998) Source of polynuclear aromatic hydrocarbons in Prince William Sound, Alaska, USA, subtidal sediments. *Environmental Toxicology and Chemistry*, 17, 1651–2.

Short, J. W., Kvenvolden, K. A., Carlson, P. R., *et al.* (1999) Natural hydrocarbon background in benthic sediments of Prince William Sound, Alaska: oil vs. coal. *Environmental Science & Technology*, 33, 34–42.

Short, J. W., Wright, B. A., Kvenvolden, K. A., *et al.* (2000) Response to comment on "Natural hydrocarbon background in benthic sediments of Prince William Sound, Alaska: oil vs. coal". *Environmental Science & Technology*, 34, 2066–7.

Silliman, J. E., Meyers, P. A. and Bourbonniere, R. A. (1996) Record of postglacial organic matter delivery and burial in sediments of Lake Ontario. *Organic Geochemistry*, 24, 463–72.

Silliman, J. E., Meyers, P. A., Ostrom, P. H., Ostrom, N. W. and Eadie, B. J. (2000) Insights into the origin of perylene from isotopic analyses of sediments from Saanich Inlet, British Columbia. *Organic Geochemistry*, 31, 1133–42.

Silliman, J. E., Meyers, P. A., Eadie, B. J. and Klump, J. V. (2001) A hypothesis for the origin of perylene based on its low abundance in sediments of Green Bay, Wisconsin. *Chemical Geology*, 177, 309–22.

Silverman, S. R. (1965) Migration and segregation of oil and gas. In: *Fluids in Subsurface Environments*, Vol. 4 (A. Young and G. E. Galley, eds.), American Association of Petroleum Geologists, Tulsa, OK, pp. 53–65.

 (1971) Influence of petroleum origin and transformation on its distribution and redistribution in sedimentary

rocks. In: *Proceedings of the Eighth World Petroleum Congress*, Applied Science Publishers, London, pp. 47–54.

Silverman, S. R. and Epstein, S. (1958) Carbon isotopic compositions of petroleums and other sedimentary organic materials. *American Association of Petroleum Geologists Bulletin*, 42, 998–1012.

Simoneit, B. R. T. (1986) Cyclic terpenoids of the geosphere. In: *Biological Markers in the Sedimentary Record* (R. B. Johns, ed.), Elsevier, New York, pp. 43–99.

 (2002) Biomass burning – a review of organic tracers for smoke from incomplete combustion. *Applied Geochemistry*, 68, 129–62.

Simoneit, B. R. T., Brenner, S., Peters, K. E. and Kaplan, I. R. (1981) Thermal alteration of Cretaceous black shale by diabase intrusions in the Eastern Atlantic – II. Effects on bitumen and kerogen. *Geochimica et Cosmochimica Acta*, 45, 1581–602.

Simoneit, B. R. T., Schoell, M., Dias, R. F. and Aquino Neto, F. R. (1993) Unusual carbon isotope compositions of biomarker hydrocarbons in a Permian tasmanite. *Geochimica et Cosmochimica Acta*, 57, 4205–11.

Simpson, I. A., Dockrill, S. J., Bull, I. D. and Evershed, R. P. (1998) Early anthropogenic soil formation at Tofts Ness, Sanday, Orkney. *Journal of Archaeological Science*, 25, 729–46.

Simpson, I. A., van Bergen, P. F., Perret, V., *et al.* (1999) Lipid biomarkers of manuring practice in relict anthropogenic soils. *The Holocene*, 9, 223–9.

Sinninghe Damsté, J. S. and de Leeuw, J. W. (1990) Analysis, structure and geochemical significance of organically-bound sulphur in the geosphere: state of the art and future research. *Organic Geochemistry*, 16, 1077–101.

Sinninghe Damsté, J. S., de Leeuw, J. W., Dalen, A. C. K., *et al.* (1987) The occurrence and identification of series of organic sulfur compounds in oils and sediment extracts. 1. A study of Rozel Point oil (USA). *Geochimica et Cosmochimica Acta*, 51, 2369–91.

Sinninghe Damsté, J. S., van Koert, E. R., Kock-van Dalen, A. C., de Leeuw, J. W. and Schenck, P. A. (1989) Characterisation of highly branched isoprenoid thiophenes occurring in sediments and immature crude oils. *Organic Geochemistry*, 14, 555–67.

Sinninghe Damsté, J. S., Kock van Dalen, A. C., Albrecht, P. A. and de Leeuw, J. W. (1991) Identification of long-chain 1,2-di-*n*-alkylbenzenes in Amposta crude oil from the Tarragona Basin, Spanish

Mediterranean: implications for the origin and fate of alkylbenzenes. *Geochimica et Cosmochimica Acta*, 55, 3677–83.

Sinninghe Damsté, J. S., de las Heras, F. X. C., van Bergen, P. F. and de Leeuw, J. W. (1993a) Characterization of Tertiary Catalan lacustrine oil shales: discovery of extremely organic sulphur-rich type I kerogens. *Geochimica et Cosmochimica Acta*, 57, 389–415.

Sinninghe Damsté, J. S., Wakeham, S. G., Kohnen, M. E. L., Hayes, J. M. and de Leeuw, J. W. (1993b) A 6,000-year sedimentary molecular record of chemocline excursions in the Black Sea. *Nature*, 362, 827–9.

Sinninghe Damsté, J. S., Schouten, S., Hopmans, E. C., van Duin, A. C. T. and Geenevasen, J. A. J. (2002a) Crenarchaeol: the characteristic core glycerol dibiphytanyl glycerol tetraether membrane lipid of cosmopolitan pelagic Crenarchaeota. *Journal of Lipid Research*, 43, 1641–51.

Sinninghe Damsté, J. S., Strous, M., Rijpstra, W. I. C., *et al.* (2002b) Linearly concatenated cyclobutane lipids form a dense bacterial membrane. *Nature*, 419, 708–12.

Slentz, L. W. (1981) Geochemistry of reservoir fluids as unique approach to optimum reservoir management. Presented at the Middle East Oil Technical Conference, March 9–12, 1981, Manama, Bahrain.

Smalley, P. C. and England, W. A. (1994) Reservoir compartmentalization assessed with fluid compositional data. *SPE Reservoir Engineering*, 8, 175–80.

Smallwood, B. J., Philp, R. P. and Allen, J. D. (2002) Stable carbon isotopic composition of gasolines determined by isotope ratio monitoring gas chromatography mass spectrometry. *Organic Geochemistry*, 33, 149–59.

Smith, J. E. (1956) Basement reservoir of La Paz-Mara oil fields, western Venezuela. *American Association of Petroleum Geologists Bulletin*, 40, 380–5.

Smith, H. M. (1968) Qualitative and quantitative aspects of crude oil composition. *US Bureau of Mines Bulletin*, 642, 1–136.

Smith, G. W., Fowell, D. T. and Melsom, B. G. (1970) Crystal structure of 18α(H)-oleanane. *Nature*, 219, 355–6.

Sofer, Z. (1980) Preparation of carbon dioxide for stable carbon isotope analysis of petroleum fractions. *Analytical Chemistry*, 52, 1389–91.

(1984) Stable carbon isotope compositions of crude oils: application to source depositional environments and petroleum alteration. *American Association of Petroleum Geologists Bulletin*, 68, 31–49.

Sofer, Z., Bjorøy, M. and Hustad, E. (1991) Isotopic composition of individual *n*-alkanes in oils. In: *Organic Geochemistry. Advances and Applications in the Natural Environment* (D. A. C. Manning, ed.), Manchester University Press, Manchester, UK, pp. 207–11.

Spies, R. B., Stegeman, J. J., Hinton, D. E., *et al.* (1996) Biomarkers of hydrocarbon exposure and sublethal effects in embiotocid fishes from a natural petroleum seep in the Santa Barbara Channel. *Aquatic Toxicology*, 34, 195–219.

Stach, E., Mackowsky, M.-T., Teichmüller, M., *et al.* (1982) *Coal Petrology*. Gebrüder Borntraeger, Berlin.

Stahl, W. J. (1977) Carbon and nitrogen isotopes in hydrocarbon research and exploration. *Chemical Geology*, 20, 121–49.

(1978) Source rock-crude oil correlation by isotopic type-curves. *Geochimica et Cosmochimica Acta*, 42, 1573–7.

(1979) Carbon isotopes in petroleum geochemistry. In: *Lectures in Isotope Geology* (F. Jager and J. C. Hunziker, eds.), Springer-Verlag, New York, pp. 274–83.

Staplin, F. L. (1969) Sedimentary organic matter, organic metamorphism, and oil and gas occurrence. *Canadian Petroleum Geologists Bulletin*, 17, 47–66.

Steen, A. (1986) Gas chromatographic/mass spectrometric (GC/MS) analysis of C_{27}–C_{30}-steranes. *Organic Geochemistry*, 10, 1137–42.

Stein, R. (1986) Organic carbon and sedimentation rate – further evidence for anoxic deep-water conditions in the Cenomanian/Turonian Atlantic Ocean. *Marine Geology*, 72, 199–209.

Steinfatt, I. and Hoffmann, G. G. (1993) A contribution to the thermochemical reduction of SO_4^{2-} in the presence of S^{2-} and organic compounds. *Phosphorus, Sulfur, Silicon and Related Elements*, 74, 431–4.

Stevens, T. O. and McKinley, J. P. (1995) Lithoautotrophic microbial ecosystems in deep basalt aquifers. *Science*, 270, 450–4.

Stinnett, J. W. (1982) The deep earth gas hypothesis: big on promises, but evidence looks thin. *Synergy*, 2, 12–20.

Stone, A. C. and Stoneking, M. (1999) Analysis of ancient DNA from a prehistoric Amerindian cemetery. *Philosophical Transactions of the Royal Society London, Biological Sciences*, 354, 153–8.

Stoneking, M. and Cann, R. L. (1989) African origin of human mitochondrial DNA. In: *The Human Revolution:*

Behavioural and Biological Perspectives on the Origins of Modern Humans (P. Mellars and C. Stringer, eds.), Edinburgh University Press, Edinburgh, pp. 17–30.

Stott, A. W. and Evershed, R. P. (1996) δ^{13}C Analysis of cholesterol preserved in archaeological bones and teeth. *Analytical Chemistry*, 68, 4402–8.

Stott, A. W., Evershed, R. P. and Tuross, N. (1997) Compound-specific approach to the δ^{13}C analysis of cholesterol in fossil bones. *Organic Geochemistry*, 26, 99–103.

Stott, A. W., Evershed, R. P., Jim, S., *et al.* (1999) Cholesterol as a new source of palaeodietary information: experimental approaches and archaeological applications. *Journal of Archaeological Science*, 26, 705–16.

Strous, M., Fuerst, J. A., Kramer, E. H. M., *et al.* (1999) Missing lithotroph identified as new planctomycete. *Nature*, 400, 446–9.

Stuermer, D. H., Peters, K. E. and Kaplan, I. R. (1978) Source indicators of humic substances and proto-kerogen. Stable isotope ratios, elemental compositions, and electron spin resonance spectra. *Geochimica et Cosmochimica Acta*, 42, 989–97.

Sugai, A., Masuchi, Y., Uda, I., Itoh, T. and Itoh, Y. H. (2000) Core lipids of hyperthermophilic archaeon, *Pyrococcus horikoshii* OT3. *Journal Japanese Oil Chemical Society*, 49, 659–700.

Suggate, R. P. (1998) Relations between depth of burial, vitrinite reflectance and geothermal gradient. *Journal of Petroleum Geology*, 21, 5–32.

Summons, R. E. and Powell, T. G. (1986) *Chlorobiaceae* in Palaeozoic sea revealed by biological markers, isotopes, and geology. *Nature*, 319, 763–5.

(1987) Identification of aryl isoprenoids in a source rock and crude oils: biological markers for the green sulfur bacteria. *Geochimica et Cosmochimica Acta*, 51, 557–66.

Summons, R. E., Brassell, S. C., Eglinton, G., *et al.* (1988) Distinctive hydrocarbon biomarkers from fossiliferous sediment of the Late Proterozoic Walcott Member, Chuar Group, Grand Canyon, Arizona. *Geochimica et Cosmochimica Acta*, 52, 2625–37.

Summons, R. E., Jahnke, L. L., Hope, J. M. and Logan, G. A. (1999) 2-Methylhopanoids as biomarkers for cyanobacterial oxygenic photosynthesis. *Nature*, 400, 554–7.

Summons, R. E., Jahnke, L. L., Cullings, K. W. and Logan, G. A. (2002a) Cyanobacterial biomarkers: triterpenoids

plus steroids? *EOS Transactions of the American Geophysical Union*, 47, Fall Meeting Supplement.

Sundararaman, P. (1985) High-performance liquid chromatography of vanadyl porphyrins. *Analytical Chemistry*, 57, 2204–6.

Swain, T. and Copper-Driver, G. (1979) Biochemical evolution in early land plants. In: *Paleobotany, Paleoecology and Evolution 1* (K. J. Niklas, ed.), Praeger Publishers, New York, pp. 103–34.

Swannell, R. P. J., Lee, K. and McDonagh, M. (1996) Field evaluations of marine oil spill bioremediation. *Microbiology Reviews*, 60, 342–65.

Sweeney, J. J. and Burnham, A. K. (1990) Evaluation of a simple model of vitrinite reflectance based on chemical kinetics. *American Association of Petroleum Geologists Bulletin*, 74, 1559–70.

Sylvester-Bradley, P. C. and King, R. J. (1963) Evidence for abiogenic hydrocarbons. *Nature*, 198, 728–31.

Szatmari, P. (1989) Petroleum formation by Fischer–Tropsch synthesis in plate tectonics. *American Association of Petroleum Geologists Bulletin*, 73, 989–98.

Taft, D. G., Egging, D. E. and Kuhn, H. A. (1995) Sheen surveillance: an environmental monitoring program subsequent to the 1989 *Exxon Valdez* shoreline cleanup. In: *Exxon Valdez Oil Spill: Fate and Effects in Alaskan Waters* (P. G. Wells, J. N. Butler and J. S. Hughes, eds.), American Society for Testing and Materials, Philadelphia, PA, pp. 215–38.

Takai, K., Moser, D. P., Deflaun, M. and Onstott, T. C. (2001) Archael diversity in waters from deep South African gold mines. *Applied and Environmental Microbiology*, 67, 5750–60.

Talbot, H. M., Watson, D. F., Murrell, J. C., Carter, J. F. and Farrimond, P. (2001) Analysis of intact bacteriohopanepolyols from methanotrophic bacteria by reversed-phase high-performance liquid chromatography-atmospheric pressure chemical ionisation mass spectrometry. *Journal of Chromatography A*, 921, 175–85.

Talukdar, S. C. and Marcano, F. (1994) Petroleum systems of the Maracaibo Basin, Venezuela. In: *The Petroleum System – From Source to Trap* (L. B. Magoon and W. G. Dow, eds.), American Association of Petroleum Geologists Tulsa, OK, pp. 463–81.

Tang, Y., Perry, J. K., Jenden, P. D. and Schoell, M. (2000) Mathematical modeling of stable carbon isotope ratios in natural gases. *Geochimica et Cosmochimica Acta*, 64, 2673–87.

Tauber, H. (1981) ^{13}C evidence for dietary habits of prehistoric man in Denmark. *Nature*, 292, 332–3.

Taylor, G. H., Teichmüller, M., Davis, A., *et al.* (1998) *Organic Petrology*. Gebrüder Borntraeger, Berlin.

Teal, J. M., Farrington, J. W., Burns, K. A., *et al.* (1992) The West Falmouth oil spill after 20 years: fate of fuel oil compounds and effects on animals. *Marine Pollution Bulletin*, 24, 607–14.

Tegelaar, E. W., de Leeuw, J. W., Derenne, S. and Largeau, C. (1989) A reappraisal of kerogen formation. *Geochimica et Cosmochimica Acta*, 53, 3103–6.

Teichmüller, M. and Durand, B. (1983) Fluorescence microscopical rank studies on liptinites and vitrinites in peat and coals and comparison with the results of the Rock-Eval pyrolysis. *International Journal of Coal Geology*, 2, 197–230.

Ten Haven, H. L. (1986) Organic and inorganic geochemical aspects of Mediterranean Late Quaternary sapropels and Messinian evaporitic deposits. Ph. D. thesis, Utrecht University, Utrecht, Germany.

(1996) Applications and limitations of Mango's light hydrocarbon parameters in petroleum correlation studies. *Organic Geochemistry*, 24, 957–76.

Ten Haven, H. L., de Leeuw, J. W., Rullkötter, J. and Sinninghe Damsté, J. S. (1987) Restricted utility of the pristane/phytane ratio as a palaeoenvironmental indicator. *Nature*, 330, 641–3.

Terken, J. M. J. and Frewin, N. L. (2000) The Dhahaban petroleum system of Oman. *American Association of Petroleum Geologists Bulletin*, 84, 523–44.

Thackeray, J. F., Van Der Merwe, N. J. and Van Der Merwe, T. A. (2001) Chemical analysis of residues from seventeenth-century clay pipes from Stratford-upon-Avon and environs. *South Africa Journal of Science*, 97, 19–22.

Thiel, V., Peckmann, J., Richnow, H. W., *et al.* (2001) Molecular signals for anaerobic methane oxidation in Black Sea seep carbonates and a microbial mat. *Marine Chemistry*, 73, 97–112.

Thiel, V., Blumenberg, M., Pape, T., Seifert, R. and Michaelis, W. (2003) Unexpected occurrence of hopanoids at gas seeps in the Black Sea. *Organic Geochemistry*, 34, 81–7.

Thiéry, R., Pironon, J., Walgenwitz, F. and Montel, F. (2000) PIT (Petroleum Inclusion Thermodynamic): a new modeling tool for the characterisation of hydrocarbon fluid inclusions from volumetric and microthermometric measurements. *Journal of Geochemical Exploration*, 69–70, 701–4.

Thomas, J. B., Mann, A. L., Brassell, S. C. and Maxwell, J. R. (1989) 4-Methyl steranes in Triassic sediments: molecular evidence for the earliest dinoflagellates. Presented at the 14th International Meeting on Organic Geochemistry, September 18–22, 1989, Paris.

Thomas, D. J., Bralower, T. J. and Zachos, J. C. (1999) New evidence for subtropical warming during the late Paleocene thermal maximum: stable isotopes from Deep Sea Drilling Project Site 527, Walvis Ridge. *Paleoceanography*, 14, 561–70.

Thompson, K. F. M. (1983) Classification and thermal history of petroleum based on light hydrocarbons. *Geochimica et Cosmochimica Acta*, 47, 303–16.

(1987) Fractionated aromatic petroleums and the generation of gas-condensates. *Organic Geochemistry*, 11, 573–90.

(1988) Gas-condensate migration and oil fractionation in deltaic systems. *Marine and Petroleum Geology*, 5, 237–46.

Thompson, K. F. M. and Kennicutt, M. C., II (1990) Nature and frequency of occurrence of non-thermal alteration processes in offshore Gulf of Mexico petroleums. In: *Gulf Coast Oils and Gases* (D. Schumacher and B. F. Perkins, eds.). Society of Economic Paleontologists and Mineralogists, Tulsa, OK, pp. 199–218.

Thorne, A., Grün, R., Mortimer, G., *et al.* (1999) Australia's oldest human remains: age of the Lake Mungo 3 skeleton. *Journal of Human Evolution*, 36, 591–612.

Timofeeff, M. N., Lowenstein, T. K. and Blackburn, W. H. (2000) ESEM-EDS: an improved technique for major element chemical analysis of fluid inclusions. *Chemical Geology*, 164, 171–82.

Tissot, B. (1969) Premières données sur les méchanismes et la cinétique de la formation du pétrole dans les sédiments. Simulation d'un schéma réactionnel sur ordinateur. *Revue de l'Insitut Français du Petrole*, 24, 470–501.

Tissot, B. P. and Welte, D. H. (1984) *Petroleum Formation and Occurrence*. Springer-Verlag, New York.

Tissot, B. P., Durand, B., Espitalié, J. and Combaz, A. (1974) Influence of the nature and diagenesis of organic matter in formation of petroleum. *American Association of Petroleum Geologists Bulletin*, 58, 499–506.

Tissot, B. P., Deroo, G. and Hood, A. (1978) Geochemical study of the Uinta Basin: formation of petroleum from the Green River Formation. *Geochimica et Cosmochimica Acta*, 42, 1469–85.

Tomczyk, N. A., Winans, R. E., Shinn, J. H. and Robinson, R. C. (2001) On the nature and origin of acidic species in petroleum. 1. Detailed acid type distribution in a California crude oil. *Energy & Fuels*, 15, 1498–504.

Tornabene, T. G. (1985) Lipid analysis and the relationship to chemotaxonomy. In: *Methods in Microbiology*, Vol. 18 (G. Gottschalk, ed.), Academic Press, London, pp. 209–234.

Torsvik, V., Ovreas, L. and Thingstad, T. F. (2002) Prokaryotic diversity – magnitude, dynamics, and controlling factors. *Science*, 296, 1064–6.

Treibs, A. (1936) Chlorophyll and hemin derivatives in organic mineral substances. *Angewandte Chemie*, 49, 682–6.

Trudinger, P. A., Chambers, L. A. and Smith, J. W. (1985) Low-temperature sulphate reduction: biological versus abiological. *Canadian Journal of Earth Science*, 22, 1910–8.

Tseng, H.-Y., Pottorf, R. J. and Symington, W. A. (2002) Compositional characterization and PVT properties of individual hydrocarbon fluid inclusions: method and application to hydrocarbon systems analysis. Presented at the Annual Meeting of the American Association of Petroleum Geologists, March 10–13, 2002, Houston, TX.

Tsuda, K., Hayatsu, R., Kishida, Y. and Akagi, S. (1958) Steroid studies. VI. Studies of the constitution of sargasterol. *Journal of the American Chemical Society*, 80, 921–5.

Tyson, R. V. (2001) Sedimentation rate, dilution, preservation, and total organic carbon: some results of a modeling study. *Organic Geochemistry*, 32, 333–9.

Tyson, R. V. and Pearson, T. H. (1991, eds.) *Modern and Ancient Continental Shelf Anoxia*. London Geological Society, London.

Uda, I., Sugai, A., Itoh, Y. H. and Itoh, T. (2001) Variations in molecular species of polar lipids from *Thermoplasma acidophilum* depend on growth temperature. *Lipids*, 36, 103–105.

Ungerer, P., Behar, F., Villalba, M., Heum, O. R. and Audibert, A. (1988) Kinetic modeling of oil cracking. *Organic Geochemistry*, 13, 235–45.

Urem-Kotsou, D., Stern, B., Heron, C. and Kotsakis, K. (2002) Birch-bark tar at Neolithic Makriyalos, Greece. *Antiquity*, 76, 962–6.

US Geological Survey, (2000) *World Petroleum Assessment 2000. Executive Summary* U.S. Geological Survey, http://greenwood.cr.usgs.gov/energy/worldenergy/dds-60/espt.html (accessed September 6, 2001).

Valisolalao, J., Perakis, N., Chappe, B. and Albrecht, P. (1984) A novel sulfur containing C_{35} hopanoid in sediments. *Tetrahedron Letters*, 25, 1183–6.

Van Aarssen, B. G. K., Cox, H. C., Hoogendoorn, P. and de Leeuw, J. W. (1990) A cadinene biopolymer in fossil and extant dammar resins as a source for cadinanes and bicadinanes in crude oils from Southeast Asia. *Geochimica et Cosmochimica Acta*, 54, 3021–31.

Van Aarssen, B. G. K., Alexander, R. and Kagi, R. I. (1996) The origin of Barrow Sub-basin crude oils: a geochemical correlation using land-plant biomarkers. *APPEA Journal*, 36, 465–76.

Van Bergen, P. F., Peakman, T. M., Leigh-Firbank, E. C. and Evershed, R. P. (1997) Chemical evidence for archaeological frankincense: boswellic acids and their derivatives in solvent soluble and insoluble fractions of resin-like materials. *Tetrahedron Letters*, 38, 8409–12.

Vance, J. E. (1998) Eukaryotic lipid-biosynthetic enzymes: the same but not the same. *Trends in Biochemical Sciences*, 23, 423–8.

Van der Berg, K. J., Pastorova, I., Spetter, L. and Boon, J. J. (1996) State of oxidation of diterpenoid *Pinaceae* resins in varnish, wax lining material, 18th century resin oil paint, and a recent copper resinate glaze. In: *Proceedings of the 11th Triennial Meeting of ICOM Committee for Conservation* (J. Bridgland, ed.), James and James, London, pp. 930–7.

Van der Doelen, G. A. (1999) Molecular studies of fresh and aged triterpenoid varnishes. Ph. D. thesis, University of Amsterdam, Amsterdam, the Netherlands.

Van der Merwe, N. J. and Vogel, J. C. (1978) ^{13}C content of human collagen as a measure of prehistoric diet in woodland North America. *Nature*, 276, 815–6.

Van Deursen, M., Beens, J., Reijenga, J., *et al.* (2000) Group-type identification of oil samples using comprehensive two-dimensional gas chromatography coupled to a time-of-flight mass spectrometer. *Journal of High Resolution Chromatography*, 23, 507–10.

Van Dorsselaer, A., Ensminger, A., Spyckerelle, C., *et al.* (1974) Degraded and extended hopane derivatives (C_{27}–C_{35}) as ubiquitous geochemical markers. *Tetrahedron Letters*, 14, 1349–52.

Van Duin, A. C. T. and Larter, S. R. (1997) Unraveling Mango's mysteries: a kinetic scheme describing the diagenetic fate of C_7-alkanes in petroleum systems. *Organic Geochemistry*, 27, 597–9.

(1998) Application of molecular dynamics calculations in the prediction of dynamical molecular properties. *Organic Geochemistry*, 29, 1043–50.

(2001). Molecular dynamics investigation into the adsorption of organic compounds on kaolinite surfaces. *Organic Geochemistry*, 32, 143–50.

Van Duin, A. C. T. and Sinninghe Damsté, J. S. (2003) Computational chemical investigation into isorenieratene cyclisation. *Organic Geochemistry*, 34, 515–26.

Van Duin, A. C. T., Bass, J. M. A. and van de Graaf, B. (1996a) A molecular mechanics force field for tertiary carbocations. *Journal Chemical Society Faraday Transactions*, 92, 353–62.

Van Duin, A. C. T., Hollanders, B., Smits, R. J. A., *et al.* (1996b) Molecular mechanics calculation of the rotational barriers of 2,2′,6-trialkylbiphenyls to explain their GC-elution behaviour. *Organic Geochemistry*, 24, 587–91.

Van Duin, A. C. T., Peakman, T. M., de Leeuw, J. W. and van de Graaf, B. (1996c) Novel aspects of the diagenesis of Δ^7-5α-sterenes as revealed by a combined molecular mechanics calculations and laboratory simulations approach. *Organic Geochemistry*, 24, 473–93.

Van Duin, A. C. T., Sinninghe Damsté, J. S., Koopmans, M. P., de Leeuw, J. W. and van de Graaf, B. (1997) A kinetic calculation method of homohopanoid maturation: applications in the reconstruction of burial histories of sedimentary basins. *Geochimica et Cosmochimica Acta*, 61, 2409–29.

Van Graas, G., Baas, J. M. A., de Graaf, V. and de Leeuw, J. W. (1982) Theoretical organic geochemistry. 1. The thermodynamic stability of several cholestane isomers calculated by molecular mechanics. *Geochimica et Cosmochimica Acta*, 46, 2399–402.

Vanko, D. A. and Stakes, D. S. (1991) Fluids in oceanic layer 3: evidence from veined rocks, Hole 735B, Southwest Indian Ridge. *Proeedings of Ocean Drilling Program, Scientific Results*, 118, 181–215.

Van Krevelen, D. W. (1961) *Coal.* Elsevier, New York.

Vaughan, D. E. W. (1988) Synthesis and manufacture of zeolites. *Chemical and Engineering Progress*, 84, 25–31.

Venkatesan, M. I. and Kaplan, I. R. (1990) Sedimentary coprostanol as an index of sewage addition in Santa Monica Basin, Southern California. *Environmental Technology*, 24, 204–13.

Venkatesan, M. I. and Mirsadeghi, F. H. (1992) Coprostanol as sewage tracer in McMurdo Sound, Antarctica. *Marine Pollution Bulletin*, 25, 328–33.

Venkatesan, M. I. and Santiago, C. A. (1989) Sterols in ocean sediments: novel tracers to examine habitats of cetaceans, pinnipeds, penguins and humans. *Marine Biology*, 102, 431–7.

Venkatesan, M. I., Linick, T. W., Suess, H. E. and Buccellati, G. (1982) Asphalt in carbon-14-dated archeological samples from Terqa, Syria. *Nature*, 295, 517–9.

Venkatesan, M. I., Ruth, E. and Kaplan, I. R. (1986) Coprostanols in Antarctic marine sediments: a biomarker for marine mammals and not human pollution. *Marine Pollution Bulletin*, 17, 554–7.

Venosa, A. D., Suidan, M. T., Wrenn, B. A., *et al.* (1996) Bioremediation of an experimental oil spill on the shoreline of Delaware Bay. *Environmental Science & Technology*, 30, 1764–75.

Venosa, A. D., Suidan, M. T., King, D. and Wrenn, B. A. (1997) Use of hopane as a conservative biomarker for monitoring the bio-remediation effectiveness of crude oil contaminating a sandy beach. *Journal of Industrial Microbiology and Biotechnology*, 18, 131–9.

Vlierboom, F. W., Collini, B. and Zumberge, J. E. (1986) The occurrence of petroleum in sedimentary rocks of the meteor impact crater at Lake Siljan, Sweden. *Organic Geochemistry*, 10, 153–61.

Vogel, J. C. and van der Merwe, N. J. (1977) Isotopic evidence for early maize cultivation in New York State. *American Antiquity*, 42, 238–42.

Volkman, J. K. (1988) Biological marker compounds as indicators of the depositional environments of petroleum source rocks. In: *Lacustrine Petroleum Source Rocks* (A. J. Fleet, K. Kelts and M. R. Talbot, eds.), Blackwell, London, pp. 103–22.

Volkman, J. K. and Maxwell, J. R. (1986) Acyclic isoprenoids as biological markers. In: *Biological Markers in the Sedimentary Record* (R. B. Johns, ed.), Elsevier, New York, pp. 1–42.

Volkman, J. K. and Nichols, P. D. (1991) Applications of thin layer chromatography-flame ionization detection to the analysis of lipids and pollutants in marine and environmental samples. *Journal of Planar Chromatography*, 4, 19–26.

Volkman, J. K., Barrett, S. M., Blackburn, S. I., *et al.* (1998) Microalgal biomarkers: a review of recent research developments. *Organic Geochemistry*, 29, 1163–79.

Voparil, I. M. and Mayer, L. M. (2000) Dissolution of sedimentary polycyclic aromatic hydrocarbons into the Lugworm's *(Arenicola marina)* digestive fluids. *Environmental Science & Technology*, 34, 1221–8.

Wachter, E. A. and Hayes, J. M. (1985) Exchange of oxygen isotopes in carbon dioxide-phosphoric acid systems. *Chemical Geology*, 52, 365–74.

Wade, W. J., Hanor, J. S. and Sassen, R. (1989) Controls on H_2S concentration and hydrocarbon destruction in the eastern Smackover trend. *Transactions – Gulf Coast Association of Geological Societies*, 34, 309–20.

Waldo, G. S., Carlson, R. M. K., Moldowan, J. M., Peters, K. E. and Penner-Hahn, J. E. (1991) Sulfur speciation in heavy petroleums: information from X-ray absorption near-edge structure. *Geochimica et Cosmochimica Acta*, 55, 801–14.

Walker, A. A. (1998) Oldest glue discovered. www.archaeology.org/online/news/glue.html (accessed February 1, 2001).

Walker, A. L., McCulloh, T. H., Petersen, N. F. and Steward, R. J. (1983) Anomalously low reflectance of vitrinite in comparison with other petroleum source-rock maturation indices from the Miocene Modelo Formation in the Los Angeles Basin, California. In: *Petroleum Generation and Occurrence in the Miocene Monterey Formation, California* (C. M. Isaacs and R. E. Garrison, eds.), Society of Econonic Paleontologists and Mineralogists, Los Angeles, pp. 185–90.

Walters, C. C. (1990) Gases and condensated from Block 511A High Island, Offshore Texas. In: *Gulf Coast Oils and Gases: Their Characteristics, Origin, Distribution, and Exploration and Production Significance* (D. Schumacher and B. F. Perkins, eds.), Society of Economic Paleontologists and Mineralogists, Tulsa, OK.

Walters, C. C. and Cassa, M. R. (1985) Regional organic geochemistry of offshore Louisiana. *Transactions: Gulf Coast Association of Geological Societies*, 35, 277–86.

Walters, C. C. and Hellyer, C. L. (1998) Multi-dimensional gas chromatographic separation of C_7 hydrocarbons. *Organic Geochemistry*, 29, 1033–41.

Walters, C. C., Chung, H. M., Buck, S. P. and Bingham, G. G. (1999) Oil migration and filling history of the Beryl and adjacent fields in the South Viking Graben, North Sea. Presented at the Annual Meeting of the American Association of Petroleum Geologists, April 11–14, 1999, San Antonio, TX.

Wang, Z. and Fingas, M. (1999) Oil spill identification. *Journal of Chromatography A*, 843, 369–411.

Wang, X. and Mullins, O. C. (1994) Fluorescence lifetime studies of crude oils. *Applied Spectroscopy*, 48, 977–84.

Wang, H. D. and Philp, R. P. (1997b) Geochemical study of potential source rocks and crude oils in the Anadarko Basin, Oklahoma. *American Association of Petroleum Geologists Bulletin*, 81, 249–75.

Wang, Z., Fingas, M. and Sergy, G. (1994) Study of 22-year-old *Arrow* oil samples using biomarker compounds by GC/MS. *Environmental Science & Technology*, 28, 1733–46.

Wang, Z., Fingas, M., Blenkinsopp, S., *et al.* (1998) Study of the 25-year-old Nipisi oil spill: persistence of oil residues and comparisons between surface and subsurface sediments. *Environmental Science & Technology*, 32, 2222–32.

Wang, Z., Fingas, M. and Page, D. S. (1999a) Oil spill identification. *Journal of Chromatography A*, 843, 369–411.

Wang, Z., Fingas, M., Shu, Y. Y., *et al.* (1999b) Quantitative characterization of PAHs in burn residue and soot samples and differentiation of pyrogenic PAHs from petrogenic PAHs – the 1994 Mobile Burn Study. *Environmental Science & Technology*, 33, 3100–9.

Wang, Z., Fingas, M. and Sigouin, L. (2000) Characterization and source identification of an unknown spilled oil using fingerprinting techniques by GC-MS and GC-FID. *LC-GC*, 18, 1058–67.

(2001a) Characterization and identification of a "mystery" oil spill from Quebec (1999). *Journal of Chromatography A*, 909, 155–69.

Wang, Z., Fingas, M. F., Sigouin, L. and Owens, E. H. (2001b) Fate and persistence of long-termed spilled *Metula* oil in the marine salt marsh environment: degradation of petroleum biomarkers. In: *Proceedings of the 2001 International Oil Spill Conference, Tampa, Florida, March 26–29, 2001*, American Petroleum Institute, Washington, DC, pp. 115–25.

Waples, D. W. (1983) A reappraisal of anoxia and organic richness, with emphasis on Cretaceous of North Atlantic. *American Association of Petroleum Geologists Bulletin*, 67, 963–78.

Warburton, G. A. and Zumberge, J. E. (1982) Determination of petroleum sterane distributions by mass spectrometry with selective metastable ion monitoring. *Analytical Chemistry*, 55, 123–6.

Watanabe, K. (2001) Microorganisms relevant to bioremediation. *Current Opinion in Biotechnology*, 12, 237–41.

Watson J. T. (1997) *An Introduction to Mass Spectrometry*, 3rd edn. Lippincott-Raven, Philadelphia, PA.

Watson, D. F. and Farrimond, P. (2000) Novel polyfunctionalised geohopanoids in a recent lacustrine sediment (Priest Pot, UK). *Organic Geochemistry*, 31, 1247–52.

Watts, S., Pollard, A. M. and Wolff, G. A. (1999) The organic geochemistry of jet: pyrolysis-gas chromatography/mass spectrometry (Py-GCMS) applied to identifying jet and similar black lithic materials – preliminary results. *Journal of Archaeological Science*, 26, 923–33.

Weitkamp, J., Schafer, K. and Ernst, S. (1991) Selective adsorption of diastereomers in zeolites. *Journal of the Chemical Society, Chemical Communications*, 1142–3.

Wellings, F. E. (1966) Geological aspects the origin of oil. *Institute of Petroleum Journal*, 52, 124–30.

Wells, P. G., Butler, J. N. and Hughes, J. S. (1995) Exxon Valdez *Oil Spill: Fate and Effects in Alaskan Waters, (3rd ASTM Environmental Toxicology and Risk Assessment Symposium)*. American Society for Testing and Materials, Philadelphia, PA.

Wellsbury, P., Goodman, K., Barth, T., *et al.* (1997) Deep marine biosphere fuelled by increasing organic matter availability during burial and heating. *Nature*, 388, 573–6.

Welte, D. H., Horsfield, B. and Baker, D. R. (1997) *Petroleum and Basin Evolution*. Springer-Verlag, New York.

Wenger, L. M., Goodoff, L. R., Gross, O. P., Harrison, S. C. and Hood, K. C. (1994) Northern Gulf of Mexico: an integrated approach to source, maturation, and migration. Presented at the *First Joint American Association of Petroleum Geologists/AMGP Research Conference*, October 2–6, 1994, Mexico, Mexico.

Weser, U., Kaup, Y., Etspüler, H., Koller, J. and Baumer, U. (1998) Embalming in the Old Kingdom of pharaonic Egypt. *Analytical Chemistry*, 70, 511–6A.

West, N., Alexander, R. and Kagi, R. I. (1990) The use of silicalite for rapid isolation of branched and cyclic alkane fractions of petroleum. *Organic Geochemistry*, 15, 499–501.

Westgate, T. D., Ward, J., Slater, G. F., Lacrampe-Couloume, G. and Sherwood Lollar, B. (2001) Abiotic formation of C_1–C_4 hydrocarbons in crystalline rocks of the Canadian Shield. Presented at the Eleventh Annual V. M. Goldschmidt Conference, May 20–24, 2001, Hot Springs, VA.

Wever, H. E. (2000) Petroleum and source rock characterization based on C_7 star plot results: examples from Egypt. *American Association of Petroleum Geologists Bulletin*, 84, 1041–54.

White, D. (1999) *The Physiology and Biochemistry of Prokaryotes, 2nd edn*. Oxford University Press, New York.

White, R. and Kirby, J. (2001) A survey of nineteenth- and early twentieth-century varnish compositions found on a selection of paintings in the National Gallery Collection. *National Gallery Technical Bulletin (London)*, 22, 64–85.

Whitehead, E. V. (1971) Chemical clues to petroleum origin. *Chemistry and Industry* 1971, 1116–8.

(1974) The structure of petroleum pentacyclanes. In: *Advances in Organic Geochemistry 1973* (B. Tissot and F. Bienner, eds.), Editions Technip, Paris, pp. 225–43.

Whiticar, M. J. and Snowdon, L. R. (1999) Geochemical characterization of selected Western Canada oils by C_5–C_8 Compound Specific Isotope Correlation (CSIC). *Organic Geochemistry*, 30, 1127–61.

Whiticar, M. J., Faber, E. and Schoell, M. (1986) Biogenic methane and freshwater environments: CO_2 reduction vs. acetate fermentation – isotope evidence. *Geochimica et Cosmochimica Acta*, 50, 693–709.

Williams, J. A. (1974) Characterization of oil types in the Williston Basin. *American Association of Petroleum Geologists Bulletin*, 58, 1243–52.

Willsch, H., Clegg, H., Horsfield, B., Radke, M. and Wilkes, H. (1997) Liquid chromatographic separation of sediment, rock, and coal extracts and crude oil into compound classes. *Analytical Chemistry*, 69, 4203–9.

Wingert, W. S. (1992) GC-MS analysis of diamondoid hydrocarbons in Smackover petroleum. *Fuel*, 71, 37–43.

Winters, J. C. and Williams, J. A. (1969) Microbiological alteration of crude oil in the reservoir. *American Chemical Society, Division of Petroleum Chemistry, New York Meeting Preprints*, 14, E22–31.

Wischmann, H., Hummel, S., Rothschild, M. A. and Herrmann, B. (2002) Analysis of nicotine in archaeological skeletons from the Early Modern Age and from the Bronze Age. *Ancient Biomolecules*, 4, 47–52.

Woese, C. R. (2002) On the evolution of cells. *Proceedings of the National Academy of Sciences, USA*, 99 8742–7.

Woese, C. R., Magrum, L. J. and Fox, G. E. (1978) Archaebacteria. *Journal of Molecular Evolution*, 11, 245–52.

Wong, K. (1999) Is out of Africa going out the door? *Scientific American*, 281, 13–4.

(2001) Shakespeare on drugs? *Scientific American News Briefs*. www.sciam.com/news (accessed March 2, 2001).

Wooley, C. (2001) The myth of the "pristine environment": past human impact on the Gulf of Alaska coast. *Spill Science & Technology Bulletin*, 7, 89–104.

Worden, R. H., Smalley, P. C. and Oxtoby, N. H. (1995) Gas souring by thermochemical sulfate reduction at 140°C. *American Association of Petroleum Geologists Bulletin*, 79, 854–63.

Xiao, Y. (2001) Modeling the kinetics and mechanisms of petroleum and natural gas generation: a first principles approach. In: *Molecular Modeling Theory and Applications in the Geosciences: Reviews in Mineralogy & Geochemistry*, Vol. 42 (R. T. Cygan and J. D. Kubicki, eds.), The Geochemical Society and Mineralogical Society of America, Washington, DC, pp. 383–436.

Xiao, Y. and James, A. T. (1997) Is acid catalyzed isomerization responsible for the invariance in the isoheptanes of petroleum. In: *Proceedings of the 18th International Meeting on Organic Geochemistry, September 22–26, 1997, Maastricht, The Netherlands*. Forschungszentrum Jülich, Jülich, Germany, pp. 769–70.

Xu, L., Reddy, C. M., Farrington, J. W., *et al.* (2001) Identification of a novel alkenone in Black Sea sediments. *Organic Geochemistry*, 32, 633–45.

Yaws, C. L., Pan, X. and Lin, X. (1993) Water solubility data for 151 hydrocarbons. *Chemical Engineering*, 100, 108–11.

Yon, D. A., Maxwell, J. R. and Rybach, G. (1982) 2,6,10-Trimethyl-7-(3-methylbutyl)-dodecane, a novel sedimentary biological marker compound. *Tetrahedron Letters*, 23, 2143–6.

Yuen, G. U., Blair, N., Des Marais, D. J. and Chang, S. (1984) Carbon isotope composition of low molecular weight hydrocarbons and monocarboxylic acids from Murchison meteorite. *Nature*, 307, 252–4.

Yuen, G. U., Pecore, J. A., Kerridge, J. F., *et al.* (1990) Carbon isotopic fractionation in Fischer–Tropsch type reactions. *Lunar and Planetary Science*, XXI, 1367–8.

Yu, Z., Peng, P., Sheng, G. and Fu, J. (2000a) The carbon isotope study of biomarkers in the Maoming and the Jianghan Tertiary oil shale. *Chinese Science Bulletin*, 45, 90–6.

Yu, Z., Sheng, G., Fu, J. and Peng, P. (2000b) Determination of porphyrin carbon isotopic composition using gas chromatography–isotope ratio monitoring mass spectrometry. *Journal of Chromatography A*, 903, 183–91.

Zelt, F. B. (1985) Natural gamma-ray spectrometry, lithofacies, and depositional environments of selected Upper Cretaceous marine mudrocks, western United States, including Tropic Shale and Tununk Member of Mancos Shale. Ph. D. thesis, Princeton University, Princeton, NJ.

Zhang, J., Quay, P. D. and Wilbur, D. O. (1995) Carbon isotope fractionation during gas–water exchange and dissolution of CO_2. *Geochimica et Cosmochimica Acta*, 59, 107–14.

Zumberge, J. E. (1987) Terpenoid biomarker distributions in low maturity crude oils. *Organic Geochemistry*, 11, 479–96.

Index

abelsonite 68
abietadiene 57
abietane-type acids 331, 334
abietic acid 320, 331, 332, 336, 337
ab initio 194, 208
abiogenic hydrocarbon gas 153, 256–9
abiogenic hypothesis 253
abiogenic methane *see* methane
abiogenic petroleum 263, 270
abiotic synthesis 36, 253
abnormal pressure *see* overpressure
Abu Dhabi 154
Abu Durba, Gulf of Suez 324–5
Abu Durba seep, Gulf of Suez 326
Acacia 329
accessory pigment *see* carotenoid
Accretionary Wedge oils, Angola 153
accuracy 125
acetate 152, 254
acetate fermentation 316
acetic acid 53, 195, 316
aceticlastic bacteria 316
acetoacetyl-ACP (acyl carrier protein) 48
Acetobacter 333
acetogenic bacteria (acetogens) 4
acetyl-CoA (acetyl-coenzyme A) 48
acetyl-CoA carboxylase (ACC) 48
acetylene 253
Acholeplasma 47, 51
acidic clay minerals 157
acritarch fluorescence 95
acritarchs 81, 95
actinomycete 47
active source rock 72, 264
acute toxicity 286
acyclic isoprenoids *see* isoprenoids
acylation 212
acyl carrier protein (ACP) 48
adamantanes 157, 158, 162, 176
adduction efficiencies for *n*-alkanes 205
adipose fats 342, 343
Aegean Sea 305
aeration 282
aerobic 7, 10, 13

aerosol particles 318, 320
age-dating
  ceramics 332
  fuel spills 313–15
  gasoline 314
age-related biomarkers 205
ages of man 322, 323
aggregation 281
Akata-Agbada(!) 84
Akhmin, Egypt 324
Alabama Embayment, Gulf of Mexico 182
Alaska Peninsula 306, 307
Alberta 10
albertite 81
algae 47, 57, 61
algaenan 45, 76
algal 188
Algeria 192, 261–2, 263
alginite 75, 76, 89
Aliambata well, Indonesia 147
aliphatic hydrocarbon 19
alkaloid 352
alkane 18
  acyclic 21–2
  adduction efficiencies 205
  bimodal *n*-alkane distribution 104
  branched (*see* isoalkane)
  cyclic 202
  monocyclic 23–4
  normal (*see* normal alkane)
alkanoic acids 47, 344
alkene (olefin) 18–19, 20, 23, 128
  acyclic 23
alkenones 213, 215
alkylaromatics 334
alkylation 130, 132, 133
alkylbenzenes 31, 128, 206, 328
alkylbenzthiophenes 334
alkylcyclohexanes 206
alkylcyclopentanes 206
alkyldiamantanes 161
alkyldibenzothiophenes 81
alkylnaphthalenes 206, 304
alkylphenanthrene 206
alkyl pyrzaines 342

alkyltoluenes 206
alkylxylenes 206
Allen Hills meteorite (ALH84001), Antarctica 298
alloxanthin *see* carotenoid
Alondra Field, California 264, 265
Alsace-Lorraine 302
Altamont-Bluebell Field, Utah 84, 259, 260
alumina 199, 200
alumina/TOC 88
aluminosilicate 203
Alzheimer's disease 51
Amazon Basin, South America 25
American Petroleum Institute (API) 120
American Society for Testing Materials (ASTM) 119
amino acid racemization 341
amino acids 79, 338, 340, 341
δ-aminolevulinate 65
ammonia 52
Amoco Cadiz 278
amphipathic 42
amplification 348
amplitude variation with offset (AVO) 111
amyrin 59
amyrone 335
Anacardiaceae trees 334
anaerobic 7, 10
  methane oxidation 59
  respiration 8, 10
Anammox 52
anammoxosome membrane 52
Anasazi 349
Andector Field, Texas 84
angiosperms 268, 321
anhydrite 80, 154, 156, 159
aniline point 127
animals 60, 61
anisic acid 346
annular depression 269
anode 216
anoxia 10–13, 16, 79–80, 88
anoxic *see* anoxia
anteisoheptane *see* 3-methylhexane

Antelope Shale, California oils 141, 152
Antes Shale *see* Utica Shale
anthracene 296
anthracene/phenanthrene 298
anthracite 91
anthropogenic markers 275
antiknock additives 314
antioxidant 71
Antrim Shale, USA 82
apatite 338, 341
API gravity 102, 120, 182
apocarotenoid 70
Appalachian Basin, USA 152
Appleton Field, Alabama 181, 189
Apsheron Peninsula, South Caspian 115
aqueous vapor extraction 280
aquifer 283
Aquinet Ouernine *see* Tanezzuft Shale
Aquitaine Basin, France 154, 303
Arabia 332
arachidic acid 49
arachidonic acid 49
aragonite 138
archaea 3, 4, 5, 6, 29, 71, 254
  lipids 46, 48, 49, 50, 60
archaebacteria *see* archaea
archeological gums and resins 329–32
archeology 322–52
Arcy-sur-Cure, France 348
Åre (Hitra) Formation, Norway 187
argillaceous source rock 264
Argo Merchant 276
aristolochene 57
aromatic hydrocarbon UCM *see* unresolved complex mixture
aromatic hydrocarbons (aromatics) 19–20, 30–1, 64, 112, 127, 128, 155, 199
aromaticity 172, 175
aromaticity ratio (B) 174
aromatics 200
aromatization 89–90
art 333–4
arteriosclerosis 51
Arun Field, Sumatra 84
aryl carotenoids 71
aryl isoprenoids (trimethylbenzenes) 149
asphalt 122, 269, 333
  deposits in Middle East 328
  floating blocks 324
  Valdez 307
asphaltene-poor oils 144
asphaltene pyrolysis 300
asphaltenes 34, 126, 128
asphaltic paint 333
Asphalt Ridge heavy oil, Utah 120

As-Sabiyah, Kuwait 329
Asuka-881458 *see* carbonaceous chondrite
asymmetric carbon atom 34, 35, 36–9, 40, 271
Athabasca heavy oil, Canada 120
Athabasca tar sand 205
Atlantic Empress 278
atmospheric bottoms 132
atmospheric distillation 122, 130, 133
atmospheric-equivalent boiling point 122
atmospheric gas oil 122
atmospheric residue 122
atomic C/N 76
atomic H/C 73, 74, 76–9, 93, 99
atomic H/C vs. O/C *see* van Krevelen diagram
atomic O/C 76–9
atomic orbitals 19
atomic S/C 79
authigenic carbonate 111
automated data inquiry for oil spills *see* oil spills
autotroph 4, 8
avenasterol 60
axial bond 37
azeotrope 199
Aztec 352

Bach Ho Field, Vietnam 261
backflush 200
background hydrocarbons 112, 115, 301–2, 308, 311, 312
background petrogenic hydrocarbons 308–12
background subtraction 244
bacteria 6
bacterial oxidation 317
bacterial sulfate reduction (BSR) 315
bacteriochlorophyll 8, 64–8
bacteriohopanepolyols 58
bacteriohopanes 42, 58
bacteriohopanetetrol 41, 42, 56, 58, 59
bacteriohopanoids, polyfunctional 58
bacteriophage 5
bacteriopheophytin 67
bacteriorhodopsin 64
bacterioruberin 71
Baghewala well, India 106
Bagnolo oil, Italy 142
Bakken Shale, USA 92, 166, 182, 184, 188
Baku, South Caspian 115
ball-and-stick projection *see* structural notation
Bampo-Peutu(!) 84
Baota Formation, China 100

Barents Sea, Norway 12
barley beer 346
baseline 104, 106, 242, 244
baseline separation (resolution) 167, 168
baseline threshold 251
basement rocks 260, 261–8
base peak 222
Base-ring Ware 350
Bazhenov Formation, Western Siberia 11, 14, 99, 143
Bazhenov-Neocomian(!) 84
Beatrice oil, North Sea 144, 235, 248
beerstone *see* calcium oxalate
beeswax 324, 333, 344–6
behenic acid 49
Belayim Formation, Gulf of Suez 325
belemnites 139
Belemnitella 137
benchtop quadrupole 227
benzene 19–20, 30, 91, 280, 291, 314, 315
benzo(a)anthracene/chrysene 298
benzo(a)pyrene 287, 298
benzocarbazoles 33, 128, 194, 199
benzo(e)pyrene/benzo(a)pyrene 298
benzo[g,h,i]perylene 311
benzohopanes 235
benzoperylene 248
benzopyrene 296
benzothiophenes 31–2, 266
Bercy (Paris), France 342
Bering River, Alaska 307, 308, 310, 311
Berkine Trend, Algeria 262, 263
Beryl Complex, North Sea 109, 170
Beryl kitchen, North Sea 109
betulins 337
B-F diagram 174–5, 176–7
bicadinanes 206, 207, 208, 307
bicarbonate 138
Big Escambia Creek Field, Alabama 181, 189
Big Horn Basin, Wyoming 141, 154
bilayer 45, 46
bile 305
bile acids 43, 347
bilins 64, 65
bioaccumulation 287, 305
bioavailability 287
biodegradation 9, 103, 104, 116, 174, 187, 263, 281–2, 294–5, 315
biodegraded oil 113, 115
biofacies 10
biogenic 252–3
biogenic hydrocarbons 301, 308
biogenic markers 275
biogeochemistry *see* organic geochemistry
biological configuration 41, 42

biological marker *see* biomarkers
biological oxygen demand (BOD) 10, 11, 12
biomacromolecule 10
biomarkers (biological markers, molecular or chemical fossils) 3, 9, 10, 252, 274, 304
    aerosol particles 318, 320
    art 333–4
    age-related 205
    concentrations in oil 64
    laboratory 198
    mass chromatograms 228, 229–30
    quantitation 240–2
    refinery products 134–5
    sulfur-bound biomarkers 80
    smoke 319–21
    wine 344
biomass 5, 9, 63
biomass burning 318
biopolymer 24, 45
bioremediation 282, 302–4
biosphere 67
biosynthesis 45–71
biotic *see* biogenic
bioturbation 10, 11, 12, 13, 16, 111
biphenyls 31, 296
biphytane (bisphytane) 25, 26–7, 28–9
biphytanyl isoprenoids 52
biphytanyl tetraether 48
birch bark tar 336–7
bisabolene 57, 330
28, 30-bisnorhopane (BNH, 28, 30-dinorhopane) 26–7, 28–9, 34, 142, 197, 245, 266, 305, 306
bison bones 350
bisphytane *see* biphytane
bitomarosite 81
bitosite 82
bitumen 10, 73, 75, 199, 252
    archeological 322, 327, 329, 351
    indigenous 100–1
    mastic 328
bitumen network 100
bitumen/TOC 73, 74, 101
bituminites 76, 81, 82, 83
bivalves 305
Black Hawk Coal, Utah 77, 79
Black Sea 11, 12, 14, 215
black shale facies *see* Green River Formation
Blacksher Field, Alabama 189
bleaching 111
Bligh Reef, Prince William Sound, Alaska 305, 306
BNH *see* 28, 30-Bisnorhopane
boat, oldest 329

boat conformation 24
BOD *see* biological oxygen demand
boghead coal 78
bone 138, 338, 341, 350
    extracts 331
Boscan Field, Venezuela 84
Bosphorus 11
Boswellia 329, 332
boswellic acids 332
Botneheia Formation, Spitsbergen 116
botryococcanes 26–7, 28–9, 104
Botryococcus 78, 81, 82, 83, 104
bottled oils 116
bottoms 132, 133
bottom-simulating reflector (BSR) 111
Brantley Jackson Field, Alabama 189
Brassica (cabbage) 343
brassicasterol 60
Brazeau River Field, Canada 155, 156
breccia 266
Brent Formation, North Sea 83
bronze (copper–tin) 322
Bronze Age 322, 323
brown algae, macrophytic 65
Brown Limestone, Gulf of Suez 325
brucite 259
Bryan Mills Field, Alabama 189
BSR *see* bacterial sulfate reduction or bottom-simulating reflector
BTEX (benzene, toluene, ethylbenzene, xylenes) 286, 314, 315
bubble point 132
Buckner Anhydrite, Gulf of Mexico 159
Bucomazi Formation, West Africa 11, 84, 154
Bucomazi-Vermelha(!) 84
Bunker C 122
burning products
    cellulose 319
    lignin 320
butane 20, 21, 22
butanoyloxyfucoxanthin *see* carotenoid
1-butene 20
butene, *cis* and *trans* 23
Buzzard's Bay, Massachusetts 274, 276

[14]C 317, 318, 332
C3 plant (Calvin pathway) 138, 139, 338, 339, 343, 344, 345
C_3-dibenzothiophenes/C_3-chrysenes 297
C_3-naphthalenes/C_3-phenanthrenes 297
C4 plant (Hatch–Slack pathway) 138, 139, 338, 339, 341, 344
C_7 hydrocarbon analysis 162–92
C_7 hydrocarbon classes 190

C_7 hydrocarbons 168, 169, 190, 212
    maturity-sensitive, 171
C_7 oil correlation star diagram (C_7-OCSD) 178, 179, 181
C_7 oil transformation star diagram (C_7-OTSD) 178, 181
C_7 quaternary/(quaternary + tertiary) 170
C_{27}–C_{28}–C_{29} distribution *see* ternary diagram
C_{30} *ent*-isocopalane *see* tricyclohexaprenane
C_{30} tetracyclic polyprenoids *see* tetracyclic polyprenoids
cabbage *see Brassica*
cacao (cocoa) 352
cadinanes 208, 331
cadinene 55, 57, 330
caffeine 352
calcite 138, 264
calcium carbide 255
calcium naphthenates 126
calcium oxalate (beerstone) 346
calibration of mass scale 220
California, offshore 278
Calvin pathway *see* C3
CAM (crassulacean acid metabolism pathway) 139
Camamu-Almada Basin, Brazil 151
Cambay-Hazad(!) 84
Cambrian-Ordovician 145
Campeche, Mexico 279
campestane 62
campestanol 342
campesterol 60, 342, 346
camphene 56, 330
Canaan, Middle East 328
Canadian Shield 257, 259
Cannabis 351
cannel coal 81, 82, 83, 329
cannibalism 349–50
canonical variable (CV) 147
Canyon Diablo Troilite (CD) 137
Cape Yakataga, Alaska 308
capillary column 210
capillary gas chromatography 210
capric acid 49
carbazole 33
carbocation 40, 160, 161, 163, 170
carbohydrates 6, 45, 256
carbon 18, 124
    quaternary 163, 170
    secondary 163
    tertiary 163, 170
carbonaceous chondrite 253, 272
carbonado 157

carbonate 138
   carbon 74
   evaporite 264
   marine 189
Carbonate Platform oils, Angola 153
carbonate source rock *see* source rock
Carbon County, Utah 79
carbon cycle 8–9
carbon dendrites 216
carbon dioxide (CO_2) 8, 138, 139, 158, 195, 253, 316, 317
   reduction 316
carbon isotope ratio ($\delta^{13}C$) 338, 339, 343
carbon isotopes 137
carbonium ion *see* carbocation
carbon preference index (CPI) 104, 294, 308
carbon residue 125
carboxyl 152
carboxylic acids 34, 63, 151–2
28-carboxyursen-12-enol 346
carcinogen 287
carene 330
Carnarvon Basin, Australia 162
Carneros oil, California 201, 227, 229–30, 268
carnivore 339, 340
β-carotane (perhydro-β-carotene) 26–7, 28–9, 149, 236–7, 238–9, 259
carotene 69, 71, 149
carotenoid 8, 29, 47, 51–2, 54, 55, 64, 69–71
Carribbean Sea 278
carrier bed 264
carrier gas 208
caryophyllene 330
casing 92
Caspian Sea 13, 278
Castillo de Bellver 278
catabolism 346
catagenesis 9, 275
Catalan oil shale, Spain 80
catalysis 166
catalyst 132
   platinum-carbon 44
catalytic cracking 130, 133
catalytic isomerization 133
catalytic reforming 130, 133
catechin 344
cathode 216
cation exchange capacity 204
cave bear 340
caving 89, 90, 92, 94
^{12}C–^{12}C 140
^{12}C–^{13}C 140

^{13}C enrichment with decreasing age 144–7
C_2-dibenzothiophene/C_2-phenanthrene versus C_3-dibenzothiophene/C_3-phenanthrene (C_2-D/C_2-P versus C_3-D/C_3-P) 296
C_2-dibenzothiophene/C_2-phenanthrene (DPI or C_2-D/C_2-P) 309, 310, 311
CDT (*see* Canyon Diablo Troilite)
cell membrane 6, 46
cellulose 10, 47, 319
cell wall 47
Celtic oppidum 338
centifolia rose 55
centipoise (cP) 121
Central Appalachian Basin, USA 141
Central Graben, North Sea 83
Central Kansas uplift 268
Cerro Negro, Venezuela 102
   heavy oil 120
Cerro Prieto, Mexico 140
cerumen (earwax) 352
Chad's Creek Well, Canada 95
chain-of-custody 287
chair conformation 24
Chalain, France 329
Chalk Group, North Sea 83
channeling contrast microscopy (CCM) 195
charcoal 93, 323
Chatom Field, Alabama 84, 189
Chattanooga Shale, USA 82
cheilanthanes *see* tricyclic terpanes
chelate 34, 68
chemical fossil *see* biomarker
chemical ionization (CI) *see* ionization
chemoautotroph 4, 6
chemocline 11
chemoheterotroph 4
chemosynthesis 7
chemosynthetic communities 278
chemotroph 4, 253
chert 86
chewing gum 336
chirality 11, 34–6, 39
chiral stationary phase 36
chitin 321
chloramphenicol 6
chlorin 67
chlorobactene 71
Chlorobiaceae (green sulfur bacteria) 11, 56, 67, 68, 149
Chlorobium 80
chloroform 199, 200
chloroform/methanol 199
Chlorogloeopsis 63

chlorophyll 8, 64–8, 71, 252
   chlorophyll a 28, 33, 65, 66, 67, 68, 271
   chlorophyll b 29, 65
Chlorophyta 82
chloroplast 3, 6, 46
CHN analyses 123
cholane 62, 201, 202, 241
5β-cholestan-3β-ol *see* coprostanol
cholestane 25, 26–7, 28–9, 34, 36, 37, 38, 62, 135, 235, 236–7, 238–9
5α-cholest-5-en-3β-ol *see* cholesterol
cholest-8(9)-en-3B-ol 63
cholesterol (5α-Cholest-5-en-3β-ol) 6, 42, 51, 60, 61, 63, 275, 276, 320, 321, 341, 346
   %cholesterol/total lipids 276
choline 48
chondrillasterol 346
C_{35} hopane index *see* hopanes
Chromatiaceae (purple sulfur bacteria) 67
chromatogram *see* gas chromatogram
chromatography *see* gas or liquid chromatography
chromosome 6
chronostratigraphic unit 15
chrysenes 31, 280, 296, 297
Chrysophyte algae 254
Chunchula Field, Alabama 189
ciliate, marine 56
cineol 56
cinnabar 316
cis-butene 23
cis,cis,cis-5, 8, 11, 24-eicosatetraenoic acid *see* arachidonic acid
cis,cis,cis-9, 12, 15-octadecanoic acid *see* linoleic acid
cis,cis-9, 12-octadecanoic acid *see* linoleic acid
cis-configuration 23
cis-9-hexadecanoic *see* palmitoleic acid
cis-9-octadecanoic *see* oleic acid
citroneliol 55, 58
Cleistosphaeridium 82
Cleopatra 324, 327
cloud point 121
coal 78, 81, 82, 138, 252, 255, 258, 298, 308, 311
coalbed methane 153
coal-source hypothesis 308–10
coal tar 298
coal versus seep hypotheses, Alaska 308–12
coaly shale 81, 82
Coast Guard Marine Safety Laboratory 299
cocaine 351–2

coccolithophorid (coccolithophore) 149
Coccus 329
cocoa tree *see Theobroma*
Code of Hammurabi 328
codeine 350
co-elution 148, 203, 233, 240
coenzyme Q *see* ubiquinone
coffee 352
Cogollo Formation, Venezuela 267
co-injection 240
coke 125, 133
coker 134
coking 130, 133
Cold Creek Field, Alabama 189
Cold Lake heavy oil, Canada 120
cold spot 248, 249
cold trapping 209
collagen 338, 341
    carnivores 338
    herbivores 338, 341
    human 339
    Neanderthal 341
    omnivores 52
collinite 89, 90
collision-activated decomposition (CAD) 227, 233
collision cell quadrupole (Q2) 218, 227, 231
collision gas 227
Columbia River basalt 254
column bleed 244
column chromatography 198, 199–200
column overload 244, 245
column temperature 251
combustion-isotope ratio mass spectrometer (combustion-IRMS) 138
commingled fluids 106–9
Commiphora 329
Como Field, Alabama 189
compartmentalization, reservoir 109
%compound depleted 295
compound-specific isotope analysis (CSIA) 148–52, 155, 190–2, 203
    alkanes 151
    archeology 345
    correlation 150–1, 272, 291, 300–1
    fatty acids 342
    gasoline 315
    paleoenvironment 149–50
    sediments 301
computational chemistry 208
Conception Bay, Newfoundland 300
condensate-associated gas 153
condensates 116, 120, 151, 203, 205, 255
condensate-to-gas ratio (CGR) 159

condensed interval 86
configurational isomerization *see* stereoisomerization
confocal scanning laser microscopy (CSLM) 195
conformation 24, 193, 207, 212
Congo copal 329, 334
congressane *see* diamantane
conifers (gymnosperms) 321
conifer resins 324, 329, 331, 332
conjugation 5
conodont alteration index (CAI) 95
Conradson carbon residue 125
conservation of hopane 294–5
contamination 113
    archeology 349, 350, 351
    bitumen and oil 100, 101, 107, 143, 203
    environmental 274
    rock 92, 94, 95, 115
    sediment or water 113, 294
Cook Inlet, Alaska 174, 306
coorongite 78
Copper Glacier, Alaska 311
coprecipitation 144
coprolites 347, 349–50
coporphyrinogen 65, 66
coprostanone 275
coprostanol (5β-cholestan-3β-ol) 275–6, 346
coprostanol/epicoprostanol (cop/e-cop) 276
cordaites *see* gymnosperms
core 113
corg *see* total organic carbon
Coriolis 8
coronene 31
correlation, stable carbon isotopes 141–4, 150–1
    see also oil–oil and oil–source rock correlation
co-sources, oil 144
cotinine 351
covalent (sigma) bond 18, 19, 20, 34
Cowboy Wash, Colorado 349
CPI *see* carbon preference index
Crassulacean acid metabolism pathway *see* CAM
Crenarchaeota 50
creosote 298, 302, 336
cresol 346
Cretaceous black shale, Cape Verde Rise 88
Cretaceous mass extinction *see* Cretaceous–Tertiary (K–T) boundary

Cretaceous–Tertiary (K–T) boundary 146
Crete 345
C-ring monoaromatic steroids *see* monoaromatic steroids
crocetane 60
Cro-Magnon 349
Crosby Creek Field, Alabama 189
cryptophyte 65
β-cryptoxanthin *see* carotenoid
crystallographic free diameter 204
CSIA *see* compound-specific isotope analysis
$^{\circ}C_{temp}$ *see* temperature of generation
C_{29} Ts, 325
cutinite 76, 89
cuttings, drill 113
cyanobacteria 7, 59, 67
Cycladic Islands 345
cycloalkane (naphthene, cycloparaffin) 23, 24, 128, 172
cycloalkanoporphyrin (CAP) 68
cycloartenol 56–8, 59, 60
cyclobutane 52
cyclodextrin 211–12
cyclohexamantane 157
cyclohexane 23, 24, 91, 192
cycloparaffin *see* cycloalkane
cyclopentane 23
cyclopropyl intermediate 166
Cypress Creek Field, Alabama 189
cytoplasmic membrane 47
cytosine 59
Czech Republic 268–9

dairy practice 343–4
Dakhleh Oasis, Western Desert 325
Dala granites 269
Dalbergia 332
Dalton 222
damascenone 58
Damborice well, Czech Republic 268, 269
dammaradienone 335
dammaranes 333, 334
dammarane-type molecules 335
dammarenolic acid 334, 335
dammar resin 55, 61, 207, 329, 334
Daphnia 286
darmarenolic acid 331
Darwinian threshold 53
data processing 220–2
data system overload 245, 246
daughter ion 218, 229, 231
daughter mode (in GCMS/MS) 227, 234
daughter quadrupole (Q3) 227, 232
DB-1 stationary phase 211, 244
DB-5 stationary phase 211

Dead Sea, Eastern Mediterranean 328, 341
Dead Sea asphalt 324–5, 326, 328, 334
Dean's splitter 212
deasphalting 133
decane 21
decision tree 287
Decorah Formation, USA 80, 105
deep-Earth gas hypothesis 253–6
deep hot biosphere hypothesis 254
deep induction log 195
Deep Sea Drilling Project (DSDP; later called Ocean Drilling Program, ODP) 74
defocusing 248
dehydroabietic acid 320, 329, 331, 332, 334, 336, 337
delta log R 83–6
Delta National Wildlife Refuge, Mississippi 289
delta-value, isotope 136
demethylated hopanes see norhopanes, 28,30-bisnorhopane, 25,28,30-trisnorhopane
dendrogram, HCA 107
density 120, 129
    floatation 75
Denver Basin, Colorado 174
deoxophylloerythroetioporphyrin (DPEP) 67, 68
deoxyribonucleic acid (DNA) 5, 347–9
    sequencing 60
Derdere Formation, Turkey 100
desalting 130
10-desmethylhopanes see norhopanes
desmethylsteranes 254
Desulfosarcina 316
Desulfovibrio 79
detection limit 125
deuterated standards 202
deuterium 138
Dexsil 211
dextral see R configuration
dextrorotatory 36, 271
diacholestanes 26–7, 28–9, 236–7, 238–9
diadinoxanthin see carotenoid
diagenesis 9, 79
diahopanes (C_{29}*–C_{34}*) 24, 206, 207
dialkylthiacyclopentane 81
diamantane (congressane) 157, 158
diamond 138, 157
diamondoids 157–62
    biodegradation 162
    maturity parameters 162
    source parameters 161–2
diapocarotenoid 70

diasterane index see diasteranes/steranes
diasteranes (rearranged steranes) 40, 266, 324, 325
diastereomer 35, 38, 39
diatomaceous 13
diatoms 8, 65, 252, 254, 268
diazotroph 4
dibenzofuran 33
dibenzothiophene/phenanthrene 12
dibenzothiophenes 32, 81, 296, 299, 308
dichloromethane see methylene chloride
dicot see angiosperm
Di-DPEP porphyrin 68
diesel 122, 132, 133, 134, 213, 214, 299
diet see paleodiet
dietary lipids 342–3
diether lipids 48
dihydro-ar-curcumene 236–7, 238–9
dihydro-bacteriochlorophyll 65
Dillinger Ranch Field, Wyoming 104
dimethylalkyl pyrophosphate (DMAPP) 53–5
dimethylallyl alcohol 55
2, 2-dimethylbutane/2, 3-dimethylbutane
4, 4-dimethylcholestene 63
dimethylcyclopentanes 23, 188
dimethyldiamantane facies ratios 161
2, 6-dimethylheptane 21
3, 4-dimethylhexane 206
dimethylnaphthalenes 311
2, 6-dimethyloctane 25, 26–7, 28–9, 30, 271
2, 4-dimethylpentane/2, 3-dimethylpentane (2, 4-DMP/2, 3-DMP) 179–83
dimethylpentanes 22
dimethylphenanthrenes 200, 207
2, 2-dimethylpropane 22
4, 4-dimethylsterols 63
Dingo Claystone, Australia 81
dinoflagellates 65, 81
    cysts (hystrichospheres) 150
28, 30-dinorhopane (DNH) see 28, 30-bisnorhopane
dinosaur bones 350
dipentene see limonene
diphtheria 6
diploptene 56, 59, 60, 61
diplopterol 56, 59, 60
α,ω-dipolar carotenoid 51
Dipterocarpaceae 207, 332, 334
direct injection see injection
dispersant 282
dispersion 280
disproportionation 143
dissociation constant 126
dissolution 280

distillation 122–3
    cuts 122, 127
    simulated 123, 129
    tower 132, 134
distributed source-rock sampling 151, 272
disulfides 31, 32, 80
diterpanes 25, 211
diterpenoid 55, 57
Djedoler 324, 327
DNA see deoxyribonucleic acid
dodecane 21
Dolni Lomna well, Czech Republic 268, 269
dolomite 258–9, 315
domains of life 3–5, 6, 46, 48, 49, 52, 64
Dominguez Field, California 265
Douala Basin, Cameroon 89
double bond 19
double-focusing mass spectrometer 217, 229
Doushantuo Formation, China 100
downstream 119
DPEP see deoxophylloerythroetioporphyrin
Draupne Shale, North Sea 83
drill-bit metamorphism 255
drilling fluid contamination 107
drilling mud 115, 255
drill stem test (DST) 107, 110
drimane 206
dry gas 166, 317
drying and sweetening 130
Dry Piney Field, Wyoming 104
dry valley 4
Dukthoth River, Alaska 311
Dunaliella 71
Duperow Formation, USA 182
Duri Field, Sumatra 84
Duvernay Formation, Canada 156, 162
dwell time 227
dyes 313
dynamic combustion 137
dysaerobic 10
dysoxic see suboxic

earwax see cerumen
East Midland Field, UK 263
Eel River Basin, California 240, 242, 248, 316
effective source rock 9, 16, 17, 72
Egemba Field, Niger Delta 84
Egret Member, Rankin Formation, Newfoundland 83, 146, 233
eicosane 21
Ekofisk Field, North Sea 84
Ekofisk Formation, North Sea 83

eladic acid 346
elastic recoil detection analysis (ERDA) 195
Eldfisk Field, North Sea 84
electric sector field strength (E) 229
electron 137
    impact (EI) *see* ionization
    multiplier (discrete dynode) 209, 216, 217, 218, 220, 227
    transport 65
    trap 215
electrostatic mass analyzer 217
electrostatic sector 217
elemental analysis 123–4
El Segundo Field, California 264, 265
embalming resin 345
Emiliania 149
emitter 216
empirical force field 193
emulsification 281
emulsified 122
emulsion 281
enantiomer 34, 35, 36, 39
endosymbiosis 5
energy consumption 10
engine knock (pre-ignition) 132
environmental chemistry procedures 287–9
environmental markers 274–6
Environmental Protection Agency (EPA) 274, 285, 312
    Center for Subsurface Modeling Support (CSMoS) 315
    ECOTOX (ECOTOXicology) AQUIRE (aquatic life) database 286
Environmental Technology Center 299
EPA *see* Environmental Protection Agency
epi-cedrol 57
epicoprostanol 276
epi-cubenol 57
epicuticular wax 47, 104
epimer *see* diastereomer
epoxy 89
equation of state (EOS) 120
equatorial bond 37
equatorial carbonate source rocks 146
equilibrium exchange 138
ergostane 25, 26–7, 28–9, 62
ergosterol 60
Erika 300, 301
eroded section 95
Eromanga Basin, Australia 207
error 251
Erthroxylon 351
Erwinia 71

Escherichia 63
essential amino acids 338, 340
estimated ultimate recovery (EUR) 260
ethane 21, 158
ethane/ethene 113
ethanol 313
ether:hexane 200
ethyl *see* functional group
ethyladamantane ratio 161
ethylbenzene 291, 314
24-ethylcholestane *see* stigmastane
24-ethyl-5β-cholestan-3β-ol 347
ethylene 19
ethylene dibromide 313, 314
ethylene dichloride 313, 314
3-ethylpentane 22
etioporphyrin 67, 68
ETR (extended tricyclic terpane ratio) *see* tricyclic terpanes
eubacteria 3, 4, 46, 49
eudesmane 26–7, 28–9
*Eugene Island, Gulf of Mexico 110, 185
eukarya *see* eukaryote
eukaryote 3, 4, 6, 43, 46, 49
euphol 59
Euphorbiaceae 25
Euphrates River, Iraq 326
Europa 254
eutrophic 8
euxinic 12, 80
evaporation 103, 280
evaporative fractionation 173–6
evaporative lithofacies 16
evaporite 189
even-carbon preference *see* carbon preference index
evolution 53
exinite 75, 89
expelled petroleum (S1) 98–100
expulsion 264
expulsion efficiency (ExEf) 98–100
Exshaw Formation, Alberta 10
extended hopanes *see* hopanes
extended tricyclic terpane ratio (ETR) *see* tricyclic terpanes
extractor plate 216
Exxon Valdez 142, 276, 278, 282, 295, 296, 297, 298, 303, 305–12

faculative aerobe 4
faculative anaerobe 4
farnesane 25, 26–7, 28–9, 271
farnesene 57, 330
farnesol 54, 55
farnesyl pyrophosphate (FPP) 54, 57
Farsund Formation, North Sea 83
fatty acids 30, 46, 47–8, 49, 342

biosynthesis 48
C$_4$–C$_{14}$ 343
C$_{16:0}$ 343
C$_{18:0}$ 343
milk 343, 344
faujasite 204
faults 107, 110–11, 112–13
    migration 110
    pemeability 110
    sealing capacity 110
FC-43 *see* perfluorotributylamine
fecal pellets 281
fecal pollution 275–6
Fedorov Field, West Siberia 84
feedstock 119, 130, 134
fenchol 56
fennel 346
Fennoscandian Shield 257, 259
Fergana Valley, Uzbekistan 278, 279
fermentation 8, 316, 317, 344
fernenes 61
ferns 60
fertilizer 302
fichtelite 26–7, 28–9
FID *see* flame ionization detector
field ionization (FI) *see* ionization
field ionization mass spectrometry (FIMS) 129
filament 215, 217
fingerprint 9, 272, 288
fingerprinting of oil spills 289–5
    non-specific methods 290–1
    source-specific methods 291
fire point 121
Fischer–Tropsch 253, 257, 258, 259, 272
fixation, carbon 7
fixation, nitrogen 7
fjord 12
flame ionization detection 220
flame ionization detector (FID) 77
flamingo 71
flash point 121
flavanol 344
flight tube 217
floral scents 58
flowering plants *see* angiosperms
fluctuating profundal lithofacies 16
fluid catalytic cracker (FCC) 134
fluid catalytic cracking 133
fluid inclusion 194–7, 258
fluid inclusion stratigraphy (FIS) 196
fluid inclusion volatiles (FIV) 195, 196
fluoranthene/pyrene 298
fluoranthenes 301, 308, 328
fluorenes 296
fluorenones 33
fluorescence intensity (I$_{max}$) 95

fluvial-lacustrine lithofacies 16
focusing plates 215
forensic environmental chemistry 313
Forties Field, North Sea 84
fossil fuel 8–9
    burning 344
Four Corners area, USA 349
Fourier transform infrared analysis
        (FTIR) 195, 291
Fourier transform-ion cyclotron
        resonance-mass spectrometry
        (FT/ICR/MS) 129
fractional conversion (f) 98–100, 117, 118,
        158
fragment ion 216
fragmentogram see mass chromatogram
fragrance 55
    see also odor
framework density 204
frankincense 329, 332
freesias 55
freezing-point depression 194
friedelane 26–7, 28–9
FTIR see Fourier transform infrared
        analysis
fucosterol 43
fucoxanthin see carotenoid
full-scan GCMS 221, 227
functional group 20
fungus 61
furan 33
furfural extraction 130
fused silica capillary column 210
fusinite 76

Galicia Coast, Spain 305
gallic acid (3, 4, 5-trihydroxybenzoic acid)
        344
gammacerane 26–7, 28–9, 36, 39, 56, 146,
        147, 211, 271, 324, 325, 334
gammacer-3β-ol see tetrahymanol
gamma-ray response 195
gamma-ray spectral log 86, 97, 255
Gandhar Field, India 84
Gardena well, Los Angeles Basin 265, 266

gas chimney 111
gas chromatogram 30, 104, 105, 180, 211
gas chromatograph 209
gas chromatograph/mass spectrometer
        209
gas chromatography (GC) 103–9, 112–13,
        208–15, 290
    chiral 211–12
    comprehensive two-dimensional GC
        212–13
    correlation 105–6

extended hold-time 212
    high-temperature 127, 342
    two-dimensional (GC × GC) 169,
        212–14, 215
gas chromatography/mass spectrometry
        see GCMS
gas chromatography/mass
        spectrometry/mass spectrometry
        see GCMS/MS
gas composition in soil 317
gas hydrate 111
gas/liquid chromatography see gas
        chromatography
gas maturity 140
gasohol 314
gas oil 132
gas oil feed (GOF) 134
gas–oil ratio (GOR) 102, 158, 159, 175
gasoline 122, 132, 133, 163, 312–13, 314,
        315
    modeling leakage 315
gasoline-naphtha (C$_4$–C$_{11}$) hydrocarbons
        163
gasoline-range hydrocarbons 156
Gas Processors Association (GPA) 83
gas-prone 79
gas wetness 113, 153
GC see gas chromatography
GC/FID see gas chromatography
GC × GC see gas chromatography
GC × GCMS 214
GC/combustion/IRMS (also called
        CSIA) 138
GCMS 195, 208–15, 222, 291
    data problems 242–51
GCMS/MS 221, 227–9
Gebel Zeit, Gulf of Suez 324–5,
        326
gelatin 333
gel-permeation chromatography
        127
gene 5, 53, 60
genetic potential (S1 + S2) 96
geochemical fossil see biomarker
geochemical log 96–7, 98
geological configuration 41, 42
geometric isomer 23
geometry optimization 193
geoporphyrin see porphyrin
geosphere 63
geothermal gradient 94
geraniol 55, 58
geranyl diphosphate 56
geranyl pyrophosphate (GPP) 54, 56
geranylfarnesol 55
geranylgeraniol 55
geranylgeranyl diphosphate 57, 69

geranylgeranyl pyrophosphate (GGPP)
        54
germacrene 57
Ghareb Formation, Jordan 324, 325
Ghawar Field, Saudi Arabia 84, 179
gilsonite 149
Gin Creek Field, Alabama 189
Gippsland Basin, Australia, 162
glial cell 51
global positioning satellite (GPS) 111
Gloeocapsomorpha 80, 81, 82, 105
glucose 6, 256
glue, oldest 341
glutamate 65
glyceraldehyde 36
glyceraldehyde-3-phosphate 53
glyceride 6
glycerol 48
glycerol dialkyl glycerol tetraethers
        (GDGTs) 50
glycerolipid 49
glycine 65
glycolipid 46, 47, 48, 49
glycoprotein 45, 47
GOR see gas–oil ratio
Gordion, Turkey 345
Gorgostane 62
Goslar, Germany 351
Gotlandien Shale see Tanezzuft Shale
gouge zone 110
Gough's Cave, UK 339
gracilicutes see Gram-negative bacteria
grain size 13
Graminaeae (grasses) 321
Gram-negative bacteria 47, 59
Gram-positive bacteria 47, 59, 316
Gram stain 47
graphite 3, 90, 91
grasses see graminaeae
Gravberg well, Sweden 253, 255, 256
gravimetric measurement 290, 291
gravity segregation 107
grease 133
grease compounding 130
Great Alaska Earthquake 306
Greenhorn Shale, USA 149
greenhouse gases 315–16
Green River(!) 84
Green River Formation, USA 11, 16, 68,
        78, 81, 82, 149, 206, 240, 259, 261,
        262
Green River paleolake 149
Green River-type oil shale see lacosite
greenschist facies (memormism) 9
green non-sulfur bacteria 67
Green sulfur bacteria see Chlorobiaceae
Green well, Iowa 104

groundwater modeling 315
group-type fractionation (SARA or PARA) 128–9
Grugan Field, Pennsylvania 152
guanine 59
guard column 200, 201
Guayuta Group oils, Venezuela 103
Gulf of California 12, 13
Gulf Coast oils 105
Gulf of Mexico 278
Gulf War, 1991 278
Gullfaks Field, North Sea 107
gum arabic 329
gum, archeological 329–32, 336
gutta-percha 52, 55
Guttenberg Member, Decorah Formation 80, 105
gymnosperm *see* conifers
gypsum 80

Hafkenscheid Collection 333
hafting (adhesive) 329
hair, human 338
Hajji Firuz, Iran 332
half-life 317
Hall Creek Field, Alabama 181, 189
halobacteria 64
halocline 11
halophile 5, 71
halophilic bacteria *see* halobacteria
Halpern parameters 177–9, 181, 187, 191
Haltenbanken area, Norway 110
Hamilton Dome, Wyoming 64, 201, 211, 227, 228, 232, 234–5, 250
Hanging Gardens of Babylon 322, 328
Hanifa-Arab(!) 84
Hanifa-Hadriya, Saudi Arabia 94
hardground 86
Harz Mountains, Germany 336, 337
hashish (tetrahydrocannabinol) 351
Hassi Messaoud Field, Algeria 192, 262, 263
Hassi R-Mel, Algeria 262
Hatch–Slack pathway *see* C4
Hatter's Pond Field, Alabama 84, 189
Haven 278, 282, 297
HBI *see* highly branched isoprenoids
H/C *see* atomic H/C
Headlee Field, Texas 84
head-to-head isoprenoids 25, 26–7, 28–9
head-to-tail isoprenoids 25, 26–7, 28–9
heart cutting 212
Heather Formation, North Sea 109, 170, 173, 181
heating oils 122
heavy coker 134
heavy oil 120

Hebron Field, Newfoundland 84
Helicobacteria 67
helium 255
Helminthoides 12
hemar 322
hematite 110–11
heme 65, 67, 68
hemipelagic sediments 14
hemiterpane 25, 26–7, 28–9, 55
hemiterpenoid 55
hen-bane 351
heptane 21, 22, 132, 164, 188
heptane ratio 171–3, 187, 188
heptylbenzene 31
herbivore 339, 340
Hercynian unconformity 262, 263
Hertfordshire, UK 347
heteroatomic molecules 31–4
    see also NSO compounds
heterotroph 4
Hevea 25
hexachlorodisilane 150
hexamantanes 157, 158
hexane 21, 167, 199, 200
hexanoyloxyfucoxanthin *see* carotenoid
HI *see* hydrogen index
Hibernia Field, Newfoundland 84
high-acid oil 126
higher hydrocarbons 167
higher plants 7, 57
High Island, Gulf of Mexico 175
high-latitude siliciclastic source rocks 146
highly branched isoprenoids (HBI) 30
    C_{25} HBI 26–7, 28–9, 232
high-performance liquid chromatography (HPLC) 68, 102, 148, 198, 200, 201
highstand systems tract (HST) 87
high-sulfur oil 80
high-temperature gas chromatography 127
histogram, reflectance 89
histone 6
Hitra Formation *see* Åre Formation
Hit-Ramadi-Abu Jir, Iraq 326, 328
homocadinanes 208
Homo erectus 347, 348
homogenization temperature 194
homohopanes *see* hopanes
homohopanoic acids 60
homolog 21, 25
homologous oils 164, 165
homologous series *see* homolog
Homo sapiens 347
hopane isomerization [22S/(22S + 22R)] 268

hopanes 34, 41–2, 64, 149, 206, 207, 211, 240, 268, 334
    C_{28} $\alpha\beta$-hopanes 271
    C_{30} hopane 24, 26–7, 28–9
    C_{35} hopanes 193, 325
    17α, 21β(H)-hopanes ($\alpha\alpha$-hopanes) 303
    17β, 21α(H)-hopanes ($\beta\alpha$-hopanes) *see* moretanes
    22S and 22R 207, 208
hopanoic (hopanoid) acids
    C_{32} hopanoid acid 63
hopanoids 42, 49, 51, 52, 55, 58–61, 63–4
hopanols 51
Hopea 61
hop-17(21)-enes 61
hop-21-ene 61
horizontal drilling 102
horizontal gene transfer (HGT) 5, 53, 71
hormones 61, 63
hot shale 255
HPLC *see* high-performance liquid chromatography
human hair 338
humic coal 79, 81, 82
Hummal, Syria 326–8
hump *see* unresolved complex mixture
humulene 57
Hunt parameters 170
Huqf Formation, Oman 147
Huron Shale, USA 141, 153
Huxford Field, Alabama 181, 189
hybrid mass spectrometer 228
hydrazine 52
hydride abstraction 216
hydride ion 40
hydrocarbon 73
    biogenic 301, 308
    input to marine environment 276
    petrogenic 301, 308
    pyrogenic 302
hydrocarbon gases 140–1, 152
hydrocarbon/TOC 101
hydrocracker 134
hydrocracking 130, 133
hydrodesulfurization 130, 133
hydrofiner 134
hydrofinishing 134
hydrogen 124, 253
hydrogen index (HI) 73, 77, 78
    original quality (HI°) 97–100, 117
hydrogen index vs. oxygen index *see* van Krevelen diagram, modified
hydrogen isotopes 150, 152–3, 317
hydrogenosome 5
hydrogen radical 40
hydrogen steam reforming 130

hydrogen sulfide (H_2S) 11, 12, 79–80, 155, 158, 195, 315
hydrolysis, acid 74
hydrolysis constant 126
hydromethylbilane 65
hydrophilic 45
hydrophobic 45
hydroprocessing 132
hydrostatic 264
hydrothermal circulation 264, 265
hydrothermal methane 259
hydrothermal vent 6
hydrotreating 130, 132, 133
hydrotroilite 80
hydrous pyrolysis 101, 259, 261, 262
hydroxyapatite 350
hydroxyl *see* functional group
hydroxylamine 52
hydroxyl wax esters 344
hydroxyoleanonic acid 333
hyodeoxycholic acid 346
Hyperion Field, California 264, 265, 266
Hyperion sewage treatment plant 276
hyperthermophile 50
hyphae 60
hystrichospheres *see* dinoflagellate cysts

Iabe-Landana source rock, Angola 153
Iatroscan 102, 103, 128, 290
Ice Man 336
ichnofossil *see* trace fossil
Idlube 255
Ill River, Alsace-Lorraine 302
illite 16
I_{max} *see* fluorescence intensity
immature 73
immunological detection assay (ELISA) 349
Indian Ocean 278
incense 329, 332
incorrect mass range 247
indeno[1, 2, 3-*cd*]pyrene/benzo [*ghi*]perylene 298
indole 33
indoline 33
inertinite 75, 76
inertodetrinite 76
infauna 13
infrared (IR) 74, 203, 290
injection 209–10
    port 209
insects 55
in situ burning 282
inspissation (evaporation) 102
interfering peaks 245
internal standard 200–2, 240
International Atomic Energy Agency 136

International Union of Pure and Applied Chemistry *see* IUPAC
intramolecular equilibrium isotope effect 139
ionization
    chemical ionization (CI) 216
    electron impact (EI) 215–16
    field ionization (FI) 216–17
ionizing chamber *see* ion source
ionizing voltage 215
ionoluminescence (IL) 195
β-ionone 55, 58
ion sampling frequency 243–4
ion source (ionizing chamber) 209, 215, 216, 217, 248
ion volume 248
Iraq oils 177
IRM/GCMS 148
Iron Age 322, 323
iron/sulfur 87
isoalkane 21, 128, 172, 202
isobutadiene *see* isoprene
isobutane *see* 2-methylpropane
isofucosterol 60
isoheptane *see* 2-methylhexane
isoheptane abundances 164
isoheptane invariance 164, 183–4, 194
isoheptane ratio *see* K_1
isohexane *see* 2-methylpentane
isolimonene 330
isomasticadienoic acid 333
isomer 21
    abundance 271
isomerization 130, 132
isooctane (2, 2, 4-trimethylpentane) 132
isoparaffin *see* isoalkane
isopentane *see* 2-methylbutane
isopentane/*n*-pentane 163
isopentenoid *see* terpenoid
isopentyl pyrophosphate (IPP) (*also called* isoprene pyrophosphate) 53–5
isopimaric acid 320, 331, 332
isoprene (methylbutadiene or isopentadiene) 24, 26–7, 28–9
isoprene rule 24–30, 271
isoprenoids 24, 206, 329
    acyclic 41, 271
    biosynthesis 54–5
    regular 30
isorenieratane 149
isorenieratene 11, 71, 149, 193
isotope 136
    see also stable isotopes
isotope fractionation 138–9, 203, 252
isotopic-age trend 146
isotopic reference samples 344
isotopic reference standards 137

isotope ratio monitoring/gas chromatography/mass spectrometry *see* IRM/GCMS
Isthmia, Greece 345
IUPAC (International Union of Pure and Applied Chemistry) 22, 204
Ixtoc well, Gulf of Mexico 278, 279

jasmine 55
Jauf Reservoir, Ghawar Field, Saudi Arabia 179
Jay Field, Alabama 189
Jebel Bichri, Syria 328
jet fuel 132, 133, 134
Jet Rock, Yorkshire, UK 116, 329
Jianghan Basin, China 68, 100, 150
Jianghan oil shale, China 150
Jobo heavy oil, Venezuela 120
Junggar (Zhungeer) Basin, China 16, 197

K_1 (Mango parameter) 156, 164, 171–3, 183–4, 187, 188
K_2 (Mango parameter) 165, 183–4, 185
K_3 (Mango parameter) 165
K_4 (Mango parameter) 165
Karababa Formation, Turkey 100
Karabogaz Formation, Turkey 100
Karalian shungite, Russia 81
Kareem Formation, Gulf of Suez 325
Katalla Beach, Alaska 307, 311
Katalla oil seep, Alaska 297, 308, 309, 310–11
kaurane 26–7, 28–9, 57
Kauri copal 329, 334
Kellis Tomb, Western Desert 325
kerogen 9, 10, 63, 75, 89, 145, 166, 167, 252, 264, 271
    kerogen 73, 76–80, 86
    see also type I, II, IIS, II/III, III, IV
kerosene 122, 132, 133
ketones 213, 342
ketonic decarboxylation 342
Khibina intrusion, Russia 258
Khuzestan, Iran 325
Kidd Creek Mine, Ontario, Canada 258
Kimmeridge-Brent(!) 83, 84
Kimmeridge Clay (Kimmeridge) Formation, North Sea 11, 86, 106, 109, 170, 181, 255
    oil 173
Kimmeridge-Hibernia(!) 83, 84
kinetic fractionation 138
kinetosome 5
King Midas 345
Kirkuk, Iraq 326, 328
Kockatea Shale, Australia 81

Kodiak, Alaska 306, 307
Kogolym oil, West Siberia 143
    isotope type-curve 143–4
KOH see potassium hydroxide
Kola Peninsula, Russia 258
Kreyenhagen Formation, California
    oils 141, 152
Kuche Depression, Tarim Basin, China
    197
kukersite 82, 83, 105
Kulthieth Formation, Gulf of Alaska 310
Kuparuk Field, Alaska 177

labdane 26–7, 28–9
laccol-based lacquer film 334
Lachine Canal, Quebec, Canada 299
lacquer films 334
lactone 334
lacosite 81–2, 83
    Green River-type 82
    Rundle-type 82
lactic acid 35, 36
lacustrine 16
    oils 187
ladderanes 52
Lagoa Feia-Carapebus(!) 84
Lagoa Feia Formation, Brazil 11
Lake Hoare, Victoria Land, Antarctica 4
Lakes 16
Lake Siljan, Sweden 269
Lake Tanganyika, East Africa 11
La Luna Formation, Venezuela 267
La Luna-Misoa(!) 84
Lama Field, Venezuela 84
lamalginite 81, 82
Lambeosaurus 350
laminae 264
lamination 10, 12, 13
lamosite 81
lampblack 333
lamps 344, 345
Landana Formation see Iabe-Landana
    source rock
land plants 47, 61, 138
lanolin 338
lanosterol 6, 56–8, 59, 60
La Paz Field, Venezuela 267
laser Raman spectroscopy 195
Latouche mine, Alaska 307
Latrobe Formation, Australia 148
lauric acid 49
Lawndale Field, California 264, 265
LC$_{50}$ 286
LD$_{50}$ 286
leaching see solubilization
leaded gasoline 313, 314
lead isotopes 314

leaf wax 302, 303, 343
least-squares fingerprint matching 288
LECO 74
Leduc Reef, Canada 195
Leiosphaeridia 95
lentil 346
levopimeric acid 331, 336
levorotatory 36, 271
Lidy Hot Springs, Idaho 253
light aromatics 286
lightered 282
light hydrocarbons 162–92
light oils 151, 203, 205
lignin 189
lignite 329
    additive 92, 94
lignoceric acid 49
like dissolves like 199
limestone 267, 324, 325
limonene 56, 329, 330
linalool 58
Lincoln Creek Formation, USA 60
Linde 5A 202, 205, 206
Linde type A zeolite 204, 206
Lindow Man 341
linearity 125
linked-scan mode 227, 229
linoleic acid 49
lipid bilayer 46
lipid membranes 45–52
lipids 45, 49, 256
lipopolysaccharide 47
lipoprotein 47
    high-density (HDL) 51
    low-density (LDL) 51
liptinite 75, 76, 89
liptodetrinite 76
liquefied natural gas 122
liquid chromatography (LC) 128
liquid petroleum gas (LPG) 133
lithoautotrophic 253
lithofacies 86
litigation 287, 288
liverwort see non-vascular plant
Lodgepole Formation, USA 184
Lodgepole Mound oil, Williston Basin
    184
Lokbatan mud volcano, South Caspian
    115
Long Beach Field, California 265
longifolene 57, 324, 329, 331
Los Angeles Basin, California 264–7
Louanne Salt, Gulf of Mexico 159
Louvre Museum 328
Lovozero intrusion, Russia 258
low-sulfur oil 80
lowstand systems tract 87

Lubna well, Czech Republic 268, 269
lubricating oil 132, 133
lubricating stock 122
lupane 206, 208, 337, 342
lupenol 59, 337
lupenone 337
lurestan, Iran 325
lutein see carotenoid
lycopa-14(E), 18(E)-diene see
    lycopadiene
lycopane 25, 26–7, 28–9
lycopene 69

maceral group 75, 76
macerals 9, 75–6, 91, 94
maceration 75
macrinite 76
Madison Group, USA 182, 188
Madre de Dios Basin, Bolivia 98
magnesium 67
magnetic mass spectrometer 217
magnetic sector field strength (B) 229
magnetite 255, 258, 259
Mahakam Delta, Borneo 13
Mahogany Member see Green River
    Formation
Maiella oil, Italy 141
Maillard condensation 342
Main Pass Field, Gulf of Mexico 108
maize 339
Majiagou Formation, China 162
malaria 54
Malaspina Glacier, Alaska 307, 311
malonyl-CoA (malonyl-coenzyme A) 48
Malossa Field, Italy 141
MA(I)/MA(I + II) see monoaromatic
    steroids
mammoth 340
Manching, Germany 338
Mandal-Ekofisk(!) 83, 84
Mandal Formation, North Sea 83
mandrake 351
manganese oxide 111
Mango parameters 156, 179–87
Mango reactions 167
Manila copal 329, 334
mantle outgassing 9
manure 347
manuring practices 346–7
Maoming Shale, China 150
Maracaibo Basin, Venezuela 267–8
Marillac, France 339
marine oils 147
marine organic matter 147–8
Marlim Field, Brazil 84
marobitosite 81
marosite 81–2, 83

Mars 254, 298
marsh gas 138
Mary Ann Field, Alabama 84, 158
Mary Rose 336, 337
mass analysis 217–20
mass analyzer 209, 216
    see also quadrupole rods
mass balance 100, 117–18
mass/charge (*m/z*) 216, 217, 219, 222, 223–4, 225–6, 235
mass chromatogram (mass fragmentogram) 211, 217, 219, 222–7, 228, 229–30, 232
    isotope peaks 235, 248
mass discrimination 210, 220
mass fragmentogram *see* mass chromatogram
mass range 245, 247
mass resolution 217
mass spectrometer 217, 218
mass spectrometric fragmentations 223–4, 225–6
mass spectrometry (MS) 215–22
mass spectrum 216, 217, 219, 220, 235–7, 238–9, 240, 241, 244
MA-steroid *see* monoaromatic steroids
mastic resin 329, 334
masticadienonic acid 333
matrix-assisted laser desorption/ionization (MALDI) 127
mature 73
maximum flooding surface (MFS) 97, 98
Maya 25, 352
McArthur River Field, Cook Inlet 84
mead 346
meat 346
Mediterranean Sea 11, 332
Mega Borg 281
meiosis 6
melting points 159
Menilitic Shale, Europe 268
mercaptan 31
mercury 316
Mesolithic Period 322, 323
mesophile 5
Mesopotamia 325
meso-pristane 38, 39
mesopyrophaeophorbide-a 68
mesotrophic 8
Messel Shale, Germany 68, 148
Messinian source rocks, Mediterranean Sea 150
metagenesis 9
metallic carbides 253
metalloporphyrin *see* porphyrins
metal catalyst 166, 167
metals 80

metamorphism 258–9
metastable reaction monitoring *see* MRM/GCMS
metaxylene *see* xylenes
metazoa 10, 11, 13
meteor impact 270
methane 3, 8, 18, 19, 21, 140, 158, 195, 315
    abiogenic 254, 272
    $\delta^{13}C$ 153, 257, 317
    δD 153, 317
    geothermal 140, 259
    microbial 316, 317
    soils 317–19
    thermogenic 316, 317
methanesulfonic acid 150
methanogen 4, 8, 59, 253, 316
methanogenic bacteria *see* methanogen
methanogenesis 316, 319
Methanosarcinales 316
methionine 6
methoxyphenols 320
methyl *see* functional group
methyladamantane index 162
methyladamantane/adamantane 162
methylalkanes 206
methylalkylcyclohexanes 206
methylbutadiene *see* isoprene
2-methylbutane 22, 26–7, 28–9, 39
24-methylcholestane *see* ergostane
methylcyclohexane 187, 188–90, 192
2-methylcyclopentadienyl manganese tricarbonyl (MMT) 313, 314
methylcyclopentane 192
methyldehydroabietate 332, 336, 337
methyldiamantane index 162
methyldiamantanes 160
2-methyldocosane 201
methylene chloride 199, 200
methyl-erthritol phosphate (MEP) pathway 53, 54
2-methyl-3-ethylheptane 25, 30, 38, 40, 271
3-methyl-3-ethylheptane 26–7, 28–9
3-methylheptadecane 201
2-methylhexane 22, 164
3-methylhexane 22, 164
2-methylhopanoids 63
methylmercury 316
methylnaphthalenes 206, 311
3-methylnonadecane 201
Methylococcus 6, 63, 150
2-methyloctadecane 45, 271
Methylosphaera 63, 150
methylotroph 4, 8, 316
methylotrophic bacteria *see* methylotroph

2-methylpentane 22
methylphenanthrenes 207, 299
2-methylpropane 22
methyl radical 19
4-methylsteranes 146, 150, 234, 246
methyl-*tert*-butyl ether (MTBE) 313, 314
methylxanthines 352
Meurhe River, Alsace-Lorraine 302
Meuse River, Alsace-Lorraine 302, 303
mevalonate (MEV) pathway 53, 54
mevalonic acid 53
micelle 282
micrinite 76
microbial gas 153
microbial methane 153, 316
microfractures 264
microscopy 75–6
Mid-Atlantic Ridge 259
Middle Ob Region, West Siberian 143
mid-ocean ridge 259
Midland Basin, USA 164, 183
migrated oil 100, 102, 112, 113, 143
migration 110
    distance 194
Mikulov marls 268
milk 343
    fluidity 48–52
    lipids 343
Minas Field, Sumatra 84, 104, 148
Mingbulak well, Uzbekistan 278, 279
Minimata disease 274
Minoan lamps 344
Minoan settlements 345
Mississippi Delta 13
Mississippi Salt Basin, Gulf of Mexico 182
Mississippi Valley-type ore deposits 262
mitigation of oil spills 282
mitochondrial DNA (mDNA) 348, 349
mitochondrial Eve 348
mitochondrion 3, 5, 6, 46
mitosis 6
mixed oils 108, 144, 160, 177
mixing, intra-reservoir 107
mixing curve 108
Moab fault 110
Moa bones 350
Moab Tongue Sandstone 110
Mobile Bay, Alabama 158, 176
mobile phase 209, 210
modeling gasoline leakage 315
modeling marine oil spills 283
modulation (slicing) 213
modulator 212
Mokattam Shale *see* Tanezzuft Shale
MOLART 334

molecular class plots *see* ring-preference plots
molecular dynamics (MD) 194
molecular fossil *see* biological marker
molecular ion (M+.) 215, 231
molecular mechanics 192–3, 268
molecular modeling 192–4
molecular sieve 128, 200
molecular weight 127–8
monoaromatic hydrocarbons 212
monoaromatic steroids (MA-steroids) 64, 201, 202, 229–30, 241, 246, 248, 334
monocot *see* angiosperm
monolayer biphytanyls 46
monomethylalkane 206
monoterpane 25, 26–7, 28–9, 55
monoterpene 329
monoterpenoid 55, 163, 336
monoterpenoid biosynthesis 55
Monte Alpi oil, Italy 142
Monterey Formation, California 150, 252
  oils 141, 142, 184
  Phosphatic Member 79
  tars 306
Moravia, Czech Republic 269
Moray Firth, UK 248
mordenite 204, 205, 206
moretanes 41, 149, 207, 300
Morichal heavy oil, Venezuela 120
morolic acid 333
moronic acid 333
morphine 350
Morro do Barro Formation, Brazil 151
mortality 286
Moselle River, Alsace-Lorraine 302
mosses 60
Mount Carmel Field, Alabama 189
Mount Sorrel granodiorite, UK 262, 264
mousse 281
Movico Field, Alabama 189
Mowry Shale, USA 11
MRM/GCMS 221, 229–31, 234–5
mtDNA *see* mitochondrial DNA
mud log 92, 93
mud volcano 115
multidimensional chromatography *see* gas chromatography
multiple ion detection (MID) *see* selected ion monitoring
multiregional evolution hypothesis 347–9
multisector mass spectrometer 227
mummia 351
mummies 323–5, 327, 329, 333, 347, 350–1
mummification 323, 351
Murchison meteorite 257, 272

murein 6, 47
mussels 289, 294, 305, 306, 308, 309
mutagen 287
Mycobacterium 47, 63
mycoplasma 6, 47
myoglobin 349
myrcene 56, 330
myristic acid 49
myrrh 329
myrrhoric acid 329
mystery oils 299–300, 301
m/z see mass/charge
m/z 191 *see* terpanes
m/z 217 *see* steranes

Nahal Heimar Cave, Israel 341
Nancy Field, Alabama 189
Nannocystis 6, 63, 150
naphtha 122
naphthalenes 31, 296, 297, 304, 311
naphthene *see* cycloalkane
  branching 172
naphthenoaromatic 31
narcotics 350–2
National Bureau of Standards see NBS
National Institute of Standards and Technology 136
natural gas 122, 138, 315–16
  microbial 316–17
  thermogenic 316–17
natural gas liquid (NGL) 17
natural gas plot 194
natural product chemistry 198
Navajo Sandstone, Utah 110
NaX (13X) molecular sieve 205–6
NBS-19 137, 139
NBS-22 140
n-decanoic *see* capric acid
n-d-M method 128
n-docosanoic *see* behenic acid
n-dodecanoic *see* lauric acid
Neanderthal 336, 339, 348, 349
Neanderthal 1856 Feldhofer type specimen 348
neat (unspiked) sample 240
negative ion electrospray (ESI) 129
negative picking 116
n-eicosanoic *see* arachidic acid
neoabietic acid 331, 336
neohop-13(18)-ene 61
Neolithic Period 322, 323
neopentane *see* 2, 2-dimethylpropane
neoxanthin *see* carotenoid
nerol 55
nerolidol 58
neuron 51

neurosporene 69
neutral loss mode (in GCMS/MS) 227, 234
neutron 136, 137
New Albany Shale, USA 136
Newtonian fluid 121
New York City 284
n-hexadecanoic *see* palmitic acid
nickel 103, 125
nickel ion 68
nickel/vanadium 88, 103
Nicotiana 351
nicotine 351–2
Niger Delta, Nigeria 96, 173
nightshade 351
Nipisi, Alberta, Canada 303–4
Nisku Formation, Alberta, Canada 155, 162
nitrate 7, 8, 10, 11, 52
nitrogen 124
  compounds 32–4
  isotopes ($\delta^{15}N$) 338, 340, 346
NMR *see* nuclear magnetic resonance
Nocardia 47
Nodular Shale, Los Angeles Basin 91, 265–7
nomenclature (structural) 21, 34
  acyclic isoprenoids 41
  steranes 34, 42–4
  terpanes 41–2
  triterpanes 34
n-nonacosane 343
nonacosan-15-ol 343
nonacosan-15-one 343
nonane 21
non-associated gas 153
non-essential amino acids 340
non-hydrocarbon gas
non-Newtonian fluid 121
non-waxy oil 148
24-nordiacholestanes 268
norhopanes
  α,β-norhopane 241
  norhopane (C_{29} norhopane) 207
  25-norhopanes (10-desmethylhopanes) 240
  30-norhopanes 236–7, 238–9
normal alkane (*n*-alkane, *n*-paraffin) 21, 30, 104, 105, 106, 113, 128, 172, 200, 202, 205, 206, 211, 248, 253, 258, 308
normal isotopic trend, hydrocarbon gases 140, 256, 272
30-normoretane 236–7, 238–9
18α-norneohopane 206, 207
Norphlet Formation, Gulf of Mexico 157, 158, 159, 161

norpristane (2, 6,
        10-trimethylpentadecane) 25,
        26–7, 28–9
North Lincolnshire, UK 341
North Sea 278
    oils 170
North Viking Graben 165
Norwegian geochemical standards (NGS)
        116
Nowruz Field, Iran 278
NSO compounds (resins) 31
n-tetracosanoic *see* lignoceric acid
n-tetradecanoic *see* myristic acid
Nubian mummy 347
nuclear magnetic resonance (NMR) 124,
        198, 203, 235
nuclear membrane 6, 46
Nugget Sandstone, Wyoming 250
nutrients 16

obligate aerobe 4
obligate anaerobe 4
O/C *see* atomic O/C
ocean
    coastal 8
    open 8
ocimene 330
ocotillone-type side chain 334
n-octadecanoic *see* stearic acid
octane 21
    rating or number 132, 313
odd-carbon preference *see* carbon
        preference index
odor 126
    see also fragrance
Odyssey 278
Oguta Field, Niger Delta 84
OI *see* oxygen index
oil 138
    assays
        advanced 127–9
        basic 120–7
oil-associated gas 153
oil cracking 159
oil deadline 158, 159
oil density 120
%oil depleted 295
oil layer thickness 288, 289
oil–oil correlation 233
oil quality, prediction 102
oil sample *see* sample quality, selection,
        storage
oil shales 81–3, 256
oil slick 280, 289
    sampling 288
oil–source rock correlation 100, 184, 233,
        272–3

Oil Spill Health Taskforce 305
oil spills 276–84
    amounts 276, 278
    Automated Data Inquiry for Oil Spills
        (ADIOS) 283, 284, 285
    compound-specific isotope analysis
        148, 151, 300–1
    diagnostic ratios 299
    fingerprinting 289–95, 300
    internal standards 292
    land 283
    largest 278
    mitigation 282
    modeling 283
    PAH parameters 295–8
    quality assurance and control
        (QA/QC) 291–4
    Nipisi pipeline 303–4
    rate of removal from sea surface 283
    removal from sediment 295
    surrogates 292
    tanker 279
    target compounds 292
    tiered approach for identification 288
    underground 283–4
    weathering 280, 294–5
oil standards 116–17
oil-to-gas conversion based on
        diamondoids (OTG$_d$) 161
oil-to-gas cracking 158–61
oil transport tonnage 279
oil–water contact (OWC) 282
Okpai Field, Niger Delta 84
Old Masters paintings 334
oleanane 26–7, 28–9, 142, 153, 196, 203,
        206, 208, 240, 254, 268,
        306, 307, 308, 311, 325, 334,
        342
oleanane/hopane *see* oleanane index
oleanane index [(oleanane/(oleanane +
        hopane)] 309, 310
18α/(18α + 18β)-oleananes 268
oleanane-type molecules 335
oleanonic acid 333
oleic acid 49
oligotrophic 8
olive oil 345, 346
olivine 259
Olmeda Formation, Argentina 185
omnivore 339, 340
on-column injection *see* injection
online combustion 138
OPEC 17
opium 350
optical activity 36, 252, 270–1
organelle 6
organic acids 125

organic carbon *see* total organic carbon
organic facies (organofacies) 13
organic geochemistry 68
organic-lean rocks 101–2, 115
organic matter
    amorphous 76
    classification 75
    gas-prone 75, 77
    marine 147–8
    non-source 75
    oil-prone 75, 77
    primary 9
    recent 294
    terrigenous 147–8
    type 77
Orgueil meteorite *see* carbonaceous
        chondrite
Orinoco Heavy Oil (Tar) Belt, Venezuela
        103
Orkney, UK 346
orthoxylene *see* xylene
Oseberg Field oil, Norwegian North Sea
        84, 116
osteocalcin 350
Otztal Alps 336
outcrop 113
"Out of Africa hypothesis" *see* recent
        replacement hypothesis
overload 210
overmature *see* postmature
overpressure 264
oxic 10–13, 16, 88, 107
    water column 10
oxidant 7
oxidation 9, 94, 134, 152, 281
oxidation–reduction reaction *see*
        redox
oxidosqualene 56, 61, 63
oxidosqualene cyclase 58
oxo-dehydroabietic acid 329
oxygen 124
    compounds 33
    free 53, 55, 57, 63
oxygenates 313, 314
oxygen index (OI) 77, 78

Pabdeh Formation, Iran 325
Pachuta Creek Field, Alabama 189
Pachyrhinosaurus 350
packed column GC 210
PAH *see* polycyclic aromatic
        hydrocarbons
PAH-RI *see* refractory PAH index
Palace of Knossos, Crete 350
paleodiet 338–44
Paleo DNA Laboratory, Lakehead
        University 349

paleoenvironment 149–50
paleo-fluid contact 196
Paleogene-Neogene 145
paleogeothermal gradient 95
paleolatitude 145–6
Paleolithic Period 322, 323
paleo-pressure 195
palmitic acid 49, 342
palmitic wax esters 344
palmitoleic acid 49
palustric acid 331, 336
palynomorph 95
Pando well, Bolivia 96, 98
Papaver 350
Parachute Creek Member 68
paraffin, normal *see* alkane
paraffin branching 172
n-paraffins *see* alkane
paraffinicity 172
   ratio (F) 174
paraxylene *see* xylene
parent ion 216, 218, 229, 231
parent–daughter ratio plots (C$_7$
   hydrocarbons) 184–6, 227
parent–daughter transition (selected
   compounds) 231, 233
parent mode (in GCMS/MS) 227, 231–4

parent quadrupole (Q1) 227, 232
partitioning, chromatographic 148
Pasenhor 324, 327
Pasteur, Louis 35
Pathfinder well, Gulf of Mexico 110
pay zone 196
PCR *see* polymerase chain reaction
PDB standard 137, 139
peak–area ratios 105
peak–height ratios 105, 108
peat 350
Peco Field, Canada 156
Peedee belemnite *see* PBD standard
Peedee Formation, South Carolina 137,
   139
pelagic sediments 14
pelagic tars 281
Pematang-Sihapas(!) 84
pentane 21, 39, 199
pentane/dichloromethane 199
pentamantane 157
2,6,10,15,19-pentamethylicosane
   (PMI) 8
pepidoglycan *see* murein
perfluorotributylamine (FC-43) 222
perhydro-β-carotane *see* β-carotane
peridinin *see* carotenoid
periodic acid 58
permeability 264, 266

perylene 31, 112, 275, 296, 308
petrogenic hydrocarbons 301, 308
petroleum 10
   in basement rocks 261–8
   contamination 294
   markers 275
   origin 252–3
   production 17
   province 17
   refining 129–34
   system 17
   toxicity 284–7, 304–5
petroleum systems 83, 84, 112, 272
   Gulf of Mexico 112
   modeling *see* thermal modeling
phase separation 173
phellandrene 330
phenanthrene/methylphenanthrene
   298
phenanthrenes 31, 206, 207, 296
phenol 33, 34
   extraction 130
pheophytin 67
phorbide 67
Phormidium 63
phosphate 7, 8
Phosphatic Member *see* Monterey
   Formation
phosphoglycolipid 47, 49
phospholipid 46, 47, 48, 49, 136
Phosphoria Formation, USA 8, 64, 79,
   104, 105
phosphosulfoglycolipid 49
phosphorite 8, 86
photic zone 7, 8, 10, 11
   anoxia 149
photoautotroph 4
photoheterotroph 4
photometer 89
photooxidation 281
photophosphorylation 64
photosynthesis 5, 71, 139
   anoxic photosynthetic bacteria 64
photosystems I and II 64, 65
phototroph 4, 7, 8
phthalates 116
phycobilin 64
phytane (Ph) 25–7, 28–2, 30, 34, 40, 41,
   67, 104, 150, 206, 211, 271
phytane/nC$_{18}$ 106, 107, 295
phytoclast 75, 89
phytoene 54, 55, 69
phytofluene 69
phytol 28, 39, 41, 65
phytoplankton 8
pi (π) bond 18–20
PI *see* production index

pimarane 26–7, 28–9
   -type acids 331
pimaric acid 320, 331
Pinaceae 334, 337
Pinda Formation, Angola 153
Pine Hill anhydrite 159
pinene 56, 329, 330
pine resins 331, 332, 336
Pineview Field, Wyoming 250
Pinus 329
Pistacia 329, 332, 334
piston (seabottom) cores 111–13, 115
pitch 334–8
plane-polarized light 270
plankton 138, 256
plasmid 5
plastid 5
Platform A, Santa Barbara Channel,
   California 276
Playa del Rey-Alondra Trend 265, 266
Playa del Rey Field, California 264, 265
plumes 313
PMI *see* 2,6,10,15,19-
   pentamethylicosane
Point Barrow, Alaska 154
poise 121
Pokachev well, West Siberia 143
polar compound 34, 200
pollen 95
pollutant 274, 275, 291, 315–16
pollution 111, 275–6, 277, 278, 312–13
Polyangium 150
polycadinenes 55, 331
polycyclic aromatic hydrocarbons (PAH,
   PNA) 31, 248, 275, 286, 291,
   295–8, 300, 305, 308–12
   carbon isotopes 301
   pyrogenic (pyrosynthetic) 298, 301, 308
   source ratios 296, 297
   weathering ratios 296, 297
polyphenols 344
polymerase chain reaction (PCR) 348,
   349
polymerization
   methane 140, 256, 257, 258, 268
   refinery 130, 133
polynuclear aromatic hydrocarbons *see*
   polycyclic aromatic hydrocarbons
polynuclear thiophenes 32
polysaccharide 7
polysiloxane stationary phases 210–11
polystyrene 127
polysulfides 79, 80
polyterpanes 25, 26–7, 28–9
Ponca City, Oklahoma 30, 271
Pool Creek Field, Alabama 189
Porifera (sponges) 61, 254

poriferastane 62
porosity 264, 265, 266
porphobilinogen 65
porphyrins 33, 34, 64–8, 200, 211, 252, 254, 271
C_{31} and C_{32} DPEP 150
porphyrin-polar fraction 199, 200
Port Ashton, Alaska 307
Posidonia Shale, Europe 171
positive picking 116
postmature 73, 100
potassium hydroxide 126
potential source rock 72
potsherds 342, 343, 344
pour point 120–1
Powder River Basin, Wyoming 174
prasinophyte algae 97
prasinoxanthin *see* carotenoid
Precambrian organic matter 146
Precambrian oils and source rocks 173
precision 125, 250
precolumn 210
preservation 9, 10, 12, 13, 14, 16
primary migration 264
primary pathway for carbon flow 149
primary production 138
primesum 186–7, 188, 189
Prince William Sound, Alaska 142, 289, 295, 296, 297, 305–12
Prinos Field, Greece 104
priority pollutants 291
pristane (Pr) 25–7, 28–9, 30, 38, 40, 41, 67, 104, 150, 206, 211, 271, 314
pristane/nC_{17} 106, 107, 294
pristane/phytane, pristane/(pristane + phytane) 106, 142, 266
processed streams 134
process models, refinery 119
producibility 121
production allocation 107
production index (PI) 73, 78, 88, 101
production string 158
productivity 12
Antarctica 12
primary 5–8, 9, 12, 13, 14, 16
secondary 8
programmed temperature vaporizing injection (PTV) *see* injection
prokaryotes (bacteria) 3–5, 46, 63
propane 21
propyl *see* functional group
24-*n*-propylcholestanes 26–7, 28–9, 203
proteins 10, 256
ancient 349–50
marine 338
proteobacteria 59, 316
protochlorophyllide 65, 66

proton 136, 137
-induced gamma-ray emission (PIGE) 195
-induced X-ray emission (PIXE) 195
transfer 216
protoporphyrin 65, 66
proximity to pay 196
Prudhoe Bay Field, Alaska 154
pseudohomologous series 220
psychrophile 4, 5, 50
PTV *see* injection
Puente Formation Schist-Conglomerate, California 265, 266–7
purity and fit 220
purple non-sulfur bacteria 67
purple sulfur bacteria *see* Chromatiaceae
purpurin 67
pycnometer 120
pyrenes 31, 308, 328
pyridine 33
pyrite 80, 87
pyrobitumen 158
Pyrococcus 54
pyrogenic hydrocarbon 302
pyrogenic index 298, 300, 301
pyrogenic PAH 298, 308
pyrogenic PAH/fossil PAH ratio 298
pyrograms *see* Rock-Eval programs
pyrolysis 336, 337
pyrolysis/gas chromatography/mass spectrometry (py/GCMS) 329, 334
pyrolysis/MSMS 334
pyrolyzate
isotope profiling 143
pyrrole 33, 329
pyruvate 53

Qasr Ibrîm, Nubian Egypt 332, 340, 342
Qiongdongnan Basin, China 162
Qixia Formation, China 100
quadratic coefficient (QC) 148
quadrupole mass filter *see* quadrupole mass spectrometer
quadrupole mass spectrometer 214
quadrupole rods 209
quality, organic matter 73
quantitation, biomarker 240–2
quantitation limit 125
quantitative structure–activity relationships (QSAR) 208
quantity 73
quantum mechanical tunneling 217
quantum mechanics 192–3
quasi-anaerobic 10
Queen Vein Coal, Alaska 307, 310
quinoline 33

Qusaiba Member, Qalibah Formation, Saudi Arabia 179

racemic 36, 271
radiogenic carbon isotopic analysis 317
see also ^{14}C
raffinate 133
rail conformation 193, 208
Rainbow Hydrothermal Field, Atlantic Ocean 259
Ramsbottom residue 125
raney nickel 80
range 125
rank *see* thermal maturity
Rastrites Shale, Sweden 270
Ratawai Shale, Iraq 176
R configuration 38
reagent gas 216
rearranged hopanes *see* diahopanes
rearranged sterane *see* diasteranes
recent replacement hypothesis 347–9
reconstructed ion chromatogram (RIC, total ion chromatogram) 220
recovery *see* estimated ultimate recovery
recycled organic matter 115
Red Desert Oil, Wyoming 148
red/green quotient (Q) 95
redox potential (oxidoreduction potential) 7, 10, 12
Red River Formation, USA 182, 184
reducing 107
reductant 7
Redwash Field, Utah 84
refinery processes 130
refinery simulator 127
see also process simulator
reflectance *see* vitrinite reflectance
refractive index (RI) 126, 128, 199, 200
refractory PAH index (PAH-RI) 296, 310
remediation 284, 318
repeat formation test (RFT) 107, 110
repeller 215
reserve growth 17
reservoir alteration vectors 174
reservoir compartments 110
reservoir continuity 105, 107, 109–10, 212
reservoir delineation 196
reservoir filling history 109–10
reservoir maturity 159
reservoir rock 10, 272
detecting petroleum 102
reservoir temperature 182
residual fuel oils 133
residual salt

residuum 122, 134
resin 34, 80, 128, 252
    archeological 329–32, 333
resinate wines 332
resinite 76, 89
resistivity curve 83
resolution 218, 242
resorcinol 105
respiration 7, 8
response factor 241, 294
retene 296, 331
retention gap technique 210
retrograde condensate 173
reversed isotopic trend, hydrocarbon
    gases 152, 256, 257, 272
reverse methanogenesis *see* anaerobic
    methane oxidation
Rhine River, Germany 302
Rhodobacter 71
Rhodoccoccus 47
rhodo-DPEP porphyrin 68
rhodo-etio porphyrin 68
ribosomal RNA (rRNA) 5
ribosome 6
RIC *see* reconstructed ion
    chromatogram
ring-preference plots 187–90
Río Desaguadero, Bolivia 280
R$_o$ *see* vitrinite reflectance
road asphalt 302, 303
road runoff 302
roasted pitch 333
robustness 125
rock-eval programs 74, 77
rock-eval pyrolysis 74, 77–8, 88–9
Rock-Eval Expert System Advisor
    (REESA) 96
rock sample *see* sample quality, selection,
    storage
    size 115
rock standards 116–17
rosin 329
Rospo Mare Oil, Italy 141
Rothamsted Experimental Station, UK
    347
rotoevaporation 142
Rozel Point, Utah 30, 80
rRNA *see* ribosomal RNA
rubber 25, 26–7, 28–9, 52, 55
Rudeis Formation, Gulf of Suez 325
ruminant 343
Rundle Oil Shale, Australia 82
Rundle-type oil shale *see* lacosite
ruthenium tetroxide (RuO$_4$) 105

S1 73, 77
S2 73, 77

S3 77, 78
Saanich Inlet, British Columbia 12
Sabine Parish, Gulf of Mexico 164, 165,
    185
sabinene 330
    hydrate 56
Sabine Pass, Gulf of Mexico 183
Sable Island, Scotian Shelf, Canada 178,
    179, 180, 181, 183, 187, 191
Sagalassos, Turkey 345
Saint John's Harbor, Newfoundland 300
Saint Lawrence River, Quebec, Canada
    299, 300
salinity 194
salt in oil 122
salt dome 112
Salym Field, West Siberia 84, 99
Salym Oil, West Siberia 143
sample clean-up and separations 199–200
sample dilution 210
sample probe 217
sample quality, selection, storage 113–16
    distributed source-rock sampling 151,
        272
    oil 113, 116
    rock 113–16
sandarach resin 329
sandaracopimaric acid 329, 331
San Joaquin Basin, California 152
San Joaquin heavy oil 134
Santa Barbara Basin, California
Santa Barbara, California 79
Santa Barbara Channel, California 274,
    276, 305
Santa Maria Basin, California 79
Santa Monica Basin, California 276
sapropelic coal 81, 82
saprophyte 4
Saratoga Chalk Formation, Louisiana 164
sargasterol *see* fucosterol
Sargelu Formation, Iran 325
saturate-aromatic fraction 199, 200
saturated hydrocarbons (saturates) 18, 64,
    128, 155, 199
saturates 200, 202, 205
saturated hydrocarbon UCM *see*
    unresolved complex mixture
Saxony Basin, Europe 171
S/C *see* atomic S/C
S configuration 38
scan analysis 219–20
scan mode 227
scanning 217
scan number 219, 240
scan rate 243, 244
scan run *see* scan analysis
Schaefer parameters 171

schist 264, 266
Schulz–Flory distribution 258
Scladina Cave, Belgium 339
sclerotinite 76
scorpion conformation 193, 208
Scotian Shelf, Canada 170, 171, 177, 179,
    181, 182, 183
screening
    crude oil 102–9
    source rock (quality and quantity)
        72–88
    source rock (thermal maturity) 73,
        88–96
Sea Empress 280
Sea Island, Kuwait 278
seal integrity 196
seal rock (cap rock) 154
sealed-tube combustion 137
seawater 10
8, 14-secohopanes 34
secondary building unit (SBU) 204
secondary cracking 9
secondary migration 263, 264
secondary pathway for carbon flow
    149
secondary porosity 151
secondary processes 105, 106
sediment
    contaminated 300
    emulsified in oil 122
    extracts 302
    Prince William Sound 309, 311, 312
    uncontaminated 113
sedimentary rock 260
sedimentation rate 13–16
    Baltic Sea 14
    Central Pacific 14
    Namibia, Africa 14
    Oregon 14
    Peru 14
seep hypothesis 310
seep oil 112
seeps 111, 269, 311
seismic amplitude anomalies 111
selected ion monitoring (SIM, multiple
    ion detection, MID) 221, 222,
    233
selected metastable ion monitoring
    (SMIM) 229
selinene 57, 330
semifusinite 76
semipermeable membrane devices
    (SPMD) 289
Senonian Limestone, Dead Sea *see*
    Ghareb Formation
Senonian Bituminous Chalk, Dead Sea
    328

separator 215
Sergipe-Alagoas Basin, Brazil 63
serine 48
serpentine 259
serpentinization 253, 257, 258–9
sesquiterpane 25
sesquiterpene 329
sesquiterpenoid 55, 57, 336
  biosynthesis 55
sesterterpane 25, 26–7, 28–9
sesterterpenoid 55
sewage pollution 275–6
Shakespeare, William 351
shale 311
Shanganning Basin, China 162
shear rate 121
shear stress 121
shellac 329
shipwrecks 336
short-column chromatography 200
shungite see Karalian Shungite
siderite 253
signal-to-noise 227, 232, 242–3, 248
Silent Spring 274
silica 264
silica gel 199
silicalite see ZSM-5
silicon tetrahedron 203
Siljan Ring, Sweden 253, 255, 257, 270
Silphium 351
SIM see selected ion monitoring
Simpson-Ellenburger(.), USA 84
simulated distillation see distillation
Sinai, Egypt 328
single bond 18
single focusing mass spectrometer 217
single quadrupole mass spectrometer 218
sinister see S configuration
sinking and sedimentation 281
sitosterol 60, 61, 342, 346
size-exclusion chromatography 127
skimmer 282
SLAP (Standard Light Antarctic
    Precipitation) 137
Sleipner area, North Sea 196
Smackover Formation, Gulf of Mexico
    146, 154, 158, 159, 160, 161, 162,
    170, 173, 176, 181, 182
Smackover-Tamman(!) 84
smectite 16
Smith Island, Alaska 289
smoke 319–21
Smørbukk North Field, Norway 109, 110
sodium borohydride 58
soft ionization 129, 216
soft-rot fungus 346
soil gas methane 317

solubilization 255, 268
solvent deasphalting 130
solvent dewaxing 130, 133
solvent extraction 130
solvent-split injection see injection
Somalia 332
sonar 111
Songliao Basin, China 11
soot 333
Soter 324, 327
source potential index (SPI) 96–7
    Niger Delta, Tertiary 96
    North Sea, Upper Jurassic 96
    Saudi Arabia, Upper Jurassic 96
source rocks 10, 72, 272
    carbonate 72–4, 264
    carbonate versus shale 74
    density 97
    quantity and quality 72–88
    screening 72–88
    shale 72–4
sour gas 154, 155
South Caspian region, Azerbaijan 115
South Cypress Creek Field, Alabama 189
South Pennine Orefield, UK 263
Soxhlet extraction 199
space-filled projection see structural
    notation
specific gravity 102, 120
Spekk Formation, Norway 187
sphingolipids 49
spiking 240
spirotriterpane 208, 240
Spirulina 71
split-flow injector 210
split injection see injection
splitless injection see injection
sponges see Porifera
spore coloration index see thermal
    alteration index
spores 95, 96
sporinite 76, 89
Sprayberry Formation, Texas 164
spreading 280
squalane (2,6,10,15,19,23-
    hexamethyltetracosane) 8, 26–7,
    28–9
squalene 54, 55–8, 59, 61, 211
squalene-hopene cyclase (squalene
    cyclase) 55, 58
stable carbon isotope ratios 271–2, 306
    C_1–C_5 hydrocarbon gases 140
    ethane 257
    lipids 339, 344
    methane 153, 257, 317
    oils 138, 142, 145, 146, 147
    propane 257

stable carbon isotope type-curve 143–4
stable isotopes 137, 235, 338–42
Standard Light Antarctic Precipitation see
    SLAP
standard temperature and pressure (STP)
stanols 43, 346, 347
star diagram 107
State Line Field, Mississippi 176, 189
Statfjord Field, North Sea 84
stationary phase 36, 209, 210
    achiral 36
steady-state kinetics 165
steady-state metal-catalysis 164
steady-state metal-catalyzed reactions 164
steam cracking 130
stearic acid 49, 342
Stenberg well, Sweden 253
stepwise aqueous oxidation 152
sterane isomerization, 20S/(20S + 20R)
    40, 242
steranes 25, 42–4, 64, 206, 211, 233,
    234–5, 245, 268
    C_{27} 233
    C_{27}/C_{29} 146
    fingerprint 326
    parent ion transitions 229, 233
    regular (4-desmethyl) 234
steranes/hopanes 146
steranes/hopanes (regular
    steranes/17α-hopanes) 146
stereochemical assignment 37, 38
stereochemistry 21, 34, 35, 37, 40–4
stereoisomer 35, 39, 42
stereoisomerization 40
steric energy 207
steroid 36, 49, 55, 61
sterols 42–4, 46, 51, 62, 63–4, 150, 254
    3β-hydroxy 63
stick projection see structural notation
stigmastanes (24-ethylcholestanes) 25,
    26–7, 28–9, 62, 160
stigmastanol 342, 346
stigmasterol 60, 61
Stockholm tar 336, 337
Stone Age 322, 323
STP see standard temperature and
    pressure
straight-run products 132
straight-run stream 134
Strange Lake, Canada 258
stratification (stratified water) 11
Streptomyces 47, 60
streptomycin 5, 6
string tubing leak 108
strontium isotopes 109–10
structural isomer 39
structural notation 20–1

sublimate 158
suboxic 10, 12
succinate 65
succinyl co-enzyme A 65
Sugar Ridge Field, Alabama 189
Sulaiy Formation, Iraq 176
sulfate 10, 11, 13, 16
  -reducing bacteria 12, 79, 316
  reduction 11, 16, 79
sulfides 79
sulfoglycolipids 49
sulfolipids 49
sulfur 79–81, 124, 142
  bacteria 7
  compounds 31–2
    aromatic 32
  sulfur isotopes 152–6
sulfurization 162
Sunnyside heavy oil, Utah 120
superacid 157
supercritical fluid chromatography 128
superoxide dimutase 4
surface geochemical exploration 111–13
surfactant 282
Susa, Iran 328
Susian culture 325
Svalbard Rock, Spitsbergen 116
Swan Hills Field, Alberta, Canada 158,
    162
Swanson River Field, Cook Inlet 84
sweet crude see low-sulfur oil
sweetening 130
sylvestrene 330
symbiont 5
synapse 51
synthetic fuels 78
synthetic oil 78
synthetic petroleum 258
syringe 208, 209
systems tracts 86

TAI see thermal alteration index
tailing 242, 244
tail-to-tail 25, 26–7, 28–9
TA/(MA + TA) see monoaromatic
    steroid aromatization
tandem mass spectrometry 68, 227–8, 233
  see also GCMS/MS
Tanezzuft Shale, North Africa 262, 263
tannins 10, 344
TAPH see total polycyclic aromatic
    hydrocarbons
taraxastanes 208, 307
taraxerenes 11
tar ball 142, 281, 306
tar, birch bark 336–7
tar coals 333

target compounds for oil spill studies
    292
Tarim Basin, China 162, 197
tar sands 81, 82, 120
tartaric acid 333, 344, 346
Tasmanite oil shale, Tasmania 82, 83
Tasmanites 75, 78, 81, 97
tasmanitid cysts 82
T-atom 204
TBR see trimethylnaphthalene ratio
TCI see transmittance color index
tea 352
teeth 338
Teflon 116
TEHRGC see thermal extraction
    high-reolusion gas
    chromatogram
telalginite 81, 82
telinite 89, 90
Tell el-Ajjul, Gaza, Palestine 350
Tell el'Oueili, Iraq 325, 328
telosite 82
temperature of generation ($^{o}C_{temp}$)
    179–83
Tenax trap 128
terebinth resin 332–3
ternary diagram ($C_{27}–C_{28}–C_{29}$) 134
  C_7 hydrocarbons 190
  steranes 134, 269, 270
terpane m/z 191 fingerprint 306, 309, 326
terpanes 41–2, 211, 250
terpenoids (isopentenoids) 24, 36, 329
  acids 329
  biosynthesis 52–3, 54, 59
  classes 26–7, 28–9, 55
terpinene 330
terpineol 56, 346
turpinolene 330
Terqa, Syria 323, 328
Terra Nova Field, Newfoundland 84
terrigenous oils 147, 187
terrigenous organic matter 16, 147–8,
    188
terrigenous versus marine 187
tetracyclic polyprenoids (TPP) 26–7,
    28–9
tetracyclic sesterterpane 26–7, 28–9
tetracyclic triterpanes 26–7, 28–9
tetraethyl lead 313, 314
tetrahymanol (gammaceran-3β-ol) 56, 59
Tetrahymena 56
tetrakishomohopanes 207
tetramantane 157
2,6,10,14-tetramethylhexadecane see
    phytane
tetramethyl-lead 313
tetramethylnaphthalenes 206

2,6,10,14-tetramethylpentadecane
    see pristane
tetrapyrrole 8, 65, 66, 67
tetraterpane 25, 26–7, 28–9
tetraterpenoid 55
TFE-fluorocarbon polymer net
    (tetrafluoroethylene polymer,
    Teflon) 288, 289
Thebes, Egypt 324
Theobroma 352
theobromine 352
theophylline 352
thermal alteration index (TAI) 77, 94–6
thermal conductivity detector (TCD) 74,
    78
thermal cracking 133, 163
thermal extraction high-reolusion gas
    chromatogram (TEHRGC) 102,
    184
thermal maturity 73, 89, 90, 103, 174
  effect on gas chromatogram 105, 106
thermal modulator 213
thermoacidophile 5
thermochemical sulfate reduction (TSR)
    80, 154–6, 159, 161, 162, 172, 173,
    176, 179, 187, 315
thermocline 11
thermogenic hypothesis 259–73
thermogenic methane 153, 316
thermophile 5
thermotaxis 67
thiaadamantanes 162
thiaalkanes 31, 32
thiacycloalkanes 32
thin-layer chromatography/flame
    ionization detection (TLC/FID)
    see iatroscan
Thiobacillus 79, 80
thiols 31, 32
thiophenes 32
  polynuclear 32
Thioploca 63
thitsiol-based lacquer film 334
Thompson B–F diagram see B–F diagram
Thompson parameters 159, 171–7,
    188
thorium/uranium 86–7
three-dimensional ion trap mass
    spectrometer 218
Thumb Bay, Alaska 307
Thur River, Alsace-Lorraine 302
thylakoid membrane 64, 71
Tia Juana Field, Venezuela 84
TIC see reconstructed ion
    chromatogram
Tilje Formation, Norway 109
Timan-Pechora Basin, Russia 144

time-of-flight mass spectrometry (TOF/MS) 214

TLC *see* thin-layer chromatography

TMI *see* 2,6,15,19-tetramethylicosane

T_{max} 73, 77, 88–9
  converting to reflectance 89
  vs. R_o, Douala Basin, Cameroon 89

TNH *see* 25,28,30-trisnorhopane

TNR *see* trimethylnaphthalene ratio

TNS *see* 25,28,30-trinorhopane

tobacco 352

TOC *see* total organic carbon

tocopherols 29

Tofts Ness, Scotland 346

toluene 20, 188–90, 199, 208, 280, 291, 314

TOM *see* total organic matter

Tomachi Formation, Bolivia 96, 98

topped 142

torbanites 78, 82, 83, 329

Toro sandstones, Papua New Guinea 196

Torque Trim 255

Torrance Field, California 265

Torrey Canyon 274, 276

total acid number (TAN) 125–6

total base number (TBN) 126

total carbon 74

total ion chromatogram *see* reconstructed ion chromatogram

total organic carbon (TOC, C_{org}) 13, 15, 16, 72–5, 78, 83–8
  Bazhenov Formation, West Siberia 14
  Black Sea sediments 13, 213
  Caspian Sea sediments, Russia 13, 73
  High TOC at the base, decreasing upward (HTB) 86
  measurement 74–5
  minimum TOC 72–4
  North Sea, Jurassic 13
  original quantity (TOC°) 97–100, 117, 118
  Paris Basin, Toarcian and Hettangian/Sinemurian 13
  Tyumen Formation, West Siberia 15
  versus uranium 87
  West Siberia, Lower-Middle Jurassic 13
  West Siberia, Upper Jurassic 13

total organic matter (TOM) 72

total polycyclic aromatic hydrocarbons (TPAH) 296, 312

total petroleum hydrocarbons 290–1, 301

total scanning fluorescence (TSF) 112, 113

Tower of Babel 322

toxicity of petroleum 284–7

toxicological indicators 304–5

toxicology 274

TPP *see* tetracyclic polyprenoids

trace fossils (ichnofossils) 12

trace metals 125

tracers
  in aerosol particles 318
  in smoke 320

Traict du Croizic, France 301

Tramutola oil, Italy 142

trans-butene 23

trans-configuration 23

transcription 6

transfer 209, 215, 248

transformation ratio 101

transgression 16, 86, 87, 97

transgressive systems tract (TST) 87

transient cavity 269

transit-time curve 83

transmission electron microscopy (TEM) 10, 75

transmittance color index (TCI) 95

transportation fuels 132

trap 111

Treibs, Alfred 68

Treibs Award 69

Treibs Medalists 69

Tretaspis Shale, Sweden 269–70

triacontane 21

triacylglycerol (TAG) 47, 343, 345

triamantane 157

triaromatic (TA) steroids 31, 64, 201, 202, 232, 241, 246

Triassic–Jurassic boundary 145

Triassic mudrocks, Barents Sea 12, 87, 88

trichodiene 57

tricyclic diterpenoid acids 332

tricyclics/17α-hopanes 98

tricyclic terpanes (cheilanthanes) 26–7, 28–9, 206, 211, 271
  C_{30} tricyclic terpane 236–7, 238–9

tricyclohexaprenane 26–7, 28–9

3,4,5-trihydroxybenzoic acid *see* gallic acid

2,2,3-trimethylbutane 165

2,6,10-trimethyldodecane (*see* farnesane)

2,6,10-trimethyl-7-(3-methylbutyl)-dodecane 26–7, 28–9, 30

1,2,7-trimethylnaphthalene 196

2,6,10-trimethylpentadecane *see* norpristane

2,2,3-trimethylpentane 22

Trinidad oils 105

triple quadrupole mass spectrometer 218, 227, 229

triple-sector mass spectrometer *see* triple quadrupole mass spectrometer

25,28,30-trisnorhopane (TNH) 142

triterpane 25, 26–7, 28–9

triterpanoid palmitate 342

triterpene, pentacyclic 55

triterpenoid 55

troilite 80

tropane alkaloids 351

trophic level 8

true boiling point (TBP) distributions 122

TSF *see* total scanning fluorescence

Ts/Tm *see* Ts/(Ts + Tm)

Tumulus Mount, Turkey 345

turbidity current 13

Turkey Creek Field, Alabama 189

turpentine 336
  tree *see* Pistacia

Tuscaloosa Formation, USA 161, 170, 173, 182

Tuxedni-Hemlock(!) 84

twisted conformation 24

two-dimensional GC *see* gas chromatography

two-dimensional NMR *see* nuclear magnetic resonance

TXL Field, Texas 84

Tynec well, Austria 269

type I 76, 77, 78, 259

type IS 78

type II 76, 77, 78–9, 107

type IIS 31, 79, 80

type II/III 79, 107

type III 76, 77, 79, 107, 189

type IV 76, 79

type curve *see* stable carbon isotope type-curve

Tyumen Formation, West Siberia 15, 143

ubiquinone 26–7, 28–9, 55

UCM *see* unresolved complex mixture

Udang oil, Borneo 148

Uinta Basin, Utah 149, 260

UK 262–4

ultralaminae 10, 75

ultramafic 253, 259

ultrastable-Y (US-Y) zeolite 148, 206, 207, 208

ultraviolet (UV) 199
  fluorescence 195, 290

Ulu Burun, Turkey 332

Umm El Tiel, Syria 326–8

unconformity 93, 262, 263, 264

undecane 21

underground leakage *see* oil spills

underground storage tanks (USTs) 283–4

unit cell 204
unresolved complex mixture (UCM) 104,
    106, 112, 113, 115, 294, 300
    aromatic hydrocarbon UCM 294
    saturated hydrocarbon UCM 294
unsaturated 19, 49
upstream 119
upwelling 8
    Peru 8
uranium 86–7
urea adduction 200, 201
uroporphyrinogen 65, 66
ursanes 334
urushiol-based lacquer film 334
US-Y molecular sieve see ultrastable-Y
Utica Shale, USA 141, 152, 153

vacuum distillation 122, 130, 133
vacuum gas oil (VGO) 132, 134
vacuum residuum or residue 122, 132
vadose zone modeling 315
Valdez, Alaska 306, 307
Valhal Field, North Sea 84
vanadium 103, 125, 295
vanadium/nickel [V/(Ni + V)] 88,
    103
vanadyl ion 68
Van-Egan Field, West Siberia 84
Van der Waals force 205
van Krevelen diagram 76, 77, 93
    maturation lines 93
    modified 77, 78
    oil shales 83
vapor extraction 318
vaporizing injection see injection
varnish 329, 334, 344
    yellowing 334
varve 12
Vendel, Uppland, Sweden 347
Venture well, Canada 177
Verrill Canyon Shale, Canada 170, 173
vetispiradiene 57
Victoria Land, Australia 5
Vienna PDB see VPDB
Vienna Standard Mean Ocean Water see
    VSMOW
Vietnam 187
Viking Graben, North Sea 83, 109, 170
Vindija Cave, Croatia 339, 340, 341
violaxanthin see carotenoid
virus 5
visbreaking 130, 133
viscometer 121
viscosity 120, 121–2, 128

dynamic 121
    kinematic 121, 127
    relative 121
vitamin A 71
vitrinite 75, 76, 89
    bituminous 91
    indigenous 90
    misidentification 93
    oxidation 93
    oxidation rims 93
    recycled 90, 94
vitrinite reflectance (R_o in oil, R_r random,
    or R_m mean) 73, 89–94, 96
    histogram 89, 90
    interpretive problems 94
    profile (R_o vs. depth) 89, 92, 94
    suppression 91, 92
    van Krevelen diagram 77
    versus stable carbon isotope differences
        among hydrocarbon gases 141
    versus Schaefer parameters 171
Vocation Field, Alabama 189
Volga River, Russia 316
Volga-Ural Basin, Russia
Vosges Mountains, Alsace-Lorraine 302
VPDB standard 137, 138
VPDB (Vienna PDB) 139
VSMOW (Vienna Standard Mean Ocean
    Water) standard 137

Wabasca heavy oil, Canada 120
Walkers Creek Field, Alabama 189
walnut hulls 116
water, emulsified in oil 122
water injection 266
water washing 174
wax 104, 120, 121, 126–7, 133
waxy oil 148, 201
weathering 103, 294–5, 308–12
weight-loss correction factor 117
Western Canada Basin 154
West Sak oils, North Slope, Alaska 177
wet gas 88, 166
whales 341
Whitby Mudstone Formation, Port
    Mulgrave area, Yorkshire, UK 116
white spirits 122
White Tiger Field, Vietnam see Bach Ho
    Field
wildcat 260
Wilmington Field, California 264, 265,
    266
Williston Basin, North America 177, 182,
    183, 184, 188, 190

Winborne Field, Canada 156
wine, biomarkers 344
Winnipegosis Formation, North America
    182, 184
Wiriagar well, Indonesia 147
wire-frame projection see structural
    notation
wood 138, 298
Woodford Shale, USA 100
wood tar see pitch
wool oil 338
workover (well maintainance) 116
World War II 305

10X zeolite 206
13X zeolite 206
Xanten-Wardt, Germany 329
xanthophyll 8, 69
X-ray absorption near-edge spectroscopy
    (XANES) 81
X-ray diffraction or crystallography 198,
    203, 240
xylenes 280, 291, 314

Yacoraite Formation, Argentina 183,
    185
Yakataga oil seep 297, 307, 309, 310–11
Yallourn lignite, Australia 63
Yellowstone National Park 140
Yem-Yegov well, West Siberia 143
Yinggehai Basin, China 162
Yorkshire, UK 346
Yusho incident 274

Zagros Mountains, Iran 332
Zdanice Well, Czech Republic 268, 269
zeaxanthin see carotenoid
Zechstein Basin, Germany 154, 176
Zemlet Field, Algeria 192
zeolite molecular sieves 202–6, 208
    adsorption of methanol 208
    concentrate compounds 205–8
    remove n-alkanes 205
    structural codes 204
zeolite structures 203–5
Zhungeer Basin, China see Junggar
    Basin
zig zag projection see structural
    notation
zingiberene 330
Zoroastrian religion 115
ZSM-5 203, 204, 205, 206
ZSM-23 206
Zymomonas 58, 61